U0314415

黄荣辉 文选

科学出版社
北京

内 容 简 介

黄荣辉院士是我国著名的气象学家，主要从事季风和大气动力学研究。他对大气中准定常行星波形成、传播和异常机理进行系统研究，提出准定常行星波在球面三维大气中沿两支波导的传播理论，研究了热带西太平洋暖池热状态及暖池上空对流活动对东亚夏季大气环流和气候异常的重要作用，提出影响中国夏季气候的大气环流异常遥相关型及其理论。

本书分两部分。第一部分为论文选编，主要选取作者有重要参考价值的部分论文，这些论文主要涉及五方面的研究：①关于准定常行星波动力学研究；②关于西太平洋暖池对东亚夏季风系统及我国气候变异的热力作用及其动力过程研究；③关于东亚冬、夏季风系统及气候变异特征研究；④关于西北太平洋台风气候学研究；⑤关于平流层重力波研究。第二部分为自述·追忆·致辞，收录作者的相关文章和大会发言稿。

本书收录了黄荣辉院士有代表性的部分重要论文，是他在大气科学领域科研成果的集中体现。本书可供大气科学、气候变化和海洋科学领域的科研人员，高等院校大气科学、气候变化等专业的研究生和教师阅读参考。

审图号：GS(2022)1074 号

图书在版编目(CIP)数据

黄荣辉文选 / 黄荣辉著. —北京：科学出版社，2022.3
ISBN 978-7-03-071875-4

Ⅰ. ①黄⋯ Ⅱ. ①黄⋯ Ⅲ. ①大气科学-文集 Ⅳ. ①P4-53

中国版本图书馆 CIP 数据核字(2022)第 042399 号

责任编辑：彭胜潮 / 责任印制：肖 兴 / 封面设计：黄华斌

科 学 出 版 社 出版
北京东黄城根北街 16 号
邮政编码：100717
http://www.sciencep.com

中国科学院印刷厂 印刷
科学出版社发行　各地新华书店经销

*

2022 年 3 月第 一 版　开本：889×1194 1/16
2022 年 3 月第一次印刷　印张：32 3/4　插页：12
字数：1 020 000

定价：398.00 元
(如有印装质量问题，我社负责调换)

黄荣辉院士

1 个人及全家照

1. 1965年在北京大学地球物理系毕业

2. 1970年在中国科学院大气物理研究所工作

3. 1980年在日本东京大学留学时在安田大讲堂前照的像片

4. 1986年获得中华全国总工会颁发的"全国优秀科技工作者"称号和"五一"劳动奖章

5. 1991年在中国科学院大气物理研究所中关村办公楼阅读文献

6. 2019年荣获庆祝中华人民共和国成立70周年纪念章

7. 1999年获何梁何利基金科学与技术进步奖

8. 2002年在中国科学院大气物理研究所中关村办公楼工作照

9. 2010年参加中国科学院院士大会

10. 2012年应邀出席中央电视台举办的教师节晚会

11. 2009年拍的全家照。前排：本人（右一）、夫人（左一）、孙子（左二）；后排：女儿（左一）、儿子（左二）、儿媳（右一）

12. 2005年6月和侄儿的合照

13. 2012年4月在美国国会大楼前的广场。本人（左二）、女儿（左一）、夫人（右一）

14. 2019年12月拍的全家照。前排：本人（左一）、夫人（右一）；后排：孙子（左一）、儿子（左二）、女婿（左三）、女儿（右三）、外孙（右二）、儿媳（右一）

2 与两位恩师的合影

← 15. 1995年陪恩师叶笃正院士（左二）访问中国台湾"中央"大学大气科学系

→ 16. 2002年陪恩师叶笃正院士（左二）和师母（左一）参加在敦煌召开的"干旱半干旱区陆-气相互作用国际研讨会"后参观月牙泉

17. 1986年陪同恩师日本东京大学岸保勘三郎教授（左一）在八达岭长城参观

18. 2002年陪恩师叶笃正院士（右二）与有关专家参观"敦煌干旱区陆-气相互作用观测试验场"

← 19. 2007年拜访恩师东京大学岸保勘三郎教授（右二）时。本人（左二）、陈文（右一）、陈光华（左一）。

20. 1992年与日本新田（Nitta）教授（右一）拜访恩师东京大学岸保勘三郎教授（左二）合影 →

21. 1996年本人（左一）与夫人（左二）拜访恩师——东京大学岸保勘三郎教授（右一）

3 参加学术会议

22. 1985年出席在日本东京召开的"东亚季风国际专家研讨会"。前排：陶诗言院士（左一）、岸保勘三郎教授（左二）；后排：本人（左一）、高由禧院士（左二）、浅井富雄教授（右三）、李崇银院士（右二）、周小平研究员（右一）

23. 1988年出席在美国洛杉矶召开的纪念J. Bjerknes学术研讨会休息时与UCLA大学的柳井迪雄（Yanai）教授（左一）及荒川昭夫（Arakawa）教授（右一）交谈

24. 1995年出席在韩国首尔召开的中、日、韩关于气候季节预测国际研讨会时与韩国气象厅长文胜义教授（左三）及其他有关韩国气象专家的合影

25. 1999年在韩国首尔召开的"韩国气象厅与中国科学院大气物理研究所合作研究讨论会"上做有关学术报告

26. 2000年主持《国家重点基础研究发展规划》第一批启动项目"我国气候灾害的形成机理和预测理论研究"项目全体会议

27. 2002年主持国家海洋局海洋-大气化学与全球变化重点实验室学术委员会第一次会议

28. 2002年参加中国科学院地学部第十届常委会第九次会议时的合影

29. 2003年参加全国政协第十届第一次会议科技界委员的合影

30. 2011年出席在湖南省张家界召开的"海峡两岸台风合作研究成果交流会"时与著名气象学家、日本全球变化研究所所长松野太郎教授（左二）及中国台湾高雄海洋科技大学陈昭铭教授（右一）和陈昭铭夫人（左一）的合影

31. 2006年出席在成都召开的"中国科学院暨两岸青年大气科学研讨会"。本人（中）、刘广英教授（右）

32. 2012年应邀访问澳门气象局时与前局长冯瑞权博士（左三）的合影。本人（左四）、陈文（左二）、王林（右二）、冯涛（右三）

4 与学生们合影

33. 1990年在参加有关学术研讨会后参观珠海有关景区时与学生合影。本人（中）、陈文（左一）、陆日宇（左二）、殷宝玉（右二）、宋玉成（右一）

← 34. 1991年在办公室指导学生科学研究。本人（中）、陈文（左）、陈金中（右）

← 35. 1992年在办公室和学生合影。本人（左三）、鹿晓丹（左一）、陈文（左二）、陈金中（右三）、殷宝玉（右二）、刘爱弟（右一）

36. 1999年在参加有关学术研讨会后与学生合影。本人（右三）、温之平（左一）、任保华（左二）、严邦良（左三）、张人禾（右二）、周连童（右一）

37. 1987年春节在北京龙潭湖和学生合影。本人（左）、薛松（右）

38. 1998年在青岛参加有关学术研讨会后与学生合影。本人（左三）、夫人（左二）、武炳义（左一）、张启龙（右二）、黄国风（右一）

39. 1999年在办公室指导学生科学研究。本人（中）、陈际龙（左）、周连童（右）

40. 2000年在安徽合肥中国科学技术大学与学生的合影。本人（右），任保华（左）。

41. 2005年考察中国科学院平凉试验站观测塔合影。本人（左）、韦志刚（右）

42. 2002年在云南参加有关学术会议后与学生们。前排：本人（左一），夫人（右一）；后排：顾雷（左一）、崔雪峰（左二）、任保华（左三）、陈文（左四）、张志华（右三）、周连童（右二）、王磊（右一）

43. 2008年在大气所中关村办公楼前与师生们合影。第一排：本人（左六）、陈文（左一）、黄刚（左二）、蔡榕硕（左三）、严邦良（左四）、Evgeny Jadin（俄罗斯，左五）；顾雷（右一）、李新荣（右二）、陈际龙（右三）、陆日宇（右四）、曹杰（右五）、陈仲良（右六）。第二排：王鹏飞（左一）、范广洲（左二）、韦志刚（左三）、温之平（左四）、周定文（左五）、任保华（左六）、周连童（左七）、王磊（左八）。后排包括：王林、黄平、鲍名、魏科、杜振彩、龚志强、胡开明、胡开喜、李超凡、林中达、刘永、屈侠等

44. 2006年在云南参加有关学术研讨会后与学生合影。本人（左三）、曹杰（左一）、黄勇勇（左二）、陈栋（右二）、顾雷（右一）

45. 2007年在日本参加有关学术研讨会后合影。本人（右）、蔡榕硕（左）

46. 2013年在杭州参加有关学术会议后与学生们的合影。本人（右二）、夫人（左二）、武亮（左一）、冯涛（右一）

47. 2011年参加有关会议后的合影。本人（左三）、陆日宇（左一）、马耀明（左二）、傅云飞（右三）、杜振彩（右二）、温之平（右一）

48. 2013年在香港参加COAA会议后的合影。本人（左二）、吴仁广（左一）、夫人（右二）、吴仁广夫人（右一）

49. 2012年在广西参加中韩季风学术交流会议后合影。本人（左二）、夫人（左一）、温之平（右二）、温之平夫人（右一）

50. 2012年在中国台湾参加海峡两岸会议后的合影。本人（左）、龚志强（右）

51. 2008年在洛阳参加有关学术会议后与学生的合影。本人（左）、黄平（右）

52. 2013年在中国台湾参加有关学术会议后与学生的合影。本人（右）、皇甫静亮（左）

53. 2013年在北京参加有关研究项目启动会后与学生的合影。本人（中）、武亮（左）、王林（右）

54. 2011年在研究所博士学位授予仪式后与学生合影。本人（左二）、武伟（左一）、王鹏飞（右二）、杜振彩（右一）

55. 2012年在研究所博士学位授予仪式后与学生的合影。本人（左二）、李超凡（左一）、陆日宇（右一）

56. 2012年在研究所博士学位授予仪式后与学生的合影。本人（左）、刘永（右）

57. 2016年在研究所博士学位授予仪式后与学生合影。本人（左三）、皇甫静亮（左一）、应俊（左二）、谭红建（右二）、王谞（右一）

58. 2020年教师节与学生合影。本人（前排右一）、夫人（前排左一）；后排：周连童、陈际龙、王磊、陈光华、周德刚、皇甫静亮、殷晓雪、张宏杰、韩永秋、汤玉莲

59. 2021年在大气所和学生合影。本人（左）、陈国森（右）

60. 2021年在研究所博士学位授予仪式后与学生合影。本人（左）、张宏杰（右）

61. 2021年在北京和学生合影。本人（右）、于非（左）

获 奖 项 目

一、科 技 奖

1. 2003年，"我国短期气候预测系统的研究"项目获国家科学技术进步奖一等奖，笔者为第二完成人。
2. 1988年，"中国卫星气象学研究"项目获国家自然科学奖三等奖，笔者为第六完成人。
3. 1991年，"大气中准定常行星波形成、传播与异常机制的研究"项目获国家自然科学奖三等奖，笔者为第一完成人。
4. 1997年，"东亚与热带大气低频变化及其气候异常机理研究"项目获国家自然科学奖三等奖，笔者为第一完成人。
5. 1986年，"大气中准定常行星波形成、传播与异常机制的研究"项目获中国科学院科学技术进步奖一等奖，笔者为第一完成人。
6. 1993年，"东亚夏季大气环流异常和短期气候变化成因的研究"项目获中国科学院科学技术进步奖二等奖，笔者为第一完成人。
7. 1997年，"我国灾害性气候的预测方法及其应用研究"项目获中国科学院科学技术进步奖二等奖，笔者为第一完成人。
8. 1999年，笔者获得何梁何利基金科学与技术进步奖。

二、荣 誉 奖

1. 1986年获得国家"五一"劳动奖章。
2. 1986年获得"国家级有突出贡献中青年科学家"称号。
3. 2017年获得"中国科学院优秀导师奖"。
4. 2018年获得中国科学院大学地球与行星科学学院"杰出贡献教师奖"。
5. 2019年获得庆祝中华人民共和国成立70周年纪念章。

出 版 专 著

1. 小仓义光著, 黄荣辉译. 大气动力学原理. 北京: 科学出版社, 1981.

2. 叶笃正, 黄荣辉, 等著. 长江黄河流域旱涝规律与成因研究. 济南: 山东科学技术出版社, 1996.

3. 黄荣辉, 郭其蕴, 孙安健, 等编. 中国气候灾害分布图集. 北京: 海洋出版社, 1997.

4. 黄荣辉, 李崇银, 王绍武, 等编. 我国旱涝重大气候灾害及其形成机理研究. 北京: 气象出版社, 2003.

5. 黄荣辉著. 大气科学概论. 北京: 气象出版社, 2005.

6. 黄荣辉, 陈文, 马耀明, 高晓清, 吕世华, 韦志刚, 张强, 卫国安, 胡泽勇, 周连童, 周德刚, 等著. 中国西北干旱区陆-气相互作用及其对东亚气候变化的影响. 北京: 气象出版社, 2011.

7. 黄荣辉, 吴国雄, 陈文, 刘屹岷, 周连童, 王林, 等编. 大气科学和全球气候变化研究进展与前沿. 北京: 科学出版社, 2014.

8. 黄荣辉, 吴国雄, 陈文, 刘屹岷, 周连童, 周德刚, 刘永, 等编. 大气科学和全球气候变化研究重大科学问题. 北京: 科学出版社, 2016.

自 序

时光流转，岁月如梭。从 1959 年考入北京大学地球物理系气象专业学习，毕业之后在中国科学院大气物理研究所当研究生，之后留所从事研究至今已有六十余载，转眼间已到退休之年。由于近年来身体欠佳，治病之余在家整理以前的论文以及著或编的书稿，不禁想起以前艰难求学历程和几十年学术研究及对学生的培养生涯。

从大学时代起，我就对大气动力学比较兴趣，大学毕业后几十年我也一直从事这方面的研究，大概做了以下几个方面的工作：①准定常行星波动力学，主要研究了地球大气中准定常行星波的形成、传播特征及规律以及异常机理；②短期气候动力学，主要研究了东亚冬、夏季风系统变异与我国旱涝气候灾害年际和年代变异成因及其机理，特别是研究了热带西太平洋暖池的热力状况与对流活动、我国西北干旱和半干旱区陆-气相互作用等对东亚夏季风系统变异的影响过程及其动力学机理，并研究了准定常行星波传播波导的变异对东亚冬季风系统年代际变异的动力作用；③西北太平洋台风气候学与生成动力学，主要研究了西北太平洋季风槽的年际、年代际变异及其对热带低压生成的影响过程以及动力学机理，特别是分析了季风槽变异通过大尺度环境场调制热带波动而使热带低压和台风生成的动力过程。此外，在 20 世纪 70 年代初还做了卫星红外遥感气温原理与反演研究。

除了学术研究外，我还乐于给学生讲课，与学生一起讨论大气动力学的一些科学问题。我与我的学生多年在中国科学院研究生院主讲了"高等大气动力学"和"热带地球流体动力学"课程。有时，遇见听过我讲课的大气和海洋科学的青年学者，他们很亲切地说"黄老师，我听过你讲的课"；听到此声音，我心里很高兴。确实，我能在课堂上给学生讲课，与学生一起讨论地球大气和海洋一些动力过程，感到很幸福。"桐花万里丹山路，雏凤清于老凤声"，看到我教过的年轻大气科学学者茁壮成长，心里感到很欣慰。

在生病之前，我与有关学者还主持和组织了我国大气科学和全球气候变化研究发展战略的研讨。经过近五年的研究与讨论，回顾了近百年来国内外大气科学的发展历程、当前的发展动向和趋势，提出了今后应开展研究的重大科学问题及应采取的战略措施。

回想我从一个农村穷苦的放牛娃成长为一名中国科学院院士，并在科研领域做出些许成绩，这除了归功于父母的养育、培养之恩以及我夫人张锦英对我生活和身体无微不至地关心外，还要特别感谢两位恩师——叶笃正院士和日本东京大学岸保勘三郎先生对我的培养；并且，我还特别感谢北京大学第一医院泌尿外科主任周利群大夫对我的病进行精心治疗，以及我的学生武亮博士和皇甫静亮博士等对我身体的关心和照顾，使我的身体得以逐渐康复，从而才能有精力来完成此书稿的收集和修改。此外，科学出版社彭胜潮编审对本书给予精心编辑，中国科学院大气物理研究所皇甫静亮副研究员对本书的校对和修改花费了很多精力，对于他们的帮助谨此表示衷心感谢。

本书主要收集了我与有关学者合作的研究论文和著作目录，选取有一定参考价值的部分论文的全文，以及反映我艰难求学历程与愉快的研究生涯的自述和缅怀我一生不能忘怀的两位恩师的文章，除此还有一些会议的致词或讲话。如果这些微不足道的研究论文或有关文章对我的学生或后人有一点点参考价值，我就心满意足了。

黄荣辉
于 2020 年 5 月

目 录

自序

第一部分 论文选编

一、准定常行星波动力学研究

The Response of a Model Atmosphere in Middle Latitude to Forcing by Topography and Stationary Heat Sources ································· 5

The Response of a Hemispheric Multi-Level Model Atmosphere to Forcing by Topography and Stationary Heat Sources Part I Forcing by Topography ································· 22

The Response of a Hemispheric Multi-Level Model to Forcing by Topography and Stationary Heat Sources Part II Forcing by Stationary Heat Sources and Forcing by Topography and Stationary Heat Sources ································· 37

关于冬季北半球定常行星波传播另一波导的研究 ································· 52

The Response of a Hemispheric Multi-Level Model Atmosphere to Forcing by Topography and Stationary Heat Sources in Summer ································· 63

夏季青藏高原上空热源异常对北半球大气环流异常的作用 ································· 77

Relationship between the Interannual Variation of Total Ozone in the Northern Hemisphere and the QBO of Basic Flow in the Tropical Stratosphere ································· 89

二、西太平洋暖池对东亚夏季风系统及我国气候变异的热力作用及其动力过程研究

The Physical Effects of Topography and Heat Sources on the Formation and Maintenance of the Summer Monsoon Over Asia ································· 101

夏季热带西太平洋上空的热源异常对东亚上空副热带高压的影响及其物理机制 ································· 112

Numerical Simulation of the Relationship between the Anomaly of the Subtropical High over East Asia and the Convective Activities in the Western Tropical Pacific ································· 122

The Influence of ENSO on the Summer Climate Change in China and Its Mechanisms ································· 135

引起我国夏季旱涝的东亚大气环流异常遥相关及其物理机制的研究 ································· 146

The East Asia/Pacific Pattern Teleconnection of Summer Circulation and Climate Anomaly in East Asia ································· 156

Impacts of the Tropical Western Pacific on the East Asian Summer Monsoon ································· 169

热带西太平洋纬向风异常对 ENSO 循环的动力作用 ································· 183

东亚夏季风爆发和北进的年际变化特征及其与热带西太平洋热状态的关系 ································· 191

Impact of Thermal State of the Tropical Western Pacific on Onset Date and Process of the South China Sea Summer Monsoon ································· 208

三、东亚冬、夏季风系统及气候变异特征研究

我国夏季降水的年代际变化及华北干旱化趋势 ································· 227

The Progresses of Recent Studies on the Variabilities of the East Asian Monsoon and Their Causes ································· 237

中国东部夏季降水的准两年周期振荡及其成因·················252
Characteristics and Variations of the East Asian Monsoon System and Its Impacts on Climate Disasters in China·················268
我国东部夏季降水异常主模态的年代际变化及其与东亚水汽输送的关系·················298
Characteristics, Processes, and Causes of the Spatio-temporal Variations of the East Asian Monsoon System·················316
关于中国西北干旱区陆-气相互作用及其对气候影响研究的最近进展·················349
20世纪90年代末中国东部夏季降水和环流的年代际变化特征及其内动力成因·················371
20世纪90年代末东亚冬季风年代际变化特征及其内动力成因·················383
Differences and Links between the East Asian and South Asian Summer Monsoon Systems: Characteristics and Variability·················401

四、西北太平洋台风气候学研究

西北太平洋热带气旋移动路径的年际变化及其机理研究·················419
台风在我国登陆地点的年际变化及其与夏季东亚/太平洋型遥相关的关系·················429
关于西北太平洋季风槽年际和年代际变异及其对热带气旋生成影响和机理的研究·················441

五、平流层重力波研究

平流层球面大气地转适应过程和惯性重力波的激发·················463

第二部分　自述·追忆·致辞

我的自述——家贫更加激励我努力拼搏·················483
缅怀我的恩师——叶笃正先生的学术成就和治学精神·················488
难忘师生情——缅怀岸保勘三郎先生·················493
上善若水——忆许惜今校长·················496
在纪念谢义炳先生诞辰百周年会上的讲话·················497
在福建省泉港二中(原惠安二中)五十周年校庆庆典上的贺词·················498
"海峡两岸大气科学研究生学术研讨会"总结·················499
《福建省惠安一中高十五组同学毕业五十周年纪念册》序言·················501
海峡两岸青年大气科学学术研讨会的开幕词·················502
《泉港区第二中学(原惠安二中)校志》序言·················503
在"首届大气科学学科建设与人才培养研讨会"上的讲话——发展大气科学　培养创新人才·················504
海峡两岸台风暴雨合作研究成果交流会的开幕词·················507
在中山大学大气科学系建系五十周年庆典会上的致词·················508
庆贺福建省泉港二中(原惠安二中)建校六十周年的贺词·················509
"东亚能量和水分循环及其与季风相互作用国际研讨会"开幕词·················510
在"海峡两岸台风和暴雨合作研究成果发表研讨会"上的致词·················511
在"第三海洋研究所海洋-大气化学与全球变化重点实验室成立十周年庆典会"上的致词·················512

附录　论文目录·················513

第一部分　论文选编

一、准定常行星波动力学研究

The Response of a Model Atmosphere in Middle Latitude to Forcing by Topography and Stationary Heat Sources*

Huang Ronghui and K. Gambo

Abstract

　　The Midlatitude standing waves forced by topography and stationary heat sources are investigated by means of a quasi-geostrophic, linear, steady-state, 34-level model with eddy viscosity and Newtonian cooling effect included. The results show that increased vertical resolution contributes significantly to the solution of the finite difference equations of motion and the thermodynamic equation, *i.e.*, our 34-level model with increased vertical resolution in both the troposphere and stratosphere, gives better results when compared to simpler models such as the "two-levels" model.

　　The computed vertical and zonal distributions of amplitude and phase for winter are in good agreement with observed standing geopotential and temperature waves. Similarly, computed and observed summer standing wave positions are also fairly consistent. However, the computed trough over the central Pacific Ocean in summer appears to be too weak.

　　The eddy viscosity, and the meridional wavelength of the topography and diabatic heating seem to have a strong influence on the response of a model atmosphere to forcing by topography and diabatic heating.

1. Introduction

　　The response of a model atmosphere to forcing by topography and stationary heat sources has been studied by many authors through the use of quasi-geostrophic, linearized, steady-state models in which the zonal basic states are assumed to be known and the heating/cooling sources are either specified or parameterized. For example, Charney and Eliassen (1949), Bolin (1950), Gambo (1956), Murakami (1963), and Sankar-Rao (1965a) investigated the stationary flow patterns in middle latitudes due to forcing by the topography. Smagorinsky (1953) and Sankar-Rao (1965c) examined the vertical distribution of stationary waves excited by diabatic heating. The Staff Members, Academia Sinica (1958), Döös (1963), Saltzman (1965), Sankar-Rao (1965b), and Derome and Wiin-Nielsen (1971) studied the effects of forcing by topography and stationary heat sources on a middle latitude model atmosphere. The response of the tropical atmosphere to local, steady forcing was also discussed by Webster (1972).

　　In these previous studies, simplified layer-models were used to discuss the vertical distribution of stationary waves. Thus, it is possible that the effect of vertically propagating, stationary planetary waves excited in the troposphere may not be treated correctly in the stratosphere, due to the poor vertical resolution in the stratosphere. Nakamura (1976) and Kirkwood and Derome (1977) showed that different results are obtained for various vertical resolutions. Their results also revealed that insufficient resolution in the stratosphere can lead to a spurious energy reflection at the upper boundary. In some cases, when disturbances are internal type in the vertical as defined by Charney and Drazin (1961), an apparent false wave structure is obtained in the troposphere. Therefore, a multi-level model with sufficient vertical resolution in the stratosphere must be used to accurately describe the vertical distribution of the amplitude and phase

---

* 本文原载于: Journal of the Meteorological Society of Japan, Vol. 59, No.2, 220-237, 1981.

of standing waves forced by topography and diabatic heating.

In this paper, our discussion focuses on the effect of vertical resolution in estimating standing wave response to forcing by topography and stationary heating/cooling sources. For simplicity, a quasi-geostrophic, linear, steady-state model is used for the discussion.* In addition, The heating/cooling sources are assumed to be given as an external forcing mechanism even though the specification of heat sources, such as the release of latent heat of condensation and the transport of sensible and latent heat from the earth's surface, is directly related to the atmospheric motions.

2. The model and parameters

We shall assume that the motion takes place on a β-plane centered at 45°N. Thus, we can use a Cartesian coordinate system in which the x and y coordinates increase to the east and north, respectively. We shall also make use of the hydrostatic approximation to introduce pressure as the vertical coordinate. As mentioned in the introduction, a quasi-geostrophic, linear, steady-state model is utilized with eddy viscosity and the effect of Newtonian cooling also incorporated in the model. To correctly describe the behavior of motion in the stratosphere, a 34-level representation of the atmosphere is used.

1) Model

The steady state, quasi-geostrophic, vorticity and thermodynamic equations in which the eddy viscosity, the effect of Newtonian cooling and the horizontal kinematic thermal diffusivity are included, may be expressed as

$$\vec{V} \cdot \nabla(\zeta+f) = f_0 \frac{\partial \omega}{\partial p} + K_m \nabla^2 \zeta, \quad (1)$$

and

$$\vec{V} \cdot \nabla\left(\frac{\partial \phi}{\partial p}\right) + \sigma \omega = -\frac{RH}{c_p p} - \alpha_R \frac{\partial \phi}{\partial p} + K_T \nabla^2 \left(\frac{\partial \phi}{\partial p}\right), \quad (2)$$

respectively. The notations used in the above equations are as follows:

$\zeta = (1/f_0)\nabla^2\phi$: vertical component of relative vorticity

$\vec{V}$: horizontal velocity vector

$\sigma = -\alpha(\partial \ln \theta)/\partial p$: static stability parameter (α:

---
* It was recently shown that the nonlinear terms in the mean-state equations are important when attempting to explain the effects of flow moving around the mountains (Ashe (1979)).

specific volume, θ : potential temperature)

H : diabatic heating per unit time and unit mass

R : gas constant (0.287 KJ·Kg^{-1}·deg^{-1})

c_p : specific heat at constant pressure (=1.004 KJ·Kg^{-1}·deg^{-1})

α_R : Newtonian cooling coefficient

K_m : horizontal eddy viscosity

K_T : horizontal kinematic thermal diffusivity

f_0 : Coriolis parameter at 45°N

ω : vertical p-velocity (dp/dt)

The structure of the model used in this study is shown in Fig. 1 with the atmosphere from $p=p_t$ to $p=p_s$ divided into N layers. The vorticity equations are specified at $n-1/2$ and $n+1/2$ levels, while the thermodynamic equations are described at n level. Thus, the model equations are

$$\vec{V}_{n-(1/2)} \cdot \nabla(\zeta+f)_{n-(1/2)} = f_0 \left(\frac{\partial \omega}{\partial p}\right)_{n-(1/2)} + \left[K_m \cdot \nabla^2\left(\frac{1}{f_0}\nabla^2\phi\right)\right]_{n-(1/2)}, \quad (3)$$

$$\vec{V}_n \cdot \nabla\left(\frac{\partial \phi}{\partial p}\right)_n + \sigma_n \omega_n = -\left(\frac{RH}{c_p p}\right)_n - (\alpha_R)_n \left(\frac{\partial \phi}{\partial p}\right)_n + (K_T)_n \cdot \nabla^2\left(\frac{\partial \phi}{\partial p}\right)_n, \quad (4)$$

$$\vec{V}_{n+(1/2)} \cdot \nabla(\zeta+f)_{n+(1/2)} = f_0 \left(\frac{\partial \omega}{\partial p}\right)_{n+(1/2)} + \left[K_m \cdot \nabla^2\left(\frac{1}{f_0}\nabla^2\phi\right)\right]_{n+(1/2)}$$

for $n=1, 2, \cdots, N$. (5)

For the upper boundary conditions, we assume that the vertical p-velocity vanishes at the top of the model, i.e.,

$$\omega = 0, \quad \text{at } p = p_t \text{ (or } z = z_t\text{)}, \quad (6)$$

where p_t (or Z_t) is the pressure (or height) at

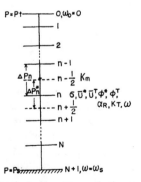

Fig. 1 Vertical structure of the model.

the top of the model.

The lower boundary condition at $p=p_s$ (p_s; surface pressure) assumes that the vertical p-velocity is caused by surface topography where the standard pressure is p_g, and also by Ekman pumping resulting from the viscosity in the Ekman layer. Thus, the vertical p-velocity at $p=p_s$ is given as (see Derome and Wiin-Nielsen, 1971)

$$\omega_s = \vec{V}_s \cdot \nabla p_g - \frac{p_s \cdot F}{2f_0} \zeta_s, \text{ at } p=p_s \text{ (or } z=0), \quad (7)$$

where $\vec{V}_s$ is the horizontal velocity vector at $p=p_s$, and $p_s=1000$ mb for simplicity. F is the friction coefficient and will be treated as a constant (4×10^{-6} s^{-1}).

To obtain the vertical distributions of standing waves, we introduce the following notations

$$(\ )_n^* = \frac{1}{2}[(\ )_{n-(1/2)} + (\ )_{n+(1/2)}], \quad (8)$$

and

$$(\ )_n^T = \frac{1}{2}[(\ )_{n-(1/2)} - (\ )_{n+1/2}]$$
$$\text{for } n=1, 2, \cdots, N. \quad (9)$$

Here, ( )* and ( )^T denote the mean state and the deviation from the mean state due to the thermal wind, respectively. Thus equations (3), (4) and (5) may be rewritten as follows:

$$(\vec{V}_n^* + \vec{V}_n^T) \cdot \nabla(\zeta_n^* + f) + (\vec{V}_n^* + \vec{V}_n^T) \cdot \nabla \zeta_n^T$$
$$= f_0 \frac{\omega_n}{\Delta p_n} - f_0 \frac{\omega_{n-1}}{\Delta p_n}$$
$$+ (K_m)_{n-(1/2)} \cdot \nabla^2 \left[\frac{1}{f_0} \nabla^2 (\phi_n^* + \phi_n^T)\right], \quad (10)$$

$$2\vec{V}_n^* \cdot \nabla \phi_n^T - \sigma_n \Delta p^* \omega_n$$
$$= \frac{RH_n}{c_p p_n} \Delta p_n^* - 2(\alpha_R)_n \cdot \phi_n^T + 2(K_T)_n \cdot \nabla^2 \phi_n^T, \quad (11)$$

$$(\vec{V}_n^* - \vec{V}_n^T) \cdot \nabla(\zeta_n^* + f) - (\vec{V}_n^* - \vec{V}_n^T) \cdot \nabla \zeta_n^T$$
$$= f_0 \frac{\omega_{n+1}}{\Delta p_{n+1}} - f_0 \frac{\omega_n}{\Delta p_{n+1}}$$
$$+ (K_m)_{n+(1/2)} \cdot \nabla^2 \left[\frac{1}{f_0} \nabla^2 (\phi_n^* - \phi_n^T)\right],$$
$$\text{for } n=1, 2, \cdots, N. \quad (12)$$

In these equations, ϕ_n^* and ϕ_n^T are regarded as the mean and thermal geopotential heights at the n'th level, respectively, while Δp_n and Δp_n^* are defined as

$$\Delta p_n = p_n - p_{n-1}$$

and

$$\Delta p_n^* = p_{n+(1/2)} - p_{n-(1/2)}$$

Hereafter, to facilitate the discussion, ϕ_n^* and ϕ_n^T are designated as the mean and thermal standing waves, respectively.

2) *The vertical difference scheme and linearization*

According to numerical tests by Nakamura (1976), vertical grid increments should be as small as $\Delta z = 1\sim 2$ km in the troposphere and $\Delta z = 2\sim 3$ km in the stratosphere, with the top of the model in the middle stratosphere. To satisfy these requirements, the vertical grid increments (Δz) used in this model are as follows:

z: 0–12 km $\quad \Delta z = 1.5$ km
12–30 km $\quad \Delta z = 2.0$ km
30–60 km $\quad \Delta z = 3.0$ km
60–92 km $\quad \Delta z = 4.0$ km

The top of our model atmosphere is defined at $z=92$ km (i.e., $p_t = 1.140 \times 10^{-3}$ mb in winter and $p_t = 8.495 \times 10^{-4}$ mb in summer). Thus, we divide the atmosphere into 34 layers from the earth's surface to $z=92$ km.

To obtain the vorticity at 1000 mb which appears in equation (7), we will extrapolate the geopotential from N level to 1000 mb ($N+1$ level) according to the hydrostatic equations, i.e.,

$$\phi_{N+1} = \phi_N^* - 1.9\phi_N^T . \quad (13)$$

Hence, the vertical p-velocity caused by Ekman pumping changes with respect to N'th level geopotential and thermal geopotential heights that are obtained from the model equations. Therefore, we can then rewrite the upper and lower boundary conditions as

$$\omega_0 = 0, \quad (14)$$

and

$$\omega_{N+1} = \vec{V}_{N+1} \cdot \nabla p_g - \frac{p_{N+1} \cdot F}{2f_0^2} \nabla^2 (\phi_N^* - 1.9\phi_N^T). \quad (15)$$

To facilitate computations, the following equations are substituted into equations (10), (11) and (12).

$$\zeta_n^* = \nabla^2 \psi_n^*, \quad \phi_n^* = f_0 \psi_n^*,$$
$$\zeta_n^T = \nabla^2 \psi_n^T, \quad \phi_n^T = f_0 \psi_n^T,$$
$$\text{for } n=1, 2, \cdots, N.$$

Here, ψ_n^* and ψ_n^T are the mean and thermal stream functions, respectively, due to forcing by topography and diabatic heating at the n'th level.

Using these equations with the upper and lower boundary conditions of equations (14) and (15), we can eliminate ω_n ($n=1, 2,\ldots N+1$) and linearize equations (10)–(12). Thus, we obtain 2×34 linear differential equations, which correspond to equations (10)–(12).

Here, we shall assume that the perturbation equations have a simple meridional structure given by $\cos(\mu y)$, i.e.,

$$[\phi_n^*, \phi_n^T, H_n, P_g]$$
$$=[\hat{\phi}_n^*(x), \hat{\phi}_n^T(x), \hat{H}_n(x), \hat{P}_g(x)]\cos(\mu y), \quad (16)$$

such that $y=0$ at 45°N, $\mu=2\pi/Ly$, and Ly is the meridional wavelength. The longitudinal dependence of the perturbation will be represented by the Fourier series

$$[\hat{\phi}_n^*(\lambda), \hat{\phi}_n^T(\lambda), \hat{H}_n(\lambda), \hat{P}_g(\lambda)]$$
$$=\sum_{k=1}^{K}\{[(A_n^*)_k, (A_n^T), (Q_n)_k, (W)_k]\cos(k\lambda)$$
$$+[(B_n^*)_k, (B_n^T)_k, (T_n)_k, (G)_k]\sin(k\lambda)\} \quad (17)$$

in which $\lambda=x/a\cos\varphi_0$ is the longitude, a represents the radius of the earth and $\varphi_0=45°N$. Here, n and k are the vertical level and the zonal wave number, respectively.

When equations (16) and (17) are substituted into the linearized differential equations (10)–(12), we obtain 4×34 linear algebraic equations. Therefore, if the forcing functions of topography and diabatic heating, i.e., $(Q_n)_k$, $(T_n)_k$, $(W)_k$, $(G)_k$, are considered to be known from observations, the terms $(A_n^*)_k$, $(A_n^T)_k$, $(B_n^*)_k$, $(B_n^T)_k$ can be computed from the 4×34 linear algebraic equations. The vertical distributions of the amplitude and phase for mean and thermal standing waves can also computed from $(A_n^*)_k$, $(B_n^*)_k$, $(A_n^T)_k$, $(B_n^T)_k$.

3) Parameters

a) Static stability σ_n: the static stability parameter σ_n is calculated from the mean temperature and density at 45°N in January and July obtained for the U.S. Standard Atmosphere (1966) by the following equation

$$\sigma_n=-\frac{1}{\rho_n}\left(\frac{\partial \ln\theta}{\partial p}\right)_n \quad \text{for } n=1, 2, \cdots, N, \quad (18)$$

where ρ is the density and θ is the potential temperature.

b) The vertical profiles of the basic zonal wind at 45°N in winter and summer are from

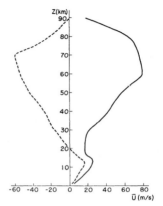

Fig. 2 The vertical profiles of the basic zonal wind at 45°N in winter (solid curves) and summer (dashed curves).

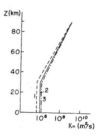

Fig. 3 The vertical distributions of the horizontal eddy viscosity coefficient. ($K_T = K_m$). The dashed, solid and dot-dashed lines show $K_m=0.7\times 10^6$ m^2 s^{-1}, 1.0×10^6 m^2 s^{-1} and 2.0×10^6 m^2 s^{-1} below 30 km height, respectively.

the computed results of R. J. Murgratroyd (1969) and are shown in Fig. 2.

c) Since the correct values of the damping coefficients in the stratosphere are not fully known, we shall assume one of plausible values, such as that shown by solid line in Fig. 3. In general, the upward propagating waves are reflected back at the top of a model. Large values of K_m are incorporated above 40 km in order to eliminate such unrealistic effects of the upper boundary condition. This value of 1.0×10^6 m^2 s^{-1} for $(K_m)_{n-1/2}$ and $(K_m)_{n+1/2}$ below the 30 km height seems to be in agreement with the eddy viscosity coefficient in the troposphere estimated from observational data (Wiin-Nielsen (1971)). Furthermore, we assume that the horizontal kinematic thermal diffusivity coefficient $(K_T)_n$ is equal to the horizontal eddy viscosity coefficient $(K_m)_n$, i.e.,

$$(K_T)_n=(K_m)_n$$

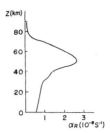

Fig. 4 The vertical profile of the Newtonian cooling coefficient.

where

$$(K_m)_n = \frac{1}{2}[(K_m)_{n-(1/2)} + (K_m)_{n+(1/2)}]$$
$$\text{for } n = 1, 2, \cdots, N.$$

d) The assumed vertical profile for the coefficient of Newtonian cooling $(\alpha_R)_n$ is shown in Fig. 4. The values of $(\alpha_R)_n$ used between 0–75 km are based on computations by Dickinson (1973). Above 75 km $(\alpha_R)_n$ is assumed to be constant and relatively small.

e) The mean zonal wind at the surface (i.e., U_s) is obtained by extrapolating from the mean zonal wind at N level, and is assumed to be 5.0 m s^{-1} in winter and 2.7 m s^{-1} in summer.

3. Computation of response to forcing by idealized topography and diabatic heating

To qualitatively understand the response to forcing by actual topography and diabatic heating, we shall first consider the response to forcing which is represented by a simple harmonic function. The amplitude of this simple forcing function is assumed to be in both winter and summer.

Let us assume a harmonic function of topographic forcing in the following form,

$$\hat{P}_g(\lambda) = W \cos(k\lambda), \quad k = 1, 2, \cdots, K. \quad (19)$$

Here, W is the amplitude of the forcing function of topography, which is assumed to be 50 mb. Similarly, the harmonic function of diabatic heating is assumed to be as follows

$$\hat{H}_n(\lambda) = Q_n \cos(k\lambda), \quad k = 1, 2, \cdots, K. \quad (20)$$

The vertical distribution of Q_n is shown in Fig. 5, which is similar to the vertical distribution used by Murakami (1972), i.e.,

$$Q_n = Q_0 \cdot \exp\left[-\left(\frac{p_n - \bar{p}}{b}\right)^2\right]$$
$$\text{for } n = 1, 2, \cdots, N, \quad (21)$$

where Q_0 is assumed to be equal to 5×10^{-6} KJ·Kg^{-1}·s^{-1}. We assume that $b = 300$ mb and

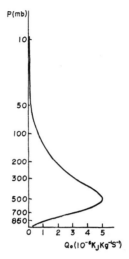

Fig. 5 The vertical profile of the amplitude of heat source.

$\bar{p} = 500$ mb, which corresponds to a vertical distribution of diabatic heating with a maximum at 500 mb, and exponential decreases above and below 500 mb.

The linear algebraic equations mentioned previously are solved only for the cases of $k = 1$ to 2 and $\mu = 0.950 \times 10^{-6}$ m^{-1} (i.e., $L_y = 60°$). The solution for the mean stream function is written in the form

$$\{\hat{\psi}_n^*(\lambda)\}_k = (\hat{C}_n^*)_k \cos k[\lambda - (\alpha_n^*)_k], \quad (22)$$

$$(\hat{C}_n^*)_k = [(A_n^*)_k^2 + (B_n^*)_k^2]^{1/2}, \quad (23)$$

$$(\alpha_n^*)_k = \frac{1}{k} \tan^{-1}[(B_n^*)_k / (A_n^*)_k],$$
$$n = 1, 2, \cdots, N, \quad k = 1, 2. \quad (24)$$

Here, $(c_n^*)_k$ and $(\alpha_n^*)_k$ are the amplitude and phase of the mean stream function for the zonal wave number k at n-level, respectively. It is convenient, for comparison with the observed values, to convert the stream function into height values, such that

$$(C_n^*)_k = \frac{f_0}{g}(\hat{C}_n^*)_k \quad (25)$$

where $(c_n^*)_k$ is the height value of mean standing waves for the zonal wave number k at n-level.

Similar expressions for the thermal stream function are written in the form

$$\{\hat{\psi}_n^T(\lambda)\}_k = (\hat{C}_n^T)_k \cdot \cos k[\lambda - (\alpha_n^T)_k], \quad (26)$$

$$(\hat{C}_n^T)_k = [(A_n^T)_k^2 + (B_n^T)_k^2]^{1/2}, \quad (27)$$

$$(\alpha_n^T)_k = \frac{1}{k} \tan^{-1}[(B_n^T)_k / (A_n^T)_k],$$
$$n = 1, 2, \cdots, N, \quad k = 1, 2. \quad (28)$$

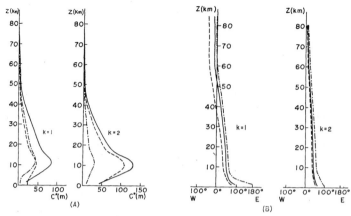

Fig. 6 The vertical distributions of amplitude (A) and phase (B) for standing waves responding to forcing by idealized topography (dashed curves), idealized diabatic heating (dot-dashed curves), and by both the idealized topography and idealized diabatic heating (solid curves) in winter at 45°N.

In these equations, $(c_n^T)_k$ and $(\alpha_n^T)_k$ are the amplitude and phase of the thermal stream function for the zonal wave number k at n-level, respectively. The term $(c_n^T)_k$ is converted into temperature values by applying the hydrostatic equation, i.e.,

$$(C_n^T) = \frac{2p_n}{R} \frac{f_0(\hat{C}_n^T)_k}{\Delta p_n^*} \qquad (29)$$

where $(c_n^T)_k$ is the temperature value of thermal standing waves for the zonal wave number k at n-level.

The results of these computations are as follows:

(I) Winter case:

a) Figures 6(A) and 6(B) show the amplitude and phase, respectively, of mean standing waves responding to forcing by topography, diabatic heating and both topography and diabatic heating.

As may be seen in Fig. 6(A), the amplitude of mean standing waves responding to forcing by either topography or diabatic heating is a maximum at 12 or 10.5 km. The reason for this maximum response is made clear by an analysis of the profile of the mean zonal wind at 45°N in winter (Fig. 2). In Fig. 2, note that there is a jet at 12 km height along 45°N, with $\partial \bar{U}/\partial p > 0$ in the 12~18 km layer and $\partial \bar{U}/\partial p < 0$ below the 12 km height. Thus, $\partial/\partial p(1/\sigma \, \partial \bar{U}/\partial p) > 0$ at around 12 km. On the other hand, it is well known that if the profile of $\bar{U}$ satisfies the equation

$$\beta - f^2 \frac{\partial}{\partial p}\left(\frac{1}{\sigma}\frac{\partial \bar{U}}{\partial p}\right)\bigg/\bar{U}$$
$$-\left\{k^2 + \mu^2 + \frac{f^2}{\sqrt{\sigma}}\frac{\partial^2}{\partial p^2}\left(\frac{1}{\sqrt{\sigma}}\right)\right\} \leq 0, \qquad (30)$$

in the case of a non-viscous atmosphere, the standing wave cannot be propagated upward (for example, refer to Nakamura, 1976). At the 12 km height, the mean zonal wind at 45°N in winter satisfies equation (30) for the case of zonal wave number 1 and $\mu = 0.950 \times 10^{-6}$ m^{-1} ($L_y = 60°$). Therefore, the standing wave for zonal wave number 1 cannot be propagated upward and the amplitude of standing wave $k = 1$ is a maximum at this height. Obviously, this height decreases as the zonal wave number k and the meridional wave number μ increase. The sensitivity of k, μ, and $\bar{U}$ to the condition described above will be examined in detail, with respect to observational data, in Section 4.

b) Figure 6(A) also shows that the amplitude of standing waves responding to forcing by topography for $k = 2$ is larger than that for $k = 1$. Conversely, the amplitude of standing waves with respect to forcing by diabatic heating is smaller for $k = 2$ than for $k = 1$. These results are in qualitative agreement with those computed by Derome and Wiin-Nielsen (1971) using a two-level model.

c) The phase of mean standing waves is tilted westward with increasing height, as shown in Fig. 6(B). This is due to the inclusion of eddy viscosity and the effect of Newtonian cooling in

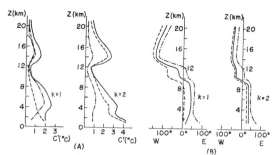

Fig. 7. As in Fig. 6, except for thermal standing waves.

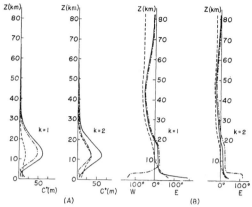

Fig. 8 As in Fig. 6, except for summer.

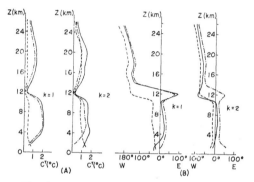

Fig. 9 As in Fig. 7, except for summer.

this model.

d) Figures 7(A) and 7(B) show the vertical distributions of amplitude and phase, respectively, for thermal standing waves in response to forcing by topography, diabatic heating and both topography and diabatic heating. In Fig. 7(A), the amplitude of thermal standing waves responding to forcing by topography is a maximum at 1.5 km, while thermal standing wave amplitude with respect to forcing by diabatic heating is a maximum near 500 mb. However, the amplitudes in both cases become minimum at around 10.5–12 km. Figure 7(B) reveals that the phase response of thermal standing waves to forcing by either the topography or the diabatic heating changes drastically, *i.e.*, the phases of thermal standing waves above 12 km and below 9 km are inversely related. Thus, the phase of thermal standing waves lags behind that of standing waves.

(II) Summer case:

a) Figures 8(A) and 8(B) are the summer counterparts of Figures 6(A) and 6(B). As shown in Fig. 8(A), the amplitude of mean standing waves responding to forcing by either topography or diabatic heating is largest at a height of about 12 km.

The mean zonal wind $\bar{U}$ at 45°N in summer (Fig. 2) is easterly (westerly) above (below) 20 km, with a subtropic jet at 12 km. Note that $\partial \bar{U}/\partial p > 0$ in the 10.5–64 km layer, $\partial \bar{U}/\partial p < 0$ below 10.5 km, and $\partial/\partial p(1/\sigma \partial \bar{U}/\partial p) > 0$ at 12 km. Hence, the mean zonal wind at 45°N in the summer may satisfy equation (30) at 12 km, and the mean standing waves at 45°N in summer cannot propagate upward above this height.

In Figure 8(A), the amplitude response of mean standing waves to forcing by topography in summer is smaller than that in winter. However, the amplitude of mean standing waves responding to forcing by diabatic heating in summer is of the same magnitude as in winter. The phase of mean standing waves responding to forcing by both topography and diabatic heating in Summer (Fig. 8(B)) shifts westward by about 50° of longitude from the winter case (see Fig. 6B).

b) Figures 9(A) and 9(B) correspond to Figs. 7(A) and 7(B) for winter described previously. As shown in Fig. 9(A), the amplitude of thermal standing waves in summer is relatively small compared with that in winter. For $k=1$, the amplitude of thermal standing wave response to forcing by diabatic heating is larger than the response to forcing by topography. As in the winter case, the phase of thermal standing waves lag behind the phase of standing wave in summer.

4. Computation of response to forcing by actual topography and observed heat sources

Derome and Wiin-Nielsen (1971) computed the amplitude and phase of mean standard pressure for zonal wave numbers 1 through 5 at

45°N (see Table 1 in their paper). Amplitude and phase values of diabatic heating at 45°N for zonal wave numbers 1 through 5 are obtained by the following method.

First, the winter and summer diabatic heating values for 30°N, 40°N 50°N and 60°N at 10° longitude intervals are obtained from the estimates of Ashe (1979), and the diabatic heating values at 45°N are determined by averaging these values. The resulting zonal distribution of diabatic heating at 45°N in winter and summer is expanded by the following Fourier series:

$$H(\lambda, \varphi) = \bar{H}(\varphi) + \sum_{k=1}^{K} [Q_k(\varphi)\cos(k\lambda) + T_k(\varphi)\sin(k\lambda)]$$
$$= \bar{H}(\varphi) + \sum_{k=1}^{K} C_k(\varphi)\cos\{k\lambda - \alpha_k(\varphi)\} \quad (31)$$

Fourier coefficients appearing in this expansion for $H(\lambda, \varphi)$ are then computed from the relations

$$\{Q_k, T_k\} = \frac{1}{\pi}\int_0^{2\pi} H(\lambda, \varphi)\{\cos(k\lambda), \sin(k\lambda)\}d\lambda,$$
$$C_k(\varphi) = [Q_k^2(\varphi) + T_k^2(\varphi)]^{1/2},$$

and

$$\alpha_k(\varphi) = \tan^{-1}[T_k(\varphi)/Q_k(\varphi)].$$

The amplitude and phase of zonal wave components 1 through 5 for standard pressure at the ground and diabatic heating are given in Table 1. Note that the largest amplitudes topography and diabatic heating in winter are found in zonal wave number 2, followed by wave number 1. However, in summer, the maximum amplitude for diabatic heating occurs in zonal wave number 1 with a secondary maximum at wave number 2.

Since we do not know the actual vertical distribution of diabatic heating in the atmosphere, we assume that it follows the distribution represented by equation (21) with the value of diabatic heating at 500 mb from Ashe (1979).

If we substitute the forcing function of actual topography, diabatic heating, and both topography and diabatic heating from Table 1, into the linear algebraic equation mentioned in section 2, we obtain the vertical distributions of amplitude and phase for mean and thermal standing waves responding to these actual forcing mechanisms.

The results of the computations are as follows:

(1) Winter case:

a) The vertical distributions of amplitude and phase response of mean standing waves to forcing by topography, diabatic heating, and both topography and diabatic heating in winter can be seen in Figs. 10(A) and (B). For comparison, the vertical distributions of amplitude and phase for standing waves during January 1964–1970, as computed by van Loon, et al. (1973) with the observed data near 45°N, are reproduced in Fig. 11. We have also calculated the mean values for the amplitude and phase of standing waves in winter (December, January, February) at 45°N by utilizing data during the period of 1971–77. (The result is not shown here.) Note that our computed results are similar to those calculated by van Loon, et al. (1973).

The computed values of standing wave amplitude for wave numbers 1–3 at 9 km, and phase from 850 to 200 mb for wave numbers 1–3, are given in Table 2. Also shown in Table 2 are the observed values.

The phases (longitude of the ridge) of standing waves for wave numbers 1–3 computed with this model are 10°, 60° and 100°E, respectively at 850 mb with shifts of 60°, 25° and 10° of longitude, respectively, to the west from 850 mb to 200 mb. On the other hand, the correspond-

Table 1. Mean amplitude and phase of the standard pressure and diabatic heating between 30°N and 60°N for zonal wave numbers 1 through 5 in winter and summer.

Zonal wave number k	Standard pressure at the ground		Diabatic heating in Winter		Diabatic heating in Summer	
	C_k (cb)	α_k (deg.)	$C_k(10^{-6}KJ\cdot Kg^{-1}\cdot s^{-1})$	α_k (deg.)	$C_k(10^{-6}KJ\cdot Kg^{-1}\cdot s^{-1})$	α_k (deg.)
1	4.22	−93.2 (266.8)	5.26	256.6	3.38	109.8
2	5.21	−17.2 (342.8)	7.27	330.5	3.37	197.8
3	2.71	−220.7 (139.3)	3.13	190.5	1.27	50.6
4	2.34	−189.6 (170.4)	2.32	213.6	1.75	21.7
5	2.28	−66.4 (293.6)	0.46	76.3	0.47	318.9

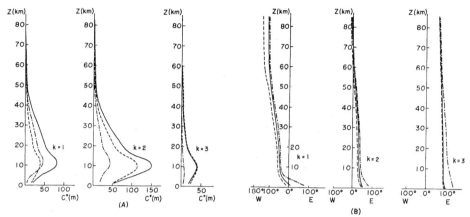

Fig. 10 The vertical distributions of amplitude (A) and phase (B) for standing waves responding to forcing by actual topography (dashed curves), actual diabatic heating (dot-dashed curves), and both the actual topography and actual diabatic heating (solid curves) in winter.

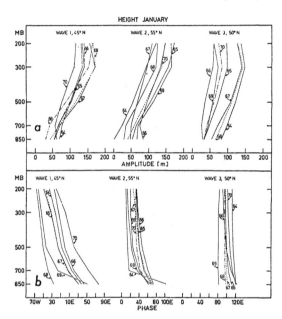

Fig. 11 The vertical distributions of amplitude (in meters) and phase for wave numbers 1–3 of geopotential height for January 1964–1970 (After van Loon et al., 1973).

ing 850 mb phases computed from the observed data are 30°, 60° and 100°E, respectively. These phases exhibit westward shifts of 80°, 30° and 0° of longitude, respectively, from 850 to 200 mb. Thus, we find that the westward tilt of standing waves computed with this model decreases as the zonal wave number increases, and the amplitude of standing waves increases from 850 to 200 mb. We may then be able to conclude from Figs. 10–11 and Table 2 that the vertical distributions of amplitude and phase for standing waves computed with this linear model are in rather good agreement with observed standing waves. However, the amplitude of wave number 3 seems to be less than the observed value.

Figure 12 shows the zonal distributions of mean standing waves responding to forcing by topography, diabatic heating and both topography and diabatic heating computed from equation (17) using $k=5$ at 700 mb, 500 mb, and 250 mb. At these three levels, the forced pattern consists of a major trough near 140°E, a ridge at about 130°W and a trough near 60°W. The major trough in Fig. 12 nearly coincides with the eastern coast of Asia, and shifts westward by 10° of longitude from 700 to 250 mb. Near the west coast of North America, a ridge tilts to the west from 130°W at 700 mb to 140°W at 250 mb. Similarly, the second major trough slopes from 60°W at 700 mb to 70°W at 250 mb in the vicinity of the eastern coast of North America. We also find that these troughs tend to occur near the regions of large-scale heating, with the ridge occurring regions of large-scale cooling, in winter. As shown in Figs. 10 and 12, the response to forcing by topography is larger than the response to forcing by diabatic heating in winter.

Figure 13 compares the results of this model with those computed by Derome and Wiin-Nielsen (1971) using their two-layer model, and

Table 2. The amplitude of standing waves for wave numbers 1–3 at 9 km, and vertical distributions of phase for wave numbers 1–3 from 850 mb to 200 mb, computed with this linear model and from the observed data.

Zonal wave number k	Amplitude at 9 km height (m)		phase from 1.5 km to 12 km height	
	Computation	observation	Computation	observation
1	75	100–160 (100–170)	10°E–50°W	30°E–50°W (25°E–35°E)
2	147	100–160 (75–150)	60°E–35°E	60°E–30°E (62°E–37°E)
3	39	50–140 (85–140)	100°E–90°E	100°E (105°E–95°E)

* Values in brackets are amplitudes at 9 km height and phase from 1.5 km to 12 km height, computed from the observational data for the period of 1971–1977.

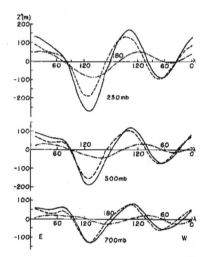

Fig. 12 The zonal distributions of standing waves ($K=5$) responding to forcing by topography (dashed curves), diabatic heating (dot-dashed curves) and both topography and diabatic heating (solid curves) in winter at 250 mb (top), 500 mb (middle), and 700 mb (bottom).

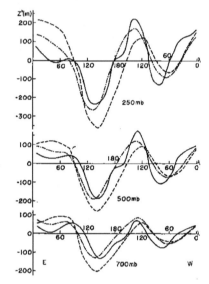

Fig. 13 The zonal distributions of standing waves ($K=5$) computed by this model (dot-dashed curves), Derome (1971) with two-levels (dashed curves), and the observed heights of standing waves for January averaged between 30°N and 60°N (solid curves), at 250 mb (top), 500 mb (middle), and 700 mb (bottom).

the observed heights of standing waves for January (averaged between 30°N and 60°N). From this figure, we find that the results from this model are far better than those of Derome and Wiin-Nielsen (1971), *i.e.*, the trough and ridge positions, as well as heights, of mean standing waves computed with this model are in good agreement with the observed standing waves. Therefore, we may be able to conclude from these results that multi-levels model are far better than simple layer models, such as the 2-level model, in the winter case.

b) Figures 14(A) and (B) are the thermal standing wave counterparts of Figs. 10(A) and 10(B), respectively, while Fig. 15 shows the vertical distributions of amplitude and phase for actual thermal standing waves computed by van Loon *et al.* (1973) with Fourier expansion from observed data in January 1964–1970. We have also calculated the mean values of amplitude and phase for thermal standing waves in winter (December, January, February) at 45°N, using data during the period of 1971–1976. (The result

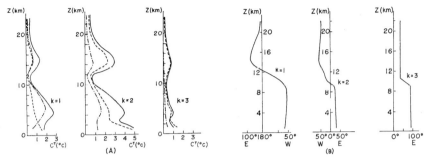

Fig. 14 The vertical distributions of amplitude (A) and phase (B) of thermal standing waves responding to forcing by topography (dashed curves), diabatic heating (dot-dashed curves), and both topography and diabatic heating (solid curves).

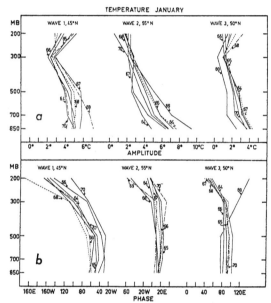

Fig. 15 As in Fig. 11, except for temperature in degrees Celsius. (After van Loon et al., 1973).

is not displayed here.) The amplitude of thermal standing waves for wave numbers 1–3 at 4.5 km, and the vertical distributions of the phase of thermal standing waves for wave numbers 1–3 at 1.5–12 km, computed from this linear model and the observational data are summarized in Table 3.

We find that the 850 to 250 mb phases (longitudes of the ridges) for wave numbers 1–3 computed by this model are 60°W–160°W, 25°E–35°W, and 90°E–35°E, respectively, while the corresponding vertical distributions of phase for observed thermal standing waves from 850 to 200 mb are 50°W–160°W, 20°E–20°W, and 100°E–60°E, respectively. From the results of this model, the largest phase change in thermal standing waves occurs near the tropopause. The amplitude of thermal standing waves decreases from 850 mb to 200 mb for wave numbers 2 and 3, but is maximum at 500 mb for wave number 1. Thus, the vertical distributions of amplitude and phase for thermal standing waves computed from this linear model are in good agreement with the observational results. However, the computed amplitudes for wave numbers 1 and 3

Table 3. The amplitude of thermal standing waves for wave numbers 1–3 in 1.5–12 km, computed with this linear model and from the observed data.

Zonal wave number k	Amplitude at 4.5 km height (°C)		phase from 1.5 km to 12 km height	
	Computation	observation	Computation	observation
1	2.8	4.0–6.0 (3.7–5.4)	60°W–160°W	50°W–160°W (55°W–88°W)
2	4.0	3.5–6.0 (2.5–3.7)	25°E–35°W	20°E–20°W (20°E–0°E)
3	1.0	1.5–3.0 (2.1–3.2)	90°E–35°E	100°E–60°E (95°E–85°E)

* Values in brackets are amplitudes at 4.5 km height and phase from 1.5 km to 12 km height, computed from the observational data for the period of 1971–1976.

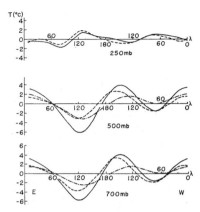

Fig. 16 As in Fig. 12, except for thermal standing waves.

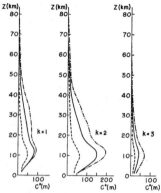

Fig. 17 The vertical distributions of amplitude for standing waves forced by both topography and diabatic heating, using three different values of μ; namely $0.856\times10^{-6}\,\mathrm{m}^{-1}$ (dot-dashed curves), $0.95\times10^{-6}\,\mathrm{m}^{-1}$ (solid curves), and $1.13\times10^{-6}\,\mathrm{m}^{-1}$ (dashed curves) corresponding to meridional wave lengths of $Ly=66°$, $60°$ and $50°$ of latitude, respectively, in winter, for $k=1$ (left), 2 (center), and 3 (right).

are relatively small compared with the observed amplitudes (especially for wave number 3).

The zonal distributions of thermal standing waves ($K=5$) in response to forcing by topography, diabatic heating and both topography and diabatic heating are shown in Fig. 16 for 700, 500 and 250 mb, respectively. Forcing effects by diabatic heating are almost the same as forcing by topography in the middle and lower troposphere. Trough or ridge positions of thermal standing waves lag behind the position of the mean standing waves, as the troughs and ridges in the troposphere are inversely related to those in the stratosphere, i.e., the ridge of a thermal standing wave in the stratosphere corresponds to the trough of a standing wave in the troposphere. This result is in good agreement with observations, but the computed amplitudes of thermal standing waves in the stratosphere are relatively small compared to the observed values.

c) Figure 17 shows the vertical distributions of amplitude for mean standing waves forced by both topography and diabatic heating, using three different values of μ; namely, 0.856×10^{-6}, 0.950×10^{-6} and $1.130\times10^{-6}\,\mathrm{m}^{-1}$. These values of μ correspond to meridional wavelengths of $Ly=66°$, $60°$ and $50°$ of latitude, respectively. The corresponding zonal distributions of mean standing waves at 700, 500 and 250 mb are reproduced in Fig. 18. As may be seen in Figs. 17 and 18, the value of μ is related to different amplitudes and phases. In the cases of $Ly=50°$ and $60°$ of latitude, the two sets of curves are very similar in phase. However, the curves of $Ly=66°$ of latitude, differ appreciably from the others.

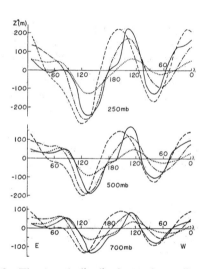

Fig. 18 The zonal distributions of standing waves forced by both topography and diabatic heating, using three different values of μ; namely, $0.856\times10^{-6}\,\mathrm{m}^{-1}$ (dashed curves), $0.950\times10^{-6}\,\mathrm{m}^{-1}$ (dot-dashed curves), and $1.13\times10^{-6}\,\mathrm{m}^{-1}$ (dot curves) at 250 mb (top), 500 mb (middle), and 700 mb (bottom). The solid curves show the respective observed standing waves in January.

Figures 17 and 18 show that the amplitude response to forcing is highly dependent upon the meridional wavelength, with $Ly=60°$ of latitude yielding the most reasonable results when com-

Table 4. The reflected height of standing waves forced by both topography and diabatic heating, using different values of μ.

μ (m^{-1}) \ k	1	2	3	4	5
0.471×10^{-6} ($L_y = 120°$)	45.0 km	39.0 km	33.0 km	10.5 km	9.0 km
0.856×10^{-6} ($L_y = 66°$)	12.0	12.0	9.0	9.0	9.0
0.950×10^{-6} ($L_y = 60°$)	12.0	10.5	9.0	9.0	7.5
1.130×10^{-6} ($L_y = 50°$)	10.5	9.0	9.0	9.0	1.5

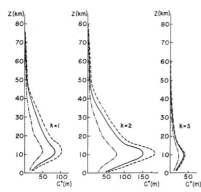

Fig. 19 The vertical distributions of standing waves in winter, forced by both topography and diabatic heating, using three different eddy viscosity coefficients, namely, $K_m = 0.7 \times 10^6$ m^2 s^{-1} (dashed curves), 1.0×10^6 m^2 s^{-1} (solid curves) and 2.0×10^6 m^2 s^{-1} (dot-dashed curves), for $k = 1$ (left), 2 (center), and 3 (right).

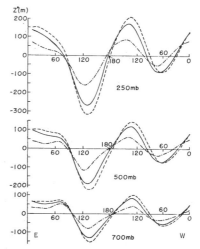

Fig. 20 The zonal distributions of standing waves in winter, forced by both topography and diabatic heating, using three different eddy viscosity coefficients; namely $K_m = 0.7 \times 10^6$ m^2 s^{-1} (dashed curves), 1.0×10^6 m^2 s^{-1} (solid curves) and 2.0×10^6 m^2 s^{-1} (dot-dashed curves), at 250 mb (top), 500 mb (middle), and 700 mb (bottom).

pared with observations. The reason why the value of $L_y = 60°$ gives reasonable results should be re-examined in the future. The dependency of μ on the estimation of response amplitude comes from the vertical propagation character of planetary waves, as discussed in section 3, [refer to equation (30)]. The reflected height of standing waves forced by both topography and diabatic heating, using different values of μ, is computed from equation (30) and shown in Table 4. Here, we find that the longer the meridional wavelength, the higher the reflected height.

d) The vertical and zonal distributions of mean standing waves forced by both topography and diabatic heating, using three different eddy viscosity coefficients (see Fig. 3) are shown in Figs. 19 and 20, respectively. Note that a larger eddy viscosity results in a smaller standing wave amplitude and steeper westward tilt. Of the three eddy viscosity coefficients which we have used, the eddy viscosity coefficient of $K_m = 1.0 \times 10^6$ m^2 s^{-1} in 1.5–30 km height yields the most reasonable results.

(II) *Summer case:*

Compared to the case of winter, the simulation of standing waves forced by topography and diabatic heating during summer has rarely been attempted. This is due to a lack of exact information pertaining to standing wave generation mechanism in summer. This itself indicates the importance of further study in this particular area. In this study, an attempt was made to compute the response of standing waves to forcing by both topography and diabatic heating in summer with this linear model, using the vertical distributions of the mean zonal wind at 45°N in summer. The computed results are as follows:

Figures 21(A) and (B) show the vertical distribution of amplitude and phase for mean standing waves responding to forcing by topography,

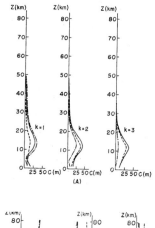

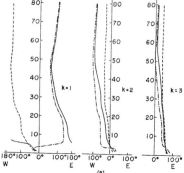

Fig. 21 As in Fig. 10, except for summer.

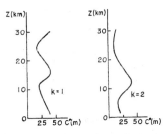

Fig. 22 The vertical distributions of amplitude of standing waves for wave numbers 1 and 2 at 45°N in summer, computed by Iwashima (1980) from the observed data during the period of 1965–1977.

diabatic heating, and both topography and diabatic heating in summer. The vertical distributions of mean standing wave amplitude for wave numbers 1 and 2 at 45°N in summer computed by Iwashima (1980) with Fourier expansion from the observed data during period of 1965–1977 are reproduced in Fig. 22. A comparison of Figs. 21 and 22 reveals that the amplitude of standing waves computed with this linear model is in rather good agreement with the observed values. Figure 21 also shows that the amplitude of mean standing waves for $k=1$ decreases with increasing height below the middle troposphere, and then increases from the middle troposphere to 12 km, while the amplitude for $k=2$ and 3 increases from 1.5 km to 12 km. These computed vertical distributions qualitatively resemble the observed distributions in summer.

Figure 23 shows the zonal distributions of mean standing waves responding to forcing by topography, diabatic heating, and both topography and diabatic heating at 700, 500, 300 and 200 mb, respectively, computed for summer from equation (17) using $K=5$. The zonal distributions of mean standing waves computed with Fourier expansion, using $K=5$, from the observed data during the period of 1951–1961 (from the Berliner Wetterkarte) are displayed in Fig. 24. The positions of three troughs (located over Baykal, North America and the west coast of Europe) and ridges (located over the Ural Mountains, the Rocky Mountains and the Atlantic Ocean) in Figs. 23 and 24 are agree rather well. However, the computed intensity of the trough over the central Pacific Ocean is especially poor compared with the observational data.

b) The thermal standing wave counterparts of Figs. 21(A) and 21(B) are shown in Figs. 25(A) and 25(B), respectively. Here, we find that thermal standing waves lag behind standing waves, with the phase of thermal standing waves above the tropopause inversely related to that in the troposphere. The zonal distributions of

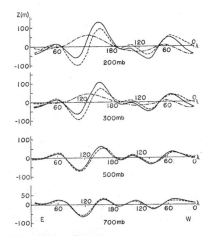

Fig. 23 As in Fig. 12, except for summer at 200 mb (top), 300 mb (upper-middle), 500 mb (lower-middle), and 700 mb (bottom).

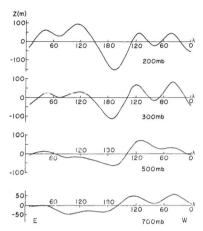

Fig. 24 The zonal distributions of standing waves in July computed with Fourier expansion, using $K=5$, from the observed data, averaged from 1951 to 1961.

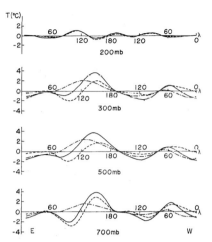

Fig. 26 As in Fig. 16, except for summer at 200 mb (top), 300 mb (upper-middle), 500 mb (lower-middle), and 700 mb (bottom).

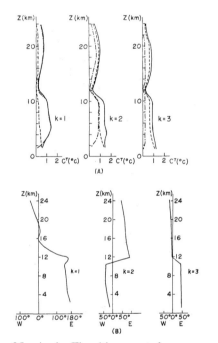

Fig. 25 As in Fig. 14, except for summer.

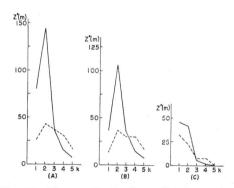

Fig. 27 The amplitude of standing waves for wave numbers 1–5 at 12 km height responding to forcing by (A) both topography and diabatic heating, (B) topography, and (C) diabatic heating in winter (solid curves) and summer (dashed curves).

thermal standing wave response to forcing by topography, diabatic heating, and both topography and diabatic heating in summer are displayed in Fig. 26 for 700, 500, 300 and 200 mb, respectively.

c) Zonal distributions of standing waves in response to forcing by both topography and diabatic heating in summer were computed by using two different values of μ; namely, 0.856×10^{-6} m^{-1}, 0.950×10^{-6} m^{-1}, which correspond to meridional wavelengths of 66° and 60° of latitude, respectively. The influence of μ on the estimation of standing waves, however, was small when compared with the winter case described earlier.

Since we have independently discussed the response of mean and thermal standing waves to forcing by topography and diabatic heating in winter and summer for $k=1\sim 3$, these results are summarized in Figs. 27 and 28 to understand the response of different wave numbers. Figures 27(A), 27(B) and 27(C) show the ampli-

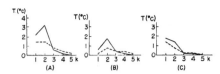

Fig. 28 As in Fig. 27, except for thermal standing waves. The amplitude of thermal standing waves is the value at 500 mb.

tude of standing waves for wave numbers 1–5 at 12 km responding to forcing by both topography and diabatic heating, topography, and diabatic heating, respectively, in winter and summer. Similarly, the amplitude response of thermal standing wave, respectively, at 500 mb in winter and summer are presented in Figs. 28(A), 28(B) and 28(C). Here, we will present a brief summary of Figs. 27 and 28, which are representative of Figs. 12, 16, 23 and 26.

a) The amplitude of mean and thermal standing wave response to forcing by topography is maximum for wave number 2 at 45°N. On the other hand, the amplitude for forcing by diabatic heating is largest for wave number 1 in both winter and summer.

b) The amplitude of mean standing waves responding to forcing by either topography or diabatic heating at 45°N is maximum in the lower layer of the stratosphere, while the amplitude of thermal standing waves is maximum within the lower layers of the troposphere in both winter and summer.

c) The amplitude of standing and thermal standing waves in response to forcing by topography and diabatic heating for $k=1$ and 2 at 45°N in winter is much larger than that in summer.

d) The amplitude of standing and thermal standing waves responding to forcing by both topography and diabatic heating is maximum for $k=2$, and decreases rather rapidly as k increases in winter. However, in summer, this amplitude decreases rather slowly with increasing k, i.e., the amplitudes of standing and thermal standing waves for wave numbers 3, 4 are not as small as in winter.

5. Conclusions

A 34-level quasi-geostrophic, linear steady-state model with eddy viscosity and the effect of Newtonian cooling included, is used to obtain the vertical and zonal distributions of mean and thermal standing waves at 45°N in response to forcing by topography, diabatic heating, and both topography and diabatic heating, respectively, in winter and summer. The results of this model indicate that the vertical and zonal distributions of mean and thermal standing waves in winter are in good agreement with the corresponding distributions of the observed mean and thermal standing waves. Furthermore, these results are far better than those computed by a two-level model (Derome and Wiin-Nielsen, 1971). However, there are some differences between the computed and observed mean standing waves in summer. Although the computed positions of troughs and ridges are in rather good agreement with the observed positions of the mean standing wave, a large difference occurs over the central Ocean, i.e., the computed trough over this region is far weaker than the observed one.

It has also been shown that eddy viscosity and the meridional wavelengths of the topography and diabatic heating seem to have a larger influence on the response of a model atmosphere to forcing by topography and diabatic heating. The best agreement with observations in winter occurs in the vertical and zonal distributions of amplitude and phase for standing waves responding to forcing with a meridional wavelength of $Ly=60°$ of latitude. However, this agreement between the computed values and observations is confined to the troposphere and lower stratosphere. In the middle and upper stratosphere, comparison of the computed results with the observations is not satisfactory. Thus, we computed the response to forcing by both the actual topography and the actual diabatic heating when the meridional wavelength is $Ly=120°$ of latitude, instead of $Ly=60°$ of latitude. These computed results are in agreement with the observations in the stratosphere. However, the vertical and the zonal distributions of mean standing waves computed with this meridional wavelength are not compatible with the observations in the troposphere. Therefore, the problem of how to select a suitable meridional wavelength remains to be solved.

The vertical and horizontal distributions of diabatic heating used in this study are based on the assumptive distribution. Although this distribution seems to be suitable over the Tibetan Plateau, the vertical and horizontal distributions of heating/cooling over the ocean should be re-examined, particularly in summer.

Some of the main deficiencies in this investigation are the use of β-plane geometry, the assumptions that the basic zonal winds do not change in the meridional direction, and the use of a fixed meridional wavelength. Therefore, further studies should be aimed at removing these restrictions by adopting a three-dimensional treatment of the subject.

Acknowledgements

The authors are grateful to Prof. T. Matsuno and Dr. H. Nakamura for their comments and suggestions during the course of this study. Thanks are due to Dr. T. Iwashima for use of his unpublished results in summer 1965–1977. Many thanks are also due to Profs. N. Saito and Takio Murakami for critical reading of the original manuscript. Thanks are also extended to Mr. Y. Fujiki for drawing the figures and Mrs. K. Kudo for her expert typing of the manuscript. This paper is sponsored in part by Grant in Aid for Scientific Research from the Ministry of Education.

References

Ashe, S., 1979: A nonlinear model of the time-average axially asymmetric flow induced by topography and diabatic heating. *Journal of the Atmospheric Sciences*, **36**, 109–126.

Bolin, B., 1950: On the influence of the earth's orography on the general character of the westerlies. *Tellus*, **2**, 184–195.

Charney, J. G., and A. Eliassen, 1949: A numerical method for predicting the perturbations in the middle-latitude westerlies. *Tellus*, **1**, 38–54.

———, and P. G. Drazin, 1961: Propagation of planetary-scale disturbances from the lower into the upper atmosphere. *Journal of Geophysical Research*, **66**, 83–109.

Derome, J. F., and A. Wiin-Nielsen, 1971: The response of a middle latitude model atmosphere to forcing by topography and stationary heat sources. *Monthly Weather Review*, **99**, 564–576.

Dickinson, R. E., 1973: Method of parameterization for infrared cooling between altitudes of 30 to 70 kilometers. *Journal of Geophysical Research*, **78**, 4451–4457.

Döös, Bo. R., 1962: The influence of exchange of sensible heat with the earth's surface on the planetary flow. *Tellus*, **14**, 133–147.

Iwashima, T., 1980: Analysis of stationary waves in summer (present at the annual meeting of the Meteorological Society of Japan in March, 1980).

Gambo, K., 1956: The topographical effect upon the jet stream in the westerlies. *Journal of the Meteorological Society of Japan*, **34**, 24–28.

Kirkwood, E., and J. Derome, 1977: Some effects of the upper boundary condition and vertical resolution on modeling forced stationary planetary waves. *Journal of the Atmospheric Sciences*, **105**, 1239–1251.

Murakami, T., 1963: The topographic effects in the three-level model of the s-coordinate, papers in meteorology and geophysics. *Tokyo Meteorological Research Institute, Japan*, **14**, 144–150.

———, 1972: Equatorial stratospheric waves induced by diabatic heat sources. *Journal of the Atmospheric Sciences*, **29**, 1129–1137.

Murgatroyd, R. J., 1969: The structure and dynamics of the stratosphere. The Global Circulation of the Atmosphere, 155–195.

Nakamura, H., 1976: Some problems in reproducing planetary waves by numerical models of the atmosphere. *Journal of the Meteorological Society of Japan*, **54**, 129–146.

Saltzman, B., 1968: Surface boundary effects on the general circulation and macroclimate: A review of the theory of the quasi-stationary perturbations in the atmosphere. *Meteorological Monographs*, **8**, 4–19.

Sankar-Rao, M., 1965a: Continental elevation influence on the stationary harmonics of the atmospheric motion. *Pure and Applied Geophysics*, **60**, 141–159.

———, 1965b: Finite difference models for the stationary harmonics of atmospheric motion. *Monthly Weather Review*, **93**, 213–224.

———, 1965c: On the influence of the vertical distribution of stationary heat sources and sinks in the atmosphere. *Monthly Weather Review*, **93**, 417–420.

Smagorinsky, J., 1953: The dynamical influence of large-scale heat sources and sinks on the quasi-stationary mean motions of the atmosphere. *Quartary Journal of the Royal Meteorological Society*, **79**, 342–366.

Staff Members, Academia Sinica 1958: On the general circulation over eastern Asia. *Tellus*, **10**, 299–312.

Van Loon, H., R. L. Jenne and K. Labitzke, 1973: Zonal harmonic standing waves. *Journal of Geophysical Research*, **78**, 4463–4471.

Webster, P. J., 1972: Response of the tropical atmosphere to local, steady forcing. *Monthly Weather Review*, **100**, 518–541.

Wiin-Nielsen, A., and J. Sela, 1971: On the transport of quasi-geostrophic potential vorticity. *Monthly Weather Review*, **99**, 447–459.

The Response of a Hemispheric Multi-Level Model Atmosphere to Forcing by Topography and Stationary Heat Sources

Part I Forcing by Topography*

Huang Ronghui and K. Gambo

Abstract

The standing waves responding to forcing by the Hemispheric topography are investigated by means of a quasi-geostrophic, steady state, 34-level model, with Rayleigh friction, the effect of Newtonian cooling and the horizontal kinematic thermal diffusivity included in a spherical coordinate system.

The results computed by this model show that the topography at high latitudes, such as Greenland Plateau, plays an important role in the standing planetary waves responding to forcing by the Hemispheric topography, and the anomaly of standing waves is connected with a zonal mean wind at the surface at 70°–85°N in winter. The computed results show also that the standing waves responding to forcing by the Hemispheric topography can be propagated vertically and laterally toward the region of larger refractive index, i.e., toward the region of weaker westerly winds at low latitudes.

The amplitudes are maxima at 38 km height, 60°N for wave number 1, and at 27 km height, 60°N for wave number 2. In addition, they have secondary peaks in the upper troposphere at 20°–30°N. This may be considered as one of reasons for the formation of standing waves in the upper troposphere at low latitudes.

1. Introduction

The response of a model atmosphere in middle latitudes to forcing by topography and stationary heat sources was investigated by authors (1981) with a β-plane approximate multi-level model. The computed results are in good agreement with observed results. In that paper, however, we assumed that the motion takes place on a β-plane centered at 45°N and the zonal mean wind is constant with respect to latitude. On the other hand, Dickinson (1968a) emphasized the role of horizontal wind shears in the vertical propagation of standing waves. Thus, the response of a hemispheric model atmosphere to forcing by topography and stationary heat sources must be discussed in a model where both the vertical and horizontal wind shears are considered.

The response of a hemispheric model atmosphere to forcing by topography has been studied by many authors. For example, the Staff Members, Academia Sinica (1958) studied the effects of forcing by topography and stationary heat sources to formation of standing troughs and ridges with a two-level quasi-geostrophic model. Egger (1976) investigated the linear response of a hemispheric model atmosphere to forcing by topography by means of a two level primitive equation model. Ashe (1979) discussed that the nonlinear terms in the mean-state equation are important when we attempt to explain the effects of the flow moving around the mountains though he used a two-level model.

In these simplified models such as the two-level model mentioned above, the effect of vertically propagating stationary planetary waves excited in the troposphere may not be treated correctly in the stratosphere, due to the poor vertical resolution in the stratosphere (Huang and Gambo, 1981). This means that the vertical

---

* 本文原载于: Journal of the Meteorological Society of Japan, Vol. 60, No.1, 78-92, 1982.

distributions of the amplitude and the phase of standing planetary waves responding to forcing by the northern hemispheric topography and stationary heat sources may not also be obtained correctly.

Matsuno (1970) showed that planetary-scale, stationary disturbances in the winter stratosphere are considered to be upward propagating internal Rossby waves forced from below. In his computation a multi-level model in the spherical coordinate system was used and the observed value of height at 500 mb level was used as the lower boundary condition. Pyle and Rogers (1980) examined the importance of chemical processes in the stratospheric transport by stationary planetary waves with the aid of Matsuno's model, but the lower boundary condition was taken at 10 km height. In these papers the use of a multi-level model was emphasized to explain the characteristic feature of stationary planetary waves in the stratosphere. However, the discussion was not extended to the process how the lower boundary condition was generated.

In this paper we assume that the stationary planetary waves in winter are induced by the forcing by topography and stationary heat sources. In the numerical computation of the response of a model atmosphere to forcing by topography and stationary heat sources we use a hemispheric 34-level model in the spherical coordinate system.

In Part I of the paper, we examine the response of a model atmosphere to forcing by topography. In Part II of the coming paper the forcing by stationary heat sources will be discussed. Generally speaking, the theoretical treatment of stationary planetary waves to the forcing by topography was confined to the discussion of the topographical effect due to the Tibetan Plateau and the Rocky Mountains. Few discussions were done about the effect of topography at high latitudes, such as the Greenland Plateau. In this circumstance we focus on the role of the Greenland Plateau.

2. The model and parameters

In order to investigate the vertical and horizontal distribution of standing planetary waves forced by the northern hemispheric topography and stationary heat sources, we adopt a spherical coordinate system, with λ as longitude and φ as latitude. In addition, we shall use a pressure coordinate p as the vertical coordinate. As mentioned in the introduction, a quasi-geostrophic, linear steady-state model is utilized with Rayleigh friction, the effect of Newtonian cooling and the horizontal kinematic thermal diffusivity incorporated in the model. To describe the behavior of motion in the stratosphere correctly, a 34-level representation of the atmosphere is used.

1) Model

The steady state, quasi-geostrophic vorticity and the thermodynamic equations in which Rayleigh friction, the effect of Newtonian cooling and the horizontal kinematic thermal diffusivity are included, may be expressed as

$$\bar{U}\frac{\partial}{a\cos\varphi\partial\lambda}(\zeta')+v'\frac{\partial}{a\partial\varphi}(\bar{\zeta}+f)$$
$$=f\frac{\partial\omega}{\partial p}-R_f\zeta', \qquad (1)$$

and

$$\bar{U}\frac{\partial}{a\cos\varphi\partial\lambda}\left(\frac{\partial\phi'}{\partial p}\right)-2\Omega_0\sin\varphi\frac{\partial\bar{U}}{\partial p}v'+\sigma\omega$$
$$=-\frac{RH}{c_p p}-\alpha_R\frac{\partial\phi'}{\partial p}+K_T\nabla^2\left(\frac{\partial\phi'}{\partial p}\right), \qquad (2)$$

respectively. The notations used in the above equations are as follows:

a : radius of the earth
$\bar{U}$: the basic zonal wind speed
v' : meridional component of perturbation motion
ϕ' : geopotential of perturbation
ζ' : vertical component of relative perturbation vorticity
$\bar{\zeta}$: vertical component of relative vorticity of the basic state
$\sigma=-\alpha\frac{\partial\ln\theta}{\partial p}$: static stability parameter (α: specific volume, θ: potential temperature)
H : diabatic heating per unit time and unit mass
R : gas constant (0.287 KJ·kg^{-1}·deg^{-1})
c_p : specific heat at constant pressure (1.004 KJ·kg^{-1}·deg^{-1})
α_R : Newtonian cooling coefficient
K_T : horizontal kinematic thermal diffusivity
R_f : Rayleigh friction coefficient of perturbation
f : Coriolis parameter
ω : vertical p-velocity (dp/dt)

$\zeta', \bar{\zeta}, \nabla^2, v'$ in a spherical coordinate system may be expressed as follows:

$$\zeta' = \frac{1}{2\Omega_0 \sin\varphi} \frac{1}{a^2} \left[\frac{\sin\varphi}{\cos\varphi} \frac{\partial}{\partial\varphi} \left(\frac{\cos\varphi}{\sin\varphi} \frac{\partial\phi'}{\partial\varphi} \right) \right.$$
$$\left. + \frac{1}{\cos^2\varphi} \frac{\partial^2\phi'}{\partial\lambda^2} \right], \tag{3}$$

$$\bar{\zeta} = \frac{1}{2\Omega_0 \sin\varphi} \frac{1}{a^2} \left[\frac{\sin\varphi}{\cos\varphi} \frac{\partial}{\partial\varphi} \left(\frac{\cos\varphi}{\sin\varphi} \frac{\partial\bar{\phi}}{\partial\varphi} \right) \right.$$
$$\left. + \frac{1}{\cos^2\varphi} \frac{\partial^2\bar{\phi}}{\partial\lambda^2} \right], \tag{4}$$

$$\nabla^2 = \frac{1}{a^2} \left[\frac{\partial^2}{\partial\varphi^2} - \tan\varphi \frac{\partial}{\partial\varphi} + \frac{1}{\cos^2\varphi} \frac{\partial^2}{\partial\lambda^2} \right], \tag{5}$$

$$v' = \frac{1}{2\Omega_0 \sin\varphi} \frac{1}{a\cos\varphi} \frac{\partial\phi'}{\partial\lambda}, \tag{6}$$

where Ω_0 is rotation rate of the earth (7.29×10^{-5} sec^{-1}).

When equations (3) and (4) are substituted into the equation (1), we obtain the following vorticity equation

$$\hat{\Omega} \frac{\partial}{\partial\lambda} \left[\frac{1}{2\Omega_0 \sin\varphi} \frac{1}{a^2} \left\{ \frac{\sin\varphi}{\cos\varphi} \frac{\partial}{\partial\varphi} \left(\frac{\cos\varphi}{\sin\varphi} \frac{\partial\phi'}{\partial\varphi} \right) \right. \right.$$
$$\left. \left. + \frac{1}{\cos^2\varphi} \frac{\partial^2\phi'}{\partial\lambda^2} \right\} \right] + \frac{1}{a} qv' = f \frac{\partial\omega}{\partial p}$$
$$- R_f \frac{1}{2\Omega_0 \sin\varphi} \frac{1}{a^2} \left[\frac{\sin\varphi}{\cos\varphi} \frac{\partial}{\partial\varphi} \left(\frac{\cos\varphi}{\sin\varphi} \frac{\partial\phi'}{\partial\varphi} \right) \right.$$
$$\left. + \frac{1}{\cos^2\varphi} \frac{\partial^2\phi'}{\partial\lambda^2} \right], \tag{7}$$

Here q is expressed as

$$q = \left[2(\Omega_0 + \hat{\Omega}) - \frac{\partial^2 \hat{\Omega}}{\partial\varphi^2} + 3\tan\varphi \frac{\partial\hat{\Omega}}{\partial\varphi} \right] \cos\varphi, \tag{8}$$

where $\hat{\Omega}$ is defined as

$$\hat{\Omega} = \frac{\bar{u}}{a\cos\varphi}, \tag{9}$$

If we include the second approximation to v' in the planetary vorticity advection term, as mentioned by Matsuno, we can get an equation from which a reasonable energy equation follows. Thus, we divide the term of $(1/a)qv'$ in (7) into the two parts

$$\frac{1}{a}qv' = \frac{2\Omega_0 \cos\varphi}{a}v' + (\text{remainder}), \tag{10}$$

The first term on the right-hand side expresses advection of planetary vorticity by the north-south wind and has been shown to be dominant in the vorticity equation of planetary scale of disturbances. If we divide the meridional component of the perturbation wind into the component of geostrophic wind and the component of nongeostrophic wind (second approximation), we can obtain the second approximation to v'

$$v' = \frac{1}{2\Omega_0 \sin\varphi} \left(\frac{1}{a\cos\varphi} \frac{\partial\phi'}{\partial\lambda} \right.$$
$$\left. - \hat{\Omega} \frac{1}{2\Omega_0 \sin\varphi} \frac{\partial^2\phi'}{\partial\varphi\partial\lambda} \right), \tag{11}$$

Thus, we obtain the following vorticity equation, by substituting (11) into (7)

$$\hat{\Omega} \frac{\partial}{\partial\lambda} \left[\frac{1}{2\Omega_0 \sin\varphi} \frac{1}{a^2} \left\{ \frac{\sin^2\varphi}{\cos\varphi} \frac{\partial}{\partial\varphi} \left(\frac{\cos\varphi}{\sin^2\varphi} \frac{\partial\phi'}{\partial\varphi} \right) \right. \right.$$
$$\left. \left. + \frac{1}{\cos^2\varphi} \frac{\partial^2\phi'}{\partial\lambda^2} \right\} \right] + \frac{1}{a} q \frac{1}{2\Omega_0 \sin\varphi}$$
$$\times \frac{1}{a\cos\varphi} \frac{\partial\phi'}{\partial\lambda} = f\frac{\partial\omega}{\partial p} - R_f \frac{1}{2\Omega_0 \sin\varphi} \frac{1}{a^2}$$
$$\times \left[\frac{\sin\varphi}{\cos\varphi} \frac{\partial}{\partial\varphi} \left(\frac{\cos\varphi}{\sin\varphi} \frac{\partial\phi'}{\partial\varphi} \right) + \frac{1}{\cos^2\varphi} \frac{\partial^2\phi'}{\partial\lambda^2} \right], \tag{12}$$

On the other hand, we get the following thermodynamic equation, by substituting (9) into (2),

$$\hat{\Omega} \frac{\partial}{\partial\lambda} \left(\frac{\partial\phi'}{\partial p} \right) - \left(\frac{\partial\hat{\Omega}}{\partial p} \right) \frac{\partial\phi'}{\partial\lambda} + \sigma\omega$$
$$= \frac{-RH}{c_p p} - \alpha_R \frac{\partial\phi'}{\partial p} + K_T \frac{1}{a^2} \left[\frac{\partial^2}{\partial\varphi^2} \right.$$
$$\left. + \tan\varphi \frac{\partial}{\partial\varphi} + \frac{1}{\cos^2\varphi} \frac{\partial^2}{\partial\lambda^2} \right] \left(\frac{\partial\phi'}{\partial p} \right) \tag{13}$$

The model equations are constituted from equations (12) and (13). The vertical structure of the model used in this study is the same as that discussed in the paper of the β-plane approximate model (Huang and Gambo, 1981),* i.e., the atmosphere from $p = p_t$ (p_t: pressure at the top of the model) to $p = p_s$ (p_s: surface pressure) is divided into N ($N = 34$) layers, such as $n = 1, 2, \ldots, N$. The level of $n = 0$ corresponds to that of $p = p_t$ while the level of $p = p_s$ to $n = N+1$. In the numerical computation, the vorticity equations included the effect of Rayleigh friction are specified at $n-1/2$ and $n+1/2$ levels, while the thermodynamic equations included the horizontal kinematic thermal diffusivity and the effect of Newtonian cooling are described at n level. Thus, the model equations are

$$\hat{\Omega}_{n-(1/2)} \frac{\partial}{\partial\lambda} \left[\frac{1}{2\Omega_0 \sin\varphi} \frac{1}{a^2} \left\{ \frac{\sin^2\varphi}{\cos\varphi} \frac{\partial}{\partial\varphi} \right. \right.$$
$$\left. \left. \times \left(\frac{\cos\varphi}{\sin^2\varphi} \frac{\partial\phi'}{\partial\varphi} \right) + \frac{1}{\cos^2\varphi} \frac{\partial^2\phi'}{\partial\lambda^2} \right\}_{n-(1/2)} \right]$$

---

* Hereafter we refer this paper as to the β-plane approximate model.

$$+\frac{1}{a}q_{n-(1/2)}\frac{1}{2\Omega_0\sin\varphi}\frac{1}{a\cos\varphi}\frac{\partial\phi'_{n-(1/2)}}{\partial\lambda}$$

$$=f\left(\frac{\partial\omega}{\partial p}\right)_{n-(1/2)}-(R_f)_{n-(1/2)}$$

$$\times\frac{1}{2\Omega_0\sin\varphi}\frac{1}{a^2}\left[\frac{\sin\varphi}{\cos\varphi}\frac{\partial}{\partial\varphi}\left(\frac{\cos\varphi}{\sin\varphi}\frac{\partial\phi'}{\partial\varphi}\right)\right.$$

$$\left.+\frac{1}{\cos^2\varphi}\frac{\partial^2\phi'}{\partial\lambda^2}\right]_{n-(1/2)},$$

$$\text{for} \quad n=1,2,\cdots,N+1 \quad (14)$$

$$\hat{\Omega}_n\left(\frac{\partial\phi'}{\partial p}\right)_n-\left(\frac{\partial\hat{\Omega}}{\partial p}\right)_n\frac{\partial\phi_n'}{\partial\lambda}+\sigma_n\omega_n$$

$$=-\left(\frac{RH}{c_pp}\right)_n-(\alpha_R)_n\left(\frac{\partial\phi'}{\partial p}\right)_n+(K_T)_n$$

$$\times\frac{1}{a^2}\left[\frac{\partial^2}{\partial\varphi^2}-\tan\varphi\frac{\partial}{\partial\varphi}+\frac{1}{\cos^2\varphi}\frac{\partial^2}{\partial\lambda^2}\right]\left(\frac{\partial\phi'}{\partial p}\right)_n$$

$$\text{for} \quad n=1,2,\cdots,N \quad (15)$$

For the upper boundary conditions, we assume that the vertical p-velocity vanishes at the top of the model, i.e.,

$$\omega=0, \text{ at } p=p_t, \text{ (or } z=z_t) \quad (16)$$

As mentioned in the paper of the β-plane approximate model, the lower boundary condition at $p=p_s$ (p_s: surface pressure) assumes that the vertical p-velocity is caused by surface topography where the standard pressure is p_G, and also by Ekman pumping resulting from the viscosity in the Ekman layer. Thus, the vertical p-velocity at $p=p_s$ is given as

$$\omega_s=\vec{V}_s\cdot\nabla p_G-\frac{p_s\cdot F}{2f}\zeta_s', \text{ at } p=p_s \text{ (or } z=0) \quad (17)$$

where $\vec{V}_s$ is the horizontal velocity vector at $p=p_s$, and $p_s=1,000$ mb for simplicity. F is the friction coefficient and will be treated as a constant (4×10^{-6} s^{-1}). ζ_s' is the vorticity of perturbation at the surface and is written as follows:

$$\zeta_s'=\frac{1}{2\Omega_0\sin\varphi}\frac{1}{a^2}\left[\frac{\sin\varphi}{\cos\varphi}\frac{\partial}{\partial\varphi}\left(\frac{\cos\varphi}{\sin\varphi}\frac{\partial\phi_s'}{\partial\varphi}\right)\right.$$

$$\left.+\frac{1}{\cos^2\varphi}\frac{\partial^2\phi_s'}{\partial\lambda^2}\right], \quad (18)$$

where ϕ_s' is the geopotential of disturbance at the surface.

2) *The vertical difference scheme*

The vertical difference scheme used in this model is the same as that discussed in the paper of the β-plane approximate model, i.e., the vertical grid increments (Δz) used in this model are as follows:

$$z: \quad 0-12 \text{ km} \quad \Delta z=1.5 \text{ km}$$
$$12-20 \text{ km} \quad \Delta z=2.0 \text{ km}$$
$$30-60 \text{ km} \quad \Delta z=3.0 \text{ km}$$
$$60-92 \text{ km} \quad \Delta z=4.0 \text{ km} \quad (19)$$

The top of this model atmosphere is defined as $z=92$ km (i.e., $p_t=1.140\times10^{-3}$ mb in winter). We divide the atmosphere into 34 layers from the earth's surface to 92 km. Thus, the following equations are obtained

$$f\left(\frac{\partial\omega}{\partial p}\right)_{n-(1/2)}=f\frac{1}{\Delta p_0}(\omega_n-\omega_{n-1}),$$

$$\text{for} \quad n=1,2,\cdots,35 \quad (20)$$

$$\left(\frac{\partial\phi'}{\partial p}\right)_n=\frac{1}{\Delta p_n*}(\phi'_{n+(1/2)}-\phi'_{n-(1/2)}),$$

$$\text{for} \quad n=1,\cdots,34 \quad (21)$$

$$\phi_n'=\frac{1}{2}(\phi'_{n+(1/2)}+\phi'_{n-(1/2)})$$

$$\text{for} \quad n=1,2,\cdots,34 \quad (22)$$

where

$$\Delta p_n=p_n-p_{n-1},$$
$$\Delta p_n*=p_{n+(1/2)}-p_{n-(1/2)}$$

To obtain the vorticity at 1,000 mb which appears in (17) we will extrapolate the geopotential from $N+1/2$ level to 1,000 mb ($N+1$ level) assuming the same temperature disturbances in both N and $N+1/2$ level atmospheres, i.e.,

$$\phi'_{N+1}=\phi'_{N+(1/2)}-0.46(\phi'_{N-(1/2)}-\phi'_{N+(1/2)}) \quad (23)$$

Substituting (20)-(22) into equations (14)-(15) and using the upper and lower boundary conditions of equations (16) and (17), we can eliminate ω_n ($n=1, 2, \ldots N+1$). Thus, we obtain 35 linear differential equations with regards to ϕ'.

All coefficients of the above equations are independent of λ, so any solution to the model equations (14)-(15) can be expressed as

$$\phi'(\lambda,\varphi,p)=\sum_{k=1}^{K}\Phi(\varphi,p)e^{ik\lambda}, \quad (24)$$

where k is the wave number in the longitudinal direction.

When (24) is substituted into the linear differential equations obtained above, we can get 35 linear differential equations to $\Phi(\varphi,p)$. In order to solve the linear algebraic equations with regard to $\Phi(\varphi,p)$, the finite-difference scheme with the grid interval of $\Delta\varphi=5°$ is used in the latitudinal direction. The number of grid points is 19×35 (19 and 35 points in the φ and p directions, respectively). The finite difference equations, thus formulated, make a system of linear equations for $\Phi_{j,n}$ at 19×35 points. Since the relaxation methods are not generally applicable for this system of linear equations, the method proposed by Lindzen and Kuo (1969) is used to solve such a system of linear equations.

In order to solve these linear algebraic equations, the lateral boundary conditions are necessary. We require that $\Phi(\varphi, p)$ should vanish at the pole and assume that $\Phi(\varphi, p)=0$ at the equator, i.e.,

$$\Phi(\varphi, p)=0 : \varphi=\pi/2 , \qquad (25)$$
$$\Phi(\varphi, p)=0 : \varphi=0 , \qquad (26)$$

Therefore, if the forcing functions of topography and diabatic heat are considered to be known from observations, the response of a model atmosphere to forcing by the hemispheric topography and stationary heat sources can be computed from these linear algebraic equations. The vertical and lateral distributions of the amplitude and the phase for standing waves in the northern hemisphere can be also computed.

3) Parameters

a) Static stability σ_n: the static stability parameter σ_n is calculated in this hemispheric model from the mean temperature and density at 45°N in January obtained for the U.S. Standard Atmosphere (1966). For simplicity, we will assume that the static stability parameter does not change with latitudes.

b) The vertical profile of the basic zonal mean wind: the observed zonal mean wind in the global atmosphere up to the mesopause level have presented by Murgatroid et al., (1965). But their results cannot be used directly in this study because the data includes small-scale features which may have significant influences on the calculation of the refractive index square of standing waves. For this reason, we shall use the vertical distribution of the zonal mean wind constructed by Matsuno (1970) up to height 60 km. This vertical distribution of the zonal mean wind is an idealized model of the zonal wind in winter, retaining only the major feature of the observed wind system. In addition, the vertical distribution from height 60 km to height 92 km is extrapolated refering the vertical distribution presented by Holton (1976). The vertical

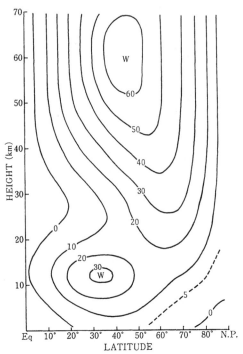

Fig. 1 The model basic state zonal mean wind distribution (ms^{-1}) in the winter Northern Hemisphere.

distribution thus obtained is similar to that constructed by Lori, et al. (1980) and is shown in Fig. 1.

c) The coefficient of Rayleigh friction R_f: The value of R_f up to 30 km height is assumed to be $0.1\times10^{-6}\,\mathrm{s}^{-1}$. We shall assume plausible values above 30 km such as shown in Fig. 2. In general, the upward propagating waves are reflected back at the top of the model. Therefore large values of R_f are incorporated above 40 km to eliminate such unrealistic effects of the upper boundary condition.

d) The horizontal kinematic thermal diffusivity coefficient K_T: we shall assume that K_T is the same as that discussed in the β-plane approximate model.

e) We shall also use the coefficient of Newtonian cooling which was adopted in the β-plane approximate model.

f) The zonal mean wind at the surface (i.e., $\bar{U}_s$) is obtained by extrapolating linearly from the zonal mean wind at $N+1/2$ level.

3. Computation of the refractive index square of standing waves

In this section we discuss the characteristic

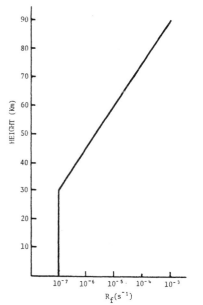

Fig. 2 The vertical distribution of the coefficient of Rayleigh friction.

features of the vertical and horizontal propagations of standing waves from the standpoint of the refractive index square introduced by Charney and Drazin (1961), in a sense of better understanding the results of numerical computation of standing waves induced by the topography in sections 4 and 5. In the discussion of the refractive index square in the vertical propagation, it is convenient to use the z-coordinate instead of the p-coordinate. Here we use the z-coordinate defined by $z = -H_0 \ln(p/p_0)$ where p_0 is the reference pressure and H_0 is the representative scale height.

If there is no heat source and Rayleigh friction, the effect of Newtonian cooling and the horizontal kinematic thermal diffusivity are not considered, we have the potential vorticity equation in the z-coordinate system from (7) as follows:

$$\hat{\Omega} \frac{\partial}{\partial \lambda} \left[\frac{\sin^2\varphi}{\cos\varphi} \frac{\partial}{\partial \varphi} \left(\frac{\cos\varphi}{\sin^2\varphi} \frac{\partial \phi'}{\partial \varphi} \right) + \frac{1}{\cos^2\varphi} \frac{\partial^2 \phi'}{\partial \lambda^2} \right.$$
$$+ 4\Omega_0^2 a^2 \sin^2\varphi \frac{\partial}{p \partial z} \left(\frac{p}{\tilde{N}^2} \frac{\partial \phi'}{\partial z} \right) \right]$$
$$+ \left[2(\Omega_0 + \hat{\Omega}) - \frac{\partial^2 \hat{\Omega}}{\partial \varphi^2} + 3 \tan\varphi \frac{\partial \hat{\Omega}}{\partial \varphi} \right.$$
$$- 4\Omega_0^2 a^2 \sin^2\varphi \frac{\partial}{p \partial z} \left(\frac{p}{\tilde{N}^2} \frac{\partial \hat{\Omega}}{\partial z} \right) \right]$$
$$\times \cos\varphi \frac{1}{\cos\varphi} \frac{\partial \phi'}{\partial \lambda} = 0, \quad (27)$$

where $\tilde{N}$ is the Brunt-Väisälä frequency.

Introducing the new variable

$$\phi(\varphi, z) = \exp\{-z/(2H_0)\} \Phi(\varphi, z)$$

We get an equation in ψ_k

$$\frac{\sin^2\varphi}{\cos\varphi} \frac{\partial}{\partial \varphi} \left(\frac{\cos\varphi}{\sin^2\varphi} \frac{\partial \psi_k}{\partial \varphi} \right) + l^2 \sin\varphi \frac{\partial^2 \psi_k}{\partial z^2}$$
$$+ \left\{ 2(\Omega_0 + \hat{\Omega}) - \frac{\partial^2 \hat{\Omega}}{\partial z^2} + 3 \tan\varphi \frac{\partial \hat{\Omega}}{\partial z} \right.$$
$$- l^2 \sin^2\varphi \left[\frac{\partial^2 \hat{\Omega}}{\partial z^2} - 2 \left(\frac{1}{2H_0} + \frac{\partial \ln \tilde{N}}{\partial z} \right) \frac{\partial \hat{\Omega}}{\partial z} \right] \right\}$$
$$\Big/ \hat{\Omega} - \frac{k^2}{\cos^2\varphi} - l^2 \sin^2\varphi \frac{1}{4H_0^2}$$
$$- l^2 \sin^2\varphi \left[\frac{1}{H_0} \frac{\partial \ln \tilde{N}}{\partial z} + 2 \left(\frac{\partial \ln \tilde{N}}{\partial z} \right)^2 \right.$$
$$\left. - \frac{1}{\tilde{N}} \frac{\partial^2 \tilde{N}}{\partial z^2} \right] \right\} \psi_k = 0 \quad (28)$$

where $l = 2\Omega_0 a / \tilde{N}$. Here we shall make a further simplification by assuming that the atmosphere is nearly isothermal. Then, $\tilde{N}$ in (28) may be considered as constant, and (28) can be written into the following simple equation (Matsuno, 1970)

$$\frac{\sin^2\varphi}{\cos\varphi} \frac{\partial}{\partial \varphi} \left(\frac{\cos\varphi}{\sin^2\varphi} \frac{\partial \psi_k}{\partial \varphi} \right) + l^2 \sin^2\varphi \frac{\partial^2 \psi_k}{\partial z^2}$$
$$+ Q_k \psi_k = 0, \quad (29)$$
$$Q_k = Q_0 - \frac{k^2}{\cos^2\varphi}, \quad (30)$$
$$Q_0 = \left[2(\Omega_0 + \hat{\Omega}) - \frac{\partial^2 \hat{\Omega}}{\partial \varphi^2} + 3 \tan\varphi \frac{\partial \hat{\Omega}}{\partial \varphi} \right.$$
$$\left. - l^2 \sin^2\varphi \left(\frac{\partial^2 \hat{\Omega}}{\partial z^2} - \frac{1}{H_0} \frac{\partial \hat{\Omega}}{\partial z} \right) \right] \Big/ \hat{\Omega}$$
$$- l^2 \sin^2\varphi \frac{1}{4H_0^2} \quad (31)$$

where H_0 is 7 km and $\tilde{N}$ is 2×10^{-2} s^{-1} in the isothermal atmosphere. As was discussed by Matsuno, (28) describes the wave propagation in the φ and z directions in an isothermal atmosphere. Here Q_k is regarded as the refractive index square of wave for zonal wave number k. If (31) is computed by the finite-difference scheme in φ and z directions, the distributions of Q_k in the $\varphi - z$ plane is obtained. Fig. 3 shows the distribution of Q_0 (the value of Q_k for $k=0$) where $\hat{\Omega}$ in (31) is computed from the zonal mean wind shown in Fig. 1. Similar distribution of Q_0 above 500 mb level was computed by Matsuno. In this paper, however, the computa-

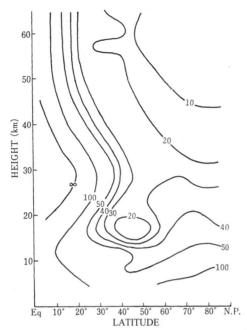

Fig. 3 The vertical distribution of the refractive index square Q_0 of wave for $k=0$ (k: zonal wave number).

tion of Q_0 in the troposphere and the lower layer of the stratosphere is carefully treated, because the discussion of wave guides (see section 4) in the upper troposphere at low latitudes through the lower troposphere at middle latitude is important in our case. We note that a minimum of Q_0 is located in the stratosphere and the mesosphere in high latitudes and another minimum of Q_0 is found at the lower stratosphere at middle latitudes. This characteristic feature of Q_0 may play an important role in the wave propagation, as described in the following sections.

4. Computation of the response to forcing by idealized topography

Before we discuss the response of a model atmosphere to forcing by the actual hemispheric topography, we examine the response of a model atmosphere to forcing by idealized topography. In this discussion we also examine the role of Q_k mentioned in section 3 from the standpoint of the vertical and horizontal propagations of standing waves induced by the topography.

In the following discussion, we consider the case where there is no stationary heat source and assume the idealized topography The basic state pressure of surface topography, p_G is generally expressed in the following form

$$p_G(\lambda, \varphi) = \sum_{k=1}^{K} \{\hat{p}_g(\varphi)\}_k e^{ik\lambda}, \tag{32}$$

Here $(\hat{p}_g(\varphi))_k$ denotes the meridional distribution of the amplitude of p_G for wave number k, and is expressed in the following complex form

$$\{\hat{p}_g(\varphi)\}_k = (p_A)_k + i(-p_B)_k, \tag{33}$$

For the sake of simplicity, the idealized topography is assumed to be expressed only by three waves of $k=1$, 2 and 3 and the form of $(\hat{p}_g(\varphi))_k$ is defined at the specified latitude such as

$$(p_A)_k = \begin{cases} 50 \text{ mb}, & \varphi = \varphi', \\ 0 \text{ mb}, & \varphi \neq \varphi', \end{cases}$$

and

$$(p_B)_k = \begin{cases} 0 \text{ mb}, & \varphi = \varphi', \\ 0 \text{ mb}, & \varphi \neq \varphi', \end{cases} \quad \text{for } k=1,2,3 \tag{34}$$

When (33) and (34) are substituted into the model equations (14)-(15) we get the vertical and latitudinal distributions of the geopotential $\Phi_k(\varphi, p)$ forced by this idealized topography. The solution to the geopotential $\Phi_k(\varphi, p)$ may be written in the form

$$\Phi_k(\varphi, p) = [\{\Phi_A(\varphi)\}_{n-(1/2)} + i\{-\Phi_B(\varphi)\}_{n-(1/2)}]_k$$
$$\text{for } n=1,2,\cdots,35, k=1,2,3. \tag{35}$$

Here n specifies the value at the pressure level. Then, the amplitude and the phase of standing waves for the zonal wave number k at latitude φ and $n-1/2$ level are written in the following forms, respectively.

$$\hat{\Phi}_k(\varphi, p) = [[\{\Phi_A(\varphi)\}_{n-(1/2)}]k^2$$
$$+ [\{\Phi_B(\varphi)\}_{n-(1/2)}]k^2]^{1/2}, \tag{36}$$

and

$$\alpha_k(\varphi, p) = \frac{1}{k} \tan^{-1}[[\{\Phi_B(\varphi)\}_{n-(1/2)}]_k$$
$$/[\{\Phi_A(\varphi)\}_{n-(1/2)}]_k],$$
$$\text{for } n=1,2,\cdots,35, \quad k=1,2,3. \tag{37}$$

It is convenient, for comparison with the observed values, to convert the geopotential into the height value, such that

$$c(\varphi, p) = \hat{\Phi}_k(\varphi, p)/g \tag{38}$$

where $c_k(\varphi, p)$ is the height value of standing wave for the zonal wave number k at $n-1/2$ level.

In this section, the response of a model atmosphere to forcing of the idealized topography is computed in the case where $\varphi' = 80°N$ or $\varphi' = 40°N$ in (34), i.e., the idealized topography is considered only at the latitude of 80°N or 40°N.

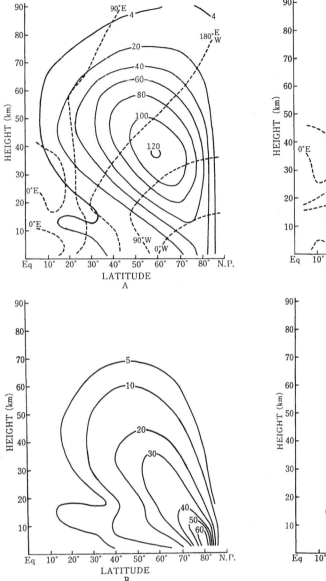
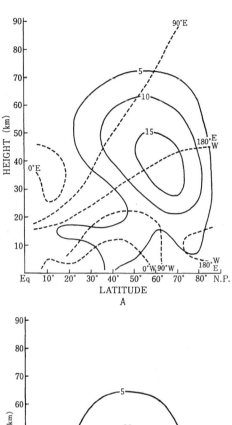
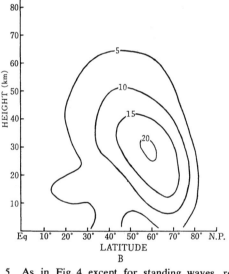

Fig. 4 The vertical distribution of amplitude (m) (solid curves) and phase (dashed curves) of standing waves, responding to forcing by idealized topography at 80°N for wave number $k=1$ (A) and $k=2$ (B), respectively. Position of ridges is shown on the right of equal phase lines.

Fig. 5 As in Fig. 4 except for standing waves, responding to forcing by idealized topography at 40°N.

In these computations we assume that the zonal mean wind at the surface, $\bar{U}_s$ in (17) is assumed as $\bar{U}_s = 4.5$ m/s in either case of $\varphi' = 80°N$ and $\varphi' = 40°N$.

Figs. 4(A) and 4(B) show the vertical distribution of the amplitude of standing waves responding to forcing by the idealized topography at 80°N for zonal wave number $k=1$ and $k=2$, respectively. On the other hand, Figs. 5(A) and 5(B) show the vertical distribution of the amplitude of standing wave responding to forcing by the idealized topography at 40°N for $k=1$ and $k=2$, respectively. In Fig. 4(A) and Fig. 5(A), the phase of standing waves is shown for the sake of reference. From these figures, we may be able to summarize the results as follows:

1) The amplitude of standing waves forced by the idealized topography at 80°N for wave number 1 is much larger than that forced by the topography at 40°N for wave number 1. The main reason for this difference may be considered in the following points;

i) The difference of the forcing: In this paper, the vertical p-velocity at the surface, ω_s is computed in a form of the geostrophic approximation of $\vec{V}_s \cdot \Delta p_G$ in (17). Thus, ω_s is proportional to $1/\cos\varphi$, i.e.,

$$\omega_s \sim \bar{U}_s \frac{ik}{a \cos \varphi} \{p_g(\varphi)\}_k ,$$

Since the zonal mean surface wind and the amplitude of the topography at 80°N are assumed to be equal to those at 40°N, the vertical velocity at the surface at 80°N is 5.8 times as large as that at 40°N because $\cos 40° \approx 5.8 \cos 80°$.

ii) The role of refractive index square of standing waves for wave number $k=1$, Q_1: Since $Q_1 = Q_0 - 1/\cos^2\varphi$, Q_1 may be considered nearly equal to Q_0 except in the vicinity of the pole. The computed value of Q_1 is shown in Fig. 6 by dashed lines. As may be seen in the figure, there is a region of small values of Q_1 above the tropospheric jet at 40°-50°N. Therefore, it may be easily understood that the standing wave for wave number 1 is propagated from the lower troposphere to the upper stratosphere in high latitudes, while the standing wave is hardly propagated upward in middle latitudes, because it is propagated through the belt of larger values of Q_1 and blocked in the region of small values of Q_1.

2) The amplitude of standing waves for wave number 1 forced by the idealized topography at 80°N has a maximum at 38 km height, 60°N and a secondary peak is found at 13 km height, 20°-30°N as shown in Fig. 4(A). The amplitude of standing waves for wave number 2 has a maximum near the surface at 80°N and also a secondary peak is found at 15 km height, 20°-30°N, as shown in Fig. 4(B). Generally speaking, the standing wave for wave number 2 in high latitudes cannot be propagated from the lower troposphere to the upper stratosphere as may be seen in Fig. 4(B), because Q_2 in high latitudes becomes negative.

On the other hand, the amplitude of standing waves for wave number 1 forced by the idealized topography at 40°N has a maximum at 38 km height, 65°N and a secondary peak is found at 15 km height, 20°-30°N, as shown in Fig. 5(A). The amplitude of standing waves for wave number 2 has a maximum at 27 km height, 60°N and also a secondary peak is found at 15 km height 20°-30°N. Although the amplitude of standing wave for wave number 3 forced by idealized topography at either 80°N or 40°N is not shown in the figure, it has a maximum in the troposphere, because it cannot be propagated into the stratosphere.

As may be seen in Figs. 4(A), 4(B), 5(A) and 5(B), the secondary peaks of the amplitude of standing waves responding to forcing by the idealized topography at 40°N and 80°N are found in the upper troposphere at low latitudes. Thus, we may be able to speculate that the propagation of standing waves responding to forcing by topography in middle and high latitudes toward low latitudes may be considered as one of main reasons for the formations of standing waves in the upper troposphere at low latitudes.

3) Since the vertical distribution of Q_1 has a small value at above 14 km height, 40°-50°N, the standing wave for wave number 1 responding to forcing by topography in high and middle latitudes has two wave guides. As may be seen in Fig. 6, a wave guide of standing waves responding to forcing by topography at 80°N for wave number 1 points from the lower troposphere at 80°N toward 38 km height, 60°N and another wave guide points from the lower troposphere in middle latitudes toward the upper troposphere at low latitudes. In the same manner, it may be seen in Fig. 7 that a wave guide of standing waves responding to forcing by the topography at 40°N for wave number 1 points from the lower troposphere at 40°N toward 38 km height, 65°N. We find that this standing wave cannot be directly propagated through the lower stratosphere in middle latitudes as mentioned before. Another wave guide points from the lower troposphere in middle latitudes toward the upper troposphere at low latitudes.

4) The phase of standing waves responding to forcing by the idealized topography at either 80°N or 40°N is tilted westward with increasing height and decreasing latitude, as shown in Fig. 4(A) or Fig. 5(A). This means that the trough and ridge axes tilt westward with height and shift westward with the decrease of latitude.

In closing the discussion of this section, we mention short comments about the result obtained by Laprise (1978). He computed the standing

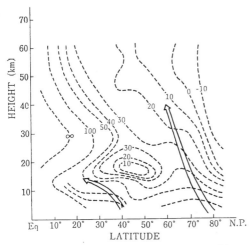

Fig. 6 Schematic picture of the wave guides of standing wave for wave number $k=1$, responding to forcing by idealized topography at 80°N. Dashed curves show the refractive index square Q_1 for wave number $k=1$.

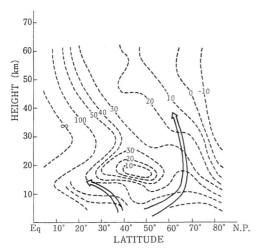

Fig. 7 As in Fig. 6 except for standing wave, responding to forcing by idealized topography at 40°N.

wave forced by the topography, by means of a linearized steady-state primitive equation model on a hemisphere. In his case, the zonal wave number 1 is treated and the maximum value of ω_s due to the topography is given at $\varphi=45°N$. The result shows that the secondary peak such as obtained in Fig. 5(A) is relatively weak and is found at about 10 km, 30°N. Although his treatment in low latitudes is more reasonable compared with ours, because of the use of a primitive model, the height and the latitude of the secondary peak seem to be unreasonable compared with the observations. Different results from ours mentioned above may be obtained from his assumption of zonal wind that there is no easterlies in low latitudes, i.e., there is no absorber over the tropics.

5. Computation of the response to forcing by the actual hemispheric topography

First, in order to compute the standing waves responding to forcing by the actual topography, we must obtain the meridional distribution of the amplitude and the phase of the actual hemispheric topography. In this paper we use the height of the topography arranged by Berkofsky and Bertoni (1955) and it is expanded into the following Fourier series

$$p_G(\lambda, \varphi) = \bar{p}_g(\varphi) + \sum_{k=1}^{K} [(p_A)_k(\varphi)\cos(k\lambda) + (p_B)_k(\varphi)\sin(k\lambda)]$$
$$= \bar{p}_g(\varphi) + \sum_{k=1}^{K} \{\hat{p}_g(\varphi)\}_k \cos\{k\lambda - \alpha_k(\varphi)\},$$
(39)

Fourier coefficients appearing in this expansion for $p_G(\lambda, \varphi)$ are then computed from the relations

$$\{(p_A)_k, (p_B)_k\} = \frac{1}{\pi}\int_0^{2\pi} p_G\{\cos(k\lambda), \sin(k\lambda)\}d\lambda,$$
$$\{\hat{p}_g(\varphi)\}_k = \{(p_A)_k^2 + (p_B)_k^2\}^{1/2},$$
$$\alpha_k(\varphi) = \tan^{-1}\{(p_B)_k/(p_A)_k\},$$
for $k=1,2,3$. (40)

The amplitudes of $(\hat{p}_g(\varphi))_k$ for $k=1$ and $k=2$ thus obtained are shown in Figs. 8(A) and 8(B), respectively. Note that the largest amplitudes of the Hemispheric topography are found at 35°N, where the Tibetan Plateau and the Rocky Mountains are found. However, we may also find the relatively large values of the amplitude in high latitudes. Thus, the forcing effects of topography in high latitudes will not be neglected compared with that in middle latitudes. If we substitute the value of $(p_A)_k$, $(p_B)_k$ of the actual topography obtained by (40) into the linear algebraic equation mentioned in section 2, we obtain the vertical distribution of the amplitude and the phase of standing waves responding to these actual forcing mechanisms.

In the studies, such as in Egger's (1976) and Ashe's investigations, the forcing effect of topography in high latitudes is almost neglected. However, according to the computed results in

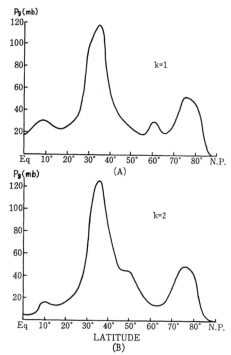

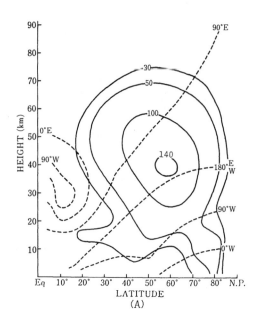

Fig. 8 The amplitude of zonal wave component 1 (A) and 2 (B) for standard pressure at ground.

section 4, the forcing effect of the topography in high latitudes is relatively large and plays an important role in the formation of standing waves. Therefore our discussion focuses on the effects of topography in high latitudes. Thus, we compute the following three experiments (A, B and C).

A) The case where the zonal mean wind at the surface, 70°-85°N is the east wind and in other regions the zonal wind is assumed such as shown in Fig. 1. In this case, the vertical distribution of the amplitude and the phase of standing waves responding to forcing by the Hemispheric topography for wave number 1 and 2 is shown in Figs. 9(A) and 9(B), respectively. In addition to Figs. 9(A) and 9(B), the vertical velocity which is computed as adding the contributions from wave numbers 1-3 is shown in Fig. 10 for the sake of reference. We may find that the rising motion occurs in the eastern coast of Greenland and sinking motion occurs in the western coast. On the other hand, a maximum sinking motion occurs in the lee side of the Tibetan Plateau, and a weaker sinking motion occurs also in the lee side of the Rocky Mountains. This is in rather good agreement with the result computed by Saltzman and Irsch (1972)

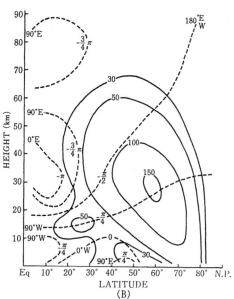

Fig. 9 The vertical distribution of amplitude (m) (solid curves) and phase (dashed curves) of standing wave, responding to forcing by Hemispheric topography for wave number $k=1$ (A) and $k=2$ (B), respectively, when the zonal mean wind at surface 70°–85°N is the east wind. Position of ridges is shown on the right of equal phase lines. The values of phase indicated with π in (B) are that computed from the equation (37).

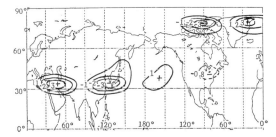

Fig. 10 The vertical velocity W_s (10^{-3}ms^{-1}) at surface induced by the Hemispheric topography.

from the observed horizontal wind. It is shown that the east wind at the surface, 70°-85°N assumed in this case is in rather good agreement with the actual zonal mean wind at 70°-85°N.

B) The case where the zonal mean wind at the surface, 70°-85°N is the west wind and in other regions the zonal wind is assumed such as shown in Fig. 1. In this case the vertical distribution of the amplitude and the phase of standing waves responding to forcing by the Hemispheric topography for wave number 1 is shown in Fig. 11.

C) The case where there is no topographical effect of the Greenland Plateau and the mean zonal wind is the same as the case (A). In this case, the vertical distribution of the amplitude and the phase of standing waves responding to forcing by the Hemispheric topography excluding the Greenland Plateau for wave number 1 is shown in Fig. 12.

Here, a brief summary of these computed results may be written as follows:

i) In the case (A) mentioned above, the phase of standing waves for wave number 1 is almost out of phase of that in the case (B) in the stratosphere. However, the difference of the phases in both cases is small in the troposphere in middle and low latitudes.

ii) In the case (C), the amplitude of standing waves for wave number 1 is appreciably small compared with that in the case (A).

These results reveal that the Greenland Plateau plays an important role for the standing waves responding to forcing by the Hemispheric topography in winter.

iii) We may find from Figs. 9(A) and 11 that the response of a Hemispheric model atmosphere is mainly due to the forcing by topography in high and middle latitudes. In addition, we may find in Figs. 9(A) and 9(B) that the amplitude of standing waves responding to forcing by the Hemispheric topography for wave number 1 has

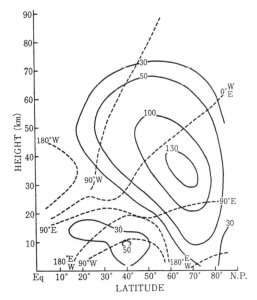

Fig. 11 As in Fig. 9 except for the zonal mean wind at surface, 70°-85°N is the west wind.

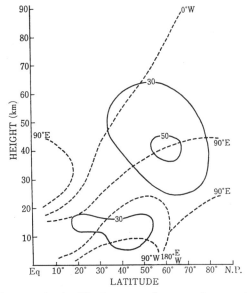

Fig. 12 As in Fig. 9 except for assuming no topography at surface, 70°-85°N.

a maximum at 38 km height, 60°N and a secondary peak is found at 15 km height, 20°-30°N, as described in section 4. The amplitude of standing waves for wave number 2 has a maximum at 27 km height, 60°N and a secondary peak where the amplitude is larger than that for wave number 1 is also found at 15 km height, 20°-30°N. As mentioned in section 4, the standing waves responding to forcing by the Hemispheric topography, especially by topography in middle latitudes, can be propagated toward low latitudes. Webster (1972) suggested that most of the time-independent circulation in low latitudes is forced by heating and orography within the tropics and the subtropics. However, we consider that the forcing effects from higher latitudes are more important in the formation of standing waves in the upper troposphere at low latitudes.

In the most part of the meridional plane, the phase of standing waves is tilted westward with increasing height and decreasing latitude and the westward tilt of standing waves decreases as the zonal wave number increases.

iv) Figs. 9(A) and 9(B) also show that the amplitude of standing waves responding to forcing by the topography for wave number 2 is larger than that for wave number 1. This result is the same as that obtained in the paper on the β-plane approximate model.

Next, we shall compute the disturbance pattern at an isobaric surface (or at a constant height level), by synthesizing wave component 1-3. Since higher wave number components decay strongly with height, inclusion of more wave components would not change substantially the result at high levels. Fig. 13 shows the disturbance pattern at 30 km level responding to forcing by the Hemispheric topography, when the zonal mean wind at the surface 70°-85°N is the east wind (case A). The forced pattern consists of a major negative anomaly over Europe, another negative anomaly over Pacific Ocean, a major positive anomaly over the eastern coast of Asia and another positive anomaly over the eastern coast of North America. Fig. 14 also shows the disturbance pattern at 30 km level responding to forcing by the Hemispheric topography, when the zonal mean wind at the surface 70°-85°N is the west wind (case B). We may find in Fig. 14 that the considerable difference occurs around Greenland as compared with Fig.

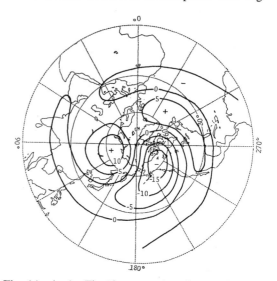

Fig. 14 As in Fig. 13 except for the zonal mean wind at surface, 70°–85°N is the west wind.

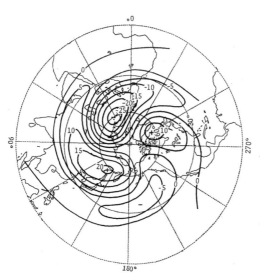

Fig. 13 The disturbance pattern (units in dm) at 30 km level, responding to forcing by the Hemispheric topography, when the zonal mean wind at surface, 70°–85°N is the east wind.

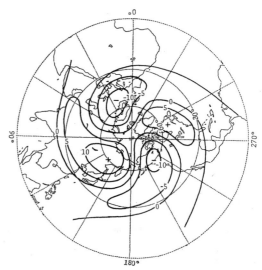

Fig. 15 As in Fig. 13 except for assuming no topography at surface, 70°–85°N.

13. In this case, a negative anomaly over Europe and a positive anomaly over the eastern coast of North America are not found. Fig. 15 shows the disturbance pattern at 30 km level responding to forcing by the Hemispheric topography, assuming that there is no topography at 70°-85°N (case C). Compared with Fig. 13, we may find that when there is no topography at 70°-85°N, the disturbance pattern responding to forcing by the Hemispheric topography becomes weak not only around Greenland, but also over Asia and Pacific Ocean. Thus, we may be able to conclude from Figs. 10-15 that the topography in high latitudes, such as the Greenland Plateau plays an important role for the standing waves responding to forcing by the Hemispheric topography.

6. Conclusions

A 34-level quasi-geostrophic linear steady-state model in a spheric coordinate with Rayleigh friction, the effect of Newtonian cooling and horizontal thermal diffusivity is used to obtain the vertical and meridional distributions of standing waves and disturbance pattern at a constant height in the response to forcing by topography in winter. The computed results are as follows:

1) The topography at high latitudes, such as Greenland, plays a considerably important role for the standing waves responding to forcing by the Hemispheric topography. Thus, the topography at high latitudes cannot be neglected in the investigation about the response of a model atmosphere to forcing by the Hemispheric topography.

2) The standing waves responding to forcing by topography at either high latitudes and middle latitudes can be vertically and laterally propagated. The amplitude of standing waves for wave number 1 is a maximum at 38 km height, 60°N, and has a secondary peak at 15 km height, 20°-30°N, while the amplitude for wave number 2 is a maximum at 27 km height, 60°N and has also a secondary peak at 15 km height, 20°-30°N.

The propagation of standing waves responding to forcing by topography at middle and high latitudes, especially at middle latitudes, toward the low latitudes may be considered as one of reasons for the formation of standing waves in the upper troposphere at low latitudes.

3) The amplitude of standing waves responding to forcing by topography for wave number 2 is larger than that for wave number 1. This is the same as the results computed by the β-plane approximate model.

In this paper, our discussion focuses only on the standing waves responding to forcing by the Hemispheric topography in winter. In a same way, the stationary heating sources play an important role in the formation of standing waves and seem to be larger than the forcing effect by topography. Therefore, further studies should be aimed at the response of a model atmosphere to forcing by stationary heat sources, and also by both topography and stationary heat sources. They will be described in the next paper of the present investigation. In that companion paper, the momentum, heat and Eliassen-Palm fluxes due to standing waves, responding to forcing by both topography and stationary heat sources, will be discussed.

Many authors discussed the problem of planetary mountain forcing in General Circulation Models, such as Kasahara, Sasamori and Washington (1973) with NCAR model, Manabe and Terpstra (1974) with GFDL model. Youngblut and Sasamori (1980) investigated the nonlinear effects of transient and stationary eddies in the winter mean circulation by means of diagnostic analysis. The results show that the nonlinear stationary eddy source generates geopotential perturbation which are comparable with the effect of the transient eddies.

The effects of topography in forcing the stationary eddy flow field of the atmosphere have been examined by Frederikson and Sawford (1981). Using nonlinear and linear spherical barotropic models, they found that the nonlinear effects are most important at low latitudes. The stationary flow field simulated by the nonlinear model is in more good agreement with the observed one than that by the linear model. However, at high levels, where the zonal flow is stronger, the linear approximation is better than at low levels. Thus, how should be nonlinear terms be included in this model have to be investigated further. By do so, we may obtain a better understanding of the nature of planetary-scale disturbances in the whole atmosphere.

Acknowledgements

The authors are grateful to Prof. T. Matsuno for his valuable comments and suggestions during the course of this study. Thanks are due to Dr. Y. Sato, Dr. Ts. Nitta and Dr. T. Satomura for discussions throughout the course of this work. Thanks are also extended to Mr. Y. Fujiki for drawing the figures and Mrs. K. Kudo for the expert typing of manuscript. This paper is sponsored in part by Grant in Aid for Scientific Research from the Ministry of Education.

References

Ashe, S., 1979: A nonlinear model of the time-average axially asymmetric flow induced by topography and diabatic heating. *J. Atmos. Sci.*, **36**, 109-126.

Berkofsky, L., and E. A. Bertoni, 1955: Mean topographic charts for the entire earth. *Bul. Amer. Meteor. Soc.*, **36**, 350-354.

Charney, J. G., and P. G. Drazin, 1961: Propagation of planetary-scale disturbances from the lower into the upper atmosphere. *J. Geophys. Res.*, **66**, 83-109.

Dickinson, R. E., 1968: Planetary Rossby waves propagating vertically through weak westerly wind wave guides. *J. Atmos. Sci.*, **25**, 984-1002.

Egger, J., 1976: The linear response of a hemispheric two-level primitive equation model to forcing by topography. *Mon. Wea. Rev.*, **104**, 351-363.

Frederikson, J. S. and B. L. Sawford 1981: Topography waves in nonlinear and linear spherical barotropic models. *J. Atmos. Sci.*, **36**, 69-86.

Holton, J. R., 1976: A semi-spectral numerical model for wave-mean flow interactions in the stratosphere: Application to sudden stratospheric warmings. *J. Atmos. Sci.*, **33**, 1639-1647.

Huang, Rong-hui and K. Gambo, 1981: The response of a model atmosphere in middle latitudes to forcing by topography and stationary heat sources. *J. Meteor. Soc. Japan*, **59**, 220-237.

Kasahara, A., T. Sasamori and W. M. Washington, 1973: Simulation experimental with a 12-layer stratospheric global circulation model. 1. Dynamical effect of the earth's orography and thermal influence of continentality. *J. Atmos. Sci.*, **30**, 1229-1251.

Laprize, R., 1978: On the influence of stratospheric conditions on forced tropospheric waves in a steady-state primitive equation model. *Atmosphere-Ocean*, **16**, 300-314.

Lindzen, R. S., and H. L. Kuo, 1969: A reliable method for the numerical integration of a large class of ordinary and partial differential equations. *Mon. Wea. Rev.*, **97**, 732-734.

Lordi, N. J., A. Kasahara and S. K. Kao, 1980: Numerical simulation of stratospheric sudden warmings with a primitive equation spectral model. *J. Atmos. Sci.*, **37**, 2746-2767.

Manabe, S., and T. B. Terpstra, 1974: Effects of mountains on the general circulation of the atmosphere as identified by numerical experiments. *J. Atmos. Sci.*, **31**, 3-42.

Matsuno, T., 1970: Vertical propagation of stationary planetary waves in the winter northern hemisphere. *J. Atmos. Sci.*, **27**, 871-883.

Murgatroyd, R. J., 1969: The structure and dynamics of the stratosphere. *The Global Circulation of the Atmosphere*, 155-195.

Pyle, J. A., and C. F. Rogers, 1980: Stratospheric transport by stationary planetary waves—the importance of chemical processes. *Quart. J. Roy. Meteor. Soc.*, **106**, 421-446.

Youngblut, C. and T. Sasamori, 1980: The nonlinear effects of transient and stationary eddies on the winter mean circulation. Part 1. diagnostic analysis. *J. Atmos. Sci.*, **37**, 1944-1957.

Staff Members, Academia Sinica 1958: On the general circulation over eastern Asia. *Tellus*, **10**, 299-312.

Saltzman, B., and F. E. Irsch III, 1972: Note on the theory of topographically forced planetary waves in the atmosphere. *Mon. Wea. Rev.*, **100**, 441-444.

Webster, P. J., 1972: Response of the tropical atmosphere to local, steady forcing. *Mon. Wea. Rev.*, **100**, 518-541.

The Response of a Hemispheric Multi-Level Model to Forcing by Topography and Stationary Heat Sources

Part II Forcing by Stationary Heat Sources and Forcing by Topography and Stationary Heat Sources[*]

Huang Ronghui and K. Gambo

Abstract

The response of a Hemispheric model atmosphere to forcing by the Hemispheric topography and the stationary heat sources in winter was examined. In this computation the model described in Part I of this paper is used.

The computed results show that the amplitude of standing wave responding to forcing by stationary heat sources at high latitudes, for zonal wave number 1, is larger than that forced by stationary heat sources at middle latitudes. The results also show that the amplitude of standing waves for zonal wave number 1 responding to forcing by the Hemispheric stationary heat sources is larger than that responding to forcing by the Hemispheric topography. In these computations, the role of refractive index square of standing waves for zonal wave number 1 or 2 is discussed in relation with the propagation of standing waves from the lower troposphere at middle and high latitudes toward the upper stratosphere at high latitudes or the upper troposphere at low latitudes.

The momentum, heat and Eliassen-Palm fluxes due to standing waves computed by this model, are qualitatively in good agreement with the observed results.

1. Introduction

In Part I of this paper (Huang and Gambo, 1982), a quasi-geostrophic, steady-state 34-level model with Rayleigh friction, Newtonian cooling effect and horizontal kinematic thermal diffusivity included in a spherical coordinate system was used to examine the standing waves responding to forcing by the Hemispheric topography in winter. In this paper, however, we discuss the standing waves responding to forcing by the Hemispheric topography and the stationary heat sources in winter. That is, the results in Part I of this paper is re-examined by considering the effect of heat sources in addition to the orographic effect.

Generally speaking, the response of a Hemispheric model atmosphere to forcing by stationary heat sources has been studied by many authors. For example, the staff member, Academia Sinica (1958) studied the effects of the forcing to formation of standing troughs and ridges. Egger (1976) investigated the response of a model atmosphere to forcing by topography and stationary heat sources by use of a conventional quasi-geostrophic two-level model with a β-plane approximation. Ashe (1978) examined the standing disturbance pattern by means of a two-level nonlinear model.

As mentioned in Part I, the simplified two level models such as mentioned above are not sufficient to describe the vertical distribution of the amplitude and phase of standing planetary waves responding to forcing by topography and heat sources, due to the poor vertical resolution. Therefore, a multi-level model with sufficient vertical resolution in the stratosphere must be used to describe accurately the vertical distribution of the amplitude and phase of standing waves by topography and stationary heat sources.

2. Model

As the same as described in Part I, a Hemi-

---

[*] 本文原载于: Journal of the Meteorological Society of Japan, Vol. 60, No.1, 93-108, 1982.

spheric, quasi-geostrophic, linear, steady-state model is utilized with Rayleigh friction, Newtonian cooling effect and the horizontal kinematic thermal diffusivity. In the vertical we divide the atmosphere into 34 layers from the earth's surface to $z=92$ km in the same way as mentioned in Part I. The vertical and meridional difference schemes for the model computation are described in Part I. Therefore we do not repeat the details of finite-difference schemes in this paper. We also use the same zonal wind as mentioned in Part I. Only the difference between Part I and Part II is that we put $H=0$ in Part I (see equations (2) or (15) in Part I). Here H is the diabatic heating per unit time and unit mass.

3. Computation of the response to forcing by idealized stationary heat sources

i) The response to forcing by idealized stationary heat sources at different latitudes

In order to investigate the forcing effects of stationary heat sources at different latitudes to a model atmosphere, we shall assume that there is no topography at the surface and the idealized stationary heat sources have the following distribution in a spherical coordinate system, with λ as longitude and φ as latitude, *i.e.*,

$$H(\lambda,\varphi,p) = \sum_{k=1}^{K} \{\hat{H}_0(\varphi,p)\}_k e^{ik\lambda}, \quad (1)$$

$$\{\hat{H}_0(\varphi,p)\}_k = (H_A)_k + i(-H_B)_k, \quad (2)$$

Here we use a pressure coordinate for the vertical one.

For the sake of simplicity, the idealized stationary heat sources are assumed to be expressed only by three zonal waves of $k=1, 2$ and 3 and the form of $(H_0(\varphi,p))_k$ is a given function specified such as

$$(H_A)_k = \begin{cases} 6.0 \times 10^{-6} \text{ KJ} \cdot \text{kg}^{-1} \cdot \text{s}^{-1}, & \varphi = \varphi' \\ 0 & \varphi \neq \varphi' \end{cases}$$

$$(H_B)_k = \begin{cases} 0, & \varphi = \varphi' \\ 0, & \varphi \neq \varphi' \end{cases}$$

$$(3)$$

Further, we assume that the vertical distribution of stationary heat sources is the same as that taken in our β-plane approximate model (Huang and Gambo, 1981), *i.e.*,

$$\{\hat{H}_0(\varphi,p)\}_k = \{\hat{H}_0(\varphi)_k \exp\left\{-\left(\frac{p-\bar{p}}{d}\right)^2\right\}, \quad (4)$$

Here we assume $d=300$ mb and $\bar{p}=500$ mb. This means that the distribution of diabatic heating has a maximum at 500 mb and decreases exponentially above and below 500 mb.

Then, the vertical and meridional distributions of the amplitude and phase of standing waves forced by idealized stationary heat sources at different latitudes are obtained by substituting idealized stationary heat sources into the model equations (14)-(15) of Part I. In this section, the response of a model atmosphere to forcing of idealized stationary heat sources is computed in the case where $\varphi'=80°$N or $\varphi'=40°$N in (3) *i.e.*, the idealized stationary heat sources are considered only at the latitude of 80°N or 40°N. In these computations, the zonal wind is assumed to be the same with that of the case (A) in Part I, *i.e.*, the zonal wind at surface, 70°-85°N is the east wind and in other regions the zonal wind is assumed such as shown in Fig. 1 of Part I.

Figs. 1(A) and 1(B) show the vertical distribution of the amplitude of standing waves responding to forcing by idealized stationary heat sources at 80°N for zonal wave number $k=1$ and $k=2$, respectively. Figs. 2(A) and 2(B) show the vertical distribution of the amplitude of standing waves responding to forcing by idealized stationary heat sources at 40°N for $k=1$ and $k=2$, respectively. We may find the following results from Figs. 1(A), 1(B) and Figs. 2(A), 2(B).

1) The amplitude of standing waves forced by idealized stationary heat sources at 80°N for $k=1$ is rather larger than that forced by idealized stationary heat sources at 40°N. However, the effect of the latitudinal difference of heat sources is not so large compared with that forced by idealized topography (refer Figs. 4(A) and 4(B) of Part I), because the geopotential responding to forcing by stationary heat source is proportional to the term $f(\partial\omega/\partial p)$, while ω is proportional to the diabatic heating, as shown in the thermodynamic equation.

The different responses of standing waves forced by the heat sources at 80°N and 40°N are easily understood from the discussion of the propagation of standing waves in connection with the meridional distribution of the refractive index square of wave mentioned in Part I. That is, the standing wave for $k=1$ at high latitudes is readily propagating from the lower troposphere to the upper stratosphere, while the standing wave for $k=1$ at middle latitudes is hardly

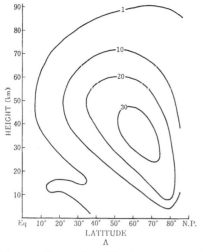 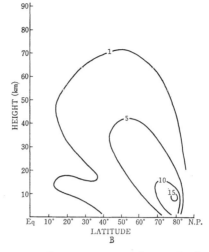

Fig. 1 The vertical distribution of amplitude (m) of standing waves, responding to forcing by idealized stationary heat sources at 80°N for wave number $k=1$ (A), and $k=2$ (B), respectively.

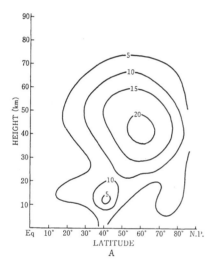 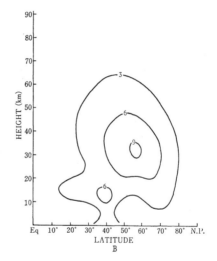

Fig. 2 As in Fig. 1 except for standing waves, responding to forcing by idealized stationary heat sources at 40°N.

propagated from the lower troposphere to the upper stratosphere, because the standing wave is propagated through the belt of the larger refractive index square for $k=0$, Q_0 and blocked in the region of small Q_0.

On the other hand, the standing wave for $k=2$ at high latitudes cannot be propagated from the lower troposphere to the upper stratosphere, because the refractive index square for $k=2$, Q_2 at high latitudes becomes negative, as shown in Fig. 6 of Part I.

2) The amplitude of standing wave responding to forcing by idealized stationary heat sources at 80°N for $k=1$ has a maximum at 38 km height, 60°N, and has a secondary peak at 13 km height, 20°-30°N, as shown in Fig. 1(A). On the other hand, the amplitude of standing wave for $k=1$ responding to forcing by idealized stationary heat sources at 40°N has a maximum at 40 km height, 60°N, and has a secondary peak at 15 km height, 20°-30°N, as shown in Fig. 2(A). The amplitude of standing waves for $k=2$ has a maximum at 31.5 km height, 55°N, and has a secondary peak at 15 km height, 20°-

30°N. The amplitude of standing waves for wave number 3 responding to forcing by idealized stationary heat sources at either 80°N or 40°N has a maximum in the troposphere, and has also a secondary peak at 15 km height, 20°-30°N, respectively, though the results are not shown in this paper. These results suggest that we have two wave guides for the propagation of standing waves responding to forcing by stationary heat sources such as explained in Part I of this paper.

As may be seen in Figs. 1(A), 1(B), 2(A) and 2(B), the secondary peak of the amplitude of standing waves responding to forcing by idealized stationary heat sources at 40°N is larger than that responding to forcing by idealized stationary heat sources at 80°N. Thus, we may be able to speculate that it may be one of reasons for the formation of standing waves in the upper troposphere at low latitudes that the standing waves responding forcing by stationary heat sources at middle latitudes can propagate toward the upper troposphere at low latitudes.

ii) *Computation of the response of forcing by the stationary heat sources with the meridional width*

It has been discussed in the β-plane approximate model (Huang and Gambo, 1981) that the meridional wavelength of the diabatic heating seems to have an influence on the response of an one-dimensional model atmosphere to forcing by the diabatic heating, *i.e.*, the reflected height of standing waves is highly dependent upon the meridional wavelength of the diabatic heating. In the case of a two-dimensional model atmosphere, however, the forced standing wave can be propagated through the two wave guides horizontally and vertically. Therefore the result obtained in the case of an one-dimensional model atmosphere will not be directly extended to the case of a two-dimensional model atmosphere. Considering this matter, we examine the standing waves responding to forcing by the stationary heat sources with the meridional width in a two-dimensional model.

The meridional distribution of idealized stationary heat sources taken in this section is similar to that used by Murakami (1972), but the two different meridional widths are assumed, *i.e.*, $(H_A)_k$ and $(H_B)_k$ in (2) are assumed as follows:

$$\begin{cases} (H_A)_k = 6.0 \times 10^{-6} \text{ KJ} \cdot \text{kg}^{-1} \text{ s}^{-1} \exp\left\{-\left(\frac{\varphi-\varphi_0}{b}\right)^2\right\}, \\ (H_B)_k = 0, \end{cases}$$
(5)

where $\varphi_0 = 40°N$, and $b = 40°$ or $20°$ in latitude, respectively.

Figs. 3(A) and 3(B) show the vertical distribution of the amplitude of standing waves responding to forcing by idealized stationary heat sources in cases where $b = 40°$ in latitude, for wave number $k=1$ and $k=2$ respectively. On the other hand, Figs. 4(A) and 4(B) show the vertical distribution of the amplitude of standing wave responding to forcing by idealized stationary heat sources in cases where $b = 20°$ in lati-

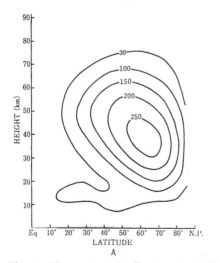
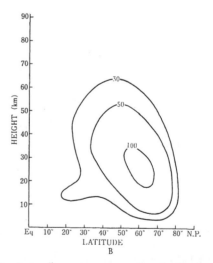

Fig. 3 The vertical distribution of amplitude (m) of standing wave, responding to forcing by idealized stationary heat sources in case where b is 40° of latitude, for wave number $k=1$ (A), and $k=2$ (B), respectively.

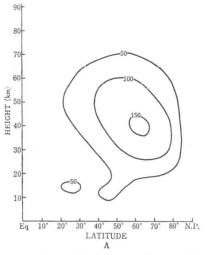
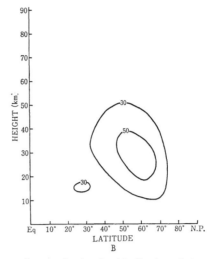

Fig. 4 As in Fig. 3 except for standing waves, responding to forcing by idealized stationary heat sources in case where b is 20° of latitude.

tude, for $k=1$ and $k=2$, respectively.

Comparing Figs. 1(A), 1(B) and Figs. 2(A), 2(B) with Figs. 3(A), 3(B) and Figs. 4(A), 4(B), we may find that the amplitude of standing waves in cases where the meridional widths ($b=40°$ or 20° in latitude) are considered is about ten times larger than that in cases of Figs. 1(A) and 1(B), because of the difference of total heating in the atmosphere. From Figs. 3(A), 3(B) and Figs. 4(A), 4(B), we also find that the maximum amplitude of standing waves in the case of $b=40°$ in latitude is about two times larger than that in the case of $b=20°$ in latitude. However, if we consider that the total heating in the atmosphere in the former case is about two times larger than that in the latter case, the discussion about the role of the meridional width of the diabatic heating in the case of an one-dimensional model atmosphere is not simply extended to the case of two-dimensional model atmosphere.

4. Computation of the response to forcing by actual Hemispheric stationary heat sources in winter

In this section, we use the result of the diabatic heating in January proposed by Ashe (1978) as the actual Hemispheric stationary heat sources in winter. Since his result denotes the mean diabatic heating between 1,000 mb and 500 mb, we assume that the vertical distribution of diabatic heating is expressed by equation (4) and $(H_0(\varphi))_k$ in (4) is given by the value estimated by Ashe. That is, $(H_0(\varphi))_k$ is computed by expanding Ashe's result into the Fourier series into the longitudinal direction. For the sake of reference, the amplitudes of $(H_0(\varphi))_k$ for zonal wave number $k=1$ and $k=2$ are shown in Fig. 5.

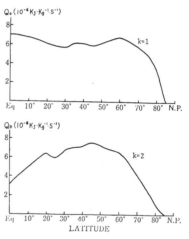

Fig. 5 The amplitude of zonal wave component 1 (upper), and 2 (bottom), for the Hemispheric stationary heat sources.

Figs. 6(A) and 6(B) show the vertical distribution of the amplitude and phase of standing wave responding to forcing by actual Hemispheric stationary heat sources for $k=1$ and $k=2$, respectively. The following results may be mentioned from Figs. 6(A) and 6(B).

1) The amplitude of standing waves responding to forcing by actual Hemispheric stationary

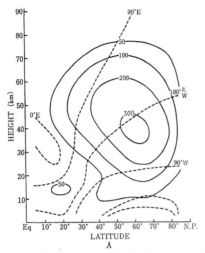
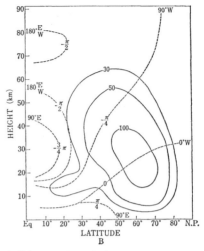

Fig. 6 The vertical distribution of amplitude (m) (solid curves) and phase (dashed curves) of standing wave, responding to forcing by the Hemispheric stationary heat sources for wave number $k=1$ (A), and $k=2$ (B), respectively. Position of ridges is shown on the right of equal phase line. The values of phase indicates with in (B) are that computed from the equation (37) in Part I.

heat sources for $k=1$ has a maximum at 38 km height, 60°N, and has a secondary peak at 13 km height, 20°N, while the amplitude of standing waves in the case of $k=2$ has a maximum at 27 km height, 60°N, and has a secondary peak at 15 km height, 20°-30°N. The phases of standing waves responding to forcing by actual Hemispheric stationary heat sources are tilted westward with increasing height in both cases of $k=1$ and $k=2$.

2) As mentioned in the case of idealized stationary heat sources, the amplitude of standing waves for $k=1$ responding to forcing by actual Hemispheric stationary heat sources is larger than that for wave number 2.

3) For $k=1$, the amplitude of standing waves responding to forcing by actual Hemispheric heat sources is larger than that responding to forcing by actual topography (see Fig. 9(A) in Part I). This result presents a contrast to the result obtained by NCAR model (Kasahara et al., 1973) where it is emphasized that the dynamical effect of the earth's topography is a dominant factor in reproducing the Aleutian high in the stratosphere.

4) The standing waves responding to forcing by actual Hemispheric stationary heat sources, especially by stationary heat sources at middle latitudes, can be propagated toward low latitudes. Webster (1972) suggested that most of the time independent circulation in low latitudes is forced by heating and orography within the tropics and subtropics. However, we consider that the forcing effects of stationary heat sources at middle and high latitudes are more important in the formation of standing waves in the upper troposphere at low latitudes.

Next, we shall compute the disturbance pattern at isobaric surface (or at constant height) by synthesizing zonal wave components of $k=1$, 2 and 3. Fig. 7 shows the disturbance pattern

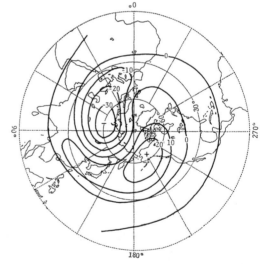

Fig. 7 The disturbance pattern (units in dm) at 30 km level, responding to forcing by the Hemispheric stationary heat sources.

at 30 km level responding to forcing by actual Hemispheric stationary heat sources. In the figure, we may find that the forced pattern consists of a negative anomaly over the eastern Hemisphere and a positive anomaly over the western Hemisphere. Similarly, it is shown in Fig. 7 that the disturbance pattern responding to forcing by actual stationary heat sources is stronger than that forced by the actual topography (see Fig. 13 in Part I).

5. Computation of standing waves responding to forcing by the actual Hemispheric topography and stationary heat sources in winter

In this section we discuss the response of standing waves to forcing by both actual topography and stationary heat sources in winter, and compare with the observed values. If we substitute the forcing function of both the actual topography and diabatic heating into the linear algebraic equations (14)-(15) in Part I, we obtain the vertical distributions of the amplitude and phase for standing waves responding to these actual Hemispheric forcing mechanisms.

The vertical distributions of the amplitude and phase response to forcing by both the actual Hemispheric topography and stationary heat sources in winter for zonal wave number $k=1$ and $k=2$ are shown in Figs. 8 and 9, respectively. For the sake of comparison, the latitude-height sections of the observed amplitude and

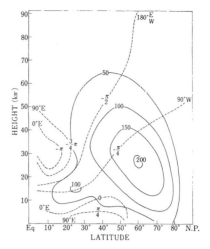

Fig. 9 As in Fig. 8 except for the wave number $k=2$.

phase for $k=1$ and $k=2$ in January, averaged over the years 1964-1970 in the troposphere and over the years 1965-1969 in the stratosphere, are reproduced from the paper of van Loon, et al. (1973) in Figs. 10 and 11, respectively.

The computed values of the maximum amplitude and secondary peak of standing waves for zonal wave numbers $k=1$-3 responding to forcing by the actual topography, actual stationary heat sources, and both the actual topography and stationary heat sources are given in Table 1. The observed values are also shown in Table 1.

Here, we may find the following results from Figs. 8-11 and Table 1.

1) The amplitude of computed standing waves responding to forcing by both the actual topography and stationary heat sources has a maximum at 38 km height, 60°N for $k=1$, and at 27 km height, 60°N for $k=2$, respectively and the computed values of the maximum amplitude are about 440 m for $k=1$ and 200 m for $k=2$, respectively. In addition, we have a secondary peak at 15 km height, 20°N for $k=1$ and at 15 km height, 25°N for $k=2$, respectively. The amplitude of standing waves for $k=3$ responding to forcing by the actual topography and stationary heat sources has a maximum in the troposphere, because it cannot be propagated into the stratosphere, but it has also a secondary peak at 13 km height, 20°-25°N.

As shown in Figs. 10 and 11 and also in Table 1, the amplitude of observed standing waves has a maximum at about 30 km height, 65°N for $k=1$ and at 30 km height, 60°N for $k=2$, re-

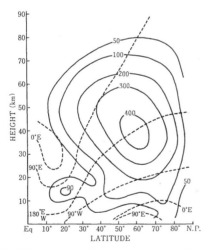

Fig. 8 The vertical distribution of amplitude (m) (solid curves) and phase (dashed curves) of standing wave, responding to forcing by both the Hemispheric topography and stationary heat sources for wave number $k=1$. Position of ridges is shown on the equal phase line.

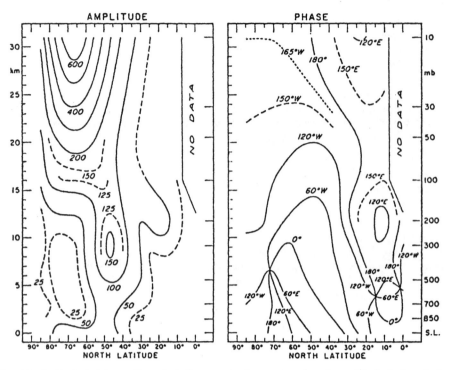

Fig. 10 Latitude-height section of stationary zonal wave number 1 amplitude (m) and phase (longitude of ridge) in January, averaged over the years 1964-1970 in the troposphere and over 1965-1969 in the stratosphere (from van Loon et al., 1973).

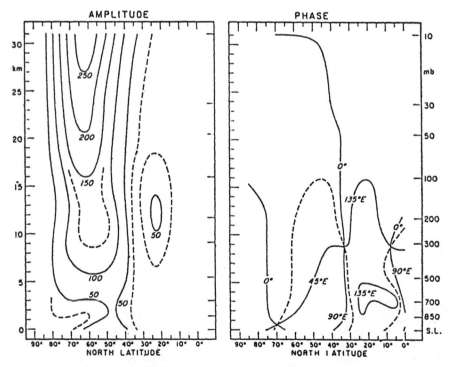

Fig. 11 As in Fig. 10 except for zonal wave number 2.

Table 1. The values and positions of maximum and secondary peak of amplitude of standing waves for wave numbers 1-3.

zonal wave number k		Computation		Observation	
		value (m)	position	value (m)	position
1	maximum	439	60°N, 38 km	600	65°N, 30 km
	secondary peak	94	20°N, 15 km	50	20°N, 12 km
2	maximum	203	60°N, 27 km	250	60°N, 30 km
	secondary peak	100	25°N, 15 km	50	20-25°N, 12 km
3	maximum	49	60°N, 10 km	100	50°N, 10 km
	secondary peak	50	20-25°N, 13 km	25	20-25N, 10 km

spectively. On the other hand, the observed vertical distribution of the amplitude of standing waves for $k=3$ differs appreciably from the other. The amplitude has a maximum at 10 km height, 50°N. It is interesting that a secondary peaks is observed in the upper troposphere at 20°-30°N for $k=1$-3.

Similarly, the phase of standing waves for $k=1$ is tilted westward with increasing height, and shifts westward in going toward the equator. Moreover, this ridges (or troughs) tilt for $k=2$ relatively little, as comparison with wave number $k=1$.

We may be able to conclude from these results that the vertical distribution of the amplitude and phase for standing waves computed from our model is in good agreement with the observational result. However, we must remark that the computed phase for $k=2$ is about 50° westward in longitude and the computed amplitude for $k=3$ is relatively small compared with the observed result.

2) A remarkable feature in the amplitude distributions computed from our model is that disturbances are almost entirely confined to high latitudes, centered at 60°N. This is one of the well-known characteristics of the stratospheric circulation. Moreover, the secondary peaks of standing waves are obtained in the upper troposphere at low latitudes. This secondary peak may be considered as one of reasons for the formation of standing waves at low latitudes.

Next, we shall compute the disturbance pattern at an isobaric surface (or at constant height), responding to forcing by the actual topography and stationary heat sources in winter, by synthesizing zonal wave components $k=1$-3. Fig. 12 shows the computed disturbance pattern at 30 km level. Fig. 13 shows the observed January plane-

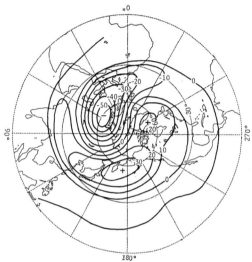

Fig. 12 The disturbance pattern (units in dm) at 30 km level, responding to forcing by both the Hemispheric topography and stationary heat sources.

tary wave height pattern due to zonal wave numbers $k=1$-3 at 10 mb (Avery, 1978). Roughly speaking, the computed pattern is in good agreement with the observed one. In both the computed and observed patterns, the positive anomaly is found over Alaska and Siberia (Aleutial high), although the computed intensity seem to be weak compared with the observed one, while the negative anomaly is found over Europe, and the computed intensity is in good agreement with the observed one.

Since the standing waves for $k=1$-3 have a secondary peak in the upper troposphere at 20°-30°N, we compute the disturbance pattern at 12 km-level, as shown in Fig. 14. For the sake of comparison, we calculate the disturbance pattern at 200 mb in January, averaged over the

Fig. 13 The observed January planetary wave height pattern (units in 100 m) due to zonal wave numbers 1-3 at 10 mb (after Avery, 1978).

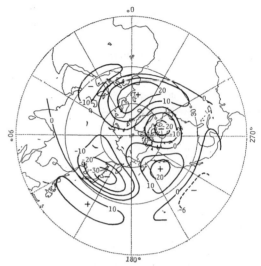

Fig. 15 The observed January planetary wave height pattern (units in *dm*) due to zonal wave number 1-3 at 200 mb.

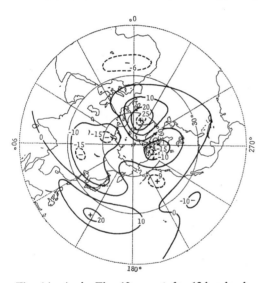

Fig. 14 As in Fig. 12 except for 12 km level.

years 1972-1977, from the observed NMC data and it is shown in Fig. 15. We may find that the computed result is in good agreement with the observed one. In both computed and observed patterns, the negative anomalies are found over the eastern coast of Asia and North America, while the positive anomalies are found near the west side of the Rocky Mountains and the Atlantic Ocean. But, the computed intensity of the negative anomaly over the eastern coast of Asia and positive anomaly over the western part of the Rocky Mountains are rather weaker than the observed ones. It is interesting to note that there are three subtropic standing disturbance patterns over the region at about 20°N in Fig. 14, i.e., the positive anomaly is found over the western coast of the Pacific Ocean, while the negative anomalies are found over the eastern coasts of the Pacific Ocean and North Africa. These computed patterns are qualitatively in good agreement with the corresponding observed patterns in Fig. 15. Thus, we may be able to speculate that the standing patterns in the upper troposphere at low latitudes, such as subtropical high over the west coast of the Pacific Ocean, are considered as the result of responding to forcing by the Hemispheric topography and stationary heat sources at middle and high latitudes. Concerning this speculation, we add some comment about the use of quasi-geostrophic model for the discussion in low latitudes. As mentioned in Part I of this paper, the similar experiment was done by Laprize (1978) by making use of a steady-state primitive equation model and he obtained a little different result from ours about the subtropic standing waves. Since there is no heat sources in his case and the zonal wind is different from ours in a sense that there is no easterlies in low latitudes, i.e., there is no absorber for standing waves in low latitudes, the direct comparison of our results with Laprize's results is not possible. However, the further study about the comparison of our results with those obtained by a primitive model will be necessary.

Next, we examine the disturbance patterns in the middle troposphere. Fig. 16 shows the computed disturbance patterns at 6 km-level. For the sake of comparison, we also calculated the disturbance patterns at 500 mb in January, averaged over the years 1972-1977, from the observed NMC data and it is shown in Fig. 17. In both the computed and observed patterns in Figs. 16 and 17, the negative anomalies are found over the eastern coasts of Asia and North America, while the positive anomalies are found over the western coasts of the Rocky Mountains and the Atlantic Ocean. We may also find that, although the computed intensities of the negative anomaly over the eastern coast of Asia and the positive anomaly over the Rocky Mountains are weaker than the observed ones, the regions of the negative and positive patterns computed by our model are in good agreement with the observed patterns. We may also find that the results obtained by our model are far better than those computed by Ashe (1978).

Thus, we may be able to conclude that the disturbance patterns at constant height responding to forcing by the Hemispheric topography and stationary heat sources in winter are relatively in good agreement with the observed patterns, and far better than the results computed with the simpler models such as the "two-levels" model.

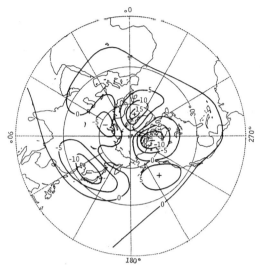

Fig. 16 As in Fig. 12 except for 6 km level.

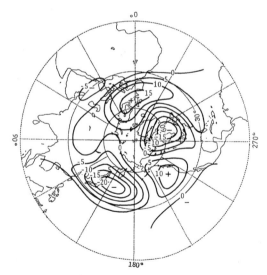

Fig. 17 As in Fig. 15 except for 500 mb.

6. Computation of the momentum and heat eddy fluxes

In the previous sections, we have computed the vertical distributions of standing waves and disturbance patterns at constant height forced by the Hemispheric topography and stationary heat sources. Therefore we can easily compute the momentum and heat fluxes of standing waves.

Considering that the geopotential ϕ' is expressed by

$$\phi'(\lambda, \varphi, p) = \sum_{k=1}^{K} \Phi_k(\varphi, p) e^{ik\lambda},$$

(k: zonal wave number), we have the momentum and heat fluxes of standing waves as follows:

$$\overline{u'v'} = \frac{1}{2} \frac{1}{4\Omega_0^2 \sin^2\varphi} \frac{1}{a^2 \cos\varphi} \sum_{k=1}^{3} k I_m$$
$$\times \left(\frac{\partial \Phi_k^*}{\partial \varphi} \Phi_k \right), \quad (6)$$

$$\overline{v'T'} = \frac{1}{2} \frac{1}{2\Omega_0 \sin\varphi} \frac{1}{a \cos\varphi} \frac{p}{R} \sum_{k=1}^{3} k I_m$$
$$\times \left(\frac{\partial \Phi_k^*}{\partial p} \Phi_k \right), \quad (7)$$

where the symbol "bar" denotes the zonal mean, Φ_k^* is the conjugate complex of Φ_k, Ω_0 is the rotation rate of the earth, p is the pressure, R is the gas constant and a is the radius of the earth, respectively. In (6) and (7), we pick up only three zonal component for the sake of simplicity.

The computed results of $\overline{u'v'}$ and $\overline{v'T'}$ are shown in Figs. 18 and 19, respectively. As may be seen in Fig. 18, the large value of poleward momentum flux is found in the stratosphere. It has a maximum at 38 km height, 45°-55°N, and has a secondary peak at 16 km height, 15°-35°N. In addition, the momentum flux toward the equator is found in the lower stratosphere over the tropics and also at high latitudes from the troposphere up to the 16 km height. This result is in qualitative agreement with the result computed by Pyle and Rogers (1980) in the stratosphere and mesosphere, and with the result obtained by Oort (1971) in the troposphere. Here we remark that the maximum value of poleward momentum flux in our case is about two times larger than that obtained Pyle and Rogers. This difference may be obtained from the use of different coefficient of Rayleigh friction, R_f in the vorticity equation and the different assumption of zonal wind in the upper stratosphere. That is, we adopt the relatively large value of R_f in the upper atmosphere compared with that of Pyle and Rogers and concerning the basic mean wind they assume that the zonal wind increases linearly with height. We may also find from Fig. 19 that the poleward heat flux in the stratosphere has a maximum at 36 km height, 65°N. This result is also in qualitative agreement with the result computed by Pyle and Rogers (1980). In addition the computed heat eddy flux has a secondary peak, in the lower stratosphere over the tropics. This result is also similar to the observed result obtained by Oort (1971).

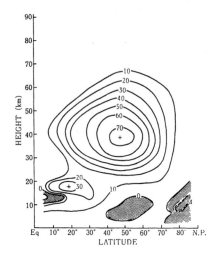

Fig. 18 The vertical distribution of horizontal momentum flux ($m^2 s^{-2}$) due to standing waves, responding to forcing by both the Hemispheric topography and stationary heat sources.

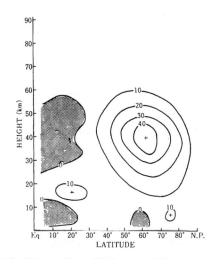

Fig. 19 The vertical distribution of horizontal heat flux (°K·ms^{-1}) due to standing waves, responding to forcing by both the Hemispheric topography and stationary heat sources.

7. Computation of Eliassen-Palm flux due to stationary planetary waves

In this section we compute the Eliassen-Palm flux due to planetary waves in the vertical and meridional plane. Concerning the estimation of the Eliassen-Palm flux from the observational data, many papers have been presented. For example, Sato (1980) made the observationary estimates of Eliassen-Palm flux due to quasi-stationary planetary waves in the troposphere. Edmon et al. (1980) also calculated Eliassen-Palm fluxes cross sections for the troposphere. Iwashima (1981) computed the Eliassen-Palm fluxes of standing waves for zonal wave numbers $k=1$-4.

The eddy momentum and heat fluxes obtained in section 6 can be directly used to compute the Eliassen-Palm flux F by means of the following relation

$$F=(F_y, F_p) \qquad (8)$$

where

$$F_y = \frac{a\cos\varphi}{\bar{U}}\overline{(\phi'v')_k} = -a\cos\varphi\overline{(u'v')_k}, \quad (9)$$

$$F_p = \frac{a\cos\varphi}{\bar{U}}\overline{(\phi'\omega')_k} = -a\cos\varphi\frac{fR}{\sigma P}\overline{(T'v')_k},$$

$$\text{for } k=1, 2, 3. \quad (10)$$

Here f is the Coriolis parameter and σ is the static stability parameter and σ is the static stability parameter defined by $\sigma = -\alpha(\partial/\partial p)\ln S$ (α: specific volume, θ: potential temperature).

It is convenient, for the sake of comparison with the observational estimates, to convert ω' in equation (10) into w', such that

$$F_z = a \cos\psi \frac{fR}{\rho g \sigma p} \overline{(T'v')}_k,$$
$$\text{for } (k = 1, 2, 3) \quad (11)$$

Figs. 20 and 21 show the Eliassen-Palm flux (F_y, F_z) due to standing wave responding to forcing by the Hemispheric topography and stationary heat sources in winter for $k=1$ and $k=2$, respectively. We may find that the Eliassen-Palm flux due to standing waves for $k=1$ has a maximum region near 40 km height, 50°-60°N, while for $k=2$ has a maximum region near 30 km height, 40°-50°N, and has a secondary peak near 16 km height over the tropics. The fluxes in Figs. 20 and 21 show that they are toward the tropics in the stratosphere, and are blocked in the regions of small values of refrac-

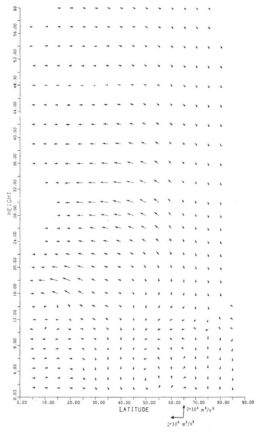

Fig. 21 As in Fig. 20 except for wave number 2.

tive index square for $k=0$, Q_0 such as in the lower stratosphere of middle latitudes, and in the easterly regions over the tropics. Moreover, a remarkable feature may be found near the tropopause at middle and high latitudes. That is, the Eliassen-Palm flux for $k=1$ near the tropopause at middle and high latitudes is toward the tropics, while the Eliassen-Palm flux for $k=2$ near the tropopause at middle and high latitudes is toward the pole. This result is in qualitative agreement with the result revealed by Iwashima (1981) from the observed data.

8. Computation of the energy flux due to stationary planetary waves

The eddy momentum and heat fluxes obtained in section 6 can be also directly used to compute the energy flux by means of the following relation

$$E = \left(\sum_{k=1}^{3} \overline{\rho(\phi'v')}_k , \sum_{k=1}^{3} \overline{\rho(\phi'w')}_k \right), \quad (12)$$

Here we compute the value of E, synthesizing

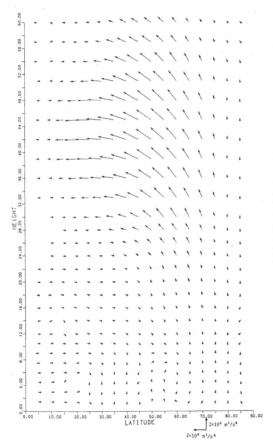

Fig. 20 The vertical distribution of Eliassen-Palm flux due to standing wave for wave number 1, responding to forcing by both the Hemispheric topography and stationary heat sources.

only three zonal components $k=1$-3, for the sake of simplicity.

Fig. 22 shows the energy flux due to standing

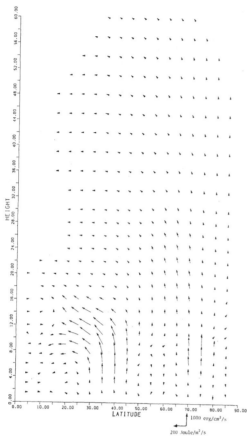

Fig. 22 The vertical distribution of energy flux due to standing waves 1-3, responding to forcing by both the Hemispheric topography and stationary heat sources.

waves responding to forcing by the Hemispheric topography and stationary heat sources in winter, computed by (12). We can clearly see that the region of the small value of refractive index square for $k=0$, Q_0 at the lower stratosphere at middle latitudes blocks the energy flux and makes it bifurcated. As a consequence the strong upward energy current appears in the upper troposphere at middle latitudes and near 70°N. This result is different from that computed by Matsuno (1970). This difference may come from the reason that the actual atmospheric static stability is used in this investigation, while Matsuno assumed the atmosphere is nearly isothermal. In Matsuno's result, a secondary peak of standing waves is not obtained. Here, as mentioned in section 6, we remark again the role of the wave guide oriented from the lower troposphere at middle latitudes toward the upper troposphere over the tropics (see Figs. 6 and 7 in Part I) to the formation of a secondary peak mentioned above. Another interesting feature is that wave energy vanishes in the line of zero zonal wind located at low latitudes which is an important energy sink for planetary waves propagating in middle and high latitudes (Dickinson, 1968).

9. Conclusion and discussion

A 34-level quasi-geostrophic, linear, steady-state model in a spherical coordinate with Rayleigh friction, the effect of Newtonian cooling and horizontal thermal diffusivity is used to obtain the vertical and meridional distributions of the amplitude and phase of standing waves responding to forcing by stationary heat sources in winter. The computed results may be summarized as follows:

1) The amplitude of standing waves forced by idealized stationary heat sources at high latitudes is rather larger than that forced by stationary heat sources at middle latitudes for zonal wave number $k=1$.

2) The standing waves responding to forcing by the actual Hemispheric stationary heat sources in winter, especially by that at middle latitudes, can be propagated toward low latitudes, and this result may be considered as one of reasons for the formation of standing waves in the upper troposphere at low latitudes.

3) For the zonal wave number $k=1$, the amplitude of standing wave responding to forcing by the Hemispheric stationary heat sources in winter is larger than that responding to forcing by the Hemispheric topography.

We computed also the vertical distributions of amplitudes and phases of standing waves, and the disturbance patterns at constant height, responding to forcing by both the Hemispheric topography and actual stationary heat sources in winter. The computed results are in good agreement with the observed ones.

We compute the momentum and heat fluxes and also the Eliassen-Palm flux due to the standing waves responding to forcing by the Hemispheric topography and stationary heat sources in winter. The computed results are qualitatively in good agreement with the observational results. The Eliassen-Palm flux for zonal wave number

$k=1$ is inverse to that for $k=2$ near the tropopause at middle and high latitudes.

As mentioned in section 4, the vertical distributions of diabatic heating used in this study are based on the assumptive distribution. Although this distribution seems to be suitable over the Tibetan Plateau, the vertical distributions of heating/cooling over the ocean should be re-examined.

As described in Part I of this paper, many authors such as Ashe (1978), and Frederikson and Sawford (1981) investigated the standing waves in nonlinear spherical model. Their results show that nonlinear effects are important. Therefore, further studies should be aimed at the incorporation of nonlinear terms in the model. The limitation of making use of a quasi-geostrophic model to the discussion of standing waves in low latitude should be also examined in the future.

Acknowledgements

The authors are grateful to Prof. T. Matsuno for his valuable comments and suggestions during the course of this study. Thanks are due to Dr. T. Iwashima for his kindness to give us the information about his unpublished results. Thanks are due to Dr. Ts. Nitta, Dr. Y. Sato, and Dr. T. Satomura for discussions throughout the course of this work. Thanks are also extended to Mr. Y. Fujiki for drawing the figures and Mrs. K. Kudo for the expert typing of manuscript. This paper is sponsored in Part by Grant in Aid for Scientific Research from the Ministry of Education.

References

Ashe, S., 1978: A nonlinear model of the time-average axially asymmetric flow induced by topography and diabatic heating. *Journal of the Atmos. Sci.*, **36**, 109-126.

Avery, S., 1978: The tropospheric forcing and vertical propagation of stationary planetary waves in the atmosphere. Ph.D. Thesis, University of Illinois at Urbana-Champaign. Available from University Microfilm, Ann Arbor, Michigan.

Dickinson, R. E., 1968: Planetary Rossby waves propagating vertically through weak westerly wind wave guides. *Journal of the Atmos. Sci.*, **25**, 984-1002.

Edmon, H. J., B. J. Hoskins and M. Z. McIntyre, 1980: Eliassen-Palm cross sections for the troposphere. *Journal of the Atmos. Sci.*, **37**, 2600-2616.

Egger, J., 1976: On the theory of the steady perturbations in the troposphere. *Tellus*, **28**, 381-389.

Eliassen, A. and E. Palm, 1961: On the transfer of energy in stationary mountain waves. *Geofysiske Publikasjoner*, **22**, 1-23.

Frederikson, J. S. and B. L. Sawford, 1981: Topography wave in nonlinear and linear spherical barotropic models. *Journal of the Atmos. Sci.*, **36**, 69-86.

Huang, Rong-hui and K. Gambo, 1981: The response of a model atmosphere in middle latitudes to forcing by topography and stationary heat sources. *Journal of the Meteor. Soc. of Japan*, **59**, 220-237.

——— and ———, 1982: The response of a hemispheric multi-level model atmosphere to forcing by topography and stationary heat sources. Part I. *Journal of the Meteor. Soc. Japan*, **60**, 78-92.

Iwashima, T., 1981: Analysis of standing wave in the atmosphere (present at the annual meeting of the Meteorological Society of Japan in March 1981).

Kasahara, A., T. Sasamori and W. M. Washington, 1973: Simulation experiments with a 12-layer stratospheric global circulation model. I. Dynamical effect of the earth's orography and thermal influence of continentality. *Journal of the Atmos. Sci.*, **30**, 1129-1137.

Laprize, R., 1978: On the influence of stratospheric conditions on forced tropospheric waves in a steady-state primitive equation model. *Atmosphere-Ocean*, **16**, 300-314.

Matsuno, T., 1970: Vertical propagation of stationary planetary waves in the winter northern hemisphere. *Journal of the Atmos. Sci.*, **27**, 871-883.

Murakami, T., 1972: Equational stratospheric waves induced by diabatic heat sources. *Journal of the Atmos. Sci.*, **29**, 1129-1137.

Oort, A. and E. M. Rasmusson, 1971: Atmospheric circulation statistics NOAA professional paper 5.

Pyle, J. A. and C. F. Rogers, 1980: Stratospheric transport by stationary planetary waves—the importance of chemical processes. *Quart. Jour. of the Royal Meteorological Society*, **106**, 421-446.

Sato, Y., 1980: Observational estimates of Eliassen and Palm flux due to quasi-stationary planetary waves. *Jour. of the Meteor. Soc. of Japan*, **58**, 430-435.

Staff Members, Academia Sinica 1958: On the general circulation over eastern Asia. *Tellus*, **10**, 299-312.

van Loon, H., R. L. Jenne and K. Labitzke, 1973: Zonal harmonic standing waves. *Journal of Geophysical Research*, **78**, 4463-4471.

Webster, P. J., 1972: Response of the tropical atmosphere to local steady forcing. *Mon. Wea. Rev.*, **100**, 518-541.

关于冬季北半球定常行星波传播
另一波导的研究*

黄荣辉

(中国科学院大气物理研究所，北京)

岸保勘三郎

(日本国东京大学地球物理系)

摘 要

本文计算了在等温大气情况下定常行星波的折射指数平方，表明了定常行星波在垂直及侧向传播中除了极地波导外，还存在着一个从中纬度对流层下层指向低纬度对流层顶附近的一支波导。

本文应用一个包括 Rayleigh 摩擦、Newton 冷却及水平涡动热量扩散，34 层定常的准地转球坐标模式计算了北半球理想地形(或定常热源)强迫所产生的定常行星波垂直及侧向传播，证明了冬季北半球定常行星波传播中存在着这一波导。

利用上述模式计算了冬季北半球实际地形及定常热源强迫所产生的定常行星波的振幅与位相的垂直分布及其在等压面上定常扰动系统的分布。说明了这一波导的存在使得中、高纬度地形与定常热源强迫所产生的定常行星波向低纬度传播，这是冬季副热带定常扰动系统形成的主要原因之一。

一、引 言

关于定常行星波的传播，许多学者已作了研究，Charney 和 Drazin 在 60 年代初期就引进了波的折射指数，来讨论具有垂直切变的基本气流中定常行星波的垂直传播[1]。Dickinson 指出了定常行星波从对流层向平流层传播中存在着一个波导，即波要从高纬度向平流层传播，这就是极地波导[2]。但是，这两个研究中所用的基本气流是理想的气流，它在波动方程中可以简单地分离成垂直方向与经向方向两个分量。因此，Matsuno 应用了一个比较符合实际的基本气流垂直分布来研究对流层的定常行星波向平流层传播，提出平流层爆发性增温机制[3]。在他的研究中，解无法表达成简单的解析表达式，而是应用数值解。

上面这些理论只说明对流层的行星尺度的波从高纬度地区向平流层传播。它解释了冬季定常行星波的振幅为什么在高纬度平流层上、中层有一个最大值，但是根据 Van Loon 的计算结果[4]，冬季纬向波数 1 与 2 的定常行星波的振幅在低纬度对流层顶附近还有一个第二峰值，关于这个峰值的形成原因至今没有人解释过。

---

* 本文原载于：中国科学(B 辑)，第 10 期，940-950，1983 年 10 月出版。

此外，近年来，冬季北半球中、高纬度与低纬度的相互作用已有许多研究。最近 Gambo 计算了十年冬季北半球 $45°N, 135°E$ 格点的 700 毫巴的高度偏差 ($z' = z - \bar{z}^\lambda$) 的月平均值与全球各网格点 200 毫巴高度偏差的月平均值的相关系数，发现冬季 $45°N, 135°E$ 700 毫巴的高度偏差值与 $25°N, 165°E$ 附近太平洋地区上空的 200 毫巴高度偏差值有很好的负相关[5]，这说明北半球中纬度对流层下层与低纬度对流层顶附近的高度偏差有很好的三维遥相关。

显然，这种三维遥相关绝不是偶然，而是有内在联系的。为了解释这种三维遥相关的机制，本文从准地转涡度方程及热力学方程入手，应用符合实际的北半球平均纬向风速剖面图，利用数值解法得出冬季北半球各波数定常行星波的折射指数平方的分布，从理论分析了定常行星波的传播除了极地波导以外，还存在一支从中纬度对流层下层指向低纬度对流层顶附近的波导。利用数值模式计算了理想强迫源其强迫所产生的定常行星波的传播情况，及其北半球实际地形及热源强迫所产生的定常行星波的传播。

二、定常行星波的折射指数平方与波传播的波导

在不考虑 Rayleigh 摩擦、Newton 冷却及水平涡动热力扩散情况下，球坐标的线性化准地转涡度方程及热力学方程分别表示成：

$$\left(\frac{\partial}{\partial t} + \bar{U}\frac{\partial}{a\cos\varphi\partial\lambda}\right)\zeta' + v'\frac{\partial}{a\partial\varphi}(\xi + f) = f\frac{\partial\omega}{\partial p}, \tag{1}$$

$$\left(\frac{\partial}{\partial t} + \bar{U}\frac{\partial}{a\cos\varphi\partial\lambda}\right)\left(\frac{\partial\phi'}{\partial p}\right) - 2\Omega_0\sin\varphi\frac{\partial\bar{U}}{\partial p}v' + \sigma\omega = 0. \tag{2}$$

这里 a 是地球半径，$\bar{U}$ 是基本气流，v' 是扰动的经向分量，ϕ' 是扰动位势，ζ' 是相对扰动涡度，ξ 是基本气流相对涡度，$\sigma = -\alpha\frac{\partial\ln\theta}{\partial p}$ 是静力稳定度参数（α 是比容，θ 是位温），f 是科氏参数，ω 是 p 坐标的垂直运动，φ 是纬度，λ 是经度，Ω_0 是地球自转角速度。

利用方程(1)与(2)可得到如下的 p 坐标位涡度方程式：

$$\left(\frac{\partial}{\partial t} + \hat{\Omega}\frac{\partial}{\partial\lambda}\right)\left\{\left[\frac{\sin\varphi}{\cos\varphi}\frac{\partial}{\partial\varphi}\left(\frac{\cos\varphi}{\sin\varphi}\frac{\partial\phi'}{\partial\varphi}\right) + \frac{1}{\cos^2\varphi}\frac{\partial^2\phi'}{\partial\lambda^2}\right]\right.$$
$$+ 4\Omega_0^2 a^2 \sin^2\varphi\frac{\partial}{\partial p}\left(\frac{1}{\sigma}\frac{\partial\phi'}{\partial p}\right)\right\} + \left[2(\Omega_0 + \hat{\Omega}) - \frac{\partial^2\hat{\Omega}}{\partial\varphi^2}\right.$$
$$\left. + 3\tan\varphi\frac{\partial\hat{\Omega}}{\partial\varphi} - 4\Omega_0^2 a^2 \sin^2\varphi\frac{\partial}{\partial p}\left(\frac{1}{\sigma}\frac{\partial\hat{\Omega}}{\partial p}\right)\right] \times \cos\varphi\frac{1}{\cos\varphi}\frac{\partial\phi'}{\partial\lambda} = 0, \tag{3}$$

其中 $\hat{\Omega} = \bar{U}/a\cos\varphi$，为方便起见，在讨论定常行星波传播时采用 z 坐标来代替 p 坐标，即 $z = -\tilde{H}_0\ln\frac{p}{p_0}$，$p_0$ 是参考压力，$\tilde{H}_0$ 是特征高度。由方程(3)可以得到 z 坐标的位涡度方程：

$$\left(\frac{\partial}{\partial t} + \hat{\Omega}\frac{\partial}{\partial\lambda}\right)\left\{\left[\frac{\sin\varphi}{\cos\varphi}\frac{\partial}{\partial\varphi}\left(\frac{\cos\varphi}{\sin\varphi}\frac{\partial\phi'}{\partial\varphi}\right) + \frac{1}{\cos^2\varphi}\frac{\partial^2\phi'}{\partial\lambda^2}\right]\right.$$
$$+ 4\Omega_0^2 a^2 \sin^2\varphi\frac{\partial}{p\partial z}\left(\frac{p}{N^2}\frac{\partial\phi'}{\partial z}\right)\right\} + \left[2(\Omega_0 + \hat{\Omega}) - \frac{\partial^2\hat{\Omega}}{\partial\varphi^2}\right.$$

$$+ 3\tan\varphi \frac{\partial \hat{Q}}{\partial \varphi} - 4\Omega_0^2 a^2 \sin^2\varphi \frac{\partial}{\partial p\partial z}\left(\frac{p}{N^2}\frac{\partial \hat{Q}}{\partial z}\right)\Bigg] \times \cos\varphi \frac{1}{\cos\varphi} \frac{\partial \phi'}{\partial \lambda} = 0, \tag{4}$$

N 是 Brunt-Väisälä 频率. 令

$$\phi'(\lambda, \varphi, z) = \text{Re} \sum_{\tilde{k}=1}^{\tilde{K}} \Phi_{\tilde{k}}(\varphi, z) e^{i\tilde{k}(\lambda - ct)}, \tag{5}$$

$\tilde{k}$ 是纬向波数，并引入新变量 $\Psi_{\tilde{k}}$

$$\Psi_{\tilde{k}}(\varphi, z) = e^{-z/2\tilde{H}_0} \Phi_{\tilde{k}}(\varphi, z). \tag{6}$$

这样由方程（4）可以得到下面波的传播方程:

$$(\hat{Q} - c)\left\{\frac{\sin\varphi}{\cos\varphi}\frac{\partial}{\partial \varphi}\left(\frac{\cos\varphi}{\sin\varphi}\frac{\partial \psi_{\tilde{k}}}{\partial \varphi}\right) + l^2 \sin^2\varphi \frac{\partial^2 \psi_{\tilde{k}}}{\partial z^2} - \left(l^2 \sin^2\varphi \frac{1}{4\tilde{H}_0^2} + \frac{\tilde{k}^2}{\cos^2\varphi}\right)\psi_{\tilde{k}}\right.$$

$$\left. - l^2 \sin^2\varphi\left[\frac{1}{\tilde{H}_0}\frac{\partial \ln N}{\partial z} + 2\left(\frac{\partial \ln N}{\partial z}\right)^2 - \frac{1}{N}\frac{\partial^2 N}{\partial z^2}\right]\psi_{\tilde{k}}\right\} + \left\{2(\Omega_0 + \hat{Q}) - \frac{\partial^2 \hat{Q}}{\partial \varphi^2}\right.$$

$$\left. + 3\tan\varphi \frac{\partial \hat{Q}}{\partial \varphi} - l^2 \sin^2\varphi\left[\frac{\partial^2 \hat{Q}}{\partial z^2} - 2\left(\frac{1}{2\tilde{H}_0} + \frac{\partial \ln N}{\partial z}\right)\frac{\partial \hat{Q}}{\partial z}\right]\right\} \times \psi_{\tilde{k}} = 0. \tag{7}$$

上式中 $l = 2\Omega_0 a/N$. 我们将进一步简化，假设大气是等温的，这样，方程（7）可写成下面简单的方程:

$$\frac{\sin\varphi}{\cos\varphi}\frac{\partial}{\partial \varphi}\left(\frac{\cos\varphi}{\sin\varphi}\frac{\partial \psi_{\tilde{k}}}{\partial \varphi}\right) + l^2 \sin^2\varphi \frac{\partial^2 \psi_{\tilde{k}}}{\partial z^2} + Q_{\tilde{k}}\psi_{\tilde{k}} = 0, \tag{8}$$

$$Q_{\tilde{k}} = Q_0 - \frac{\tilde{k}^2}{\cos^2\varphi}, \tag{9}$$

$$Q_0 = \left[2(\Omega_0 + \hat{Q}) - \frac{\partial^2 \hat{Q}}{\partial \varphi^2} + 3\tan\varphi \frac{\partial \hat{Q}}{\partial \varphi} - l^2\sin^2\varphi\left(\frac{\partial^2 \hat{Q}}{\partial z^2} - \frac{1}{\tilde{H}_0}\frac{\partial \hat{Q}}{\partial z}\right)\right]\Big/$$

$$(\hat{Q} - c) - l^2\sin^2\varphi \frac{1}{4\tilde{H}_0^2}. \tag{10}$$

在定常情况下，$c = 0$, 这样 Q_0 变成下式:

$$Q_0 = \left[2(\Omega_0 + \hat{Q}) - \frac{\partial^2 \hat{Q}}{\partial \varphi^2} + 3\tan\varphi \frac{\partial \hat{Q}}{\partial \varphi} - l^2\sin^2\varphi\left(\frac{\partial^2 \hat{Q}}{\partial z^2} - \frac{1}{\tilde{H}_0}\frac{\partial \hat{Q}}{\partial z}\right)\right]\Big/\hat{Q}$$

$$- l^2\sin^2\varphi \frac{1}{4\tilde{H}_0^2}, \tag{11}$$

在等温大气情况下，$\tilde{H}_0$ 是 7 公里，N 是 $2 \times 10^{-2} s^{-1}$. 方程（8）描述了在等温大气中定常行星波在 $\varphi - z$ 面上的传播。只要有基本气流的分布，由（11）式可以求出 Q_0 的分布，并由（9）式求出 $Q_{\tilde{k}}$ 的分布。

令 $y = \tanh^{-1}(\sin\varphi)$, 方程（8）可变成 y, z 坐标的方程，即

$$\frac{1}{\cos^2\varphi}\frac{\partial^2 \psi_{\tilde{k}}}{\partial y^2} + l^2 \sin^2\varphi \frac{\partial^2 \psi_{\tilde{k}}}{\partial z^2} + Q_{\tilde{k}}\psi_{\tilde{k}} = 0. \tag{12}$$

令

$$\psi_{\tilde{k}}(y, z) = \bar{\phi}_0 e^{i(\tilde{m}y + \tilde{n}z)}, \tag{13}$$

并把方程（13）代入方程（12），令 $m = \dfrac{\tilde{m}}{\cos\varphi}$, $\tilde{a}^2 = l^2\sin^2\varphi$, $\tilde{a}\tilde{n} = n$, 可得到下面频率方

程,即
$$m^2 + n^2 = Q_{\tilde{k}}. \tag{14}$$

从方程(14)可以看到,$Q_{\tilde{k}}$ 就是波数为 $\tilde{k}$ 的定常行星波在 y-z 面上的折射指数平方. 从方程得知,波要在 y-z 面上传播,m 与 n 必须是实数,$m^2 + n^2 > 0$,从方程(14)得到 $Q_{\tilde{k}}$ 必须大于零. 反之,在 $Q_{\tilde{k}} < 0$ 的地方,如在 $\bar{Q} < 0$（也就是东风带）的地方,波是不容易往上传播的.

我们定义定常行星波在 y-z 面上群速度的分布为波的传播路径. 但是,由于冬季北半球平均纬向风速分布在垂直与经向均有切变,并且由于地球曲率的关系,波在传播中要发生折射,使得波的传播集中在某一、二个"通道"（称为波导）进行. 下面将推导定常行星波在传播中的折射情况.

假设在 y-z 面上行星波的频散方程是
$$\tilde{\omega} = F(y, z, K^2), \tag{15}$$

上式 $K^2 = k^2 + m^2 + n^2$,设波在 y-z 面上的群速度 $c'_g = (c_{gy}, c_{gz})$ 及波在传播中其路径与水平方向所成的角 $\tilde{\alpha}$,则

$$\tan \tilde{\alpha} = c_{gz}/c_{gy} = \frac{\partial \tilde{\omega}/\partial n}{\partial \tilde{\omega}/\partial m}. \tag{16}$$

由(15)式,对于某一纬向波数所对应的 K,记为 $K_{\omega'}$,(16)式可变成:

$$\tan \tilde{\alpha} = \frac{\partial F/\partial (K^2) \cdot \dfrac{\partial K^2_{\omega'}}{\partial n}}{\partial F/\partial (K^2) \cdot \dfrac{\partial K^2_{\omega'}}{\partial m}} = \frac{2n \partial F/\partial (K^2)}{2m \partial F/\partial (K^2)} = \frac{n}{m} \tag{17}$$

并且有下式关系:

$$\frac{\partial F}{\partial y} = -\frac{\partial F}{\partial (K^2)} \cdot \frac{\partial K^2_{\omega'}}{\partial y}, \quad \frac{\partial F}{\partial z} = -\frac{\partial F}{\partial (K^2)} \cdot \frac{\partial K^2_{\omega'}}{\partial z}. \tag{18}$$

从曾庆存的著作[6]中 §11.5 可以知道:

$$\frac{d_g m}{dt} = -\frac{\partial \tilde{\omega}}{\partial y}, \quad \frac{d_g n}{dt} = -\frac{\partial \tilde{\omega}}{\partial z}, \tag{19}$$

上式中 $\dfrac{d_g}{dt} = \dfrac{\partial}{\partial t} + c_g \cdot \nabla$ 是随路径的个体微商. 因此

$$\begin{cases} n \dfrac{d_g m}{dt} = -n \dfrac{\partial F}{\partial y} = n \dfrac{\partial F}{\partial (K^2)} \cdot \dfrac{\partial K^2_{\omega'}}{\partial y} = c_{gz} \dfrac{\partial K^2_{\omega'}}{\partial y}, \\ m \dfrac{d_g n}{dt} = -m \dfrac{\partial F}{\partial z} = m \dfrac{\partial F}{\partial (K^2)} \cdot \dfrac{\partial K^2_{\omega'}}{\partial z} = c_{gy} \dfrac{\partial K^2_{\omega'}}{\partial z}. \end{cases} \tag{20}$$

对(17)式沿波的传播路径取微商,则得

$$\frac{d_g \tan \tilde{\alpha}}{dt} = \sec^2 \tilde{\alpha} \frac{d_g \tilde{\alpha}}{dt} = m^{-2} \left(m \frac{d_g n}{dt} - n \frac{d_g m}{dt} \right). \tag{21}$$

把(20)式代入上式,则可得

$$\frac{d_g \tilde{\alpha}}{dt} = \frac{1}{m^2 + n^2} \left(c_{gy} \frac{\partial K^2_{\omega'}}{\partial z} - c_{gz} \frac{\partial K^2_{\omega'}}{\partial y} \right) = \frac{1}{K^2_\omega - k^2} i \cdot c'_g \times \nabla K^2_{\omega'}.$$

如果把上式中 k 令为 $\dfrac{\tilde{k}}{\cos\varphi}$，并利用（9）式与（14）式，则

$$\frac{d_g \hat{a}}{dt} = \frac{1}{Q_{\tilde{k}}} i \cdot c'_g \times \nabla Q_0 \tag{22}$$

i 是 λ 方向的单位矢量． 这就是波在传播中其路径折射所要遵守的方程，波的传播路径是与 $Q_{\tilde{k}}$ 与 Q_0 的梯度有关．虽然我们研究的是定常情况，即 $\dfrac{\partial \hat{a}}{\partial t}=0$．但是从（22）式可以看到定常行星波在 y-z 面上的折射情况．若 $Q_{\tilde{k}} > 0$，波的传播路径总是向着 Q_0 的梯度方向折射．

从上面结论可知，Q_0 与 $Q_{\tilde{k}}$ 的分布对于定常行星波传播的路径是非常重要的，因此，我们对（11）式取 φ, z 方向的有限差分形式并利用图 1 所示的冬季北半球平均纬向风速的垂直分布，计算出 Q_0 的分布（见图 2）． 并由（10）式计算出 $Q_{\tilde{k}}$ 的分布． 图 3.a 中虚线就是 Q_1 的分布． 下面我们分析 $\tilde{k}=1$ 定常行星波的传播．

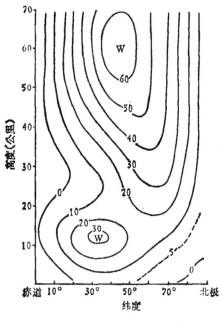 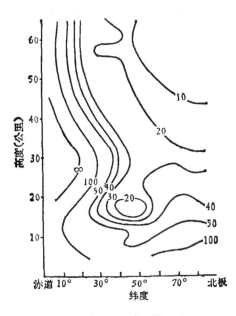

图 1　冬季北半球平均纬向风速分布
（单位：米/秒）

图 2　Q_0 分布图

从图 2 可以看到在高纬度平流层上层有一个 Q_0 最小值的区域，并且在中纬度平流层下层也存在一个最小值的区域，而在高纬度及低纬度地区分别是 Q_0 值相对大的区域． 我们把这个分布简单画出一个示意图（见图 3b）．在图中极地是一刚壁，假设有一强迫源所产生的定常行星波向上传播时，波沿 Q_0 的梯度方向折射，即它要向高纬度 Q_0 大的区域折射，但它受极地刚壁的反射，这样波将会聚焦在高纬度地区而往平流层传播．因此，高纬度地区上空就是定常行星波向平流层传播的一个通道．这与 Dickinson 所指出的极地波导是一致的[2]．当然，波传播到平流层时，它同样向着 Q_0 大的方向折射，即波将向赤道方向传播．此外，我们从中纬度与低纬度之间来看，在中纬度平流层下层是一个 Q_0 最小值的区域，而冬季低纬度对流层是一个 Q_0 大的区域，按照（22）式所示的折射规律，波同样也会向着低纬度方向折射，但是，冬季在低

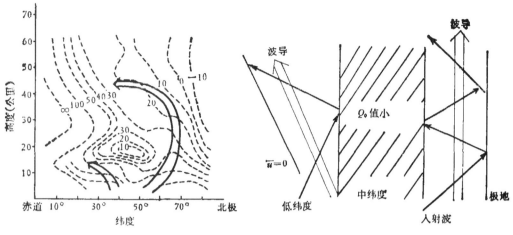

图 3a 位于40°N理想地形强迫所产生的 $\tilde{k}=1$ 定常行星波传播波导示意图(实箭号)
(图中虚线表示 $\tilde{k}=1$ 定常行星波折射指数平方 Q_1 的分布图)

图 3b 波导示意图

纬度地区有一东西风交界处,即 $\overline{U}=0$,波不会越过此线传播到东风区. 因此在低纬度地区也存在着一支波导,即波要从中、高纬度的对流层下层通过对流层向低纬度的对流层顶附近传播. 图 3a 的粗箭号就是从理论上提出的二支波导的示意图.

从图 3a 我们可以看到定常行星波它不能直接从中纬度对流层下层传播到平流层,它必须绕开位于中纬度平流层下层 Q_0 值小的区域.

我们同样计算了 Q_2,根据公式 (9),在极圈附近高纬度的 Q_2 值为负值,这样,在极圈附近的高纬度地区强迫源强迫所产生的定常行星波不能往平流层传播.

四、北半球理想地形及定常热源强迫所产生的定常行星波的传播

为了证明另一波导的存在,我们利用数值模式计算了北半球模式大气对理想地形(或定常热源)的响应,研究北半球理想(或定常热源)强迫所产生的定常行星波的传播.

本文所用的模式是一个包括 Rayleigh 摩擦、Newton 冷却及水平涡动热力扩散,34 层定常准地转球坐标模式,其模式方程如下:

$$\hat{Q}_{n-\frac{1}{2}} \frac{\partial}{\partial \lambda} \left\{ \frac{1}{2\Omega_0 \sin\varphi} \frac{1}{a^2} \left[\frac{\sin^2\varphi}{\cos\varphi} \frac{\partial}{\partial \varphi} \left(\frac{\cos\varphi}{\sin^2\varphi} \frac{\partial \phi'}{\partial \varphi} \right) + \frac{1}{\cos^2\varphi} \frac{\partial^2 \phi'}{\partial \lambda^2} \right] \right\}_{n-\frac{1}{2}}$$
$$+ \frac{1}{a} q_{n-\frac{1}{2}} \times \frac{1}{2\Omega_0 \sin\varphi} \times \frac{1}{a\cos\varphi} \frac{\partial \phi'_{n-\frac{1}{2}}}{\partial \lambda} = f\left(\frac{\partial \omega}{\partial p}\right)_{n-\frac{1}{2}} - (R_f)_{n-\frac{1}{2}}$$
$$\times \frac{1}{2\Omega_0 \sin\varphi} \frac{1}{a^2} \left[\frac{\sin\varphi}{\cos\varphi} \frac{\partial}{\partial \varphi} \left(\frac{\cos\varphi}{\sin\varphi} \frac{\partial \phi'}{\partial \varphi} \right) + \frac{1}{\cos^2\varphi} \frac{\partial^2 \phi'}{\partial \lambda^2} \right]_{n-\frac{1}{2}} \quad (23)$$

$$\hat{Q}_n \frac{\partial}{\partial \lambda} \left(\frac{\partial \phi'}{\partial p}\right)_n - \left(\frac{\partial \hat{Q}}{\partial p}\right)_n \frac{\partial \phi'_n}{\partial \lambda} + \sigma_n \omega_n = -\left(\frac{RH}{c_p p}\right)_n - (\alpha_R)_n \left(\frac{\partial \phi'}{\partial p}\right)_n$$
$$+ (K_T) \times \frac{1}{a^2} \left[\frac{\partial^2}{\partial \varphi^2} - \tan\varphi \frac{\partial}{\partial \varphi} + \frac{1}{\cos^2\varphi} \frac{\partial^2}{\partial \lambda^2} \right] \left(\frac{\partial \phi'}{\partial p}\right)_n \quad (24)$$

$$\hat{\Omega}_{n+\frac{1}{2}} \frac{\partial}{\partial \lambda} \left\{ \frac{1}{2\Omega_0 \sin\varphi} \frac{1}{a^2} \left[\frac{\sin^2\varphi}{\cos\varphi} \frac{\partial}{\partial\varphi}\left(\frac{\cos\varphi}{\sin^2\varphi}\frac{\partial\phi'}{\partial\varphi}\right) + \frac{1}{\cos^2\varphi}\frac{\partial^2\phi'}{\partial\lambda^2}\right]\right\}_{n-\frac{1}{2}} + \frac{1}{a}q_{n+\frac{1}{2}}$$

$$\times \frac{1}{2\Omega_0 \sin\varphi} \times \frac{1}{a\cos\varphi} \frac{\partial\phi'_{n+\frac{1}{2}}}{\partial\lambda} = f\left(\frac{\partial\omega}{\partial p}\right)_{n+\frac{1}{2}} - (R_f)_{n+\frac{1}{2}} \times \frac{1}{2\Omega_0 \sin\varphi} \frac{1}{a^2}$$

$$\times \left[\frac{\sin\varphi}{\cos\varphi}\frac{\partial}{\partial\varphi}\left(\frac{\cos\varphi}{\sin\varphi}\frac{\partial\phi'}{\partial\varphi}\right) + \frac{1}{\cos^2\varphi}\frac{\partial^2\phi'}{\partial\lambda^2}\right]_{n+\frac{1}{2}}. \tag{25}$$

上面公式中，H 是单位时间与单位质量的非绝热加热，R 是气体常数（0.287 千焦·公斤$^{-1}$·度$^{-1}$），c_p 是定压比热(1.004 千焦·公斤$^{-1}$·度$^{-1}$)，α_R 是 Newton 冷却系数，K_T 是水平涡动扩散，R_f 是扰动的 Rayleigh 摩擦系数，n 表示模式层次. 在上述方程中，q 为下式，

$$q = \left[2(\Omega_0 + \hat{\Omega}) - \frac{\partial^2 \hat{\Omega}}{\partial\varphi^2} + 3\tan\varphi \frac{\partial\hat{\Omega}}{\partial\varphi}\right]\cos\varphi,$$

在推导模式方程时，其涡度平流中的行星涡度平流项的 v' 引进非地转分量.

我们假设在模式顶的垂直速度为零，即

$$\omega = 0, \quad 在 \ p = p_t \ (\text{或} \ z = z_t), \tag{26}$$

作为上边界条件. 并假设地表面垂直速度由气流爬山及 Ekman 层粘性所产生的 Ekman 抽吸所引起，即

$$\omega_s = \boldsymbol{V}_s \cdot \nabla p_G - \frac{p_s \cdot \tilde{F}}{2f}\zeta'_s, \quad 在 \ p = p_s \ (\text{或} \ z = 0), \tag{27}$$

作为下边界条件. $\boldsymbol{V}_s$ 是在 $p = p_s$ 的水平风速矢量. 为了简单起见，$p_s = 1000$ 毫巴，$\tilde{F}$ 是摩擦系数，取为常数 $(4 \times 10^{-6} s^{-1})$，$\zeta'_s$ 是地表面的扰动涡度. p_G 为地形高度(换算为气压).

模式的差分方案与差分方程中所出现的系数均与文献[7]相同. 即垂直方向取 34 层，经向差分距取 5° 纬距. 静力稳定度由美国出版的标准大气的 $\bar{p}, \bar{T}, \bar{\rho}$ 计算而得到.

上面所得到的模式方程一般不能应用张驰方法，故应用 Lindzen 和 Kuo 所提出的方法来解这样线性方程组[8].

为了解这些线性代数方程组，还需要侧边界条件，我们需要 $\Phi_k(\varphi, p)$ 在极地及赤道为零，即

$$\Phi_k(\varphi, p) = 0, \quad \varphi = \frac{\pi}{2}; \quad \Phi_k(\varphi, p) = 0, \quad \varphi = 0. \tag{28}$$

因而，如果地形与非绝热加热的强迫函数已知，由模式就可以得到北半球地形及定常热源强迫所产生的定常行星波的传播情况.

我们假设强迫源为一理想地形(或理想热源)，其地表面地形的标准压力为 p_G，它可写成

$$p_G(\lambda, \varphi) = \text{Re} \sum_{k=1}^{K} (\hat{p}_g(\varphi))_k e^{ik\lambda}, \tag{29}$$

$(\hat{p}_g(\varphi))_k$ 是波数 k 地形的振幅. 为简单起见，φ 分别取 80°N 与 40°N，其值取 50 毫巴. 并把它代入模式方程，就能得到理想地形强迫所产生的定常行星波的垂直及经向分布情况. 图 4 与图 5 分别表示位于 80°N 与 40°N 理想地形强迫所产生的纬向波数 $k = 1$ 的定常行星波的振幅与位相的垂直分布.

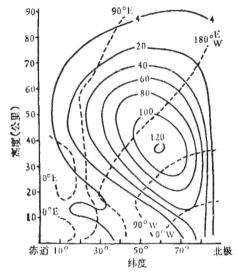 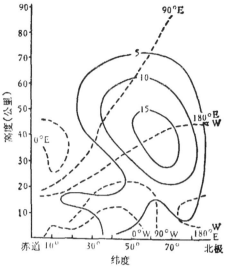

图4 位于 80°N 理想地形强迫所产生的纬向波数 $\tilde{k}=1$ 定常行星波的振幅（单位是米，实线）与位相（虚线）的垂直分布

图5 位于 40°N 理想地形强迫所产生的 $\tilde{k}=1$ 定常行星波的振幅（单位是米，实线）与位相（虚线）的垂直分布

从图4可以看到，位于 80°N 理想地形强迫所产生的纬向波数 $\tilde{k}=1$ 定常行星波振幅的最大值，位于40公里高度，60°N附近，并且在 25°N，15公里高度附近存在着第二峰值。从图5同样可以看到，位于 40°N 理想地形强迫所产生的纬向波数 $\tilde{k}=1$ 定常行星波的振幅最大值位于 40°公里高度，65°N附近，并且在 25°N，15公里高度也有一个第二峰值。

从振幅及位相的分布可以看到，位于 80°N 地形及定常热源强迫所产生的纬向波数1定常行星波的一支波的传播路径，是高纬度的对流层通过极地波导向平流层传播。而另一支波的传播波导是从高纬度对流层下层指向低纬度对流层顶附近。同样，位于 40°N 所产生的纬向波数1定常行星波传播中，也存在两支波导：由于中纬度平流层存在一个 Q_0 值小的区域，因此，中纬度地形（或定常热源）强迫所产生的定常行星波不能直接向中纬度平流层传播，而必须通过对流层向北传播。然后通过极地波导再向平流层传播。另一波导从中纬度对流层低层向低纬度对流层顶附近传播。

从上面分析，我们可以看出中、高纬度地形（或定热源）强迫所产生的定常行星波传播中确实存在一支从中纬度对流层下层向低纬度对流层顶附近传播的波导。

五、波导在中、低纬度相互作用中的作用

我们应用 Berkofsky 和 Bertoni 所计算的北半球地形资料作为北半球实际地形[9]，并应用 Ashe 所计算的1月份非绝热加热的结果，作为冬季北半球500毫巴实际定常热源[10]。我们对实际地形与定常热源进行富氏展开，并且，定常热源的垂直分布与 Murakami 所用的垂直分布一样[11]。从而我们就可以得到北半球实际地形与定常热源强迫所产生的定常行星波的振幅与位相的垂直分布。

图6表示纬向波数1冬季北半球实际地形与定常热源强迫所产生的定常行星波的振幅与

位相的垂直分布. 从这个图可以看到,我们的计算结果与 Van Loon 从观侧资料计算的结果一致[4]. 令人兴趣的是我们所计算的纬向波数 1 的定常行星波在 25°N, 15 公里高度附近存在着振幅第二峰值,这也与 Van Loon 从实际观侧资料所计算的结果一致. 因此,位于中、高纬度地形与定常热源强迫所产生的定常行星波,通过我们上述所提出的第二波导向低纬度对流层上层传播,这是形成低纬度对流层上层与平流层低层定常行星波的主要原因. Webster 提出低纬度平均环流是由于热带与副热带地形及定常热源强迫所造成[12]. 然而,我们认为冬季从中、高纬度强迫所产生的定常行星波通过波导向低纬度传播是更重要的. 这是中、低纬度相互作用的机制.

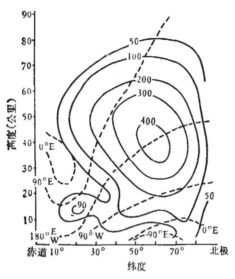

图 6 冬季北半球实际地形与定常热源强迫所产生的纬向波数 $k=1$ 定常行星波的振幅(单位是米,实线)与位相(虚线)的垂直分布

下面,我们把纬向波数 1—3 分量合成来计算冬季实际北半球地形与定常热源强迫所形成的等压面上的定常扰动系统. 图 7 表示由此模式所计算的 12 公里高度的定常扰动分布,与我们由观测资料所计算的结果一致. 令人感兴趣的是图 7 中在 20°N 附近大约有三个副热带定常扰动系统. 正距平位于太

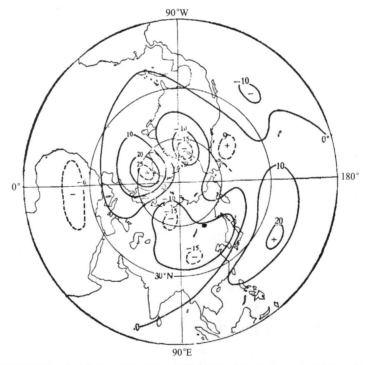

图 7 由模式计算所得到冬季北半球实际地形与定常热源强迫所产生的 12 公里高度上的定常扰动系统
(单位是十米)

平洋的西部,这就是副热带高压,它与实际位置比较一致. 此外还有两个负距平区域,分别位于太平洋的东部与北美. 这都是中、高纬度(主要是中纬度)地形与定常热源强迫所产生的定常行星波向低纬度传播所形成的. 这可以解释引言中所述 45°N, 135°E 附近的 700 毫巴高度偏差值与低纬度太平洋地区的 200 毫巴高度偏差值有一个很好的三维遥相关.

六、能量通量的计算

我们应用 Eliassen-Palm 所推导的公式来计算北半球实际地形与定常热源强迫所产生的定常行星波的能量通量矢量[13],即

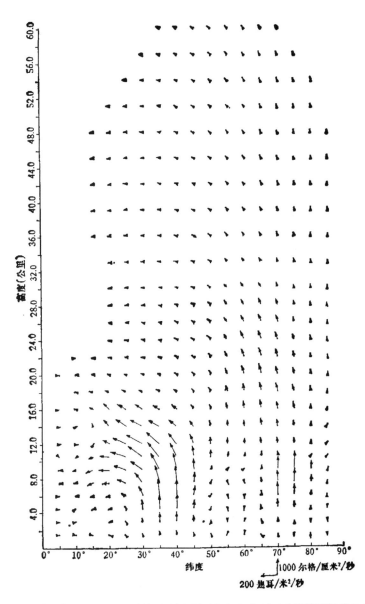

图 8　冬季北半球实际地形与定常热源强迫所产生的定常行星波的能量通量矢量分布图

$$E = \left(\sum_{k=1}^{3} \bar{\rho} \, \overline{(\phi'v')}_k, \sum_{k=1}^{3} \bar{\rho} \, \overline{(\phi'w')}_k \right), \tag{30}$$

$\bar{\rho}$ 是基本态的密度。"——"表示沿纬圈的平均。

图 8 表示冬季北半球地形与定常热源强迫所产生的定常行星波的能量通量的分布。从图 8 可以看到，在中纬度平流层低层 Q_0 值小的区域，波的能量通量显著减弱，强的上升能流出现在中纬度及 70°N 附近的高纬度地区。因此，从图 8 也可以看出，北半球地形与定常热源强迫所产生的定常行星波的传播中，存在两支波导：一支是从高纬度对流层指向平流层中、上层，另一支波导从中纬度对流层下层指向低纬度对流层顶附近。并且我们也可以看到，中纬度地形与定常热源强迫所产生的定常行星波的能量通过我们所指出的另一波导向低纬度对流层顶附近传播，这是低纬度对流层上层与平流层下层定常扰动重要的能量来源。

七、结 论

本文从理论上讨论了定常行星波传播中除了极地波导以外，还存在着一支从中纬度对流层下层指向低纬度对流层顶附近传播的波导。

本文还应用上述模式，计算了北半球地形与定常热源强迫所产生的定常行星波的振幅与位相的垂直分布及其在等压面上的定常扰动系统的分布，说明了波导在中、低纬度相互作用中所起的作用。

本文还计算了北半球地形与定常热源强迫所产生的定常行星波的能量通量矢量。从能量通量矢量的分布再一次证明了另一支波导的存在，并且说明了中纬度地形与定常热源强迫所产生的定常行星波的能量向低纬度传播是低纬度定常扰动主要的能量来源。

本文承蒙叶笃正先生提出宝贵意见，在此表示感谢。

参 考 文 献

[1] Charney, J. G. & Drazin, P. G., *Journal of Geophysical Research*, **66**(1961), 83—109.
[2] Dickinson, R. E., *Journal of the Atmospheric Sciences*, **25**(1968), 984—1002.
[3] Matsuno, T., *ibid.*, **27**(1970), 871—883.
[4] Von Loon, H., Jenne, R. L. & Labitzke, K., *Journal of Geophysical Research*, **78**(1973), 4463—4471.
[5] Gambo, K. & Kudo, K., *Journal of the Meteorological Society of Japan*, **61**(1983), 1.
[6] 曾庆存，数值天气预报的数学物理基础，科学出版社，1979.
[7] 黄荣辉，大气科学，**7**(1983),4期.
[8] Lindzen, R. S. & Kuo, H. L., *Monthly Weather Review*, **97**(1969), 732—734.
[9] Berkofsky, L. & Bertoni, E. A., *Bulletin of the American Meteorological Society*, **36**(1955), 350—354.
[10] Ashe, S., *Jouranl of the Atmospheric Sciences*, **36**(1979), 109—126.
[11] Murakami, T., *ibid.*, **29**(1972), 1129—1137.
[12] Webster, P. J., *Menthly Weather Review*, **100**(1972), 518—541.
[13] Eliassen, A. & Palm, E., *Geofysiske Publikasjoner*, **22**(1961), 1—23.

The Response of a Hemispheric Multi-Level Model Atmosphere to Forcing by Topography and Stationary Heat Sources in Summer*

Huang Ronghui

Institute of Atmospheric Physics, Academia Sinica

and

K. Gambo

Geophysical Institute, Tokyo University, Japan

Abstract

The stationary planetary waves responding to forcing by topography and stationary heat sources in summer are investigated by means of a steady-state, linear, quasi-geostrophic, 34-level model, with Rayleigh friction, the effect of Newtonian cooling and the horizontal kinematic thermal diffusivity included in a spherical coordinate system.

The results show that the main stationary planetary waves responding to forcing by both topography and heat sources in summer are confined to the troposphere over the subtropics. A secondary peak of the maximum amplitude is also found in the upper troposphere at high latitudes for zonal wave numbers $k=1$ and $k=2$. The amplitude of stationary planetary waves responding to forcing by heat sources is larger than that responding to forcing by topography. In this computation, the role of refractive index square of stationary planetary waves in summer is also discussed in order to make clear the differences of response of a model atmosphere to forcing in cases of winter and summer.

1. Introduction

The response of a northern hemispheric model atmosphere to forcing by topography and stationary heat sources in winter was investigated by Huang and Gambo (1982)* with a steady state, quasi-geostrophic, linear 34-level model in which Rayleigh friction, effect of Newtonian cooling and horizontal kinematic thermal diffusivity are included. Generally speaking, the simulation of stationary planetary waves forced by topography and diabatic heating during summer has rarely been attempted until the present time.

Huang and Gambo (1981) had studied the response of a model atmosphere in middle latitudes to forcing by topography and stationary heat sources in summer by means of a β-plane approximate multi-level model. In that paper, however, we assumed that the motion takes place on a β-plane centered at 45°N and the zonal mean wind is constant with respect to latitude. The stationary planetary waves responding to forcing by topography and stationary heat sources can be propagated only vertically, but cannot be propagated laterally. Thus, it was shown from the computed results that there are some differences between the calculation and the observation, especially, a large difference occurs over the central Pacific Ocean, i.e., the computed trough over this region is far weaker than the observed one. Ashe (1979) discussed the response of a northern hemispheric model atmosphere to forcing by topography and stationary heat sources in summer with a nonlinear two-level model. But his results were not in agreements with the observation, due to the poor vertical resolution in his model.

Considering these situations, we re-examine three-dimensional response of a model atmos-

---

* 本文原载于: Journal of the Meteorological Society of Japan, Vol. 61, No.4, 495-509, 1983.

phere to forcing by topography and heat sources in summer. As is well known, the simulation of stationary flow pattern in summer is to make clear the large-scale monsoon circulation over the tropics. This means that the simulation of wind field over the tropics is important. In this paper, however, we do not directly discuss the stationary flow pattern over the tropics. Due to the restriction of quasi-gestrophic model which we use, our discussion is concentrated on the large-scale height fields over the subtropics, and middle and high latitudes. The model which we use in this paper is a steady-state, quasi-geotrophic, linear, 34-level model which is the same model as described in the paper of HG.

2. Model

The steady state, linear, quasi-geostrophic vorticity and thermodynamic equations in which Rayleigh friction, effect of Newtonian cooling and horizontal kinematic thermal diffusivity are included, may be expressed in spherical coordinates (λ, φ, p) as

$$\bar{U}\frac{\partial}{a\cos\varphi\partial\lambda}(\zeta') + v'\frac{\partial}{a\partial\varphi}(\bar{\zeta}+f) = f\frac{\partial\omega}{\partial p} - R_f\zeta' \quad (1)$$

$$\bar{U}\frac{\partial}{a\cos\varphi\partial\lambda}\left(\frac{\partial\phi'}{\partial p}\right) - 2\Omega_0\sin\varphi\frac{\partial\bar{U}}{\partial p}v' + \sigma\omega$$
$$= -\frac{RH}{c_p p} - \alpha_R\frac{\partial\phi'}{\partial p} + K_T\nabla^2\left(\frac{\partial\phi'}{\partial p}\right) \quad (2)$$

respectively. The notation used in the above equations is as follows;

- a : radius of the earth
- Ω_0 : rotation rate of the earth.
- $\bar{U}$: the basic zonal wind speed
- v' : meridional component of perturbation motion
- ϕ' : geopotential of perturbation
- ζ' : vertical component of relative perturbation vorticity
- $\bar{\zeta}$: vertical component of relative vorticity of the basic state
- $\sigma = -\alpha\dfrac{\partial\ln\theta}{\partial p}$: static stability parameter (α: specific volume, θ: potential temperature)
- H : diabatic heating per unit time and unit mass
- R : gas constant (0.287 KJ Kg^{-1} deg^{-1})
- c_p : specific heat at constant pressure (1.004 KJ Kg^{-1} deg^{-1})
- α_R : Newtonian cooling coefficient
- K_T : horizontal kinematic thermal diffusivity
- R_f : Rayleigh friction coefficient
- f : Coriolis parameter
- ω : vertical p-velocity (dp/dt)

Since the details of the finite-difference scheme for equations (1) and (2) is written in the paper of HG, we do not explain about the finite-difference scheme in this paper. However, we mention about the upper and lower boundary conditions, the latitudinal boundary conditions and the parameters which we use in the model atmosphere, for the sake of better understanding of the model atmosphere.

2.1 Boundary conditions

For the upper boundary conditoins, we assume that the vertical p-velocity vanishes at top of the mode, i.e.,

$$\omega = 0, \quad \text{at } p = p_t \text{ (or } z = z_t) \quad (3)$$

For the boundary condition at $p = p_s$ (p_s: surface pressure), we assume that the vertical p-velocity is caused by surface topography where the standard pressure is p_s, and also by Ekman pumping resulting from the viscosity in the Ekman layer. Thus, the vertical p-velocity at $p = p_s$ is given as follows

$$\omega_s = V_s\cdot\nabla p_G - \frac{p_s F}{2f}\zeta_s' \quad \text{at } p = p_s \text{ (or } z = z_s) \quad (4)$$

Here V_s is the horizontal velocity vector at $p = p_s$, and $p_s = 1000$ mb for simplicity. F is the friction coefficient and will be treated as a constant (4×10^{-6} s^{-1}). ζ_s' is the vorticity of perturbation at the surface. The vertical difference scheme used in this paper is the same as that discussed in the paper of HG. The top of this model atmosphere is defined as $Z = 92$ km (i.e., $p_t = 8.459\times10^{-4}$ mb) and we divide the atmosphere into 34 layer from the earth's surface to 92 km.

For the latitudinal boundary conditions, we assume that

$$\phi' = 0 \quad \text{at } \varphi = \frac{\pi}{2} \text{ and } \varphi = 0 \quad (5)$$

Here ϕ' is the perturbation of geopotential.

2.2 Parameter

a) Static stability σ: the stability parameter σ is calculatd in this hemispheric model from the mean temperature and density at 45°N in July obtained from the U.S. standard atmosphere (1966). For the simplicity, we will assume that the static stability paramter does not change with latitudes.

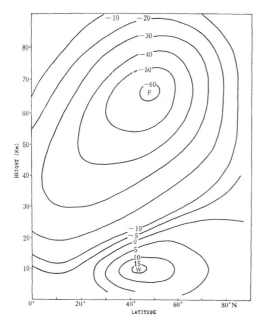

Fig. 1 The model basic state zonal mean wind distribution (ms^{-1}) in the summer Northern Hemisphere.

b) The vertical profile of the basic zonal mean wind: the observed zonal mean wind in the global atmosphere up to the mesopause level is obtained by Murgatroyd et al. (1969). However, we did not use their results directly in this study, because the data include small-scale features which may have significant influence on the calculation of the refractive index square of stationary planetary waves. For this reason, we smoothed their results to obtain an idealized model of the zonal wind in summer, retaining the major feature of the observed wind. The distribution of the basic zonal wind used in this paper is shown in Fig. 1.

c) The coefficient of Rayleigh friction: the coefficient of Newtonian cooling and the horizontal kinematic thermal diffusivity coefficient are assumed as the same as that in the paper of HG.

3. Computation of the refractive index square of stationary planetary waves

In order to discuss the characteristics of the vertical and horizontal propagations of stationary planetary waves, we compute the refractive index square of stationary planetary wave, proposed by Charney and Drazin (1961), in summer.

As mentioned in the paper of HG, if there is no heat sources and Rayleigh friction, effect of Newtonian cooling and horizontal kinematic thermal diffusivity are not considered, and assume that the atmosphere is nearly isothermal, we can obtain the potential vorticity equation in Z-coordinate system from equations (1) and (2),

$$\frac{\sin^2\varphi}{\cos\varphi}\frac{\partial}{\partial\varphi}\left(\frac{\cos\varphi}{\sin^2\varphi}\frac{\partial\psi_k}{\partial\varphi}\right)+l^2\sin^2\varphi\frac{\partial^2\psi_k}{\partial z^2}$$
$$+Q_k\psi_k=0 \quad (6)$$

$$Q_k=Q_0-\frac{k^2}{\cos^2\varphi} \quad (7)$$

$$Q_0=\left[2(\Omega_0+\hat{\Omega})-\frac{\partial^2\hat{\Omega}}{\partial\varphi^2}+3\tan\varphi\frac{\partial\hat{\Omega}}{\partial\varphi}\right.$$
$$\left.-l^2\sin^2\varphi\left(\frac{\partial^2\hat{\Omega}}{\partial z^2}-\frac{1}{H_0}\frac{\partial\hat{\Omega}}{\partial z}\right)\right]\bigg/\hat{\Omega}$$
$$-l^2\sin^2\varphi\frac{1}{4H_0^2} \quad (8)$$

Here, $Z=-H_0\ln(p/p_0)$ (p_0: reference pressure, H_0: representative scale height), the geopotential height ϕ' is expressed by

$$\phi'(\lambda,\varphi,z)=R_e\sum_{k=1}^{K}e^{z/2H_0}\psi_k(\varphi,z)e^{ik\lambda}$$

and $l=2\Omega_0 a/\tilde{N}$ ($\tilde{N}$: Brunt-Vaisälä frequency). In this paper we assume that H_0 is 7 km and $\tilde{N}$ is 2×10^{-2} s^{-1}. The equation (6) is the two-dimensional wave equation. It describes the wave propagation in the φ and Z directions in an isothermal atmosphere. Here, Q_k is regarded as the refractive index square of wave for zonal wave number k.

The distribution of Q_k in summer is computed from the zonal mean wind shown in Fig. 1 by equations (6) and (7), and the refractive index square Q_0 for wave number $k=0$ is shown in Fig. 2. We note that a minimum of Q_0 is located in the lower stratosphere at middle and high latitudes. The values of Q_0 in the tropo-

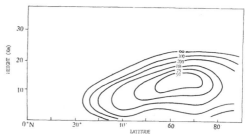

Fig. 2 The vertical distribution of the refractive index square Q_0 for $k=0$ (k: zonal wave number) in summer.

sphere and stratosphere at low latitudes, and in the stratosphere at middle and high latitudes are negative. Therefore, the stationary planetary waves in summer cannot propagate in the stratosphere. This situation is different from that in winter.

4. Computation of the response to forcing by topography

Before we discuss the response of a model atmosphere to forcing by the actual hemispheric topography, we examine the response of a model atmosphere to forcing by idealized topography. In this discussion, we also examine the role of Q_k in the vertical and horizontal propagations of stationary planetary waves induced by the topography.

In the following discussion, we consider the case where there is no stationary heat source and assume the idealized topography. The basic state pressure of surface topography, P_G is generally expressed in the following form

$$P_G(\lambda, \varphi) = R_e \sum_{k=1}^{K} (\hat{P}_g(\varphi))_k e^{ik\lambda}, \quad k=1, 2, \ldots K$$

Here $(\hat{P}_g(\varphi))_k$ denotes the meridional distribution of the amplitude of P_G for wave number k and is expressed in the following complex form

$$\{\hat{P}_g(\varphi)\}_k = (P_A)_k + i(-P_B)_k \quad (9)$$

For the sake of simplicity, the idealized topography is assumed to be expressed only by three waves of $k=1, 2, 3$ and the form of $(\hat{P}_g(\varphi))_k$ is defined at the specific latitude such as

$$(P_A)_k = \begin{cases} 50 \text{ mb}, & \varphi = \varphi', \\ 0 \text{ mb}, & \varphi \neq \varphi', \end{cases}$$
$$(P_B)_k = \begin{cases} 0 \text{ mb}, & \varphi = \varphi', \\ 0 \text{ mb}, & \varphi \neq \varphi', \end{cases} \quad \text{for } k=1, 2, 3. \quad (10)$$

In this section, the response of a model atmosphere to forcing by the idealized topography is computed not only at 40°N and 80°N, but also at 30°N, because there is the interface of westerly and easterly in summer around 30°N. In following computations we assume that the zonal mean wind at the surface $\bar{U}_s$ in (4) is assumed as $\bar{U}_s = 2.3$ m/s in all cases of $\varphi' = 80°$N, 40°N and 30°N.

Figs. 3(A) and 3(B) show the vertical distribution of the amplitude and phase of stationary planetary waves responding to forcing by the idealized topography at 80°N for zonal wave number $k=1$ and $k=2$, respectively. On the other hand, Figs. 4(A) and 4(B) show the verti-

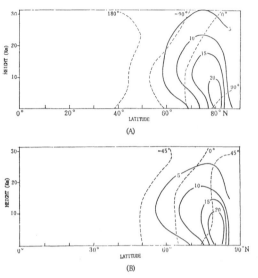

Fig. 3 The vertical distribution of amplitude (m) (solid curves) and phase (dashed curves) of stationary planetary waves, responding to forcing by idealized topography at 80°N for wave number $k=1$ (A) and $k=2$ (B) respectively. Position of ridges is shown on the right of equal phase lines.

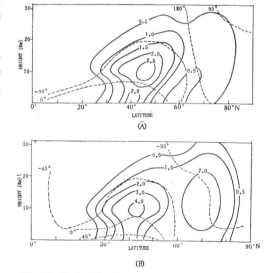

Fig. 4 As in Fig. 3 except for stationary planetary waves, responding to forcing by idealized topography at 40°N.

cal distribution of the amplitude and phase of stationary planetary waves responding to forcing by the idealized topography at 40°N for $k=1$ and $k=2$, respectively. From these figures, we may be able to summarize the results as follows:

1) The amplitude of stationary planetary waves by the idealized topography at 80°N is much larger than that forced by the topography at 40°N for wave number 1. The main reason for this difference may be considered in the following point. The vertical p-velocity at the surface ω_s is computed in a form of the geostrophic approximation of $V_s \cdot \nabla P_G$ in (4). Thus, ω_s is proportional to $1/\cos\varphi$ because the zonal mean surface wind and the amplitude of the topography at 80°N are assumed to be equal to those at 40°N.

2) The amplitude of stationary planetary waves for wave number 1 forced by the idealized topography at 80°N has a maximum near ground, and a secondary peak is not found in low latitudes such as was found in the case of winter. It will be also pointed from Fig. 3 that the polar wave guide in the stratosphere is not clearly seen in the case of summer compared with the case of winter.

3) The amplitude of stationary planetary waves for $k=1$ and $k=2$ forced by the idealized topography at 40°N has a maxima in the upper troposphere at 45°N, and a secondary peak is found in the upper troposphere at high latitudes. This is considered as the result of the propagations of stationary planetary waves responding to forcing by the idealized topography in middle latitude. Fig. 5 is the schematic picture of the wave guides of stationary planetary wave for wave number $k=1$, responding to forcing by the idealized topography at 40°N.

For the sake of comparison, the vertical and meridional distributions of amplitude and phase of stationary planetary waves for $k=1$ and $k=2$ responding to forcing by the idealized topography at 30°N, are also shown in Figs. 6(A) and 6(B), respectively. As may be seen in the figure, the waves can not be propagated to the upper troposphere over the subtropics.

In this section, we discuss the response of a model atmosphere to forcing by the actual hemispheric topography in summer based on the results mentioned above. For the computation we use the same meridional distribution of the amplitude and phase of the actual hemispheric topography which we used in the case of winter (refer to HG). Figs. 7(A) and 7(B) show the vertical distribution of the amplitude and phase responding to forcing by the northern hemispheric topography for zonal wave number $k=1$ and $k=2$, respectively. We may find that the amplitude of stationary planetary waves responding to forcing by topography for $k=1$ and $k=2$ in summer is much less than that in winter.

In order to examine the effect of Greenland Plateau. The results for $k=1$ and $k=2$ are shown in Figs. 8(A) and 8(B), respectively. Comparing Figs. 8(A) and 8(B) with Figs. 7(A) and 7(B), we may be able to mention that the effect of Greenland Plateau seems to be small except at high latitudes. This result may be understood from the result shown in Figs. 3(A) and 3(B), that is, the stationary planetary waves in summer responding to forcing by the topography at high latitudes can not be propagated to middle latitudes.

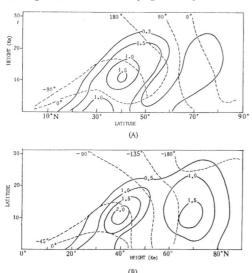

Fig. 5 As in Fig. 3 except for stationary planetary waves, responding to forcing by idealized topography at 30°N.

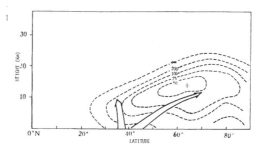

Fig. 6 Schematic picture of the wave guides of stationary planetary wave for wave number $k=1$, responding to forcing by idealized topography at 40°N. Dashed curves show the refractive index square Q_1 for wave number $k=1$.

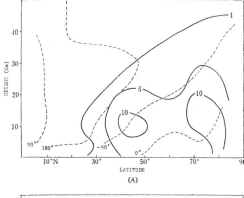

(A)

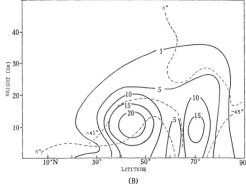

(B)

Fig. 7 The vertical distribution of amplitude (m) (solid curves) and phase (dashed curves) of stationary planetary wave, responding to forcing by Northern Hemispheric topography for wave number $k=1$ (A) and $k=2$ (B), respectively. Position of ridges is shown on the right of equal phase lines.

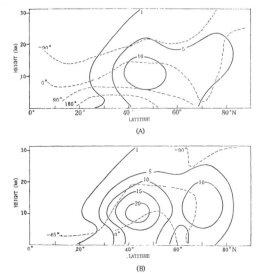

Fig. 8 As in Fig. 7 except for assuming no topography at surface, 70°–85°N.

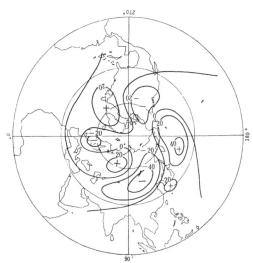

Fig. 9 The disturbance pattern (units in m) at 12 km level, responding to forcing by the Northern Hemispheric topography.

Next, we shall compute the disturbance pattern at an isobaric surface (or at a constant height level), by synthesizing wave component of $k=1-3$. Fig. 9 shows the disturbance pattern at 12 km level responding to forcing by the northern hemispheric topography. A remarkable feature is that disturbances are mainly confined to middle latitudes. The disturbance pattern consists of a major negative anomaly over Asia, another negative anomalies over North America and Western Europe, while a major positive anomaly over Pacific Ocean and other positive anomalies over the eastern coast of Atlantic Ocean and eastern Europe. Fig. 10 shows the disturbance pattern at 12 km level responding to forcing by the northern hemispheric topography assuming that there is no topographical effect of Greenland Plateau. Compared with Fig. 9, the disturbance pattern in Fig. 10 is almost same as that in Fig. 9. This will be explained from the result that the topographical effect of Greenland Plateau for the stationary planetary waves in middle latitudes seems to be weak. This result is different from that in the case of winter. As explained in the paper of HG, the Greeland Plateau plays an important role for the stationary waves in middle latitudes in winter.

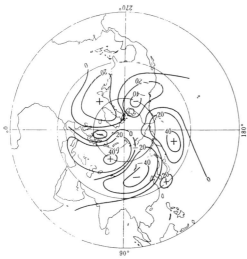

Fig. 10 As in Fig. 10 except for assuming no topography at surface 70°–85°N.

5. Computation of the response to forcing by stationary heat sources

As the first step to discuss the response of a model atmosphere to forcing by the actual northern hemispheric stationary heat sources, we examine the response of a model atmosphere to forcing by idealized heat sources at specified latitudes. As the idealized heat sources, we assume the following distribution in a spherical coordinate system:

$$H(\lambda, \varphi, p) = R_e \sum_{k=1}^{K} (\hat{H}_0(\varphi, p))_k e^{ik\lambda} \quad (11)$$

Here, H is the diabatic heating per unit time and unit mass and we use a pressure coordinate for the vertical one. $\hat{H}_0(\varphi, p)$ is a complex function expressed in the form of

$$(\hat{H}_0(\varphi, p))_k = (\hat{H}_A(\varphi, p))_k + i(-\hat{H}_B(\varphi, p))_k \quad (12)$$

For the sake of simplicity, the idealized stationary heat sources are assumed to be expressed only by three zonal waves of $k=1, 2$ and 3 and the form of $(H_0(\varphi, p))_k$ is a given function specific such as

$$(H_A)_k = \begin{cases} 6.0 \times 10^{-6} \text{ KJ Kg}^{-1} \text{s}^{-1}, & \varphi = \varphi', \\ 0, & \varphi \neq \varphi', \end{cases}$$

$$(H_B)_k = \begin{cases} 0, & \varphi = \varphi', \\ 0, & \varphi \neq \varphi', \end{cases}$$

$$k = 1, 2, 3. \quad (13)$$

Further, we assume that the vertical distribution of stationary heat sources is the same as that taken in the case of winter (refer to HG), i.e.,

$$(\hat{H}_0(\varphi, p))_k = (\hat{H}_0(\varphi))_k \exp\left\{-\left(\frac{p-\bar{p}}{d}\right)^2\right\}. \quad (14)$$

Here we assumed $d=300$ mb and $\bar{p}=500$ mb. This means that the distribution of diabatic heating has a maximum at 500 mb and decreases exponentially above and below 500 mb.

Then, the vertical and meridional distributions of the amplitude and phase of stationary planetary waves forced by idealized stationary heat sources at differente latitudes are obtained by substituting idealized stationary heat sources into the model equations described in section 2. In this section, the response of a model atmosphere to forcing by idealized stationary heat sources is computed not only in cases where $\varphi'=80°$N and $\varphi'=40°$N, but also in the case where $\varphi'=30°$N. The case where $\varphi'=30°$N is picked up since there is the interface of westerly and easterly in summer around 30°N.

Figs. 11(A) and 11(B) show the vertical distribution of the amplitude and phase of stationary planetary waves responding to forcing by idealized stationary heat sources at 80°N for zonal wave number $k=1$ and $k=2$, respectively. Figs. 12(A) and 12(B) show the vertical distribution of the amplitude and phase of stationary

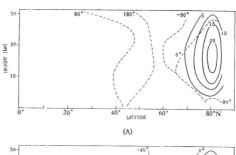

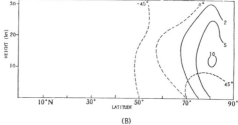

Fig. 11 The vertical distribution of amplitude (m) (solid curves) and phase (dashed curves) of stationary wave, responding to forcing by idealized stationary heat sources at 80°N for wave number $k=1$ (A), $k=2$ (B), respectively.

planetary waves responding to forcing by idealized stationary heat sources at 40°N for $k=1$ and $k=2$, respectively. We may find the following results from Figs. 11(A) 11(B) and Figs. 12(A), 12(B):

(1) The considerable difference between the winter and summer responses to forcing by stationary heat sources may be found in the vertical distribution of amplitude and phase of waves. This is closely connected with difference of the zonal mean wind and static stability.

(2) The amplitude of stationary planetary wave to forcing by idealized stationary heat sources at 80°N for $k=1$ has a maximum in the upper troposphere at 80°N, and a secondary peak is not found in low latitude. On the other hand, the amplitude of stationary planetary wave for $k=1$ responding to forcing by idealized stationary heat sources at 40°N has a maximum in the upper troposphere at 40°N and have a secondary peak in the upper troposphere at high latitudes for $k=1$ and $k=2$. The secondary peak will be considered as the results due to the propagationary planetary waves to forcing by stationary heat sources.

Figs. 13(A) and 13(B) show the vertical and meridional distribution of amplitude and phase of stationary planetary waves responding to forcing by idealized stationary heat sources at 30°N for $k=1$ and $k=2$, respectively. We may also find that they have a maximum in the upper

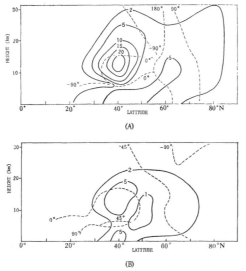

Fig. 12 As in Fig. 11 except for stationary planetary wave, responding to forcing by idealized stationary heat sources at 40°N.

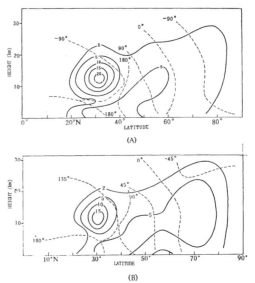

Fig. 13 As in Fig. 11 except for stationary planetary wave, responding to forcing by idealized stationary heat sources at 30°N.

troposphere at 30°N, and also have a secondary peak in the upper troposphere at high latitudes (55°N for $k=1$ and 70°N for $k=2$).

We may find from the above mentioned results that the stationary planetary waves responding to forcing by stationary heat sources in the subtropics may influence on the circulation in middle and high latitudes. Here we remark that Simmons (1981) and Webster (1972) suggested that the forcing by diabatic heating in the tropics influences on the circulation in middle and high latitudes in winter. However, we consider that the influence of the subtropical heat sources to middle and high latitudes seems to be more effective in summer, because the diabatic heating over the Tibetan Plateau will be intensified in summer.

Based on the results mentioned above, we examine the response of a model atmosphere to forcing by actual hemispheric heat sources in summer. In the following discussion, we assume that the vertical distribution of diabatic heating is expressed by equation (14) and $(\hat{H}_0(\varphi))_k$ is given by the value estimated by Ashe. However, this assumption is inappropriate for the low latitudes in summer. Lindzen et al. (1982) used the thermal forcing function in which the height of maximum value increases with descreasing latitude. Their assumption seems to be suitable to the observed vertical distribution of diabatic heating in the northern hemisphere. For the sake of simplicity, we assume that the vertical distribution of actual diabatic heating is also expressed by equation (14), but $p=500$ mb for $\varphi=40°N$, and $p=400$ mb for $\varphi\leqq40°N$.

$(\hat{H}_0(\varphi))_k$ is computed by expanding Ashe's result into the Fourier series in the longitudinal direction. For the sake of reference, the amplitudes of $(H_0(\varphi))_k$ for zonal wavenumber $k=1$ and $k=2$ are shown in Fig. 14. We find maximum amplitude of diabatic heating for $k=1$ and $k=2$ around 30°N, while the secondary peak is found at high latitudes.

Figs. 15(A) and 15(B) show the vertical distribution of the amplitude and phase of stationary planetary waves responding to forcing by actual northern hemispheric stationary heat sources for $k=1$ and $k=2$, respectively. The following results may be mentioned from Figs. 15(A) and 15(B):

(1) The amplitudes of stationary planetary waves responding to forcing by actual stationary heat sources for $k=1$ and $k=2$ have a maximum at 12 km height, 30°N and a secondary peak at 13 km height, 70°N.

(2) As mentioned in the case of idealized stationary heat sources, the amplitude of stationary planetary wave for $k=1$ responding to forc-

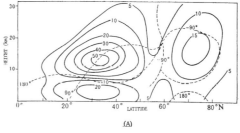

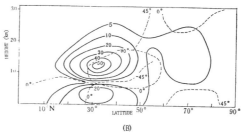

Fig. 15 The vertical distribution of amplitude (m) (solid curves) and phase (dashed curves) of stationary planetary wave, responding to forcing by the Northern Hemispheric stationary heat sources for wave number $k=1$ (A) and $k=2$ (B), respectively. Position of ridges is shown on the right of equal phase line.

ing by actual stationary heat sources is larger than that for wave number 2.

(3) For $k=1$, the amplitude of stationary wave responding to forcing by actual northern hemispheric heat sources is larger than that responding to forcing by actual topography (see Fig. 7(A)). From this result, we may be able to emphasize that the thermal effect is a dominant factor in reproducing the Tibetan high in summer.

(4) The stationary planetary waves responding to forcing by stationary heat sources at middle latitudes can be propagated not only toward the upper troposphere around 30°N, but also toward the upper troposphere at high latitudes.

(5) The amplitudes of stationary planetary waves for $k=1$ and $k=2$ decreases with increasing height around the middle troposphere, and then increases from the middle troposphere to 12 km. These computed vertical distributions are qualitatively in good agreement with the observed distributions in summer.

Next, we shall compute the disturbance pattern at an isobaric surface (or at a constant height) by synthesizing zonal wave components of $k=1$, 2 and 3. The computed disturbance pattern

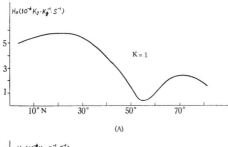

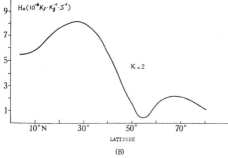

Fig. 14 The amplitude of zonal wave component 1 (upper), and 2 (bottom) for the Hemispheric stationary heat sources.

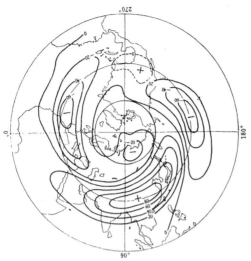

Fig. 16 The disturbance pattern (units in m) at 12 km level, responding to forcing by the Northern Hemispheric stationary heat sources.

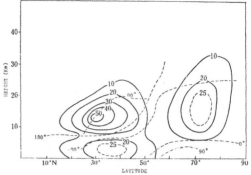

Fig. 17 The vertical distribution of amplitude (m) (solid curves) and phase (dashed curves) of stationary wave, responding to forcing by both the Northern Hemispheric topography and stationary heat sources for wave number $k=1$. Position of ridge is shown on the equal phase line.

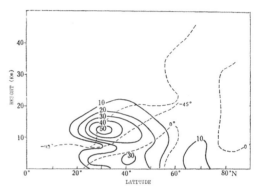

Fig. 18 As in Fig. 8 except for the wave number $k=2$.

at 12 km height is shown in Fig. 16. In the figure, we may find that the stationary pattern is found mainly in the subtropics. The forced pattern consists of a major positive anomaly over South Asia, centered at Tibetan Plateau, another positive anomaly over North America, and two negative anomalies over the Pacific Ocean and the Atlantic Ocean, respectively. In addition, a negative anomaly is found over North Asia. Similarly, it is shown in Fig. 16 that the disturbance pattern responding to forcing by actual stationary heat sources is stronger than that forced by the actual topography (see Fig. 9).

6. Computation of stationary planetary wave responding to forcing by the actual northern hemispheric topography and the stationary heat sources

In this section, we discuss the response of stationary planetary waves to forcing by both the actual topography and stationary heat sources in summer. If we substitute the forcing function of both the actual topography and diabatic heating into the model equations described in section 2, we obtain the vertical distribution of the amplitude and phase of stationary waves responding to forcing by both the actual hemispheric topography and the stationary heat sources in summer. The results are shown in Figs. 17 and 18 for $k=1$ and $k=2$, respectively. For the sake of comparison, the observed ones for $k=1$ and $k=2$ in July are shown in Figs. 19(A), 19(B) and 20(A) 20(B), respectively.* In Fig. 19(A), the maximum of amplitude at 30°N is found around 100 mb–150 mb level according to Iwashima's result. Comparing Figs. 17 and 18 with Figs. 19 and 20, we may find the following results:

1) The amplitudes of computed stationary planetary waves responding to forcing by both the actual topography and stationary heat sources have a maximum at 13 km height, 30°N for $k=1$ and $k=2$, and the computed values of the

---

* The observed amplitude and phase in July are computed from the NMC data during the period of 1963-1979. The results are quite similar with those obtained independently from the summer data during the period of 1973-1977, by Iwashima (1980).

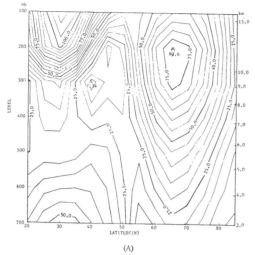

Fig. 18(A) Latitude-height section of stationary zonal wave number 1 amplitude (m) in July, averaged over the years 1963-1979. Contour interval is 5 m.

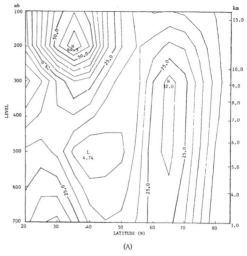

Fig. 20(A) As in Fig. 19(A) except for zonal wave number 2.

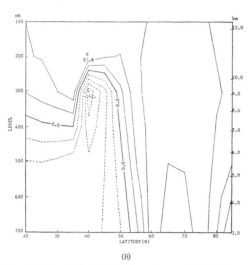

Fig. 19(B) Latitude-height section of stationary zonal wave number 1 phase (longitude of ridge) in July, averaged over the years 1963-1979. Contour interval is 30° longitude.

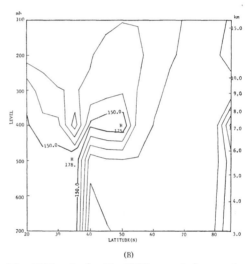

Fig. 20(B) As in Fig. 19(B) except for zonal wave number 2.

maximum amplitude are about 60 m for $k=1$ and $k=2$. In addition, we may find a secondary peak at 13 km height, 70°N for $k=1$, while the amplitude of a secondary peak for $k=2$ is small. The latitudes and heights of maximum amplitude and a secondary peak seem to be in good agreement with the observed results for $k=1$ and $k=2$, respectively. However, we remark that the computed values of maximum amplitude and a secondary peak for $k=1$ and $k=2$ are small compared with the observed ones. The ratio of the computed amplitude to the observed one is about 1/2–1/3. It may be possible to speculate that the value of diabatic heating in summer computed by Ashe is rather small.

2) The computed main disturbances around 30°N are found in the upper and lower troposphere, while the disturbance is relatively weak in middle troposphere. This is qualitatively in good agreement with the observed result.

Next, we shall compute the disturbance pattern at an isobaric surface (or at constant height) responding to forcing by the actual topography and stationary heat sources in summer, by synthesizing zonal wave components $k=1-3$. Fig. 21 shows the computed disturbance pattern at 12 km height. For the sake of comparison, we calculated the disturbance pattern at 200 mb in July, averaged over the years 1972–1977 from the observed MMC data and it is shown in Fig. 22. In the computed patterns, the major positive anomaly is found over the South Asia, centered in Tibetan Plateau. This major disturbances is so called Tibetan high. Another two positive anomalies are found over America and Europe. The major negative anomaly is found over the central Pacific Ocean. These negative anomalies are so called T.U.T.T. (Tropical Upper Tropospherical Trough). Compared with Fig. 22, the major disturbance patterns at 200 mb in summer, i.e., Tibetan high and T.U.T.T. may be considered as the results forced by stationary heat sources. In high latitudes, the computed pattern is not in good agreement with the observed pattern.

7. Computation of the momentum and heat eddy fluxes

In the previous sections, we have computed the vertical distribution of stationary planetary waves. Therefore, we can easily compute the eddy momentum and heat fluxes due to the stationary planetary waves. Considering that the geopotential ϕ' is expressed by

$$\phi'(\lambda,\varphi,p)=R_e\sum_{k=1}^{K}\Phi_k(\varphi,p)e^{ik\lambda}$$

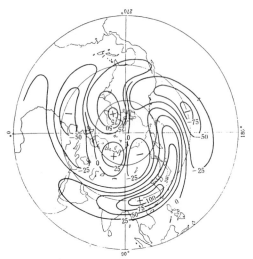

Fig. 21 The disturbance pattern (units in m) at 12 km level, responding to forcing by both the Northern Hemispheric topography and stationary heat sources.

(k: zonal wave number), we have the momentum and heat fluxes due to the stationary planetary waves as follows:

$$\overline{u'v'}=\frac{1}{2}\frac{1}{4\Omega_0^2\sin^2\varphi}\frac{1}{a^2\cos\varphi}\sum_{k=1}^{3}kI_m\left(\frac{\partial\Phi_k^*}{\partial\varphi}\Phi_k\right) \quad (15)$$

$$\overline{v'T'}=\frac{1}{2}\frac{1}{2\Omega_0\sin\varphi}\frac{1}{a\cos\varphi}\frac{p}{R}\sum_{k=1}^{3}kI_m\left(\frac{\partial\Phi_k^*}{\partial p}\Phi_k\right) \quad (16)$$

Here the symbol "bar" dennotes the zonal mean, Φ_k^* is the conjugate complex of Φ_k, Ω_0 is the rotation rate of the earth, p is the pressure, R is the gas constant and a is the radius of the earth, respectively. In (15) and (16), we pick up only three zonal components for the sake of simplicity.

The computed results of $\overline{u'v'}$ and $\overline{v'T'}$ are shown in Fig. 23 and Fig. 24 respectively. As may be seen in Fig. 23, the large value of poleward momentum flux is found in the upper troposphere at 30°N and the equatorward momentum flux is found at high latitudes. The result is in qualitative agreement with the result computed by Newell *et al.* (1974) from the observed data. In the same way, the maximum

Fig. 22 The observed July planetary wave height pattern (units in mb) due to zonal wave number 1–3 at 200 mb.

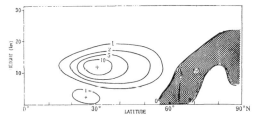

Fig. 23 The vertical distribution of horizontal momentum (m² s⁻²) due to stationary planetary waves, responding to forcing by the Northern Hemispheric topography and stationary heat sources.

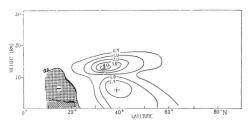

Fig. 24 The vertical distribution of horizontal heat flux (°K ms⁻¹) due to stationary planetary waves, responding to forcing by both the Northern Hemispheric topography and stationary heat sources.

poleward heat flux is found at 13 km height, 35°N, while equatorward heat flux is found at low latitudes. However, this computational result is not in good agreement with the observed one (Oort and Rasmusson, 1971). Concerning this problem, it may be pointed from Fig. 17 and Figs. 19(a) and 19(b) that the negative contribution of computed zonal wave number 1 (tilted eastward) to the value of $\overline{v'T'}$ around 13 km height, 35°N is less estimated compared with the observed one. This problem is remained to be solved in the future.

8. Computation of the energy flux due to the stationary planetary waves

The eddy momentum and heat fluxes obtained in section 7 can be also directly used to compute the energy flux by means of the following relation

$$E = \left(\sum_{k=1}^{3} \bar{\rho}(\overline{\phi'v'})_k, \sum_{k=1}^{3} \bar{\rho}(\overline{\phi'w'})_k \right) \quad (17)$$

Here, we compute the value of E, synthesizing only three zonal components $k = 1-3$ for the sake of simplicity.

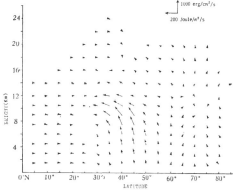

Fig. 25 The vertical distribution of energy flux due to stationary planetary waves 1–3, responding to forcing by both the Hemispheric topography and stationary heat sources.

Fig. 25 shows the energy flux due to the stationary planetary waves responding to forcing by the northern hemispheric topography and the stationary heat sources in summer, computed by (17). We find the major upward energy flow at middle latitudes. This upward energy flow can be propagated toward the upper troposphere around 30°N and also toward the upper troposphere in high latitudes, respectively.

We may also find that the energy flux in the lower troposphere over the tropics is horizontally toward the subtropics. As a consequence, the wave energy convergence occures in the interface between the westerly and eastrly, i.e., the line of zero zonal wind located in the subtropics is the energy sink for planetary waves. However, we must mention here that there is no energy flow toward the equator in the troposphere over the tropics. This unreasonable result may come from the insufficient treatment of wind field over the tropics due to the quasi-geostrophic model.

9. Summary and discussion

The stationary planetary waves responding to forcing by the northern hemisphere topography and heat sources in summer are investigated by means of a steady-state, linear, quasi-geostrophic, 34-level model, with Rayleigh friction, effect of Newtonian cooling and horizontal kinematic thermal diffusivity included in a spherical coordinate system. In this computation, the role of refractive index square of stationary planetary waves is discussed in relation with the propagation of stationary planetary waves from the lower troposphere at middle latitudes toward the upper troposphere at high latitudes or the upper troposphere around 30°N.

The vertical distribution of the amplitude and phase of stationary planetary waves, and the standing disturbance pattern at constant height responding to forcing by both the topography and stationary heat sources are computed. The computed results show that the standing disturbance pattern responding to forcing by both topography and stationary heat sources is mainly confined to the troposphere over the subtropics.

Compared with the computed results of the winter response, there are considerable differences between the stationary planetary waves in winter and that in summer. These differences are as follows:

1) The amplitude of stationary planetary waves responding to forcing by both topography and stationary heat sources for $k=1$ and $k=2$ in winter is much larger than that in summer. The amplitude in winter has a maximum in the stratosphere around 65°N, and have a secondary peak in the upper troposphere at low latitudes for $k=1$ and $k=2$. However, the amplitude in summer has a maximum in the upper troposphere around 30°N for $k=1$ and $k=2$, and have a secondary peak in the upper troposphere at high latitudes for $k=1$.

2) The stationary planetary-scale disturbances are mainly confined to high latitudes, centered at 60°N, in winter, while they are mainly confined in middle-low latitudes, centered at 30°N, in summer.

3) In winter, the maximum amplitude of stationary planetary waves responding to forcing by both topography and stationary heat sources is found for $k=1$ and decreases rather rapidly as k increases. However, in summer, the maximum amplitude decreases rather slowly with increasing k, i.e., the amplitudes of stationary planetary waves for wave numbers 3, 4, ... are not as small as that in winter. These differences may be considered mainly due to the difference of wave guides in summer and winter. In this paper, we have discussed the response of a model atmosphere to forcing only for zonal wave numbers $k=1-3$. However, the response for zonal wave numbers $k=4-6$ seems to be important in the case of summer, because the resonance wave number in middle latitudes will be found at $k=4-6$. This problem should be discussed in the future.*

Acknowledgements

The authors are grateful to Prof. T. Matsuno for his valuable comments and suggestions during the course of this study. Thanks are due to Dr. T. Satomura for discussion throughout the course of this work and also to Prof. T. Iwashima for his kindness to give us the information about his unpublished results. Thanks are also extended to Mrs. Wang Wan-wen for drawing the figures and Mrs. K. Kudo for the expert typing of manuscript.

References

Ashe, S., 1979: A nonlinear model of the time average axially asymmetric flow induced by topography and diabatic heating. *J. Atmos. Sci.*, **36**, 109-126.

Charney, J. G. and P. G. Drazin, 1961: Propagation of planetary-scale disturbances from the lower into the upper atmosphere. *J. Geophys. Res.*, **66**, 83-109.

Huang, Rong-hui and K. Gambo, 1981: The response of a model atmosphere in middle latitudes to forcing by topography and stationary heat sources. *J. Meteor. Soc. Japan*, **59**, 220-237.

——— and ———, 1982: The response of a hemispheric multi-level model atmosphere to forcing by topography and stationary heat sources. Part I & II. *J. Meteor. Soc. Japan*, **60**, 78-108.

Iwashima, T., 1980: Analysis of stationary planetary waves in summer (Presented at the annual meeting of the Meteorological Society of Japan in March 1980).

Lindzen, R. S., T. Aso and D. Jacqmin, 1982: Linearized calculations of stationary waves in the atmosphere. *J. Meteor. Soc. Japan*, **60**, 66-77.

Murgatroyd, R. J., 1969: The structure and dynamics of the stratosphere. *The Global Circulation of the Atmosphere*, 155-195.

Newell, R. E., J. W. Kidson, D. G. Vincent and G. J. Boer, 1972, 1974: *The General Circulation of the Tropical Atmosphere and Integrations with Extratropical Latitudes.* Vol. 1 (1972), 258 pp., Vol. 2 (1974), 371 pp., The MIT press.

Oort, A. and E. M. Rasmusson, 1971: Atmospheric circulation statistics. NOAA professional paper 5.

Simmons, A. J., 1981: Tropical influences on stationary wave motion in middle and high latitudes Technical report No. 26, ECMWF.

Webster, P. J., 1972: Response of the tropical atmosphere to local steady forcing. *Mon. Wea. Rev.*, **100**, 518-541.

夏季青藏高原上空热源异常对北半球大气环流异常的作用*

黄荣辉

(中国科学院大气物理研究所)

提 要

本文应用缓变媒质中行星波的传播理论来研究夏季基本气流中定常行星波的三维传播，解释夏季北半球大气环流三维遥相关的物理机制。

本文还应用一个包括Rayleigh摩擦、Newton冷却及水平涡旋热力扩散的准地转34层球坐标模式来研究夏季青藏高原热源异常对北半球中、高纬度定常扰动系统的影响。计算结果表明夏季青藏高原上空的热源异常将产生北半球中、高纬度大气环流的异常，若青藏高原上空热源增强就会引起南亚高压增强，我国东北受槽控制而产生冷夏，并且将引起鄂霍茨克海上空高压加强，同时在阿拉斯加将产生槽，这与实际资料所得到的结果一致。

一、引 言

许多研究表明夏季青藏高原是北半球一个强大的热源，似如"大气海洋"中的一个热岛。叶笃正与高由禧先生根据长时间地面观测资料计算了青藏高原的感热输送与潜热输送[1]，其计算结果表明了从5月到7月份高原西部存在一个很大的感热输送通量，而从6月到8月份在高原的东部存在一个很大的潜热输送。Ashe也利用气候资料计算了北半球夏季行星尺度的热源分布[2]，计算结果表明夏季最大热源位于青藏高原的东南部。最近Nitta与Luo and Yanai分别计算了夏季青藏高原上空的热源分布[3,4]，他们的计算结果表明平均加热率最大位于青藏高原的东南部，并且最大加热率位于400—500 mb。

至今还没有人计算夏季青藏高原上空热源的多年变动情况。陈烈庭统计了青藏高原上空的积雪面积、厚度、积雪天数与我国初夏季风的关系[5]，统计结果表明，若冬季青藏高原积雪面积大、厚度深，天数多，则我国初夏的西南季风就弱，反之就强。这表明冬季青藏高原的积雪情况对夏季青藏高原上空的热源有很大影响。这是因为地表面积雪会大大增大太阳辐射的反射率，从而使得地表面对短波辐射吸收大大减少，并且由于融雪需要热量，这也使得热源大大减弱，因此，可以推理，若青藏高原积雪面积大、厚度深、天数多，则夏季青藏高原上空的热源可能弱，反之就强。此外，根据Luo and Yanai的计算结果，高

---

* 本文原载于：气象学报，第43卷，第2期，208—220，1985年5月出版.

原上空的水汽对高原的热源影响很大,干燥期其最大加热率只有 1~2°K/天,而在湿润期可以达到 5°K/天,因此,高原上空的水汽变化可以直接引起高原热源的异常。

由于高原热源的强弱将直接影响南亚高压的强弱,因此,高原热源的异常肯定会影响北半球大气环流。Asakura 与 Gambo and Kudo 分别计算了夏季北半球大气环流的遥相关[6,7],他们的计算结果都表明夏季中、高纬度环流的异常与青藏高压有很大关系。关于北半球冬季大气环流三维遥相关机制已有不少解释,如作者与 Shukla et al. 从不同角度分别解释了冬季北半球大气环流三维遥相关的物理机制[8,9]。但是,关于夏季大气环流三维遥相关的物理机制至今还没有人研究过,本文利用缓变媒质中波的传播理论及数值试验来说明夏季青藏高原热源的异常对北半球大气环流异常的作用,从而说明夏季北半球大气环流的三维遥相关的物理机制。

二、夏季北半球大气环流三维遥相关的观测事实

Asakura 利用 1946—1965 年 20 年资料计算了夏季西藏高原地区 500 mb 高度场与

图 1 北半球 7 月份西藏高原地区 500 mb 高度场与北半球 500 mb 高度场的遥相关。

北半球 500 mb 高度场的遥相关。Gambo 和 Kudo 利用 1963—1979 年 17 年资料计算了夏季 30°N，115°E 为基点的 700 mb 高度偏差值与北半球 700 mb 高度偏差场及 500 mb 高度偏差场的三维遥相关。他们的计算结果表明了北半球夏季大气环流中存在着很好的三维遥相关，为了便于下面从理论上及数值试验上讨论，我们把他们的计算结果表示在图 1。

从图 1 我们可以看到，夏季当青藏高压加强，我国北方的低压槽要发展，这将带来我国东北与日本北部的冷夏，并且鄂霍次克海高压加强，而且在阿拉斯加要产生槽，北美地区高压要加强；此外，太平洋副热带高压减弱。这种大气环流的遥相关可以用来作长期预报的依据。

三、夏季定常行星波的传播理论

为了解释上面所述的大气环流异常的三维遥相关，我们将应用缓变媒质中波的传播理论。

假设在大气中没有热源与粘性，并且不考虑 Newton 冷却，球坐标的线性化涡度方程与热力学方程分别可写成

$$\left(\frac{\partial}{\partial t} + \bar{U}\frac{\partial}{a\cos\varphi\partial\lambda}\right)\zeta' + v'\frac{\partial}{a\partial\varphi}(\bar{\zeta}+f) = f\frac{\partial\omega}{\partial P} \tag{1}$$

$$\left(\frac{\partial}{\partial t} + \bar{U}\frac{\partial}{a\cos\varphi\partial\lambda}\right)\left(\frac{\partial\phi'}{\partial P}\right) - 2\Omega_0\sin\varphi\frac{\partial\bar{U}}{\partial P}v' + \sigma\omega = 0 \tag{2}$$

这里 a 是地球半径，φ 是纬度，λ 是经度，$\bar{U}$ 是基本气流，v' 是扰动风速的经向分量，ζ' 是相对扰动涡度的垂直分量，$\bar{\zeta}$ 是基本态的相对涡度的垂直分量，f 是科氏参数，ω 是垂直速度，Ω_0 是地球自转角速度。

在讨论定常行星波传播时，应用 Z 坐标代替 p 坐标是方便的。这里我们所用的 Z 坐标定义为 $Z = -H_0 \ln\left(\frac{p}{P_0}\right)$，$P_0$ 是参考压力，H_0 是特征高度，这样可得到 Z 坐标系的位涡度方程：

$$\left(\frac{\partial}{\partial t} + \hat{\Omega}\frac{\partial}{\partial\lambda}\right)\left\{\left[\frac{\sin\varphi}{\cos\varphi}\frac{\partial}{\partial\varphi}\left(\frac{\cos\varphi}{\sin\varphi}\frac{\partial\phi'}{\partial\varphi}\right) + \frac{1}{\cos^2\varphi}\frac{\partial^2\phi'}{\partial\lambda^2}\right]\right.$$
$$\left. + 4\Omega_0^2 a^2 \sin^2\varphi\frac{\partial}{p\partial Z}\left(\frac{p}{\tilde{N}^2}\frac{\partial\phi'}{\partial Z}\right)\right\} + \left[2(\Omega_0+\hat{\Omega}) - \frac{\partial^2\hat{\Omega}}{\partial\varphi^2} + 3\tan\varphi\frac{\partial\hat{\Omega}}{\partial\varphi}\right.$$
$$\left. - 4\Omega_0^2 a^2 \sin^2\varphi\frac{\partial}{p\partial Z}\left(\frac{p}{\tilde{N}^2}\frac{\partial\hat{\Omega}}{\partial Z}\right)\right]\cos\varphi\frac{1}{\cos\varphi}\frac{\partial\phi'}{\partial\lambda} = 0 \tag{3}$$

上式中 $\hat{\Omega}$ 定义为

$$\hat{\Omega} = \frac{\bar{U}}{a\cos\varphi}$$

是基本气流的角速度，$\tilde{N}$ 是 Brunt-Väisälä 频率。

我们设方程 (3) 的任何一个解可表达成

$$\phi'(\lambda, \varphi, Z, t) = \text{Re} \sum_{k=1}^{K} \Phi_k(\varphi, Z, t) e^{ik\lambda} = \text{Re} \sum_{k=1}^{K} e^{Z/2H_0} \Psi_k(\varphi, Z) e^{ik(\lambda - ct)}$$

(4)

k 是纬向波数。把方程(4)代入方程(3)可得到下列方程，

$$(\hat{\Omega} - c) \left\{ \left[\frac{\sin\varphi}{\cos\varphi} \frac{\partial}{\partial \varphi} \left(\frac{\cos\varphi}{\sin\varphi} \frac{\partial \Psi_k}{\partial \varphi} \right) + l^2 \sin^2\varphi \frac{\partial^2 \Psi_k}{\partial Z^2} - \left(l^2 \sin^2\varphi \frac{1}{4H_0^2} + \frac{k^2}{\cos^2\varphi} \right) \Psi_k \right. \right.$$
$$\left. - l^2 \sin^2\varphi \left[\frac{1}{H_0} \frac{\partial \ln \tilde{N}}{\partial Z} + 2\left(\frac{\partial \ln \tilde{N}}{\partial Z} \right)^2 - \frac{1}{\tilde{N}} \frac{\partial^2 \tilde{N}}{\partial Z^2} \right] \right] \Psi_k + \left\{ 2(\Omega_0 + \hat{\Omega}) - \frac{\partial^2 \hat{\Omega}}{\partial \varphi^2} \right.$$
$$\left. + 3\tan\varphi \frac{\partial \hat{\Omega}}{\partial \varphi} - l^2 \sin^2\varphi \left[\frac{\partial^2 \hat{\Omega}}{\partial Z^2} - 2\left(\frac{1}{2H_0} + \frac{\partial \ln \tilde{N}}{\partial Z} \right) \frac{\partial \hat{\Omega}}{\partial Z} \right] \right\} \Psi_k = 0$$

(5)

上式中

$$l = 2\Omega_0 a / \tilde{N},$$

(6)

我们将进一步简化，假设大气是近似等温的，$\tilde{N}$ 可认为是常数，这样方程(5)可写成

$$\frac{\sin\varphi}{\cos\varphi} \frac{\partial}{\partial \varphi} \left(\frac{\cos\varphi}{\sin\varphi} \frac{\partial \Psi_k}{\partial \varphi} \right) + l^2 \sin^2\varphi \frac{\partial^2 \Psi_k}{\partial Z^2} + Q_k \Psi_k = 0,$$

(7)

$$Q_k = Q_0 - \frac{k^2}{\cos^2\varphi}$$

(8)

$$Q_0 = \left[2(\Omega_0 + \hat{\Omega}) - \frac{\partial^2 \hat{\Omega}}{\partial \varphi^2} + 3\tan\varphi \frac{\partial \hat{\Omega}}{\partial \varphi} - l^2 \sin^2\varphi \left(\frac{\partial^2 \hat{\Omega}}{\partial Z^2} - \frac{1}{H_0} \frac{\partial \hat{\Omega}}{\partial Z} \right) \right] / (\hat{\Omega} - c)$$
$$- l^2 \sin^2\varphi \frac{1}{4H_0^2}$$

(9)

在等温大气 H_0 是 7 公里，$\tilde{N}$ 是 2×10^{-2} 秒$^{-1}$，方程 (7) 描述了波在等温大气中 φ 与 Z 方向的传播。

对于定常行星波 $c \approx 0$，因此，方程可写成

$$Q_0 = \left[2(\Omega_0 + \hat{\Omega}) - \frac{\partial^2 \hat{\Omega}}{\partial \varphi^2} + 3\tan\varphi \frac{\partial \hat{\Omega}}{\partial \varphi} - l^2 \sin^2\varphi \left(\frac{\partial^2 \hat{\Omega}}{\partial Z^2} - \frac{1}{H_0} \frac{\partial \hat{\Omega}}{\partial Z} \right) \right] / \hat{\Omega}$$
$$- l^2 \sin^2\varphi \frac{1}{4H_0^2}。$$

(10)

这样我们可以从方程(10)计算出 Q_0 以后，再从方程(8)计算出 Q_k 来。Q_k 是波数 k 的定常行星波的折射指数平方，Q_0 可以认为是波数 0 的折射指数平方，它的分布特点对于波的传播起着重要作用。

然而，由于在实际基本气流中存在着垂直及经向切变以及地球曲率的原因，定常行星波的传播路径在传播中将被折射。若假设波的传播路径与水平方向之间的夹角是 $\hat{\alpha}$，我们应用波数与频率之间的运动学关系，并假设基本气流是缓变的，利用 WKBJ 方法就可得到如下关系式[8]

$$\frac{d_g \hat{\alpha}}{dt} = \frac{1}{Q_k} \vec{i} \vec{C}_g' \times \nabla Q。$$

(11)

$\vec{i}$ 是 λ 方向的单位矢量，$\vec{C}_g'$ 是群速度在经圈方向的投影，并且我们定义 $\frac{d_g}{dt}$ 为下式，

$$\frac{d_g}{dt} = \frac{\partial}{\partial t} + \vec{C}'_g \cdot \nabla$$

它表示以群速度 $\vec{C}'_g$ 移动的随体微商。方程(11)描述了行星波传播路径的变化特点，它说明了定常行星波传播路径的变化是由 Q_1 与 Q_0 的梯度所决定。显然，传播路径的变化总是向着 Q_0 的梯度方向折射。

由夏季的基本气流并利用(9)式可以算得波折射指数平方 Q_0 与 Q_1 的分布。图2中的虚线表示 Q_1 的分布。除了极地附近外，Q_0 值与 Q_1 值相近，因此我们可以看到夏季正的 Q_0 最小值位于中、高纬度平流层低层及对流层上层，而在热带及平流层的 Q_0 值是负的。因此，夏季定常行星波不能传播到平流层，然而，定常行星波能够在对流层内传播。

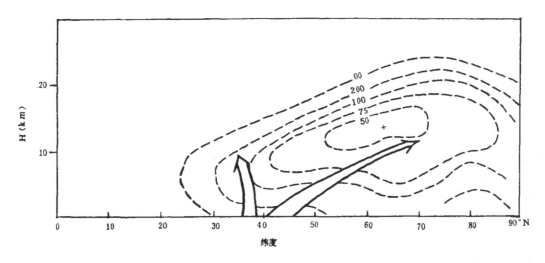

图 2　北半球夏季定常行星波传播波导示意图（虚线表示 Q_1，实箭号表示波导）

按照方程(11)，一旦由于某种强迫机制强迫所产生定常行星波入射到位于中纬度对流层上层 Q_0 值小的区域，则波将向着副热带上空折射，但夏季东西风交界线位于副热带，故波不能通过此临界面而传播，这个区域就像"波导管"一样，定常行星波能够沿着此波导向着对流层上层传播。从图2也可以看到，在中纬度 ∇Q_0 的方向是向北的，而在高纬度的 ∇Q_0 的方向是向南的，波能够聚焦在 50—70°N 的对流层。因此，在副热带由于某种强迫源强迫所产生的定常行星波能够向中高纬度地区传播，图2中的实箭号表示夏季 $k=1$ 定常行星波传播波导示意图。

四、模 式 与 参 数

1. 模式

上面我们已在理论上分析了夏季受迫定常行星波的传播。为了实际考察这种受迫定常行星波的三维传播，本文将采用一个包括 Rayleigh 摩擦、Newton 冷却及水平涡旋热力扩散的准地转34层球坐标模式来模拟夏季定常行星波的传播，其模式方程是

$$\hat{\Omega}_{n-\frac{1}{2}}\frac{\partial}{\partial \lambda}\left\{\frac{1}{2\Omega_0\sin\varphi}\frac{1}{a^2}\left[\frac{\sin^2\varphi}{\cos\varphi}\frac{\partial}{\partial\varphi}\left(\frac{\cos\varphi}{\sin^2\varphi}\frac{\partial\phi'}{\partial\varphi}\right)+\frac{1}{\cos^2\varphi}\frac{\partial^2\phi'}{\partial\lambda^2}\right]\right\}_{n-\frac{1}{2}}+\frac{1}{a}q_{n-\frac{1}{2}}\times\frac{1}{2\Omega_0\sin\varphi}$$

$$\frac{1}{a\cos\varphi}\frac{\partial\phi'}{\partial\lambda}_{n-\frac{1}{2}}=f\left(\frac{\partial\omega}{\partial p}\right)_{n-\frac{1}{2}}-(R_f)_{n-\frac{1}{2}}\times\frac{1}{2\Omega_0\sin\varphi}\frac{1}{a^2}\left[\frac{\sin\varphi}{\cos\varphi}\frac{\partial}{\partial\varphi}\left(\frac{\cos\varphi}{\sin\varphi}\frac{\partial\phi'}{\partial\varphi}\right)\right.$$

$$\left.+\frac{1}{\cos^2\varphi}\frac{\partial^2\phi'}{\partial\lambda^2}\right] \tag{12}$$

$$\hat{\Omega}_n\frac{\partial}{\partial\lambda}\left(\frac{\partial\phi'}{\partial p}\right)_n-\left(\frac{\partial\hat{\Omega}}{\partial p}\right)_n\frac{\partial\phi_n'}{\partial\lambda}+\sigma_n\omega_n=-\left(\frac{RH}{C_p p}\right)_n-(a_R)_n\left(\frac{\partial\phi'}{\partial p}\right)_n+(K_T)_n\times$$

$$\times\frac{1}{a^2}\left[\frac{\partial^2}{\partial\varphi^2}-\tan\varphi\frac{\partial}{\partial\varphi}+\frac{1}{\cos^2\varphi}\frac{\partial^2}{\partial\lambda^2}\right]\left(\frac{\partial\phi'}{\partial p}\right)_n \tag{13}$$

$$\hat{\Omega}_{n+\frac{1}{2}}\frac{\partial}{\partial\lambda}\left\{\frac{1}{2\Omega_0\sin\varphi}\frac{1}{a^2}\left[\frac{\sin^2\varphi}{\cos\varphi}\frac{\partial}{\partial\varphi}\left(\frac{\cos\varphi}{\sin^2\varphi}\frac{\partial\phi'}{\partial\varphi}\right)+\frac{1}{\cos^2\varphi}\frac{\partial^2\phi'}{\partial\lambda^2}\right]\right\}_{n+\frac{1}{2}}+\frac{1}{a}q_{n+\frac{1}{2}}\times\frac{1}{2\Omega_0\sin\varphi}$$

$$\frac{1}{a\cos\varphi}\frac{\partial\phi'_{n+\frac{1}{2}}}{\partial\lambda}=f\left(\frac{\partial\omega}{\partial p}\right)_{n+\frac{1}{2}}-(R_f)_{n+\frac{1}{2}}\times\frac{1}{2\Omega_0\sin\varphi}\frac{1}{a^2}\left[\frac{\sin\varphi}{\cos\varphi}\frac{\partial}{\partial\varphi}\left(\frac{\cos\varphi}{\sin\varphi}\frac{\partial\phi'}{\partial\varphi}\right)+\right.$$

$$\left.\frac{1}{\cos^2\varphi}\frac{\partial^2\phi'}{\partial\lambda^2}\right] \tag{14}$$

上面公式中，H 是单位时间与单位质量的非绝热加热，R 是气体常数（0.287 千焦/公斤·度），C_P 是定压比热（1.004 千焦/公斤·度），a_R' 是 Newfon 冷却系数，K_T 是水平涡动热力扩散，R_f 是扰动的 Rayleigh 摩擦系数，n 表示模式层次，q 为下式

$$q=\left[2(\Omega_0+\hat{\Omega})-\frac{\partial^2\hat{\Omega}}{\partial\varphi^2}+3\tan\varphi\frac{\partial\hat{\Omega}}{\partial\varphi}\right]\cos\varphi$$

在推导模式方程时，其涡度平流中的行星涡度平流项的 v' 引进了非地转分量，这样可得到合理的能量方程。

我们假设在模式顶的垂直速度为零，即

$$\omega=0,\ \text{在}\ p=p_t\ (\text{或}\ Z=Z_t) \tag{15}$$

作为上边界条件。并假设地表面垂直速度由气流爬山及 Ekman 层粘性所产生的 Ekman 抽吸所引起，即

$$\omega_s=\vec{V}_s\cdot\nabla P_G-\frac{p_s\cdot F}{2f}\zeta_s',\ \text{在}\ p=p_s(\text{或}\ Z=0) \tag{16}$$

作为下边界条件。$\vec{V}_s$ 是在 $p=p_s$ 的水平风速矢量，为简单起见，取 $p_s=1000$ mb，F 是摩擦系数并取为常数（4×10^{-6}秒$^{-1}$），ζ_s' 是地表面的扰动涡度。

在此模式所用的垂直差分方案与在 β 平面近似模式所用的方案相同[10]，即从地面到 92 公里高度（$p_t=8.459\times10^{-4}$mb）分成 34 层，在经向取 5°纬距为差分间隔。

上面所得到的模式方程中一般不能应用张驰方法，故应用 Lindzen and kuo 所提出的方法来解模式方程[11]。

为了解这些线性代数方程组，还需要侧边界条件，故假设 $\Phi_k(\varphi,p)$ 在极地及赤道为零，即

$$\Phi_k(\varphi, p) = 0, \ \varphi = \frac{\pi}{2}; \ \Phi_k(\varphi, p) = 0, \ \varphi = 0 \tag{17}$$

2. 参数

a) 静力稳定度参数 σ_R：此模式所用的静力稳定度是从美国出版的标准大气 7 月份 45°N 的平均温度与密度计算而得到。为简单起见，我们假设静力稳定度参数将不随纬度而改变。

b) 纬向平均风场的垂直廓线：我们采用 Murgatrod 所计算的分布[12]，但加于光滑，消去"小尺度"特征。

c) Rayleigh 摩擦系数 R, 与 Newton 冷却系数 α_R 取与 β 平面近似模式相同。

d) 水平涡旋热力扩散系数 K_T：由于夏季热带及副热带纬向平均风速都很小，故在热量平衡公式中的热量扩散系数要适当取比冬季小，故取 0.1×10^6 米^2/秒。

这样，若强迫源已知，则从模式方程就可以得到定常热源强迫所产生的定常行星波的传播情况。

五、青藏高原热源异常对北半球大气环流异常的影响

为了说明夏季青藏高原热源异常对北半球大气环流异常的影响，我们利用上述数值模式分别计算模式大气对地形与实际热源的气候平均值的响应及模式大气对地形与热源异常的响应。

1. 模式大气对地形与北半球实际热源气候平均值的响应

我们利用北半球实际地形与 Ashe 所计算的夏季北半球热源的气候分布作为强迫源，可以计算出夏季北半球实际地形与热源气候平均值强迫所产生的各等压面上定常扰动系统的分布。

图 3 表示夏季北半球实际地形与热源气候值强迫所产生的 300 mb 面上定常扰动分布。可以看到它与由实测资料所计算的分布（见文献 13 中的图 22）比较一致。主要的扰动系统有南亚高压，太平洋中部的热带对流层上层槽。

2. 模式大气对地形与青藏高原热源异常时的热源强迫的响应

由于我们没有夏季青藏高原热源异常的实际分布值，故在本试验中采用一个理想热源异常时的距平（相对于气候平均值的偏差），其垂直分布如下：

$$H'(\lambda, \varphi, p) = \hat{H}'(\varphi, p) \exp\left[-\left(\frac{p-\bar{p}}{d}\right)^2\right] \tag{18}$$

上式中 $d = 300$ mb，$\bar{p} = 500$ mb，这表示非绝热加热的异常在 500 mb 为最大，这也是与高原热源的实际分布比较符合的。我们还假设 500 mb 热源异常时的距平分布如下：

$$\hat{H}'(\varphi, p) = \begin{cases} \hat{H}'_0 \left(\sin\frac{\pi(\varphi-\varphi_1)}{(\varphi_2-\varphi_1)} \sin\frac{\pi(\lambda-\lambda_1)}{(\lambda_2-\lambda_1)} \right)^2, & \lambda_1<\lambda<\lambda_2, \ \varphi_1<\varphi<\varphi_2 \\ 0 & \text{其它区域} \end{cases} \tag{19}$$

$\lambda_1 = 45°$E，$\lambda_2 = 135°$E，$\varphi_1 = 15°$N，$\varphi_2 = 45°$N。根据 Luo and yanai 的计算结果，取 $\frac{1}{c_p}\hat{H}'_0 = 1.38°$K/天。图 4 表示 500 mb 面上热源异常时的距平分布。热源距平中心位于

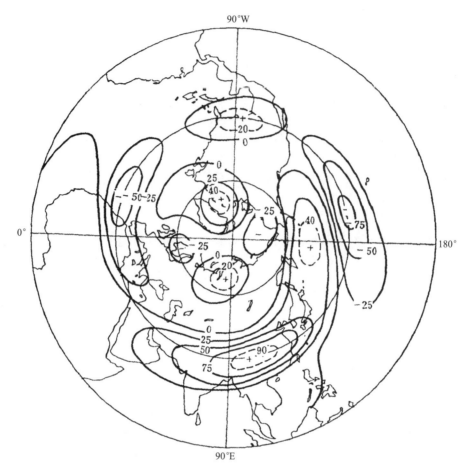

图 3　北半球夏季实际地形与热源气候值强迫所产生的 300 mb 面上定常扰动分布（单位是米）

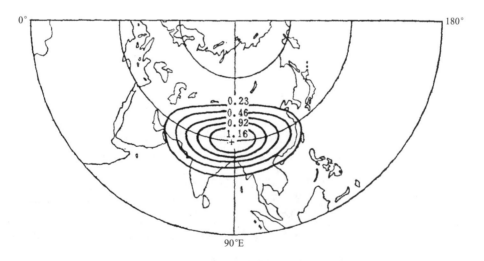

图 4　500 mb 面上热源异常的距平分布图

30°N, 90°E, 也就是位于青藏高原的上空, 并且热源异常是正距平。

我们把这个热源异常的理想距平分布迭加在北半球热源的气候分布, 就可以得到异常热源分布, 即

$$\bar{H}(\lambda, \varphi, p) + H'(\lambda, \varphi, p) = H(\lambda, \varphi, p) \tag{20}$$

$\bar{H}$ 是热源的气候平均值。我们把此异常热源代入上述模式就可以计算出当青藏高原热源异常时各等压面的定常扰动系统的分布及其各波数振幅与位相的垂直分布。

为了便于比较当青藏高原热源异常时与正常年份各等压面上定常扰动的差别以及各波数振幅、位相等等的差别, 我们把当青藏高原热源异常时强迫所产生的各波数定常行星波的振幅与位相及各等压面上定常扰动的高度值减去正常年份(气候平均值)的各波数定常行星波的振幅与位相值及各等压面上定常扰动的高度值。即

$$(Z^*)' = Z' - Z'_0 \tag{21}$$

Z' 是当青藏高原热源异常时强迫所产生的某等压面上某格点的扰动高度值, 而 Z'_0 是地形与热源气候平均值强迫所产生的某等压面上某格点的扰动高度值, 故 $(Z^*)'$ 为当青藏高原热源异常时所产生的扰动高度的异常值。

图 5 中的实线与虚线分别表示由于青藏高原上空热源异常而引起的波数 1 振幅与位相的距平。从图 5 可以发现波数 1 准定常行星波的最大距平位于 30°N 时近的对流层上层, 并且在高纬度对流层上层存在着距平的第二峰值, 而位相距平说明了位相异常是从亚热带随着纬度增加向西倾斜, 因此, 可以看到定常行星波从副热带向中、高纬度地区传播, 其异常可以从副热带向中、高纬度地区传播, 这个计算结果不仅可以证实上述理论分析所得的结论, 而且说明了位于青藏高原的热源异常可以影响到北半球中、高纬度行星波的异常。

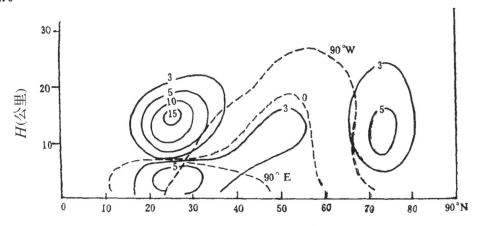

图 5 当青藏高原上空热源异常时, 波数 1 准定常行星波振幅与位相的异常情况分布。图中虚线表示位相, 实线表示振幅。

下面, 我们将计算各等压面上由于热源异常而产生高度场的偏差情况。图 6 与图 7 分别表示由于青藏高原上空热源异常所产生的 300 mb 与 500 mb 等压面上扰动高度场的偏差情况, 图中的实箭号表示由理论上所得到的准定常行星波的传播路径。从图 6 与图 7 可以发现, 当青藏高原上空热源发生异常时, 北半球定常扰动系统不仅在副热带地区

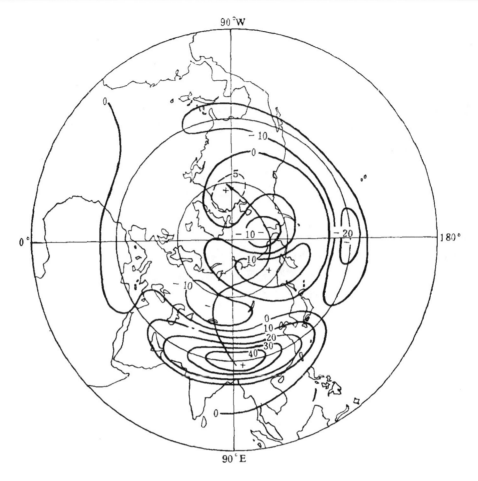

图 6 当青藏高原热源异常时，地形与热源强迫所产生 300 mb 上扰动高度场的偏差（单位是米）。

发生异常，而且在中、高纬度地区发生异常。当青藏高原热源加强时，青藏高压要加强（300mb上南亚高压要加强），而我国北方低压槽要加深，这容易造成我国东北与日本北部冷夏；并且鄂霍茨克海高压要加强，阿拉斯加地区产生槽；此外，500 mb 面上副热带高压减弱。这些计算结果完全与 Asakura 等人由实际观测资料所得的结果是一致的。

上面计算结果可以说明夏季青藏高原热源不仅会引起东亚地区环流的异常，而且可以引起太平洋副热带高压的异常，甚至可以影响北美环流的异常。这种环流的遥相关正是由于准定常行星波的传播而造成，而传播路径是由夏季的基本气流所决定，当基本气流一定时，波传播总是沿着特定波导而进行的，故它的异常也沿着特定波导而传播，这对于长期天气预报是有参考意义。如1979年高原地面热源与大气热源偏弱，根据许致远等人的统计，我国黑龙江省夏季8月份偏暖[14]，因此，夏季高原上空热源异常与我国东北夏季气温有直接关系。

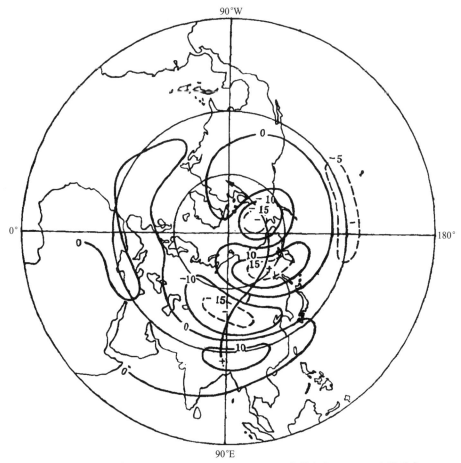

图 7 当青藏高原热源异常时，地形与热源强迫所产生 500 mb 上扰动高度场的偏差（单位是米）。

六、结 论 与 讨 论

本文应用缓变媒质中波动的传播理论研究了夏季基本气流中强迫定常行星波的三维传播规律。夏季定常行星波不仅能向亚热带对流层上层传播，而且可以由亚热带向中、高纬度地区对流层上层传播。

本文应用一个包括 Rayleigh 摩擦、Newton 冷却及水平涡旋 热力扩散准地转 34 层球坐标模式研究了夏季由于青藏高原热源异常对北半球中、高 纬 度 定常扰动系统的影响。计算结果表明：

1) 夏季青藏高原热源异常将影响北半球中、高纬度对流层大气环流的异常。

2) 若夏季青藏高原热源增强，青藏高压(300 mb 面上的南亚高压)就增强，随后我国北方的槽要加深，这样使我国东北与日本北部将产生冷夏，并且鄂霍茨克海高压加强，阿拉斯加西部将产生槽，北美地区的高压要加强，此外，副热带高压将减弱，这是与实际资料所得的结论一致。

本文所讨论的青藏高原热源的异常是理想情况,实际的异常情况有待于进一步研究,并且引起异常的原因更是需要研究的。

参 考 文 献

[1] 叶笃正,高由禧,青藏高原气象学,科学出版社,1979年。
[2] Ashe, S., A nonlinear model of the time average axially asymmetric flow induced by topography and diabatic heating, *J. Atmos. Sci.*, 36, 109—126, 1979.
[3] Nitta, T., Observational study of heat sources over the eastern Tibetan Plateau during the summer monsoon, *J. Meteor. Soc. Japan*, 61, 590—605, 1983.
[4] Luo, H. and M. Yanai, The large-scale circulation and heat sources over the Tibetan Plateau and surrounding areas during the early summer of 1979, *Mon. Wea. Rev.*, 112 (to be published) 1984.
[5] 陈烈庭,青藏高原冬春季异常雪盖影响初夏季风的统计分析,中长期水文气象预报文集,185—194,1978年。
[6] Asakura, T., Dynamical climatology of atmospheric circulation over East Asia centered in Japan, Papers in Meteorology and Geophysics, 19, 1—68, 1968.
[7] Gambo, K. and K. Kudo, Teleconnections in the zonally asymmetric height field during the northern hemispheric summer, *J. Meteor. Soc. Japan*, 61, 829—837, 1983.
[8] 黄荣辉、岸保勘三郎,关于冬季北半球定常行星波传播另一波导的研究,中国科学,1983年第10期,940—950。
[9] Shukla, J. and J. M. Wallace, Numerical simulation of the atmospheric response to equatorial Pacific sea surface temperature anomalies, *J. Atmos. Sci.*, 40, 1613—1630.1983.
[10] Huang, Rong-hui and K. Gambo, The response of a model atmosphere in middle latitudes to forcing by topography and stationary heat sources, *J. Meteor. Soc. Japan*, 59, 220—237, 1981.
[11] Lindzen, R. S., and H. L. Kuo, A reliable method for the numerical integration of a large class of ordinary and partial differential equations, *Mon. Wea. Rew.*, 96, 732—734, 1969.
[12] Murgatroyd, R. J., The structure and dynamics of the stratosphere, the global circulation of the atmosphere, 155—195, 1969.
[13] Huang, Rong-hui and K. Gambo, The response of a hemispheric multilevel model atmosphere to forcing by topography and stationary heat sources in summer, *J. Meteor Soc. Japan*, 61, 495—505, 1983.
[14] 许致远、白人海、魏松林,黑龙江省夏季低温与北太平洋海温异常的联系及其长期预报,东北夏季低温长期预报文集,219—237,1983年。

Relationship between the Interannual Variation of Total Ozone in the Northern Hemisphere and the QBO of Basic Flow in the Tropical Stratosphere[*]

Huang Ronghui(黄荣辉) and Wang Lianying(王连英)

Institute of Atmospheric Physics, Academia Sinica, Beijing 100080

ABSTRACT

The harmonic analyses of monthly mean total ozone in the atmosphere over the Northern Hemisphere for 26 years (1960–1985) are made by using the Fourier expansion. The analysed results show that there is obviously a quasi-biennial oscillation (QBO) in the interannual variations of the amplitudes of total ozone. Generally, the amplitudes of wavenumber 1 and 2 during the westerly of the equatorial QBO are larger than those during the easterly. In the early winter, the amplitude of wavenumber 1 during the easterly phase is larger, and in the late winter, it is larger during the westerly phase. These are in good agreement with the observational distributions.

I. INTRODUCTION

Although the total ozone is very small in the atmosphere, it is one of the important compositions of atmosphere. It absorbs ultraviolet radiation in short wave radiation from the sun. The ultraviolet radiation is largely related to the human life, and is directly related to the cancer of human skin. On the other hand, since a photochemical reaction can be caused in the ozone due to absorption of ultraviolet, the heating sources in the middle atmosphere may be affected. Thus, it would play an important role for the circulation of middle atmosphere and for the climatic change.

Owing to the Dobson's observational stations of the total ozone increase continuously, a global Dobson's observational network of the ozone has been set up. Therefore, a good monitoring of interannual and intraseasonal variation of the total ozone in an air column can be made on the globe. Dutch (1971) analysed the seasonal variation of the total ozone in the atmosphere by using the data of global Dobson's observational network. The analysed results show that there is an obvious seasonal variation of total ozone either in the Southern Hemisphere or in the Northern Hemisphere. London et al. (1976) also analysed the variation of total ozone with the help of the observational data of satellite. His results show that the zonal mean features of total ozone obtained from the data of satellite are analogous to that from the data of global Dobson's observational network.

As far back as 1940's, it was found that ozone is a tracer, which can be used to trace the variation of the atmospheric circulation. Thus, recently, more investigations focus attention on the relationship between the variation of the total ozone and the atmospheric circulation. For example, Gushein (1976) analysed the relationship between the circulation at 100 hPa and the variation of total ozone. He pointed out that when a zonal circulation is prevailing in the Northern Hemisphere, the zonal circulation can cause large total ozone in high latitudes to be

---

[*] 本文原载于: Advances in Atmospheric Sciences, Vol.7, No.1, 47–56, 1990.

separated from small total ozone in low latitudes, and the poleward meridional transfer of ozone may be blocked. Therefore, the variation of total ozone is not obvious; however, a meridional circulation is prevailing, the poleward meridional transfer becomes evident, in this case, total ozone tends to have obvious variation. Recently, Tung et al. (1986) investigated the causes of ozone home over the Antarctica. Their investigations show that because large part of the Southern Hemisphere is covered by ocean, the quasi-stationary planetary waves forced by the large-scale topography and the difference between ocean and land, are weak, the polar vortex surrounding the Antarctica is much stable than that in the Northern Hemisphere. Therefore, the transfer to the Antarctica due to disturbance is weak, the ozone home is easily formed over the Antarctica. All these can explain that atmospheric circulation may play an important role in the distribution of ozone. Thus, the distribution of ozone would in turn well reflect the variation of atmospheric circulation.

In this paper, the interannual and seasonal variations of total ozone in the Northern Hemisphere for every wavenumber are analysed by using the Fourier expansion with observational data of monthly mean total ozone for 26 years from 1960 to 1985 (from Ozone Data for the World, January, 1960–December, 1985, Atmospheric Environment Service, Canada).

A quasi-geostrophic, 34-level, 2-dimensional model with Rayleigh friction, Newtonian cooling and horizontal eddy thermal diffusion is used to calculate forced quasi-stationary planetary waves during different phases of the equatorial QBO in late winter. The relationship between the interannual variation of total ozone and the QBO of basic flow in the tropical stratosphere is explained.

II. THE INTERANNUAL VARIATION OF AMPLITUDE OF TOTAL OZONE FOR VARIOUS WAVENUMBERS

The interannual variation of the atmospheric circulation in the equatorial lower stratosphere exhibits a quasi-biennial oscillation. This phenomenon was discovered by Reed et al. (1961), Vergand and Ebdon (1961) from the observational data. Recently, Holton and Tan (1982) have pointed out from the analyses of observational data that in late winter, the amplitudes of planetary waves for wavenumbers 1 and 2 in high and middle latitudes are larger during the westerly phase of the equatorial QBO; while in early winter, they are smaller. Angell (1980) investigated the interannual variation of ozone at Resolute station near the North Pole, and pointed out that there is an obvious quasi-biennial oscillation in the amount of total ozone. Moreover, Kulbarni (1980) analysed the variation of ozone at some stations in the Northern Hemisphere, and also discovered that there is a quasi-biennial oscillation in total ozone. However, the variations from those investigations are analysed by using the data only at one station. Hasebe (1980) studied the interannual variations of the zonal and meridional mean total ozone using the data of total ozone for 14 years from 1962 to 1976. He discussed mainly the interannual variation of phase of total ozone in the Northern Hemisphere for various wavenumbers and not the interannual and seasonal variations of amplitude of total ozone. Thus, it is a significant subject to study how the interannual variations of amplitude of total ozone in the Northern Hemisphere for various wavenumbers would be.

Fig.1 shows the interannual variation of amplitude of annual mean total ozone in the Northern Hemisphere for wavenumber 1 during 1960–1985. We can see from Fig.1 that there is a distinct difference of the interannual variation of amplitude for wavenumber 1 between in the south of 30°N and in the north of 30°N. In the south of 30°N, the amplitude of total ozone for wavenumber 1 decreases obviously from 1960's to the present. The amplitude of total ozone for wavenumber 1 is larger in the early and middle 1960's, while in the late 1960's, it decreases suddenly. It seems to experience a catastrophic event.

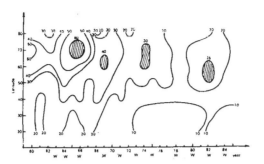

Fig.1. Interannual variation of amplitude of annual-mean total ozone in the Northern Hemisphere for wavenumber 1 during 1960–1985 (Unit in Dobson). W indicates the westerly phase of basic flow in the tropics.

We also see from Fig.1 that in the north of 40°N, especially in the area of 60°–70°N, there is an obvious quasi-biennial oscillation in the distribution of amplitudes of total ozone for wavenumber 1. Generally, during the westerly phase of the equatorial QBO, the amplitude of total ozone for wavenumber 1 is larger.

We have also analysed the interannual variation of amplitude of total ozone for wavenumber 2 for 26 years from 1960 to 1985. As shown in Fig.2, there is a distinct difference of the interannual variation of amplitude of total ozone for wavenumber 2 between in the south of 30°N and in the north of 30°N. In the south of 30°N, the amplitude of total ozone for wavenumber 2 is larger in 1960's, while in the period from the late 1960's to the present, the amplitude of total ozone for wavenumber 2 decreases obviouly. This is analogous to the variation of amplitude for wavenumber 1. In the north of 30°N, especially in the area of 40°–50°N, there is a quasi-biennial oscillation in the distribution of amplitude of wavenumber 2. Generally, during the westerly phase of the equatorial QBO, the amplitude of wavenumber 2 is slightly larger.

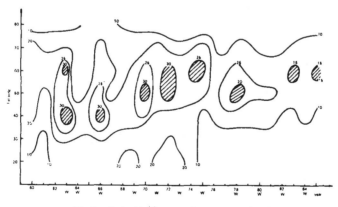

Fig.2. As in Fig.1 except for wavenumber 2.

In order to see clearly the differences between the amplitude of total ozone for various wavenumbers for different phase of the equatorial QBO, the latitude-time cross sections of amplitude of total ozone for wavenumbers 1 and 2 in easterly and westerly categories of the equatorial QBO in the early and late winter are shown in Figs. 3a,b, and Figs. 4a,b. From Figs.3 and 4 it can be seen that the amplitude of total ozone for wavenumber 1 is larger in the easterly category than that in the westerly category for the early winter. Moreover, the total ozone for wavenumber 2 exhibits slightly stronger amplitudes in the easterly category for the early winter, but the difference between categories is not statistically significant. However, As shown in Figs. 4a,b, in the late winter, the amplitude of wavenumber 1 is obviously stronger in the westerly category, and the amplitude of wavenumber 2 seems to be slightly larger.

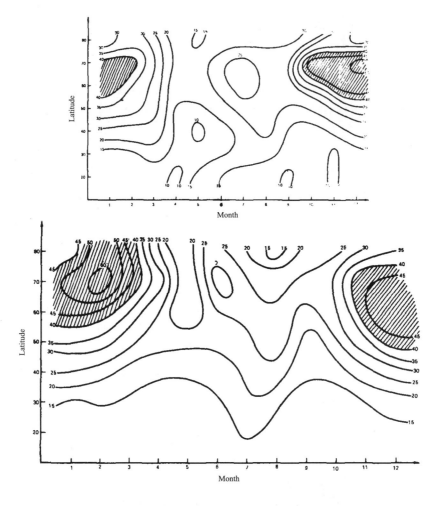

Fig.3. Latitude–time sections of the amplitude (unit in Dobson) of the total ozone for wavenumber 1 (a) in the easterly category; (b) in the westerly category.

From the above–mentioned harmonic analyses of the total ozone, we can see that there is an obvious quasi–biennial oscillation in the interannual variation of the amplitudes of wavenumbers 1 and 2. Generally, when the basic flow is westerly in the tropical stratosphere, their amplitudes are larger for the late winter. This is in good agreement with the distribution of amplitudes of planetary wavenumbers 1 and 2 obtained by Halton and Tan (1982).

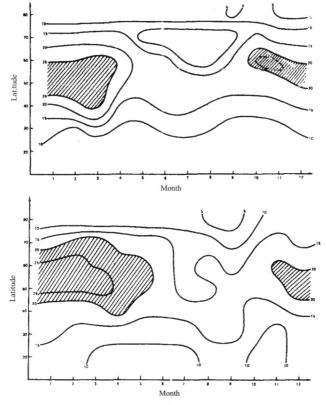

Fig.4. As in Fig.3 except for wavenumber 2.

III. MODEL AND PARAMETERS

Above, we have analysed the relationship between the distribution of the amplitude of total ozone and the QBO of basic flow in the tropical stratosphere in winter from the observational facts. In order to make this relationship much clear, a quasi-geostrophic, 34-level, two-dimensional model with the Rayleigh friction, the Newtonian cooling and the horizontal eddy thermal diffusion is used to make numerical simulation of this problem.

1. Model

As described in Huang and Gambo's (1982) paper, the model equations used in this paper are

$$\hat{\Omega}_{n-\frac{1}{2}}\frac{\partial}{\partial \lambda}\left\{\frac{1}{2\Omega_0 \sin\varphi}\frac{1}{a^2}\left[\frac{\sin^2\varphi}{\cos\varphi}\left(\frac{\cos\varphi}{\sin^2\varphi}\frac{\partial \varphi'}{\partial \varphi}\right)+\frac{1}{\cos^2\varphi}\frac{\partial^2 \varphi'}{\partial \lambda^2}\right]\right\}_{n-\frac{1}{2}}+\frac{1}{a}q_{n-\frac{1}{2}}\frac{1}{2\Omega_0 \sin\varphi}$$

$$\times \frac{1}{a\cos\varphi}\frac{\partial \varphi'_{n-\frac{1}{2}}}{\partial \lambda}=f\left(\frac{\partial \omega}{\partial p}\right)_{n-\frac{1}{2}}-(R_f)_{n-\frac{1}{2}}\times \frac{1}{2\Omega_0 \sin\varphi}\frac{1}{a^2}\left[\frac{\sin\varphi}{\cos\varphi}\frac{\partial}{\partial \varphi}\left(\frac{\cos\varphi}{\sin\varphi}\frac{\partial \varphi'}{\partial \varphi}\right)\right.$$

$$\left.+\frac{1}{\cos^2\varphi}\frac{\partial^2 \varphi'}{\partial \lambda^2}\right]_{n-\frac{1}{2}}, \qquad (1)$$

$$\hat{\Omega}_n \frac{\partial}{\partial \lambda}\left(\frac{\partial \varphi'}{\partial p}\right)_n - \left(\frac{\partial \hat{\Omega}}{\partial p}\right)_n \frac{\partial \varphi'_n}{\partial \lambda} + \sigma_n \omega_n = -\left(\frac{RH}{C_p p}\right)_n - \left(\alpha_R\right)_n \left(\frac{\partial \varphi'}{\partial p}\right)_n + \left(K_T\right)_n$$
$$\times \frac{1}{a^2}\left[\frac{\partial^2}{\partial \varphi^2} - \tan\varphi \frac{\partial}{\partial \varphi} + \frac{1}{\cos^2\varphi}\frac{\partial^2}{\partial \lambda^2}\right]\left(\frac{\partial \varphi'}{\partial p}\right)_n, \quad (2)$$

$$\hat{\Omega}_{n+\frac{1}{2}} \frac{\partial}{\partial \lambda}\left\{\frac{1}{2\Omega_0 \sin\varphi}\frac{1}{a^2}\left[\frac{\sin^2\varphi}{\cos\varphi}\frac{\partial}{\partial \varphi}\left(\frac{\cos\varphi}{\sin^2\varphi}\frac{\partial \varphi'}{\partial \varphi}\right) + \frac{1}{\cos^2\varphi}\frac{\partial^2 \varphi'}{\partial \lambda^2}\right]\right\}_{n+\frac{1}{2}} + \frac{1}{a}q_{n-\frac{1}{2}}\frac{1}{2\Omega_0 \sin\varphi}$$
$$\times \frac{1}{a\cos\varphi}\frac{\partial \varphi'_{n+\frac{1}{2}}}{\partial \lambda} = f\left(\frac{\partial \omega}{\partial p}\right)_{n+\frac{1}{2}} - (R_f)_{n+\frac{1}{2}} \times \frac{1}{2\Omega_0 \sin\varphi}\frac{1}{a^2}\left[\frac{\sin\varphi}{\cos\varphi}\frac{\partial}{\partial \varphi}\left(\frac{\cos\varphi}{\sin\varphi}\frac{\partial \varphi'}{\partial \varphi}\right)\right.$$
$$\left. + \frac{1}{\cos^2\varphi}\frac{\partial^2 \varphi'}{\partial \lambda^2}\right]_{n+\frac{1}{2}}, \quad (3)$$

where $\hat{\Omega} = \frac{U}{a\cos\varphi}$ is the angle velocity of basic flow. H is diabatic heating per unit time and unit mass. R is gas constant ($0.287 KJ/Kg \cdot K$), C_p is specific heat at constant pressure ($1.004 KJ/Kg \cdot K$), α_R is Newtonian cooling coefficient, K_T is horizontal eddy thermal diffusion, R_f stands for Rayleigh friction coefficient of perturbation, n represents level in the model, and is expressed:

$$q = \left[2(\Omega_0 + \hat{\Omega}) - \frac{\partial^2 \hat{\Omega}}{\partial \varphi^2} + 3\tan\varphi \frac{\partial \hat{\Omega}}{\partial \varphi}\right]\cos\varphi,$$

is derived from the model equations, the component of non-geostrophic wind v' is included in the planetary vorticity advection term in the vorticity advection, so that we can get a reasonable energy equation.

For the upper boundary condition, we assume that the vertical p-velocity vanishes at the top of model,
$$\omega = 0, \quad \text{at} \quad p = p_t \quad (\text{or} \quad Z = Z_t). \quad (4)$$

For the lower boundary condition, we assume p-velocity at P_s is caused by surface topography and by Ekman pumping resulting from the viscosity in the Ekman layer:

$$\omega = \vec{V}_s \cdot \nabla P_G - \frac{P_s F}{2f}\zeta'_s, \quad \text{at} \quad P = P_s \quad (\text{or} \quad Z = 0), \quad (5)$$

where $\vec{V}_s$ is horizontal velocity vector at $P = P_s$ and $P_s = 1000$ hPa, for simplicity. F is friction coefficient and will be treated as a constant ($4 \times 10^{-6} s^{-1}$), ζ'_s is vorticity of perturbation at the surface. P_G is height of topography.

The vertical finite difference scheme used in this model is the same as that discussed in the paper of plane approximate model (See Huang and Gambo (1981)), i. e., we divide the atmosphere into 34 layers from the earth's surface to the top of this model atmosphere, the finite-difference scheme with an interval $\Delta\varphi = 5^0$ is used in latitudinal direction.

Since the relaxation methods are not, generally, applicable for the model equations obtained above, the method proposed by Lindzen and Kuo (1969) is used to solve the model equations.

In order to solve those algebraic equations, we assume

$$\varphi'(\lambda,\varphi,p) = Re \sum_{k=1}^{k} \Phi_R(\varphi,p)e^{ik\lambda}, \qquad k = 1,2,......K \tag{6}$$

and assume

$$\Phi_k(\varphi,p) = \begin{cases} 0, & \varphi = \frac{\pi}{2}, \\ 0, & \varphi = 0, \end{cases} \tag{7}$$

that $\Phi_k(\varphi,p)$ should vanish at the pole and equator.

2. Parameters

(1) Static stability parameter: The static stability used in this model is calculated from the mean temperature and density at 45°N in July for the U.S. Standard Atmosphere. For simplicity, we assume that the static stability parameter does not change with latitude.

(2) The vertical profile of the basic zonal-mean wind: In order to calculate the relationship between amplitude of planetary waves and basic flow in the tropical stratosphere, two different profiles of basic zonal-mean wind are used in this paper. One is easterly in the middle and lower stratosphere of the tropics; another is westerly. Moreover, these two profiles of basic flows are the same in areas of middle and high latitudes and tropical troposphere. They may represent the two categories of the equatorial QBO for the late winter.

(3) The coefficients of Rayleigh friction R_f and Newtonian cooling α_R and coefficients of horizontal eddy diffusion K_T are the same as those in Huang and Gambo's (1982) paper.

IV. RELATIONSHIP BETWEEN THE FORCED QUASI-STATIONARY PLANETARY WAVES AND THE BASIC FLOW IN THE TROPICAL STRATOSPHERE

The model equations (1)–(3) are used to compute the distribution of quasi-stationary planetary waves forced by topography and heat sources in the easterly and westerly categories of the equatorial QBO.

Fig.5a is the distribution of amplitude and phase of planetary wavenumber 1 forced by the Northern Hemispheric topography and heat sources in the easterly category, while Fig.5b is the same as Fig. 5a but in the westerly category. Figs. 6a,b, are the distributions of amplitude and phase of quasi-stationary planetary wave for wavenumber 2 forced by the Northern Hemispheric topography and heat source in the easterly and westerly categories, respectively. Comparing Fig.5 with Fig.6, we can find that the amplitudes of planetary waves for wavenumbers 1 and 2 forced by topography and heat sources are larger in the westerly category than in the easterly category.

In order to see clearly the influence of the basic flow in the tropical stratosphere on the quasi-stationary planetary waves in middle and high latitudes, we calculate the quasi-stationary disturbance pattern at isobaric levels by synthesizing the components of zonal wavenumbers 1–3. Fig.7a is the stationary disturbance pattern at 30 Km height forced by the topography and heat sources in the easterly category. The disturbance pattern exhibits disturbance about wavenumber 1. The Aleutian high is located over the Aleutian islands and the polar vortex is located over Europe. Fig.7b is the same as Fig.7a but in the westerly category. We can find from Fig.7b that the Aleutian high located over the Aleutian area intensifies and the polar vortex located over Europe also intensifies.

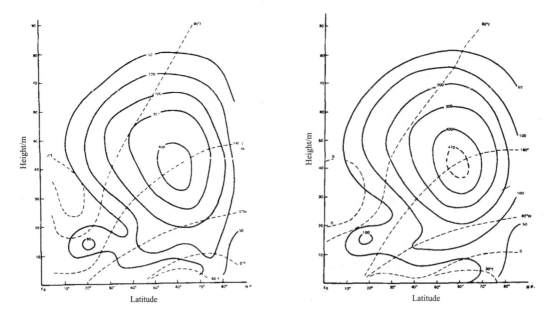

Fig.5. The distribution of quasi-stationary planetary wave for wavenumber 1 forced by the Northern Hemispheric topography and heat sources in the different categories of the equatorial QBO. The solid curves denote amplitude (Unit in m), the dashed curves denote phase. (a) in the easterly category; (b) in the westerly category.

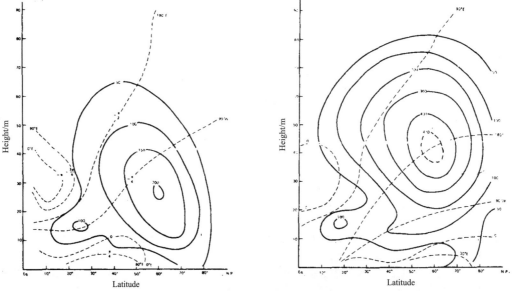

Fig.6. As in Fig.5 except for wavenumber 2.

As mentioned in the introduction, ozone is a tracer of the atmospheric circulation in the stratosphere. Variations of the stratospheric circulation, especially the planetary-scale circulation, would influence greatly on the distribution of ozone. As a result, the larger amplitude planetary-scale disturbances, forced by topography and heat sources, may cause larger amplitudes of total ozone distribution during the westerly category of the equatorial QBO.

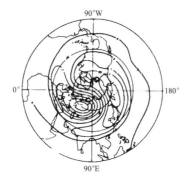 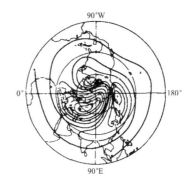

Fig.7. The disturbance patterns (Unit in 100 m) at 30 Km height forced by topography and heat sources in the different categories, (a) in the easterly category; (b) in the westerly category.

V. CONCLUSIONS

The harmonic analyses of monthly mean total ozone in the Northern Hemisphere for 26 years during 1960–1985 are made by using the Fourier expansion in this paper. The analysed results show that there is an obvious quasi–biennial oscillation in the interannual variation of amplitude for wavenumbers 1 and 2 in the north of 30°N. Amplitudes of total ozone for wavenumbers 1 and 2 are larger in the westerly category of the equatorial QBO during the late winter. These are in good agreement with the observational distributions of planetary waves.

In this paper, a quasi–geostrophic 34–level, two–dimensional model with the Rayleigh friction, the Newtonian cooling and the horizontal eddy thermal diffusion is used to calcutate the forced planetary waves for the late winter during the different categories of the equatorial QBO. The results show that the amplitudes of wavenumbers 1 and 2 in the westerly category are larger than that in the easterly category. This can demonstrate that the planetary–scale disturbances forced by topography and heat sources may cause larger amplitudes of total ozone distribution during the westerly category of the equatorial QBO.

REFERENCES

Angell, J.K., (1980), Temperature and ozone variation in the stratosphere, *Pure and Appl. Geophys.*, **118**: 378–386.

Dutch, H.U., (1971), Photochemistry of Atmospheric ozone, *Advances in Geophysics*, **15**: 219–322.

Gushein, G. P. (1976) Oscillation of total ozone and air circulation in the stratosphere during wintertime, G. G. O. Pub. 357.

Huang, Ronghui and K. Gambo, (1981), The response of a model atmosphere in middle latitudes to forcing by topography and stationary heat sources, *J. Meteor. Soc. Japan*, **59**: 220–237.

Hasebe, F., (1980), A global analysis of the fluctuation of total ozone, *J. Meteor. Soc. Japan*, **58**: 95–117.

Holton, J.R., and H, C, Tan, (1982), The quasi–biennial oscillation in the Northern Hemisphere lower stratosphere, *J. Meteor, Soc, Japan*, **60**: 140–148.

Huang Ronghui and K, Gambo (1982), The response of a Hemisphere multi–level model atmosphere to forcing by topography and heat sources, Part. II, *J. Meteor. Soc. Japan*, **60**: 93–108.

Kulbarni, R. N., (1980), Atmospheric ozone over Australia A review, *Pure and Appl. Geophys.* **118**: 387–400.

Lindzen, R. S., and H. L. Kuo, (1969), A reliable method for the numerical integration of a large class of ordinary and partial differential equations, *Mon. Wea. Rev.*, **97**: 732–734.

London, J., Frederick, J. and Anderson, G., (1976), Satellite observations of the global distribution of stratospheric ozone, *J.Geophys. Res.*, **82**: 2523–2556.

Ozone data for the world, January, 1960–December, 1985, Atmospheric Environment Service, Canada.

Reed, R, J., et al., (1961), Evidence of the downward-propagating annual wind reversal in the equatorial stratosphere, *J. Geophys. Res.*, **66**: 813–818.

Tung, K. K., M.K.W.Ko, J.M.Rodgiguez and N.D. Sze (1986), Are Antarctic ozone variations a manifestation of dynamics or chemistry? *Nature*, **322**: 811–814.

Vergand, R, G., and R, A. Ebdon, (1961), Fluctuations in tropical stratospheric winds, *Meteor. Mag.*, **90**: 125–143.

Von Lonn, H., R. L. Janne and K. Labitzke, (1973), Zonal harmonic standing waves, *J. Geophys. Res.*, **78**: 4463–4471.

二、西太平洋暖池对东亚夏季风系统及我国气候变异的热力作用及其动力过程研究

The Physical Effects of Topography and Heat Sources on the Formation and Maintenance of the Summer Monsoon Over Asia*

Huang Ronghui(黄荣辉) and Yan Bangliang(严邦良)

Institute of Atmospheric Physics, Academia Sinica, Beijing

ABSTRACT

The physical effects of topography and heat sources on the formation and maintenance of the summer monsoon over Asia are discussed in this paper by using the transformed Eularian-mean equations and a quasi-geostrophic 34-level spherical coordinate model.

The computed results of the divergence of the E-P flux, the induced meridional circulation and the perturbation geostrophic wind speed induced by the forcing of topography and heat sources show that the diabatic heating effect over the Tibetan Plateau may play an important role for the formation and maintenance of the summer monsoon over Asia, which is much greater than the dynamical effect of topography.

The computed results also show that, of the physical effects of topography and heat sources on the formation and maintenance of the summer monsoon over Asia, the effect of forced meridional circulation is larger than that of the divergence of E-P flux of the induced waves.

I. INTRODUCTION

Monsoon is a climatic system in which wind systems change with seasons. Over South Asia and East Asia, in particular, it has obviously different wind systems in wintertime and summertime. Generally speaking, the monsoon circulation prevaling over South Asia and East Asia has three specific features: (1) there is an anticyclone in the upper troposphere over the Tibetan Plateau, i.e. the so-called South Asian high, and there is a very strong easterly jet over South Asia and East Aisa; (2) a southwest flow prevails in the lower troposphere over South Asia, while over East Asia the flow from South Asia changes into southerly due to the influence of southeast flow from the west of the West Pacific subtropical high; and (3) this monsoon circulation brings the monsoon rainfall in India and Bengladesh and causes the plum rains in China and Japan.

Recently, many investigations show that the said monsoon circulation is in connection with the strong diabatic heating over the Tibetan Plateau. Yeh and Gao (1979) show that the heat source over the southeast of the Tibetan Plateau is the largest heat source during the Northern Hemispheric summer. It is not only a heat source over Eurasian continent that is in contradiction to the cold source over the ocean, but also like a "heat island" directly heating the atmosphere, as the Plateau is very high. Using a numerical model, Ji and Tibaldi (1984) calculated the effect of diabatic heating over the Tibetan Plateau for the onset of the monsoon over Asia. Kuo and Qian (1982) made some numerical experiments on the maintenance of the monsoon circulation over Asia by GCM. The computed results show that the diabatic heating over the Tibetan Plateau plays an important role in the maintenance of the summer monsoon circulation over Asia. However, the physical mechanism of the heat source over the Tibetan

Plateau for the formation and maintenance of the summer monsoon over Asia has not been deeply discussed so far. In this paper, the transformed Eularian-mean equations are used to discuss the influence of planetary-scale eddies induced by the Plateau topography and heat source on the flow, and thus explain the effect of the heat source over the Tibetan Plateau on the formation of the mean monsoon circulation in summer. In addition, a geostrophic 34-level model is used for studying the thermal effect of the diabatic heating over the Tibetan Plateau on the mean monsoon circulations over South Asia and East Asia.

II. THE TRANSFORMED EULARIAN-MEAN EQUATIONS AND THE INDUCED MERIDIONAL CIRCULATION

According to Edmon's (1980) derivation, taking the Eularian-mean of the equations of motion, continuity equation, and thermodynamic equation in the spherical coordinates, and neglecting the small terms, we can obtain the following equations:

$$\frac{\partial \bar{U}}{\partial t} - f\bar{v}^* - \bar{D} = \frac{1}{a\cos\varphi}\nabla \cdot F, \tag{1}$$

$$f\bar{U}_p - \frac{R}{a}\bar{\theta}_\varphi = 0, \tag{2}$$

$$\frac{1}{a\cos\varphi}(\bar{v}^*\cos\varphi)_\varphi + \bar{\omega}^*_p = 0, \tag{3}$$

$$\frac{\partial \bar{\theta}}{\partial t} + \bar{\theta}_p \bar{\omega}^* - \bar{S} = 0. \tag{4}$$

Eqs. (1)—(4) are the so-called transformed Eularian-mean equations. R is the gas constant, $\bar{D}, \bar{S}$ is the Eularian-means of friction diabatic heating, respectively. $\bar{\omega}$ is the Eularian-mean of vertical velocity. $(\bar{v}^*, \bar{\omega}^*)$ is the residual meridional circulation and can be written as

$$\bar{v}^* = \bar{v} - \frac{\partial(\overline{\theta'v'}/\bar{\theta}_p)}{\partial p}, \tag{5}$$

$$\bar{\omega}^* = \bar{\omega} + \frac{\partial(\overline{\theta'v'}/\bar{\theta}_p \times \cos\varphi)}{a\cos\varphi\partial\varphi}. \tag{6}$$

From Eqs. (5) and (6) we can see that the so-called residual meridional circulation consists of the meridional circulation of the basic state and that induced by topography and heat source. Because $\bar{v}=0$ and $\bar{\omega}=0$ for the meridional circulation of the basic state in this investigation, $(\bar{v}^*, \bar{\omega}^*)$ is the induced meridional circulation. Thus (5) and (6) can be written as

$$\bar{v}^* = -\frac{\partial}{\partial p}\left(\frac{\overline{\theta'v'}}{\bar{\theta}_p}\right), \tag{7}$$

$$\bar{\omega}^* = \frac{1}{a\cos\varphi}\frac{\partial}{\partial\varphi}\left(\frac{\overline{\theta'v'}}{\bar{\theta}_p}\cos\varphi\right). \tag{8}$$

Moreover, F in Eq. (1) is the divergence of the Eliassen-Palm flux and may be expressed as

$$\nabla \cdot F = \frac{1}{a\cos\varphi}\frac{\partial}{\partial\varphi}[F(\varphi)\cos\varphi] + \frac{\partial}{\partial p}[F(p)], \tag{9}$$

$$F = (F(\varphi), F(p)) = \left(-a\cos\overline{u'v'}, fa\cos\varphi\frac{\overline{\theta'v'}}{\overline{\theta}_P}\right). \tag{10}$$

In the above equations, θ' is the perturbation potential temperature. From Eq. (1) we can obtain the following results:

(1) If $\overline{D}=0$, $\overline{S}=0$, and there is no heat transfer, the basic flow will not vary when the E-P flux due to perturbation vanishes.

(2) If there is heat transfer due to perturbation, the westerly flow will be accelerated, when the E-P flux due to perturbation is divergent and the induced meridional circulation is northward.

(3) When the E-P flux due to perturbation is convergent and the induced meridional circulation is southward, the westerly flow will decrease, while the easterly flow accelerates.

III. MODEL AND PARAMETERS

1. *Model*

In order to discuss the effects of divergence of E-P flux and the induced meridional circulation due to perturbation by the forcing of topography and heat source on the flow, we have to use a numerical model to compute the planetary waves and quasi-stationary perturbations at isobaric surfaces induced by topography and heat source. In this paper, a quasi-geostrophic 34-level spherical coordinate model is used, in which Rayleigh friction, the effect of Newtonian cooling and the horizontal thermal diffusivity are included (Huang and Gambo, 1982). For the planetary-scale motion, the divergent component of motion may play an important role in the meridional transfer of potential vorticity. Therefore, the effect of non-geostrophic wind on the meridional transfer of potential vorticity is considered and we can obtain the following model equations,

$$\hat{\Omega}_{n-\frac{1}{2}}\frac{\partial}{\partial\lambda}\left\{\frac{1}{2\Omega_0\sin\varphi}\frac{1}{a^2}\left[\frac{\sin^2\varphi}{\cos\varphi}\frac{\partial}{\partial\varphi}\left(\frac{\cos\varphi}{\sin^2\varphi}\frac{\partial\phi'}{\partial\varphi}\right)+\frac{1}{\cos^2\varphi}\frac{\partial^2\phi'}{\partial\lambda^2}\right]\right\}_{n-\frac{1}{2}}$$
$$+\frac{1}{a}q_{n-\frac{1}{2}}\frac{1}{2\Omega_0\sin\varphi}\frac{1}{a\cos\varphi}\frac{\partial\phi'_{n-\frac{1}{2}}}{\partial\lambda}=f\left(\frac{\partial\omega}{\partial p}\right)_{n-\frac{1}{2}}-(R_f)_{n-\frac{1}{2}}\frac{1}{2\Omega_0\sin\varphi}\frac{1}{a^2}$$
$$\times\left[\frac{\sin\varphi}{\cos\varphi}\frac{\partial}{\partial\varphi}\left(\frac{\cos\varphi}{\sin\varphi}\frac{\partial\phi'}{\partial\varphi}\right)+\frac{1}{\cos^2\varphi}\frac{\partial^2\phi'}{\partial\lambda^2}\right]_{n-\frac{1}{2}}, \tag{11}$$

$$\hat{\Omega}_n\frac{\partial}{\partial\lambda}\left(\frac{\partial\phi'}{\partial p}\right)_n-\left(\frac{\partial\hat{\Omega}}{\partial p}\right)_n\frac{\partial\phi'_n}{\partial\lambda}+\sigma_n\omega_n=-\left(\frac{RH}{C_p p}\right)_n-(\alpha_R)\left(\frac{\partial\phi'}{\partial p}\right)_n+(K_T)_n$$
$$\times\frac{1}{a^2}\left[\frac{\partial^2}{\partial\varphi^2}-\tan\varphi\frac{\partial}{\partial\varphi}+\frac{1}{\cos^2\varphi}\frac{\partial^2}{\partial\lambda^2}\right]\left(\frac{\partial\phi'}{\partial\varphi}\right)_n, \tag{12}$$

$$\hat{\Omega}_{n+\frac{1}{2}}\frac{\partial}{\partial\lambda}\left\{\frac{1}{2\Omega_0\sin\varphi}\frac{1}{a^2}\left[\frac{\sin^2\varphi}{\cos\varphi}\frac{\partial}{\partial\varphi}\left(\frac{\cos\varphi}{\sin^2\varphi}\frac{\partial\phi'}{\partial\varphi}\right)+\frac{1}{\cos^2\varphi}\frac{\partial^2\phi'}{\partial\lambda^2}\right]\right\}_{n+\frac{1}{2}}+\frac{1}{a}q_{n+\frac{1}{2}}$$
$$\times\frac{1}{2\Omega_0\sin\varphi}\frac{1}{a\cos\varphi}\frac{\partial\phi_{n-\frac{1}{2}}}{\partial\lambda}=f\left(\frac{\partial\omega}{\partial p}\right)_{n+\frac{1}{2}}-(R_f)_{n+\frac{1}{2}}\frac{1}{2\Omega_0\sin\varphi}\frac{1}{a^2}\left[\frac{\sin\varphi}{\cos\varphi}\right.$$
$$\left.\times\frac{\partial}{\partial\varphi}\left(\frac{\cos\varphi}{\sin\varphi}\frac{\partial\phi'}{\partial\varphi}\right)+\frac{1}{\cos^2\varphi}\frac{\partial^2\phi'}{\partial\lambda^2}\right]_{n+\frac{1}{2}}, \tag{13}$$

where H is the diabatic heating per unit time per unit mass,

R the gas constant, C_P the specific heat at constant pressure, α_R the Newton cooling coefficient, K_T the horizontal eddy thermal diffusivity, R_f the Rayleigh friction coefficient of perturbation, n indicates levels in the model, and q is expressed as

$$q=\left[2(\Omega_0+\hat{\Omega})-\frac{\partial^2\hat{\Omega}}{\partial\varphi^2}+3\tan\varphi\frac{\partial\hat{\Omega}}{\partial\varphi}\right]\cos\varphi$$

where Ω_0 is the angle velocity of the earth rotation and $\hat{\Omega}$ is the angle velocity of the basic flow, i.e.,

$$\hat{\Omega}=\frac{U}{a\cos\varphi}$$

2. *Parameters*

(1) Static stability parameters σ_n: The static stability used in this model is calculated from the mean temperature and density at 45°N in July of the U.S. Standard Atmosphere and is assumed not to change with latitudes.

(2) The vertical profile of the basic zonal mean flow: The profile computed by Murgatrogid is used. In order to eliminate small-scale features, a smooth operator is used. The result is shown in Fig. 1

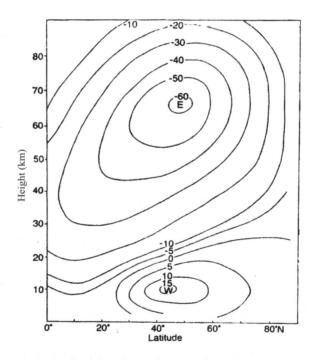

Fig. 1. The vertical distribution of zonal mean flow (m/s) in summer.

The coefficients of Rayleigh friction R_f and Newtonian cooling α_R are the same as these in Huang and Gambo (1982).

(4) The horizontal eddy thermal diffusivity coefficient K_T: Because the zonal mean winds in the tropics and subtropics in summer are very weak, the thermal diffusivity coefficient in the thermal balance equation should be smaller than that in winter and will be taken as a constant of 0.1×10^6 m² s⁻¹.

Thus, if the forcing source is given, the vertical distribution of stationary planetary waves and the stationary perturbation patterns at isobaric surface induced by the forcing source can be computed by using the model equations.

IV. DIVERGENCE OF E-P FLUX AND FORCED MERIDIONAL CIRCULATION INDUCED BY TOPOGRAPHY AND HEAT SOURCE

In order to investigate the physical roles of the diabatic heating over the southeast of the Tibetan Plateau in the formation and maintenance of the summer monsoon, we have made two numerical calculations: one is the divergence of E-P flux and the induced meridional circulation of the quasi-stationary planetary waves due to the forcing by topography during the northern summer; the other is the divergence of E-P flux and the induced meridional circulation of the quasi-stationary planetary waves due to the forcing by topography and heat source during the northern summer.

1. *A Case Considering Only Topographical Forcing*

The Northern Hemispheric topography arranged by Berkofsky and Bertni (1955) is used as the actual topography. It is expanded into Fourier series and is substituted into Eqs. (11)—(13). In this way we can obtain the distributions of amplitude and phase of quasi-stationary planetary waves due to the forcing by actual topography. Then, we can calculate the divergence of E-P flux and the forced meridional circulation of the guasi-stationary planetary waves induced by the forcing of topography.

Fig. 2 (a), (b), (c) shows the divergence of E-P flux of the quasi-stationary planetary waves due to the forcing by topography for wavenumbers 1, 2 and for composing wave components of $K=1-3$, resepectively. From Fig. 2 we can see that the E-P flux of the quasi-stationary planetary wave due to the forcing by topography is convergent in the subtropic troposphere, but it is divergent in the upper troposphere in middle latitudes. From Eq. (1) we can find that the forcing effect of topography is favorable for the easterly strengthening in the subtropics. However, in the case of middle latitudes, it is favorable for the westerly strengthening in the upper troposphere, but for the easterly strengthening or the westerly weakening in the lower troposphere.

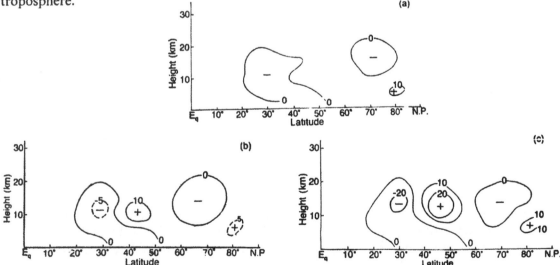

Fig. 2. The divergence (units, m/s) of E-P flux of the quasi-stationary planetary waves due to the forcing by topography (a) for wavenumber 1, (b) for wavenumber 2, (c) for composing wave components of $k=1-3$.

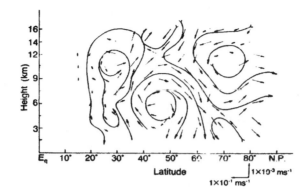

Fig. 3. The Induced meridional circulation due to the topographical forcing, for composing wave components of $k = 1-3$.

In order to see clearly the influence of the forcing effect of topography on the flow, we have computed the meridional circulation induced by topography. Fig. 3 that shows the meridional circulation induced by topography, we can see that there is a direct circulation to the north of 30°N. Moreover, there is a very weak meridional circulation in the middle and upper troposphere in the region of 20°—30°N, and in the middle and lower troposphere in the region of 20°—25°N, while a downward flow occurs in the region near 20°N. In order to investigate the effect of this induced meridional circulation on the flow, we have also computed the value of $f\bar{v}^*$ in this circulation. Fig. 4 shows the Coriolis force $f\bar{v}^*$ in the mean meridional circulation induced by topography in summer. From Fig. 3 and Fig. 4 we can see that the topographical forcing effect makes the induced meridional circulation flow southward and the value of $f\bar{v}^*$ become negative in the upper subtropical troposphere. The flow accelerates westward due to the Coriolis force, and the easterly enhances, while the westerly weakens. However, in the lower subtropical troposphere, the induced meridional circulation flows northward, $f\bar{v}^*$ has a positive value, the flow accelerates eastward, and the westerly enhances, while the easterly weakens. In middle-latitude areas it is in contrary to that in the subtropics.

So far we have discussed theoretically the role of plateau topography in the formation of the Asian monsoon. Now we shall discuss the physical roles of topography and heat source for the formation of summer monsoon in Asia, using model Eqs. (11)—(13).

2. A Case Considering Both Topographical and Thermal Forcing

As mentioned in the introduction, the southeast part of the Tibetan Plateau is a region in which the diabatic heating rate is the largest in the atmosphere over the Northern Hemisphere. In order to compute the roles of heat source over the Tibetan Plateau in the formation and maintenance of the mean monsoon circulation over Asia, we first assume that the vertical distribution of the heat source over the Northern Hemisphere is given by

$$H_0(\lambda, \varphi, p) = \hat{H}_0(\lambda, \varphi) \exp\left(-\left(\frac{p-\bar{p}}{d}\right)^2\right), \tag{14}$$

where $d = 300$ hPa. Observations show that the height of maximum heating rate decreases with increasing latitude. At middle latitudes, the maximum heating rate is located at 500 hPa, while over the Tibetan Plateau it is at 400 hPa. Thus, we take $\bar{p} = 500$ hPa for $\varphi \geq 40°N$, and

$\bar{p} = 400$ hPa for $\varphi < 40°$N, i.e., at middle and high latitudes, the maximum heating rate is located at 500 hPa, while in the subtropics it is at 400 hPa.

We expand the actual heat sources computed by Ashe (1979) into Fourier series and substitute them into the model equations to compute the quasi-stationary planetary waves and quasi-stationary disturbance pattern at constant height level, induced by both topography and heat source.

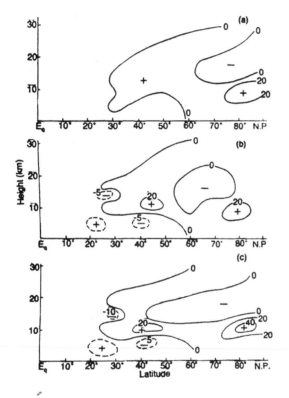

Fig. 4. $f\bar{v}^*$ in the meridional circulation induced by topography (units: 10^{-6} m s^{-2}) (a) for wavenumber 1; (b) for wavenumber 2; (c) for composing wave components of $k = 1-3$.

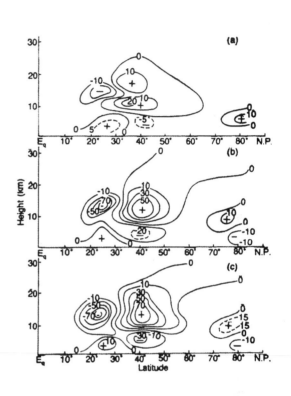

Fig. 5. As in Fig. 2, except that these are for the forcing by both topography and heat source.

Fig. 5 a, b, c is the divergence of E-P flux of the quasi-stationary planetary waves due to the forcing by topography and heat source for wavenumber 1, wavenumber 2, and composing wave component of $K = 1-3$. Comparing Fig. 5 with Fig. 2, we can see that in the subtropics, the divergence of E-P flux taking into account both topography and heat source is obviously different from that taking into account only topography. When the forcing effect of heat source is considered, the E-P flux is convergent in the upper subtropical troposphere and much larger than that due to topography. However, it is divergent in the lower subtropical troposphere. Moreover, the convergence of E-P flux in the upper troposphere and the divergence in the lower troposphere in middle-latitude areas are much stronger than those due to the topographical effect. That is to say, in summer, the forcing effect of heat source over the Tibetan Plateau is largely favorable for the easterly acceleration in the upper subtopical troposphere, while it is so for the westerly strengthening or the easterly weakening in the lower subtropical troposphere.

In order to see clearly the forcing effect of heat source on the zonal mean flow, we have computed the meridional circulation induced by topography and heat source. Fig. 6 shows the meridional circulation induced by the forcing of both topography and heat source for composing wave component of $K=1—3$. Comparing Fig. 6 with Fig. 3, we can see that, when considering both topography and heat source, the monsoon circulation in the subtropics extends southward significantly, the upward flow is over 30°N, and the downward flow is over 10°N. Moreover, the meridional circulation is stronger than that induced only by topography. This induced meridional circulation is obviously divided into two circulations: one is located in the upper and middle troposphere; the other in the middle and lower troposphere. The northerly flow in the upper troposphere and the southerly flow in the lower troposphere induced by topography plus heat source are stronger than those induced only by topography. In order to further study the effect of the induced meridional circulation due to topography and heat source on the flow, we have also computed the Coriolis force $f\bar{v}^*$ in the meridional circulation. Fig. 7 a,b,c show the values of $f\bar{v}^*$ due to topography and heat source for wavenumbers 1, 2, and the composing wave component of $K=1—3$. Comparing Fig. 7 with Fig. 4, we find a great difference between them. $f\bar{v}^*$ is negative in the upper subtropical troposphere and it is positive in the lower subtropical troposphere, and its values are larger than those in Fig. 4. Moreover, $f\bar{v}^*$ is positive in the upper troposphere and it is negative in the lower troposphere at middle latitudes. Its values are also larger than those in Fig. 4. Thus, because of the Coriolis torque, the induced meridional circulation due to the thermal effect of plateau will create a strong easterly acceleration in the upper subtropical troposphere and the westerly acceleration in the lower subtropical troposphere. This is favorable for the formation and maintenance of the southwest monsoon.

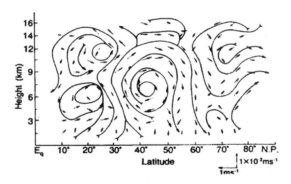

Fig. 6. As in Fig. 3, except that it is for the forcing by both topography and heat source.

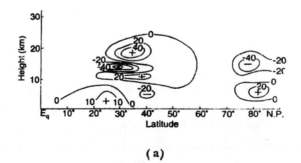

(a)

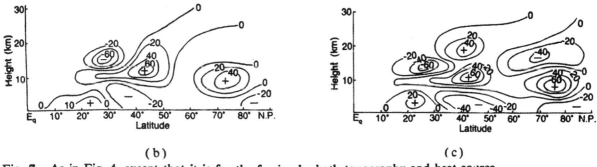

Fig. 7. As in Fig. 4, except that it is for the forcing by both topography and heat source.

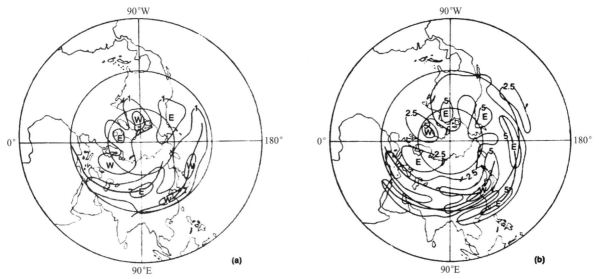

Fig. 8. The perturbation geostrophic wind speed (units: m s^{-1}) induced by topography. (a) at the 3 km height level; (b) at the 12 km height level.

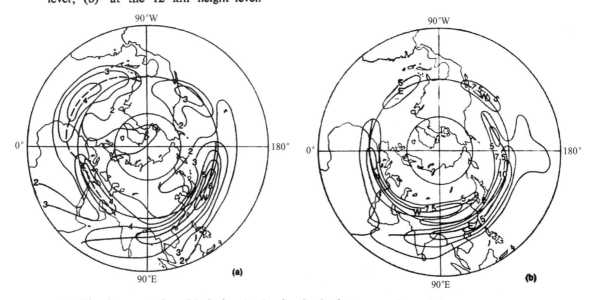

Fig. 9. As in Fig. 8, except that this is for the forcing by both topography and heat sources.

V. THE PERTURBATION FLOW INDUCED BY TOPOGRAPY AND HEAD SOURCES

In the above we have discussed theoretically the forcing effect of topography and heat sources on the zonal mean flow. Now we shall compute the stationary disturbance patterns and perturbation flow at constant height level, using the numerical model. Fig. 8 (a) shows the perturbation geostrophic wind speed induced only by topography at the 3 km height level; Fig. 8(b) shows the pertubation geostrophic wind speed induced only by topography at the 12 km height level. In Fig. 8 we can see that the southwest perturbation flow, due to only topographical forcing, is weak in the lower troposphere over South Asia. Moreover, the east perturbation flow is also weak in the upper troposphere over South Asia. Thus, if only the topographical effect is considered, the monsoon circulation should be very weak over South Asia.

In order to see clearly the effect of heat source over the plateau on the summer monsoon over Asia, we have also computed the quasi-stationary disturbance patterns and perturbation flow at constant-height levels using the numerical model. Fig. 9 (a) shows the perturbation geostrophic wind induced by both topography and heat sources at the 3 km height level; Fig. 9(b) shows the perturbation geostrophic wind induced by both topography and heat sources at the 12 km height level. Comparing Fig. 9 with Fig. 8 we find a large difference between them. When both topography and heat sources are considered, a strong southwest wind belt passes through the India Peninsula, the Bay of Bengal, and the south of China to the south of Japan. This strong southwest wind belt plays an important role in the maintenance of the summer monsoon over Asia. Moreover, in Fig. 9 we can also see that a strong easterly belt passes through the West Pacific Ocean, the south of China, the Indian Peninsula to West Africa in the upper troposphere. The maximum strength of this flow may reach 15 m s^{-1} and even larger. This easterly also plays an important role for the maintenance of the monsoon circulation in the upper troposphere over Asia. Besides, in Fig. 9 we can find a strong westerly belt from North Africa along the Mediterranean Sea to the Caspian Sea.

The computed results show that the heat sources over the Plateau play an important role for the formation and maintenance of the summer monsoon over Asia.

VI. CONCLUSIONS AND DISCUSSIONS

The physical effects of topography and heat sources on the formation and maintenance of the summer monsoon over Asia are discussed by computing the divergence of the E-P flux of the planetary waves, induced meridional circulation, and perturbation geostrophic wind speed due to the forcing by topography and heat source. The computed results show that in summer the divergence of E-P flux of the quasi-stationary waves and the induced meridional circulation due to the forcing by the plateau topography are very weak. Moreover, in the distribution of perturbation geostrophic wind speed induced by only topography, the strongest southwest flow is located in the lower troposphere over East Asia at middle latitudes, but it is very weak; in the upper troposphere the strongest easterly flow is located over the south of China and the West Pacific Ocean and it is also very weak. Thus one can see that the forcing effect of the Plateau's topography alone can not form the monsoon circulation. The computed results also show that, when the forcing effect of diabatic heating over the plateau are considered, both the divergence of E-P flux of the quasi-stationary planetary waves and the induced meridional circulation due to the forcing by topography and heat source are very strong in the upper and lower subtropical troposphere. These make the easterly accelerate largely in the upper subtropical troposphere and the westerly accelerate in the lower subtropical troposphere.

The results computed by the model also show that, in the distribution of perturbation geostrophic wind speed induced by both topography and heat source, a strong southwest wind belt passes through the Indian Peninsula, the Bay of Bengal and the south of China to the West Pacific Ocean, while a strong easterly passes the West Pacific Ocean, the south of China, and

India to West Africa. Thus, the heat source over the Tibetan Plateau may play an important role in the formation and maintenance of the monsoon circulation over South Asia and East Asia.

From the computed results it can be seen that, for the forcing effects by topography or by topography plus heat source, the order of magnitude of Coriolis torque $f\bar{v}^*$ in the induced meridional circulation in the same as that of the divergence of E-P flux, i.e., $\nabla \cdot F/a\cos\varphi$, in the upper troposphere. However, the Coriolis torque $f\bar{v}^*$ in the meridional circulation is much larger than the divergence of E-P flux in the lower troposphere. Thus, in the discussion of wave-flow interaction in the lower troposphere the effect of the induced meridional circulation should be considered in addition to the divergence of E-P flux.

Because the model used in this paper is a hemispheric one, we can only discuss the effects of topography and heat sources over the Northern Hemisphere on the formation and maintenance of the Asia monsoon. The effect of the cross-equator flow from the Southern Hemisphere on the formation of the Asian monsoon can not be addressed here.

REFERENCES

Ashe, S. (1979), A nonlinear model of the time-average axially asymmetric flow induced by topography and diabatic heating, *J. Atmos. Sci.*, 36: 109—126.

Berkofsky, L. and Bertoni, E.A. (1955), Mean topographic charts for the entire earth, *Bul. Amer. Meteor. Soc.*, 36: 350—354.

Edmon, H.L., Hoskins, B.J. and McIntyre, M.E. (1980), Eliassen-Palm cross sections for the troposphere, *J. Atmos. Sci.*, 37: 2600—2616.

Huang, R.H. and Gambo, K. (1982), The response of a hemispheric multi-level model atmosphere to forcing by topography and stationary heat sources, Part I, II, *J. Meteor. Soc. Japan*, 60: 78—108.

Huang, R.H. and Gambo, K. (1983), The response of a hemispheric multi-level model atmosphere to forcing by topography and stationary heat sources in summer, *J. Meter. Soc. Japan*, 61, 495—505.

Ji, Liren and Tibaldi, R. (1984), Numerical experiments on the seasonal transition of the general circulation over Asia in summer, *Adv. Atmos. Sci.*, 1: 128—139.

Kuo, H.L. and Qian, Y.F. (1982), Numerical simulation of the development of mean monsoon circulation in July, *Mon. Wea. Rev.*, 110:1879—1897.

Yeh Tucheng and Gao Youxi (1980), Meteorology of Qinghai-Xizang Plateau, Science Press, Beijing, China, 278pp. (in Chinese).

夏季热带西太平洋上空的热源异常对东亚上空副热带高压的影响及其物理机制*

黄荣辉

(中国科学院大气物理研究所)

李维京

(中国科学院兰州高原大气物理研究所)

提　要

许多观测事实表明，当热带西太平洋海表面温度增高时，热带西太平洋上空，特别是我国南海及菲律宾周围的对流活动则增强。并且表明东亚上空副热带高压的强弱与此对流强弱有很大相关，并将严重地影响东亚季风降水的异常。

本文从理论上及数值计算方面研究了热带太平洋上空，特别是我国南海与菲律宾周围的对流活动所引起的热源加强对北半球大气环流异常的影响。结果表明：热带西太平洋上空特别是我国南海与菲律宾上空的热源加强，则位于东亚上空的副热带高压将加强。

一、序　言

最近几年来，人们发现低纬度海温异常不仅影响低纬度的纬向-垂直环流，而且可以影响北半球中高纬度的大气环流。Wallace等人从北半球多年观测资料发现北半球大气环流的异常会出现一种太平洋-北美型，即所谓PNA型遥相关[1]。后来，Shukla和Wallace，Tokioka通过数值试验说明了冬季热带东太平洋海温异常可以引起北半球PNA型大气环流的异常[2,3]。最近，黄荣辉从理论上及数值计算方面说明了冬季热带上空的热源强迫所产生的准定常行星波可以准水平地通过对流层传播到北半球中、高纬度的对流层上层。因此，热带东太平洋海温异常将引起中、高纬度大气环流的异常，并使得我国东北上空的高度场出现负距平，引起我国东北地区的冷夏。

关于夏季大气环流的遥相关研究还不多。最近黄荣辉已从观测事实及数值模拟说明了夏季青藏高原热源的异常不仅将影响南亚环流的异常，而且将影响东亚环流的异常[4]。近来，人们已开始注意热带西太平洋的海温异常对北半球大气环流异常的影响。Nitta的研究来表明了西太平洋上空的云量长期变化与热带西太平洋的海面温度有很大关系[5]，当热带西太平洋海温高，就会引起我国南海与菲律宾东北部上空的云量增加。Kurihara也得出同样结果[6]。

夏季西太洋副热带高压对于我国东部、日本、朝鲜半岛等东亚地区的气候有着重要的影响。它的位置及强度的异常直接影响着东亚夏季气候的异常，若在盛夏，它的强

---

* 本文原载于：大气科学, 12(特刊): 107-116, 1988年11月出版.

度偏强、位置偏北,则在我国东部,朝鲜半岛及日本等东亚地区会出现高温少雨的天气。在这方面已有不少研究。但是,关于西太平洋副热带高压的年际变化与季节内变化及其物理机制至今还不太清楚。为此,有必要从观测事实、理论及数值计算方面来探讨夏季西太平洋副热带高压发生异常的物理机制。

二、夏季西太平洋副热带高压与我国南海、菲律宾周围对流活动的关系

Kurihara 和 Kawahara 从观测事实说明了 1978,1984 年由于热带西太平洋海温偏高,则在我国南海、菲律宾周围的上空对流活动加强[6]。这些对流活动直接影响着东亚上空副热带高压的位置与强度。图1是1984年8月月平均500hPa高度与菲律宾周围平均的TBB(黑体温度)之间的关系,其中斜线区域表示TBB在 $-10℃$ 以下,平均的TBB越低,则表示对流活动愈强。因为从卫星所测得的黑体温度在有云的情况是云体的温度,故其所测得的TBB低;而在没有云盖的情况,卫星所测得的是海面与陆地的温度,故所得的TBB就高。因此,平均的TBB低说明该区域云盖出现的频数大,即对流活动强。从图1可以看到强的对流区位于我国南海与菲律宾周围区域。而强的西太平洋副热带高压位于日本及我国东部上空,它使日本、朝鲜半岛及我国的盛夏平均气温比常年高 0.5—1.0℃,而在这些地区的降水是常年的60—80%,这反映梅雨锋是弱的。

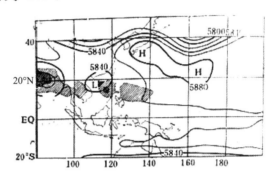

图1 1984年8月月平均500hPa高度场(实线)与菲律宾周围平均的TBB分布(斜线区域)

其中斜线阴影区表示TBB低于-10℃

图2是1985年8月月平均500hPa高度场与距平分布图,其中虚线表示该月相对于常年平均的异常分布图。我们可以看到西太平洋副热带高压异常偏北、强度偏强,它控制着日本岛的上空。并且,我们还可以看到其异常距平分布出现像一波列的遥相关型分布。一个负距平区域出现在我国南海与菲律宾上空;而正距平区域出现在日本上空;并且另

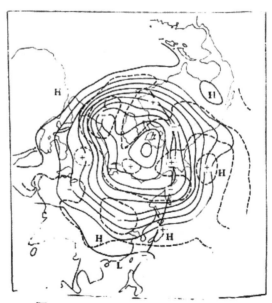

图2 1985年8月月平均500hPa高度场(实线)与距平(虚线)分布图

一个负距平区域与一个正距平区域分别出现在鄂霍次克海及阿拉斯加的上空。与1984年盛夏一样,1985年我国华南、华中、黄淮流域相继出现干旱、高温天气,在江淮地区的夏季降水只是常年的50%左右。

为了更清楚地看到西太平洋副热带高压的异常与菲律宾周围对流活动的关系,我们分别作沿某经度的高度值与TBB值的纬度-时间剖面图。图3是1984年8月份沿130—150°E平均而得到的每日高度场与TBB的纬度-时间剖面图。其中阴影区域表示TBB低于-10℃,即表示对流活动频繁而强盛的区域。图4是1985年8月份沿140°E的5天平均的500hPa高度场与沿130°E的5天平均TBB的纬度-时间剖面图。从这两图可以看到东亚上空的西太平洋副热带高压的异常发展是与我国南海、菲律宾周围的西太平洋上空的对流活动有很紧密的关系。并且可以看到在菲律宾周围对流活动活跃10天以后,东亚上空的西太平洋副热带高压就会异常发展,位置偏北、强度偏强。

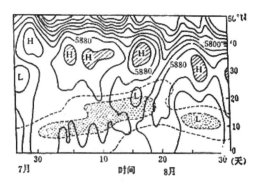

图3 1984年8月份沿130—150°E平均而得到的500hPa高度(实线)与TBB的纬度-时间剖面图 其中阴影区域表示TBB低于-10℃

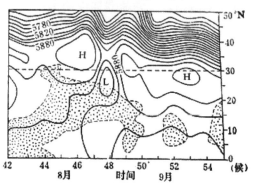

图4 1985年8月沿140°E的5天平均的500 hPa高度(实线)与沿130°E的5天平均TBB的纬度-时间剖面图 其中阴影区域表示TBB低于-10℃

三、夏季强迫行星波的传播特征

上述从观测事实得到的关系决不是偶然的,而是有它们的内在联系的。下面,我们从夏季强迫行星波的传播特性来讨论东亚上空的副热带高压与菲律宾周围的对流活动所引起的热源的关系。

当热带西太平洋海表温度发生异常增暖,则不仅使得海洋向大气的感热输送大大增加,并且根据Nitta, Kurihara和Kawahara的研究,将会在西太平洋,特别是菲律宾周围的积云对流大大加强,这个积云对流将会使潜热释放大大增加,产生很强的热源。根据Maruyama等人的估计,其云量异常可达到±2.5,这相应于降水可达到±9mm/d的异常[7]。因此,在热带西太平洋的海温增暖时,在西太平洋上空,特别在我国南海与菲律宾周围会形成一个很强的热源。由于这个热源的强迫作用,将产生大气的准定常行星波。

作者曾讨论了夏季准定常行星波在球面斜压大气中的传播[8],证明了夏季副热带准定常行星波可以准水平地传到中、高纬度对流层地区。因此，我们可以利用二维的涡度方程近似地讨论夏季强迫行星波在球面上的传播。

在球面上的涡度方程是

$$\left(\frac{\partial}{\partial t} + \widehat{\Omega}\frac{\partial}{\partial \lambda}\right)\left\{\frac{1}{f}\left[\frac{\sin\varphi}{\cos\varphi}\frac{\partial}{a\partial\varphi}\left(\frac{\cos\varphi}{\sin\varphi}\frac{\partial \phi'}{a\partial\varphi}\right) + \frac{1}{a^2\cos^2\varphi}\frac{\partial^2 \phi'}{\partial \lambda^2}\right]\right\}$$

$$+ \frac{\partial \bar{q}}{a\partial\varphi} \cdot \frac{1}{f}\frac{\partial \phi'}{a\cos\varphi\partial\lambda} = 0 \quad (1)$$

其中

$$\frac{\partial \bar{q}}{a\partial\varphi} = \frac{1}{a}\left[2(\Omega_0 + \widehat{\Omega}) - \frac{\partial^2 \widehat{\Omega}}{\partial \varphi^2} + 3\tan\varphi\frac{\partial \widehat{\Omega}}{\partial \varphi}\right]\cos\varphi$$

$$\widehat{\Omega} = \frac{\bar{U}}{a\cos\varphi}$$

引入流函数 Ψ'，利用Marcator投影方法，令

$$\begin{cases} dx = ad\lambda \\ dy = ad[\tanh^{-1}(\sin\varphi)] \end{cases} \quad (2)$$

在 x, y 坐标系的波数 k 与 (λ, φ) 坐标系的波数 $\tilde{k}$ 有如下关系

$$k = \frac{\tilde{k}}{a}$$

这样，方程(1)可表示成

$$\left(\frac{\partial}{\partial t} + a\widehat{\Omega}\frac{\partial}{\partial \lambda}\right)\left[\frac{1}{\cos^2\varphi}\left(\frac{\partial^2 \Psi'}{\partial y^2} + \frac{\partial^2 \Psi'}{\partial x^2}\right)\right] + \frac{\bar{q}_y}{\cos^2\varphi}\cdot\frac{\partial \Psi'}{\partial x} = 0 \quad (3)$$

其中

$$\bar{q}_y = \frac{\partial \bar{q}}{a\partial\varphi}\cos\varphi$$

因为我们所研究的对象是准定常行星波的传播，所以 Ψ' 是缓变的，这样，可利用WKBJ方法来解方程(3)，引入缓变坐标系 X, Y, T，即

$$X = \varepsilon x \quad Y = \varepsilon y \quad T = \varepsilon t$$

ε 是一个正的小参数，设方程(3)为一波包解，即

$$\Psi'(x,y,t) = \widehat{\Psi}(X,Y,T)e^{i\Theta/\varepsilon} \quad (4)$$

把振幅 $\widehat{\Psi}(X,Y,T)$ 按小参数 ε 展开，可得

$$\widehat{\Psi}(X,Y,T) = \widehat{\Psi}_0(X,Y,T) + \varepsilon\widehat{\Psi}_1(X,Y,T) + \varepsilon^2\widehat{\Psi}_2(X,Y,T) + \cdots\cdots \quad (5)$$

把(4)，(5)式代入方程(3)，可得

$$\left[-i(\widehat{\omega} - a\widehat{\Omega}k) + \varepsilon\frac{\partial}{\partial T} + \varepsilon a\widehat{\Omega}\frac{\partial}{\partial X}\right] \times \left\{\varepsilon^2\left[\frac{1}{\cos^2\varphi}\left(\frac{\partial^2 \widehat{\Psi}_0}{\partial X^2} + \frac{\partial^2 \widehat{\Psi}_0}{\partial Y^2}\right)\right]\right.$$

$$+ \varepsilon i\left[\frac{1}{\cos^2\varphi}\left(2k\frac{\partial \widehat{\Psi}_0}{\partial X} + 2m\frac{\partial \widehat{\Psi}_0}{\partial Y}\right) + \frac{1}{\cos^2\varphi}\left(\widehat{\Psi}_0\frac{\partial k}{\partial X} + \widehat{\Psi}_0\frac{\partial m}{\partial Y}\right)\right]$$

$$\left. - \frac{1}{\cos^2\varphi}(k^2 + m^2)\widehat{\Psi}_0\right\} + \frac{1}{\cos^2\varphi}\bar{q}_y ik\widehat{\Psi}_0 + O(\varepsilon^2) = 0 \quad (6)$$

取方程(6)的 ε^0 级近似解，可得下式：

$$i(\hat{\omega} - a\hat{\Omega}k)\left[\frac{1}{\cos^2\varphi}(k^2+m^2)\right]\hat{\psi}_0 + \bar{q}_y\frac{ik}{\cos^2\varphi}\hat{\psi}_0 = 0 \qquad (7)$$

从(7)式可得行星波的频散关系

$$\hat{\omega} = a\hat{\Omega}k - \frac{\bar{q}_y \cdot k}{k^2+m^2} \qquad (8)$$

$\hat{\omega}$ 是局地瞬时频率，k, m 分别是 X, Y 的局地瞬时波数。于是可得准定常行星波在 X, Y 方向的群速度分量方程

$$\begin{cases} C_{gX} = \dfrac{\partial \hat{\omega}}{\partial k} = \dfrac{2\bar{q}_y \cdot k^2}{(k^2+m^2)^2} \\ C_{gY} = \dfrac{\partial \hat{\omega}}{\partial m} = \dfrac{2\bar{q}_y \cdot k \cdot m}{(k^2+m^2)^2} \end{cases} \qquad (9)$$

并且由(8)式可得

$$(k^2+m^2) = \frac{\bar{q}_y}{a\hat{\Omega}} = K_s^2 \qquad (10)$$

我们由基本气流的分布，利用(10)式可算得 K_s^2 的分布。我们定义波的传播路径与 X 方向的夹角是 θ，这样就有

$$\tan\theta = \frac{C_{gY}}{C_{gX}} = \frac{m}{k} = \sqrt{\left(\frac{K_s}{k}\right)^2 - 1} \qquad (11)$$

从(10)式可以看到，当 $K_s > k$ 时，波向极地传播，当 $K_s = k$ 时，波的传播路径将转向，这点也称转向点，从这点之后，波从高纬度向低纬度传播。因此，波在球面上的传播路径是

$$\frac{d\varphi}{d\lambda} = \begin{cases} \cos\varphi\sqrt{\left(\dfrac{aK_s}{k}\right)^2 - 1} & \text{在转向点之前} \\ -\cos\varphi\sqrt{\left(\dfrac{aK_s}{k}\right)^2 - 1} & \text{在转向点之后} \end{cases} \qquad (12)$$

由(12)式可以得到夏季准定常行星波在500hPa面上的传播路径。图5是夏季位于菲律宾周围热源强迫所产生的准定常行星波的传播路径。可以看到大约纬向波数为1—3的行星尺度的波列可以由西太平洋上空经东亚、北美上空向大西洋上空传播。而纬向波数大约为4,5,6的长波尺度的波列将向北美落基山脉方向传播。

下面，我们利用群速度可以大概估算行星尺度波列传播到极地所需的时间，

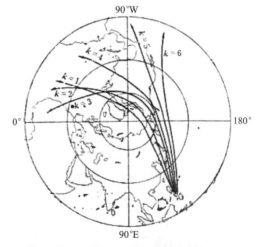

图5 位于菲律宾周围热源强迫所产生的准定常行星波的传播路径

$$|C_g| = \sqrt{C_{g\lambda}^2 + C_{g\varphi}^2} = \frac{2\bar{q}_\varphi \cdot k}{(k^2+m^2)^2}\sqrt{k^2+m^2}$$

$$= \frac{2a\widehat{\Omega}\cdot k}{K_s} = \frac{2U\cdot \hat{k}}{\cos\varphi(aK_s)} \tag{13}$$

表1给出了波数大约为1的行星尺度波列在500hPa各纬度的群速度。从表1可以估算西太平洋上空500hPa面上波数大约为1的行星尺度波列从20°N传播到转向点附近所需要的时间是16天左右，而从20°N传播到40°N附近则需要10天左右。这个结果与从观测所得的结论相一致。

表 1

纬度 °N	20	25	30	35	40	45	50	55	60	65	70	75	80
$|C_g|$	0.4	1.0	2.0	4.6	4.7	10.6	5.6	8.4	9.2	10.4	11.4	14.3	18.6

取方程(6)的$O(\varepsilon)$近似解，则可得到波作用守恒方程

$$\frac{\partial}{\partial t}A + \nabla\cdot(\mathbf{C}_g\cdot A) = 0 \tag{14}$$

上式 $A = \frac{1}{2}\frac{K_s^4|\widehat{\Psi}_0|^2}{\bar{q}_\varphi}$，对于准定常波，$\frac{\partial}{\partial t}A \approx 0$，因而，我们有方程

$$\nabla\cdot(\mathbf{C}_g\cdot A) \approx 0 \tag{15}$$

对于一般实际气流行星波的传播路径并不是一大圆，其$\nabla\cdot\mathbf{C}_g \neq 0$，因而，一般说来$A$的局地分布不容易从(15)式估计出。这就是说，波在实际气流中传播，其振幅的局地分布是不容易估计的，更不是如Hoskins和Karoly所指出的波在传播中其振幅是向极增加[9]。

四、夏季热带西太平洋上空热源异常对北半球中、高纬度地区大气环流的影响

上面，我们已从理论上分析了北半球夏季位于菲律宾周围强迫源强迫所产生的准定常行星波的传播规律。下面，我们应用一个包括Rayleigh摩擦、Newton冷却及水平热力扩散的34层球坐标模式来模拟由于在我国南海、菲律宾周围上空的热源异常所产生的环流异常。关于模式的结构与参数已在文献[10,11]中阐述，本文不再重复。

正如在第二节所述，当在我国南海与菲律宾周围上空有异常对流活动产生，那么在菲律宾上空的热源就会加强，按照Maruyama的研究结果[7]，这个热源异常大约可达到± 5 K/d。因此，我们分别假设两个理想的热源异常距平分布，一个中心位于热带西太平洋上空；另一个中心位于我国南海及菲律宾周围上空。这两个热源的水平尺度不同。我们假设热源异常距平的水平分布为

$$\Delta \widehat{H}_0(\lambda, \varphi) = \begin{cases} \Delta H_0 \left[\sin\dfrac{\pi(\varphi-\varphi_1)}{(\varphi_2-\varphi_1)} \sin\dfrac{\pi(\lambda-\lambda_1)}{(\lambda_2-\lambda_1)} \right]^2 & \lambda_1<\lambda<\lambda_2 \\ & \varphi_1<\varphi<\varphi_2 \\ 0 & \text{其它区域}, \end{cases} \quad (16)$$

上式中加热率 $\dfrac{1}{C_p}\Delta H_0 = 5.0\,\text{K/d}$，在该区域中平均加热率大约为 $3.0\,\text{K/d}$。并且，我们假设热源的垂直分布为

$$\Delta \widehat{H}(\lambda, \varphi, P) = \Delta \widehat{H}_0(\lambda, \varphi)\exp\left[-\left(\frac{p-\overline{P}}{d}\right)^2\right] \quad (17)$$

上式中 $d = 300\,\text{hPa}$，$\overline{P} = 400\,\text{hPa}$，这表示最大热源异常位于 $400\,\text{hPa}$，这是与实际比较符合的。这样，我们就可以计算由于热源异常而引起准定常行星波振幅与位相的异常及各压高面上准定常扰动型异常的距平分布。

1. 热带西太平洋上空热源异常所引起的东亚大气环流的异常

首先，我们取一热源异常距平位于西太平洋上空，其中心位于 $135°\text{E}$，$15°\text{N}$，即取 $\lambda_1 = 105°\text{E}, \lambda_2 = 165°\text{E}$；$\varphi_1 = 0°\text{N}, \varphi_2 = 30°\text{N}$。图6是这个理想热源异常的距平分布。我们把这个热源异常的距平分布叠加在北半球热源的气候分布 $\overline{H}(\lambda,\varphi,p)$ 上，即异常的热源分布是

$$\widetilde{H}(\lambda,\varphi,p) = \overline{H}(\lambda,\varphi,p) + \Delta\widehat{H}(\lambda,\varphi,p)$$

我们把此异常热源代入模式就可以计算出西太平洋热源异常时各等压面定常扰动系统的分布及其各波数振幅与位相的分布。并且，把这些分布减去正常年份各等压面上定常扰动系统的分布及各波数振幅与位相的分布就可以得到西太平洋热源异常所造成北半球大气环流的异常情况。

图7是由于热带西太平洋上空热源异常所引起的 $500\,\text{hPa}$ 扰动高度场异常的距平分布。从图7可以看到热带西太平洋热源加强时，从中印半岛到我国南部上空的扰动高度场将出现负距平，而从北太平洋、日本到我国长江流域以北的广大地区上空的扰动高度场将出现正距平。这将引起副热带高压位置异常偏北，控制日本及我国长江流域以北

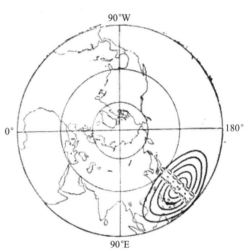

图6 一个位于 $400\,\text{hPa}$ 假想的热源异常距平分布

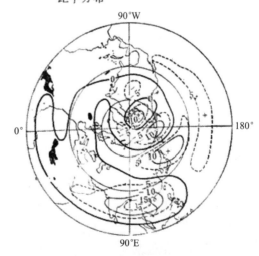

图7 由图6中所示的热带西太平洋上空热源异常所引起的北半球 $500\,\text{hPa}$ 扰动高度场异常的距平分布（单位：m）

地区，并且其强度偏强，它将使得我国长江流域以北、日本等地区出现干旱与热夏。上面我们所述的1984，1985年我国长江流域中、下游地区出现干旱及热夏就是由于这种环流异常所造成。此外，另一个负距平及正距平区域分别出现在阿留申及阿拉斯加地区的上空。由模式所计算的结果不仅与图2中所示的东亚大气环流的异常分布比较一致，而且与我们由理论上所得的结果也比较一致。

2. 我国南海与菲律宾上空热源异常所引起的东亚大气环流的异常

在第二小节中，我们已简述了当热带西太平洋海表温度偏高时，在我国南海及菲律宾周围的上空将引起强盛的对流活动。因此，我们将取一热源异常距平位于我国南海及菲律宾周围的上空。如图8所示，一个假想的热源异常距平中心位于 125°E，15°N，即位于菲律宾上空。并取 $\lambda_1 = 110°E$，$\lambda_2 = 140°E$；$\varphi_1 = 0°N$，$\varphi_2 = 30°N$。同样，把此热源的异常距平迭加在北半球热源的气候分布上，同上一样也可以得到异常热源分布，并且，由模式可以计算出当我国南海与菲律宾周围上空热源异常所引起的东亚大气环流的异常分布。

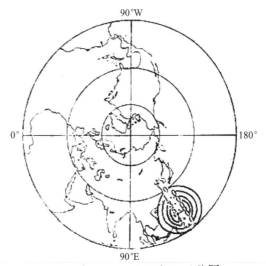

图8 一个位于400hPa假想的热源异常距平分布

图9是由于菲律宾上空热源异常所引起的500hPa扰动高度场异常的距平分布。从图9可以看到，当热源或积云对流活动在菲律宾周围的上空加强后，则在中印半岛和我国华南地区上空的高度场将出现负距平，而在我国江淮地区及日本上空的高度场将出现正距平。上述区域正是夏季西太平洋副热带高压在盛夏季节的位置。这说明当菲律宾周围对流加强后，西太平洋副热带高压主要位于我国江淮流域及日本本州上空，强度偏强，在我国华南盛行偏东气流。因此，我们可以从数值试验看到东亚上空的西太平洋副热带高压是与我国南海、菲律宾周围上空的热源强弱紧密相关的。这种关系也是由于准定常行星波传播的结果。

我们把图9与图7相比较，也可以看到热源异常距平的水平尺度影响着环流

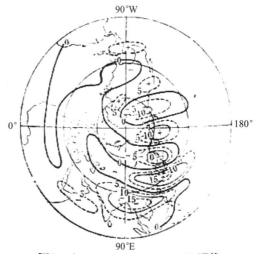

图9 由图8中所示的菲律宾周围热源异常所引起的北半球500hPa扰动高度场异常的距平分布（单位：m）

异常型的分布。图 8 中所示热源异常距平的纬向尺度比图 6 中所示的热源异常距平尺度小，故两者所引起的环流异常距平的分布也有所不同。前者由于热源异常的水平尺度大，故由它所引起的准定常波的异常类似于图 5 中波数小的传播路径往大西洋方向传播，因此，扰动高度场的异常距平的分布如图 5 中所示的波数小的路径由南亚经东亚向大西洋方向分布。而后者由于热源异常的水平尺度小，故它所引起的准定常行星波的异常类似于图 5 中所示的波数大的路径往北美西部落基山脉一带传播，故扰动高度场的异常距平的分布也如图 5 中所示的波数大的路径由南亚经东亚向北美西部方向传播。

五、结论与讨论

由上面所述的观测事实、理论分析及数值计算可以得到如下结论：

（1）在东亚上空的西太平洋副热带高压的异常发展是与热带西太平洋海面温度的异常增暖，特别与我国南海、菲律宾周围上空的对流活动的活跃紧密相关。

（2）这种遥相关可能是由于菲律宾周围上空的热源强迫所产生的准定常行星波传播的结果。

从上面所讨论的结果可以看到：当 El Niño 事件发生，如1987年，热带西太平洋的海面温度出现负异常，这样在热带西太平洋上空，特别在菲律宾周围上空的对流活动偏弱，即在西太平洋上空的热源比正常年份偏弱。根据上面的结论可以类推：在 El Niño 年的夏季，在中印半岛及我国南方上空的扰动高度场将出现正距平；而在长江以北及日本一带上空的扰动高度场将出现负距平。实际观测事实与此相符，1987 年夏季 500hPa 高度场上在中印半岛及我国长江以南出现负距平，而长江以北及日本一带出现正距平。由此可以看到，在 El Niño 年的夏季，东亚上空的西太平洋的副热带高压偏弱。

本文曾在东亚大气环流国际讨论会上报告过。

参 考 文 献

[1] Wallace, J.M. and D.S. Gutzler, 1981, Teleconnections in the geopotential height field during the Northern Hemisphere winter, Mon. Wea. Rev., Vol. 109, p. 784—812.

[2] Shukla, J. and J.M. Wallace, 1983, Numerical simulation of the atmospheric response to equatorial sea surface temperature anomalies, J. Atmos. Sci., Vol. 40, p. 1613—1630.

[3] Tokioka, T., Yamazaki, K., and M. Chiba, 1985, Atmospheric response to the sea surface temperature anomalies observed in early summer of 1983, J. Meteor. Soc. Japan, Vol. 63, p. 565—588.

[4] Huang, Ronghui, 1985, Numerical simulation of the three-dimensional teleconnections in the summer circulation over the Northern Hemisphere, Adv. Atmos. Sci. Vol. 2, p. 81—92.

[5] Nitta, T., 1986, Long-term variations cloud amount in the Western Pacific region, J. Meteor. Soc. Japan, Vol 64. p. 373—390.

[6] Kurihara, K., and M. Kawahara, 1986, Extremes of East Asia weather during the post ENSO years of 1983/84, J. Meteor. Soc. Japan, Vol. 64, p.493—503.

[7] Maruyama, T., T. Nitta and Y. Tsuneoka, 1986, Estimation of monthly rainfall from satellite-observed cloud amount in the tropical Western Pacific, J. Meteor. Soc. Japan,

Vol. 64, p. 149—155.

[8] Huang, Ronghui, 1984, The characteristics of the forced stationary planetary wave propagations in summer Northern Hemisphere, Adv. Atmos. Soc., Vol. 1, p. 84—94.

[9] Hoskins, B. J. and D.J. Karoly, 1981, The steady linear response of a spherical atmosphere to thermal and orographic forcing, J. Atmos. Soc., Vol. 34, p. 1179—1196.

[10] Huang, Ronghui and K. Gambo, 1982, The response of a hemisphere multi-level model atmosphere to forcing by topography and stationary heat sources, Part I, II, J. Meteor. Soc., Vol. 60, p. 78—108.

[11] Huang, Ronghui and K. Gambo, 1983, The response of a hemispheric multi-level model atmosphere to forcing by topography and stationary heat sources in summer, J. Meteor Soc. Japan, Vol. 61, p. 495—505.

Numerical Simulation of the Relationship between the Anomaly of the Subtropical High over East Asia and the Convective Activities in the Western Tropical Pacific*

Huang Ronghui(黄荣辉) and Lu Li(卢里)

Institure of Atmospheric Physics, Academia Sinca, Beijing

ABSTRACT

In this paper, a close relationship between the intraseasonal variation of subtropical high over East Asia and the convective activities around the South China Sea and the Philippines is analysed from OLR data.

This relationship is studied by using the theory of wave propagating in a slowly varying medium and by using a quasi-geostrophic, linear, spherical model and the IAP-GCM, respectively. The results show that when the SST is warming around the western tropical Pacific or the Philippines, the convective activities are intensified around the Philippines. As a consequence, the subtropical high will be intensified over East Asia. The computed results also show that when the anomaly of convective activities are caused around the Philippines, a teleconnection pattern of circulation anomalies will be caused over South Asia, East Asia and North America.

I. INTRODUCTION

Huang and Li (1987), Huang and Wu (1988) have studied the influence of ENSO on the summer climate change in China and its mechanism from the observed data. It is discovered that in the developing stage of ENSO, the SST in the western tropical Pacific is colder in summer, the convective activities may be weak around the South China Sea and the Philippines. As a consequence, the subtropical high will shift southward. Therefore, a drought may be caused in the Indo-China Peninsula and the southern China. Moreover, in midsummer the subtropical high is weak over the Yangtze River valley and Huaihe River valley, the summer monsoon rainfall belt easily forms and maintains over the Yangtze River valley and Huaihe River valley, and the flood may be caused in the area between the Yangtze River and the Huaihe River. On the contrary, in the decaying stage of ENSO, the convective activities may be strong around the Phillippines, and the subtropical high shifts northward, and a drought may be caused in the area between the Yangtze River and the Huaihe River.

In the above-mentioned papers, the relationship between the interannual variation of subtropical high and the convective activities around the Philippines and the South China Sea is clear. In order to explain the relationship in more detail, the relationship between the intraseasonal variation of subtropical high over East Asia and the convective activites around the South China Sea and the Phillippies in summer is analysed in this paper by using the observed data. The theory of wave propagating characteristics of quasi-stationary planetary waves is used to explain this relationship. The anomaly of disturbance height field over the Northern Hemisphere due to SST anomaly in the western tropical Pacific is simulated by using the general circulation model of I. A. P. Moreover, in order to compare with the results

---

* 本文原载于: Advances in Atmospheric Sciences, Vol.6, No.2, 202-214, 1989.

simulated by using GCM, a linear, quasi-geostrophic, 34-level spherical coordinate model is utilized to calculate the influence of heat source anomaly around the Philippines on the atmospheric circulation over East Asia.

II. RELATIONSHIP BETWEEN THE INTRASEASONAL VARIATION OF SUBTROPICAL HIGH AND THE CONVECTIVE ACTIVITIES AROUND THE PHILIPPINES

In order to investigate the relationship between the intraseasonal variation of subtropical high over East Asia and the convective activities around the Philippines, the latitude-time cross-sections of 500 hPa height averaged for 125°-145°E and 5-day mean TBB along 130°E are made. Fig. 1 indicates the latitude-time cross-section of 500 hPa height averaged for 125°-145°E and 5-day mean TBB below -10℃ along 130°E. The lower the TBB in an area is, the stronger the convective activities are. Obviously, in the summer of 1984 the convective activities are very strong around the Philippines and the South China Sea. Fig.2 shows the latitude-time cross-section of 5 day mean 500 hPa height along 130°E and 5-day mean TBB below -10℃ along 130°E. We can see that in the summer of 1985, the convective activities are strong around the Philippines and the South China Sea. Moreover, in both summer of 1984 and 1985, the subtropical high locates further north than the normal (dashed line in Fig.2). Therefore, there is a close relation between the anomalous development of subtropical high over East Asia and the intensified convective activities around the Philippines and the South China Sea. The appearance of the intensified convective activities around the Philippines is earlier about 10 days than the development of subtropical high over East Asia.

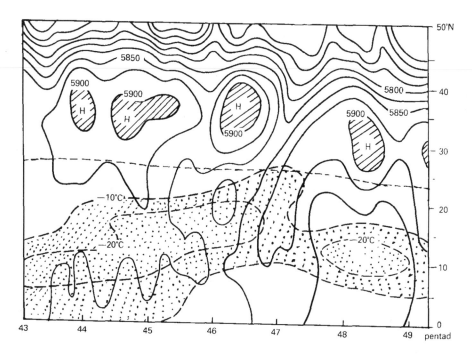

Fig.1. Latitude-time cross-section of 500 hPa height averaged for 125°-145°E and 5-day mean TBB below -10℃ (dotted areas) along 130°E in August 1984.

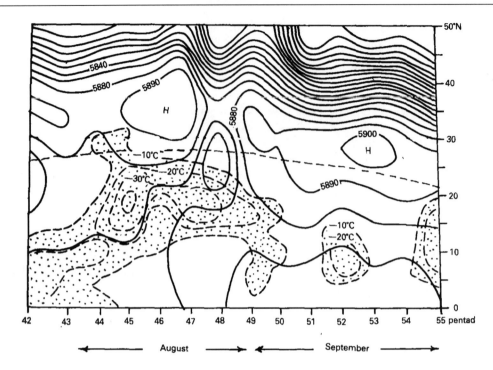

Fig.2. Latitude–time cross–section of 5–day mean 500 hPa height along 130 ° E and 5–day mean TBB below −10℃ (dotted areas) along 130 ° E in August 1985.

Fig.3 shows the latitide–time cross–section of 5–day mean 500 hPa height along 120 ° E and TBB betow −10℃ along 130 ° E during August 1987.

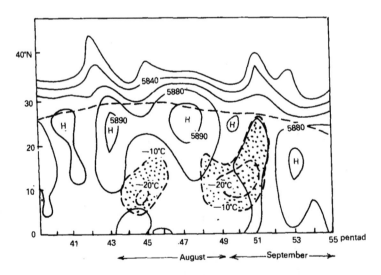

Fig.3. Latitude–time cross–section of 5–day mean 500 hPa height along 120 ° E (solid curves) and TBB below −10℃ along 130 ° E during August 1987.

The dotted area indicates the TBB below −10℃. Obviously, the convective activities are very weak around the Philippines. From Fig.3 we can see that in the summer of 1987, the subtropical high over East Asia shifts southward. This may be due to the weak convective activities around the Philippines.

From the above-mentioned results, in the summer, when the convective activities are intensified around the Philippines, after 10 days the subtropical high will be intensified and shifted northward.

III. THE PROPAGATING CHARACTERISTICS OF QUASI-STATIONARY PLANETARY WAVES FORCED BY HEAT SOURCE AROUND THE PHILIPPINES

The above-mentioned relation between the unusual development of the subtropical high and the intensified convective activities around the Philippines is by no means accidental, but inevitable. It results from the propagation of quasi-stationary planetary waves forced by the heat source around the Philippines.

Huang (1984) has discussed the propagating laws of quasi-stationary planetary waves in the summer basic flow, and pointed out that during the Northern Hemisphere summer, the quasi-stationary planetary waves can quasi-horizontally propagate from the subtropics to the middle and high latitudes. As a consequence, we may discuss the propagations of planetary waves forced at the subtropics in a sphere using a two-dimensional vorticity equation and thermodynamical equation.

The vorticity equation in a spherical coordinates is as follows:

$$\left(\frac{\partial}{\partial t} + \hat{\Omega}\frac{\partial}{\partial \lambda}\right)\left\{\frac{1}{f}\left[\frac{\sin\varphi}{\cos\varphi}\frac{\partial}{a\partial\varphi}\left(\frac{\cos\varphi}{\sin\varphi}\frac{\partial\varphi'}{a\partial\varphi}\right) + \frac{1}{a^2\sin^2\varphi}\frac{\partial^2\varphi'}{\partial\lambda^2}\right]\right\}$$
$$+ \frac{1}{a}\frac{\partial\bar{q}}{\partial\varphi} \times \frac{1}{f}\frac{\partial\varphi'}{a\cos\varphi\partial\lambda} = 0, \tag{1}$$

where $\bar{q}$ is the vorticity of the basic flow, and its gradient in a spherical coordinates may be expressed as

$$\frac{\partial\bar{q}}{a\partial\varphi} = \frac{1}{a}\left[2(\Omega_0 + \hat{\Omega}) - \frac{\partial^2\hat{\Omega}}{\partial\varphi^2} + 3\mathrm{tg}\varphi\frac{\partial\hat{\Omega}}{\partial\varphi}\right]\cos\varphi.$$

Ω_0 is the rotation angle of the earth, and $\hat{\Omega}$ is the angle velocity of the basic flow. Introducing the streamfunction and using Mercator projection

$$\begin{cases} dx = ad\lambda, \\ dy = ad[\tan h^{-1}(\sin\varphi)], \end{cases} \tag{2}$$

The relation between the zonal wavenumber k in x, y coordinate system and the wavenumber $\tilde{k}$ in λ, φ coordinate system is

$$k = \tilde{k}/a,$$

and, Eq. (1) can be expressed as

$$\left(\frac{\partial}{\partial t} + a\hat{\Omega}\frac{\partial}{\partial x}\right)\left\{\frac{1}{\cos^2\varphi}\left[\frac{\partial^2\Psi'}{\partial x^2} + \frac{\partial^2\Psi'}{\partial y^2}\right]\right\} + \frac{1}{\cos^2\varphi}\bar{q}_v \times \frac{\partial\Psi'}{\partial x} = 0. \tag{3}$$

Where, $\bar{q}_y$ is

$$\bar{q}_y = \frac{\partial \bar{q}}{a\partial \varphi}\cos\varphi.$$

Because the problem, investigated in this paper, is the propagations of quasi-stationary planetary waves, and the amplitude of wave is the slowly varying function of x, y, t, we can use WKBJ method to solve Eq. (3). A slowly varying coordinate system $(X, Y, T,)$ is introduced, i. e.,

$$X = \varepsilon x, \quad Y = \varepsilon y, \quad T = \varepsilon t. \tag{4}$$

Where ε is a positive small parameter. Assuming the solution of Eq. (3) is a solution of wavepacket, i. e.,

$$\Psi'(x, y, t) = \hat{\Psi}'(X, Y, T)e^{i\Theta/\varepsilon} \tag{5}$$

and

$$\frac{\partial \Theta}{\partial T} = -\hat{\omega}, \quad k = \frac{\partial \Theta}{\partial X}, \quad m = \frac{\partial \Theta}{\partial Y}. \tag{6}$$

k, m are the local zonal and meridional wavenumber, respectively, $\hat{\omega}$ is the local frequency. The amplitude $\hat{\Psi}(X, Y, T)$ may be expanded with a small parameter, then, we have

$$\hat{\Psi}(X,Y,T) = \hat{\Psi}_0(X,Y,T) + \varepsilon\hat{\Psi}_1(X,Y,T) + \varepsilon^2\hat{\Psi}_2(X,Y,T) + \ldots \tag{7}$$

Substituting Eqs. (4)–(7) into Eq. (3), we can obtain the following equation

$$\left[-i(\hat{\omega} - a\Omega k) + \varepsilon\frac{\partial}{\partial T} + \varepsilon a\Omega\frac{\partial}{\partial X}\right] \times \left\{\varepsilon^2\left[\frac{1}{\cos^2\varphi}\left(\frac{\partial^2 \hat{\Psi}_0}{\partial X^2} + \frac{\partial^2 \hat{\Psi}_0}{\partial Y^2}\right)\right]\right.$$
$$\left. + \varepsilon i\left[\frac{1}{\cos^2\varphi}\left(2k\frac{\partial \hat{\Psi}_0}{\partial X} + 2m\frac{\partial \hat{\Psi}_0}{\partial Y}\right) + \frac{1}{\cos^2\varphi}\left(\hat{\Psi}_0\frac{\partial k}{\partial X} + \hat{\Psi}_0\frac{\partial m}{\partial y}\right)\right] - \frac{1}{\cos^2\varphi}(k^2 \right.$$
$$\left. + m^2)\hat{\Psi}_0\right\} + \frac{1}{\cos^2\varphi}\bar{q}_y ik\hat{\Psi}_0 + 0(\varepsilon^2) = 0. \tag{8}$$

The variations of propagating ray, propagating speed of waves and the variations of wave amplitude during the propagating of waves can be obtained from Eq. (8)

Taking the zero-order approximation of Eq. (8),

$$\left\{i(\hat{\omega} - a\Omega k)\left[\frac{1}{\cos^2\varphi}(k^2 + m^2)\right]\right\}\hat{\Psi}_0 + \frac{ik\bar{q}_y}{\cos^2\varphi}\hat{\Psi}_0 = 0,$$

the dispersion of planetary waves can be obtained, i.e.,

$$\hat{\omega} = a\Omega k - \frac{\bar{q}_y \cdot k}{k^2 + m^2}. \tag{9}$$

The components of group velocity of planetary waves in the X, Y directions C_{gX} and C_{gY} can be obtained from Eq. (10). For quasi-stationary planetary waves, they are

$$\begin{cases} C_{gX} = \dfrac{\partial \hat{\omega}}{\partial k} = 2\overline{q}_y \cdot k^2 / (k^2 + m^2)^2, \\ C_{gY} = \dfrac{\partial \hat{\omega}}{\partial m} = 2\overline{q}_y \cdot k \cdot m / (k^2 + m^2)^2, \end{cases} \quad (10)$$

respectively. From Eq. (10), we have the following relation

$$(k^2 + m^2) = \dfrac{\overline{q}_y}{a\hat{\Omega}} = K_s^2. \quad (11)$$

Using the distribution of the basic flow and Eq.(11), the distribution of K_s may be obtained. We define that the angle between propagating ray and X direction in the horizontal is θ, and we have

$$\mathrm{tg}\theta = \dfrac{C_{gY}}{C_{gX}} = \dfrac{m}{k}. \quad (12)$$

Substituting (11) into (12), we may obtain the equation of propagating ray path of waves.

$$\mathrm{tg}\theta = \begin{cases} \sqrt{\left(\dfrac{K_s}{k}\right)^2 - 1}, & \text{before turning point} \\ -\sqrt{\left(\dfrac{K_s}{k}\right)^2 - 1}, & \text{after turning point.} \end{cases} \quad (13)$$

From Eq. (13), we may see that when $K_s > k$, the waves can propagate poleward. When $K_s = k$, there is a turning point in propagating ray path. Transforming X, Y coordinate system into the spherical coordinate system, the variation of the propagating ray of waves is as follows.

$$\dfrac{d\varphi}{d\lambda} = \begin{cases} \cos\varphi \left[\left(\dfrac{aK_s}{\tilde{k}}\right)^2 - 1\right]^{\frac{1}{2}}, & \text{before turning point} \\ -\cos\varphi \left[\left(\dfrac{aK_s}{\tilde{k}}\right)^2 - 1\right]^{\frac{1}{2}}, & \text{after turning point.} \end{cases} \quad (14)$$

From Eq. (14) we may obtain the propagating ray path of quasi-stationary planetary waves in summer. Fig.4 is the propagating ray path of quasi-stationary planetary waves by a forcing source around the Philippines in a realistic summer current. We may see that the planetary-scale disturbance for zonal wavenumber 1–3 propagates toward the Atlantic Ocean, while the long wave-scale disturbance for zonal wavenumber about 4–6 propagates toward North America and Rocky Mountains.

We may also estimate the propagating period from the area around the Philippines using the group velocity. For the disturbance of wavenumber 1, the propagating period from the area around the Philippines to the turning point of its propagating ray path is about 20 days; for the disturbance of wavenumber 2, this period is about 10 days. Besides, the velocity of propagation from the subtropics to middle latitudes is slower than that from middle latitudes to the vicinity of the pole.

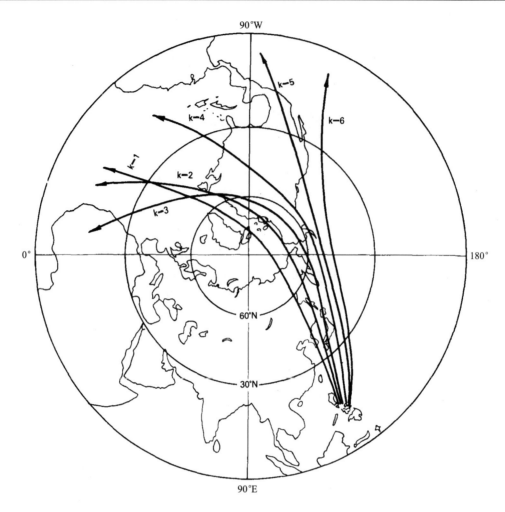

Fig. 4. Propagating ray path of quasi-stationary planetary waves forced by a forcing source around the Philippines.

IV. NUMERICAL SIMULATION OF THE INFLUENCE OF SST ANOMALY IN WESTERN TROPICAL PACIFIC ON THE SUMMER CIRCULATION ANOMALY OVER THE NORTHERN HEMISPHERE

In Section II and Huang and Wu (1987)'s paper, the close relation between the development of the subtropical high over East Asia and the convective activities around the Philippines has been shown. According to the results investigated by Kurihara (1986) and Nitta (1986), the covective activities around the Philippines are closely associated with the SST in the western tropical Pacific. When the SST in the western tropical Pacific is above the normal, the convective activities are strong around the Philippines and the South China Sea. In order to investigate the influence of SST anomaly in the western tropical Pacific on the atmospheric circulation anomaly during the Northern Hemisphere summer, we use the GCM of Institute of Atmospheric Physics, designed by Zeng Qingcun et al. (1986). This is a two-level global circulation model, in which the finite difference scheme of calculating the dynamics is unique, but the radiation heating, sensible and latent heat transfer from the sea and land surface are similar to the two-level GCM of OSU (see Chan, et al., 1982) The distribution of climatological mean SST used in this investigation is the same as Shukla and Wallace (1983).

In this model, we use an idealized distribution of SST anomaly shown in Fig. 5. We may see that an average SST anomaly is about 1℃. We add the summer SST anomaly to the climatological distribution of SST in the globe, and can obtain the distribution of anomaly SST in the Pacific. The climatological wind field and height field are used as initial wind field and height field in the model, respectively. Integrations of this model, then, are performed for one month from the initial wind field and height field for July. Moreover, we subtract the height values at isobaric surfaces and the precipitations in the case obtained from the climatological mean distribution of SST from those in the anomaly SST case. Thus, the height anomalies at isobaric surfaces and the precipitations caused by the SST anomaly in the western tropical Pacific are obtained.

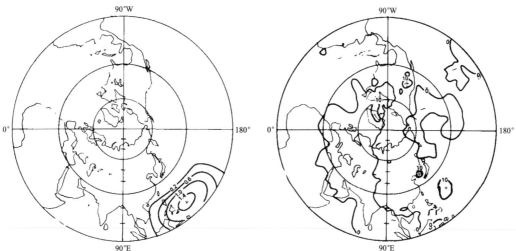

Fig.5. An idealized distribution of SST anomaly in the western tropical Pacific.

Fig.6. Distribution of 500 hPa disturbance height anomaly at the 6th model day.

First, let us see the propagation of 500 hPa disturbance height anomaly caused by the SST anomaly in the western tropical Pacific. Fig.6 shows the results of the 6th day integrated by the GCM. We can see that a positive anomaly of disturbance height is found over the western tropical Pacific, and a negative anomaly of disturbance height over Indo-China Peninsula. By the 12th day, as shown in Fig.7, a positive anomaly is found over East Asia, and another positive anomaly over Aleusian region. While over the western coast of North America, a negative anomaly can be found. We can clearly see the propagation of the anomaly pattern. This propagation of 500 hPa disturbance height anomaly is due to the propagations of planetary-wave trains forced by heat source over the western tropical Pacific.

Fig. 8 shows the distribution of 500 hPa disturbance height anomaly averaged for 30 days of the model integrations.

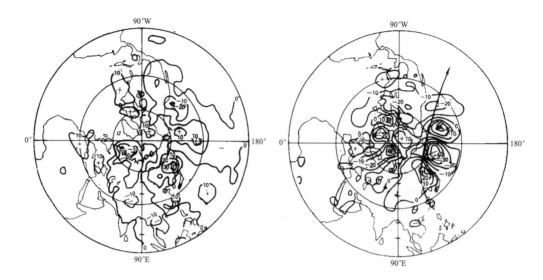

Fig. 7. As in Fig.6, but for the 12th-day result.

Fig. 8. Anomaly distribution of 500 hPa height field in July due to the SST anomaly in the western tropical Pacific, averaged for 30 days of the model integration of the IAP-GCM. (units in m).

We can find that when the SST anomaly appears in the western tropical Pacific, a negative anomaly of disturbance height field is found over Indo-China Peninsula and the southern China, while a positive anomaly locates over the northern China and Siberia. Besides, another negative anomaly is located over the Okhotsk Sea and Kamchaka, and another positive anomaly is found over Alaska, while a negative anomaly is located over the west coast of the United States of America.

Fig.9 is the anomaly distribution of precipitation in July due to the SST anomaly in the western tropical Pacific, averaged for 30 days of the model integration. Obviously, a center of positive precipitation anomaly is located around the South China Sea, the Philippines and Indonesia. Maximum precipitation anomaly may reach 15 mm d^{-1}. The mean precipitation anomaly in the region is about 8 mm d^{-1} and this is in good agreement with the observed results obtained by Maruyama et al (1986). Moreover, we may find that a negative anomaly of precipitation is located in Japan and the Huaihe River valley. Besides, other positive anomalies are located in the Aleutian region, the west coast of North America, and the west coast of Mexico respectively. This computed pattern is very analogous to the teleconnection pattern obtained by Nitta (1987) from the satellite data of OLR. Tokioka, et al. (1985) have simulated the atmospheric circulation anomaly pattern due to the SST anomaly observed in early summer of 1983 with the MRI-GCM. Our results seem to be in better agreement with the observed fact than that obtained by Tokioka.

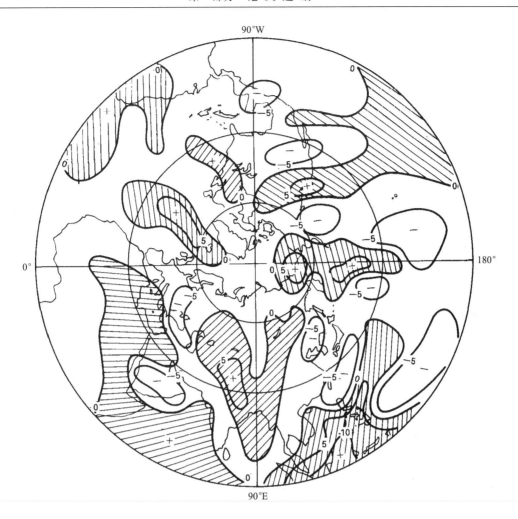

Fig.9. Anomaly distribution of precipitation in July due to the SST anomaly in the western tropical Pacific, averaged for 30 days of the model integration of IAP-GCM. (Units in mm)

V. THE STATIONARY MODEL COMPUTATION OF THE INFLUENCE OF HEAT SOURCE ANOMALY OVER THE PHILIPPINES ON THE ATMOSPHERIC CIRCULATION OVER EAST ASIA

In order to compare with the results simulated by using GCM, a quasi-geostrophic, 34-level spherical coordinate model with Rayleigh friction, Newtonian cooling effect and horizontal thermal diffusivity is utilized to compute the influence of heat source anomaly over the Philippines on the atmospheric circulation over East Asia. The model and parameters are the same as those in Huang's (1984) paper.

As mentioned in Section II, when anomaly convective activities are caused around the South China Sea and the Philippines, the heat sources anomaly will be caused around there. According to the result investigated by Maruyama et al. (1986), maximum values of the cloud anomaly may attain ±2.5. This corresponds to about 9.0 mm d^{-1} of precipitation anomaly.

Thus, we assume that an idealized heat source anomaly at 400 hPa is located around the South China Sea and the Philippines. Its horizontal distribution is given as follows.

$$\Delta \hat{H}_0(\lambda, \varphi) = \begin{cases} \Delta \hat{H}_0 \left(\sin\frac{\pi(\varphi - \varphi_1)}{(\varphi_2 - \varphi_1)} \sin\frac{\pi(\lambda - \lambda_1)}{(\lambda_2 - \lambda_1)} \right)^2, & \varphi_1 < \varphi < \varphi_2, \lambda_1 < \lambda < \lambda_2 \\ 0, & \text{otherwise}, \end{cases} \quad (15)$$

where the anomaly of heating rate $\frac{1}{c_p}\Delta \tilde{H}_0 = 5.0 \text{ K d}^{-1}$ and the averaged anomaly of heating rate is about 2.5 K d^{-1}. Moreover, we assume that the vertical distribution of this anomaly of heat source is

$$\Delta \hat{H}_0(\lambda, \varphi, p) = \Delta \hat{H}_0(\lambda, \varphi) \exp\left[-\left(\frac{p - \bar{p}}{d}\right)^2 \right]. \quad (16)$$

where $d = 3000$ hPa, $\bar{p} = 400$ hPa. Thus, the vertical distribution of amplitude and phase of quasi-stationary planetary waves and quasi-stationary disturbance pattern at isobaric surfaces can be computed.

An anomaly distribution of heat source centered at 125°E, 15°N is taken, and $\lambda_1 = 110°$ E, $\lambda_2 = 140°$ E, $\varphi_1 = 0°$ N, $\varphi_2 = 30°$ N are taken. We add the idealized anomaly of heat source to the summer climatological distribution of heat source over the Northern Hemisphere $\bar{H}_0$ (λ, φ, p),

$$H_0(\lambda, \varphi, p) = \bar{H}_0(\lambda, \varphi, p) + \Delta \hat{H}_0(\lambda, \varphi, p). \quad (17)$$

We can obtain the distribution of anomaly heat source around the Philippines. Moreover, in order to obtain the anomaly pattern of atmospheric circulation due to the anomaly of heat source, we subtract the height values of stationary disturbance at isobaric sources in the normal case from those in the anomaly case of heat source.

Fig.10 shows the anomaly distribution of 500 hPa disturbance height field caused by the heating anomaly around the Philippines. From Fig.11, We can find that when heat source or convective activities intensifies around the Philippines, a negative anomaly of disturbance height field is found over South Asia including southern China, while a positive anomaly of disturbance height field is found over the Yangtze River valley, the Huaihe River valley and Japan. In midsummer, the subtropical high is located over there. Thus, we may consider that in the summer circulation the subtropical high over East Asia is related to the heat source over the South China Sea and the Philippines, especially, to the convective activities around the Philippines. We may suggest that the anomaly of subtropical high over East Asia may be caused by the convective activities around the Philippines. Moreover, another negative anomaly is found around Hokkaido of Japan, and an anticyclone will be intensified over Aleutian region, a negative anomaly will appear over Alaska. This computed teleconnection pattern of circulation anomaly is in good agreement, not only with the observed facts, but also with the theoretical results. The computed distribution of wave train is in good agreement with the propagation ray path obtained from the theoretical analyses. This may explain that in summer, due to the anomaly of heat source around the Philippines, a teleconnection pattern of circulation anomalies may be caused from South Asia through East Asia to North America.

This is due to the propagation of quasi-stationary planetary waves.

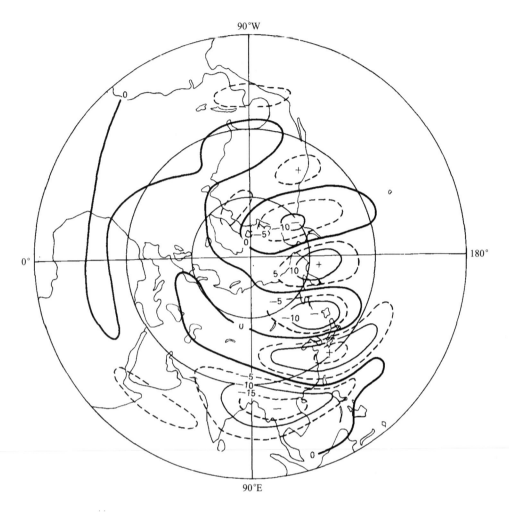

Fig.10. Anomaly distribution of the summer 500 hPa disturbance height field due to a heating anomaly around the Philippines computed by a linear model.

VI. CONCLUTIONS AND DISCUSSIONS

In this paper, the relationship between the intraseasonal variations of subtropical high over East Asia and the convective activities around the South China Sea and the Philippines is analysed from latitude–time cross–section of OLR data. The analysed results show that there is a close relationship between the intraseasonal variations of subtropical high over East Asia and the convective activities around the Philippines. When convective activities are intensified around the South China Sea and the Philippines, the subtropical high will shift northward and will be intensified.

This relationship is also studied by using the theory of wave propagating in a slowly varying medium. It shows that the close relationship between the intraseasonal variations of subtropical high over East Asia and the convective activities around Philippines is due to the propagation of quasi-stationary planetary waves forced by heat source around the Philippines. A quasi-geostrophic, linear, spherical model and the IAP-GCM are used to simulate this relationship, respectively. The results show that when the SST is warming around the western tropical Pacific or the Philippines, the convective activities are intensified around the Philippines. As a consequence, the heat source is stronger than the normal. Due to the propagation of quasi-stationary planetary waves around the Philippines, the subtropical high will intensify over East Asia in midsummer. The computed results also show that when the anomaly of convective activities are caused around the Philippines, a teleconnection pattern of circulation anomalies will be caused over South Asia, East Asia and North America.

The SST anomaly used in this study is an idealized distribution. The circulation anomalies caused by a real SST anomaly needs to be investigated in future.

The authors wish to express thanks to Prof. J. M. Wallace of the University of Washington, Prof. E. M. Rasmusson and Prof. J. Shukla of the University of Maryland and Dr. M. Blackmon of NCAR for their valuable discussions in the Joint Japan–U.S. Workshop on the EL Nino–Southern Oscillation Phenomenon. Thanks are also due to Prof. C. R. Mechoso, and Prof. M. Yanai of UCLA, and Prof. J. Horel of the University of Utah for their helpful comments in the Jacob Bjerknes Symposium on Air–Sea Interactions.

REFERENCES

Chan, S. J., J. W. Lingaas, M. E. Schlesinger, R. L. Mobley and W. L. Gates (1982), A documentation of the OSU two-level atmospheric general circulation model, Climate Research Institute of Oregon State University, Report No. 35

Huang, R. H (1984), The characteristics of the forced stationary wave propagations in summer Northern Hemisphere, Adv. Atmos. Sci., 1; 85–94.

Huang, R. H. and Li, W. J.(1987), Influence of the anomaly of heat source over the northwestern tropical Pacific for the subtropical high over East Asia, published in Proceedings of International Conference on the General Circulation of East Asia, April 10–15, 1987, Chengdu, China, 40–45.

Huang, R. H. and Wu, Y. F.(1988), The influence of ENSO on the summer climate change in China and its mechanism, Submitted to Adv. Atmos. Sci.

Kurihara, K., and M. Kawahara (1986), Extremes of East Asia weather during the post ENSO years of 83/84–severe cold winter and hot dry summer, J. Meteor. Soc. Japan, **64**: 493–503.

Maruyama, T., Ts. Nitta and Y. Tsuneoka (1986), Estimation of monthly rainfall from satellite-observed cloud amount in the tropical western Pacific, J. Meteor. Soc. Japan, **64**: 149–155.

Nitta, Ts. (1986), Long-term variations of cloud amount in the western Pacific region, J. Meteor. Soc. Japan, **64**: 373–390.

Nitta, Ts. (1987), Convective activities in the tropical western Pacific and their impact on the Northern Hemisphere summer circulation, J. Meteor. Soc. Japan, **65**: 373–390.

Tokioka, T., K. Yamazaki and M. Chiba(1985), Atmospheric response to the sea surface temperature anomalies observed in early summer of 1983; A numerical experiment, J. Meteor. Soc. Japan, **63**: 565–588.

Zeng, Qingcun et al. (1986), A global grid point general circulation model, Short- and Medium-Range Numerical Weather Prediction, collection of papers presented at the WMO, IUGG NWP Symposium, Tokyo, **4–8** August, 1986, 424–430.

The Influence of ENSO on the Summer Climate Change in China and Its Mechanisms*

Huang Ronghui (黄荣辉) and Wu Yifang (吴仪芳)

Institute of Atmospheric Physics, Academia Sinica, Beijing

ABSTRACT

The influence of ENSO on the summer climate change in China and its mechanism from the observed data is discussed. It is discovered that in the developing stage of ENSO, the SST in the western tropical Pacific is colder in summer, the convective activities may be weak around the South China Sea and the Philippines. As a consequence, the subtropical high shifted southward. Therefore, a drought may be caused in the Indo-China peninsula and in the South China. Moreover, in midsummer the subtropical high is weak over the Yangtze River valley and Huaihe River valley, and the flood may be caused in the area from the Yangtze River valley to Huaihe River valley. On the contrary, in the decaying stage of ENSO, the convective activities may be strong around the Philippines, and the subtropical high shifted northward, a drought may be caused in the Yangtze River valley and Huaihe River valley.

I. INTRODUCTION

It is well-known that the ENSO event is a most important phenomenon in the tropical air-sea interaction. When the SST in the eastern equatorial Pacific shows an anomalous warming phenomenon, an anomalous climate may be caused in the global. Early in 1960s Bjerknes (1969) put forward that an equatorial zonal circulation in the Pacific Ocean called the Walker circulation, is associated with the widespread changes SST in the tropical Pacific, and made a hypothesis of teleconnection between the cold winter in the eastern of America and the anomalous warming phenomenon of SST in the eastern equatorial Pacific. Namias (1978) pointed out that since the 1976/1977 EL Nino event showed a trough over the Aleutian region, a ridge developed over the area near the Rocky Mountain and another trough intensified over the eastern of the United States. Later, Horel and Wallace (1981), Rasmusson and Carpenter (1982, 1983), Huang Ronghui (1986) documented in detail the global features of the ENSO and its impact on the global atmospheric circulations.

The above-mentioned investigations mainly focused on the influence of the ENSO on the winter circulation. Recently, Fu (1987), Wang (1986) have investigated the influence of the ENSO event on the summer climate in China. Their investigations showed that the ENSO events may also bring about the summer climate anomalies in China. The summer climate anomalies are very important to the agriculture in China. The prediction of summer climate anomalies, especially the prediction of drought and flood in summer, is an important topic in the long-range weather forecasting in China.

II. OBSERVATIONAL FACTS OF INFLUENCE OF ENSO ON THE SUMMER CLIMATE CHANGE IN CHINA

The results investigated by Wang (1986) and Fu (1987) show that there is a good negative

---

* 本文原载于: Advances in Atmospheric Sciences, Vol.6, No.1, 21-32, 1989.

correlation between the summer mean surface temperature in the northeastern of China and the anomalous warming of SST in the eastern equatorial Pacific. When the EL Nino event occurs, the summer mean surface temperature is anomalously cold in the northeastern of China. This may bring about a cooling summer over there.

The ENSO events also heavily influence the drought and flood in the eastern and southern part of China. We investigated the correlation between the rainfall in China and SST averaged from $180°-80°W$, $5°N-5°S$ in the developing stage of an ENSO event and the decaying stage of an ENSO event, respectively. The correlation coefficients are computed from the observational data of 8 ENSO events during 1950–1980 by using the point correlation method. The correlation coefficients $\tau_{i,j}$ are

$$\tau_{i,j} = \frac{S_{x,y}}{\sqrt{S_{xx} \cdot S_{yy}}}, \quad S_{x,y} = \frac{1}{N}\sum_{j=i}^{N}\Delta R(i,j)\Delta T_s(j),$$

$$S_{yy} = \frac{1}{N}\sum_{j=1}^{N}[\Delta R(i,j)]^2, \quad S_{xx} = \frac{1}{N}\sum_{j=1}^{N}[\Delta T_s(j)]^2.$$

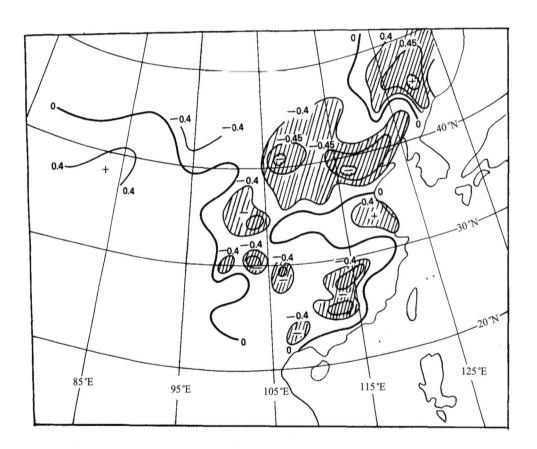

Fig.1. The correlation between the summer rain fall in China and the SST in the eastern equatorial Pacific in the developing stage of ENSO. Shaded areas indicate the coefficient of correlation above 0.4.

Where $R(i,j)$ is the rainfall anomalies in China during the summer, $T_s(j)$ is the SST anomalies averaged from 180°–80°W, 5°N–5°S. i indicates the ordinal number of grid point; j indicates the ordinal number of ENSO event; N is the number of ENSO events during 1950–1980. Fig.1 is the correlation between the summer rainfall in China and the SST in the eastern equatorial Pacific in the developing stage of ENSO. The confidence levels of all correlations have been up to above 95%, and the confidence level of correlation coefficient 0.5 to about 99%. A positive correlation may be found in the area from the Yangtze River valley to the Huaihe River valley, and a negative correlation may be found to the South of Yangtze River and in the Yellow River valley, i. e., when the ENSO event is in a developing stage, the flood may be caused in the area from the Yangtze River valley to the Huaihe River valley, but the drought will be caused tothe South of the Yangtze River and in the Yellow River valley.

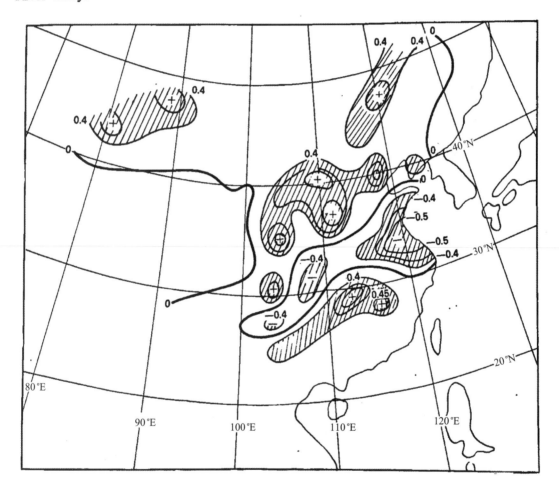

Fig.2. The correlation between the winter SST in the eastern equatorial Pacific in the decaying stage of ENSO and the rainfall in China in next summer.

The influence of SST anomaly in the eastern equatorial Pacificon the summer rainfall in China is very different in the different stage of the ENSO event. Fig. 2 shows the correlation between the winter SST in the eastern equatorial Pacific in the decaying stage of ENSO and the rainfall in China in next summer. We have also made the tests of confidence level for this correlation. The confidence levels of all correlations have also achieved about 95%, and the confidence level of correlation coefficient 0.5 have achieved about 99%. We can see that a good negative correlation is located in the Yangtze River valley and Huaihe River valley. Moreover, a positive correlation is located to the south of Yangtze River, and another positive correlation is located in the Yellow River valley. This means that when the ENSO event is in the decaying stage, in next summer a hot and drought will be caused in the Yangtze River valley and the Huaihe River valley, while a flood may be caused to the south of Yangtze River and a positive anomaly of precipitation may appear in the Yellow River valley.

As studied by Wallace and Gutzler (1981), Shukla and Wallace (1983), Huang Ronghui (1986), the heavy cold winter in the north-eastern part of China may be caused by the PNA pattern of atmospheric circulation anomaly due to the ENSO event. However, the drought and flood in the Yangtze River valley and Huaihe River valley in summer, perhaps, may not be explained by the PNA pattern because the PNA pattern is weak in summer. Thus, it is necessary to investigate the causes of the influence of ENSO on the summer climate change in East Asia.

III. RELATIONSHIP BETWEEN THE SST IN THE WESTERN TROPICAL PACIFIC AND THE SST IN THE EASTERN EQUATORIAL PACIFIC

In order to investigate the causes of the influence of ENSO events on the summer climate change in East Asia, first, we analyse the relationship between the SST anomaly in the western tropical Pacific and that in the eastern equatorial Pacific in the period of the ENSO events. Fig.3 shows the lagged correlation between the SST in the North Pacific in June and the winter SST averaged from $180°-80°W, 5°N-5°S$ in the period of the ENSO events. We have also made the tests of confidence level for this correlation. The confidence levels of all correlations have also been up to above 95%, and the confidence level of correlation coefficient 0.5 to about 99%. Thus, this correlation is confident. From this figure we can see that when ENSO event occurs, a negative anomaly of SST may appear in the western tropical Pacific. Besides, we analysed the SST anomalies in the Pacific during ENSO event by using the EOF expansion. We may find that there is a seasaw effect between the SST anomaly in the western tropical Pacific and that in the eastern equatorial Pacific.

According to the results investigated by Kurihara (1986) and Nitta (1986), there is a closely relationship between the convective activities around the Philippines and the SST anomaly in the western tropical Pacific. When the SST in the western tropical Pacific is colder than the normal, for example, in the developing stage of the ENSO events, the convective activities are weak around the Philippines and the South China Sea. On the contrary, when the SST in the weastern tropical Pacific is warmer than the normal, the convective activities are strong around the Philippines and the South China Sea.

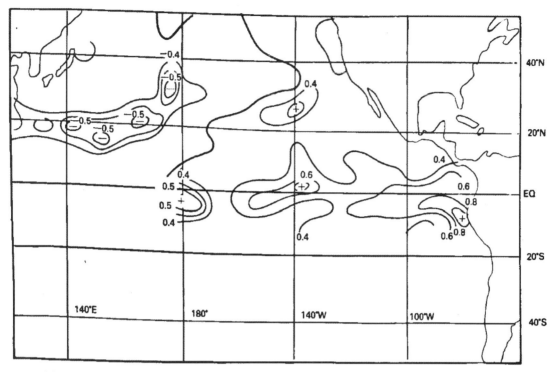

Fig.3. The lagged correlation lag between the SST in the North Pacific in June and the winter SST averaged from 180°–80°W, 5°N–5°S.

IV. RELATIONSHIP BETWEEN THE SUBTROPICAL HIGH OVER EAST ASIA AND THE CONVECTIVE ACTIVITIES AROUND THE PHILIPPINES AND THE SOUTH CHINA SEA

In summer, the subtropical high over the northwestern Pacific plays an important role for the climate change over East Asia including China, Korean peninsula and Japan. The anomalies of its location and intensity will largely cause the climate anomalies, such as the air temperature and precipitation in East Asia. For example, the location of the subtropical high is unusually northward shifted, a hot and dry weather will be caused in the East China, Korean peninsula and Japan, Recently there have been some observational studies on the behaviors of the subtropical high and its association with the climate anomalies over East Asia. However, It has not been understood how the subtropical high is maintained, or what causes the interannual and intraseasonal variability of the subtropical high. Thus, it is necessary to investigate the interannual and intraseasonal variability of the subtropical high. This paper mainly forcus on the interannual variability of the subtropical high.

As mentioned in Section III, the SST in the western tropical Pacific largely influences on the convective activities around the Philippines and the South China Sea. The SST in the western tropical Pacific is warmer than the normal in next summer of the decaying stage of ENSO. Thus, the convective activities may be intensified around the Philippines. In the following, we will analyse the interannual variation of the subtropical high and its relation to the convective activities around the Philippines.

Fig. 4 shows the relationship between the monthly mean 500 hPa height and TBB over East Asia in August 1984, the shaded areas indicate TBB below $-10°$C, i.e., indicate the intensified convective activities. From Fig. 4 it can be seen that the intensified convective activities were located around the South China Sea and the Philippines. Moreover, the strong subtropical high was over Japan and the East China. This unusually northward shifted subtropical high had caused a hot and dry monsoon in Japan, the South China and the Yangtze River valley.

Similarly, during the midsummer of 1985, the subtropical high was also unusually northward shifted and covered wide areas including the East China, Korean peninsula and Japan, as shown in Fig.5. Moreover, we may also find that there is a teleconnection pattern. A negative anomaly may be found around the South China Sea and the Philippines, and a positive anomaly may be found over Japan, another negative and positive anomaly may be found over Okhotsk Sea and Alaska, respectively. This is analogous to a Rossby-wave train.

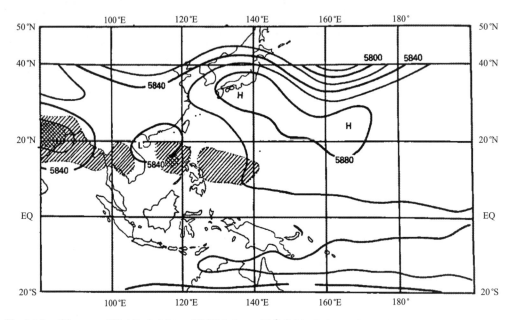

Fig.4. Monthly mean 500 hPa height and TBB below −10 °C (shaded areas) over East Asia in August 1984.

The anomalously northward shifted subtropical high had brought some anomalies of the climate there. As shown in Fig. 6, in the summer of 1985, the monsoon rainfall was 50% to 80% of the normal in the area between the Yangtze River valley and Huaihe River valley. Moreover, the mean temperature anomaly were 1 °C above the normal. This explains that in this summer, Mei-Yu front was very weak.

As shown in Fig. 5, the shaded areas indicate the intensified convective activities. The strong convective activities may be found over the Philippines.

From the above-mentioned results we may conclude that when the convective activities are intensified around the Philippines, the subtropical high will shift northward, the summer in the Yangtze River valley and Huaihe River valley may be a hot and drought summer. Recently, Nitta (1987) investigated the interannual variations of convective activities over the western tropical Pacific and their impact on the Northern Hemisphere circulation using satellite cloud amount for 7 years. His results show that during summers when SST in the tropical western Pacific is about 1.0 °C warmer than the normal, active convection regions are shifted northeastward from the normal to the subtropical western Pacific, and a high pressure anomaly predominates in middle latitudes extending from East China, through Japan to North Pacific. Our results analysed from 1984, 1985 are in good agreement with his results. Both Nitta's results and our results show that there is a good relationship between the subtropical high over East Asia and the convective activities around the Philippines and the South China Sea.

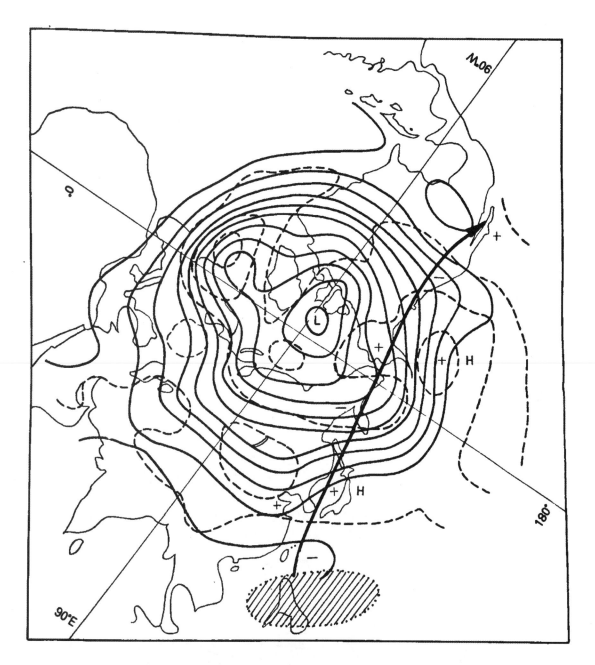

Fig.5. Monthly mean 500 hPa height in August 1985 and the summer anomaly (dotted curves) of mean 500 hPa height in 1985. The shaded areas indicate the intensified convective activities.

In the following, we will discuss the cases of weak convective activities, in the spring of 1987. According to the distribution of SST in the western tropical Pacific, Huang (1987) we predicted that the convective activities may be weak around the Philippines and the South China Sea in the summer of 1987, due to the SST anomaly is negative in the western tropical Pacific. Moreover, we also predicted that the subtropical high may shift southward. Our prediction is in good agreement with the observational facts. In order to compare with the case of strong convective activities, we present the convective activities and the anomaly of 500 hPa height field in this paper. Fig. 7 is the monthly mean 500 hPa height in August and the summer anomaly of 500 hPa height field (dotted curves) in 1987. We can see that the subtropical high anomalously shifted southward, a positive anomaly is found over the South China and Indo-China peninsula, while a negative anomaly may be found over the Yangtze River valley, Huaihe River valley and Japan. Moreover, another positive anomaly and negative anomaly may be found over Okhotsk Sea and Alaska, respectively. This pattern is just contrary to the previous pattern, but it is also analogous to a Rossby-wave train.

(图略)

Fig.6. Anomaly (%) distribution of precipitation in China during the summer of 1985.

注：本书中部分论文的地图插图有删减，读者可参阅原刊物中对应插图。下同。

The southward shifted subtropical high is favorable to the formation and maintenance of Mei-yu front over the Yangtze River valley and Huaihe River valley. As shown in Fig. 8 in

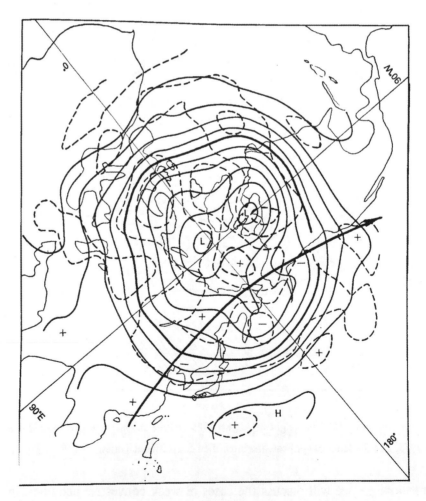

Fig.7. As in Fig. 5 except for the summer of 1987.

the summer of 1987, the summer monsoon rainfall anomaly was 50% in the Yangtze River valley and Huaihe River valley, while in the Yellow River valley and the South China the summer rainfall anomaly was about −30%. In this summer the large flood was caused in the Yangtze River valley and Huaihe River valley, and has made the economy suffer heavy losses.

The southward shifted subtropical high is closely related with the weak convective activities around the Philippines. Fig. 9 is the anomaly distribution of cloud amount during the summer of 1987. Obviously, a positive anomaly of cloud amount may be found over the equatorial area near to the dateline, a negative anomaly of cloud amount may be found over the Philippines and Indonesia. This may explain that the convective activities is weak in the summer of 1987.

(图略)

Fig.8. As in Fig. 6 except for the summer of 1987.

From SST anomaly distribution, we may see that since the EL Nino event developed again, the SST in the eastern tropical Pacific is warming, while it is colder in the western tropical Pacific. As a consequence, the convective activities was weak around the Philippines. This is in agreement with the results obtained by Kurihara and Nitta.

From the above-mentioned results we may also conclude that when the convective activities are weak around the Philippines, the subtropical high will shift southward, the flood may be caused in the Yangtze River valley and Huaihe River valley.

V. CONCLUSIONS

From the above-mentioned analysis, the following conclusions can be made.

1) In the developing stage of ENSO, the SST in the western tropical Pacific is cooling in summer, the convective activities around the South China Sea and the Philippines are weak. As a consequence, the subtropical high may shift southward; a drought may be caused in the Indo-China peninsula and the South China. Moreover, in midsummer, the subtropical high is weak over the Yangtze River valley and Huaihe River valley, the summer monsoon rainfall belt is easily formed and maintained over the Yangtze River valley and Huaihe River valley. Therefore, a flood may be caused in the area from the Yangtze River valley to Huaihe River valley, while a drought may be caused in the Yellow River valley and in the South China;

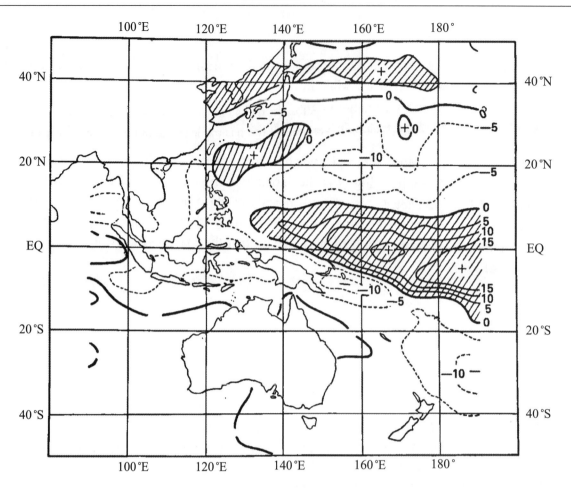

Fig.9. The anomaly (%) distribution of cloud amount during the summer of 1987. Shaded areas indicate positive anomalies.

2) In the decaying stage of ENSO, in next summer the SST in the western tropical Pacific is warming, the convective activities around the South China Sea and the Philippines will be strong. As a consequence, the subtropical high shift northward, and a flood may be caused in South China. Moreover, in midsummer, the subtropical high is strong over the Yangtze River valley and Huaihe River valley; the summer monsoon rainfall belt is weak and is not easily maintained over the Yangtze River valley and Huaihe River valley; a drought may be caused over the Yangtze River valley and Huaihe River valley. Moreover, the rainfall may appear a positive anomaly in the Yellow River valley.

As shown in above-discussed, the convective activities around the Philippines is closely related not only with the SST anomaly in the weastern tropical Pacific, but also with the atmospheric circulation over there. Generally, the convective activities is strong in the case of the warming SST in the western tropical Pacific. However although the SST is warm in the weastern tropical Pacific, the convective activities is not strong in some time.

The authors are grateful to Prof. T. Matsuno of Tokyo University for giving chance of reporting this paper in Joint Japan-U. S. Workshop on the EL Nino-Southern Oscillation Phenomenon. Thanks are due to Prof. J. M. Wallace of the University of Washington, Prof. E. M. Rasmusson, Prof. J. Shukla of the University of Maryland and

Dr. M. Blackmon of NCAR for their valuable discussions in this workshop.

REFERENCES

Bjerknes, J. (1969), Atmospheric teleconnection from the equatorial Pacific, *Mon. Wea. Rev.*, **97**: 163–172.

Fu, C. B. and Ye. D. Z. (1988), The tropical very-low frequency oscillation on interannual scale, *Adv. Atmos. Sci.*, **5**: 369–388.

Horel, J. D. and J. M. Wallace (1981), Planetary-scale atmospheric phenomena associated with the Southern Oscillation, *Mon. Wea. Rev.*, **109**: 813–829.

Huang, R. H. (1986), Physical mechanism of influence of heat source anomaly over low latitudes on general circulation over the Northern Hemisphere in winter, *Scientia sinica (Series B)*, **29**: 970–985.

Huang, R. H. (1987), Influence of the heat source anomaly over the western tropical Pacific on the subtropical high over East Asia, proceedings in International Conference on the General Circulation of East Asia, April 10–15, 1987, Chengdu, China.

Kurihara, K. and M. Kawahara (1986), Extremes of East Asian weather during the post ENSO years of 83/84-severe cold winter and hot dry summer, *J. Meteor. Soc., Japan*, **64**: 493–503.

Namias, J. (1976), Negative ocean-air feedback system over the North Pacific in the transition from warm to seasons, *Mon. Wea. Rev.*, **104**: 1107–1114.

Nitta, Ts (1986), Long-term variations of cloud amount in the weastern Pacific region, *J. Meteor. Soc., Japan*, **64**: 373–390.

Nitta, Ts, (1987), Convective activities in the tropical Western Pacific and their impact on the Northern Hemisphere summer circulation, *J. Meteor. Soc., Japan*, **62**: 165–171.

Rasmusson, E. M. and T. H. Carpenter (1982), Variations in tropical sea surface temperature and surface wind fields associated with the Southern Oscillation EL Nino, *Mon. Wea. Rev.*, **110**: 354–384.

Rasmussom, E. M., and J. M. Wallace (1983), Meteorological aspects of the EL Nino-Southern Oscillation, *Science*, **222**: 1195–1202.

Shukla, J., and J. M. Wallace (1983), Numerical simulation of the atmospheric response to equatorial sea surface temperature anomalies, *J. Atmos. Sci.*, **40**: 1613–1630.

Wallace, J. M. and D. S. Gutzler (1981), Teleconnection in the geopotential height field during the Northern Hemisphere winter, *Mon. Wea. Rev.*, **109**: 784–812.

Wang Shaowu and Zhu Hong (1986), El Nino and cooling summer in East Asia, *Kexue Tongbao*, **31**: 474–478.

引起我国夏季旱涝的东亚大气环流异常遥相关及其物理机制的研究*

黄荣辉

（中国科学院大气物理研究所）

提　　要

本文综述了引起我国夏季旱涝的东亚大范围、持续性大气环流异常遥相关方面的国内外研究状况.本文特别强调了关于夏季东亚大气环流异常遥相关物理机制的研究,其中不少研究是作者多年努力的成果.本文还指出在这方面应进一步研究的问题.

一、引　言

近年来,世界许多地方发生了大范围的干旱与洪涝,给世界人民带来严重的灾害.我国地处东亚,是强季风区域,由于季风的年际变异很大,使我国不断遭受干旱与洪涝灾害,特别是长江、黄河流域不断发生严重的洪涝与干旱.这些洪涝与干旱都是大气环流发生持续性异常所造成的.近年来的研究表明,某区域大气环流的异常不仅仅是由于该区域大气的动力、热力异常所造成,它也可能是由于别的区域环流异常所造成.许多观测事实已表明一个区域的大气环流异常可以引起另一些区域大气环流的异常,这种区域之间环流异常的相关性或由它所引起的要素异常的相关性就称为遥相关.

由于遥相关现象的发现,许多动力学家进行了遥相关动力学的研究,提出了低频变化动力学,这给旱涝预测带来一定的物理依据.本文就国内外关于引起大范围、持续性旱涝的大气环流异常遥相关观测事实及其物理机制的研究现状作一综述,特别是作者本人在这方面的研究成果.本文着重综述与我国旱涝有关的大气环流异常遥相关方面的研究.

二、北半球夏季大气环流异常遥相关的观测研究

关于北半球冬季大气环流遥相关的研究是比较多的,也是比较早的.早在本世

---

* 本文原载于：大气科学, 第14卷, 第1期, 108—117, 1990年3月出版.

纪30年代,Walker和Bliss就系统地分析了地面气象要素,他们发现大西洋与欧洲之间的海平面气压及地表面附近空气的温度存在着明显的遥相关现象[1].后来Wallace与Gutzler利用海平面气压与500hPa高度场资料系统地计算了北半球高度场的点相关[2],他们发现大气环流异常的遥相关存在着几种型:太平洋北美型(简称为PNA型)、西大西洋型(简称为WA型)、东大西洋型(简称为EA型)、欧亚型(简称为EU型)及西太平洋型(简称为WP型).Gambo与Kudo利用扰动高度场分析了北半球大气环流异常的遥相关型[3],他们的结果说明了用扰动高度场更能够表现大气环流的三维遥相关,特别是欧亚型与太平洋北美型遥相关.

关于北半球夏季大气环流遥相关迄今还没有详细的研究,而夏季大气环流的变化对于雨带分布、梅雨的强弱起着决定性的作用.也就是说,北半球大气环流的变异对于我国旱涝有着十分重要的作用.因此,研究北半球夏季大气环流异常的遥相关型是十分重要的.

1. 东亚夏季环流和降水异常与菲律宾周围对流活动的关系

夏季西太平洋副热带高压对于我国东部、日本、朝鲜半岛等东亚地区的气候有着重要的影响,它的位置及强度的异常直接影响着东亚夏季气候的异常.若在盛夏,它的强度偏强,位置偏北,则在我国东部、朝鲜半岛及日本等东亚地区会出现高温少雨的天气.在这方面已有不少研究.但是,关于西太平洋副热带高压的年际变化与季节内变化及其物理机制至今还不太清楚.

最近,作者分析了夏季东亚上空的西太平洋副热带高压与菲律宾周围对流活动的关系,提出一个夏季菲律宾周围的对流活动与东亚环流异常遥相关的概念图[4-6].

图1是夏季菲律宾周围对流活动加强后,北半球大气环流异常距平分布的示意图.从图1可以看到,当夏季菲律宾周围对流活

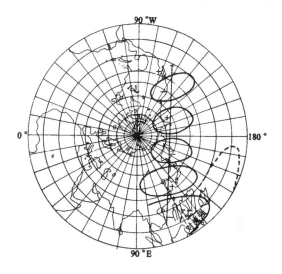

图1 夏季菲律宾周围对流活动加强后北半球大气环流异常距平分布示意图

动加强后,在500hPa高度场,一个负距平区域出现在我国南海与菲律宾上空;一个正距平区域出现在我国江淮流域与日本上空,并且另一个负距平区域与正距平区域分别出现在鄂霍茨克海及阿拉斯加的上空.这种异常距平的分布宛如一个Rossby波列的传播.最近,陈烈庭从北方涛动指数与北半球夏季500hPa异常距平的相关中也得到这个遥相关型[7],我们把这个遥相关型称为东亚太平洋型.Nitta从卫星观测的多年高云量的异常也得出同样的结果[8].

1984,1985年夏季菲律宾周围的对流活动强,使得西太平洋副热带高压位置偏北.图2是1984年8月500hPa高度场与菲律宾周围平均TBB之间的关系,其中斜线区表示TBB温度在-10℃以下,即对流活动活跃的区域.从图2可以看到强的对流活动区位于我国南海与菲律宾一带,而强的副热带高压位于日本及我国江淮流域上空,使我国东部及日本本州出现高温少雨天气.这种现象同样在1985年盛夏出现,图3是1985年8月北半球500hPa高度场(实线)与异常距平(虚线)的分布.从图2与图3可以看到这两年盛夏

500hPa 高度场的距平有类似的分布.1985 年盛夏西太平洋副热带高压异常偏北,在东亚的强度偏强,我国华中,江淮流域及黄淮流域的雨量比常年少 50%.

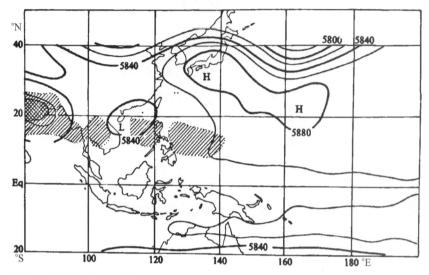

图 2　1984 年 8 月 500hPa 高度场(实线)与菲律宾周围 TBB(斜线区)的分布

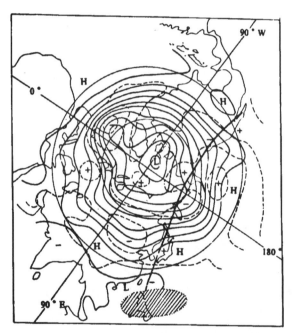

图 3　1985 年 8 月北半球 500hPa 高度场(实线)与距平(虚线,单位:50GPM)的分布

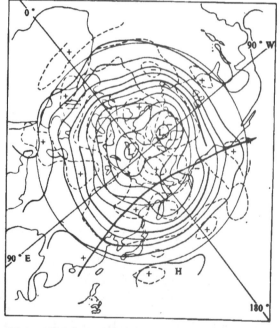

图 4　1987 年 8 月北半球 500hPa 高度场(实线)与距平(虚线,单位:50GPM)的分布

与 1984,1985 年相反,当夏季菲律宾周围的对流活动弱时,则西太平洋副热带高压偏南.图 4 是 1987 年 8 月北半球 500hPa 高度场(实线)与异常距平(虚线)的分布图.从图 4 可以看到,1987 年盛夏北半球环流异常的距平分布与 1984,1985 年的距平分布相反.一个负距平区域出现在我国江淮流域及日本上空,西太平洋副热带高压的位置偏南,它控制着我国江南及日本的南部;在我国南海与菲律宾上空是正的距平区域;并且另一个正距平与负距平区域分别位于鄂霍茨克海及阿拉斯加的上空.在 1987 年盛夏,我国江淮流域的雨量比常年多了 50% 以上,造成江淮流域很大的洪涝.1987 年盛夏我国南海与菲律宾周

围的对流活动比较弱.

Nitta 与 Kurihara 研究了热带西太平洋上空,特别是菲律宾周围云量的长期变化[9,10].他们的研究表明菲律宾周围的对流活动与热带西太平洋海表温度有很大关系,当热带西太平洋海表温度高,则菲律宾周围的对流活动就强.如 1977,1978,1981,1984,1985 等年菲律宾周围盛夏的对流就较强;而在 1980,1982,1983,1987 等年盛夏菲律宾周围的对流就较弱.

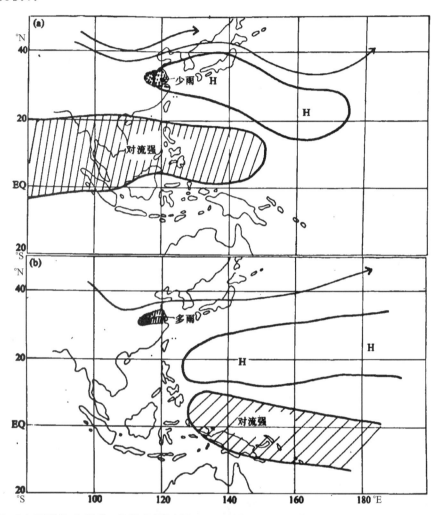

图 5 热带西太平洋海表温度、菲律宾周围的对流、西太平洋副热带高压、江淮流域降水之间的关系
(a)热带西太平洋海表温度高的情况;(b)热带西太平洋海表温度低的情况.

从我国夏季降水距平图可以看到,在 1977,1978,1981,1984,1985 年江淮流域偏旱;而在 1980,1982,1983,1987 年江淮流域偏涝.因此,我们可以作图(见图 5)表示热带西太平洋海表温度、菲律宾周围的对流活动、西太平洋副热带高压、我国江淮流域降水之间的关系.图 5a 表示热带西太平洋海表温度高的情况;而图 5b 则表示偏低的情况.可以看到,当热带西太平洋海表温度偏高时,在盛夏,菲律宾、我国南海与中印半岛的对流偏强,西太平洋副热带高压位置偏北偏西,我国江淮流域、朝鲜半岛及日本的降水偏少;相反,当热带西太平洋海表温度偏低时,在盛夏,菲律宾、我国南海的对流偏弱,西太平洋副热带高压位置偏南,并且呈条状结构,我国江淮流域的降水偏多.

2. 夏季大气环流遥相关型的年际变化

上面我们从观测事实指出在北半球夏季大气环流的遥相关中存在着东亚太平洋型,它对东亚的环流及降水有很大影响.黄荣辉与孙凤英分析了1971—1977年夏季北半球大气环流遥相关型的年际变化[11].发现在ENSO事件发生的年份或ENSO事件的第二年上述环流遥相关型比较典型,而别的年份则不明显.1972,1973年夏季环流异常的东亚太平洋型比较明显,而其它年份不很明显.并且1972与1973年夏季造成东亚太平洋型的准定常行星波列的传播路径也不一样.1972年夏季南亚地区与我国南方出现正相关区;在我国东北、苏联西伯利亚上空出现负相关区;在鄂霍茨克海及勘察加半岛上空出现另一个正相关区;在阿拉斯加地区出现另一个负相关区.波列由菲律宾经东亚传播到美洲的加拿大.而1973年夏季在阿留申群岛上空出现负相关区,在北美西部落基山脉一带出现一片正相关区.这说明1973年夏季准定常行星波波列由菲律宾周围经东亚、阿留申群岛向北美西部落基山脉一带传播.

从上述可以看到夏季大气环流的东亚太平洋型遥相关是有很大的年际变化,这是由于基本气流对行星尺度扰动的传播有很大作用所造成.

三、我国江淮流域旱涝与ENSO现象的关系

黄荣辉与吴仪芳利用30年观测资料分析了ENSO现象的不同阶段对我国旱涝分布的影响[12].研究表明ENSO现象的不同阶段对我国旱涝的分布有不同的影响.如图6所示,在ENSO现象的发展阶段,我国江淮流域的降水异常与赤道东太平洋的海温异常有一个较大的正相关;而黄河流域、华北地区及江南、华南地区的降水异常与赤道东太平

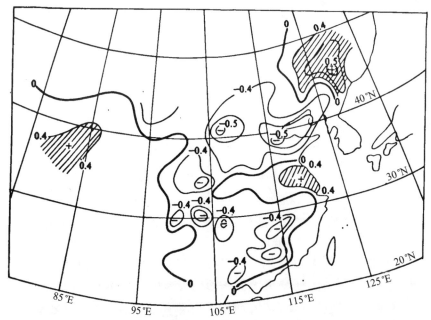

图6 ENSO发展阶段我国夏季降水异常与赤道东太平洋海温异常的相关分布

洋的海温异常有一个较大的负相关.这就是说,当ENSO现象处于发展阶段,该年夏季我国江淮流域降水将会偏多,可能发生洪涝;而黄河流域、华北地区的降水将会偏少,发生干旱;江南地区的降水也会偏少,也可能发生干旱.

然而,当ENSO处于恢复阶段,如图7所示,我国江淮流域的降水异常与前冬赤道东太平洋的海温异常有一个较大的负相关区;而黄河流域、华北地区及江南、华南地区的降

水异常与前冬赤道东太平洋海温异常有一个较大的正相关区.这就是说,当赤道东太平洋海温异常处于恢复阶段,若海温距平还在正距平范围,则夏季我国江淮流域的降水将会偏少而发生干旱;而黄河流域、华北地区及江南、华南地区的降水可能偏多.但是,当赤道东太平洋处于冷水年,特别是反 ENSO 年,则夏季江淮流域的降水偏多,而黄河流域、华北地区及华南地区的降水可能偏少而发生干旱.

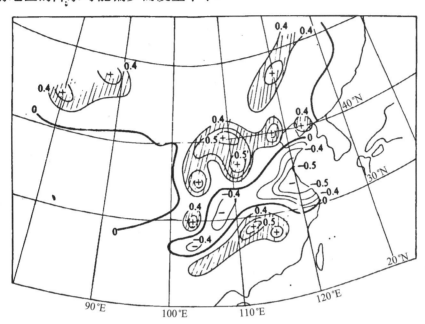

图 7　ENSO 衰减阶段我国夏季降水异常与赤道
东太平洋海温异常的相关分布

这个观测事实是与第二节中所阐述的事实相一致的.黄荣辉与吴仪芳的研究表明了赤道东太平洋的海温与热带西太平洋的海温之间存在着一个"跷跷板"结构.当 ENSO 现象处于发展期,一般热带西太平洋的海温偏低;相反,当 ENSO 现象处于恢复期,一般热带西太平洋的海温就偏高,菲律宾周围的对流活动就强,从而引起东亚上空副热带高压的位置偏北,引起东亚地区夏季高温少雨天气.

上述这些大气环流的遥相关可以作为短期气候预测的依据,有的已成为长期天气预报的很有用的方法.

四、夏季大气环流遥相关物理机制的研究

为了利用大气环流遥相关来作旱涝的短期气候预测(季度与年际预测),许多动力学家都在探讨这种遥相关的物理机制.这几年来的研究表明,引起大气环流遥相关有如下几种物理机制.

1. 准定常行星波在三维空间中的传播

Hoskins 与 Karoly 把叶笃正所提出的罗斯贝波频散理论推广到球面大气中[13],他们的研究表明罗斯贝波的波列在球面大气中是沿大圆路径传播的,并且波列向极地方向传播,其振幅要增加.黄荣辉利用波的折射指数平方与 E-P 通量系统地研究了北半球冬夏准定常行星波的传播规律[14-16],指出准定常行星波在球面上传播路径的变化遵从下列方程式:

$$\frac{d_g \alpha}{dt} = \frac{1}{2Q_k} \vec{i} \cdot \vec{C_g} \times \nabla Q_o \qquad (1)$$

其中 Q_k 是波数为 k 的准定常行星波的折射指数平方,Q_0 可以看成是波数为 0 的波的折射指数平方. $\vec{C_g}$ 是波的群速度在经圈面上的投影,$\vec{i}$ 是纬圈方向上的单位矢量,d_g/dt 表示沿群速度方向的随体微商,α 是波传播路径与水平方向的夹角. 黄荣辉利用上述理论及数值模拟的结果提出冬季准定常行星波在三维传播中有两支波导.

黄荣辉还研究了夏季准定常行星波的传播规律[17],指出夏季准定常行星波不能传播到平流层,但它能在对流层中传播. 如图 8 所示,准定常行星波能够准水平地从副热带地区向高纬度对流层传播.

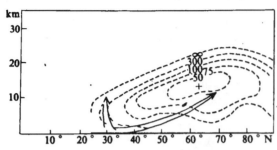

图 8 北半球夏季准定常行星波传播示意图 图中虚线表示 Q_0 值

黄荣辉还利用 WKBJ 方法求出了准定常行星波在球面大气中的传播路径是:

$$\frac{d\varphi}{d\lambda} = \begin{cases} \cos\varphi \sqrt{\left(\frac{ak_s}{k}\right)^2 - 1} & \text{在转向点之前} \\ -\cos\varphi \sqrt{\left(\frac{ak_s}{k}\right)^2 - 1} & \text{在转向点之后.} \end{cases} \qquad (2)$$

其中 k 是纬向波数,a 是地球半径,k_s 可以看作波在球面上传播的折射指数,它是:

$$K_s^2 = k^2 + m^2 = \frac{\overline{q_y}}{a\hat{\Omega}} \qquad (3)$$

$\overline{q_y}$ 是基本态位涡度的南北梯度,$\hat{\Omega} = U/a\cos\varphi$,它是基本气流的角速度. 并且,从群速度可以估算出西太平洋上空 500hPa 面上波数大约为 1 的行星尺度波列从菲律宾周围传播到大西洋约需 1 个月时间.

黄荣辉从实际夏季基本气流求出行星波在球面的传播路径(见图 9)以及在传播中波振幅的局地变化[18]. 对于实际夏季基本气流而言,$\nabla \cdot \vec{C_g} \neq 0$,波在传播中振幅的局地变化是不容易估计的,只有当行星波的传播路径是一大圆时,波在传播中的振幅才是极向增加的.

由于准定常行星波在球面大气中的传播,某区域大气环流的异常会引起另一些区域大气环流的异常.

2. 下垫面热力状况的异常引起行星尺度扰动的异常

近年来,许多研究表明了下垫面热力状况的异常是引起大气环流异常的重要原因. Moura 与 Shukla,Keshavamurly,Shukla 与 Wallace,Tokioka 模拟了 PNA 型遥相关的产生[19-22],黄荣辉从理论上及数值模拟方面说明了低纬度东太平洋热源异常将会引起 PNA 型环流的异常[23],Gambo,卢里与李维京模拟了热带大西洋海温异常对北半球冬季大气环流欧亚型异常的作用[24].

黄荣辉研究了夏季模式大气对热源强迫的响应[25]. 但是,对于夏季大气环流异常的遥相关型模拟至今还不多. 黄荣辉与李维京,黄荣辉与卢里分别利用一个动力机制模式及大气环流模式模拟了菲律宾周围对流活动加强后东亚上空环流的异常[26,27].图10表示夏季菲律宾周围对流活动加强后所引起的北半球夏季500hPa环流异常的距平分布.从图10可以看到,当菲律宾周围的对流活动加强后,我国华南及南亚地区上空的高度场出现负距平;在江淮流域及日本上空的高度场出现正距平,这个地区正是夏季西太平洋副热带高压控制的地区,这将引起我国江淮地区及日本上空的副热带高压增强.因此,可以认为夏季东亚上空西太平洋副热带高压的强弱是与菲律宾周围上空的对流活动强弱有密切的关系.

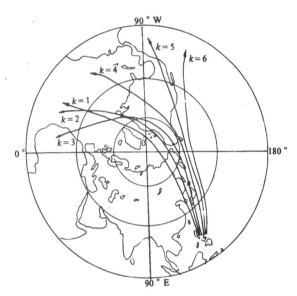

图9 位于热带西太平洋的准定常行星波的传播路径

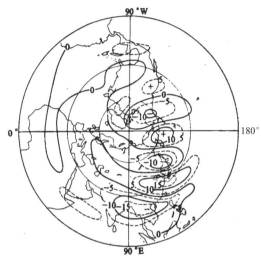

图10 由一个动力机制模式计算得到的夏季菲律宾周围对流活动加强后所引起的北半球夏季500hPa环流异常的距平分布 单位:GPM

此外,由大气环流模式计算的结果与图10相类似.

大气的下垫面不仅是海洋,还有陆地,特别是青藏高原,由于它的地势高,它的热状况直接关系到对流层中层的加热.叶笃正与高由禧,Nitta,罗会邦与Yanai的研究都表明了北半球夏季最大热源位于青藏高原的东南部[28-30].而青藏高原冬春积雪状况,如积雪天数、积雪厚度与积雪面积等对夏季青藏高原的热状况有很大影响.

黄荣辉利用数值模式模拟了夏季青藏高原上空热源异常对北半球大气环流异常的作用[31].如图11所示,若夏季青藏高原热源增强,则青藏高压就增强,我国北方的槽要加深;鄂霍茨克海上空的高压要加强;阿拉斯加地区的槽要加深;北美地区的脊要加强.这些模拟结果与实际资料分析所得到的结果一致

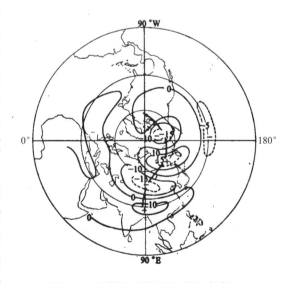

图11 由模式计算所得到的青藏高原上空热源增强后北半球夏季大气环流异常的距平分布 单位:GPM

五、应进一步研究的问题

从八十年代初到现在,大气环流异常的遥相关现象吸引着许多气象学家,他们对遥相关现象及机理作了不少研究.但是,目前国际上许多研究主要集中在冬季大气环流遥相关方面,对影响夏季旱涝的大气环流遥相关研究还比较少.这一方面是由于夏季大气环流异常比冬季弱,其遥相关型比冬季不明显;另一方面是由于夏季大气环流异常的成因比较复杂.

作者认为要弄清楚我国夏季旱涝的物理机制就必须进一步研究东亚夏季大气环流异常的遥相关及其机制,要研究热带西太平洋海温、ENSO现象、从孟加拉湾经中印半岛到菲律宾周围的对流、西太平洋副热带高压与东亚夏季旱涝之间的关系等.

从现在的研究表明,菲律宾周围的对流活动与东亚夏季旱涝有密切关系,而菲律宾周围的对流活动不仅与热带西太平洋的海表温度有关,而且与沃克环流及跨赤道气流有关.因此,必须进一步研究热带西太平洋地区的纬圈环流、经圈环流与菲律宾周围对流活动的关系.

总之,引起我国夏季旱涝的东亚大气环流异常的物理成因是很复杂的,本文所述的遥相关及其物理机制是很初浅的,还有许多课题有待进一步研究.

参 考 文 献

[1] Walker, G.T., and E.W.Bliss, 1932, World Weather, *V.Mem.Roy.Meteor.Soc.* Vol.4, 53–84.

[2] Wallace, J.M., and D.S.Gutzler, 1981, Teleconnections in the geopotential height field during the Northern Hemisphere winter, *Mon.Wea.Rev.*, Vol.109, 784–812.

[3] Gambo, K., and K.Kudo, 1983, Three-dimensional teleconnection in the zonally asymmetric height field during the Northern Hemisphere winter, *J.Meteor.Soc.Japan*, Vol.61, 36–50.

[4] Huang, R.H.(黄荣辉), and Li, W.J.(李维京), 1987, Influence of the heat source anomaly over the western tropical Pacific on the Subtropical High over East Asia, International Conference on the General Circulation of East Asia, April 10–15, 1987, Chengdu, China.

[5] 黄荣辉、李维京,1988,夏季热带西太平洋上空的热源异常对东亚上空副热带高压的影响及其物理机制,大气科学(特刊),107–116.

[6] Huang, R.H.(黄荣辉), and Wu, Y.F.(吴仪芳), 1987, The influence of the ENSO on the summer climate change in China and its mechanism, Japan–U.S.Workshop on the El Niño Southern Oscillation Phenomenon, November 3–7, 1987, Tokyo, Japan.

[7] 陈烈庭、吴仁广,北方涛动对北半球温带环流的影响(待发表)

[8] Nitta, Ts., 1986, Long-term variation of cloud amount in the western Pacific region, *J.Meteor, Soc.Japan*, Vol.64, 373–390.

[9] Nitta, Ts., 1987, Convective activities in the tropical western Pacific and their impact on the Northern Hemisphere summer circulation, *J.Meteor.Soc.Japan*, Vol. 64, 373–390.

[10] Kurihara, K.and M.Kawahara, 1986, Extremes of East Asian weather during the post ENSO years of 1983/84 —Severe cold winter and hot dry summer, *J.Meteor.Soc.Japan*, Vol.64, 493–503.

[11] 黄荣辉、孙凤英,北半球夏季遥相关型的年际变化及其数值模拟(待发表)

[12] Wu, Y.F.(吴仪芳) and Huang, R.H.(黄荣辉), 1988, A possible approach to increasing the accuracy of long-range weather forecast, *Annual Report of Institute of Atmospheric Physics, Academia Sinica*, Vol.7, 138–143.

[13] Hoskins, B.J.and D.J.Karoly, 1981, The steady linear response of a spherical atmosphere to thermal and orographic forcing, *J.Atmos.Sci.*, Vol.38, 1179–1196.

[14] Huang, R.H.(黄荣辉) and K.Gambo, 1982, The response of a hemispheric multi-level model atmosphere to forcing by topography and stationary heat sources, Part Ⅰ, Ⅱ, *J.Meteor.Soc.Japan*, Vol.60, 78–108.

[15] 黄荣辉, 岸保勘三郎, 1983, 关于冬季北半球定常行星波传播另一波导的研究, 中国科学B辑, 第10期, 940–950.

[16] 黄荣辉, 1984, 球面大气中行星波的波作用守恒方程及用波作用通量所表征的定常行星波传播波导, 中国科学B辑, 第8期, 766–775.

[17] Huang, R.H.(黄荣辉), 1984, The characteristics of the forced stationary planetary wave propagations in summer Northern Hemisphere, *Adv.Atmos.Sci.*, Vol.1, 85–94.

[18] Huang, R.H.(黄荣辉), 1986, The physical mechanism of the three-dimensional teleconnection in the summer circulation and application in the long-range weather forecasting, First WMO Conference on Long-Range Forecasting, Sep.29–Oct.3, Sofia.

[19] Moura, A.D.and T.Shukla, 1981, On the dynamics of droughts in northeast Brazil: Observations, theory and numerical experiments with a general circulation model, *J.Atmos.Sci.*, Vol.38, 2653–2675.

[20] Keshavamurty, R.N., 1982, Response of the atmosphere to sea surface temperature anomalies over the equatorial Pacific and teleconnections of the Southern Oscillation, *J.Atmos.Sci.*, Vol, 39, 1241–1259.

[21] Shukla, J.and J.M.Wallace, 1983, Numerical simulation of the atmospheric response to equatorial sea surface temperature anomalies, *J.Atmos.Sci.*, Vol.40, 1613–1630.

[22] Tokioka, T.Yamazaki, K.and M.Chiba, 1985, Atmospheric response to the sea surface temperature anomalies observed in early summer of 1983, *J.Meteor.Soc.Japan*, Vol.63, 565–588.

[23] 黄荣辉, 1985, 冬季低纬度热源异常在北半球对流层大气环流异常中的作用, 气象学报, Vol.43, 411–423.

[24] Gambo, K., Lu, L(卢里), and Li, W.J.(李维京), 1987, Numerical simulation of Eurasian pattern teleconnection in the atmospheric circulation during the Northern Hemisphere winter, *Adv.Atmos. Sci.*, Vol.4, 385–394.

[25] Huang, R.H.(黄荣辉) and K.Gambo, 1983, The response of a hemispheric multi-level model atmosphere to forcing by topography and stationary heat sources in summer, *J.Meteor.Soc.Japan*, Vol. 61, 495–509.

[26] Huang, R.H.(黄荣辉) and Li, W.J.(李维京), 1988, Anomaly of the Subtropical High and its association with the propagations of the summer quasi-stationary planetary waves over East Asia, Summer School on Large-scale Dynamics of the Atmosphere, 5–20 August, 1988.

[27] Huang, R.H.(黄荣辉) and Lu, L.(卢里), 1989, Numerical simulation of the relationship between the anomaly of the Subtropical High over East Asia and the convective activities in the western tropical Pacific, *Adv.Atmos.Sci.*, Vol. 6, 202–214.

[28] 叶笃正、高由禧, 1979, 青藏高原气象学, 科学出版社.

[29] Nitta, T., 1983, Observational study of heat sources over the eastern Tibetan Plateau during the summer monsoon, *J.Meteor.Soc.Japan*, Vol.61, 590–605.

[30] Luo, H.(罗会邦) and M.Yanai, 1980, The large-scale circulation and heat sources over the Tibetan Plateau and surrounding areas during the early summer of 1979, *Mon.Wea.Rev.* Vol.108, 1840–1853.

[31] Huang, R.H.(黄荣辉), 1985, The numerical simulation of the three-dimensional teleconnections in the summer circulation over the Northern Hemisphere, *Adv.Atmos.Sci.*, Vol.2, 81–92.

The East Asia/Pacific Pattern Teleconnection of Summer Circulation and Climate Anomaly in East Asia[*]

Huang Ronghui(黄荣辉)

Institute of Atmospheric Physics, Academia Sinca, Beijing

ABSTRACT

In this paper, many observations show that the thermal states including the SST, the convective activities in the western Pacific warm pool largely influence the interannual and intraseasonal variations of summer circulation and the climate anomalies in East Asia. Moreover, it is pointed out that there is a teleconnection pattern of summer circulation anomalies in the Northern Hemisphere, the so-called East Asia / Pacific pattern.

The cause of the teleconnection pattern is studied by using the theory of quasi-stationary planetary wave propagation, and it may be due to the propagation of quasi-stationary planetary waves forced by heat source around the Philippines. Moreover, this pattern is well simulated by using a quasi-geostrophic, linear, spherical model and the IAP-GCM, respectively.

I. INTRODUCTION

East Asia is a strong monsoon region. The summer monsoon in East Asia is influenced by not only the Indian monsoon, but also the western Pacific subtropical high. Because of the large interannual and intraseasonal variations of the western Pacific subtropical high, the summer climate anomaly is large in East Asia. The summer droughts and floods are frequently caused there, especially in the area from the Changjiang River valley to the Huanghe River valley.

Huang (1985) has pointed out that the anomaly of heat source over the Tibetan Plateau influences the anomaly of the summer monsoon. As is well-known, the western Pacific warm pool is a region of maximum SST in the world's sea. Many observations show that the thermal states including the SST, the sea level and the convective activities in the warm pool largely influence the summer circulation and climate anomalies in East Asia. In this paper, the relationship between the circulation and climate anomalies in East Asia and the thermal states and convective activities in the warm pool is analysed by using the observed data. Moreover, it is pointed out that there is a teleconnection pattern of summer circulation anomalies in the Northern Hemisphere, the so-called East Asia / Pacific pattern. Besides, the physical mechanism of the teleconnection pattern is studied theoretically and numerically.

---

[*] 本文原载于: Acta Meteorological Sinica, Vol.6, No.1, 25-37, 1992.

II. RELATIONSHIP BETWEEN THE SUMMER CIRCULATION AND CLIMATE ANOMALIES IN EAST ASIA AND THE THERMAL STATES AND CONVECTIVE ACTIVITIES IN THE WESTERN PACIFIC WARM POOL

The western Pacific warm pool is an accumulated region of very warm water. It has the highest SST and the most intense atmospheric convection. Its impacts on the summer circulation and climate anomalies in East Asia are large. In order to make the impacts clear, the relationship between the summer circulation and climate anomalies in East Asia and the thermal states and convective activities in the warm pool is analysed by using the observed data including SST anomalies, high cloud amounts, 500hPa height fields and precipitations for 12 summers from 1978 to 1989. The analysed results are shown in Fig.1.

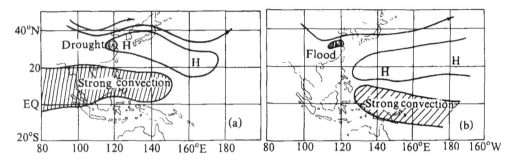

Fig.1. Relationship between the SST in the western tropical Pacific, the convective activities around the Philippines, the western Pacific subtropical high, the precipitation in East Asia. (a) The case of warmer SST in the western Pacific warm pool. (b) The case of colder SST in the western Pacific warm pool.

Fig.1 indicates the relationship between the SST and convective activities in the western Pacific warm pool, the subtropical high and the precipitation anomalies in East Asia, and Figs. 1a and 1b are the cases of the warmer and colder SST in the warm pool, respectively. The figure shows that when the SST in the warm pool is above normal, the warm water is accumulated in the western tropical and equatorial Pacific, and a cold tongue extends westward from the Peru coast along the equatorial Pacific, generally in the post ENSO years, such as in the summers of 1978, 1981, 1984, 1985, 1988, etc., the convective activities are intensified from the Indo-China Peninsula to the area around the Philippines, the western Pacific subtropical high may be unusually northward, the summer rainfall may be below normal in the Changjiang River valley and the Huaihe River valley of China and south Japan, and droughts may be caused there. On the contrary, when the SST in the warm pool is below normal, the warm water extends eastward from the warm pool along the western equatorial Pacific, generally in the developing stage of the ENSO events, such as in the summers of 1980, 1982, 1983 and 1987, the convective activities are weak around the Philippines, but they are intensified around the dateline, the western Pacific subtropical high shifts southward, the summer rainfall may be above normal in the Changjiang River valley and the Huaihe River valley of China and south Japan, and floods may be caused there, while it may be below normal in the Huanghe River valley and north China.

The warm pool may influence not only the circulation and climate anomalies in East Asia, but also the circulation anomalies in North Pacific and North America. The teleconnection

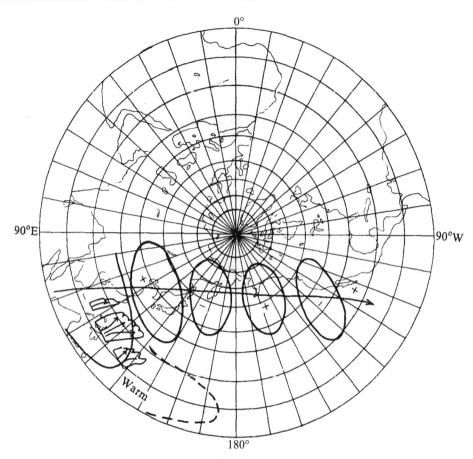

Fig.2. Sketch map of the summer circulation anomaly distribution over the Northern Hemisphere after the enhancement of the convective activities around the Philippines.

patterns of summer circulation in the warmer and colder SST in the warm pool are also analysed with the daily data, respectively.

Fig.2 explains the convective activities intensified around the Philippines: a negative anomaly of 500hPa height field is located over the South China Sea and the Philippines, and a positive anomaly is located over north China, i. e., the subtropical high shifted northward, and another negative and positive anomalies are located over the Okhotsk Sea and Alaska, respectively. Moreover, negative and positive anomalies may be found over the northwestern and southwestern part of the U. S. A. respectively. The time evolution of the daily 500hPa height anomalies along the characteristic points (whose correlation coefficients are maximal) during the summers of 1978, 1981, 1984, 1985 and 1988 is analysed. It shows that the propagating period of 500hPa height anomalies from the area around the Philippines to the western coast of North America is about one month. This anomaly distribution pattern is analogous to the propagation of planetary-wave train (see Huang and Li, 1987). This teleconnection pattern may be called the East Asia / Pacific pattern. Nitta (1987) also got the same results from the high-cloud amount anomalies.

III. RELATIONSHIP BETWEEN THE INTRASEASONAL VARIATION OF THE SUBTROPICAL HIGH AND THE CONVECTIVE ACTIVITIES AROUND THE PHILIPPINES

Generally, the rain band of summer monsoon rainfall is located in the south of the Changjiang River valley in May and the first 10 days of June, and it moves abruptly northward and is located in the north to the Changjiang River and the Huaihe River valley in mid-June. This is the beginning of the Meiyu season. Obviously, the abrupt change of the rain band is associated with the abrupt northward shift of the western Pacific subtropical high. In later July, the rain zone moves abruptly northward again and reaches north China. This means the end of the Meiyu season. This is also associated with the abrupt northward shift of the subtropical high. Thus, the early and later nothward shift of the subtropical high influences largely the summer rainfall in east China.

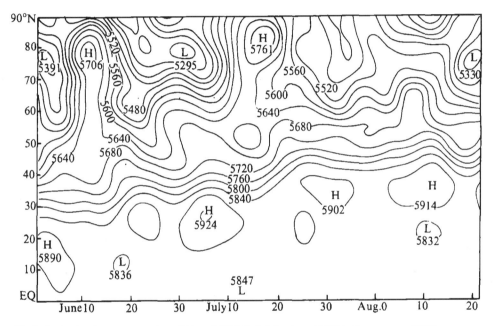

Fig.3. Latitude-time cross-section of filtered 500 hPa height along 135°E in the summer of 1985.

In order to investigate the causes of the abrupt northward shift of the western Pacific subtropical high, the intraseasonal variation of it has been analysed by using the filtered daily 500hPa data for 12 summers from 1978 to 1989. In order to obtain the filtered data, a low-pass filter is used to eliminate high-frequency oscillations with the period less than 10 days. It may be discovered that the intraseasonal variation of the western Pacific subtropical high in the drought summers is not same as that in the flood summer in the Changjiang River valley and the Huaihe River valley. In the drought summer, the abrupt northward shift of the subtropical high is obvious, while it is not obvious in the flood summer. Fig.3 is the latitude-time cross-section of filtered 500hPa height along 135°E in the summer of 1985. It can be seen From Fig.3 that the subtropical high shifted abruptly northward in 18 June 1985, the ridge of it shifted from 15°N to 25°N. This made the monsoon rain band move to the Changjiang River valley and

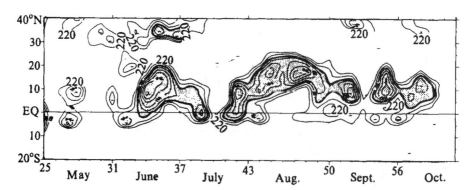

Fig.4. Latitude-time cross-section of 5-day mean OLR, the shaded area indicates OLR below 200 W / m².

Huaihe River valley. However, the subtropical high shifted abruptly northward again in 18 July 1985 and the ridge of it shifted to 35°N. This resulted in the drought during July and August 1985 in the Changjiang River valley and the Huaihe River valley.

The abrupt northward shift of the western Pacific subtropical high is closely associated with the convective activities around the Philippines. Fig.4 is the latitude-time cross-section of 5-day mean OLR. From Fig.4, we can see that the convective activities intensified around the Philippines from early June to later June, and intensified again around the Philippines from mid-July to later August. Therefore, there is a close relation between the anomalous northward shift of the subtropical high and the intensified convective activities around the Philippines. Moreover, the appearance of the intensified convective activities around the Philippines seems to be about 10 days earlier than the northward shift of the subtropical high.

However, the intraseasonal variation of the subtropical high is not obvious and the abrupt northward shift of it is not obvious either in the summers of weak convective activities around the Philippines. In the summers, the position of the subtropical high shifts southward and the rain band of the summer monsoon may be maintained in the Changjiang River valley and Huaihe River valley for long time, and floods are frequently caused there.

IV. THE PROPAGATING CHARACTERISTICS OF QUASI-STATIONARY PLANETARY WAVES FORCED BY HEAT SOURCE AROUND THE PHILIPPINES IN SUMMER

The above-mentioned relation between the unusual northward shift of the subtropical high and the intensified convective activities around the Philippines is by no means accidental, but inevitable. It results from the propagation of quasi-stationary planetary waves forced by the heat source around the Philippines.

Huang (1984) has discussed the propagating laws of quasi-stationary planetary waves in the summer basic flow, and pointed out that during the Northern Hemisphere summer, the quasi-stationary planetary waves can quasi-horizontally propagate from the subtropics to middle and high latitudes. As a consequence, we may discuss the propagations of planetary waves forced at the subtropics in a sphere using a two-dimensional vorticity equation and thermodynamical equation.

The vorticity equation in spherical coordinates is as follows:

$$\left(\frac{\partial}{\partial t}+\hat{\Omega}\right)\left\{\frac{1}{f}\left[\frac{\sin\varphi}{\cos\varphi}\frac{\partial\Phi'}{a\partial\varphi}\left(\frac{\cos\varphi}{\sin\varphi}\frac{\partial\Phi'}{a\partial\varphi}\right)+\frac{1}{a^2\cos^2\varphi}\frac{\partial^2\Phi'}{\partial\lambda^2}\right]\right\}+\frac{1}{a}\frac{\partial\bar{q}}{\partial\varphi}\times\frac{1}{f a\cos\varphi}\frac{\partial\Phi'}{\partial\lambda}=0, \quad (1)$$

where $\bar{q}$ is the vorticity of the basic flow, and its gradient in spherical coordinates may be expressed as

$$\frac{\partial\bar{q}}{a\partial\varphi}=\frac{1}{a}\left[2(\Omega_0+\hat{\Omega})-\frac{\partial^2\hat{\Omega}}{\partial\varphi^2}+3\text{tg}\varphi\frac{\partial\hat{\Omega}}{\partial\varphi}\right]\cos\varphi,$$

Ω_0 is the rotating angular velocity of the earth and $\hat{\Omega}$ is the angular velocity of the basic flow. Introducing the streamfunction and using Mercator projection

$$\begin{cases} dx = ad\lambda, \\ dy = ad[\tanh^{-1}(\sin\varphi)], \end{cases} \quad (2)$$

the relation between the zonal wavenumber k in x, y coordinate system and the wavenumber $\tilde{k}$ in λ, φ coordinate system can be

$$k=\frac{\tilde{k}}{a},$$

and, Eq. (1) can be expressed as

$$\left(\frac{\partial}{\partial t}+a\hat{\Omega}\frac{\partial}{\partial x}\right)\left[\frac{1}{\cos^2\varphi}\left(\frac{\partial^2\Psi}{\partial x^2}+\frac{\partial^2\Psi}{\partial y^2}\right)\right]+\frac{1}{\cos^2\varphi}\bar{q}_y\times\frac{\partial\Psi'}{\partial x}=0, \quad (3)$$

where

$$\bar{q}_y=\frac{\partial\bar{q}}{a\partial\varphi}\cos\varphi.$$

Because the problem investigated in this paper is the propagations of quasi-stationary planetary waves, and the amplitude of wave is the slowly varying function of x, y, t, we can use WKBJ method to solve Eq. (3).

The dispersion of planetary waves can be obtained

$$\hat{\omega}=a\hat{\Omega}k-\frac{\bar{q}_y k}{k^2+m^2}. \quad (4)$$

The components of group velocity of planetary wave in the x and y directions C_{gx} and C_{gy} can be obtained from Eq. (4). For quasi-stationary planetary waves, they are

$$\begin{aligned} C_{gx} &= \frac{\partial\hat{\Omega}}{\partial k}=\frac{2\bar{q}_y k^2}{(k^2+m^2)^2}, \\ C_{gy} &= \frac{\partial\hat{\omega}}{\partial m}=\frac{2\bar{q}_y km}{(k^2+m^2)^2}. \end{aligned} \quad (5)$$

From Eq. (4), we have the following relation

$$(k^2 + m^2) = \frac{\overline{q}_y}{a\hat{\Omega}} = K_s^2. \tag{6}$$

Using the distribution of the basic flow and Eq. (6), the distribution of K_s may be obtained.

Defining the angle between propagating ray and X direction in the horizontal as θ, we have

$$\mathrm{tg}\theta = \frac{C_{gy}}{C_{gx}} = \frac{m}{k}. \tag{7}$$

Substituting (6) into (7), we may obtain the equation of propagating ray path of waves:

$$\mathrm{tg}\theta = \begin{cases} \sqrt{(K_s/k)^2 - 1}, & \text{before turning point} \\ -\sqrt{(K_s/k)^2 - 1}. & \text{after turning point} \end{cases} \tag{8}$$

From Eq. (8), we may see that when $K_s > k$, the waves can propagate poleward; when $K_s = k$, there is a turning point in propagating ray path. Transforming x, y coordinate system into the spherical coordinate system, the variation of the propagating ray of waves is as follows:

$$\frac{d\varphi}{d\lambda} = \begin{cases} \cos\varphi\left[\left(\frac{aK_s}{k}\right)^2 - 1\right]^{1/2}, & \text{before turning point} \\ -\cos\varphi\left[\left(\frac{aK_s}{k}\right)^2 - 1\right]^{1/2} & \text{after turning point} \end{cases} \tag{9}$$

From Eq. (9) we may obtain the propagating ray path of quasi-stationary planetary waves in summer. Fig.5 is the propagating ray path of quasi-stationary planetary waves forced by a forcing source around the Philippines in a realistic summer current. We may see that the planetary-scale disturbance for zonal wavenumbers 1—3 propagates toward the Atlantic Ocean, while the long wave-scale disturbance for zonal wavenumbers about 4—6 propagates toward North America and Rocky Mountains.

We may also estimate the propagating period from the area around the Philippines using the group velocity. For the disturbance of wavenumber 1, the propagating period from the area around the Philippines to the turning point of its propagating ray path is about 20 days; for the disturbance of wavenumber 2, this period is about 10 days. Besides, velocity of propagation from the subtropics to middle latitudes is slower than that from middle latitudes to the vicinity of the pole.

V. NUMERICAL SIMULATION OF THE INFLUENCE OF SST ANOMALY IN WESTERN TROPICAL PACIFIC ON THE SUMMER CIRCULATION ANOMALY OVER THE NORTHERN HEMISPHERE

In Section III, the close relation between the northward shift of the subtropical high over East Asia and the convective activities around the Philippines has been shown. According to the results investigated by Kurihara (1986) and Nitta (1986), the convective activities around the Philippines are closely associated with the SST in the western tropical Pacific. When the SST in the western tropical Pacific is above normal, the convective activities are strong around the Philippines and the South China Sea. In order to investigate the influence of SST anomaly in the western tropical Pacific on the atmospheric circulation anomaly during the Northern Hemisphere summer, we use the GCM of Institute of Atmospheric Physics, designed by Zeng et al. (1986). This is a two-level global circulation model, in which the finite difference scheme of calculating the dynamics is unique, but the radiation heating, sensible and latent heat transfer from the sea and land surface are similar to the two-level GCM of OSU (see Chan et al., 1982). The distribution of climatological mean SST used in this investigation is the same as Shukla and Wallace (1983).

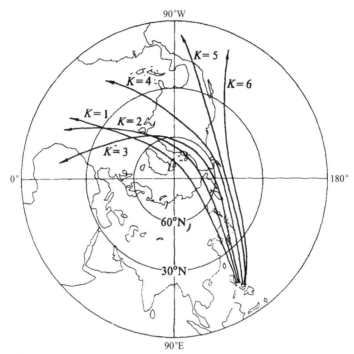

Fig.5. Propagating ray path of quasi-stationary planetary waves forced by a forcing source around the Philippines.

In this model, an idealized distribution of SST anomaly shown in Fig. 6 is used. We may see that an average SST anomaly is about 1°C. The summer SST anomaly is added to the climatological distribution of SST in the globe, and the distribution of anomaly SST in the Pacific can be obtained. The climatological wind field and height field are used as initial wind field and height field in the model, respectively. Integrations of this model, then, are performed for one month from the initial wind field and height field for July. Moreover, we subtract the height values at isobaric surfaces and the precipitations in the case obtained from the climatological mean distribution of SST from those in the anomaly SST case. Thus, the height anomalies at isobaric surfaces and the precipitations caused by the SST anomaly in the western tropical Pacific are obtained.

First, let us see the propagation of 500hPa disturbance height anomaly caused by the SST anomaly in the western tropical Pacific. Fig.7 shows the result of the 6th day integrated by the IAP-GCM. We can see that a positive anomaly of disturbance height is found over the western tropical Pacific, and a negative anomaly of disturbance height over Indo-China Peninsula. By the 12th day, as shown in Fig.8, a positive anomaly is found over East Asia, and another positive anomaly over Aleutian region. While over the western coast of North America, a negative anomaly can be found. We can clearly see the propagation of the anomaly pattern. This propagation of 500 hPa disturbance height anomaly is due to the propagations of planetary-wave trains forced by heat source over the western tropical Pacific.

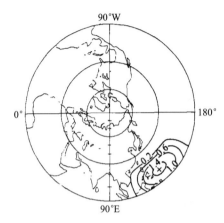

Fig. 6. An idealized distribution of SST anomaly in the western tropical Pacific.

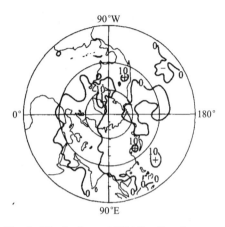

Fig. 7. Distribution of 500 hPa disturbance height anomaly at the 6th model day.

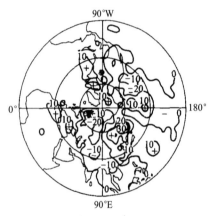

Fig. 8. As in Fig.7, but for the 12th-day result.

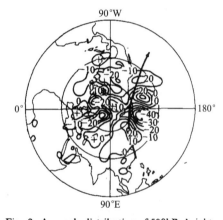

Fig. 9. Anomaly distribution of 500hPa height field in July due to the SST anomaly in the western tropical Pacific averaged for 30 days of the model integration of the IAP-GCM (units in m).

Fig.9 shows the distribution of 500 hPa disturbance height anomaly averaged for 30 days of the model integrations.

We can find that when the SST anomaly appears in the western tropical Pacific, a negative anomaly of disturbance height field is found over Indo-China Peninsula and south China, while a positive anomaly locates over the north China and Siberia. Besides, another negative anomaly is located over the Okhotsk Sea and Kamchaka, and another positive anomaly is found over Alaska, while a negative anomaly is located over the west coast of the United States of America.

Fig.10 is the anomaly distribution of precipitation in July due to the SST anomaly in the western tropical Pacific, averaged for 30 days of the model integration. Obviously, a center of positive precipitation anomaly is located around the South China Sea, the Philippines and Indonesia. Maximum precipitation anomaly may reach 15mm d^{-1}. The mean precipitation anomaly in the region is about 8 mm d^{-1} and this is in good agreement with the observed results obtained by Maruyama et al. (1986). Moreover, we may find that a negative anomaly of precipitation is located in Japan and the Huaihe River valley. Besides, other positive anomalies are located in the Aleutian region, the west coast of North America, and the west coast of Mexico, respectively. This computed pattern is very analogous to the teleconnection pattern obtained by Nitta (1987) from the satellite data of OLR. Tokioka et al. (1985) have simulated the atmospheric circulation anomaly pattern due to the SST anomaly observed in early summer of 1983 with the MRI-GCM. Our results seem to be in better agreement with the observed fact than that obtained by Tokioka.

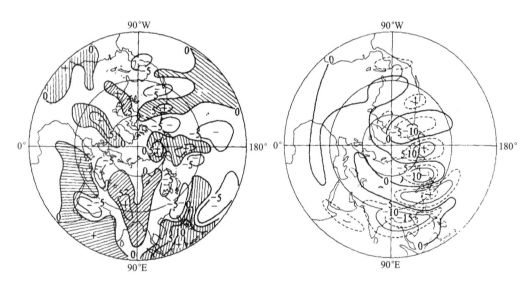

Fig. 10. Anomaly distribution of precipitation in July due to the SST anomaly in the western tropical Pacific, averaged for 30 days of the model integration of IAP-GCM(units in mm d^{-1}).

Fig. 11. Anomaly distribution of the summer 500 hPa disturbance height field due to a heating anomaly around the Philippines computed by a linear model.

VI. THE STATIONARY MODEL COMPUTATION OF THE INFLUENCE OF HEAT SOURCE ANOMALY OVER THE PHILIPPINES ON THE ATMOSPHERIC CIRCULATION OVER EAST ASIA

In order to compare with the results simulated by using GCM, a quasi-geostrophic, 34-level spherical coordinate model with Rayleigh friction, Newtonian cooling effect and horizontal thermal diffusivity is utilized to compute the influence of heat source anomaly over the Philippines on the atmospheric circulation over East Asia. The model and parameters are the same as those in Huang's (1984) paper.

As mentioned in Section II, when anomaly convective activities are caused around the South China Sea and the Philippines, the heat source anomaly will be caused around there. According to the result investigated by Maruyama et al. (1986), maximum values of the cloud anomaly may attain ± 2.5. This corresponds to about 9.0 mm d^{-1} of precipitation anomaly. Thus, we assume that an idealized heat source anomaly at 400 hPa is located around the South China Sea and the Philippines. Its horizontal distribution is given as

$$\Delta \hat{H}_0(\lambda, \varphi) = \begin{cases} \Delta \hat{H}_0 \left(\sin\frac{\pi(\varphi - \varphi_1)}{\varphi_2 - \varphi_1} \sin\frac{\pi(\lambda - \lambda_1)}{\lambda_2 - \lambda_1} \right)^2, & \varphi_1 < \varphi < \varphi_2, \lambda_1 < \lambda < \lambda_2 \\ 0, & \text{otherwise} \end{cases} \quad (10)$$

where the anomaly of heating rate $\Delta \hat{H}_0 / C_p = 5.0$K d^{-1}, and the averaged anomaly of heating rate is about 2.5 K d^{-1}. Moreover, we assume that the vertical distribution of this anomaly of heat source is

$$\Delta \hat{H}_0(\lambda, \varphi, p) = \Delta \hat{H}_0(\lambda, \varphi) \exp\left[-\left(\frac{p - \bar{p}}{d}\right)^2 \right], \quad (11)$$

where $d = 300$ hPa, $\bar{p} = 400$ hPa. Thus, the vertical distribution of amplitude and phase of quasi-stationary planetary waves and quasi-stationary disturbance pattern at isobaric surfaces can be computed.

An anomaly distribution of heat source centered at 125°E, 15°N is taken, with $\lambda_1 = 110$°E, $\lambda_2 = 140$°E, and $\varphi_1 = 0$°N, $\varphi_2 = 30$°N. We add the idealized anomaly of heat source to the summer climatological distribution of heat source over the Northern Hemisphere $\overline{H}_0(\lambda, \varphi, p)$:

$$H_0(\lambda, \varphi, p) = \overline{H}_0(\lambda, \varphi, p) + \Delta \hat{H}_0(\lambda, \varphi, p).$$

We can obtain the distribution of anomaly heat source around the Philippines. Moreover, in order to obtain the anomaly pattern of atmospheric circulation due to the anomaly of heat source, we subtract the height values of stationary disturbance at isobaric sources in the normal case from those in the anomaly case of heat source.

Fig. 11 shows the anomaly distribution of 500 hPa disturbance height field caused by the heating anomaly around the Philippines. From Fig.11, we can find that when heat sources or convective activities intensify around the Philippines, a negative anomaly of disturbance height field is found over South Asia including south China, while a positive anomaly of disturbance height field is found over the Changjiang River valley, the Huaihe River valley and Japan. In midsummer, the subtropical high is located over there. Thus, we may consider that in the summer circulation the subtropical high over East Asia is related to the heat source over the South China Sea and the Philippines, especially, to the convective activities around the Philippines. We may suggest that the anomaly of subtropical high over East Asia may be caused by the convective activities around the Philippines. Moreover, another negative anomaly is found around Hokkaido of Japan, and an anticyclone will be intensified over Aleutian region, a negative anomaly will appear over Alaska. This computed teleconnection pattern of circulation anomaly is in good agreement, not only with the observed facts, but also with the theoretical results. The computed distribution of wave train is in good agreement with the propagation ray path obtained from the theoretical analyses. This may explain that in summer, due to the anomaly of heat source around the Philippines, a teleconnection pattern of circulation anomalies may be caused from South Asia through East Asia to North America. This is due to the propagation of quasi-stationary planetary waves.

VII. CONCLUSIONS AND DISCUSSIONS

In this paper, many observations show that the thermal states including the SST, the sea level height and the convective activities in the western Pacific warm pool largely influence summer circulation and climate anomalies. Moreover, it is pointed out that there is a teleconnection pattern of summer circulation anomalies in the Northern Hemisphere, the so-called East Asia / Pacific pattern.

In this paper, the relationship between the intraseasonal variation of subtropical high over East Asia and the convective activities around the South China Sea and the Philippines is analysed from latitude-time cross-section of OLR data. The analysed results show that there is a close relationship between the intraseasonal variation of subtropical high over East Asia and the convective activities around the Philippines. When convective activities are intensified around the South China Sea and the Philippines, the subtropical high will shift northward.

This relationship is also studied by using the theory of wave propagating in slowly varying media. It shows that the close relationship between the intraseasonal variation of subtropical high over East Asia and the convective activities around Philippines is due to the propagation of quasi-stationary planetary waves forced by heat source around the Philippines.

A quasi-geostrophic, linear, spherical model and the IAP-GCM are used to simulate this relationship, respectively. The results show that when the SST is warming around the western tropical Pacific or the Philippines, the convective activities are intensified around the Philippines. As a consequence, the heat source is stronger than the normal. Due to the propagation of quasi-stationary planetary waves caused around the Philippines, the subtropical high will shift northward over East Asia in midsummer. The computed results also show that when the anomaly of convective activities is caused around the Philippines, a teleconnection pattern of circulation anomalies will be caused over South Asia, East Asia and North America.

The SST anomaly used in this study is an idealized distribution. The circulation anomalies caused by a real SST anomaly in the summer of post ENSO year has also been investigated, the simulated results are the same as the above-mentioned results.

REFERENCES

Chan, S, J., Lingaas, J. W., Schlesinger, M. E., Mobley R. L. and Gates, W. L. (1982), A documentation of the OSU two-level atmospheric general circulation model, Climate Research Institute of Oregon State University, Report No. 35.

Huang Ronghui (1984), The characteristics of the forced stationary wave propagations in the summer Northern Hemisphere, *Adv. Atmos. Sci.*, **1**:85—94.

Huang Ronghui (1985), The numerical simulation of the three-dimensional teleconnections in the summer circulation over the Northern Hemisphere, *Adv. Atmos. Sci.*, **2**:81—92.

Huang Ronghui and Li Weijing (1987), Influence of the anomaly of heat source over the northwestern tropical Pacific for the subtropical high over East Asia, published in *Proceedings of International Conference on the General Circulation of East Asia*, April 10—15, 1987, Chengdu, China, pp.40—45.

Kurihara, K. and Kawahara, M. (1986), Extremes of East Asia weather during the post ENSO years of 83 / 84-severe cold winter and hot dry summer, *J. Meteor. Soc. Japan*, **64**:493—503.

Maruyama, T., Nitta, Ts and Tsuneoka, Y. (1986), Estimation of monthly rainfall from satellite-observed cloud amount in the tropical western Pacific, *J. Meteor. Soc, Japan*, **64**:149—155.

Nitta, Ts. (1986), Long-term variations of cloud amount in the western Pacific region, *J. Meteor. Soc. Japan*, **64**: 373—390.

Nitta, Ts. (1987), Convective activities in the tropical western Pacific and their impact on the Northern Hemisphere summer circulation, *J. Meteor. Soc. Japan*, **65**:373—390.

Tokioka, T., Yamazaki, K. and Chiba, M. (1985), Atmospheric response to the sea surface temperature anomalies observed in early summer of 1983; a numerical experiment, *J. Meteor. Soc. Japan*, **63**:565—588.

Zeng Qingcun et al. (1986), A global grid point general circulation model, short- and medium-range numerical weather predicition, collection of papers presented at the WMO / TUGG NWP symposium, Tokyo, 4—8 August, 1986, pp.424—430.

Impacts of the Tropical Western Pacific on the East Asian Summer Monsoon*

Huang Ronghui and Sun Fengying

(Institute of Atmospheric Physics, Chinese Academy of Sciences)

Abstract

In this paper, the impacts of the convective activities in the western Pacific warm pool on the interannual and intraseasonal variations of the summer monsoon in East Asia are analyzed by using the observed data for 12 summers from 1978 to 1989. The analyzed results show that both interannual and intraseasonal variabilities of the East Asian summer monsoon are greatly influenced by the convective activities in the warm pool. Generally, the monsoon rainfall is below normal in East Asia and the abrupt change of the monsoon circulation is obvious in the summer of strong convective activities around the Philippines.

The impacts of the convective activities in the warm pool on the summer monsoon in East Asia and the East Asia/Pacific teleconnection pattern of summer circulation anomalies due to the convection are discussed by using the theory of planetary wave propagation and the numerical modelling by the IAP-GCM, respectively.

1. Introduction

East Asia is a part of the monsoon climate region in the world. There are many characteristic weather systems in different seasons, such as Mei-Yu in China (Baiu in Japan) in summer, tropical cyclones, persisting northwest and northeast winds and cold surges in winter. According to Tao (1985)'s suggestion, the main components of the East Asian summer monsoon system are: the monsoon trough in the South China Sea and the western Pacific, the cross-equatorial flow to the east of 100°E, the subtropical high in the western Pacific, the upper-level easterly flow, the Mei-Yu (or Baiu) frontal zones in East Asia and the mid-latitude disturbances. Therefore, the summer monsoon in East Asia is influenced not only by the Indian monsoon, but also by the western Pacific subtropical high. Because of large interannual and intraseasonal variations of the western Pacific subtropical high, the anomalies of summer monsoon rainfall and circulation are large in East Asia. Summer drought and floods frequently occurred there, especially in the area from the Yangtze River valley to south Japan.

Concerning the causes of the Asian summer monsoon, previously, many investigations emphasized the thermal effect of the Tibetan Plateau (See Ye and Gao, 1979, Nitta, 1983, Luo and Yanai, 1984, Huang, 1982). Huang (1984, 1985) pointed out that the heating anomaly over the Tibetan Plateau can cause the anomaly of the Asian summer monsoon. Recently, many observational facts have shown that the tropical western Pacific is a region of the highest SST in the world's sea surface. Thus, this region is known as "Warm pool". Because the warm pool has the highest SST, the air-sea interaction is very strong in the region. Moreover, due to the ascending branch of the Walker circulation over this area, the convergence of air and moisture leads to strong convective activities and heavy precipitation. According to the results studied by Cornejo-Garrido and Stone (1977), Hartmann *et al.* (1984), the heating of the atmosphere due to the convective activities over the warm pool also supplies the energy to drive the Walker circulation in the zonal direction.

The thermal states of the warm pool and the convective activities over the warm pool may play an important role in the variability of summer circulation over the Northern Hemisphere. Huang and Li (1987) have discussed the effect of a heat source due to the convective activities over the warm pool on the circulation anomalies over the Northern Hemisphere from the observed data and the theoretical analysis. They pointed out that there may be a teleconnection pattern of the summer circulation anomalies from the area around the Philippines to North America through East Asia and explained that the effect may be due to the propagations of quasi-stationary planetary waves forced by

ⓒ1992, Meteorological Society of Japan

---

* 本文原载于: Journal of the Meteorological Society of Japan, Vol.70, No.1, 343-356, 1992.

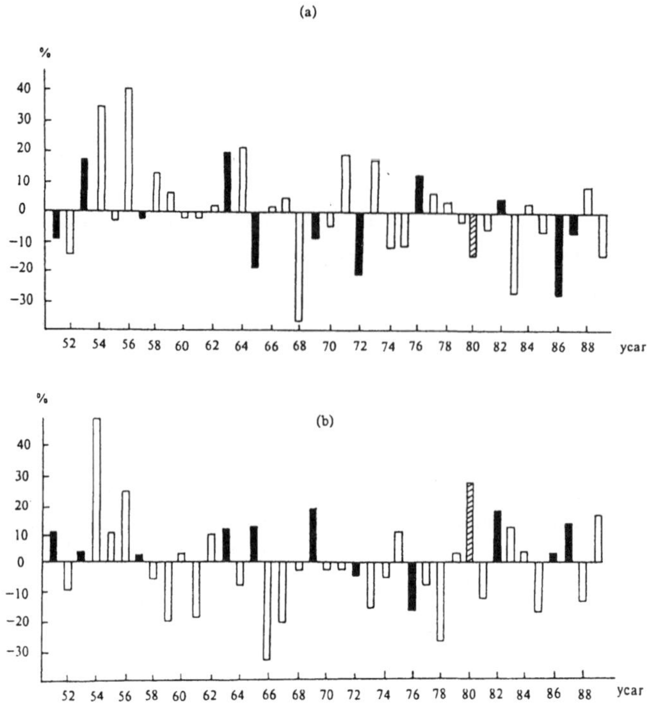

Fig. 1. Summer monsoon rainfall anomaly percentage in some regions of China during June–August. (a) in the Yellow River valley, (b) in the Yangtze River and the Huaihe River valley. The black shaded pillars indicate ENSO events. Because the SSTA in the equatorial eastern Pacific in 1980 was above normal, but it was not an ENSO year, 1980 also is marked by a shaded pillar.

the heat source over the warm pool. Nitta (1987) investigated the effect of the convective activities over the tropical western Pacific on the Northern Hemispheric circulation from the observed data of high-cloud amount by satellite and suggested the so-called Pacific-Japan oscillation. Kurihara and Kawahara (1986), Kurihara (1989) analyzed the relationship between the summer climate anomalies in Japan and the thermal states in the tropical western Pacific and pointed out that there is a good positive correlation between the summer temperature in Japan and the thermal states in the tropical western Pacific.

Although many studies have been made by meteorologists regarding the summer monsoon in East Asia, there remain many unsolved problems. Particularly, the physical mechanisms of interannual and intraseasonal variabilities of the East Asian summer monsoon should be studied further. Therefore, in this paper, the impacts of convective activities in the western Pacific warm pool on the East Asian monsoon are analyzed by using the observed data. Moreover, the physical mechanism of the impacts are simply discussed by using the theory of planetary wave propagation and the numerical modelling of the IAP-GCM.

2. Impacts of the western Pacific warm pool on the interannual variations of the East Asian summer monsoon

Because the western Pacific warm pool has the highest SST and the most intense convective activities, it is a strong heat source for the atmospheric general circulation, and the impacts of the heat source on the summer monsoon circulation and monsoon rainfall in East Asia may be large. We have analyzed the relationships between the summer monsoon circulation in East Asia and the thermal states of the warm pool and convective activities over the warm pool by using the observed data including the SST data for 1951–1989, based on COADS, OLR and High Cloud Amount data for 1978–1989, edited by the Long-range Forecast Division J.M.A., the ST anomaly data for 1972–1989 in the subsurface layer of the western Pacific warm pool along 137°E, collected by the Marine Department, J.M.A. and daily 500 hPa height fields. A data set of daily precipitations in summer for 1951–1989 at 336 stations in China is used to analyze the interannual variations of the summer precipitation in some regions of China.

Figure 1a is the rainfall anomaly percentage in summer in the Yellow River valley and north China during June–August, and Fig. 1b is that in the

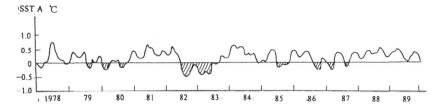

Fig. 2. Interannual variations of SST anomaly averaged for the area of 125°E–145°E, Eq.–15°N.

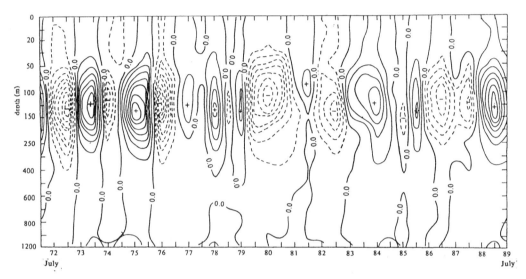

Fig. 3. Vertical-time section of sea water temperature anomaly averaged for the 137°E line from 2°N to 10°N. The contour interval is 0.5°C and dashed lines show negative values (After Yasunari, 1990).

Yangtze River and the Huaihe River valley. The Huaihe River is located between the Yellow River and the Yangtze River, as shown in Figs. 4 and 6. Because the Mei-Yu front is frequently maintained in the region of the Yangtze River and the Huaihe River valley in summer (*e.g.* Gao and Xu, 1962), the interannual variations of summer monsoon rainfall in the region are emphasized in this paper. It may be seen from Fig. 1b that in the summers of 1978, 1981, 1985, 1988, the summer monsoon rainfalls were below normal, and there were droughts in the Yangtze River and the Huaihe River valley, while in 1981, 1982, 1983, 1987, the summer rainfalls were above normal, and floods occurred there.

The summer rainfall anomalies are closely related to the thermal states of the western Pacific warm pool. In order to explain this relationship, the interannual variations of SST anomalies in some regions are analyzed from the COADS data set. Figure 2 is the interannual variation of SST anomaly averaged for the area of 125°E–145°E, Eq.–15°N. We found that in years when there were above- (below-) normal summer monsoon rainfall in the Yangtze River and the Huaihe River valley, such as in the summers of 1980, 1982, 1987, (1978, 1981, 1988) the SST anomalies of the western Pacific warm pool were negative (positive).

Although the western Pacific warm pool is characterized by a deep thermocline compared to the eastern Pacific, the heat content anomalies in the western Pacific warm pool are in close association with the SST (Yamagata, 1991). Moreover, the heat content anomalies are well related to the sea temperature (ST) anomalies in the subsurface layer of the depth about from 50 m to 300 m, as shown in Fig. 3 (after Yasunari, 1990). Obviously, in the period from 1978 to 1989 the sea temperatures in the subsurface layer of the warm pool in 1980, 1982, 1986, 1987 were below normal. In the summers of these years, the monsoon rainfalls were above normal in the Yangtze River and the Huaihe River valley, but the sea temperatures in 1978, 1988 were above normal and the summer monsoon rainfalls were below normal in the Yangtze River and the Huaihe River valley. Moreover, comparing Fig. 2 with Fig. 3, it may be seen that the sea temperature anomaly in the subsurface layer is larger and more obvious than the SST anomaly in the western Pacific warm pool. Therefore, the ST anomaly in the subsurface layer of the warm pool may be a better signal of a strong or weak summer monsoon in East Asia.

In order to show the relationship between the

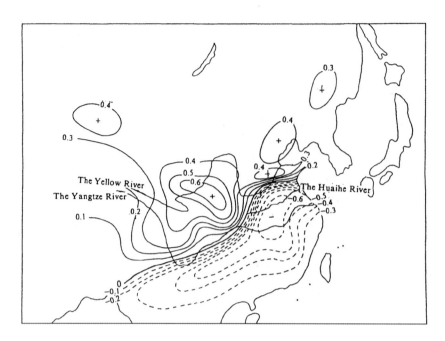

Fig. 4. Simultaneous correlation between the July monsoon precipitation in China and the July Sea temperature averaged for 137°E line from 2°N to 10°N in the 150 m depth of the western Pacific warm pool.

summer monsoon rainfall in East Asia and the sea temperature in the subsurface layer of the western Pacific warm pool, the monthly precipitation anomalies in China and the sea temperature anomalies in the subsurface layer of the warm pool in July for 18 months from 1972 to 1989 are used to calculate the one-point correlation between them. Figure 4 is the distribution of one-point correlation coefficients between the monsoon rainfall in China and the ST in the 150 m depth averaged form 2°N to 10°N along 137°E in July. The negative correlation coefficients in the Yangtze River and the Huaihe River valley are statistically significant at the 95 % confidence level, and the positive correlation coefficients are also statistically significant in the Yellow River valley. Figure 4 shows that there is a remarkable negative correlation between the summer monsoon rainfall in the Yangtze River valley and the ST in the subsurface layer of the warm pool and also a positive correlation between the summer monsoon rainfall in the Yellow River valley and the ST in the subsurface layer of the warm pool. Therefore, when the ST in the subsurface layer of the western Pacific warm pool is below normal, there is an above normal summer monsoon rainfall in the Yangtze River valley and a below normal rainfall in the Yellow River valley.

The thermal states in the western Pacific warm pool greatly affect the convective activities there. The active area of convection in a warm year of the warm pool is different from that in a cold year. Fig-

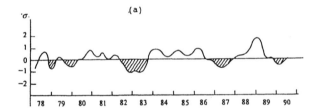

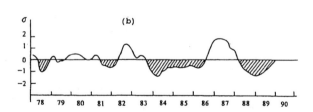

Fig. 5. Interannual variations of the normalized anomaly 3-month running mean high-cloud amount averaged for the area around the Philippines (110°E–140°E, 10°N–20°N) (a) and the equatorial area near the dateline (170°E–170°W, 5°N–5°S) (b), respectively.

ures 5a and 5b indicate interannual variations of the normalized anomalies of monthly mean high-cloud amount around the Philippines (110°E–140°E, 10°N–20°N) and over the equatorial area near the dateline, respectively. Comparing Fig. 5 with Figs. 2 and 3, we see that when the SST in the warm pool

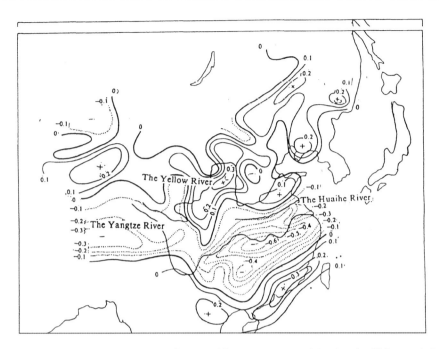

Fig. 6. Simultaneous correlation between the monthly monsoon precipitation in China and the monthly mean convective activities around the Philippines in summer.

is above normal, *i.e.* when warm sea water is accumulated in the western Pacific warm pool, such as in 1978, 1981, 1984, 1988, the convective activities are intensified around the Philippines. On the other hand, when the SST in the warm pool is below normal, *i.e.*, the warm sea water shifts eastward from the warm pool along the equatorial western Pacific, such as in 1982, 1986, 1987, the convective activities are weak around the Philippines, but may be intensified over the equatorial area near the dateline. However, there are some exceptions to the relationship. For example, the ST in the subsurface layer of the warm pool in the summer of 1985 was below normal, but the convective activities were intensified around the Philippines. On the contrary, the ST in the subsurface layer of the warm pool in the summer of 1983 was above normal, but the convective activities were weak, and it also was above normal in the summer of 1989, but the convective activities were normal.

Comparing Fig. 5 with Fig. 1, it is obvious that in the summers of strong convective activities around the Philippines, the summer monsoon rainfalls are below normal in the Yangtze River and the Huaihe River valley. On the contrary, in the summers of weak convective activities around the Philippines, the summer monsoon rainfalls are above normal in the Yangtze River and Huaihe River valley, while the summer rainfalls are below normal in the Yellow River valley. Figure 6 is the simultaneous correlation between the monthly precipitation anomalies in China and the monthly mean anomaly of high-cloud amount around the Philippines during June, July and August from the summer of 1978 to the summer of 1989. It can be seen that there is a high negative correlation in the Yangtze River and the Huaihe River valley. The correlation coefficients are statistically significant at the 95 % confidence level in the middle and lower reaches of the Yangtze River. This analyzed result is in good agreement with the result analyzed by Huang (1987) using the observed data of 1985 and 1987.

The western Pacific subtropical high is a component of the East Asian summer monsoon system. If its location shifted northward, a hot summer and drought may occur in the region from the Yangtze River valley, Korean Peninsula to south Japan. On the contrary, if its location shifted southward, a flood summer may occur there. The interannual variations of the subtropical high are closely related to the convective activities over the warm pool. Figures 7a and 7b show monthly mean 500 hPa height field over East Asia averaged for the summers of strong and weak convective activities around the Philippines, respectively. In years when there were strong convective activities around the Philippines, the location of the western Pacific subtropical high shifts northward, and its ridge line is located at 30°N in August. On the contrary, in the years of weak convective activities around the Philippines, the subtropical high shifts southward and its ridge line is located to the south of 30°N in August.

Figure 8 shows the relations between the thermal states of the warm pool, the atmospheric convec-

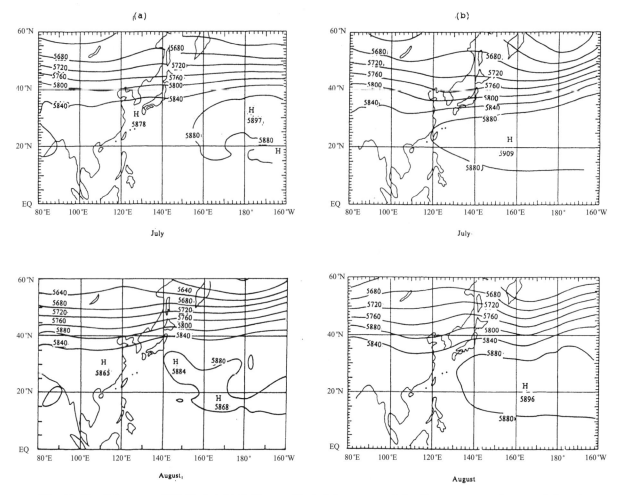

Fig. 7. Monthly mean 500 hPa height field over East Asia, averaged for the summers of strong (a) and weak (b) convective activities around the Philippines, respectively. Units are GPM.

tions over the warm pool, the western Pacific subtropical high and the East Asian summer monsoon rainfall. When the SST in the warm pool is above normal, i.e., the warm sea water is accumulated in the western Pacific warm pool, and when a cold tongue extends westward from the Peruvian coast along the equatorial eastern Pacific, the convective activities are intensified from the Indo-China Peninsula to the area around the Philippines, the western Pacific subtropical high may shift unusually northward, and the East Asian summer monsoon rainfall may be below normal (Fig. 8a). When the SST in the warm pool is below normal, i.e., the warm sea water extends eastward from the warm pool along the equatorial western Pacific, the convective activities are weak around the Philippines and are intensified over the equatorial central Pacific near the dateline, the western Pacific subtropical high may shift southward, and the East Asia summer monsoon rainfall may be above normal (Fig. 8b).

3. Impacts of the convective activities over the warm pool on the intraseasonal variation of the East Asian summer monsoon

Generally, the rain band of the East Asian summer monsoon rainfall is located to the south of the Yangtze River valley during May and in the first 10 days of June, and it moves abruptly northward and is located to the north of the Yangtze River valley, the Huaihe River valley and Japan in mid-June. This is the beginning of the Mei-Yu season in China or the Baiu season in Japan. Obviously, the abrupt movement of the rain band is closely associated with the abrupt northward shift of the western Pacific subtropical high. Yeh and Tao (1959) discovered the abrupt change of circulation over East Asia during June. This abrupt change of planetary-scale circulation will bring the onset of the East Asian summer monsoon. Later on, Krishnamuti and Ramanathan (1982), McBride (1987) also pointed out the abrupt change of the Indian summer monsoon and Australian summer monsoon. In late July, the rain band moves abruptly northward again and ar-

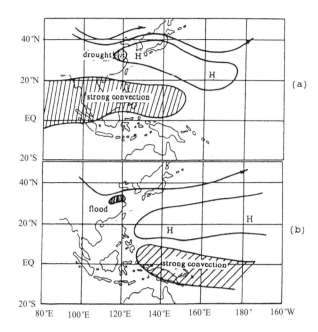

Fig. 8. Schematic map of relationship between the SST in the western Pacific warm pool, the convective activities around the Philippines, the western Pacific subtropical high, the summer monsoon rainfall in the Yangtze River valley and the Huaihe River valley. (a) in the warming case of the western Pacific warm pool. (b) in the cooling case of the western Pacific warm pool.

rives in north China, and the Mei-Yu season in the Yangtze river valley or the Baiu season in Japan ends. This is also associated with the abrupt northward shift of the subtropical high. Thus, the early and late northward shift of the subtropical high may greatly influence the summer monsoon rainfall in the Yangtze River and the Huaihe River valley.

Although the change of the East Asian summer circulation during June is abrupt in a climatic sense, it is different in different years, and in some years the change is not obvious. This may be seen from the intraseasonal variations of the low-pass filtered daily 500 hPa height field for 12 summers from 1978 to 1989. By latitude-time cross sections of low-pass filtered 500 hPa along 135°E, it is found that the intraseasonal variations of the East Asian summer monsoon circulation in the drought years are not the same as those in the flood years of the Yangtze River and the Hauaihe River valley. In the drought summers, such as in the summers of 1978, 1981, 1985 and 1988, the abrupt change of the East Asian summer monsoon circulation and the northward shift of the western Pacific subtropical high were very obvious, while they were not obvious in the flood summers, such as in the summers of 1980, 1982, 1983, 1986, 1987.

Figure 9 is a latitude-time cross section of the low-pass filtered 500 hPa height along 135°E during the summer of 1985. It can be seen that the western Pacific subtropical high shifted abruptly from 18°N to 28°N in mid-June, in which the monsoon rain band moved to the Huaihe River valley. The subtropical

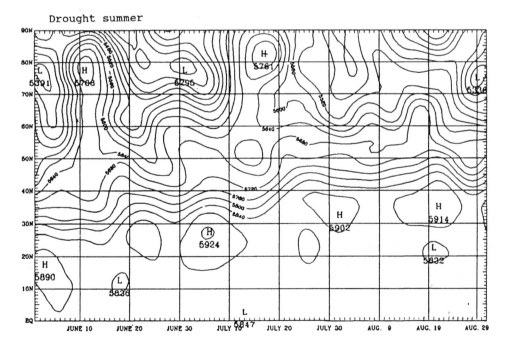

Fig. 9. Latitude-time cross section of low-pass filtered 500 hPa height along 135°E during the summer of 1985. Units are GPM.

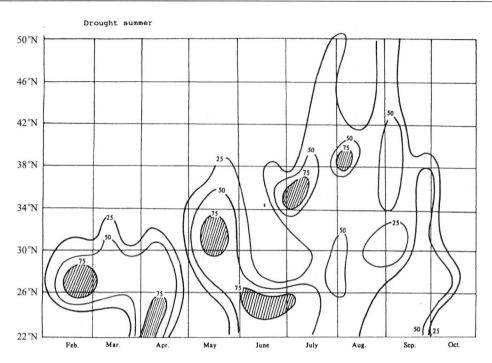

Fig. 10. Latitude-time cross section of 10-day precipitation along 115°E during the summer of 1985. Units are mm.

high shifted abruptly northward again in mid-July and its ridge shifted to 35°N, and the monsoon rain band moved to the Yellow River valley and north China. Figure 10 is a latitude-time cross section of 10-day's precipitation along 115°E during the summer of 1985. The monsoon rainfall was very weak and was 30–50 % less than the normal in the Yangtze River and the Huaihe River valley in the summer of 1985. Due to the anomalous northward movement of the rain band, a hard drought occurred in the Yangtze River and the Huaihe River valley in this summer. Similar cases also occurred in the summers of 1981 and 1988.

In the flood years of the Yangtze River and the Huaihe River valley, the abrupt change of the East Asian summer monsoon circulation and the abrupt northward shift of the western Pacific subtropical high were not remarkable. It can be seen from Fig. 11 that in 1987 the abrupt northward shift of the western Pacific subtropical high was not obvious. The ridge line of the subtropical high remained at about 20°N from early June to mid-July and it moved to 30°N in late July. This made the monsoon rain band stay in the Yangtze River and the Huaihe River valley (see Fig. 12). The precipitation anomaly percentage was +30 % and more in the Yangtze River and the Huaihe River valley, and a severe flood occurred in the region. The same cases also occurred in 1989, 1982, 1983 *etc.*.

What causes the difference between the intraseasonal variations of the western Pacific subtropical high in the drought summers and in the flood summers of the Yangtze River and the Huaihe River valley? This difference may be caused by the difference of the intraseasonal variations of the convective activities around the Philippines.

Kawahara and Hayashi (1987) showed that the interannual and intraseasonal variations are significant around the Philippines. Kurihara (1989) also pointed out that the standard deviations of the 10-day mean OLR are large in the eastern equatorial Indian Ocean and in the region from the South China Sea to the east to the Philippines along 15°N–20°N and the standard deviation is the largest around the north of the Philippines. We have analyzed the intraseasonal variations of the OLR around the Philippines and their effects on the intraseasonal variations of the OLR around the Philippines and their effects on the intraseasonal variations of the subtropical high. The analyzed results show that the intraseasonal variations of the convective activities were remarkable around the Philippines in the summers of strong convective activities around the Philippines, such as in the summers of 1978, 1981, 1985 and 1988 *etc.* Figure 13 is a latitude-time cross section of the 5-day mean OLR around the Philippines from May to October 1985. The convective activities were intensified around the Philippines and their intraseasonal variations were remarkable. The convective activities intensified from early June to late June and intensified again from mid-July to late August. Similar cases also occurred in the summers of 1981, 1988. Comparing Figs. 9 and 10 with Fig. 13 it is very clear that there is a close relationship

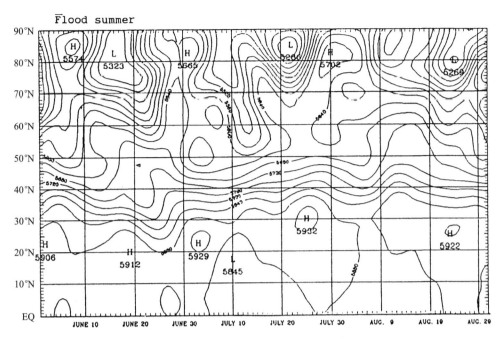

Fig. 11. Same as in Fig. 9 but for the summer of 1987.

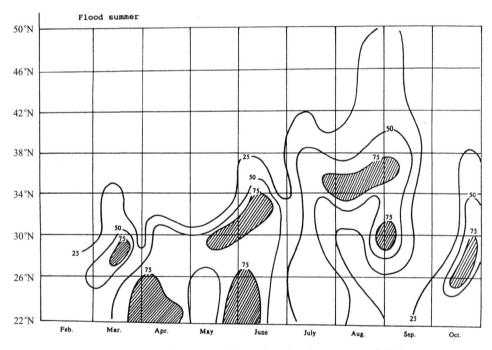

Fig. 12. Same as in Fig. 10 but for the summer of 1987.

between the intraseasonal variations of the East Asian summer monsoon circulation and the intraseasonal variations of the convective activities around the Philippines, and the abrupt northward shift of the subtropical high is closely associated with the intensified convective activities around the Philippines. The appearance of the intensified convective activities around the Philippines seems to be about 10 days earlier than the northward shift of the subtropical high.

In the years of weak convective activities around the Philippines, such as in the summers of 1980, 1982, 1983, 1986, 1987, their intraseasonal variations were weak, as shown in Fig. 14. In these years the abrupt northward shift of the subtropical high was not obvious. The subtropical high remained at the normal position or shifted southward.

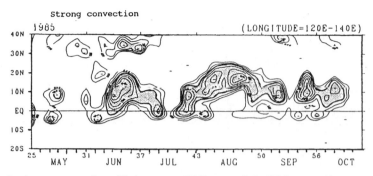

Fig. 13. Latitude-time cross section of 5-day mean OLR around the Philippines from May to October 1985. The shaded area indicates OLR below 200 W/M². (After Monthly Report on Climate System, J.M.A.)

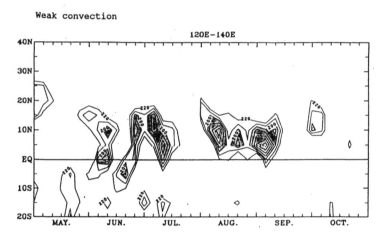

Fig. 14. Same as in Fig. 13 but for the summer of 1987.

4. Teleconnection pattern of the summer circulation anomalies

The convective activity anomalies in the western Pacific warm pool may influence not only the East Asian summer monsoon circulation, but also the circulation anomalies over North Pacific and North America. The one-point correlation of low-pass filtered daily 500 hPa disturbance height is computed with the basic point 20°N, 120°E for the summers of strong and weak convective activities around the Philippines, respectively. The computed results show that the teleconnection pattern of summer circulation in the summer of strong convective activities around the Philippines is different from that in the summers of weak convective activities around the Philippines.

Figure 15 is the distribution of correlation coefficients between the filtered daily 500 hPa disturbance height at various grid-points of the Northern Hemisphere and that at the basic point 20°N, 120°E in the summer of strong convective activities around the Philippines. One-point correlation coefficients in Fig. 15 are calculated by using the low-pass filtered daily 500 hPa disturbance height fields from 1 June to 31 August for 5 summers of strong convective activities around the Philippines, such as 1978, 1981, 1984, 1985 and 1988. The low-pass filter is designed to pass the low-frequency modes whose periods are larger than 10-days. All maximum correlation coefficients in various regions from Southeast Asia to North America are statistically significant. There is an obvious teleconnection pattern of the summer circulation anomalies from East Asia to North America through the North Pacific. Generally, since the negative anomaly of the 500 hPa height field is frequently located around the Philippines in the summer of strong convective activities around the Philippines, an area of positive correlation in Fig. 15 may mean that a negative anomaly of 500 hPa height field occurred frequently over this area. It may be seen from Fig. 15 that in the summer of strong convective activities around the Philippines, negative anomalies are located over the South China Sea and the Philippines, positive height anomalies are located over north China, Japan and the Okhotsk Sea and the subtropical high shifted northward. Another negative anomaly is located over the Aleutian area, and positive and negative anomalies are found over the western part of the U.S.A. and the southern part of North America. This anomaly distribution

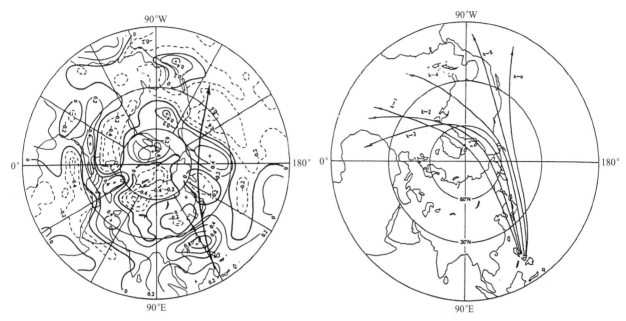

Fig. 15. Distribution of one-point correlation coefficients between low-pass filtered daily 500 hPa height values at every grid-point of the Northern Hemisphere and that at the basic point 120°E, 20°N for the summers of strong convective activities around the Philippines.

Fig. 16. Propagating ray path of quasi-stationary planetary waves forced by a source around the Philippines in summer.

is like the propagation of planetary-wave train (See Huang and Li, 1987). This teleconnection pattern may be called the East Asia/Pacific pattern (EAP pattern). In general, the correlation between the anomalies of the 500 hPa height field over the Northern Hemisphere and any localized anomaly around the Philippines in summer is not very obvious on the interannual time scale. The obvious teleconnection pattern shown in Fig. 15 may explain that the correlation between the circulation anomalies in the middle and high latitudes of the Northern Hemisphere and tropical circulation anomaly can be enhanced by focusing on the intraseasonal variation in the summers of strong convective activities around the Philippines. This pattern was also studied by Nitta (1987), Kurihara and Tsuyuki (1987). However, comparing Fig. 15 with Fig. 13b in Nitta's (1987) paper, there is a difference between both Figs. This may be due to the fact that Fig. 13b in Nitta's (1987) paper is obtained from the calculation of one-point correlation between the 500 hPa height fields and the convective activities around the Philippines.

5. Numerical simulation of the influence of a thermal anomaly in the western Pacific warm pool on the summer circulation over the Northern Hemisphere

Huang and Gambo (1983) investigated the quasi-stationary planetary waves forced by heat source and topography in summer. Huang (1984) has also discussed the propagating laws of quasi-stationary planetary waves in the summer basic flow, and pointed out that during the Northern Hemisphere summer, quasi-stationary planetary waves can quasi-horizontally propagate from the subtropics to middle and high latitudes.

Huang and Lu (1989) has calculated the propagating ray path of quasi-stationary planetary waves by a source around the Philippines in a realistic summer current. From the propagating ray of the quasi-stationary planetary waves shown in Fig. 16, it may be seen that the planetary-scale disturbances forced by the heat source around the Philippines with zonal wavenumbers 1–3 can propagate toward the Atlantic Ocean through East Asia, while the long wave-scale disturbances with zonal wavenumbers about 4–6 can propagate toward the western coast of North America and Rocky Mountains.

In order to investigate the influence of the thermal state of the western Pacific warm pool on the East Asian monsoon circulation and Northern Hemisphere circulation, a numerical experiment was made using the IAP-GCM designed by Zeng Qingcun et al. (1986). The IAP-GCM is a two-level global circulation model, in which the finite difference scheme of calculating the dynamics is unique, but the radiation heating, sensible and latent heat transfers from the sea and land surfaces are similar to the two-level GCM of OSU (See Chen, et al. 1982). The distribution of climatological mean SST used in this investigation is the same as that in Shukla and Wallace's (1983) paper. In this cal-

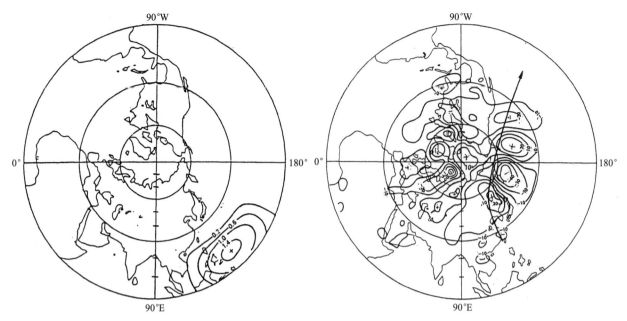

Fig. 17. An idealized distribution of the SST anomaly in the western Pacific warm pool.

Fig. 18. Anomaly distribution of the 500 hPa height field in July due to the idealized SST anomaly in the western Pacific warm pool shown in Fig. 17, averaged for 30 days of the model integrations of the IAP-GCM. Units are GPM.

culation, an idealized distribution of positive SST anomaly located in the western Pacific warm pool shown in Fig. 17 is used. This idealized summer SST anomaly distribution is added to the climatological distribution of SST in the globe, and an anomalous SST distribution in the Pacific can be obtained. The climatological wind field and height field in June are used as initial wind field and height field in the calculation, respectively. An integration is performed for one month with an anomalous SST distribution. This calculation is called an "anomalous run". Another integration of this model is also performed for one month with the climatological SST distribution and this is called a "control run". The height values of 500 hPa and precipitation obtained with the control run are subtracted from that obtained in the anomalous run, and the height anomalies of 500 hPa and the precipitation anomalies caused by the idealized SST anomaly distribution in the tropical western Pacific can be obtained.

Figure 18 shows the anomaly distribution of the 500 hPa height in July averaged for 30 days day of the model integrations, in response to a positive SST anomaly appearing in the western Pacific warm pool. It is seen from Fig. 18 that a negative anomaly of the 500 hPa height occurs in the Indo-China Peninsula and south China, while a positive anomaly appears over the west coast of Canada and the west coast of the U.S.A., is located over the Okhotck Sea and Kamchatka, another positive and negative anomaly respectively. Besides, a positive anomaly is located over Mexico. The pattern shown in Fig. 18 is in agreement with the theoretical analysis shown in Fig. 16, and it seems to be analogous to Fig. 13b in Nitta's (1987) paper. Moreover, the simulated locations of the maximum height anomalies are analogous to the observed result shown in Fig. 15, except for the eastward shift of the simulated result in North America. Nikaidou (1989) have also simulated the EAP pattern by using a GCM with a large positive SSTA in the western subtropical Pacific. His result is in good agreement with the observed facts. Huang and Lu (1989) has calculated the response of a 34-level model atmosphere to a forcing from a heat source anomaly in the tropical western Pacific. Their result is also in good agreement with the result shown in Fig. 18. These calculations show that the cause of the EAP telconnection pattern may be due to the propagation of quasi-stationary planetary waves forced by a source around the Philippines. Besides, Tsuyuki and Kurihara (1989) showed that the EAP pattern may be caused by a barotropic unstable mode in a zonally asymmetric flow. Therefore, the physical mechanism of the EAP pattern should be investigated further.

6. Discussions and conclusions

In this paper, observations show that the thermal states including the sea temperatures in the surface and subsurface layer of the western Pacific warm pool and the convective activities over the warm pool largely influence summer monsoon circulations and summer precipitations in East Asia.

Moreover, it is pointed out that there is a teleconnection pattern of summer circulation anomalies, *i.e.*, the so-called East Asia/Pacific teleconnection pattern, from Southeast Asia to the western coast of North America through East Asia during the Northern Hemisphere summer.

The relationship between the intraseasonal variations of the western Pacific subtropical high which may affect the onset and maintenance of the East Asia summer monsoon during June and the convective activities around the Philippines is analyzed by using the low-pass filtered daily 500 hPa height and latitude-time cross section of OLR data. It is found that there is a close relationship between the anomalous northward shift of the western Pacific subtropical high and the intensified convective activities around the Philippines. In summer, when there are intensified convective activities around the Philippines, there is an abrupt northward shift of the western Pacific subtropical high in early June or mid-June, and the summer monsoon rainfall may be weak in the Yangtze River and the Huaihe River valley. On the contrary, in summer when there are weak convective activities around the Philippines, the abrupt northward shift of the subtropical high is not obvious, and the summer monsoon rainfall may be heavy in the Yangtze River and the Huaihe River valley.

The two-level IAP-AGCM with an idealized distribution of SST anomaly in the tropical western Pacific are used to simulate this relationship. The result shows that when the SST in the tropical western Pacific warm pool is above normal, the western Pacific subtropical high will shift northward, and there is a teleconnection pattern of circulation anomalies in Southeast Asia, East Asia, the North Pacific and the western coast of North America.

We have used the above-mentioned results in the seasonal prediction of the East Asian summer monsoon rainfall and found that it improved the accuracy of our forecasting. We have successfully predicted the anomaly distributions of the summer monsoon rainfall in 1990 and 1991 using the thermal state in the surface and the subsurface layer of the western Pacific warm pool and the convective activities around the Philippines, particularly for the summer of 1991. According to the observed SST and ST in the subsurface layer of the western Pacific warm pool and the convective activities around the Philippines in the spring of this year, we predicted that, in the summer of 1991, the monsoon rainfall may be larger than normal by 30% or more in the Huaihe River valley and the middle and lower reaches of the Yangtze River, and a severe flood may occur in these regions, but it may be below normal in the Yellow River valley and south China. Observations showed that there was a very heavy monsoon rainfall which was more than double normal, and severe flood occurred in the Huaihe River valley and in the lower reaches of the Yangtze River from later May to mid-July, and there was a drought in the Yellow River valley and south China. It seems to us that the thermal state of the western Pacific warm pool may be a good predictor for the forecasting of the summer monsoon rainfall anomaly in East Asia, especially in the long-range prediction of droughts and floods in East Asia.

Acknowledgements

The authors are grateful to Prof. Tao Shiyan for his valuable comments and suggestions during the course of this study. The authors also are grateful to Prof. Ts. Nitta for valuable discussions and his kindness to provide us with the OLR data. Thanks are due to Profs. T. Yamagata and T. Yasunari for giving me the information about ST in the subsurface layer along 137°E. Thanks are also extended to the Long-range Forecast Division J.M.A. for providing the Monthly Report on Climate System. This paper is sponsored in part by the National Natural Science Foundation of China under grant 9488009 of the Climate Research Program.

References

Chan, S.J., J.W. Lingaas, Me.E. Schlesinger, R.L. Mobley and W.L. Gates, 1982: A documentation of the OSU two-level atmospheric general circulation model. Climate Research Institute of Oregon State University, Report No. 35.

Cornejo-Garrido, A.G. and P.H. Stone, 1977: On the heat balance of the Walker circulation. *J. Atmos. Sci.*, **34**, 1155–1162.

Gao, Y.X. and S.Y. Xu, 1962: Problem on monsoon over East Asia. In collected paper of the Institute of Geophysics and meteorology. Academia Sinica, No. 1, 1–106 (in Chinese).

Hartmann, D., H. Hendon and R.A. Houze, 1984: Some implications of the mesoscale circulations in tropical cloud clusters for large-scale dynamics and climate. *J. Atmos. Sci.*, **41**, 113–121.

Huang, R.H., 1982: The comparison between winter and summer response of a Northern Hemispheric model atmosphere to forcing by topography and stationary heat sources. *Proc. the First Sino-American Workshop on Mountain Meteorology*, 175–203.

Huang, R.H. and K. Gambo, 1983: The response of a hemisphere multi-level model atmosphere to forcing by topography and stationary heat sources in summer. *J. Meteor. Soc. Japan*, **61**, 494–509.

Huang, R.H., 1984: The characteristics of the forced planetary wave propagations in the summer Northern Hesiphere. *Adv. Atmos. Sci.*, **1**, 85–94.

Huang, R.H., 1985: Numerical simulation of the three-dimensional teleconnections in the summer circulation over the Northern Hemisphere, *Adv. Atmos. Sci.*, **2**, 81–92.

Huang, R.H. and W.J. Li, 1987: Influence of the heat source anomaly over the western tropical Pacific on the subtropical high over East Asia. *Proc. International Conference on the General Circulation of East Asia*, April 10–15, 1987, Chengdu, 40–51.

Huang, R.H. and L. Lu, 1989: Numerical simulation of the relationship between the anomaly of the subtropical high over East Asia and the convective activities in the western tropical Pacific. *Adv. Atmos. Sci.*, **6**, 202–214.

Kawahara, M. and K. Hayashi 1987: Convective activities and circulation in the tropics. *Prec. the Annual Meeting for Technical Development of Long-Range Forecast in Fiscal 1986, Forecast Department of Japan Meteorological Agency*, 3–39 (in Japanese).

Krishnamurti, T.N. and Y. Ramanathan, 1982: Sensitivity of monsoon onset of differential heating. *J. Atmos. Sci.*, **39**, 1290–1306.

Kurihara, K. and M. Kawahara, 1986: Extremes of East Asian weather during the post ENSO years of 1983/84 severe cold winter and hot dry summer. *J. Meteor. Soc. Japan*, **64**, 494–503.

Kurihara, K., 1989: A climatological study on the relationship between the Japanese summer weather and the subtropical high in the western Northern Pacific. *Geophy. Mag.*, **43**, 45–104.

Luo, H.B. and M. Yanai, 1984: The large-scale circulation and heat sources over the Tibetan Plateau and surrounding areas during the early summer of 1979. *Mon. Wea. Rev.*, **108**, 1849–1853.

McBride, J.L., 1987: The Australian summer monsoon, *Monsoon Meteorology*, 203–232. Edited by C.P. Chang and T.N. Krishnamurti, Oxford University Press.

Nikaidou, Y., 1989: The PJ-like north-south oscillation found in 4-month integrations of the global spectral model T42. *J. Meteor. Soc. Japan*, **67**, 587–604.

Nitta, Ts., 1983: Observational study of heat sources over the eastern Tibetan Plateau during the summer monsoon. *J. Meteor. Soc. Japan*, **61**, 590–605.

Nitta, Ts., 1987: Convective activities in the tropical western Pacific and their impact on the Northern Hemisphere summer circulation. *J. Meteor. Soc. Japan*, **64**, 373–390.

Shukla, J. and J.M. Wallace, 1983: Numerical simulation of atmospheric response to equatorial sea surface temperature anomalies. *J. Atmos. Sci.*, **41**, 1613–1630.

Tao, S.Y. and L.X. Chen, 1985: Summer monsoon in East Asia, *Prec. International Conference on Monsoons in the Far East*, Tokyo, 5–8 Nov. 1985, 1–11.

Tsuyuki, T. and K. Kurihara, 1989: Impact of convective activity in the western tropical Pacific on the East Asian summer circulation. *J. Meteor. Soc. Japan*, **67**, 231–247.

Yamagata, T., 1991: Asian monsoon and ocean circulation in the West Pacific. To be published in *Proc. of the International Conference on Oceans, Climate and Man*, Turin, 15–17 April 1991.

Yasunari, T., 1990: Impact of Indian monsoon on the coupled atmosphere/ocean system in the tropical Pacific. *Meteor. Atmos. Phys.*, **44**, 29–41.

Ye, D.Z. and Y.X. Gao, 1979: *Tibetan Plateau meteorology* (in Chinese). Meteorological Press, 279 pp.

Yeh, T.C., S.Y. Tao and M.C. Li, 1959: The abrupt change of circulation over the Northern Hemisphere during June and October. *Atmosphere and the Sea in Motion*, 249–267.

Zeng, Q.C., et al., 1986: A global grid point general circulation model, in Short-and Medium-Range Numerical Weather Prediction, *Collection of papers presented at the WMO/IUGG NWP Symposium*, Tokyo, 4–8 Aug. 1986, 424–430.

热带西太平洋纬向风异常对 ENSO 循环的动力作用*

黄荣辉　张人禾　严邦良

(中国科学院大气物理研究所，北京 100080)

摘要　根据观测资料，分析了1982/1983年，1986/1987年，1991/1992年和1997/1998年El Niño 事件发展和衰减以及 La Niño 事件发生过程中赤道西太平洋对流层下层环流和纬向风异常及其作用. 结果表明，在 El Niño 事件发展阶段前，在热带西太平洋上空对流层下层产生气旋性环流异常，从而使印度尼西亚和赤道西太平洋上空产生西风异常；而当 El Niño 事件发展到成熟阶段，在热带西太平洋上空对流层下层产生反气旋性环流异常，从而使印度尼西亚和赤道西太平洋上空产生东风异常. 还利用一个简单的热带海洋动力学模式，计算了 20 世纪最强的 1997/1998 ENSO 循环过程中赤道海洋波动对实际海表风应力距平的响应. 结果表明，热带西太平洋海表附近的纬向风异常，通过激发 Kelvin 波与 Rossby 波对 El Niño 事件的发展与衰减和 La Niño 事件的发生起到重要的动力作用.

El Niño 事件是热带太平洋地区海气相互作用最重要的现象. 它的发生会在全球引起严重的气候异常，从而在世界许多地区造成严重的旱涝与低温冷害，使许多国家的工农业生产受到很大损失. El Niño 事件的发生也给我国带来严重的灾害，在 El Niño 事件发展的夏季，我国华北往往发生干旱；而在 El Niño 事件的衰减期，我国长江流域往往发生洪涝. 为此，目前我国和世界各国的气象与海洋学家非常重视这一现象的发生规律及其机理的研究，以达到有朝一日可以预测这一现象的发生，从而给气候灾害的预测提供可靠的信息与物理依据.

Bjerkness[1]首先提出 El Niño 事件是赤道东太平洋海气相互作用结果的假说. 到目前为止，对 El Niño 产生机制及海气相互作用进行了大量的研究. McCreay[2], McCreay 和 Anderson[3], Anderson 和 McCreay[4]较系统地研究了 ENSO 循环的物理机制，他们从理论上提出赤道海洋波动在 ENSO 循环中的作用. Schopf 和 Suarez[5]从不稳定海气相互作用和赤道波系传播的观点解释了 ENSO 循环，表明西太平洋暖池(Warm Pool)处于暖的状态是 El Niño 事件发生必不可少的条件，只有西太平洋暖池热容量处于异常大的状态，随后才有可能发生 El Niño 事件. 然而，有的年份西太平洋暖池的温度异常高，其热容量异常大，但这些年的第二年并不发生 El Niño 事件. 这表明，西太平洋暖池处于异常暖的状态只不过是 El Niño 事件发生的必要条件之一，还应有热带西太平洋上空大气状态的条件. 因此，还应分析在 El Niño 事件发生过程中热带太

---

* 本文原载于：中国科学(D 辑)，第 31 卷，第 8 期，697-704，2001 年 8 月出版.

平洋大气环流和海面风应力的异常情况.

Philander[6]利用数值试验说明了赤道中、东太平洋信风的减弱对赤道东太平洋增温的动力作用. Tang 和 Weisberg[7]利用简单线性重力约化模式的数值模拟结果讨论了 1982/1983 年海表附近西风应力异常对赤道中、东太平洋增温的作用. 这些研究只是从数值试验来说明赤道太平洋海表附近西风应力异常对赤道东太平洋增温的作用, 因此, 很有必要从理论上进一步从赤道太平洋海面附近实测风应力异常对赤道海洋波动的激发作用来说明实际海表附近纬向风应力异常对 ENSO 循环的动力作用. 本研究应用 1980~1998 年 NCEP/NCAR 热带太平洋海表温度以及 850 hPa 风场的再分析资料和 FSU 热带太平洋海表风场的观测资料, 分析和讨论 1980~1998 年中所发生的 4 次 El Niño 事件中热带太平洋上空对流层下层环流和风场的演变及其作用, 特别是分析了 1997/1998 年热带太平洋纬向风应力异常对 1997/1998 ENSO 循环的动力作用.

1 热带西太平洋上空对流层下层纬向风异常对 ENSO 循环的影响

图 1 是 1980~1998 年季节平均的赤道东太平洋 NINO3 区海表温度距平(图 1(a))和赤道西太平洋上空 850 hPa 纬向风距平(图 1(b))的年际变化图. 从图 1(a)可以清楚看到, 从 1980~1998 年期间, 除了在 1993 年和 1994 年也发生了较弱的增温事件, 赤道中、东太平洋在 1982/1983 年, 1986/1987 年, 1991/1992 年和 1997/1998 年发生了 4 次很明显的 El Niño 事件, 其中 1997/1998 年 El Niño 事件为 20 世纪最强; 并且还可以看到, 1984/1985 年, 1988/1989 年, 1995/1996 年和 1998 年冬季赤道中、东太平洋发生了明显的冷事件, 即发生了 La Niña 事件.

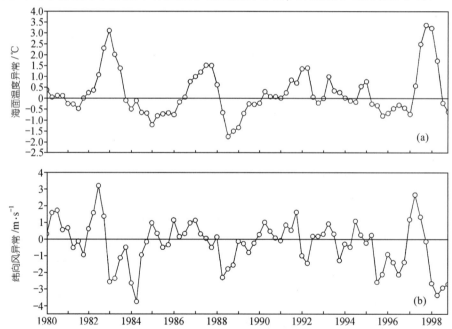

图 1　1980~1998 年季节平均的赤道东太平洋 NINO3 区(90°~150°W, 5°N~5°S)海表温度距平
(a)和赤道西太平洋(120°~160°E, 5°N~5°S)上空 850 hPa 纬向风距平(b)的年际变化

从图 1(b)可以看到, 热带西太平洋上空对流层下层纬向风距平年际变化呈波动振荡. 在

1982/1983 年 El Niño 事件发展阶段之前的 1981 年冬季到 1982 年秋季, 在 1986/1987 年 El Niño 事件发展阶段之前的 1985 年冬季到 1986 冬季, 1991/1992 年 El Niño 事件发展阶段之前的 1991 年春季到秋季和 1997/1998 年 El Niño 事件发展阶段之前的 1996 年秋季到 1997 年夏季, 均出现较强的西风异常. 此外, 与较弱的 1993 和 1994 年 El Niño 事件相联系, 1992 年冬季到 1993 年春季和 1994 年夏季也出现了明显的西风异常. 同时从图 1(b)还可看到, 在 La Niña 事件发生之前, 既从 1982 年冬季到 1983 年夏季, 1988 年的春季到秋季, 1995 年夏、秋、冬季和 1997 年冬季到 1998 年秋季, 均出现明显的东风异常. 这说明了 ENSO 循环与赤道西太平洋上空对流层下层的纬向风异常有很大关系.

比较图 1(a)和图 1(b), 我们可以清楚看到, 这 4 次 El Niño 事件发生前, 赤道西太平洋 850 hPa 纬向风异常均有较大的西风异常; 而在这 4 次 La Niña 事件发生前, 赤道西太平洋均有较大的东风异常. 黄荣辉和张人禾[8]分析了 1980~1994 年所发生的 3 次 ENSO 循环的赤道太平洋上空对流层下层纬向风的变化情况, 表明了在 El Niño 事件的发展阶段赤道西、中太平洋有强的西风异常; 并且在紧随西风异常的西部为东风异常区, 它随西风异常的东传而向东伸展. 这里的分析结果与他们所发现的现象一致.

2 热带西太平洋上空对流层下层环流异常对 ENSO 循环的影响

利用 NCEP/NCAR 资料, 我们分析了 1980~1998 年 4 次 El Niño 事件发展与衰减阶段前热带西太平洋上空 850 hPa 环流异常的季度平均距平风场分布. 图 2(a)和(b)分别是 1982 年春季、以及 1996 年冬季热带西太平洋上空 850 hPa 距平风场分布. 可以看出, 在 El Niño 事件发生之前, 热带西太平洋上空对流层下层有一个明显的气旋性异常风场分布, 在印度尼西亚和菲律宾以东的热带西太平洋上空有明显的西风异常. 这个西风异常将有利于西太平洋暖池海域产生暖 Kelvin 波, 使暖池暖水向东输送, 从而使 El Niño 事件产生[9]. 图 2(c)~(d)分别是 1982 年冬季和 1997 年秋季 El Niño 事件成熟期热带西太平洋上空 850 hPa 季度平均的距平风场分布. 可以明显地看到, 在 El Niño 衰减之前, 在热带西太平洋上空对流层下层有一明显的反气旋式的距平风场分布, 这样, 从巴布亚新几内亚沿印度尼西亚到苏门答腊一带有明显的东风异常, 这个东风异常使得西太平洋暖池产生冷 Kelvin 波, 从而使 El Niño 事件衰减、消亡[9], 并使 La Niña 事件产生.

上述热带西太平洋环流异常是与对流活动相联系. Ren 和 Huang[10]的研究表明, 当西太平洋暖池暖, 则它上空对流活动强; 反之, 则它对流活动弱. 对于大尺度环流, 由涡度方程, 即

$$\frac{d}{dt}(f+\varsigma) \approx -fD,$$

可以推断出环流异常. 其中 f 是柯氏参数, ς 是环流的相对涡度, H 是大气的标高, D 是环流的散度. 当热带西太平洋对流活动强, 低层气流辐合加强, 则 $D<0$, 因此, $\frac{d}{dt}(f+\varsigma)>0$, f 不变化, 则有 $\frac{d}{dt}\varsigma>0$, 故在热带西太平洋上空气旋性环流要增强; 相反, 热带西太平洋对流活动弱, 则 $D>0$, $\frac{d}{dt}(f+\varsigma)<0$, 同样, 有 $\frac{d}{dt}\varsigma<0$, 故在热带西太平洋上空反气旋性环流要

增强.

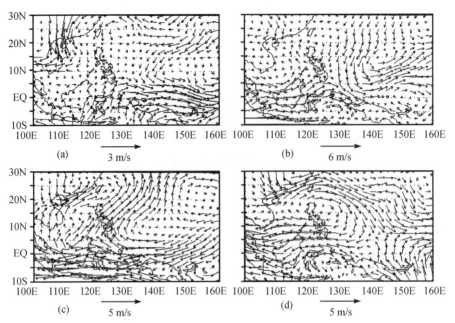

图 2　1980~1998 年 4 次 El Niño 事件发展阶段前(图(a), (b))以及成熟期(图(c), (d))时热带西太平洋上空 850 hPa 距平风场分布

(a) 1982 年春季和(b)1996 年冬季; (c) 1982 年冬季和(d)1997 年秋季

从以上分析可以看到, 热带太平洋海温、对流、环流之间是相关的. 下面将分析环流异常与赤道海洋波动之间的关系.

3　一个简单的热带太平洋海洋动力学模式

为了研究赤道太平洋海表附近的纬向风应力异常对赤道太平洋海洋波动的动力作用, 按照 Gill 的研究[11], 可给出赤道 β 平面上的无量纲线性浅水波方程组, 即

$$\frac{\partial u}{\partial t} - yv = -\frac{\partial h}{\partial x} + X, \qquad (1)$$

$$yu = -\frac{\partial h}{\partial x} + Y, \qquad (2)$$

$$\frac{\partial h}{\partial t} + \left(\frac{\partial u}{\partial x} + \frac{\partial v}{\partial y}\right) = 0. \qquad (3)$$

上述方程中无量纲变量的定义与一般的定义相同. X 和 Y 为无量纲纬向和经向风应力异常, 它们分别为 $X = \tau^x/[\rho H_0(c^3\beta)^{1/2}]$, $Y = \tau^y/[\rho H_0(c^3\beta)^{1/2}]$. $\rho = 1.026 \times 10^3 \text{kg/m}^3$; $c = (g'H_0)^{1/2}$, g' 是约化重力, 为 $5.6 \times 10^{-2} \text{m/s}^2$, H_0 是海洋混合层平均厚度, 取为 150 m; $\beta = 2.28 \times 10^{-11} \text{s}^{-1}\text{m}^{-1}$, 是赤道附近科里奥利力参数随纬度的变化. τ^x 和 τ^y 分别是热带太平洋海表附近的纬向和经向风应力异常.

对于方程组(1)~(3), 经向边界条件取为自然边界条件; 在纬向东边界 $x = x_E$ 和西边界 $x = x_W$, 边界条件分别取为[5]:

$$u(y, x = x_E) = 0, \quad \int_{-\infty}^{+\infty} u(y, x = x_W) \mathrm{d}y = 0. \tag{4}$$

引入如下新变量 $q = h + u, r = h - u$,并将变量按照韦伯函数[10] $\Psi_m(y)$ 展开,即

$$\begin{pmatrix} q(x,y,t) \\ r(x,y,t) \\ v(x,y,t) \end{pmatrix} = \sum_{m=0}^{\infty} \begin{pmatrix} q_m(x,t) \\ r_m(x,t) \\ v_m(x,t) \end{pmatrix} \Psi_m(y). \tag{5}$$

若只考虑纬向风应力异常强迫,并对展开(5)式只取到 4 阶,这样方程(1)~(3)变成下式:

$$\frac{\partial q_0}{\partial t} + \frac{\partial q_0}{\partial x} = \int_{-\infty}^{+\infty} X \psi_0 \mathrm{d}y, \tag{6}$$

$$\frac{\partial q_2}{\partial t} - \frac{1}{3}\frac{\partial q_2}{\partial x} = \frac{1}{3}\int_{-\infty}^{\infty} X\{\psi_2 - \sqrt{2}\psi_0\}\mathrm{d}y, \tag{7}$$

$$\frac{\partial q_4}{\partial t} - \frac{1}{7}\frac{\partial q_4}{\partial x} = \frac{1}{7}\int_{-\infty}^{\infty} X\{3\psi_4 - 2\sqrt{3}\psi_2\}\mathrm{d}y, \tag{8}$$

$$r_2 = 2\sqrt{\frac{1}{3}}q_4,$$

其中(6)式代表Kelvin波型的响应,(7)式和(8)式分别为 2 阶和 4 阶 Rossby 波型的响应. 因此, 根据热带太平洋海表附近实际纬向风应力距平值,就可以由(6)~(8)式求出对赤道太平洋海表附近纬向风应力距平强迫响应的 Kelvin 波与 Rossby 波.

4 热带太平洋纬向风距平对1997/1998年赤道太平洋ENSO循环的动力作用

在 1,2 节中已从观测资料分析了 1980~1998 年 4 次 ENSO 循环中热带太平洋纬向风应力异常与 El Niño 事件发展和衰减的关系,本节将通过由方程(6)~(8)和 FSU 热带太平洋海表风应力距平资料所计算的赤道海洋 Kelvin 波和 2 阶、4 阶 Rossby 波,来说明热带太平洋海表上空纬向风应力异常对 1997/1998 年 ENSO 循环的动力作用.

图 3(a)~(c)上图分别是 1997 年 3, 5 和 10 月份由热带太平洋实际海表纬向风应力距平强迫所产生的热带太平洋 Kelvin 波、2 阶 Rossby 波,而下图分别是实测的赤道太平洋 5° S~5° N 平均的海表纬向风应力距平与海表温度距平(SSAT). 由图 3(a)可以看到, 1997 年 3 月, 由于在西太平洋暖池区海表附近有很大的西风应力异常, 因此在赤道西太平洋与中太平洋激起暖 Kelvin 波, 使得赤道西、中太平洋 SST 有正的异常;正如图 3(b)所示, 到 1997 年 5 月, 随着西风应力距平的东传, 暖 Kelvin 波很快东传到赤道东太平洋, 并且在东边界反射出暖 Rossby 波, 这使得赤道东太平洋 SST 很快增高, 从而爆发 El Niño 事件. 同时, 还可以看到, 在赤道中、西太平洋激起冷 Rossby 波, 这个冷 Rossby 波将向西传播; 到了 1997 年 7 月, 东传的暖 Kelvin 波在赤道东太平洋发展, 并且被赤道东边界反射的暖 Rossby 波也继续加强, 这使得 El Niño 事件加强, 赤道东太平洋 SST 继续升高(图略); 到了 1997 年 10 月, 如图 3(c)所示, 暖 Kelvin 波在赤道东太平洋继续发展, 并且被东边界反射的暖 Rossby 波也继续加强, 这使得赤道中、东

太平洋 SST 发展到顶点，即 El Niño 事件达到成熟期. 并且，从图 3(c)还可以看到西传的冷 Rossby 波传播到西太平洋暖池区并发展，这使得西太平洋暖池变冷，为 El Niño 事件的衰减提供先行条件.

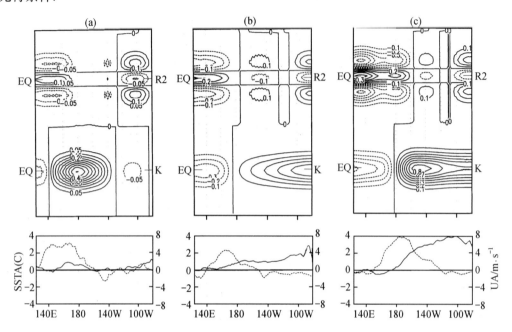

图 3　赤道海洋 Kelvin 波与 Rossby 波对赤道太平洋海表纬向风应力异常强迫的响应(上图)及观测的赤道太平洋海表温度距平(实线)和纬向风应力距平(虚线)(下图)的时空分布图
(a) 1997 年 3 月；(b) 1997 年 5 月；(c) 1997 年 10 月

图 4(a)和(b)上图分别是 1998 年 3, 5 月由热带西太平洋实际风应力距平强迫所产生的热带太平洋 Kelvin 波与 2 阶 Rossby 波，而下图分别是实测的赤道太平洋 $5°$ S~$5°$ N 平均海表纬向风应力距平与海表温度距平. 由图 4(a)可以看到，到了 1998 年 3 月，由于在西太平洋暖池区海温降低并出现较大东风应力距平，这使得赤道西太平洋出现冷 Kelvin 波，并且此冷 Kelvin 波向东传播，同时，还可以看到在赤道中东太平洋出现暖 Rossby 波；到了 1998 年 5 月，正如图 4(b)所示，由于冷 Kelvin 波东传并发展，使得赤道中太平洋海表 SST 下降，El Niño 事件逐渐衰亡，同时，还可以看到暖 Rossby 波已传到西太平洋暖池区，这将使西太平洋暖池 SST 又将逐渐回升；到了 1998 年 8 月，冷 Kelvin 波已传播到赤道中、东太平洋(图略)，这使 El Niño 事件完全衰亡，这样，此次 El Niño 事件经历了形成、发展、成熟和衰减阶段. 另外，还可以看到暖 Rossby 波又在西太平洋暖池区继续发展，使暖池区 SST 再度升高，这为下一次 El Niño 事件提供先行条件.

从上面所述的热带太平洋海表附近纬向风应力距平强迫所产生的 Kelvin 波和 Rossby 波的时空演变可以看到，热带西、中太平洋海表西风应力异常对于赤道太平洋 El Niño 事件的发生、发展起到重要作用；而东风应力异常对于 El Niño 事件的衰减起到重要的动力作用. 因此，可以说热带西、中太平洋纬向风应力异常对于热带太平洋 ENSO 循环起到重要的动力作用.

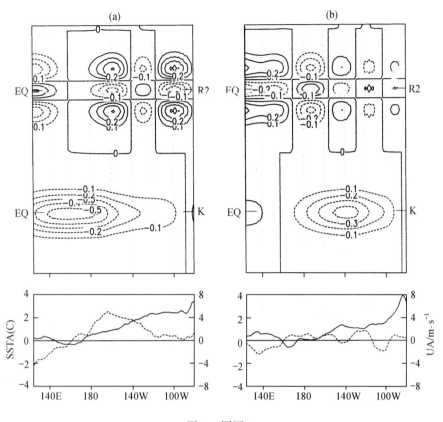

图 4 同图 3
(a) 1998 年 3 月; (b) 1998 年 5 月

5 结论

本研究利用 1980~1998 年实际观测的热带太平洋海表温度和 850 hPa 环流场和纬向风异常, 分析了 1980~1998 年 4 次 ENSO 循环过程中热带西太平洋对流层下层环流异常和纬向风异常对 El Niño 事件发展和衰减的作用. 结果表明, 在 El Niño 事件发展阶段前, 热带西太平洋上空对流层下层有明显的气旋性环流异常, 并在赤道西太平洋和印度尼西亚一带有西风异常; 而在 El Niño 事件衰减前(即在 El Niño 事件成熟期), 热带西太平洋上空对流层下层有明显的反气旋性环流异常, 并在赤道西太平洋和印度尼西亚一带有东风异常.

本文还利用一个简单的热带海洋动力模式, 讨论了 1997/1998 年 ENSO 循环过程中赤道 Kelvin 波和 Rossby 波对实际观测的热带太平洋海表风应力距平的响应. 计算结果表明, 在 El Niño 事件发展阶段, 赤道西太平洋海表附近的西风应力异常激发了东传暖 Kelvin 波和西传冷 Rossby 波; 而在 El Niño 事件衰减前, 赤道西太平洋海表附近的东风应力异常激发了东传冷 Kelvin 波和西传暖 Rossby 波. 这些波动对 1997/1998 年 ENSO 循环起到重要的动力作用.

参 考 文 献

1 Bjerknes J. A possible response of the atmospheric Hadley circulation to equatorial anomalies of ocean temperature. Tellus,

1969, 18: 820~829
2. McCreary J P. A model of tropical ocean-atmosphere interaction. Mon Wea Rev, 1983, 111: 370~387
3. McCreay J P, Anderson D L T. A simple model of El Niño and the Southern Oscillation. Mon Wea Rev, 1984, 112: 934~946
4. Anderson D L T, McCreay J P. Slowly propagating disturbances in a coupled ocean-atmosphere model. J Atmos Sci, 1995, 42: 615~628
5. Schopf P S, Suarez M J. Vacillations in a coupled ocean-atmosphere model. J Atmos Sci, 1998, 45: 549~566
6. Philander S G H. The response of equatorial ocean to a relaxation of trade winds. J Phys Oceanogr, 1981, 11: 176~189
7. Tang T Y, Weisberg R H. On the equatorial Pacific response to the 1982~1983 El Niño and Southern Oscillation event. J Mar Res, 1984, 42: 809~829
8. 黄荣辉, 张人禾. ENSO 循环与东亚季风环流相互作用过程的诊断研究. 见: 叶笃正主编. 赵九章纪念文集. 北京: 科学出版社, 1997. 93~109
9. 张人禾, 黄荣辉. El Niño 事件发生和消亡中热带太平洋纬向风应力的动力作用. Ⅰ资料诊断和理论分析. 大气科学, 1998, 22: 587~599
10. Ren B H, Huang R H. Interannual variability of the convective activities associated with the East Asian summer monsoon seem from TBB variability. Adv Atmos Sci, 1999, 15: 77~90
11. Gill A E. Some simple solutions for heat-induced tropical circulation. Q J R Meteor Soc, 1980, 106: 447~462

东亚夏季风爆发和北进的年际变化特征及其与热带西太平洋热状态的关系[*]

黄荣辉[1]　顾雷[1]　徐予红[1]　张启龙[2]　吴尚森[3]　曹杰[4]

(1 中国科学院大气物理研究所，北京 100080；2 中国科学院海洋研究所，青岛 266071；
3 中国气象局广州热带海洋气象研究所，广州 510080；4 云南大学大气科学系，昆明 650091)

摘　要　利用我国测站的降水资料、卫星测得的 OLR 和高云量资料、SST 和 137°E 次表层海温资料以及 NCEP/NCAR 再分析资料，分析了东亚夏季风的爆发和北进的年际变化特征及其与热带西太平洋热状态的关系。分析结果表明：当春季热带西太平洋处于暖状态，菲律宾周围对流活动强，在这种情况下，南海上空对流层下层有气旋性距平环流，西太平洋副热带高压偏东，从而使得南海夏季风爆发早；并且，当夏季热带西太平洋也处于暖状态，菲律宾周围对流活动也很强，在这种情况下，西太平洋副热带高压北进时，在 6 月中旬和 7 月初存在明显的突跳，从而使得东亚季风雨带在 6 月中旬明显由华南北跳到江淮流域，并于 7 月初由江淮流域北跳到黄河流域、华北和东北地区。这将引起江淮流域和长江中、下游夏季风降水偏少，并往往发生干旱，而黄河流域、华北和东北地区的夏季降水正常或偏多。相反，当春季热带西太平洋处于冷状态，菲律宾周围对流活动弱，在这种情况下，南海上空对流层下层有反气旋性距平环流，西太平洋副热带高压偏西，从而使得南海夏季风爆发晚；并且，当夏季热带西太平洋也处于冷状态，菲律宾周围对流活动也很弱，在这种情况下，西太平洋副热带高压北进时，在 6 月中旬或 7 月初向北突跳并不明显，而是以渐进式向北移动，从而使得东亚季风雨带一直维持在长江流域和淮河流域。这将引起此两流域夏季风降水偏多，并往往发生洪涝，而黄河流域、华北和东北地区的夏季降水偏少，发生干旱。作者还从非线性多平衡态动力理论说明了菲律宾周围对流活动强弱对西太平洋副热带高压北进时以突跳或渐进式向北移动起到重要作用。

Characteristics of the Interannual Variations of Onset and Advance of the East Asian Summer Monsoon and Their Associations with Thermal States of the Tropical Western Pacific

HUANG Rong-Hui[1], GU Lei[1], XU Yu-Hong[1], ZHANG Qi-Long[2],
WU Shang-Sen[3], and CAO Jie[4]

1　*Institute of Atmospheric Physics, Chinese Academy of Sciences, Beijing*　100080
2　*Institute of Oceanography, Chinese Academy of Sciences, Qingdao*　266071
3　*Guangzhou Institute of Tropical and Marine Meteorology, China Meteorological Administration, Guangzhou*　510080
4　*Department of Atmospheric Sciences, Yunnan University, Kunming*　650091

Abstract　Characteristics of the interannual variations of onset and advance of the East Asian summer monsoon and their associations with thermal states of the tropical western Pacific are analyzed by using the precipitation data at observational stations of China, data of OLR and high cloud amount observed by satellite, data set of SST and ST in subsurface of the

---

[*] 本文原载于：大气科学，第 29 卷，第 1 期，20-36，2005 年 1 月出版.

western Pacific along 137°E and the NCEP/NCAR reanalysis data. The results show that when the tropical western Pacific is in a warming state in spring, convective activities are intensified around the Philippines. In this case, there is a cyclonic anomaly circulation in the lower troposphere over the South China Sea, and the western Pacific subtropical high shifts eastward, thus, the early onset of the South China Sea summer monsoon (SCSM) can be caused. Moreover, when the tropical western Pacific is also in a warming state, convective activities around the Philippines are also strong in summer, since the western Pacific subtropical high abruptly shifts northward in mid-June and early July during its northward advance in this case, the abrupt northward-shift of the East Asian summer monsoon rainband from South China to the Yangtze River and the Huaihe River valleys is obvious in mid-June and this monsoon rainband again jumps northward from the Yangtze River and the Huaihe River valleys to the Yellow River valley, North China and Northeast China in early July. This can cause that summer monsoon rainfall is below normal and drought may occur in the Yangtze River and the Huaihe River valleys and the middle and lower reaches of the Yangtze River, but summer rainfall is normal or above normal in the Yellow River valley, North China and Northeast China. On the other hand, when the tropical western Pacific is in a cooling state in spring, convective activities are weakened around the Philippines. In this case, there is an anticyclonic anomaly circulation in the lower troposphere over the South China Sea, and the western Pacific subtropical high shifts westward, thus, the late onset of the SCSM can be caused. Moreover, when the tropical western Pacific is also in a cooling state, convective activities around the Philippines are also weak in summer, since the abrupt northward-shift of the western Pacific subtropical high is not obvious in mid-June and early July and it gradually shifts northward during its northward advance in this case, the East Asian summer monsoon rainband can be maintained in the Yangtze River and the Huaihe River valleys. This can cause that the summer monsoon rainfall is above normal and flood may occur in these two valleys, but summer rainfall is below normal and drought may occur in the Yellow River valley, North China and Northeast China. It is also explained by using the dynamical theory of nonlinear multiple equilibrium that the strong or weak convective activities around the Philippines play an important role in the abrupt or gradual northward shift of the western Pacific subtropical high.

Key words summer monsoon, interannual variation, the tropical western Pacific subtropical high, convective activity

1 引言

今年 2 月 21 日是叶笃正院士的 90 华诞。叶笃正院士是世界著名的气象学家，也是中国现代大气科学主要奠基人之一。他在 Rossby 波的能量频散、大气运动的适应过程、东亚大气环流的演变特征与机理、青藏高原的动力、热力作用以及全球气候变化和可持续性发展等研究领域作出了系统而开创性贡献。

我国地处东亚季风区，东亚季风的变异严重影响着我国旱涝气候灾害的发生，从而给我国带来严重的经济损失，为此，中国气象学者很早就重视对东亚季风的研究。早在 60 年前，我国著名气象学家竺可桢[1]首先提出东亚夏季风对中国降水的影响，之后，涂长望和黄仕松[2]研究了东亚夏季风的进退对中国降水季节内变化的影响。为了改进中国的天气预报，提高其预报水平，叶笃正院士继竺可桢、涂长望和赵九章先生之后，对东亚大气环流的变化特征和机理作了系统而开创性的研究[3~11]。叶笃正等[12]早在 20 世纪 50 年代就指出东亚上空在 6 月中旬存在着行星尺度环流的突变，这个突变将带来东亚夏季风在长江流域和淮河流域的爆发。在叶笃正发现东亚季风季节转换存在突变之后，在 20 世纪 80 年代，Krishnamurti 和 Ramanatahan[13]以及 McBride[14]也发现印度夏季风和澳大利亚季风环流也同样存在着这种突变。叶笃正和他的合作者[10]在 20 世纪 60 年代初所撰写的《大气环流的若干基本问题》，系统地总结了他和合作者关于东亚和全球大气环流的研究成果，被国际气象学家公认为世界上最早的大气环流动力学的著作之一。此书不仅系统地讨论了北半球大气环流的特征、演变过程和基本原理，而且指出：大气环流的基本要素都不是独立的，它们是相互关联的，并形成一个内在的整体。这个概念不仅对于当时大气环流动力学的发展起到重要作用，而且对于当今气候系统动力学的研究也有重要的启迪。

作为叶笃正院士的学生，长期受到叶笃正院士亲自指导和启发，我和我的学生从 20 世纪 80 年代中后期起，对于东亚夏季风的季节内、年际和年代际变化作了一些研究[15~19]，其部分研究结果也已总结在叶笃正院士和我以及其他专家共同撰写的《长江黄河流域规律和成因研究》的专著[20]中。近几年来，我们在东亚季风爆发和进退的年际变化作了进一步的研究，表明东亚夏季风的爆发和进退有很大的年际变化，并且从观测事实和动力理论分析了东亚夏季风爆发和进退与热带西太平洋热力状态的关系。因此，本文就作者最近关于东亚季风爆发和北进的年际变化特征及其与热带西太平洋热力状态的关系方面的研究进展作一概述，并以此文来庆贺我的恩师叶笃正院士的90华诞。

2 南海夏季风爆发日期的年际变化及其对长江、淮河流域梅雨的影响

Tao 和 Chen[21]指出亚洲季风首先在南海地区爆发，一般平均在5月中旬，比印度季风要早一个月。在中国，往往把发生在南海地区的夏季风称为南海季风。南海季风的爆发标志着东亚季风的来临和中国华南前汛期即将开始，因而它具有重要的预报意义[22,23]。亚洲季风在南海爆发之后，它将经过二次阶段性北进和二次停滞，最后在7月中旬，夏季风可以到达华北和东北一带，并在这两地区停滞一段时间，于8月中旬又南撤。

如何衡量亚洲夏季风在南海地区的爆发？这是近年来许多学者所关注的问题，因此，不同研究者从不同角度出发，从而定义出不同的季风指数。正如 Wang 等[24]所回顾的，约有20种之多，有的指数从动力要素出发，有的从动力和热力要素相结合出发来定义亚洲季风的爆发。从不同角度出发来定义的季风指数，其所描述的南海夏季风爆发日期有很大不同。正如 Wang 等[24]所指出，利用南海区域候平均850 hPa的纬向风出现西风来定义季风爆发，这不仅简单，而且与南海区域降水有一定的相关性。但 Wang 等[24]所定义的南海季风爆发日期只能是候，因此，本文应用梁建茵和吴尚森[25]所提出的南海季风爆发的定义：在（5°N～20°N，110°E～120°E）区域平均的850 hPa面上连续出现西风的天数大于5天，并且，之后在此区域出现的西南风中断的天数不得超过前西南风持续时期的三倍。

图1是按照梁建茵和吴尚森[25]所定义的南海夏季风1950～1999年的爆发日期。从图1可以看到，南海夏季风的爆发日期有很显著的年际变化，它最早爆发于4月下旬，最晚爆发于6月初，平均为5月19日（即5月第4候）。考虑到从1978年起才能得到高云量的卫星观测资料，因此，本文重点分析20世纪80～90年代南海夏季风的爆发及其与热带西太平洋热力状态的关系。若把南海季风爆发日期早于5月19日2天以上的年份定义为南海季风早爆发年，这样，1979，1981，1984，1985，1986，1994和1999年为南海季风早爆发年；而把爆发日期晚于5月19日2天以上的年份定义为南海季风晚爆发年，这样，1982，1983，1987，1988，1991，1993和1998年为南海季风晚爆发年。

南海季风爆发的早晚不仅对位于中国的长江、淮河流域、日本和朝鲜半岛的东亚夏季风爆发的早晚和夏季风降水有很大影响，而且对印度夏季风爆发早晚和季风降水也有重要影响[26]。图2是1950～1999年中国夏季（6～8月）降水和南海夏季风爆发日期的相关系数分布图。从图2可以清楚地看到，一个超过95%显著性检验的正相关区位于江淮流域和长江下游，而一个弱的负相关位于东北南部、华北东部以及黄河的上游地区。这表明：若南海季风爆发是早的（即早于5月第4候），则江淮流域、长江上游地区的夏季风降水偏弱，如1981，1984，1985，1994和1999年的春季南海夏季风爆发日期偏早，则这些年的夏

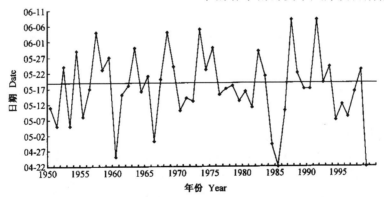

图1 1950～1999年南海夏季风爆发日期的年际变化（南海季风爆发日期引自文献[25]）
Fig. 1 The onset dates of the South China Sea summer monsoon (SCSM) during 1950–1999. The onset dates of the SCSM are quoted from reference [25]

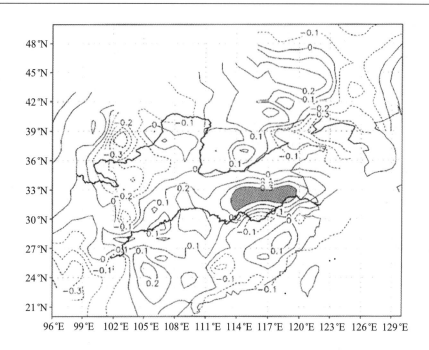

图 2 1950~1999 年中国夏季降水和南海夏季风爆发日期的相关系数分布图（南海季风爆发日期引自文献 [25]）实线：正相关，虚线：负相关；阴影区：超过 95% 显著性检验的正相关区域

Fig. 2 Distribution of the correlation coefficients between the onset dates of the South China Sea summer monsoon and the summer rainfall in China. The onset dates of the SCSM are quoted from reference [25]. The solid and dashed lines indicate positive and negative correlation, respectively, and the positive correlation areas over the 95% significance level are shaded

季，江淮流域的降水偏少，而东北南部、华北东部和黄河上游地区降水偏多；相反，若南海夏季风爆发是晚的（即晚于 5 月第 4 候），则江淮流域、长江上游地区的夏季风降水偏强，如 1982，1983，1987，1991 和 1998 年，南海夏季风爆发偏晚，则这些年的夏季，江淮流域和长江流域上游地区夏季风降水偏多，并引起严重的洪涝灾害。这些事实表明：在南海夏季风爆发较早的年份，亚洲夏季风一般偏强，并可以推进到华北和东北一带；相反，在南海夏季风爆发较晚的年份，亚洲夏季风一般偏弱，东亚夏季风一般较迟爆发，并且往往不能推进到华北和东北地区，而在江淮流域停滞较长时间。这说明在南海夏季风爆发晚的夏季，由于北方得不到充足的水汽供应而南方又频繁受到夏季风扰动的影响，这就很容易使江淮流域夏季风降水偏多，并引起洪涝。然而，虽然有个别年份南海夏季风爆发偏早，但由于北方冷空气活动较强，东亚夏季风也较迟推进到北方，从而引起南涝北旱的异常降水分布，如 1999 年夏季，虽然南海夏季风爆发较早，但由于东亚夏季风在 8 月中旬才到达华北且停滞在此区域很短时间就迅速南撤了，使得从江淮流域到华北地区发生了严重干旱[27]。这些事实都表明南海季风爆发的早晚可以作为我国东部夏季降水的季节预测的一个物理因子。

3 热带西太平洋的热力状态对南海夏季风爆发的影响

热带西太平洋是全球高海温区域，故此海域又称西太平洋暖池（The warm pool）。许多研究已表明热带西太平洋海域的热状态对于热带和东亚地区气候的年际变化有很大影响[28,29]。由于南海临近于热带西太平洋，热带西太平洋的热力变化直接影响着南海上空的大气环流和对流活动，因此，热带西太平洋的热力状态将严重影响着南海季风的爆发。为此，本节利用 NCEP/NCAR 再分析的 SST 资料和日本气象厅"凌风丸"海洋考察船所观测的 137°E 的次表层海温资料，分析热带西太平洋的热力状态对南海夏季风爆发的影响。图 3 是 NINO.west（0°~14°N，130°E~150°E）区域春季（3~5 月）平均的 SST 距平和南海夏季风爆发日期的年际变化曲线。比较图 3 中虚线所示的南海季风爆发日期与实线所示的 NINO.west 区域的春季 SST 距平的年际变化，可以发现这两者是反位相的。图 3 表明，一般当春季西太平洋暖池处于暖状态，南海夏季风爆发早，而当春季西太平洋暖池处于冷状态，则南海夏季风爆发日期就晚。

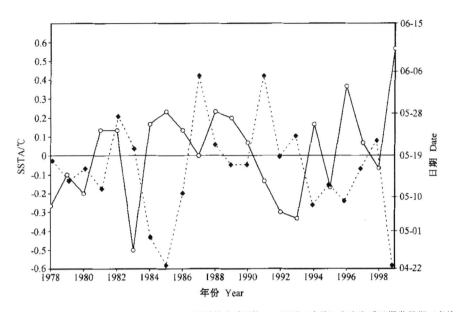

图3 1978~1999年NINO.west（0°~14°N，130°E~150°E）区域的春季平均SST距平（实线）和南海季风爆发日期（虚线）的年际变化曲线（SST距平资料来源于NCEP/NCAR再分析资料，南海季风爆发日期引自文献[25]）

Fig. 3 The SST anomalies in NINO.west (i.e., 0°–14°N, 130°E–150°E) (solid line) and the onset dates of the South China Sea summer monsoon (dashed line) during 1978–1999. The data of SST are derived from NCEP/NCAR reanalysis data, and the onset dates of the SCSM are quoted from reference [25]

为了更进一步说明西太平洋暖池春季的热力状态对南海夏季风爆发的影响，本研究分别对南海季风早爆发与晚爆发年份的春季热带太平洋SST距平作合成分析。图4a和图4b分别是对1978~1999年南海夏季风早爆发年份（1979，1981，1984，1985，1986，1994，1996和1999年）和晚爆发年份（1982，1983，1987，1988，1991，1993和1998年）春季热带太平洋SST距平的合成图。从图4a可以看到：在南海季风爆发早的春季，在热带西太平洋有正的SST距平，而在热带中、东太平洋有负的SST距平，这类似于La Niña事件发展阶段的SST距平分布；相反，在南海季风爆发晚的春季（图4b），在热带西太平洋有负的SST距平，而在热带中、东太平洋有正的SST距平，这类似于El Niño事件发展阶段的SST距平分布。因此，从热带太平洋春季的SST距平合成分布可以清楚地看到，一般当春季热带西太平洋处于暖状态，南海季风早爆发，而当春季热带西太平洋处于冷状态，则南海季风爆发就晚。

春季，西太平洋暖池的热状态还可以进一步从次表层海温的状态来表述。与热带东太平洋海温垂直结构相比，热带西太平洋的斜温层顶的深度是深的，因此，在此海域的热容量异常是与此海域次表层的海温异常密切相关[28]，因此，热带西太平洋春季次表层海温的热状态可能是南海夏季风爆发早晚的很好信号。图5是1967~1999年期间南海夏季风爆发早和爆发晚年的1月份沿137°E热带和副热带太平洋次表层海温距平的合成图。把图5a和图5b相比较，可以明显看到，在热带海域两者有很大差别。正如图5a所示，对于那些南海夏季风早爆发的年份，1月份在2°N~15°N区域沿137°E的次表层的合成海温距平是正的，最大合成距平值可达到0.6℃，它位于5°N附近赤道西太平洋的150 m深处；相反，对于那些南海夏季风晚爆发的年份（图5b），1月份在2°N~15°N区域沿137°E的次表层的合成海温距平是负的，最大合成距平值可达到-1.0℃，它同样位于5°N附近赤道西太平洋的150 m深处。

上述分析事实充分表明热带西太平洋的热状态对南海夏季风爆发有很大的影响。

4 热带西太平洋对流活动对南海夏季风爆发的影响

为什么热带西太平洋热力状态对南海夏季风爆发会有如此大的影响，这是值得进一步深入研究的问题。正如上节所述，热带西太平洋是全球高海温的区域，由于此海域很暖，这里的海气相互作用非常强，并且Walker环流的上升支也位于此海域上空[28]，因此，在此海域上空强的气流和水汽的辐合导致了强的对流活动和强降水。正如Nitta[30]所指出，此海域上

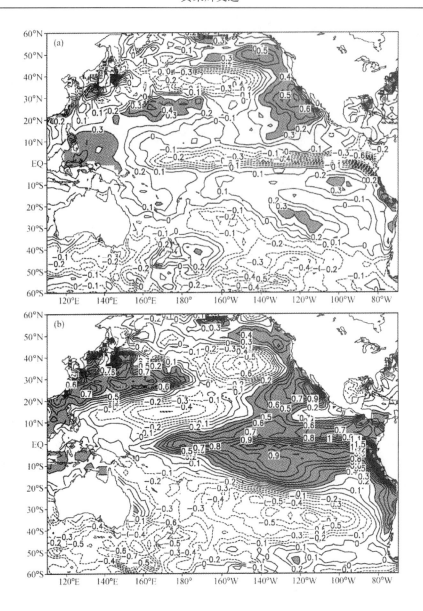

图 4 1978~1999 年期间南海季风爆发早（a）和爆发晚（b）年份的春季（3~5 月）热带太平洋 SST 距平的合成图（单位：℃，SST 资料来源于 NCEP/NCAR 再分析资料）。实线：正距平；虚线：负距平；阴影区：SST 距平大于 0.3℃ 的正距平区

Fig. 4 The composite distributions of the SST in the tropical western Pacific in spring (March – May) for the cases of early onset (a) and the cases of late onset (b) of the SCSM, respectively. Units: ℃. The data of SST is from NCEP/NCAR reanalysis data. The solid and dashed lines indicate the positive and negative anomalies, respectively, and the positive SST anomalies above 0.3 ℃ are shaded

空的对流活动与此海域的海温密切相关，当热带西太平洋［特别是（0°~14°N，130°E~150°E）海域］处于暖状态，则菲律宾周围的对流活动就强；相反，当热带西太平洋处于冷状态，则菲律宾周围的对流活动就弱。因此，菲律宾周围的对流活动可以很好地表征热带西太平洋的海气相互作用状态。并且，菲律宾周围的 TBB，OLR 或高云量变化方差要比 SST 变化的方差大得多，所以可以用菲律宾周围的对流活动来讨论热带西太平洋对南海季风爆发的影响过程会更清楚。为此，本研究应用菲律宾周围（10°N~20°N，110°E~140°E）的卫星测得的高云量（HCA）来分析热带海气相互作用对南海夏季风爆发的影响。图 6 中实线与虚线分别表示春季菲律宾周围的高云量与南海季风爆发日期的年际变化，它清楚地表明了这两者有很好的反相关，它们的相关系数可以达到 -0.76，这超过了 0.99 的显著性检验。从这相关系数可以充分说明

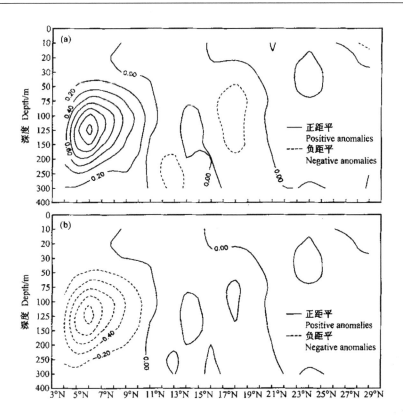

图 5 对 1967~1999 年期间南海夏季风爆发早（a）和爆发晚（b）年份的 1 月份沿 137°E 热带和副热带太平洋次表层海温距平（单位：℃）合成的纬度—时间剖面图［资料来源：日本气象厅"凌风丸"考察船沿 137°E 剖面的观测资料（见 Monthly Report on Climate System，JMA）］

Fig. 5 The composite distributions of latitude – time cross section of the subsurface sea temperature anomalies in the tropical and subtropical Pacific along 137°E in January for the cases of early onset (a) and the cases of late onset (b) of the SCSM, respectively. The subsurface sea temperature data are derived from the Oceanographic Research Vessel. RYOFU-MARU' of JMA

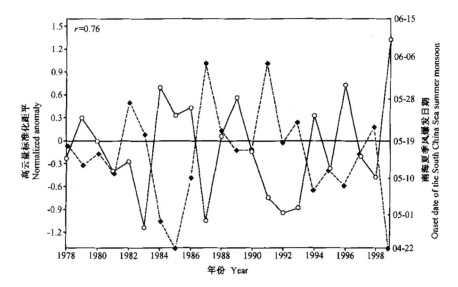

图 6 1978~1999 年春季（3~5 月）菲律宾周围（10°N~20°N，110°E~140°E）平均高云量（HCA）的标准化距平（实线）与南海夏季风爆发日期（虚线）的年际变化曲线［高云量资料来源于日本气象厅气候系统月报（Monthly Report on Climate System，JAM）］

Fig. 6 The interannual variations of normalized anomaly of the high cloud amount (HCA) around the Philippines (i.e., 10°N – 20°N, 110°E – 140°E) in spring (solid line) and the onset dates of the South China Sea summer monsoon (dashed line) during 1978 – 1999. The data of the HCA are derived from Monthly Report on Climate System, JMA

菲律宾周围春季的对流活动强弱对南海夏季风的爆发有重要的影响。在菲律宾周围对流活动强的春季，南海夏季风爆发是早的；相反，在菲律宾周围对流活动弱的春季，南海夏季风爆发是晚的。

为了更清楚地表达热带西太平洋上空对流活动对南海季风爆发的影响，本研究分别对热带西太平洋海温处于暖状态和冷状态的 5～6 月份对流活动 (OLR) 和 700 hPa 气流距平作合成分析 (图 7)。由于 OLR 是往外长波辐射，在无云情况，OLR 值是地表面的往外的长波辐射，因此，在对流活动弱的区域，OLR 值将偏高；反之，在云量多的区域，OLR 值是云顶的长波辐射值，因此，在对流活动强的区域，OLR 值偏低。从图 7a 可以看到，在热带西太平洋处于偏暖状态的 5～6 月份，从热带西太平洋经南海到中印半岛其 OLR 值偏低，这说明此区域对流活动是强的；并且，从 700 hPa 风场距平分布可以看到：在热带东印度洋与苏门答腊岛上空有西风距平气流，一个气旋性异常环流位于中印半岛、南海和华南上空的对流层低层，在南海中部和北部上空有强的西南风的距平气流，而反气旋性异常环流偏东。这些都表明：在热带西太平

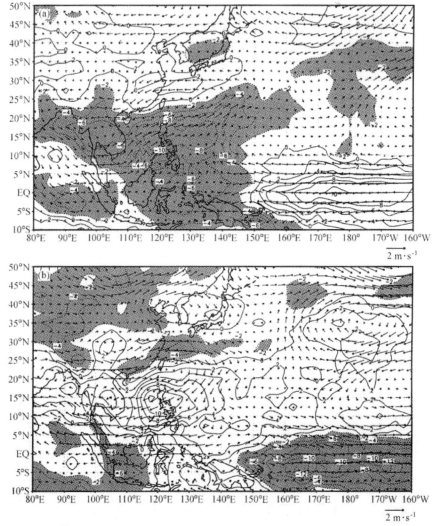

图 7 1978～1999 年期间热带西太平洋处于偏暖状态 (a) 和偏冷状态 (b) 的 5～6 月份热带西太平洋东亚和东亚上空 OLR 和 700 hPa 风场距平的合成分布 (OLR 取之于 NOAA 卫星资料集，风场取于 NCEP/NCAR 再分析资料)。实线：正 OLR 距平，虚线：负 OLR 距平 (单位：$W \cdot m^{-2}$)；阴影：低于 $-2\ W \cdot m^{-2}$ 的 OLR 距平值 (即强对流活动区)

Fig. 7 The composite distributions of OLR and wind anomaly field at 700 hPa averaged for the period from May to June for the warming states (a) and the cooling states (b) of the tropical western Pacific during 1978–1999. The OLR data are derived from NOAA, and the 700 hPa wind data are derived from NCEP/NCAR reanalysis data. The solid and dashed lines indicate positive and negative anomalies, respectively. Units: $W \cdot m^{-2}$, the negative anomalies lower than $-2\ W \cdot m^{-2}$ (i.e., strong convection area) are shaded

洋处于偏暖状态，由于菲律宾对流活动偏强，强烈的上升运动引起了热带东印度洋、中印半岛上空西风气流很强，出现强西风距平气流，南海上空出现了气旋性环流，而太平洋副热带高压偏东，从而导致了南海夏季风早爆发。相反，正如图7b所示，在热带西太平洋处于偏冷状态的5～6月份，从热带西太平洋经南海到中印半岛，其OLR距平为正，即OLR值偏高，这说明此区域对流活动是弱的，而强对流活动区域位于印度尼西亚以东的赤道中太平洋上空；并且，在热带东印度洋与印度尼西亚的苏门答腊岛上空有东风距平气流，一个反气旋性异常环流位于中印半岛、南海和华南上空的对流层低层，在南海中部和北部上空有强的东北风的距平气流。这些都表明：在热带西太平洋处于偏冷状态，由于菲律宾周围对流活动偏弱，引起了热带东印度洋、中印半岛上空西风气流很强，并出现东风距平环流，南海上空的反气旋环流很强，从而导致了南海季风晚爆发。

5 东亚夏季风北进的年际变化特征及其与热带西太平洋热状态的关系

亚洲夏季风在南海爆发以后，它将经过阶段性的北进与停滞，在7月初或中旬到达华北与东北地区。正如图8所示，一般在5月份和6月上旬，由于西太平洋副热带高压偏南，因此，季风达到华南和江南地区，这正是华南和南岭一带的前汛期；到了6月上、中旬，西太平洋副热带高压的脊线北移到20°N附近，季风雨带北跳到长江、淮河流域，这时长江和淮河流域的梅雨、日本的"Baiu"和韩国的"Changma"就开始了；雨带在这些地区停滞一段以后，到了7月中旬，由于西太平洋副热带高压脊线又北移到30°N，使得季风雨带又北跳到华北、东北一带，这时我国北方雨季开始，而江淮流域梅雨结束；到了8月中旬左右，此雨带又迅速南撤到长江流域。然而，由于西太平洋副热带高压的演变有很大的年际变化，因而，东亚夏季风的进退也有很大的年际变化。若西太平洋副热带高压的脊线在6月上旬就北跳到20°N附近，到了7月中旬它还不北移，这样，季风雨带就长期维持在江淮流域，因而引起此地区严重洪涝灾害的发生，而黄河流域、华北地区季风降水很弱，往往发生干旱；相反，若西太平洋副热带高压在6月中旬之后才北跳20°N附近，而在7月中旬之前，它又迅速北跳，这样，季风雨带在江淮流域停滞的时间就很短，因而，此地区就会发生少梅或空梅，夏季风降水偏少，从而造成此地区的高温、少雨天气，并且，由于季风

较早到达黄河流域、华北和东北地区，使得这些地区降水正常或偏多。

上面所述的我国东部夏季风降水的季节内变化，正如叶笃正等[12]所指出的，在6月中旬发生突变，这对气候平均意义上而言是正确的。然而，正如作者在以前的研究[15～19]中所指出，夏季西太平洋副热带高压和东亚季风雨带的北进有很大的年际变化，它依赖于热带西太平洋的热状态，特别是依赖于菲律宾周围的对流活动的情况。由于以前研究所用的观测资料的时期较短（1978～1990年），故本文应用较长时期观测资料（1978～2000年），进一步深入分析东亚夏季风进退的年际变化特征。

为了讨论热带西太平洋热状态、特别是菲律宾周围对流活动强弱对东亚夏季风北进年际变化的影响，有必要分析菲律宾周围对流活动的年际变化。图9是菲律宾周围（10°N～20°N，110°E～140°E）春、夏、秋、冬各季平均的标准化高云量距平的年际变化。正如上面所述，若某夏季高云量距平是正的，则可表示该夏季菲律宾周围对流活动是强的；反之，则表示该夏季菲律宾周围对流活动是弱的。从图9可以看到，1978，1981，1984，1985，1988，1994和1999年的夏季，菲律宾周围的对流活动是强的，而1980，1982，1983，1991，1992，1996和1998年的夏季，菲律宾周围的对流活动是弱的。

为了讨论东亚夏季风进退的年际变化特征对热带西太平洋热力状态，特别是对菲律宾周围对流活动强弱的依赖，下面分别分析在菲律宾周围对流活动强和对流活动弱的年份，东亚3～10月份沿115°E（即110°E～120°E平均）的5天降水量季节内变化的合成情况。

5.1 在菲律宾周围对流强的夏季

图10是对1978～2000年期间菲律宾周围对流强的年份，东亚3～10月份沿115°E（即110°E～120°E平均）的5天降水量合成的纬度—时间剖面图。从图10可以看到，在6月中旬季风雨带从江南北跳到江淮流域，并在7月初北跳到华北地区，在8月中旬雨带就迅速南撤到江淮流域。在这种情况下，江淮流域夏季降水偏少，发生干旱灾害。为了更清楚地说明在菲律宾对流活动强时夏季风北进的季内变化特征，本研究分别取1985年和1994年东亚3～10月份，沿115°E（即110°E～120°E平均）的5天降水量的季节内变化作为例子。如图11a和b所示，在5月和6月初季风雨带位于华南和江南，并在6月中旬北跳到江淮流域，7月初又北跳到华北和东北一带，此雨带于8月

中旬南撤到江淮流域。在这两个夏季，由于季风很快北进到华北，江淮流域夏季降水偏少，比正常值偏低 30%～50%，出现了高温干旱的气候灾害。

从上述分析可以看到，在热带西太平洋处于偏暖状态时，菲律宾周围对流活动偏强，从而使南海夏季风爆发就偏早，它将于5月19日之前爆发；随后，

在5月和6月初夏季风北进并停滞在华南一带，并于6月中旬很快地从华南、江南一带北进到江淮流域，使得江淮流域梅雨开始。然而，在这种情形下，季风在江淮流域停滞时间不长，于7月初季风又很快地从江淮流域北进到黄河流域、华北和东北地区，使北方雨季较早开始；随后，季风在华北和东北地区停滞一

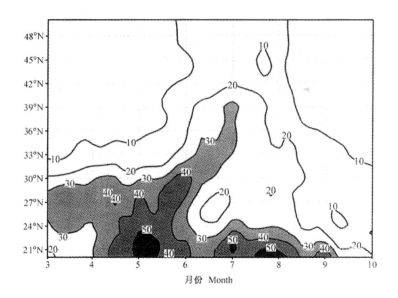

图 8 1961～1990 年东亚沿 115°E（即 110°E～120°E 平均）候降水量的 30 年气候平均值的纬度—时间剖面图（单位：mm）。阴影：候降水量大于 30 mm

Fig. 8 Latitude – time cross section of climatological mean of 5 days precipitation along 115°E (i.e., averaged between 110°E – 120°E) for 30 years from 1961 to 1990. Units: mm. The rainfalls above 30 mm are shaded

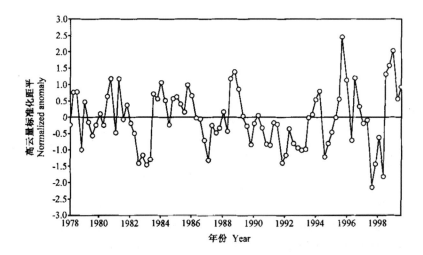

图 9 1978～2000 年菲律宾周围（10°N～20°N，110°E～140°E）春、夏、秋、冬各季平均的标准化高云量距平的年际变化［资料引自日本气象厅气候系统月报（Monthly Report on Climate System，JMA）］

Fig. 9 The interannual variations of normalized anomaly of HCA in spring, summer, autumn and winter around the Philippines (i.e., 10°N – 20°N, 110°E – 140°E) from 1978 – 2000. The data of HCA are quoted from Monthly Report on Climate System of JMA

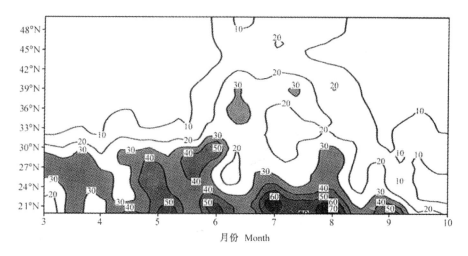

图 10 对 1978～2000 年期间菲律宾周围（10°N～20°N，110°E～140°E）对流活动强的年份东亚 3～10 月份沿 115°E（即 110°E～120°E 平均）的 5 天降水量合成的纬度—时间剖面图（单位：mm）。阴影：降水量大于 30 mm

Fig. 10 The composite distribution of latitude – time cross section of 5 days rainfall along 115°E (i.e., averaged between 110°E – 120°E) for the cases of strong convective activities around the Philippines (i.e., 10°N – 20°N, 110°E – 140°E) from March to October. Units: mm. The rainfalls above 30 mm are shaded

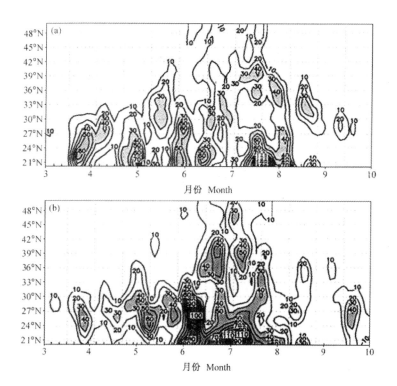

图 11 1985 年（a）和 1994 年（b）东亚 3～10 月份沿 115°E（即 110°E～120°E 平均）的 5 天降水量的纬度—时间剖面图（单位：mm）。阴影：降水量超过 30 mm

Fig. 11 Latitude – time cross section of 5 days rainfall along 115 °E (i.e., averaged between 110°E – 120 °E) from March to October in 1985 (a) and 1994 (b), respectively. Units: mm. The rainfalls over 30 mm are shaded

段时间，于 8 月中旬迅速南撤到江淮流域。因此，在这种情况下，东亚夏季风经历了两次阶段性北进和三次停滞过程，并于 8 月中旬迅速从华北地区南撤到江淮流域，之后又南撤到华南地区。

5.2 在菲律宾对流活动弱的夏季

图 12 是对 1978～2000 年期间菲律宾周围对流活动弱的年份（1980，1982，1983，1987，1991，1992，1993，1996 和 1998 年）东亚 3～10 月份沿 115°E（即

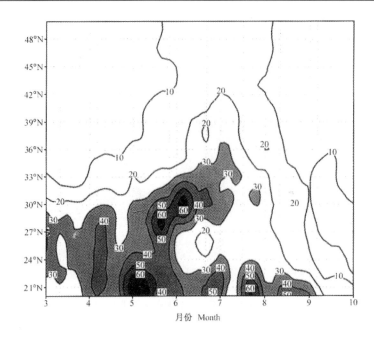

图 12 对 1978~2000 年期间菲律宾周围（10°N~20°N，110°E~140°E）对流动弱的年份东亚 3~10 月份沿 115°E（即 110°E~120°E 平均）的 5 天降水量合成的纬度–时间剖面图（单位：mm）。阴影：降水量大于 30 mm

Fig. 12 The composite distributions of latitude – time cross section of 5 days rainfall along 115°E (i.e., averaged between 110°E – 120°E) for the cases of weak convective activities around the Philippines (i.e., 10°N – 20°N, 110°E – 140°E) from March to October. Units: mm. The rainfalls above 30 mm are shaded

110°E~120°E 平均）的 5 天降水量合成的纬度—时间剖面图。从图 12 可以看到，东亚夏季风雨带并不经历二次北跳和三次停滞，而是从 5 月到 6 月份一直维持在长江和淮河流域并逐渐北移，到了 7 月初才北移到黄淮流域。由于雨带长时间维持在长江流域和江淮流域，造成这两流域季风降水偏强，从而致使此两流域发生严重洪涝，并且，由于在这种情况下，季风到达华北偏晚并停滞很短时间，在 8 月初就南撤到江淮流域，故华北地区夏季降水偏弱，从而导致华北地区发生干旱。为了更清楚地说明在菲律宾周围对流活动偏弱的年份季风北进的季内变化特征，本研究分别取 1980 年和 1998 年东亚 3~10 月份沿 115°E（即 110°E~120°E 平均）的 5 天降水量的季内变化作为例子。如图 13a 和 b 所示，在 5~6 月份，季风雨带位于长江流域，之后渐渐北移到江淮流域，并一直维持到 7 月中旬；到了 7 月中旬，弱的季风降水才在华北出现，并于 8 月初季风雨带迅速南撤到江淮流域。在这两个夏季，由于季风长期停滞在长江流域和江淮流域，使得这两流域季风降水很强。1980 年夏季，长江流域降水比正常值偏多了 30%~50%，而 1998 年夏季，长江流域降水比正常值偏多了 50%~100%，造成了严重洪涝，江淮流域降水比正常值也偏多了 30%。

从上述分析事实可以看到，在热带西太平洋处于偏冷状态，这引起了菲律宾周围对流活动偏弱，从而使得南海夏季风爆发就偏晚，它将于 5 月 19 日之后爆发，随后季风渐渐北移到华南、长江流域和江淮流域一带。在这种情况下，季风并不经历二次阶段性北进和三次停滞过程，而是渐进式地北进，并长时间维持在长江流域和江淮流域，只是到了 7 月中旬，弱的季风才北进到华北。因此，在这种情形下，在长江流域或江淮流域夏季风降水偏强，并往往发生严重洪涝，而华北地区降水偏少，易发生干旱。

6 西太平洋副热带高压的进退及其与热带西太平洋热状态的关系

前面所述的分析事实清楚地表明，热带西太平洋的热力状态对东亚夏季风及其雨带的进退有重要影响，其机理是值得进一步深入研究的。黄荣辉和孙凤英[17~19]指出，夏季西太平洋副热带高压的北进严重地影响着季风雨带的季节内变化，并且它的北进特征是与热带西太平洋热力状态有关，特别是与菲律宾对

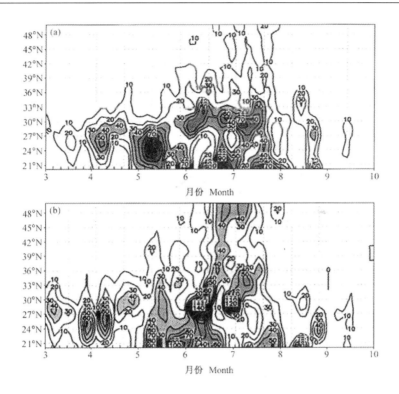

图 13　1980 年（a）和 1998 年（b）东亚 3～10 月份沿 115°E（即 110°E～120°E 平均）的 5 天降水量的纬度—时间剖面图（单位：mm）。阴影：降水量大于 30 mm

Fig. 13　Latitude – time cross section of 5 days rainfall along 115°E (i.e., averaged between 110°E – 120°E) from March to October in 1980 (a) and 1998 (b), respectively. Units: mm. The rainfalls over 30 mm are shaded

流活动有密切相关。Nitta[30]也指出菲律宾周围对流活动的年际变化严重影响着东亚夏季气候。由于以前研究所用的资料时间较短，有必要进一步利用更长时间的资料来分析夏季西太平洋副热带高压的季节内变化特征，因此，本研究分析了 1978～2000 年沿东亚 135°E 500hPa 高度的纬度—时间剖面图（图14）。

6.1　在菲律宾周围对流强的夏季

图 14a 是对菲律宾周围对流活动强的年份（1978，1981，1984，1985，1988，1994 和 1999 年）东亚沿 135°E 500 hPa 高度合成的纬度—时间剖面图，从图中可以看到，在 6 月初，西太平洋副热带高压的脊线从 18°N 突跳到 25°N，这引起了季风雨带突然北跳到江淮流域，并且在 7 月初，它又从 25°N 北跳到 33°N，这引起了季风雨带北跳到黄河流域、华北和东北地区，并使江淮流域梅雨结束。此外，从图 14a 还可以看到，从 8 月中旬以后，西太平洋副热带高压又迅速南撤到长江以南，使得雨带又南撤到江淮流域。

6.2　在菲律宾周围对流弱的夏季

图 14b 是对菲律宾周围对流活动弱的年份（1980，1982，1983，1987，1991，1992，1993，1996 和 1998 年）东亚沿 135°E 500 hPa 高度合成的纬度—时间剖面图，如图所示，西太平洋副热带高压无论在 6 月中旬或 7 月上旬，其突然北跳并不明显。在 5 月份西太平洋副热带高压脊线一直停滞在 15°N 附近，于 6 月初渐渐北移到 20°N 附近，并一直维持到 7 月中旬，在这种情形下，西太平洋副热带高压的脊线于 7 月下旬南撤，于 8 月初才南撤到华南，这使得季风雨带长时间维持在长江流域和淮河流域。

从以上分析可以看到，在菲律宾周围对流活动弱的夏季，西太平洋副热带高压的季内变化特征与在菲律宾周围对流活动强的夏季的变化特征不同。这表明，夏季西太平洋副热带高压的北进特征是与菲律宾对流活动紧密相关。当菲律宾对流活动强，则西太平洋副热带高压在向北移动时，于 6 月上旬和 7 月中旬突跳就明显。这不仅使东亚季风雨带在长江、淮河流域停滞时间短，造成长江流域和淮河流域季风降水偏少，而且使季风较早到达华北、东北一带，并较晚从华北南撤到江淮流域，引起华北、东北地区降水正常或偏多；相反，当菲律宾活动弱，则西太平洋副热带高压在北进时，在 6 月中旬和 7 月上旬突跳并不明显，而是以渐进式向北移动。这不仅使东亚季风雨带长时间停滞在长江流域和江淮流域，造成此两流域季

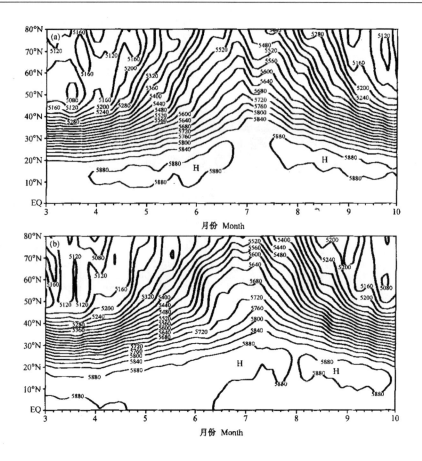

图 14 对于 1978～2000 年期间菲律宾周围对流活动强的年份（a）和对流弱的年份（b）东亚沿 135°E 500 hPa 高度合成的纬度—时间剖面图（单位：gpm，高度场资料来源于 NCEP/NCAR 再分析资料）。等值线间隔：20 gpm

Fig. 14 The composite distributions of latitude – time cross section of 500 hPa along 135°E for the cases of strong convective activities (a) and weak convective activities (b) around the Philippines, respectively. Units: gpm. The data are from NCEP/NCAR reanalysis data, and contour interval is 20 gpm

风降水偏强，引起洪涝灾害，而且造成季风较晚到达华北、东北一带，而又较早就南撤到江淮流域，从而导致华北、东北地区季风降水偏少，引起干旱灾害。

6.3 菲律宾周围对流活动所产生的热力强迫对西太平洋副热带高压非线性演变的动力作用

从以上分析可以看到，标志着东亚季风的北进和南撤的东亚季风雨带的季节内演变，与西太平洋副热带高压位置的季节内变化有密切关系，而夏季西太平洋副热带高压位置的季节内变化是与热带西太平洋的热力状态，特别是与菲律宾周围的对流活动紧密相关。当菲律宾对流活动强，西太平洋副热带高压在北进时，于 6 月中旬和 7 月上旬时发生突跳式北进，而在菲律宾周围对流活动弱时，西太平洋副热带高压在北进时，并不发生突跳，而是以渐进式向北移动，这是什么原因引起？为此，我们应用 Charney 和 Devore 所提出的大气环流演变多平衡态理论[31]，研究了西太平洋副热带高压的非线性演变过程[32]。研究结果[32]表明了西太平洋副热带高压北进时在 6 月中旬是否发生突跳，这主要依赖于菲律宾周围对流活动引起的热力强迫。当菲律宾对流活动所引起的热力强迫是强的，并超过某一临界值，在这种情况下，对此热力强迫所响应的波动之间以及波—流之间的相互作用相当强，加之外部热力的作用，从而引起西太平洋副热带高压在 6 月中旬发生突跳式的北进；相反，当菲律宾对流活动所引起的热力强迫是弱的，在这种情况下，对此热力强迫所响应的波动之间和波流之间的相互作用很弱，甚至不存在，其结果是西太平洋副热带高压只是随着外部热力强迫的振荡而振荡，从而引起西太平洋副热带高压在 6 月中旬并不发生突跳式的北进，而是以渐进式北进。

7 结论

本研究利用了我国测站的降水资料、NOAA 卫星测得的 OLR 资料、日本 GMS 卫星测得的高云量资料、日本"凌风丸"考察船测得的 137°E 次表层海温剖面资料以及 NCEP/NCAR 风场与 SST 的再分析资料,分析了东亚夏季风的爆发和进退的年际变化特征及其与热带西太平洋热状态的关系。分析结果表明:当春季热带西太平洋处于暖状态(图 15a),菲律宾周围对流活动强,在这种情况下,南海上空对流层下层有气旋性距平环流,西太平洋副热带高压偏东,从而使得南海夏季风爆发早;并且,当夏季热带西太平洋也处于暖状态,菲律宾周围对流活动也很强,在这种情况下,西太平洋副热带高压北进时,在 6 月中旬和 7 月初存在着明显的突跳,从而使得东亚季风雨带在 6 月中旬明显由华南北跳到江淮流域,并于 7 月初由江淮流域北跳到黄河流域、华北和东北地区。这将引起江淮流域和长江中、下游夏季风降水偏少,往往发生干旱,而黄河流域、华北和东北地区的夏季降水正常或偏多。相反,当春季热带西太平洋处于冷状态(图 15b),菲律宾周围对流活动弱,在这种情况下,南海上空对流层下层有反气旋性距平环流,西太平洋副热带高压偏西,从而使得南海夏季风爆发晚;并且,当夏季热带西太平洋处于冷状态,菲律宾周围对流活动也很弱的情况下,西太平洋副热带高压北进时,在 6 月中旬或 7 月初向北突跳并不明显,而是以渐进式向北移动,从而使得东亚季风雨带一直维持在长江流域和淮河流域。这将引起长江流域和江淮流域夏季风降水偏多,往往发生洪涝,而黄河流域、华北和东北地区的夏季降水偏少,发生干旱。

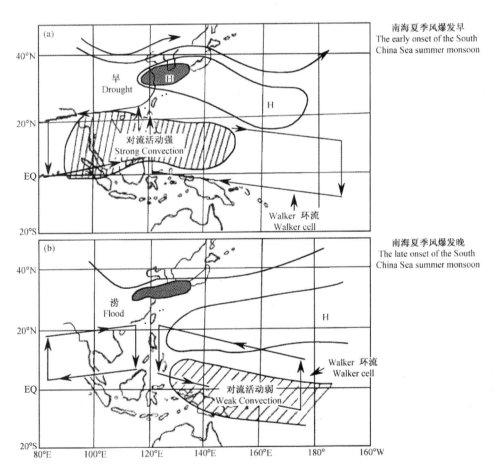

图 15 热带西太平洋(暖池)的热力状态、菲律宾周围对流活动、南海季风爆发早晚、西太平洋副热带高压与江淮流域旱涝分布的关系示意图:(a)暖池处于暖状态;(b)暖池处于冷状态

Fig. 15 Schematic map of the relationships among the SST in the tropical western Pacific (TWP), the convective activities around the Philippines, the western Pacific subtropical high, the onset of the South China Sea summer monsoon (SCSM) and the summer rainfall in the Yangtze River and the Huaihe River valleys. (a) In the warming state of the TWP; (b) In the cooling state of the TWP

本文还指出，在菲律宾周围不同对流活动所产生的热力强迫对西太平洋副热带高压演变的动力作用，从非线性动力理论说明了菲律宾周围对流活动强弱对西太平洋副热带高压北进以突跳或渐进式向北移动起到重要作用。

参考文献

[1] 竺可桢. 东南季风与中国之雨量. 地理学报, 1934, **1**: 1~26
Zhu Kezhen. Southeast monsoon and rainfall in China. *J. Chinese Geogr. Soc.* (in Chinese), 1934, **1**: 1~26

[2] 涂长望, 黄仕松. 夏季风进退. 气象杂志, 1944, **18**: 1~20
Tu Changwang, Huang Shisong. The advance and retreat of the summer monsoon. *Meteor. May.* (in Chinese), 1994, **18**: 1~20

[3] 叶笃正, 杨大升. 北半球大气中角动量的年变化和输送的机制. 气象学报, 1955, **26**: 281~294
Yeh Tu-Cheng, Yang Da-Sheng. The mechanism of the annual variation and transfer of the angular momentum of the atmosphere in the Northern Hemisphere. *Acta Meteor. Sinica* (in Chinese), 1955, **26**: 281~294

[4] 叶笃正, 朱抱真. 从大气环流看远东地区过渡季节的开始. 气象学报, 1955, **26**: 71~87
Yeh Tu-Cheng, Chu Pao-Chen. The onset of transtioal seasons of Far East from the view-point of general circulation. *Acta Meteor. Sinica* (in Chinese), 1955, **26**: 71~87

[5] Yeh Tu-Cheng. On the mechanism of maintenance of zonal circulation. *Geophysica* (Finland), 1957, **6**: 607~620

[6] Yeh Tu-Cheng, Staff Members of Academia Sinica. On the general circulation over Eastern Asia, (I). *Tellus*, 1957, **9**: 432~446

[7] Yeh Tu-Cheng, Staff Members of Academia Sinica. On the general circulation over Eastern Asia, (II). *Tellus*, 1958, **10**: 58~75

[8] Yeh Tu-Cheng, Staff Members of Academia Sinica. On the general circulation over Eastern Asia, (III). *Tellus*, 1958, **10**: 299~312

[9] 叶笃正, 陶诗言, 李麦村. 在六月和十月大气环流的突变现象. 气象学报, 1958, **29**: 249~263
Yeh Tu-Cheng, Dao Shih-Yen, Li Mei-Tsun. The abrupt change of the atmospheric circulation during June and October. *Acta Meteor. Sinica* (in Chinese), 1958, **29**: 249~263

[10] 叶笃正, 朱抱真. 大气环流的若干基本问题. 北京: 科学出版社, 1958, 159pp
Yeh Tu-Cheng, Chu Pao-Chen. *Some Fundamental Problem of the General Circulation of the Atmospheree* (in Chinese), Beijing: Science Press, 1958, 159pp

[11] 叶笃正, 陶诗言, 朱抱真, 等. 北半球冬季阻塞形势的研究. 北京: 科学出版社, 1962,
Yeh Tu-Cheng, Dao Shih-Yen, Chu Pao-Chen, et al. *Studies on the Blocking Situation in the Northern Hemisphere Winter* (in Chinese), Beijing: Science Press, 1962, 128pp

[12] Yeh Tu-Cheng, Dao Shih-Yen, Li Mei-Tsun. The abrupt change of circulation over the Northern Hemisphere during June and October. *The Atmosphere and the Ocean in Motion*, New York: Rockefeller Institute Press in Association with Oxford University Press, 1959, 249~267

[13] Krishnamurti T N, Ramanathan Y. Sensitivity of monsoon onset to differential heating. *J. Atmos. Sci.*, 1982, **39**: 1290~1306

[14] McBride J J. The Australian summer monsoon. *Monsoon Meteorology*. Chang C P, Krishnamurti T N, Eds. Oxford University Press, 1987, 203~232

[15] 黄荣辉, 李维京. 夏季热带西太平洋上空的热源异常对东亚上空副热带高压的影响及其物理机制. 大气科学, 1988, 特刊: 95~107
Huang Ronghui, Li Weijing. Influence of the heat source anomaly over the tropical western Pacific on the subtropical high over East Asia and its physical mechanism. *Chinese J. Atmos. Sci.* (in Chinese), 1988, special Issue: 95~107

[16] Huang Ronghui, Sun Fongying. Impact of the tropical western Pacific on the East Asian summer monsoon. *J. Meteor. Soc.* Japan, 1992, **70** (1B): 243~256

[17] 黄荣辉, 孙凤英. 热带西太平洋暖池的热状态及其上空的对流活动对东亚夏季气候异常的影响. 大气科学, 1994, **18**: 141~151
Huang Ronghui, Sun Fengying. Impacts of the thermal state and the convective activities in the tropical western Pacific warm pool on the summer climate anomalies in East Asia. *Chinese J. Atmos. Sci.* (in Chinese), 1994, **18**: 141~151

[18] 黄荣辉, 孙凤英. 热带西太平洋暖池上空对流活动对东亚夏季风季节内变化的影响. 大气科学, 1994, **18**: 456~465
Huang Ronghui, Sun Fengying. Impact of the convective activities over the West Pacific warm pool on the intraseasonal variability of summer climate in East Asia. *Chinese J. Atmos. Sci.* (in Chinese), 1994, **18**: 456~465

[19] Huang Ronghui, Zhou Liantong, Chen Wen. The progresses of recent studies on the variabilities of the East Asian monsoon and their causes. *Adv. Atmos. Sci.*, 2003, **20**: 55~69

[20] 叶笃正, 黄荣辉, 等. 长江黄河流域旱涝规律和成因研究. 济南: 山东科技出版社, 1992, 387pp
Ye Duzheng, Huang Ronghui, et al. *Studies on the Regularity and cause of Droughts and Floods in the Yangtze River Valley and the Yellow River Valley* (in Chinese). Jinan: Shandong Science and Technology Press, 1992, 387pp

[21] Tao Shiyan, Chen Longxun. A review of recent research on the East Asian summer monsoon in China. *Monsoon Meteorology*. Chang C P, and Krishnamurti T N, Eds. Oxford University Press, 1987, 60~92

[22] Ding Yihui. Summer monsoon rainfalls in China. *J. Meteor. Soc. Japan*, 1992, **70**: 373~396

[23] Ding Yihui, Wang Qiyi, Yan Junyue. Some aspects of climatology of the summer monsoon over the South China Sea. *From Atmospheric Circulation to Global Change*. Edited by IAP/CAS, China Meteorological Press, 1996, 107~117

[24] Wang B, Lin H, Zhang Y S, et al. Definition of South China sea monsoon onset and commencement of the East Asia summer monsoon. *J. Climate*, 2004, **17**: 699~710

[25] 梁建茵, 吴尚森. 南海西南季风爆发日期及其影响因子. 大气科学, 2002, **26**: 829~844
Liang Jianyin, Wu Shangsen. A study of southwest monsoon onset date over the South China Sea and its impact factors. *Chinese J. Atmos.*

Sci. (in Chinese), 2002, **26**: 844~855

[26] 丁一汇, 马鹤年. 东亚季风的研究现状. 见: 何金海编. 亚洲季风研究的新进展. 北京: 气象出版社, 1996, 1~14
Ding Yihui, Ma Henian. The present status and future of research of the East Asian monsoon. *The Recent Advances in Asian monsoon Research* (in Chinese). He Jinhai, Ed. Beijing: China Meteorological Press, 1996, 237pp

[27] Ding Yihui, Sun Ying. A study on anomalous activities of East Asian summer monsoon. *J. Meteor. Soc. Japan*, 2001, **79**: 1119~1137

[28] Cornejo-Garrido A G, Stone P H. On the heat balance of the Walker circulation. *J. Atmos. Sci.*, 1977, **34**: 1155~1162

[29] Huang Ronghui, Lu Li. Numerical simulation of the relationship between the anomaly of the subtropical high over East Asia and the convective activities in the western tropical Pacific. *Adv. Atmos. Sci.*, 1989, **6**: 202~214

[30] Nitta Ts. Long-term variations of cloud amount in the western Pacific region. *J. Meteor. Soc. Japan*, 1986, **64**: 373~380

[31] Charney J G, Devore J G. Multiple flow equilibria in the atmosphere and blocking. *J. Atmos. Sci.*, 1979, **36**: 1205~1216

[32] 曹杰, 黄荣辉, 谢应齐, 等. 西太平洋副热带高压演变物理机制的研究. 中国科学 (B辑), 2002, **32**: 659~666
Cao Jie, Huang Ronghui, Xie Yingqi, et al. Research on the evolution mechanism of the western Pacific subtropical high. *Science in China* (Series B, in Chinese), 2002, **45**: 659~666

Impact of Thermal State of the Tropical Western Pacific on Onset Date and Process of the South China Sea Summer Monsoon*

Huang Ronghui*[1] (黄荣辉), Gu Lei[1] (顾雷), Zhou Liantong[1] (周连童), and Wu Shangsen[2] (吴尚森)

[1] State Key Laboratory of Numerical Modeling for Atmospheric Sciences and Geophysical Fluid Dynamics (LASG), Institute of Atmospheric Physics, Chinese Academy of Sciences, Beijing 100080
[2] Guangzhou Institute of Tropical and Oceanic Meteorology, Guangzhou 510080

ABSTRACT

Since the early or late onset of the South China Sea summer monsoon (SCSM) has a large impact on summer monsoon rainfall in East Asia, the mechanism and process of early or late onset of the SCSM are an worthy issue to study. In this paper, the results analyzed by using the observed data show that the onset date and process of the SCSM are closely associated with the thermal state of the tropical western Pacific in spring. When the tropical western Pacific is in a warming state in spring, the western Pacific subtropical high shifts eastward, and twin cyclones are early caused over the Bay of Bengal and Sumatra before the SCSM onset. In this case, the cyclonic circulation located over the Bay of Bengal can be early intensified and become into a strong trough. Thus, the westerly flow and convective activity can be intensified over Sumatra, the Indo-China Peninsula and the South China Sea (SCS) in mid-May. This leads to early onset of the SCSM. In contrast, when the tropical western Pacific is in a cooling state, the western Pacific subtropical high anomalously shifts westward, the twin cyclones located over the equatorial eastern Indian Ocean and Sumatra are weakened, and the twin anomaly anticyclones appear over these regions from late April to mid-May. Thus, the westerly flow and convective activity cannot be early intensified over the Indo-China Peninsula and the SCS. Only when the western Pacific subtropical high moves eastward, the weak trough located over the Bay of Bengal can be intensified and become into a strong trough, the strong southwesterly wind and convective activity can be intensified over the Indo-China Peninsula and the SCS in late May. Thus, this leads to late onset of the SCSM. Moreover, in this paper, the influencing mechanism of the thermal state of the tropical western Pacific on the SCSM onset is discussed further from the Walker circulation anomalies in the different thermal states of the tropical western Pacific.

1. Introduction

The South China Sea (SCS) and its surrounding areas are a key region of the Asian monsoon. They are not only the vital linkage between the East Asian monsoon region and the South Asian monsoon region, but also a linkage between the Eurasian continent and the tropical western Pacific. According to the studies made by Tao and Chen (1987) and He et al. (1987), the earliest onset of the Asian summer monsoon is found in the region over the SCS and the Indo-China Peninsula. The appearance of strong convective activity and the southwesterly flow over the SCS signals the onset of the Asian summer monsoon. Generally, the summer monsoon over the SCS is called as the South China Sea summer monsoon (SCSM) in China. Since the SCSM onset has a significant influence on the mei-yu in the Yangtze River valleys and the Huaihe River valleys, the Baiu in Japan and the Changma in Korea, Chinese meteorologists and many scholars in the world pay much effort to investigate onset and its physical processes of the SCSM (e.g., Zhu et al., 1986; Huang and Tao, 1992; Ding et al., 1996; Wang et al., 2004). In order to understand the regularity

---

* 本文原载于: Advances in Atmospheric Sciences, Vol. 23, No.6, 909-924, 2006.

and cause of the summer monsoon over the SCS and its surrounding areas, and to improve the prediction of the SCSM variability, an integrated large-scale experiment on the South China Sea Monsoon (SCSMEX) had been implemented during the period from May to August, 1998 (e.g., Ding et al., 2002). This experiment has provided an observed dataset of atmosphere and ocean for the study of the SCSM onset.

In order to investigate the interannual variability of onset date and process of the SCSM, it is necessary to define an index for measuring the SCSM onset. However, the definition of the SCSM onset is complex. Recently, many scholars pay attention to this problem and proposed many ideas. Summarizing these ideas, there are two viewpoints: One emphasized the important role of the sea temperature of the SCS in the SCSM onset. For example, He et al. (1992) pointed out that the SCSM onset may be associated with the appearance of the high sea temperature in the southern part of the SCS. And He and Luo (1999) proposed that the SCSM onset may be due to the reversion of meridional temperature gradient in the Indo-China Peninsula and the SCS. Besides, Ose et al. (1997) showed that the sea temperature in the SCS is very important for Asian monsoon and may be an index for Asian monsoon and ENSO system. Another mainly considered the effect of circulation variations over the tropical western Pacific, the Indo-China Peninsula and the tropical eastern Indian Ocean on the SCSM onset. For example, Murakami et al. (1986) put forward that the transition of pressure gradient between Asia and Australia from winter pattern to summer pattern can cause the SCSM onset. Lau and Peng (1990), and Miao and Lau (1991) suggested that the actions of the Madden-Julian Oscillation (MJO) in the circulation over the tropical eastern Indian Ocean and the Indo-China Peninsula can trigger the SCSM onset. Liang and Wu (2002), and Wang et al. (2004) considered the effect of the westerly wind over the Bay of Bengal and the tropical eastern Indian Ocean on the SCSM onset.

The SCSM onset should be not a local phenomenon. It may not be only due to the local thermal contrast, such as the contrast of the thermal state between the SCS and the Indo-China Peninsula (e.g., He and Luo, 1999), but also be a large scale phenomenon of the transition from winter circulation to summer circulation over the tropical western Pacific, the SCS and the Indo-China Peninsula. Numerous studies showed that in boreal summer, there are three major heat sources, which are located over the SCS and the tropical western Pacific, the Bay of Bengal and the Tibetan Plateau, respectively (e.g., Chen and Li, 1981; Luo and Yanai, 1984; Yanai et al., 1992). Moreover, Chen et al. (1983) pointed out that the transition from a heat sink to a heat source first occurs over the SCS and its surrounding areas, and onset of the Asian summer monsoon is closely related to the heating transition over this region. The formation of Asian monsoon circulation over the SCS may be considered as the adaptation of atmospheric motion to the heating transition by semi-geostrophic or geostrophic balance (e.g., Chao and Lin, 1996; Xie and Saiki, 1999). However, this heating transition is closely associated with the thermal state of the tropical western Pacific because the heating transition depends on convective activity over the tropical western Pacific and the SCS (e.g., Huang and Sun, 1992; Ren and Huang, 1999; Huang et al., 2003). This shows that the SCSM onset may be closely associated with the thermal state of the tropical western Pacific. Thus, the impact of the thermal state of the tropical western Pacific on the SCSM onset is examined in this study.

In order to study the impact of the thermal state of the tropical western Pacific on the onset date and process of the SCSM, in this study we used the observed data including the sea surface and subsurface emperature in the tropical Pacific and high cloud amount around the Philippines for 1978–1999 provided by JMA, OLR for 1979–1999 provided by National Oceanic and Atmospheric Administration (NOAA), and the National Centers for Environmental Prediction/National Center for Atmospheric Research (NCEP/NCAR) reanalysis data of wind fields for 1950–1999 (e.g., Kalnay et al., 1996), especially NCEP/NCAR reanalysis II for 1979–1999. Moreover, the physical mechanism for the impact of thermal state of the tropical western Pacific on the SCSM onset is investigated further by examining the influence of the thermal state of the tropical western Pacific on the Walker circulation over the area from the tropical eastern Indian Ocean to the tropical western Pacific.

2. Influence of the SCSM onset on summer monsoon rainfall in China

There are many definitions of the SCSM onset in the studies of the interannual variations of the SCSM. Briefly, these definitions may be summarized into two kinds: One is defined by using rain or OLR (or TBB and BT), which may describe the characteristic of strong rainfall in the region of the SCSM. For example, Tao and Chen (1985) suggested that the abrupt increase of rainfall in the SCSM region may be considered as a criteria for measuring SCSM onset. Jin (1999) pointed out that the TBB$\leqslant$274 K, which may describe the characteristic of strong convective activity over the SCS, can be considered as a criteria for measuring the SCSM onset. Qian and Yang (2000) defined a criteria for measuring the SCSM onset using precipitation (R), brightness temperature (BT), OLR and lower-tropospheric winds, and suggested that the OLR< 240 W m^{-2}, $R > 4$–6 mm d^{-1}, BT<244 K may be used as the threshold values of SCSM onset. Another is defined by using wind. For instance, Li and Qu (1999) proposed the large difference between the divergence in the upper troposphere and that in the lower troposphere over the SCS may be as criteria for measuring the SCSM onset. And Liang and Wu (2002) proposed the following definition of the SCSM onset:

(1) The continuous days of the westerly wind at 850 hPa averaged for the area (5°–20°N, 110°–120°E) is longer than 5 days;

(2) The westerly flow is mainly from the Bay of Bengal;

(3) The following continuous break days of the southwesterly flow are not longer than the three times of the preceding continuous days of the southwesterly flow.

This definition is different from that suggested by Wang et al. (2004), who proposed a definition of the SCSM onset with pentad mean zonal wind at 850 hPa averaged over the region of 5°–15°N, 100°–120°E. According to the above-mentioned definitions of the SCSM onset suggested by Liang and Wu (2002) and Wang et al. (2004), the interannual variations of the SCSM onset date from 1950 to 1999 are shown in Fig. 1, respectively. It may be seen from Fig. 1 that the SCSM onset is generally in mid-May, which is in agreement with the result studied by Tao and Chen (1987). Moreover, Fig. 1 features a large intreannual variations in the SCSM onset date, the earliest onset could be found in late April, but the latest onset was in early June. Moreover, it may be also found from Fig. 1 that there are some differences between the SCSM onset date decided by using the definition of the SCSM onset by Liang and Wu (2002) and that cited from Wang et al. (2004).

Li and Zhang (1999) showed that the SCSM onset has a great influence on the Asian summer monsoon circulation over East Asia. Thus, there may be a good relationship between onset date of the SCSM and the mei-yu in the Yangtze River and the Huaihe River valleys (i.e., Jianghuai valley in Chinese), the Baiu in Japan and the Changma in Korea. In order to study this relationship, the correlations between onset date of the SCSM and summer rainfall in China are analyzed by using the observed data of summer precipitation and the SCSM onset date for the period of 1951–1999 shown in Fig. 1, since the observed data of summer precipitation in China are available from 1951. Figure 2 is the distribution of correlation coefficients between the summer (June–August) precipitation in China and the SCSM onset date defined by Liang and Wu (2002) for 49 summers from 1951 to 1999. It may be clearly seen from Fig. 2 that there is an obvious positive correlation in the Yangtze River and the Huaihe River valleys, the correlation coefficient between them is larger than 0.3 and exceeds the 95% confidence level in this region. This shows that when the SCSM onset is early, such as in 1994, summer rainfall is below normal in the Yangtze River and the Huaihe River valleys. In contrast, when the SCSM onset is late, such as in 1987 and 1991, summer rainfall is above normal, and severe flooding may occur in this region. This correla-

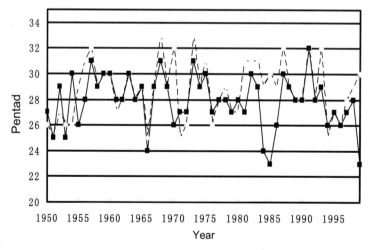

Fig. 1. Interannual variations of the SCSM onset date (solid lines) during 1950–1999 decided by using Liang and Wu's (2002) definition and those (dashed line) cited from Wang et al. (2004).

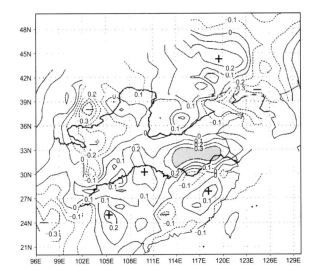

Fig. 2. Distribution of the correlation coefficients between the summer (June–August) rainfall anomalies in China and the SCSM onset dates during the period of 1951–1999. The solid and dashed contours indicate positive and negative correlations, respectively, and the correlations over the 95% confidence level are shaded. Adopted from Huang et al. (2005).

tion has been used in the seasonal prediction of summer monsoon rainfall anomalies in the Yangtze River and the Huaihe River valleys.

Besides, as shown in Fig. 2, there are also negative correlations between the SCSM onset dates and the summer rainfall anomalies in the southern part of Northeast China and the eastern part of Northwest China, and the correlation coefficients also exceed the 95% confidence level in these regions. Thus, when the SCSM onset is early, there may be positive summer rainfall anomalies in these regions.

3. Impact of the thermal state of the tropical western Pacific on the SCSM onset date

As shown by many studies, the tropical western Pacific is a region of the highest SST in the global sea surface and is known as "the warm pool" (e.g., Cornejo-Garrido and Stone, 1977; Nitta, 1986, 1987; Huang and Li, 1987, 1988). Since the SCS is located near this region, onset and maintenance of the SCSM may be greatly influenced by the thermal state of the tropical western Pacific. Therefore, impact of the thermal state of the tropical western Pacific on the SCSM onset is analyzed by using the observed data in this section. Moreover, the impact of the thermal state of the tropical western Pacific and convective activity over this region on the SCSM onset from 1978 are emphasized in the following analyses, because the observed data of high cloud amount over East Asia and the tropical western Pacific from 1978 are available, and because the interdecadal variation of Asian summer monsoon from the late 1970s can be avoided

As shown in Fig. 1, the climatological-mean onset date of the SCSM is about 19th May. Therefore, if the onset date of the SCSM is more than four days earlier than 19th May, the SCSM onset is considered as a case of early onset in this study, such as in 1979, 1981, 1984,

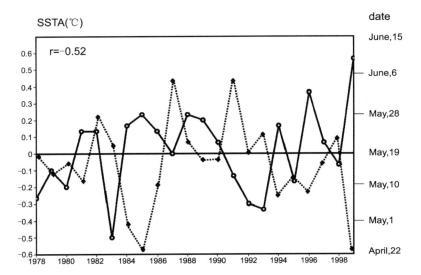

Fig. 3. Interannual variations of the SST anomaly averaged for the area of El Niño.west (i.e., 0°–14°N, 130°–150°E) in spring (March–May) (solid line, Unit: °C) and the SCSM onset date (dashed line) from 1978 to 1999. Data of SST are obtained from Monthly Report on Climate System, JMA, and the SCSM onset dates are provided by Liang and Wu (2002).

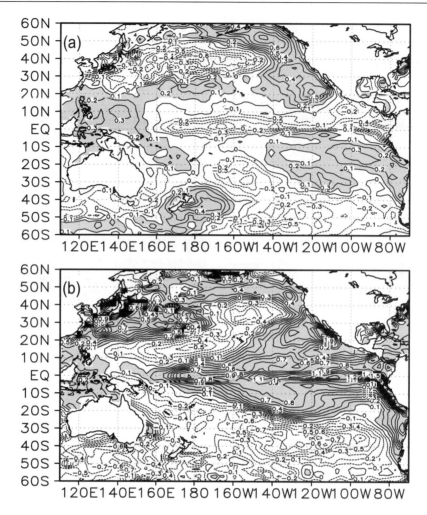

Fig. 4. Composite distributions of the SST anomalies in the Pacific Ocean in spring (March–May) for the early cases (a) and the late cases (b) of the SCSM onset. Unit: °C. The solid and dashed contours (contour interval: 0.1°C) in Figs. 4a and 4b denote positive and negative SST anomalies, respectively, and the positive anomalies over 0.3 are shaded. Data of SST are obtained from Monthly Report of Climate System, JMA.

1985, 1986, 1994, 1996 and 1999. On the other hand, if the onset date of the SCSM is more than four days later than 19th May, the SCSM onset is considered as a case of late onset, such as in 1982, 1987, 1991, 1993 and 1998. By this way, the SCSM onset is considered as normal in 1978, 1980, 1983, 1988, 1989, 1990, 1992 and 1997. The early or late onset of the SCSM may be closely associated with the thermal state of the tropical western Pacific. Figure 3 shows the interannual variations of the SST anomaly averaged over the area of El Niño.west (i.e., 0°–14°N, 130°–150°E) in spring (March–May) and the SCSM onset dates from 1978 to 1999, respectively. Compared the SCSM onset dates (dashed line) with the SST anomalies in the tropical western Pacific (solid line) shown in Fig. 3, it can be found that there is an out-phase relationship between them. The correlation coefficient between them is -0.52 and exceeds the 95% confidence level, which is -0.423. As shown in Fig. 3, the early onset of the SCSM generally occurs in a warming period of the tropical western Pacific, while in a cooling period of this region, the SCSM onset is late.

In order to show clearly the relationship between the SCSM onset date and the thermal state of the tropical western Pacific, composite distributions of the SST anomalies in the tropical Pacific for the springs

(March–May) for the early and late onsets of the SCSM are analyzed using the observed data of SST anomalies in the Pacific Ocean from 1978 to 1999, respectively (see Fig. 4). As shown in Fig. 4a, positive SST anomalies can be found in the tropical western Pacific and negative SST anomalies appear in the equatorial eastern Pacific for the early case of the SCSM onset. This is similar to a distribution of the SST anomalies in developing stage of a La Niña event. This indicates that early onset of the SCSM generally occurs in a warming period of the tropical western Pacific, i.e., generally in developing stage of a La Niña event. On the other hand, it can be seen from Fig. 4b that for the late case of the SCSM onset, negative SST anomalies distribute in the tropical western Pacific and positive SST anomalies appear in the equatorial eastern Pacific, which is similar to a distribution of the SST anomalies in developing stage of an El Niño event. This indicates that late onset of the SCSM generally occurs in a cooling period of the tropical western Pacific, i.e., generally in developing stage of an El Niño event. Therefore, there is also good correlation between the SST anomalies averaged for the Niño-3 area (i.e., 5°N–5°S, 90°–150°W) in spring and the SCSM onset dates from 1978 to 1999. The correlation coefficient between them is 0.48, which also exceeds the 95% confidence level. However, it is smaller than the correlation coefficient between the SST anomalies in the El Niño.west and the SCSM onset dates.

Since the tropical western Pacific is characterized by a deep thermocline and the heat content anomalies in this region are in close association with the sea temperature (ST) anomalies in the subsurface layer of the tropical western Pacific (e.g., Yasunari, 1990; Huang and Sun, 1992), the ST anomaly in the subsurface layer of the tropical western Pacific may be a better signal of early or late onset of the SCSM. Thus, data of the sea temperature anomalies in the subsurface layer of the Pacific along 137°E in January from 1967 to 1999, which are observed by the Oceanographic Research Vessel, JMA, (i.e., JMA, Monthly Report on Climate System 1978–2000) are also used to analyze the relationship between the SCSM onset and the ST anomaly in the subsurface layer of the tropical western Pacific (Figure is omitted). The result shows that there is also a close relationship between the SCSM onset and the thermal state in the subsurface layer of the tropical western Pacific. Early onset of the SCSM occurs generally in a warming period of the tropical western Pacific, but late onset of the SCSM is generally caused in a cooling period of this region.

4. Influence of the convective activity over the tropical western Pacific on the SCSM onset date

As described in Section 3, the tropical western Pacific is a region of the highest SST in the global sea surface. Due to the warm state of this region, the air-sea interaction is very strong, and the ascending branch of the Walker circulation is over the region. Thus, strong convergence of the air and moisture leads to strong co-

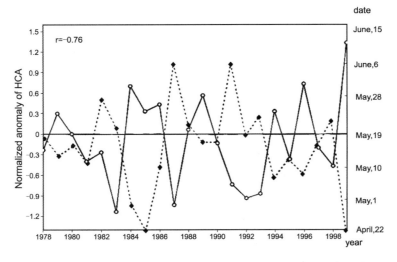

Fig. 5. Interannual variations of normalized high cloud amount (HCA) averaged for the area around the Philippines (i.e., 10°–20°N, 110°–140°E) in spring (solid line) and the SCSM onset dates (dashed line) (from Liang and Wu, 2002). The observed data of HCA is obtained from Monthly Report on Climate System, JMA.

nvective activity and heavy rainfall there (e.g., Connejo-Garrido and Stone, 1977; Hartmann et al., 1984). Nitta's (1986) study showed the convective activity over the tropical western Pacific is closely associated with the sea surface temperature of this region. When the tropical western Pacific is in a warming state, i.e., the SST anomaly in the area of El Niño.west (i.e., $0°–14°N$, $130°–150°E$) is positive, convective activity is strong around the Philippines, and vice versa. Therefore, the thermal state of the tropical western Pacific can be well described by the convective activity around the Philippines.

In order to explain well the impact of convective activity over the tropical western Pacific on the SCSM onset, the relationship between the SCSM onset date and convective activity over the tropical western Pacific in spring is analyzed using the observed data of high cloud amount (HCA) around the Philippines (see Fig. 5). The solid line in Fig. 5 denotes the interannual variations of normalized HCA anomaly around the Philippines (i.e., $10°–20°N$, $110°–140°E$) in spring from 1978 to 1999. As shown in Fig. 5, there is also an out-phase relationship between the SCSM onset dates (dashed line in Fig. 5) and the HCA anomalies around the Philippines in spring. The correlation coefficient between them reaches to -0.76, which exceeds the 99% confidence level. This can explain well that the convective activity over the tropical western Pacific in spring has a great influence on the SCSM onset. In a spring with strong convective activity around the Philippines, which are generally found in a warming period of this region, the SCSM onset is early. On the other hand, the SCSM onset is late in a spring with weak convective activities over the tropical western Pacific, which is generally found in a cooling period of the tropical western Pacific.

From the analyses mentioned in Sections 3 and 4, it may be concluded that the thermal state of the tropical western Pacific and the convective activity around the Philippines have a larger influence on the SCSM onset than the thermal state of the equatorial eastern Pacific.

5. Influence of the thermal state of the tropical western Pacific on the SCSM onset process

The thermal state of the tropical western Pacific has important influences on not only the SCSM onset date described in Section 3, but also the SCSM onset process. Lau et al. (2000), and Ding and Liu (2001) analyzed the onset process of the SCSM in 1998, and pointed out the important role of the twin cyclones over the tropical eastern Indian Ocean and the Bay of Bengal in the SCSM onset. However, the SCSM onset process in different thermal states of the tropical western Pacific may be different. Thus, the onset processes of the SCSM in the warming and cooling states of the tropical western Pacific will be discussed, respectively, as follows.

5.1 *The onset process of the SCSM in a warming state of the tropical western Pacific*

In order to study the influence of the thermal state of the tropical western Pacific on the SCSM onset process, composite distributions of pentad mean OLR and wind field at 850 hPa for various pentads of the springs with the warming and cooling states of the tropical western Pacific during 1979–2000 are given, respectively. Figures 6a–6f are the composite distributions of pentad mean OLR and wind at 850 hPa for the springs with the warming state of the tropical western Pacific i.e., for the springs of 1981, 1984, 1985, 1986, 1988, 1989, 1990, 1994, 1996 and 1999. As shown in Fig. 6a, on April 21–25, there are twin weak cyclones, which are symmetric to the equator, over the Bay of Bengal and the west of Sumatra, and convective activity is intensified over the Indo-China Peninsula and Sumatra. On April 26–30, as shown in Fig. 6b, while the western Pacific subtropical high shifts eastward, the twin cyclones move close to the Indo-China Peninsula, and the convective activity over the Indo-China Peninsula and Sumatra is further intensified. In early May, as shown in Figs. 6c and 6d, the twin cyclones become stronger, which leads to the stronger westerly flow along the equatorial eastern Indian Ocean. As shown in Fig. 6e, on May 11–15, the western Pacific subtropical high shifts eastward to the east of the Philippines. The twin cyclones are further intensified, and the cyclone located over the Bay of Bengal changes into a strong trough. This may cause the strong westerly flow over the equatorial eastern Indian Ocean to the west of Sumatra and the stronger southwesterly flow over the Indo-China Peninsula, and the convective activity is further intensified over the Indo-China Peninsula and the equatorial eastern Indian Ocean. On May 16–20, as shown in Fig. 6f, the trough located over the Bay of Bengal is further intensified with a strong cyclonic circulation appearing over the area to the west of Sumatra, the strong westerly winds prevail over the tropical eastern Indian Ocean, and the southwesterly flow becomes stronger over the Indo-China Peninsula and the SCS. This leads to early onset of the SCSM.

In order to emphasize the role of the twin cyclones in the onset process of the SCSM, the composite distributions of pentad mean wind anomaly at 850 hPa for the springs with the warming state of the tropical western Pacific are also analyzed. From Figs. 7a–7f, it

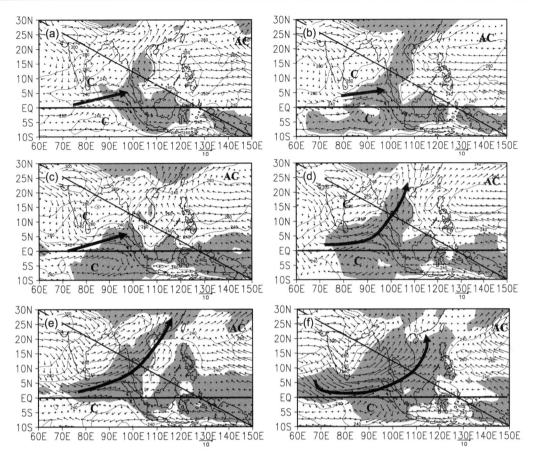

Fig. 6. Composite distributions of pentad mean OLR (Units: W m^{-2}, dashed contours) and wind field at 850 hPa (Units: m s^{-1}) for the springs with the warming state of the tropical western Pacific on (a) April 21–25, (b) April 26–30, (c) May 1–5, (d) May 6–10, (e) May 11–15, and (f) May 16–20. The areas of OLR<230 W m^{-2} are shaded. The data of wind are from the NCEP/NCAR reanalysis II for the period of 1978–2000 (e.g., Kalnay et al., 1996).

may be seen that the twin anomaly cyclones appear over the equatorial eastern Indian Ocean from late April to mid-May. When the twin anomaly cyclones move eastward close to Sumatra and the Indo-China Peninsula, they become greatly intensified. This leads to the strong westerly anomalies over the Indo-China Peninsula, Sumatra and the SCS. Thus, this can cause early onset of the SCSM.

The above-mentioned results indicate that intensification of the twin cyclones over the equatorial eastern Indian Ocean and Sumatra and the eastward shift of the western Pacific subtropical high play a significant role in the onset process of the SCSM.

5.2 *The onset process of the SCSM in the cooling state of the tropical western Pacific*

Figures 8a–8h are the composite distributions of pentad mean OLR and wind field at 850 hPa for the springs with the cooling state of the tropical western Pacific, i.e., for the springs of 1982, 1983, 1987, 1988, 1991, 1993 and 1998. As shown in Figs. 8a and 8b, in late April, there are twin cyclones over the tropical Indian Ocean to the west of Sumatra, and the western Pacific subtropical high shifts westward and is located over the SCS and the tropical western Pacific. And strong convective activity appears over the Indonesia including Sumatra. However, as shown in Figs. 8c–8f, the western Pacific subtropical high continuously shifts westward from late April to mid-May and an anticyclonic circulation appears over the SCS in mid-May, which causes the weakening of the twin cyclones. Wang et al. (2000), and Wang and Zhang (2002) analyzed the evolution of circulation over the tropical western Pacific around the Philippines during

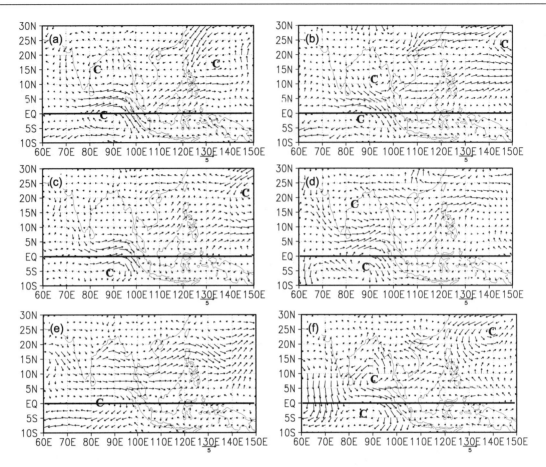

Fig. 7. Composite distributions of pentad mean wind anomaly (Units: m s^{-1}) at 850 hPa for the springs with the warming state of the tropical western Pacific on (a) April 21–25, (b) April 26–30, (c) May 1–5, (d) May 6–10, (e) May 11–15, and (f) May 16–20. The data of wind field are from the NCEP/NCAR reanalysis II for the period of 1978–2000 (e.g., Kalnay, et al., 1996).

developing stage of El Niño events, in which the SST anomalies are similar to the distribution shown in Fig. 4, i.e., the tropical western Pacific is in a cooling state. They pointed out that an anticyclonic circulation appears over the area around the Philippines during development of El Niño. The present result is in agreement with the results by Wang et al. (2000) and Wang and Zhang (2002). On May 21–25, as shown in Fig. 8g, a cyclone appears over the equatorial eastern Indian Ocean and the west of Sumatra, and the trough located over the Bay of Bengal becomes stronger, while the western Pacific subtropical high shifts eastward to the east of the Philippines. These cause the intensification of the convective activity over the equatorial eastern Indian Ocean, Sumatra and the Indo-China Peninsula. As shown in Fig. 8h, the trough located over the Bay of Bengal and the cyclone located to the west of Sumatra are further intensified in late May, which cause the strong southwesterly flow over the Indo-China Peninsula and the SCS. This leads to late onset of the SCSM.

In order to compare the difference between the onset process of the SCSM in the warming state of the tropical western Pacific and that in the cooling state, the composite distribution of pentad mean anomaly wind field at 850 hPa for the springs with the cooling state of the tropical western Pacific, as show in Figs. 9a–9h. Compared Figs. 9a–9f with Figs. 7a–7f, it can be found that the evolution of anomaly wind field at 850 hPa in the cooling state of the tropical western Pacific is significantly different from that in the warming state. It can be clearly seen from Figs. 9a–9f that in the cooling state of the tropical western Pacific, the twin anomaly anticyclones always appear over the equatorial eastern Indian Ocean and the Bay of Bengal, which can cause the strong easterly wind

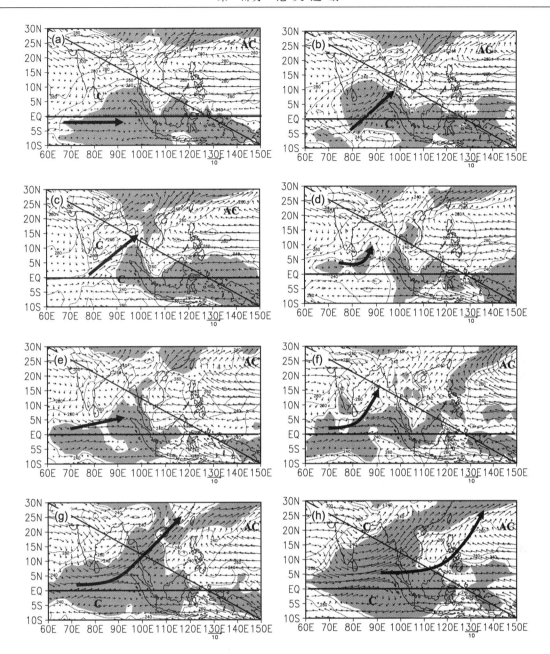

Fig. 8. Same as Fig. 6 but for the cooling state of the tropical western Pacific.

anomalies over the Indo-China Peninsula and the equatorial eastern Indian Ocean from late April to mid-May. When the twin anomaly anticyclones are weakened in late May, the twin weak anomaly cyclones appear over the Indo-China Peninsula, Sumatra and the equatorial eastern Indian Ocean, which induces the southerly wind anomalies over the Indo-China Peninsula and the southwesterly wind anomalies over the northern part of the SCS and South China. Thus, this leads to late onset of the SCSM.

Compared the onset process of the SCSM in the warming state with that in the cooling state of the tropical western Pacific, it can be clearly seen that there is an obvious difference between them. When the tropical western Pacific in a warming state in spring, since the western Pacific subtropical high early moves

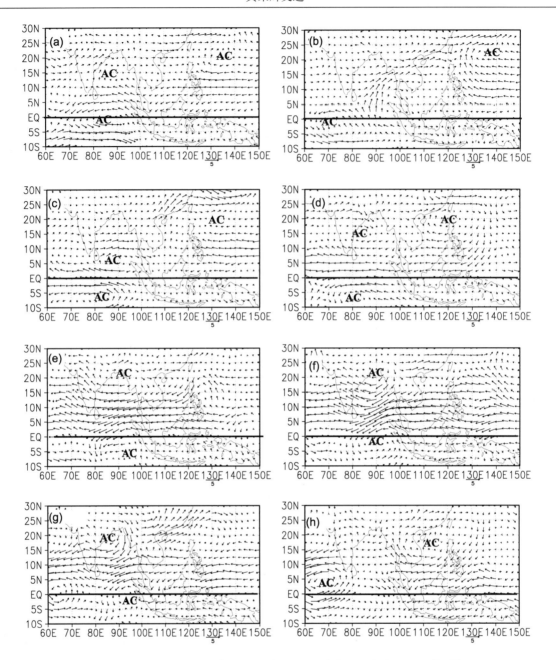

Fig. 9. Same as Fig. 7 but for the cooling state of the tropical western Pacific.

eastward to the east of the Philippines, the twin anomaly cyclones appear over the equatorial eastern Indian Ocean from late April to mid-May. This can cause that the twin cyclones being symmetric to the equator are early formed over the Bay of Bengal and the west of Sumatra before the SCSM onset. Moreover, since the cyclone located over the Bay of Bengal is greatly intensified and becomes a strong trough in mid-May, the convective activity is early intensified over the Indo-China Peninsula and Sumatra, and the strong southwesterly flow early appears over the equatorial eastern Indian Ocean, the Indo-China Peninsula and the SCS. This leads to the early onset of the SCSM. On the other hand, when the tropical western Pacific is in a cooling state in spring, since the twin anomaly anticyclones always appear over the equatorial eastern Indian Ocean and the Bay of Bengal and the western Pacific subtropical high shifts westward from late April to mid-May, and an anticyclonic circulation appears over the SCS in mid-May. This leads to the weakening of the twin cyclones over these regions. In late May, the western Pacific subtropical high shifts eastward to the east of the Philippines, which leads to the appearance of a strong trough over the Bay of Bengal. This strong trough can cause the intensification of convective activity over the Indo-China Peninsula and the SCS. Thus, the strong southwesterly flow can be caused over the Indo-China Peninsula, the northern part of SCS and South China. This leads to late onset of the SCSM.

From the influences of the thermal state and convective activity over the tropical western Pacific on the SCSM onset mentioned in Sections 3 and 4, and from the impact of the SCSM onset on the summer monsoon rainfall in China described in Section 2, it may be found that there are consistent links of the summer monsoon rainfall in China to the thermal state and convective activity over the tropical western Pacific. Huang and Li (1987, 1988), and Huang and Sun (1992) investigated the relations among the thermal states of the tropical western Pacific, the convective activity over the tropical western Pacific, the western Pacific subtropical high and the summer monsoon rainfall in China. Their results showed as follows: When the tropical western Pacific is in a warming state, the convective activity is intensified from the Indo-China Peninsula to the area around the Philippines, the western Pacific subtropical high may shift unusually northward, and the summer monsoon rainfall in the Yangtze River and Huaihe River valleys may be below normal. In this case, the SCSM onset is early. Oppositely, when the tropical western is a cooling state, the convective activity is weakened around the Philippines and is intensified over the equatorial central Pacific near the dateline, the western Pacific subtropical high may shift southward and westward, and the summer monsoon rainfall may be above normal in the Yangtze River and Huaihe River valleys of China. In this case, the SCSM onset is late. Moreover, Huang et al. (2004), and Huang et al. (2005) investigated further the impact of the thermal state and convective activity in the tropical western Pacific on the summer monsoon rainfall in East Asia by examing the intraseasonal evolution of the East Asian summer monsoon. These results showed the effect of thermal state and the convective activity over the tropical western Pacific on the East Asian summer monsoon circulation through the so-called East Asian/Pacific pattern (EAP pattern) (e.g., Nitta, 1987; Huang and Li, 1987, 1988). Therefore, the impact of the thermal state and convection over the tropical western Pacific on the summer monsoon rainfall in China may be also through their impact on the SCSM onset. This may also suggest that the significant impact of the early or late onset of SCSM on the summer monsoon rainfall in China is also through the EAP pattern.

6. The possible physical mechanism of the impact of the thermal state of the tropical western Pacific on the SCSM

The impact of thermal state of the tropical western Pacific and the convective activity around the Philippines on the SCSM onset may be interpreted from the relationship between the thermal state of the tropical western Pacific and the Walker circulation over the tropical Pacific and Indian Ocean. Since the convective activity generally is very strong over the tropical western Pacific as shown in Section 4, the associated heating supplies the energy to drive the strong ascending branch of the Walker circulation (e.g., Connejo-Garrido and Stone, 1977; Hartmann et al., 1984). In order to explain the possible physical mechanism of the impact of thermal state of the tropical western Pacific on the SCSM onset, the wind data from NCEP/NCAR reanalysis II from 1979 to 1999 are also used to analyze the Walker circulation over the tropical western Pacific and the Indian Ocean. As shown in Section 2, the SCSM onset generally occurs in mid-May, but the latest onset can be found in early June. Therefore, the impact of thermal state of the tropical western Pacific in spring on the Walker circulation over the tropical western Pacific and the tropical eastern Indian Ocean during May and June is focused in this section.

Figures 10a and 10b are the composite distributions of zonal-altitude circulation anomaly over the tropical western Pacific, the SCS and the tropical Indian Ocean (i.e., average between $5°–15°N$) averaged during the period of 1 May–30 June for the warming and cooling states of the tropical western Pacific in spring for 1979–1999, respectively. Figure 10a features a stronger ascending branch of the Walker circulation over the tropical western Pacific and stronger westerly anomalies in the lower troposphere over the tropical eastern Indian Ocean and the SCS. As shown in Figs. 7e and 7f, an anomalous westerly wind at 850 hPa appears over the area from the tropical eastern Indian Ocean to Sumatra from mid-May, and then the southwesterly flow can early enter into the area over the SCS from the Indo-China Peninsula. As a consequence, the SCSM onset is early in a warming stage of the tropical western Pacific.

On the other hand, when the tropical western Pacific is in a cooling stage, the composite distribution of zonal-altitude circulation anomaly shown in Fig. 10b is opposite to that shown in Fig. 10a. In this case, an anomalous descending flow appears over the tropical western Pacific, which shows a weaker ascending branch of the Walker circulation over this region, and easterly anomaly appears in the lower troposphere over the tropical eastern Indian Ocean and the SCS. As shown in Figs. 9a 9f, the anomaly easterly wind at 850 hPa appears over the area from the tropical eastern Indian Ocean to Sumatra from late April to mid-May. Thus, the southwesterly flow can not enter early into the SCS. As shown in Figs. 8g and 8h, the southwesterly flow can be found over the Indo-China Peninsula and the SCS in late May. As a consequence, the SCSM onset is late in a cooling stage of the tropical western Pacific.

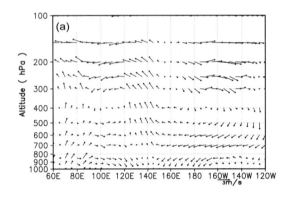

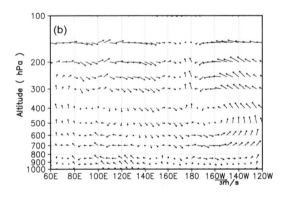

Fig. 10. Composite distributions of zonal-altitude circulation anomaly averaged over 5°–15°N during the period of May 1–June 30 for (a) the warming and (b) the cooling states of the tropical western Pacific in spring for 1979–1999. The climatological mean zonal-altitude circulation averaged for 1979–1999 is taken as the normal. The anomalies of vertical velocity and the anomalies of zonal velocity in the figures are multiplied by 150 and 10, respectively. Wind data from the NCEP/NCAR reanalysis II (e.g., Kalnay et al., 1996) are used in this analysis.

7. Summaries

Since the early or late onset of the SCSM has an important influence on the East Asian summer monsoon, which seriously affects flooding and drought disasters in the Yangtze River and the Huaihe River valleys, in this study, the influence of thermal states of the tropical western Pacific on the onset date of the SCSM is analyzed by using the observed data including the sea surface and subsurface temperature in the tropical Pacific, the high cloud amount obtained by GMS satellite, JMA, and the NCEP/NCAR reanalysis data of wind fields. From analyzed results, the impacts of thermal states of the tropical western Pacific on the SCSM onset date may be schematically summarized in Fig. 11. As shown in Fig. 11a, when the tropical western Pacific is in a warming state, convective activity are intensified from the Indo-China Peninsula to the east of the Philippines, and the western Pacific subtropical high shifts unusually northward and eastward. In this case, the SCSM onset is early, and the summer monsoon rainfall generally is below normal, and drought may occur in the Yangtze River and the Huaihe River valleys of China, South Korea and Japan. On the other hand, as shown in Fig. 11b, when the tropical western Pacific is in a cooling state, convective activity is weak around the Philippines and is intensified over the equatorial central Pacific near the dateline, and the western Pacific subtropical high shifts southward and westward. In this case, the SCSM onset is late, and the summer monsoon rainfall generally is above normal, and flood may occur in the Yangtze River and the Huaihe River valleys of China, South Korea and Japan.

Moreover, the influence of thermal state of the tropical western Pacific on the SCSM onset process is also analyzed by using OLR and the NCEP/NCAR reanalysis data of wind fields. The results show that there is an obvious difference in the onset process of the SCSM between warming and cooling states of the tropical western Pacific. When the tropical western Pacific is in a warming state in spring, the western Pacific subtropical high early moves eastward to the east of the Philippines, the twin cyclones, especially the twin anomalous cyclones, are early formed over the Bay of Bengal and the west of Sumatra before the SCSM onset. Moreover, since one of the twin cyclones located over the Bay of Bengal is intensified and becomes into a strong trough, and other cyclone is also intensified west to Sumatra, convective activity is early intensified over the Indo-China Peninsula and Sumatra, and the strong westerly flow early appears over the equatorial eastern Ocean, Sumatra and the SCS in mid-May. These lead to the early onset of the

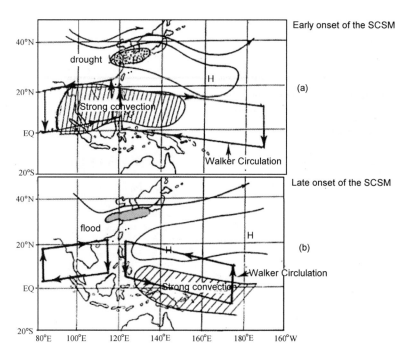

Fig. 11. Schematic diagram of the relationships among the SST in the tropical western Pacific (TWP) convective activities around the Philippines, the western Pacific subtropical high, the onset of the South China Sea summer monsoon (SCSM), and the summer rainfall in China and its surrounding regions. (a) In the warming state of the TWP; (b) In the cooling state of the TWP.

SCSM. However, when the tropical western Pacific is in a cooling state, the western Pacific subtropical high extends westward from late April, and an anticyclonic circulation appears over the SCS in mid-May, the twin cyclones located over the equatorial eastern Indian Ocean, the Bay of Bangal and Sumatra are weakened, especially, the twin anomaly anticyclones always appear over these regions from late April to mid-May. When the western Pacific subtropical high moves eastward from late May, the trough located over the Bay of Bengal is greatly intensified. Thus, convective activity is late intensified over the equatorial eastern Indian Ocean, the Indo-China Peninsula, and Sumatra, and the strong southwesterly flow later appears over the Indo-China Peninsula and the SCS. These lead to late onset of the SCSM.

Furthermore, the physical mechanism for influence of thermal state of the tropical western Pacific on the SCSM onset date and process is analyzed further from the Walker circulation anomalies over the tropical western Pacific and eastern Indian Ocean in the different thermal states of the tropical western Pacific. The results show that when the tropical western Pacific is in warming state, a strong ascending branch of the Walker circulation is found over the tropical western Pacific, and westerly anomaly can be found over the tropical eastern Indian Ocean and west to Sumatra. This can cause the westerly flow to enter early into the Indo-China Peninsula and the SCS. Thus, the SCSM onset may be early in this case. In contrast, when the tropical western Pacific is in a cooling state, an ascending branch of the Walker circulation is weaker over the tropical western Pacific, and strong westerly flow late appears over the tropical eastern Indian Ocean and west to Sumatra. Thus, the southwesterly flow enters late into the Indo-China Peninsula and the SCS, and the SCSM onset may be late in this case.

Acknowledgments. This study was supported by the National Natural Science Foundation of China grant No.40575026 and "National Key Programme for Developing Basic Science" Projects 2004CB418303 and 2006CB403600.

REFERENCES

Chao Jiping, and Lin Yonghui, 1996: The motion of tropical semi-geostrophic adaptation. *From Atmospheric Circulation to Global Change*, edited by Institute of Atmospheric Physics, Chinese Academy of Sciences, China Meteorological Press, 237–246.

Chen Longxun, and Li Weiliang, 1981: The atmospheric heat budget during summer in the Asian monsoon region. *Proc. Symposium on the Summer Monsoon in Southeast Asia*, August 15–21, 1980, Hangzhou, The People's Press of Yannan Province, 86–101. (in Chinese)

Chen Longxun, Li Weiliang, and He Jinhai, 1983: On the atmospheric heat source over Asia and its relation to the formation of the summer circulation. *Proc. First Sino-American Workshop on Mountain Meteorology*, Amer. Meteor. Soc., 265–290.

Connejo-Garrido, A. G., and P. H. Stone, 1977: On the heat source of the Walker circulation. *J. Atmos. Sci.*, **34**, 1155–1162.

Ding Yihui, and Y. Liu, 2001: Onset and the evolution of the summer monsoon over the South China Sea during SCSMEX field experiment in 1998. *J. Meteor. Soc. Japan*, **79**, 255–276.

Ding Yihui, Wang Qiyi, and Yan Junyue, 1996: Some aspects of climatology of the summer monsoon over the South China Sea. *Atmospheric Circulation to Global Change*, China Meteorological Press, 107–117.

Ding Yihui, Li Congyin, Liu Yanju, Zhang Jin, and Song Yafang, 2002: South China Sea monsoon experiment. *Climatic and Environmental Research*, **7**, 202–208. (in Chinese)

Hartmann, D., H. Hendon, and R. A. Houze, 1984: Some implications of the meso-scale circulations in tropical cloud clusters for large-scale dynamics and climate. *J. Atmos. Sci.*, **41**, 113–121.

He, H., J. W. McGinnis, Z. Song, and M. Yanai, 1987: Onset of the Asian monsoon in 1979 and the effect of the Tibetan Plateau. *Mon. Wea. Rev.*, **114**, 594–604.

He Youhai, Guan Cuihua, and Gan Zijun, 1992: Heat oscillation in the upper ocean of the South China Sea. *Acta Oceanologica Sinica*, **11**, 375–388. (in Chinese)

He Jinhai, and Luo Jingjia, 1999: Features of the South China Sea monsoon onset and Asian summer monsoon establishment sequence along with its individual mechanism. *The Recent Advances in Asian Monsoon Research*, He et al., Eds., China Meteorological Press, 74–81. (in Chinese)

Huang Ronghui, and Li Weijing, 1987: Influence of the heat source anomaly over the tropical western Pacific on the subtropical high over East Asia. *Proc. Interal Conference on the General Circulation of Eat Asia*, Chengdu, April 10–15, 1987, 40–50.

Huang Ronghui, and Li Weijing, 1988: Influcence of the heat source anomaly over the tropical western Pacific on the subtropical high over East Asia and its physical mechanism. *Chinese J. Atmos. Sci.*, **14** (Special Issue), 95–107. (in Chinese)

Huang, R., and F. Sun, 1992: Impact of the tropical western Pacific on the East Asian summer monsoon. *J. Meteor. Soc. Japan*, **70**(1B), 243–256.

Huang Ronghui, Zhou Liantong, and Chen Wen, 2003: The progresses of recent studies on the variabilities of the East Asian monsoon and their causes. *Adv. Atmos. Sci.*, **20**, 55–69.

Huang Ronghui, Huang Gang, and Wei Zhigang, 2004: Climate variations of the summer monsoon over China. *East Asian Monsoon*, C. P. Chang, Ed., World Scientific Publishing Co. Pte. Ltd., 213–270.

Huang Ronghui, Gu Lei, Xu Yufong, Zhang Qilong, Wu Shangsen, and Cao Jie, 2005: Characteristics of the interannual variations of onset and advance of the East Asian summer monsoon and their association with thermal states of the tropical western Pacific. *Chinese J. Atmos. Sci.*, **29**, 20–36. (in Chinese)

Huang Zun, and Tao Shiyan, 1992: Diagnostic study of the bursting processes of Asian summer monsoon in 1983. *Acta Meteorologica Sinica*, **50**, 210–217. (in Chinese)

Jin Zuhui, 1999: The Climatological characteristics of the South China Sea summer monsoon onset revealed by using the observed data TBB. *The Onset and Evolution of the South China Sea Summer Monsoon and Their Interaction with the Ocean*, Ding Yihui and Li Congyin, Eds., China Meteorological Press, 57–65. (in Chinese)

Japan Meteorological Agency: Monthly Report on Climate System 1978–2000.

Kalnay, E. M., and Coauthors, 1996: The NCEP/NCAR 40-year reanalysis project. *Bull. Amer. Meteor. Soc.*, **77**, 437–471.

Lau, K.-M., and L. Peng, 1990: Origin of low frequency (intraseasonal) oscillation in the tropical atmosphere. Part III: Monsoon dynamics. *J. Atmos. Sci.*, **47**, 1443–1462.

Lau, K.-M., and Coauthors, 2000: A report of the field operations and early results of South China Sea monsoon experiment (SCSMEX). *Bull. Amer. Meteor. Soc.*, **81**, 1261–1270.

Li Chongyin, and Qu Xin, 1999: The evolution characteristics of atmospheric circulation in the onset of the South China Sea summer monsoon. *The Onset and Evolution of the South China Sea Summer Monsoon and Their Interaction with the Ocean*, Ding Yihui and Li Chongyin, Eds., China Meteorological Press, 5–12. (in Chinese)

Li Chongyin, and Zhang Liping, 1999: Activities of the South China Sea summer monsoon and their influence. *Chinese J. Atmos. Sci.*, **23**, 257–266. (in Chinese)

Li Chongyin, and Qu Xin, 2000: Large Scale atmospheric circulation evolutions associated with summer monsoon onset in the South China Sea. *Chinese J. Atmos. Sci.*, **27**, 518–535. (in Chinese)

Liang Jianyin, and Wu Shangsen, 2002: A Study of Southwest monsoon onset date over the South China Sea and its impact factors. *Chinese J. Atmos. Sci.*, **26**, 844–855. (in Chinese)

Luo Huibang, and M. Yanai, 1984: The large-scale circulation and heat sources over the Tibetan Plateau and surrounding areas during the early summer of 1979. Part II: Heat and moisture budges. *Mon. Wea. Rev.*, **112**, 966–989.

Miao Jinhai, and K. M. Lau, 1991: Low-frrequency (30–60 day) oscillation of summer monsoon rainfall over East Asia. *Chinese J. Atmos. Sci.*, **15**, 63–71. (in Chinese)

Murakami, T., Chen, L. X., and A. Xie., 1986: Relationship among seasonal cycle, low-frequency oscillations and transient disturbances as revealed from OLR data. *Mon. Wea. Rev.*, **114**, 1456–1465.

Nitta, Ts., 1986: Long-term variations of cloud amount in the western Pacific region. *J. Meteor. Soc. Japan.*, **64**, 373–300.

Nitta, Ts., 1987: Convective activities in the tropical western Pacific and their impact on the Northern Hemisphere summer circulation. *J. Meteor. Soc. Japan*, **64**, 373–390.

Ose, T., Y. K. Song, and A. Kitoh, 1997: Sea surface temperature in the South China Sea-an index for the Asian monsoon and ENSO system. *J. Meteor. Soc. Japan*, **75**, 1091–1107.

Qian, W. F., and S. Yang, 2000: Onset of the regional monsoon over Southeast Asia. *Meteorology and Atmospheric Physics*, **75**, 29–38.

Ren Baohua, and Huang Ronghui, 1999: Interannual variability of the convective activities associated with the East Asian summer monsoon obtained from TBB variability. *Adv. Atmos. Sci.*, **16**, 77–90.

Tao Shiyan, and Chen Longxun, 1985: The East Asian summer monsoon. *Proc. Int. Conf. on Monsoon in the Far East*, Tokyo, 1–11.

Tao, S. Y., and L. X. Chen, 1987: A review of recent research on the East Asian summer monsoon in China. *Monsoon Meteorology*, C. P. Chang and T. N. Krishnamurti, Eds., Oxford University Press, 60–92.

Wang, B., R. G. Wu, and X. H. Fu, 2000: Pacific-East Asian teleconnection: Does ENSO affect East Asian Climate. *J. Climate*, **13**, 1517–1536.

Wang, B., and Q. Zhang, 2002: Pacific-East Asian teleconnection. Part II: How the Philippine Sea anticyclone established during development of El Niño. *J. Climatc*, **15**, 3252–3265.

Wang, B., H. Lin, and Y. S. Zhang and M. Mu, 2004: Definition of South China Sea monsoon onset and commencement of the East Asia summer monsoon. *J. Climate*, **17**, 699–710.

Xie, S. P., and N. Saiki, 1999: Abrupt onset and slow seasonal evolution of summer monsoon in an idealized GCM simulation. *J. Meteor. Soc. Japan*, **77**, 949–968.

Yanai, M., C. Li, and Z. S. Song, 1992: Seasonal heating of the Tibetan Plateau and its effects on the evolution of the Asian summer monsoon. *J. Meteor. Soc. Japan*, **70**(1B), 319–351.

Yasunari, T., 1990: Impact of Indian monsoon on the coupled atmosphere/ocean system in the tropical Pacific. *Meteorology and Atmospheric Physics*, **44**, 29–41.

Zhu Qiangen, He Jinhai, and Wang Panxun, 1986: A study of the circulation differences between East Asian and Indian summer monsoons with their interaction. *Adv. Atmos. Sci.*, **3**, 466–477.

三、东亚冬、夏季风系统及气候变异特征研究

我国夏季降水的年代际变化及华北干旱化趋势[*]

黄荣辉　徐予红　周连童

(中国科学院大气物理研究所　北京　100080)

摘　要　利用1951~1994年全国336个测站夏季(6~8月)降水和太平洋海表面温度资料来分析我国夏季降水的年代际变化及华北地区干旱化趋势。分析结果表明,我国夏季降水在1965年前后发生了一次气候跃变,华北地区从1965年后夏季降水明显减少,干旱化的趋势明显,这种趋势与西非萨赫尔地区干旱化的趋势相似;分析结果还表明我国80年代的气候与70年代的气候有较大差别,这种差别表现在长江、淮河流域从70年代末起降水增多,涝灾明显增多,而华南和华北在80年代降水明显比70年代少,干旱趋势加重。然而,从90年代中期开始,华北地区北部的降水有增多的趋势。上述所发生的气候变化可能主要是由于60年代中期和从80年代到90年代初赤道东、中太平洋海表温度明显增加,而在70年代明显降低所造成。这种现象似如年代际的"ENSO循环"现象,它对全球和我国气候变化有较大影响,特别对华北地区干旱化趋势有很大影响。然而从90年代中期开始,赤道中、东太平洋海温有下降的趋势,这有利于华北降水的增多。

1　引　言

众所周知,全球增暖及其对环境、水资源和粮食生产等的影响已是世界科学界日益重视的重要科学问题之一,它是各国政府作经济和社会决策时必须认真考虑的重要问题。华北地区是我国人口密集、土地、矿产资源丰富、经济发达的地区,也是我国工农业主要产地之一。然而,华北地区又是我国水资源十分贫乏的地区之一。水资源的缺乏致使华北地区许多河流断流干涸,特别是黄河断流不仅时间延长,而且河段增长。水资源的缺乏已严重妨碍华北地区工农业生产的进一步发展,也严重影响着华北地区城乡人民的生活。可以说,水资源缺乏是华北今后经济发展中亟待解决的重大问题。

关于全球气温变化,Hansen[1]已有较详细的分析,他发现在最近100年全球气温平均

---

[*] 本文原载于:高原气象,第18卷,第4期,465-476,1999年11月出版.

上升了 0.8 ℃,其中从 19 世纪 70 年代末到现在,大约气温上升了 0.3 ℃。Bradley 等[2]比较了我国气温变化与北半球气温变化的异同,指出我国气温变化总的趋势是与北半球的趋势一致,但从年代际的变化来看,这两者也有差别,他们的结论与《中国气候蓝皮书》中的分析结果是一致的。最近,陈隆勋等[3]利用全国 160 个站的气候资料分析了我国气温的变化,他们的结果表明:我国北方包括华北、东北和西北最近 39 年来气温上升,但在江淮地区和四川等地区气温下降。

关于气候变化,正如上面所述,许多研究主要集中讨论全球和我国气温的变化,但由于我国是季风气候,气候变化更多表现在降水的变化。黄荣辉等[4]从我国夏季降水的变化来讨论我国 80 年代以前的气候变化及对水资源的影响,但只用了我国 60 个站的地面气候资料。为了更深入和详细地讨论我国夏季气候的变化,我们应用全国 336 个测站 1951～1994 年夏季 6～8 月的旬降水来讨论我国旱涝分布的变化及华北地区的干旱化趋势,并且利用 1951～1996 年太平洋海表面温度(SST)来讨论我国夏季气候变化的可能成因,特别是华北地区干旱化趋势的可能成因。

2 我国夏季降水异常的年代际分布

由于我国处于东亚季风区,降水的气候振荡要比气温的振荡复杂得多。为了研究我国和各地区夏季降水异常的年代际分布和详细说明我国各区域夏季降水的变化,本研究把全国 336 个测站按图 1 所示,根据地理环境和气候特征划分成 7 个区域。这 7 个区域如下:1 区是黄河流域和华北地区;2 区是江淮流域;3 区是长江中、下游流域;4 区是华南地区;5 区是东北和内蒙东部地区;6 区主要是西北和内蒙西部地区;7 区是西南地区。

为了研究我国夏季降水距平的年代际分布,我们分别制作了我国从 50 年代到 90 年代初夏季降水距平各年代际的分布。图 2a～e 分别是我国 50,60,70,80 和 90 年代初夏季降水距平百分率的分布。从图 2a 可以看到,在 50 年代,华北和黄河流域、江淮流域、华南和东北中部以及西南地区夏季降水偏多,而西北和江南地区降水偏少,全国降水以偏多为

(图略)

图 1 我国气候(降水)分区

Fig. 1 Climatic regions of China

(图略)

图 2 我国各年代际的夏季降水距平百分率的分布

Fig. 2 Interdecadal distributions of the summer precipitation anomaly percentage in China.

主；到60年代，如图2b所示，我国东部出现了一个与50年代大致相反的降水距平分布，华北地区、江淮地区、东北地区降水从偏多变成偏少，而江南地区从偏少变成偏多；到70年代，如图2c所示，黄河流域和华北地区、江淮流域的降水继续偏少，长江中、下游地区、华南的北部、西南和西北降水由偏多变成偏少，也就是说，在70年代由50年代的"丰水期"变成"枯水期"；到了80年代，如图2d所示，华北和华南地区的降水偏少加剧，但是，长江中下游和江淮地区、西北和东北地区北部的降水却由偏少变成偏多，西北地区的降水依然偏多；90年代初，如图2e所示，华北南部和东部的降水依然偏少，长江流域特别是江南、两湖流域降水偏多，并且还可以看到，华北东北部，特别是滦河流域，降水有所回升。因此，80年代到90年代初的气候与70年代相比，发生了较大变化。

从上面分析结果还可以看到，我国降水距平年代际分布的变化具有如下特征：华南地区的变化与华北地区的变化特征比较一致，而江淮流域的变化与东北北部的变化比较一致，这种变化似如一波列的分布。

3 我国各区域夏季降水异常的年代际变化特征

我国地处东亚季风区，气候变化在降水变化方面的反映尤其明显，夏季降水不仅有年际变化，而且有年代际的变化。图3是我国各区域夏季6～8月区域平均的降水距平百分率的变化情况。

从图3可以看到各区域降水的年代际变化如下：

(1) 黄河流域和华北地区：如图3a所示，在50年代降水偏多，而从1965年以后降水明显减少，这种干旱趋势在70年代虽有所缓和，但从70年代末干旱加剧，并一直延续到80年代末和90年代初。在80年代华北干旱相当严重，区域总平均降水量比50年代约减少近30%。然而，华北北部的降水从90年代初有增加的趋势。

(2) 江淮流域：如图3b所示，在50年代降水偏多，从50年代末到60年代中降水明显减少，整个70年代降水偏少，约比50年代减少了20%，但从70年代末到80年代末，区域总降水量明显偏多，与50年代相近；90年代初，从江淮流域到汉江流域的降水比80年代有所减少。

(3) 长江中、下游地区：如图3c所示，在50年代降水明显偏多，但从50年代末到60年代中降水明显减少，从1965年后虽有增加，但直到70年代末，降水还是比正常偏少；从70年代末到80年代和90年代初明显增加，比正常偏多。

(4) 华南地区：如图3d所示，从50年代初到60年代中降水比正常偏多，并逐渐增加，从1965年起到70年代末降水明显减少，但仍比正常偏多，从70年代末到80年代，降水比正常偏少，比50年代约减少20%以上，呈干旱趋势。然而，华南地区降水从90年代初有增加的趋势。

(5) 东北地区：如图3e所示，从50年代到60年代中降水比正常偏多，从1965年后降水明显减少，在70年代降水比正常偏少，约比50年代偏少20%以上。从80年代起到90年代初降水逐渐增加，它的变化趋势与江淮流域的降水变化趋势相近。

(6) 西北地区：如图3f所示，50年代降水偏多，但从60年代起一直到70年代中降水明显减少，从70年代中以后，特别是80年代降水逐渐增加，这种增加趋势在90年代初

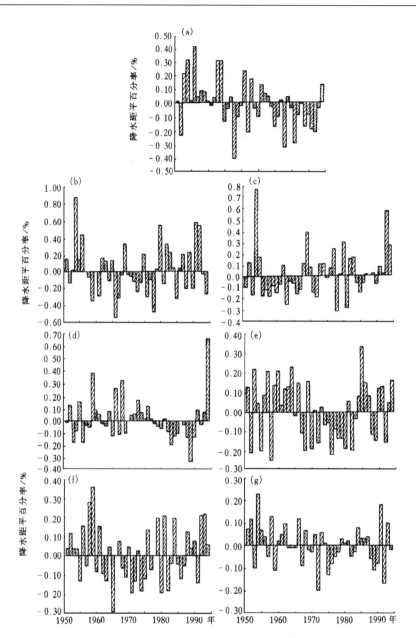

图 3 我国各区域夏季 6~8 月降水距平百分率的年际变化
(a) 黄河流域和华北地区, (b) 江淮流域, (c) 长江中、下游地区,
(d) 华南地区, (e) 东北地区, (f) 西北地区, (g) 西南地区

Fig. 3 Interannual variation of the summer precipitation anomaly percentages in various regions of China.
(a) The Yellow River valley and North China, (b) The Yangtze River-the Huai River valley,
(c) The middle and lower reaches of the Yangtze River, (d) South China,
(e) Northeast China, (f) Northwest China, (g) Southwest China

尤其明显。

(7) 西南地区：如图 3g 所示，降水的气候振荡比我国其它地区小，50 年代降水偏多，60 年代降水有所减少，70 年代降水明显减少，且比正常偏少较多；而从 80 年代起到 90

年代初,降水有所增加,在正常范围摆动。

从上面各区域降水的变化趋势可以看到如下几个特点:

(1) 我国夏季降水在1965年前后有一次明显的跃变,表现为从多变少的干旱趋势,特别在黄河流域和华北、华南地区这种干旱趋势一直延续到90年代初。

(2) 我国在70年代和80年代旱涝分布有较大的不同,从降水看,可以说在70年代末又发生了一次气候跃变。在70年代,江淮流域和长江中、下游地区夏季降水比正常偏少,而华南地区降水比正常偏多;从70年代末到80年代末和90年代初,江淮流域和长江中、下游地区的降水明显增多,比正常偏多;而华北、华南地区降水明显减少,降水比正常偏少,特别是华北地区的降水比70年代继续减少,干旱化趋势严重。

(3) 我国夏季降水年代际变化,从南到北呈现出一遥相关型。华南地区降水的变化与黄河流域的降水变化比较一致,而江淮流域的降水变化与东北地区的降水变化也比较一致。

Yamamoto等[5]指出:50年代北半球气温变化发生了一次气候跃变,但是,从我国降水的变化来看,似乎在60年代中期和70年代末发生了二次跃变。这两次跃变在我国华北地区的降水变化的表现尤其明显。

4 华北地区干旱化趋势和90年代初的变化趋势

4.1 华北地区干旱化趋势

从图3a可以看到:华北地区降水在1965年发生了一次跃变,降水由偏多明显变成偏少;并且在70年代末又发生了一次跃变,降水再一次减少,使华北地区干旱化趋势进一步加剧,80年代的降水约比50年代减少了近30%,干旱化日趋严重,使水资源严重减少。图4是华北地区河川、湖泊和土壤所含的实测水资源距平百分率的年代际变化,可以看到80年代华北地区水资源约比50年代减少了一半。由于华北地区水资源严重缺乏,人均水资源占有量只有全国的1/6,而耕地亩均水资源占有量只有全国的1/10。

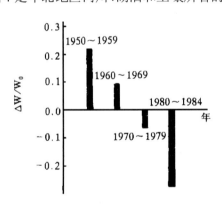

图4 华北地区水资源距平百分率的年代际变化

Fig. 4 Interdecadal variation of the water resource anomaly percentage in North China

我国华北地区干旱化的趋势与西非萨赫尔地区的干旱化趋势相类似。图5是西非萨赫尔地区降水标准化距平的年际变化曲线。从图5可以看到,在50年代萨赫尔地区的降水偏多,而从60年代中期到80年代末降水一直处于负距平,最大负距平可达到1.5个方差。图3a与图5相比较,可以看到,我国华北地区降水减少,干旱化趋势与萨赫尔地区干旱化趋势相一致。严中伟等[6]也曾提出我国华北与萨赫尔地区的降水变化之间有一个正相关关系。

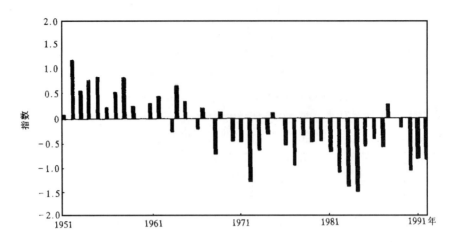

图 5 西非萨赫尔地区降水标准化距平($\Delta R/\sigma$)的年际变化(引自文献〔7〕)

Fig. 5 Interannual variation of the normalized precipitation anomalies in the Sahel area of West Africa (from reference 〔7〕)

4.2 90年代华北地区降水的变化趋势

从图 2e 所示 90 年代初华北地区降水异常的分布可看到,在华北北部 90 年代初降水异常分布与 80 年代的分布很不一样。80 年代华北和黄河中、下游地区降水偏少,干旱严重,而 90 年代初,华北北部包括滦河流域、内蒙南部和河套地区,降水异常由偏少变成偏多,然而华北中部、南部和东部降水依然偏少。从图 3a 所示的华北地区平均降水距平的演变趋势可看到,华北地区降水虽在 90 年代还是干旱时期,但与 80 年代相比已有缓慢回升的趋势。因此,可以预测:从 90 年代中后期起华北地区降水可能由偏少变成偏多。

5 热带太平洋海温的年代际变化对华北地区降水的影响

上面已阐述了我国夏季降水的年代际变化情况和华北干旱化的趋势,为什么会引起我国夏季降水的年际变化呢?这可能是由全球气候系统变化所引起的,特别是热带太平洋海温较大的年代际变化对我国降水可能有较大影响。

5.1 年代际的 ENSO 循环现象

图 6 是 $0°\sim 10°N$ 平均 SST 距平的 5 年滑动平均变化曲线。从图 6 可以看到,热带太平洋海温距平的变化存在着一个年代际的"El Niño"现象。在本世纪 10 年代中期以及 20 年代末到 30 年代初,热带中、东太平洋的海表温度较高;40 年代初热带中太平洋的海表温度也较高;50 年代中期到 50 年代末的热带东太平洋的海表温度比正常海温偏高;特别在 60 年代中期,赤道中、东太平洋的海表温度明显偏高,比正常海温高出 0.6 ℃;而从 70 年代初到 70 年代后期,赤道中、东太平洋的海温明显偏低,比正常海温偏低 0.6 ℃;而到了 70 年代末到 80 年代中后期,赤道中、东太平洋海温又升高,比正常海温升高了 0.4 ℃左右。

从图 6 可清楚看到,在热带中、东太平洋海表温度的变化呈现出一个年代际的"ENSO 循环"。在 60 年代中期与 80 年代热带中、东太平洋发生明显的年代际"El Niño 现象",而在70年代热带中、东太平洋发生了年代际的"La Niña现象"。为了更详细地研究70年代与80年代这个年代际的"ENSO循环",下面分析这两个海域平均的SST距平的5年滑动平均的变化情况。

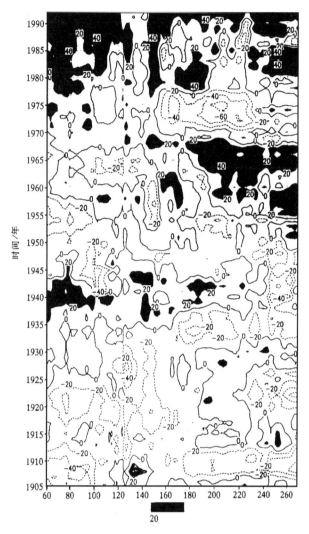

图6 热带太平洋 $0°\sim10°N$ 平均SST距平的5年滑动平均经度-时间剖面变化(单位:℃)

Fig. 6 The longitude-time cross section of 5-years running mean SST anomalies averaged for the area of $0°\sim10°N$ of the tropical Pacific. Unit: ℃

从70年代后半期到80年代后半期,全球平均气温约升高0.3 ℃,但在赤道太平洋海表面温度约上升了0.6 ℃以上,比全球平均气温上升的幅度大。图7分别是赤道中太平洋与东太平洋海表面温度距平(SSTA)冬、夏季5年滑动平均的变化曲线。从图7清楚看到:70年代赤道中、东太平洋SST偏冷,低于气候平均海表面温度;而到80年代,赤道中、东太平洋SST升高、海温偏暖,高于气候平均海表面温度,冬季赤道中太平洋的SST比70年代SST约上升1.0 ℃,夏季约上升0.5 ℃;而冬季赤道东太平洋80年代的SST比70年代SST约上升0.8 ℃,夏季约上升0.5 ℃。这种年代际的海面温度的变化似如年际变化的"ENSO 循环"一样,因此,我们可以定义一个年代际的"ENSO 循环",即 70 年代赤道中、东太平洋海温似如"La Niño"时期,即处于反 ENSO 特征,80 年代赤道中、东太平洋海温似如"El Niño"时期。

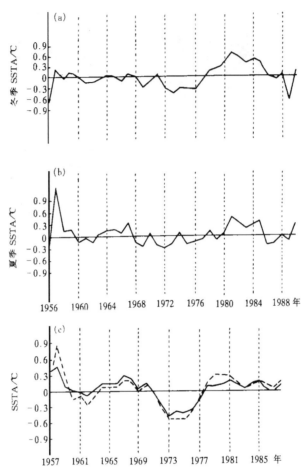

图 7　5 年滑动平均的冬季(a)和夏季(b)赤道东太平洋 SSTA 以及赤道中太平洋 SSTA(c)的分布
图中实线是夏季,虚线是冬季
Fig. 7　The 5-years running mean SST anomalies in the equatorial eastern Pacific (a),
(b) and in the equatorial central Pacific (c) in summer (solid line) and winter (dashed line)

5.2　热带太平洋年代际"ENSO 循环"对我国降水年代际变化的影响

黄荣辉等[8]分析了我国夏季季风降水年际变化与赤道中、东太平洋海温的关系并指出:当 ENSO 现象处于发展阶段,该年夏季我国江淮流域降水将会偏多,而黄河流域、华北地区降水将会偏少,往往发生干旱,江南地区的降水也会偏少,也可能发生干旱;相反,当赤道东太平洋海温处于下降阶段,即 ENSO 现象处于衰减阶段,江淮流域降水将会偏少而发生干旱,而黄河流域、华北地区和江南、华南地区的降水可能偏多。

最近我们计算了 1950~1996 年我国夏季降水和夏季赤道东太平洋(NINO.3)区的夏季与上一年秋季的 SST 之差的相关数分布(见图 8),其结果与文献[8]的结果相似。图 8 表明我国夏季降水与 ENSO 循环不同阶段赤道东太平洋的 SST 距平存在很好的相关。在

ENSO现象处于发展到成熟阶段,该年夏季我国江淮流域降水可能偏多,而黄河流域、华北和江南地区降水可能偏少,往往发生干旱;相反,当ENSO现象处于衰减阶段,江淮流域降水往往偏少;而黄河流域、华北和江南地区的降水可能偏多,特别在洞庭湖、鄱阳湖以及沅江流域往往发生洪涝,这种现象发生在1997/1998年ENSO循环尤其明显。

图3a~e与图8相比较,上面结论也适合于年代际降水异常与"ENSO循环"阶段的关系。在年代际"El Niño现象"发生阶段,我国江淮流域多雨,黄河流域和华北地区降水偏少并发生干旱,江南地区也易发生干旱,如在60年代中期华北地区少雨,干旱严重,而到80年代华北地区降水再度减少,发生了持续干旱,江淮流域降水偏多;而在70年代的年代际"La Niña"发生阶段,我国江淮流域降水偏少,发生干旱;而黄河下游和华北东部和南部地区降水偏多,比60年代有所增加,并且江南地区,特别是洞庭湖、鄱阳湖、沅江流域夏季季风降水偏多,发生洪涝。

(图略)

图8 我国夏季降水距平和赤道东太平洋NINO.3区夏季与上一年秋季海温差的相关系数分布

Fig. 8 Distribution of the correlation coefficients between the summer precipitation anomalies in China and the differences between the SST in the NINO.3 of the Equatorial Pacific in summer and those in last Autumn

5.3 90年代初热带中、东太平洋海温变化及其对我国华北降水的可能影响

从图6可以看到,从90年代初热带中、东太平洋海温又再度升高,可能又发生年代际的"El Niño现象"。正如图8所示,90年代初连续发生了3次"El Niño"现象,致使华北中、南部降水继续减少,干旱化更加严重,而华北、东北部降水有所增加。

根据赤道中、东太平洋海温的年代际变化情况可以推测:从90年代后半期开始年代际的"El Niño"现象将要结束,而年代际的"La Niña现象"将开始,也就是说,热带中、东太平洋的海表温度将降低。这意味着我国华北地区降水从90年代后半期开始将可能增加。

6 结 论

本研究利用1951~1994年全国336个测站夏季(6~8月)降水资料和太平洋海表温度分析了我国夏季降水的年代际变化以及华北地区干旱化趋势。分析结果表明:我国夏季降水在1965年前后发生了一次气候跃变,从1965年后我国华北夏季降水明显减少,干旱化趋势明显,这种趋势与西非萨赫尔地区干旱化的趋势相似;分析结果还表明了我国80年代的气候与70年代的气候有较大差别,这种差别表现在长江、淮河流域从70年代末起降水明显增多,涝灾增多,而华北和华南地区降水明显比70年代少,干旱趋势加重。此外,分析还表明从90年代中期开始华北北部的降水增多。

本研究通过我国降水与热带中、东太平洋海表温度变化的关系,分析了我国降水的年代际变化和华北地区干旱化趋势的可能成因。分析结果表明:上述我国降水的年代际变化可能主要是由于60年代中期和从80年代到90年代初赤道中、东太平洋海表温度明显增加,而70年代明显降低所造成。这种现象似如年代际的"ENSO循环"现象,它对全球和我国气候变化都有较大的影响,特别对华北地区干旱化趋势有很大的影响。分析结果还表明,从90年代中期开始,热带中、东太平洋海温有下降的趋势,这可能有利于华北降水的增多。

参考文献

1. Hansen J, S Lebedeff. Global surface air temperature: Update through 1987. Geophy Res Letters, 1998, 15: 323~326
2. Bradley R S et al. Secular fluctuations of temperature over the northern hemisphere land. The Climate of China and Global Climate. Proceedings of the Beijing International Symposium on Climate, Oct. 30~Nov. 3, 1984, Beijing, China, 76~87
3. Chen L X, Shao Yongning, Dong Min. Preliminary analysis of climatic variation during the last 39 years in China. Adv Atmos Sci, 1991, 8(3): 279~288
4. 黄荣辉, 张庆云. 华北降水的年代际变化及其对经济的影响. 关于华北地区水资源合理开发利用论文集. 北京: 水利电力出版社, 1990. 95~101
5. Yamamoto R et al. An analysis of climatic jump. J Meteor Soc Japan, 1986, 64: 273~281
6. Yan Z H, J J Ji, Q E Ye. Northern hemispheric summer climatic jump in 1960's. Science in China (Series B), 1990, 33: 61~70
7. Fourth Annual Climate Assessment. 1992
8. Huang Ronghui, Wu Yifang. The influence of ENSO on the summer climate change in China and its mechanism. Adv Atmos Sci, 1989, 6: 21~30

The Progresses of Recent Studies on the Variabilities of the East Asian Monsoon and Their Causes

Huang Ronghui (黄荣辉), Zhou Liantong (周连童), and Chen Wen (陈文)

LASG, Institute of Atmospheric Physics, Chinese Academy of Science, Beijing 100080

ABSTRACT

The variabilities of the East Asian summer monsoon are an important research issue in China, Japan, and Korea. In this paper, progresses of recent studies on the intraseasonal, interannual, and interdecadal variations of the East Asian monsoon, especially the East Asian summer monsoon, and their causes are reviewed. Particularly, studies on the effects of the ENSO cycle, the western Pacific warm pool, the Tibetan Plateau and land surface processes on the variations of the East Asian summer monsoon are systematically reviewed.

1. Introduction

A monsoon is a kind of climatic phenomenon in which the dominant wind system changes with the seasons. Many studies have shown that the Asian monsoon plays an important role in the global climate variability (Tao and Chen, 1987; Yasunari, 1990; Ding, 1994; Huang et al., 1996; Chang et al., 2000). Asian monsoons can bring large amounts of water vapour from the Pacific and Indian Oceans to the mainland. As a consequence, a large amount of rainfall can be formed in the monsoon region. Because of the close relationship between monsoon and rainfall, monsoon variability influences economy, industry, agriculture and daily life of the people in the monsoon region, especially considering the droughts and floods caused by the monsoon which have brought heavy economic losses to China, Japan, and Korea (Huang and Zhou, 2002).

The East Asian monsoon is a submonsoon system of the Asian monsoon. In East Asia, there are many characteristic weather systems in different seasons, such as the Meiyu in China, the Changma in Korea, and the Baiu in Japan in summer, persisting northwesterly and northeasterly winds and cold surges in winter. Moreover, the interannual and intraseasonal variabilities of the East Asian monsoon are very large and can cause drought and flooding in the eastern part of China, in Korea, and in Japan. Therefore, the variability of the East Asian monsoon has been an important scientific issue in China, Korea, and Japan during the last sixty years, and many advances in the study of the East Asian summer monsoon have been obtained in last century. In 1987, Tao and Chen (1987) gave a systematic review of these advances. Since their review, studies on the interdecadal, interannual and intraseasonal variations of the East Asian monsoon and their causes have been extended and in particular, the physical mechanism of the interannual and intraseasonal variabilities of the monsoon has been greatly studied recently. In order to summarize the advances made, recent studies on the East Asian monsoon are systematically reviewed in this paper.

2. Intraseasonal variability of the East Asian summer monsoon

2.1 *Intraseasonal variability of the summer monsoon rainband over East Asia*

Many studies have shown that the intraseasonal variations of the summer monsoon rainband over East Asia are closely associated with the advance and retreat of the East Asian summer monsoon (Zhu, 1934; Tu and Huang, 1944). Therefore, the intraseasonal variability of the East Asian summer monsoon (EASM) may be well seen from the intraseasonal variations of summer monsoon rainband over East Asia. Figure 1a shows the latitude–time cross section of 5-day precipitation along 115°E (110°–120°E) averaged over the last 40 years from 1961 to 2000. Figure 1a clearly shows that the summer monsoon rainband is located over the area to the south of the Yangtze River during the period from May to the first 10 days of June, and then it abruptly moves northward to sit over the Yangtze and Huaihe River valleys of China. This is the beginning of the Meiyu season in the Yangtze River valley. Moreover, it may also be seen from Fig. 1a that the monsoon rainband again moves northward to North China in July. Due to this northward movement of the rainband, the Meiyu season ends in the Yangtze River valley and the rainy season begins in North and Northeast China.

The northward movement of the summer monsoon rainband over East Asia is dependent on the onset and maintenance of the East Asian summer monsoon and it is closely associated with the northward advance of the EASM front (see Wang et al., 2001). Figure 1b shows the average dates of the onset of EASM suggested by Tao and Chen (1987). Comparing Fig. 1a with Fig. 1b, the northward movement of the summer monsoon rainband over East Asia is in good agreement with the onset of the Asian summer monsoon in East Asia. According to the result shown in Fig. 1b, the beginning date of the Baiu season in Southwest Japan is almost the same as the beginning of the Meiyu season in the Yangtze River valley; following the start of the Meiyu season in the Yangtze River valley, the Changma season begins in South Korea. Lu et al. (1995) showed that there is a good relationship between the summer rainfall anomalies in the Huaihe River valley of China and those in South Korea.

The abrupt movement of the rainband is closely associated with the abrupt northward shift of the western Pacific subtropical high (see Huang and Sun, 1992; Ding, 1992). Yeh et al. (1959) first discovered the abrupt change of the circulation over East Asia during early and mid-June. This abrupt change of planetary-scale circulation brings about the onset of EASM. Later, Krishnamurti and Ramanathan (1982) and McBride (1987) further pointed out the abrupt change of Indian summer monsoon circulation and Australian summer monsoon circulation. However, recent studies have shown that the abrupt change of the circulation over East Asia during early or mid-June is closely associated with the convective activities around the Philippines (Huang and Sun, 1992).

2.2 Relationship between the intraseasonal variation of the summer monsoon rainband and the thermal state of the western Pacific warm pool

The intraseasonal variability of summer monsoon rainfall in East Asia is closely associated with the thermal state of the western Pacific warm pool. Kurihara (1989), Huang et al. (1992), and Huang and Sun (1992) showed that there is a close relationship between the anomalous northward shift of the western Pacific subtropical high and the convective activities around the Philippines. In summers with strong convective activity around the Philippines, the abrupt northward shift of the western Pacific subtropical high is obvious in early or mid-June, and the summer monsoon rainfall may be weak in the Yangtze and Huaihe River valleys of China, in South Korea, and in Japan. On the other hand, in summers with weak convective activity around the Philippines, the abrupt northward shift of the western Pacific subtropical high is not obvious, and the summer monsoon rainfall may be strong in the Yangtze and Huaihe River basins of China, in South Korea, and in Japan.

As mentioned above, the intraseasonal variability of EASM is greatly influenced by the thermal state of the western Pacific warm pool. Huang and Sun (1992) showed that there are close relationships among the thermal states of the warm pool, the convective activities around the Philippines, the western Pacific subtropical high, the onset of the South China Sea monsoon, and the summer monsoon rainfall in East Asia. As shown in Fig. 2a, when the SST in the warm pool is above normal, i.e., the warm sea water is accumulated in the western Pacific warm pool, and the cold tongue extends westward from the Peruvian coast along the equatorial eastern Pacific, the convective activities are intensified from the Indo-China Peninsula to the east of the Philippines, and the western Pacific subtropical high may shift unusually northward. In this case, the onset of the South China Sea summer monsoon (SCSM) may be early and the summer monsoon rainfall may be below normal in East Asia, especially in the Yangtze and Huaihe River basins of China, in South Korea, and in Japan. On the other hand, as shown in Fig. 2b, when the SST in the warm pool is below normal, i.e., the warm sea water extends eastward from the warm pool along the equatorial western Pacific, the convective activities are weak around the Philippines and are intensified over the equatorial central Pacific near the dateline, and the western Pacific subtropical high may shift southward. In this case, the onset of SCSM may be late and the summer monsoon rainfall may be above normal in the Yangtze and Huaihe River valleys of China, in South Korea, and in Southwest Japan.

The influence of the thermal state in the tropical western Pacific and the convective activities over the tropical western Pacific on the intraseasonal variability of EASM may be through the 30–60 day oscillation. Huang (1994), Sun and Huang (1995), and Li (1996) showed the intraseasonal variability of EASM is associated with the 30–60 day oscillation propagated from the tropical eastern Indian Ocean, the Indo-China Peninsula, and the South China Sea to the East Asian monsoon region.

3. Interannual variability of EASM

3.1 Interannual variability of summer monsoon rainfall in East Asia

The monsoon index is a criterion measuring the strength of the monsoon, necessary for the study of the interannual variability of the monsoon. So far, two kinds of definitions of the Asian monsoon index are being used. One is defined from the thermodynamic elements. For example, Tao and Chen (1987) defined the monsoon index with the strength of monsoon rainfall, while Murakami and Matsumoto (1994) defined the monsoon index from outgoing longwave

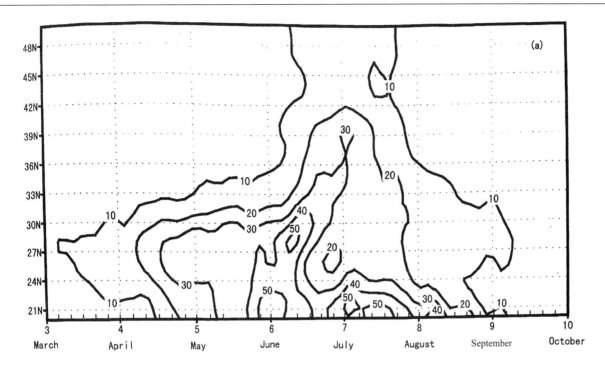

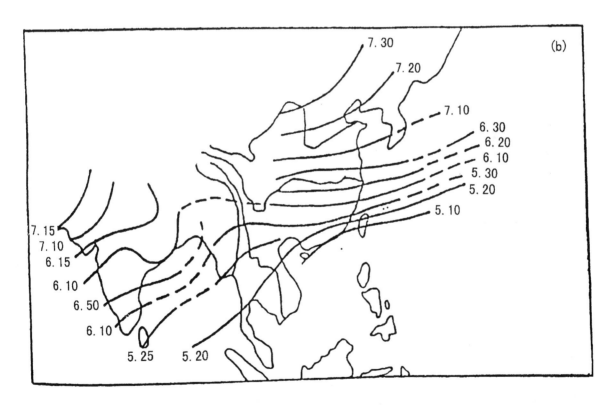

Fig.1. (a) Latitude-time cross section of 5-day precipitation along 115°E (110°–120°E) averaged over the last 40 years from 1961 to 2000 (units: mm), (b) Average dates of the onset of the East Asian summer monsoon (from Tao and Chen, 1987).

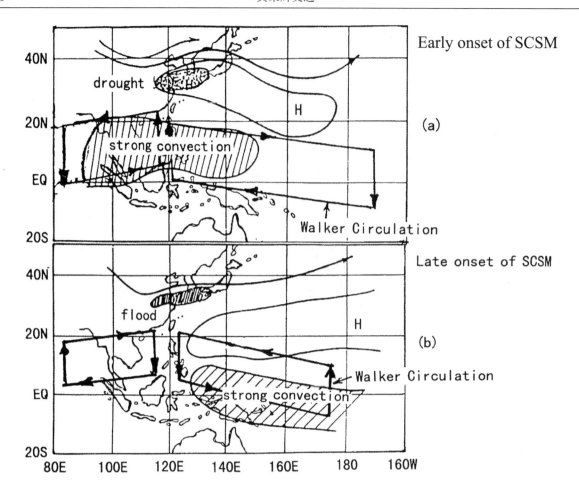

Fig. 2. Schematic map of the relationships among the SST in the western Pacific warm pool (WPWP), the convective activities around the Philippines, the western Pacific subtropical high, the onset of SCSM and the summer monsoon rainfall in East Asia. (a) in the warming state of the WPWP; (b) in the cooling state of the WPWP.

radiation (OLR). The other defined from the dynamic elements. For example, Webster and Yang (1992) and Zeng et al. (1994) defined the monsoon index from the difference between the zonal winds in the high layer and those in the lower layer of the atmosphere over a monsoon region. There are some advantages and disadvantages to these definitions. The former is easily influenced by local thermodynamic conditions, but the latter is only suitable for the South Asian monsoon region. This is mainly due to the differences between the South Asian monsoon and the East Asian monsoon. Since the East Asian summer monsoon is a kind of subtropical monsoon and its meridional component is larger, it may not be suitable to define the strenth of EASM with only the zonal component of the wind field. Moreover, the summer monsoon rainfall tends to be weak in the Yangtze and Huaihe River valleys of China, in South Korea, and in Japan when the South Asian summer monsoon is strong. Recently, Huang and Yan (1999) showed that the interannual variability of EASM can be well described using the EAP index, which is defined with the 500 hPa height anomalies in summer according to the EAP (East Asia/Pacific) teleconnection pattern of the summer circulation anomalies suggested by Nitta (1987) and Huang and Li (1987).

Figure 3 shows the interannual variation of the EAP index defined by Huang and Yan (1999). The negative (positive) index indicates above (below) normal amounts of summer monsoon rainfall in the Yangtze and Huaihe River valleys. It may be clearly seen from the figure that the interannual variability of EASM is large, and summer drought and flood disasters frequently occur there, especially in the area from the Yangtze River valley to Southern Japan through South Korea. For example, a particularly severe flood

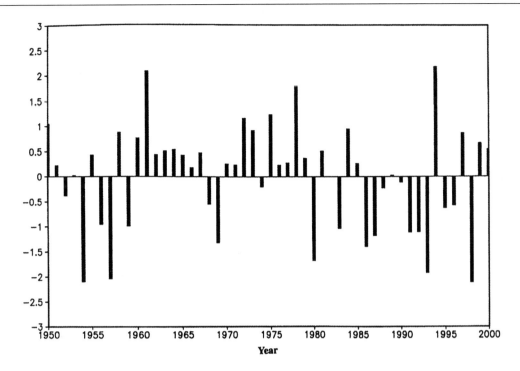

Fig. 3. Interannual variation of the EAP index for the summers from 1950 to 2000 (from Huang and Yan 1999, with additional data added for 1999 to 2000).

disaster occurred in the Yangtze River valley in the summer of 1998 (Huang et al., 1998), and prolonged and severe drought disasters have continuously appeared in North China from the summer of 1997 to the summer of 2001 (Huang and Zhou, 2002). Moreover, recent studies have shown that there is an obvious biennial oscillation in the summer monsoon rainfall in East Asia, especially in the Yangtze, Huaihe, and Yellow River valleys and North China (Miao and Lau, 1990; Yasunari, 1991; Lau and Shen, 1992; Lu, 1995; Yin et al., 1996).

3.2 *Causes of the interannual variability of EASM*

The causes of the interannual variability of EASM are complex, since it is influenced by many factors as shown in Fig. 4. It may be seen from the figure that the interannual variability of EASM is influenced by not only the Indian monsoon, the western Pacific subtropical high, and disturbances in the atmosphere in middle latitudes, but also by the West Pacific warm pool, the ENSO cycle in the tropical Pacific and the Tibetan Plateau, polar ice and Eurasian snow cover, and land-surface processes etc. Influenced by these physical factors, the interannual variability of EASM is very obvious.

(1) Dynamic and thermal effects of the Tibetan Plateau

The Tibetan Plateau has an important dynamic effect on the interannual variability of EASM. Using the GFDL 9-layer GCM simulations, Hahn and Manabe (1975) showed that due to the dynamic effect of the Tibetan Plateau, strong southwest flow can extend from the Bay of Bengal to the eastern part of China including South China, the Yangtze and Huaihe River valleys and North China. In the late 1970s, Ye and Gao (1979) first pointed out the thermal effect of the Tibetan Plateau on the Asian monsoon. Later, many investigators also emphasized the thermal effect of the Tibetan Plateau on the Asian summer monsoon (Nitta 1983; Luo and Yanai, 1984; Huang, 1984, 1985) and pointed out that the heating anomaly over the Tibetan Plateau has a large impact on the anomalies of the Asian summer monsoon circulation. Wu and Zhang (1997) explained that the Tibetan Plateau is an air-pump for the Asian summer monsoon onset through sensible heating and plays a triggering role in the monsoon onset. Recently, Zhang et al. (2002) pointed out that the heating over the Tibetan Plateau has a large and important effect on the east-west oscillation of the South Asian high, which has a significant influence on

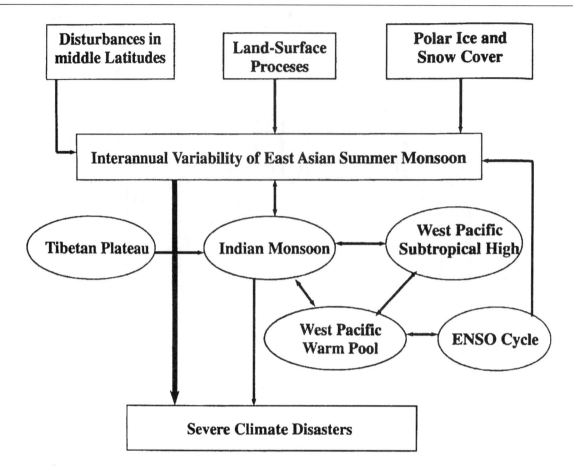

Fig. 4. Schematic map of the factors affecting the interannual variability of EASM.

the summer climate anomalies in the Yangtze and Huaihe River valleys.

(2) The thermal effect of the West Pacific warm pool

Studies by many scholars (Nitta, 1987; Huang and Li 1987; Kurihara, 1989; Huang and Sun, 1992) have shown that the thermal states in the West Pacific warm pool and the convective activities over the warm pool also play important roles in the interannual variability of EASM, as shown in Fig.2. Huang and Li (1987) and Huang and Sun (1992, 1994) made systematic investigations on the influence of the thermal states of the warm pool and the convective activities around the Philippines on the interannual anomalies of East Asian monsoon circulation from observed data and the theories of dynamics. These studies showed that there is a teleconnection pattern of the summer circulation anomalies over the Northern Hemisphere, namely, the so-called East Asia/Pacific teleconnection pattern (EAP pattern), which is also called the East Asia/North America teleconnection pattern by Lau and Shen (1992). This teleconnection pattern shows that the quasi-stationary planetary wavetrain can propagate from Southeast Asia to the western coast of North America through East Asia during boreal summer. This teleconnection pattern can greatly influence the interannual variability of EASM. Huang and Lu (1989) and Nikaido (1989) demonstrated with numerical simulations that the EAP pattern teleconnection is associated with the thermal states of the tropical western Pacific. Recently, Lu (2001) and Lu and Dong (2001) also showed that the convective activities over the West Pacific warm pool have a significant impact on the zonal shifts of the western Pacific subtropical high, and it was pointed out that if the convective activities are strong over the warm pool, then the western Pacific subtropical high shifts eastward, but it extends westward if the convective activities are weaker over the warm pool.

(3) Impact of the ENSO cycle

It is well known that the ENSO cycle is one of the most striking phenomena in the tropics and has a great influence on the Asian monsoon. The weak Indian summer monsoon tends to occur in El Niño years. Huang and Wu's study (1989) first showed that the summer monsoon rainfall anomalies in East Asia may depend on the stages of the ENSO cycle. Recently, Huang and Zhou (2002) pointed out from the composite analyses of the summer monsoon rainfall anomalies in different stages of the ENSO cycle that during the developing stage of an ENSO event, flooding tends to occur in the Yangtze and Huaihe River basins of China, in South Korea and in Southwest Japan, but drought may result in North China, as shown in Fig. 5a. On the other hand, during the decaying stage of an ENSO event, as shown in Fig. 5b, drought tends to occur in the Yangtze and Huaihe River valleys of China, in South Korea, and in Japan, but the summer monsoon rainfall may be normal or above normal in North China. Also, large positive monsoon rainfall anomalies appear to the south of the Yangtze River and severe flooding tends to occur there. Zhang et al. (1996) and Zhang (2001) also pointed out that southerly wind anomalies can appear in the lower troposphere along the coast of East Asia during the mature phase of an ENSO event, and the intensified southerly winds will favour the transport of water vapor from the Bay of Bengal and the tropical western Pacific to the eastern part of China.

Because of the interaction between the Asian monsoon and the ENSO cycle, the physical process of the influence of the ENSO cycle on the interannual variation of the East Asian monsoon is very complex and needs to be studied further.

(4) Impact of Eurasian snow cover

The interannual variability of Asian monsoon is also influenced by the Eurasian snow cover and the land-sea thermal contrast. Hahn and Shukla (1976) and Dickinson (1984) investigated the relationship between Indian monsoon rainfall and Eurasian snow cover. Their investigations showed that there is an inverse relationship between the two quantities. This inverse relationship has been demonstrated by Khandeker (1991) using detailed data. Moreover, Chen and Yan (1979, 1981) and Wei and Luo (1996) studied the influence of Tibetan plateau snow cover on EASM and pointed out that the Tibetan Plateau snow cover greatly influences the summer monsoon rainfall in the middle and lower reaches of the Yangtze River.

From the studies mentioned above, it may be seen that the interannual variation of EASM is closely associated with the thermal states of the western Pacific warm pool and the convective activities around the Philippines, the different stages of the ENSO cycle, the snow cover on the Eurasian continent, and the thermal effect of Tibetan Plateau, etc.

4. Interannual variability of the East Asian winter monsoon and its relation to the EASM

East Asia is also a region of strong winter monsoons. The East Asian winter monsoon (EAWM) features strong northwesterlies over North, Northeast China, Korea, and Japan and strong northeasterlies along the coast of East Asia (Academia Sinica, 1957; Chen et al., 1991; Ding, 1994). The strong winter monsoon can not only bring disasters such as low temperatures and severe snow storms in Northwest and Northeast China, North Korea, and North Japan in winter and severe sand dust-storms in North and Northwest China, Korea, and Japan in spring etc., but can also cause strong convective activities over the maritime continent of Borneo and Indonesia. Chang et al. (1979) and Lau and Chang (1987) pointed out that the tropical convective activities over the maritime continent may be closely associated with EAWM. Besides, strong and frequent activities of cold waves caused by strong EAWM may trigger the occurrence of El Niño (Li, 1988). Tomita and Yasunari (1996) also pointed out that EAWM might play a key role in the biennial oscillation of the ENSO/monsoon system. Therefore, the interannual variability of EAWM is also an interesting scientific issue.

4.1 *Interannual variability of EAWM*

Chen and Graf (1998) and Chen et al. (2000) systematically investigated the interannual variability of EAWM and its relation to EASM with a new definition of the EAWM index. Generally, the temperature difference or the pressure difference between the continent and ocean may be used to define the EAWM index, but the winds along the coast of East Asia seem to be more simple for defining the EAWM index. Therefore, Chen et al. (2000) chose the normalized meridional wind anomalies at 10 m over the East China Sea (25°–40°E, 120°–140°E) and the South China Sea (10°–25°N, 110°–130°E), averaged from November to March of the next year, as the EAWM index. This index is not only simple, but also has a higher correlation with the 500 hPa geopotential height fields around East Asia. Figure 6 indicates the interannual variation of the EAWM index.

From Fig. 6, it may be clearly seen that the interannual variability of EAWM is very significant. If the index shown in Fig. 6 is negative, the EAWM is stronger, corresponding to the strong northerly winds along the coast of East Asia and a cold East Asian continent and surrounding seas. In this case, the high pressure in the lower troposphere is strong over the East Asian continent and the East Asian trough at 500 hPa is also strong over Northeast China. On the other hand, if the index is positive, it corresponds to the weak northerly winds along the coast of East Asia

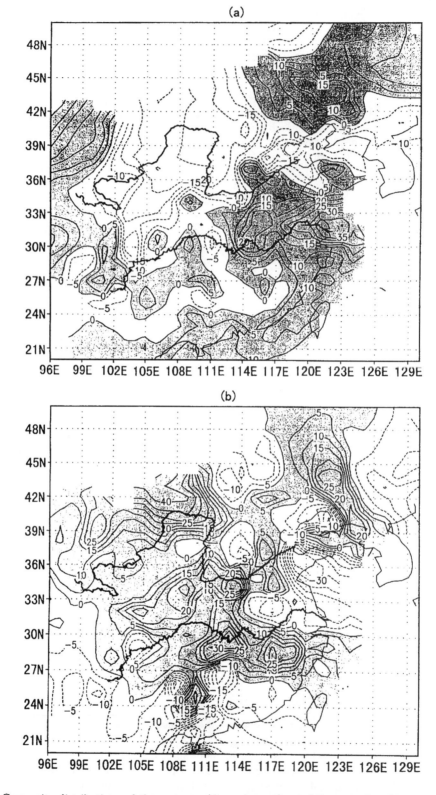

Fig. 5. Composite distributions of the summer (June–August) rainfall anomalies (percentage) in (a) the developing stage and (b) the decaying stage of ENSO cycles from 1951–1999. Shaded areas indicate positive rainfall anomalies.

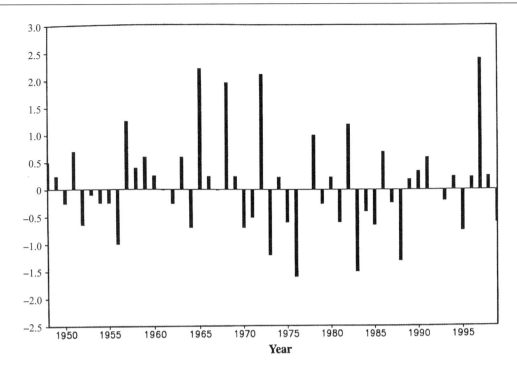

Fig. 6. Interannual variation of the EAWM index for the winters from 1948 to 1999.

and a warm East Asian continent and surrounding seas, and the high pressure in the lower troposphere over East Asia and the East Asian trough at 500 hPa are just opposite to those for the strong EAWM. Moreover, Fig. 6 shows that the EAWM was stronger from the mid 1970s to the late 1980s, but tended to be weaker from the early 1990s.

4.2 Influence of EAWM variability on EASM variability

Sun et al. (1996) showed the EAWM intensity can influence the subsequent EASM intensity. The investigation made by Chen et al. (2000) also showed that after strong EAWM, the western Pacific subtropical high will shift northward in the following summer, thus, a hot and drought-prone summer may occur in the Yangtze and Huaihe River valleys. On the other hand, a flood-prone summer may occur in the Yangtze and Huaihe River valleys following a weak EAWM. For example, in the summer of 1998, the severe flood occurred in the Yangtze River valley following the weak EAWM in the winter of 1997. This can confirm the connection between the EAWM and the following EASM.

4.3 Impact of the ENSO cycle on the interannual variability of EAWM

The ENSO cycle has a large influence not only on the interannual variability of EASM, but also on the interannual variability of EAWM. Tao and Zhang (1998) pointed out that during the winters of El Niño years, the circulation anomaly pattern over East Asia is not favorable for the outbreak of a cold surge, which may lead to the occurrence of a weak winter monsoon over East Asia. The result studied by Chen et al. (2000) also showed that the interannual variability of EAWM is greatly influenced by the processes associated with the SST anomalies in the tropical Pacific.

4.4 Effect of the East Asian monsoon on the ENSO cycle

The interaction between the Asian monsoon and the ENSO cycle is very obvious. Both diagnostic and modelling studies have revealed that the variabilities of the Asian summer monsoon activity have a significant effect on the atmosphere/ocean coupled system in the equatorial Pacific. Yamagata and Matsumoto (1989), Yasunari (1990), and Yasunari and Seki (1992) pointed out that a weaker (stronger) Asian summer monsoon seems to lead an anomalous state of the atmosphere/ocean system, which is favorable for an El Niño (anti-El Niño or La Niña event in the equatorial eastern Pacific. Li (1988, 1990), Huang et al. (1992), and Huang and Fu (1996a, b) suggested from the analysis of observed data that the anomalous East Asian monsoon may play an important role for the occurrence of an ENSO event. As shown in Fig. 7, the westerly anomalies over the tropical western Pacific have a significant dynamical effect on the formation of an El Niño event in the equatorial central and eastern Pacific through the excitation of equatorial Kelvin waves and Rossby waves in the tropical Pacific (Huang et al. 1998; Zhang and Huang, 1998; Huang and Zhang, 2001). Moreover, the analyzed results have shown that the westerly anomalies over the equatorial western Pacific may originate not only from the South Asian monsoon region, but also from the East Asian monsoon region (Huang et al., 1996; Chao and Chao, 2001). The southward propagation of the westerly wind anomalies in the lower troposphere over the East Asian monsoon region may lead to the westerly wind anomalies over the tropical Pacific through the EU pattern teleconnection.

Li (1988) and Li and Mu (1998) studied further the triggering effect of EAWM on ENSO events in the equatorial Pacific through data analysis and numerical simulation. They found that a strong EAWM will frequently bring a strong cold surge and can intensify the convective activities over the equatorial western Pacific. This, in turn, may strengthen the 30–60 day oscillation in the atmosphere over the tropical western Pacific, and the intensified low-frequency oscillation may trigger an ENSO event.

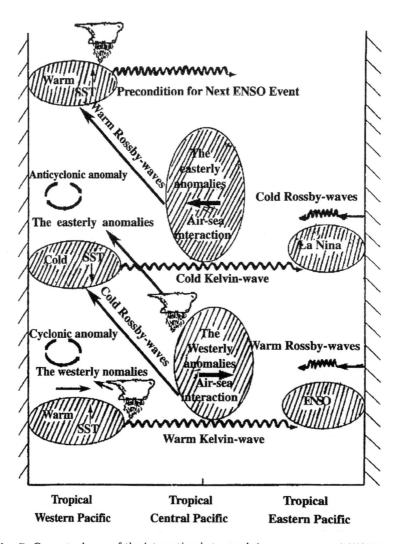

Fig. 7. Conceptual map of the interaction between Asian monsoon and ENSO cycle.

5. Interdecadal variability of EASM

5.1 *Interdecadal variation of EASM*

Recently, Huang et al. (1999) and Huang (2001) systematically studied the interdecadal variation of summer monsoon rainfall in East Asia. The results showed that the interdecadal fluctuation of summer (June–August) monsoon rainfall is more obvious than the decadal variation of surface temperature in East Asia. From the decadal-mean anomaly distributions of summer precipitation in the 1950s, to the 1990s, it can be seen that the anomaly distributions of summer monsoon rainfall in the period of the 1980s and the 1990s were obviously different from those in the 1970s in East Asia. during the 1980s and 1990s, campared with those in the 1970s, the summer monsoon rainfall obviously increased in the Yangtze River valley, but opposite phenomenon appeared in North China and the Yellow River valley, and prolonged severe droughts occurred in North China.

The interdecadal variation of summer monsoon rainfall may be seen from the interannual variations of summer precipitation in various regions of China. The analyzed results show that the interannual variations of summer precipitation in North China, the Yangtze River valley, and South China before 1976 were very different from those after the late 1970s. Thus, the differences between the summer precipitation anomalies (percentage) averaged for 1977–2000 and those averaged for 1967–1976 at 160 stations in China are analyzed (see Fig. 8). As shown in the figure, there are large differences between the summer precipitation anomalies averaged for 1977–2000 and those averaged for 1967–1976 in North China and the Yangtze River valley. From 1977 to 2000, the summer precipitation continuously decreased in North China and prolonged, severe droughts occurred frequently in this region, but in the Yangtze River valley, the summer precipitation increased obviously and serious floods were frequently caused there.

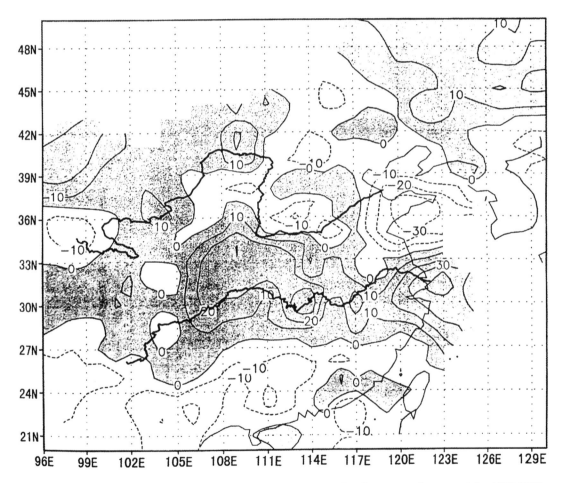

Fig. 8. Difference between the summer monsoon rainfall anomalies (percentage) averaged for 1997–2000 and those averaged for 1967–1976 in China. The shaded areas indicate positive values, dashed lines the negative values.

5.2 *Relationship between the decadal variability of EASM and the decadal variation of SST in the tropical Pacific*

In order to investigate the cause of the interdecadal variability of summer precipitation in East Asia, the ten-year running mean of SST anomalies in the tropical Pacific and their impact on the summer monsoon rainfall in East Asia are analyzed. The analyzed result shows that there is an obvious decadal variability in the SST anomalies of the equatorial central and eastern Pacific. This region of the Pacific was cooling in the 1970s and warming remarkably in the 1980s and 1990s (see Huang, 2001). Thus, the difference between the SST anomalies averaged for 1977–1999 and those averaged for 1967–1976 in the Pacific is also analyzed (see Fig. 7). As shown in the figure, an obvious El Niño–like SST anomaly pattern appeared in the tropical central and eastern Pacific from 1977 to 1999. This may explain the fact that the "decadal El Niño event" seems to occur from the 1980s to now, while the "decadal La Niña event" occurred in the 1970s. Therefore, there may be a dedacal El Niño–like cycle" in the interdecadal variability of SST anomalies of the tropical Pacific.

The above-mentioned analyses show that the decadal variability of summer monsoon rainfall in East Asia may be closely associated with the decadal variation of SST anomalies in the tropical Pacific. When the equatorial central and eastern Pacific is in a decadal warming episode, the summer monsoon rainfall is strong in the Yangtze River valley but weak in North China. Thus, droughts occurred in North China and floods frequently appeared in the Yangtze River valley in the period of the 1980s and 1990s because this period was in the warming episode of the equatorial central and eastern Pacific on the interdecadal time scale.

6. Problems for further study

It can be seen from the above review that great progress has been made in recent research on the East Asian monsoon variabilities and their causes. However, many problems on the variabilities of EASM and their causes are still not clear and need to be studied further:

(1) As mentioned in section 2, many investigations showed that the low-frequency oscillation in the tropics and mid-latitudes has an important impact on the East Asian monsoon, while the monsoon actions including the onset, maturity, and decay of the monsoon also influence the low-frequency oscillation. However, the interaction between them is not yet clear. Therefore, study of the interaction between the monsoon activity and the low-frequency oscillation in the atmosphere is also an important research issue, and is currently being faced by the Climate Variability and Predictability Programme (CLIVAR).

(2) There are complex dynamic and thermal processes in the East Asian monsoon system. The monsoon system, especially the East Asian monsoon rainband, includes systems with different spatial and temporal scales, and the interactions among these systems are important feedbacks to the monsoon. Moreover, the dynamical stability of the monsoon circulation has

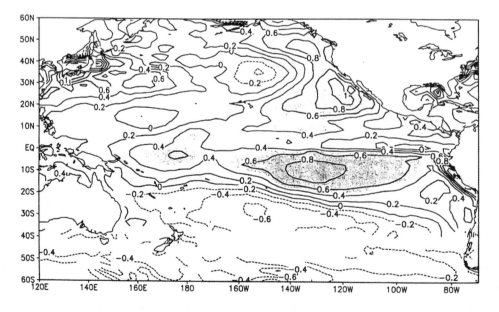

Fig. 9. Difference between the winter SST anomalies averaged for 1977–1999 and those averaged for 1967–1976 in the Pacific (units: °C).

an important effect on the monsoon activity and maintenance. However, there is little study on these problems. Therefore, the internal-dynamical process of the Asian monsoon system should be studied further.

(3) The different stages of the ENSO cycle influence the East Asian summer monsoon circulation and rainfall. However, through what process is the East Asian summer monsoon influenced? The answer is not yet clear. Moreover, the anomalous East Asian monsoon also influences the occurrence of the ENSO event. These influences are through the low-frequency oscillation or other physical process. The problem should be studied further through the analysis of observed data, dynamical theories, and numerical simulations in detail.

(4) The East Asian monsoon has a large interdecadal variability. However, the cause of this variability is not yet clear, and studies on it are scarce. This problem needs to be studied further from the air-land-ocean interaction on the interdecadal time-scale.

The above review shows that the East Asian monsoon variabilities on the intraseasonal, interannual, and interdecadal time scales are still important research issues. Therefore, understanding the interannual and interdecadal variabilities of the Asian monsoon system is the main objective of CLIVAR. Thus, after the implementation of research program of CLIVAR, it will be possible to further understand the physical mechanism of the interannual and interdecadal variabilities of the East Asian monsoon.

Because the authors' knowledge about the studies of East Asian monsoon variabilities is finite, and hence limited, this review mainly emphasizes the results studied by Chinese meteorologists and includes only a part of these results.

Acknowledgments. This paper was supported by the "National Key Programme for Developing Basic Sciences" Project G1998040900(I) and Project ZKCXZ-SW-210 of the Chinese Academy of Sciences.

REFERENCES

Chang, C. P., J. Erickson, and L. M. Lau, 1979: Northeasterly cold surges and near-equatorial disturbances over the winter-MONEX area during 1974. Part I: Synoptic aspects. *Mon. Wea. Rev.*, 107, 812–829.

Chang, C. P., Y. S. Zhang, and T. Li, 2000: Interannual and interdecadal variations of the East Asian summer monsoon and tropical Pacific SSTs. I, II. *J. Climate*, 13, 4310–4340.

Chao Q. C., and Chao J. P., 2001: The impact of the tropical wesern Pacific and the East Indian Ocean on the developing of ENSO. *Progress in Natural Sci.*, 11, 1293–1200.

Chen L. T., and Yan Z. X., 1979: Impact of the snow cover over the Tibetan Plateau in winter and spring on the atmospheric circulation and the precipitation in South China during the rainy season. *Collected Papers of Medium-and Long-Range Hydrological-Meteorological Forecast*, I, Water Resources and Electric Power Press, Beijing, 185–195. (in Chinese)

Chen L. T., and Yan Z. X., 1981: The statistical analysis of the impact of the anomalous snow cover over the Tibetan Plateau on the early-summer monsoon. *Collected Papers of Medium-and Long-Range Hydrological-Meteorological Forecast*, Water Resources and Electric Power Press, Beijing, 133–141. (in Chinese)

Chen L. X., Zhu Q. G., and Luo H. B., 1991: *East Asian Monsoon*, China Meteorological Press, Beijing, 362pp. (in Chinese)

Chen, W., and H. F. Graf, 1998: The interannual variability of East Asian winter monsoon and its relationship to global circulation. Max-Planck-Institute fur Meteorologie, Report No. 250.

Chen, W., H. F. Graf, and Huang R. H., 2000: Interannual variability of East Asian winter monsoon and its relation to the summer monsoon. *Advances in Atmospheric Sciences*, 17, 48–60.

Dickinson, R. E., 1984: Eurasian snow cover versus Indian rainfall—An extension of the Hahn-Shukla results. *J. Clim. Appl. Meteor.*, 23, 171–173.

Ding, Y. H., 1992: Summer monsoon rainfall in China. *J. Meteor. Soc. Japan*, 70, 373–396.

Ding, Y. H., 1994: *Monsoon over China*. Kluwer Academic Publishers, 420pp.

Gadgil, S., and G. Asha, 1992: Intraseasonal variation of the summer monsoon, I. Observation aspects. *J. Meteor. Soc. Japan*, 70, 517–527.

Hahn, D. C., and S. Manabe, 1975: The role of mountains in the South Asian monsoon circulation. *J. Atmos. Sci.*, 32, 1515–1541.

Hahn, D. J., and J. Shukla, 1976: An apparent relationship between snow cover and the Indian monsoon rainfall. *J. Atmos. Sci.*, 33, 2461–2462.

Huang G., and Yan Z. W., 1999: The East Asian summer monsoon circulation anomaly index and its interannual variation. *Chinese Sci. Bull.*, 44, 1325–1328.

Huang R. H., 1984: The characteristics of the forced planetary wave propagations in the summer Northern Hemisphere. *Advances in Atmospheric Sciences*, 1, 85–94.

Huang R. H., 1985: Numerical simulation of the three-dimensional teleconnections in the summer circulation over the Northern Hemisphere. *Advances in Atmospheric Sciences*, 2, 81–92.

Huang R. H., and Li W. J., 1987: Influence of the heat source anomaly over the tropical western Pacific on the subtropical high over East Asia. Proceedings of the International Conference on the General Circulation of East Asia, Chengdu, 10–15 April, 1987, 40–51.

Huang R. H., and Lu L., 1989: Numerical simulation of the relationship between the anomaly of the subtropical high over East Asia and the convective activities in the western tropical Pacific. *Advances in Atmospheric Sciences*, 6, 202–214.

Huang R. H., and Wu Y. F., 1989: The influence of ENSO on the summer climate change in China and its mechanisms. *Advances in Atmospheric Sciences*, 6, 21–32.

Huang R. H., Yin B. Y., and Liu A. D., 1992: Intraseasonal variability of the East Asian summer monsoon and its association with the convective activities in the tropical western Pacific. *Climate Variability*, Ye Duzheng et al., Eds., China Meteorological Press, Beijing, 134–155.

Huang, R. H., and F. Y. Sun, 1992: Impact of the tropical western Pacific on the East Asian summer monsoon. *J. Meteor. Soc. Japan*, **70** (1B), 243–256.

Huang R. H, and Sun F. Y., 1994: Impact of the thermal state and convective activities over the western Pacific warm pool on summer climate anomalies in East Asia. *Chinese J. Atmos. Sci.*, **18**, 262–272.

Huang R. H., 1994: Interaction between the 30-60 day oscillation, the Walker circulation and the convective activities over the tropical western Pacific and their relations to the interannual oscillation. *Advances in Atmospheric Sciences*, **11**, 367–384.

Huang R. H., and Fu Y. F., 1996a: Some progresses and problems in the study on the ENSO cycle dynamics. *Collected Papers of the Project "Disastrous Climate Prediction and Its Impact on Agriculture and Water Resources" II. Processes and Diagnosis of Disatrous Climate*, Huang Ronghui et al., Eds., China Meteorological Press, Beijing, 1972–188. (in Chinese)

Huang R. H., and Fu Y. F., 1996b: The interaction between the East Asian monsoon and ENSO cycle. *Climate Environ. Res.*, **1**, 38–54. (in Chinese)

Huang R. H., Xu Y. H., Wang P. F., and Zhou L. T., 1998: The features of the particularly severe flood over the Changjiang (Yangtze River) basin during the summer of 1998 and exploration of its cause. *Climate Environ. Res.* **3**, 300–313. (in Chinese)

Huang, R. H., 2001: Decadal variability of the summer monsoon rainfall in East Asia and its association with the SST anomalies in the tropical Pacific. CLIVAR Exchange, **2**, 7–8.

Huang R. H., Zang X. Y., and Zhang R. H., 1998: The westerly anomalies over the tropical western Pacific and their dynamical effect on the ENSO cycles during 1980–1994. *Advances in Atmospheric Sciences*, **15**, 135–151.

Huang R. H., Xu Y. H., and Zhou L. T., 1999: The interdecadal variation of summer precipitations in China and the drought trend in North China. *Plateau Meteor.*, **18**, 465–476. (in Chinese)

Huang R. H., Zhang R. H., and Yan B. L., 2001: Dynamical effect of the zonal wind anomalies over the tropical western Pacific on ENSO cycles. *Science in China (Series D)*, **44**, 1089–1098.

Huang R. H., and Zhou L. T., 2002: Research on the characteristic formation mechanism and prediction of severe climate disasters in China. *J. Nature Disasters*, **11**, 1–9. (in Chinese)

Khandekar, M. L., 1991: Eurasian snow cover, Indian monsoon and El Niño/Southern Oscillation-A synthesis. *Atmos. Ocean*, **29**, 636–647.

Krishnamurti, T. N., and Y. Ramanathan, 1982: Sensitivity of monsoon onset to differential heating. *J. Atmos. Sci.*, **39**, 1290–1306.

Kurihara, K., 1989: A climatological study on the relationship between the Japanese summer weather and the subtropical high in the western northern Pacific. *Geophys. Mag.*, **43**, 45–104.

Lau, K. M., and C. P. Chang, 1987: Planetary scale aspects of the winter monsoon and atmospheric teleconnections. *Monsoon meteorology*, C. P. Chang and T. N. Krishnamurti, Eds., Oxford University Press, 161–201.

Lau, K. M., and S. H. Shen, 1992: Biennial oscillation associated with the East Asian summer monsoon and tropical sea surface temperature. *Climate Variability*, Ye Duzheng et al., Eds., China Meteorological Press, Beijing, 53–58.

Li C. Y., 1988: The frequent actions of the East Asian trough and the occurrences of El Niño. *Scientia Sinica (Series B)*, 667–674. (in Chinese)

Li C. Y., 1990: Interaction between anomalous winter monsoon in East Asia and El Niño events. *Advances in Atmospheric Sciences*, **7**, 36–46.

Li C. Y., 1996: The precipitation during the rainy season in the Yangtze River and Huaihe River valley and the action of interseasonal oscillation in the tropical atmosphere. *Collected Papers of the Project Disastrous Climate Prediction and Its Impact on Agriculture and Water Resources II. Processes and Diagnosis of Disatrous Climate*, Huang Ronghui et al., Eds. China Meteorological Press, Beijing, 67–71. (in Chinese)

Li C. Y., and Mu M. Q., 1998: Numerical simulation of anomalous winter monsoon in East Asia exciting ENSO. *Chinese J. Atmos. Sci.*, **22**, 481–490.

Lu, R. Y., Y. S. Chuang, and R. H. Huang, 1995: Interannual variations of the precipitation in Korea and the comparison with those in China and Japan. *J. Korean Environ. Sci. Soc.*, **4**, 345–365.

Lu, R. Y., 2001: Interannnual variability of the summertime North Pacific subtropical high and its relation to atmospheric convection over the warm pool. *J. Meteor. Soc. Japan*, **79**, 771–783.

Lu, R. Y., and B. W. Dong, 2001: Westward extension of North Pacific subtropical high in summer. *J. Meteor. Soc. Japan*, **79**, 1229–1241.

Luo, H. B., and M. Yanai, 1984: The large-scale circulation and heat sources over the Tibetan Plateau and surrounding areas during the early summer of 1979. *Mon. Wea. Rev.*, **108**, 1849–1853.

McBride, J. J., 1987: The Australian summer monsoon. *Monsoon Meteorology*, C. P. Chang and T. N. Krishnamurti, Eds., Oxford University Press, 203 232.

Miao, J. H., and K. M. Lau, 1990: Interannual variability of the East Asian monsoon rainfall. *Quart. J. Appl. Meteor.*, **1**, 377–382. (in Chinese)

Murakami, T., and J. Matsumoto, 1994: Summer monsoon over the Asian continent and western North Pacific. *J. Meteor. Soc. Japan*, **72**, 745–791.

Nikaido, Y., 1989: The P-J like north-south oscillation found in 4-month integration of the global spectral model T42. *J. Meteor. Soc. Japan*, **67**, 587–604.

Nitta, Ts., 1983: Observational study of heat sources over the eastern Tibetan Plateau during the summer monsoon. *J. Meteor. Soc. Japan*, **61**, 590–605.

Nitta, Ts., 1987: Convective activities in the tropical western Pacific and their impact on the Northern Hemisphere summer circulation. *J. Meteor. Soc. Japan*, **64**, 373–390.

Palmer, T. N. et al., 1992: Modelling of interannual variations of summer monsoon. *J. Climate*, **5**, 399 417.

Academia Sinica, 1957: On the general circulation over Eastern Asia (I). *Tellus*, **9**, 432–446.

Sun, A. J., and Huang R. H., 1994: Low frequency oscillation characteristics of 500 hPa geopotential height fields in summer of 1983 and 1985 over the Northern Hemisphere. *Chinese J. Atmos. Sci.*, **18**, 365 375.

Sun, B. M., and Sun S. Q., 1994: The analysis on the features of the atmospheric circulation in preceding winter for the summer drought and flooding in the Yangtze and Huaihe River valley. *Advances in Atmospheric Sciences*, **11**, 79–90.

Tao S. Y., and Chen L. X., 1987: A review of recent research on the East Asian summer monsoon in China. *Monsoon Meteorology*, C. P. Chang and T. N. Krishnamurti, Eds., Oxford University Press, 60–92.

Tao S. Y., and Zhang Q. Y., 1998: Response of the Asian winter and summer monsoons to ENSO events. *Chinese J. Atmos. Sci.*, **22**, 399–407.

Tomita, T., and T. Yasunari, 1996: Role of the northeast winter monsoon on the biennial oscillation of the ENSO/Monsoon system. *J. Meteor. Soc. Japan*, **74**, 399–413.

Tu, C. W., and S. S. Huang, 1944: The advance and retreat of the summer monsoon. *Meteor. Mag.*, **18**, 1–20.

Wang, A. Y., S. F. Fong, Y. H. Ding, C. S. Wu, K. H. Lam, I. P. Hao, Q. Fan, C. M. Ku, W. S. Lin, and C. M. Tam, 2001; Onset, maintenance and retreat of Asian summer monsoon, *Climatological ATLAS for Asian Summer Monsoon*, S. K. Fong, and A. Y. Wang, Eds., Macau Foundation, Macau, 251–318.

Webster, P. J., and S. Yang, 1992: Monsoon and ENSO: Selectively interactive system. *Quart. J. Roy. Meteor. Soc.*, **118**, 877–926.

Wei Z. G., and Luo S. W., 1996: Impact of the snow cover in western China on the precipitation in China during the rainy season, *Collected Papers of the Project "Disastrous Climate Prediction and Its Impact on Agriculture and Water Resources", II. Processes and Diagnosis of Disatrous Climate*, Huang Ronghui et al., Eds., China Meteorological Press, Beijing, 40–45. (in Chinese)

Wu, G. X., and Y. S. Zhang, 1997: Tibetan Plateau forcing and the timing of the monsoon onset over South Asia and the South China Sea. *Mon. Wea. Rev.*, **126**, 917–927.

Yamagata, T., and Y. Matsumoto, 1989: A simple ocean-atmosphere coupled model for the origin of a warm El Niño/Southern Oscillation event. *Philos. Trans. Roy. Soc. London, A*, **329**, 225–236.

Yasunari, T., 1990: Impact of Indian monsoon on the coupled atmosphere/ocean systems in the tropical Pacific. *Meteor. Atmos. Phys.*, **44**, 29–41.

Yasunari, T., 1991: The monsoon year-A new concept of the climatic year in the tropics. *Bull. Amer. Meteor. Soc.*, **72**, 133–1338.

Yasunari, T., and Y. Seki, 1992: Role of the Asian monsoon on the interannual variability of the global climate system. *J. Meteor. Soc. Japan*, **70**, 179–189.

Ye D. Z., and Gao Y. X., 1979: *Tibetan Plateau Meteorology*, Science Press, Beijing, 279pp. (in Chinese)

Yeh, T. C., (Ye Duzheng), S. Y. Tao, and M. C. Li, 1959: The abrupt change of circulation over the Northern Hemisphere during June and October. *Atmosphere and the Sea in Motion*, 249–267.

Yin B. Y., Wang L. Y., and Huang R. H., 1996: Quasi-biennial oscillation of summer monsoon rainfall in East Asia and its physical mechanism. *Collected papers of the Project "Disastrous Climate Prediction and Its Impact on Agriculture and Water Resources" II. Processes and Diagnosis of Disatrous Climate*, Huang Ronghui et al., Eds., China Meteorological Press, Beijing, 196–205. (in Chinese)

Zhang, Q., G. X. Wu, and Y. F. Qian, 2002: The bimodality of the 100 hPa South Asian high and its association on the climate anomaly over Asia in summer. *J. Meteor. Soc. Japan* (to be published).

Zhang, R. H., A. Sumi, and M. Kimoto, 1996: Impact of El Niño the East Asian monsoon: A diagnostic study of the 86/87 and 91/92 events. *J. Meteor. Soc. Japan*, **74**, 49–62.

Zhang R. H., and Huang R. H., 1998: Dynamical effect of zonal wind stresses over the tropical Pacific on the occurring and vanishing of El Niño, Part I: Diagnostic and theoretical analyses. *Chinese J. Atmos. Sci.*, **22**, 587–599.

Zhang R. H., 2001: Relations of water vapor transports from Indian monsoon with those over East Asia and the summer rainfall in China. *Advances in Atmospheric Sciences*, **18**, 1005–1017.

Zeng, Q. C., B. L. Zhang, and Y.L. Liang, 1994: East Asian summer monsoon-A case study. *Proc. Indian Nation. Sci. Acad.*, **60**(1), 81–96.

Zhu K. E., 1934: Southeast monsoon and rainfall in China. *J. Chinese Geogr. Soc.*, **1**, 1–27.

中国东部夏季降水的准两年周期振荡及其成因*

黄荣辉[1]　陈际龙[1]　黄　刚[1]　张启龙[2]

(1 中国科学院大气物理研究所季风系统研究中心，北京 100080；
2 中国科学院海洋研究所，青岛 266071)

摘　要　应用中国160测站降水资料和ERA-40再分析资料以及EOF和熵谱分析方法，分析了中国夏季（6~8月）降水和东亚水汽输送通量的年际变化，表明中国（特别是华南、长江流域和淮河流域以及华北等地区）夏季降水具有2~3 a周期变化特征，即准两年周期的振荡特征，并表明中国降水的这种周期振荡与东亚上空夏季风水汽输送通量的准两年周期振荡密切相关；并且，还利用NCEP/NCAR的海表温度和日本气象厅的沿137°E海温剖面观测资料，分析了热带西太平洋表层与次表层海温的年际变化，揭示了热带西太平洋热力状态的变化也有显著的准两年周期的变化特征。作者利用相关和集成分析来讨论热带西太平洋热状态的准两年周期振荡对中国夏季降水和东亚水汽输送的影响，表明了热带西太平洋海温的准两年周期振荡对东亚夏季风及其所驱动的水汽输送都有很大影响。此外，作者还利用东亚/太平洋型（EAP型）遥相关理论，简单地讨论了热带西太平洋热力状态的准两年周期振荡影响中国夏季风降水准两年周期变化的物理机制。

The Quasi-Biennial Oscillation of Summer Monsoon Rainfall in China and Its Cause

HUANG Rong-Hui[1], CHEN Ji-Long[1], HUANG Gang[1], and ZHANG Qi-Long[2]

1 *Center for Monsoon System Research, Institute of Atmospheric Physics, Chinese Academy of Sciences, Beijing*　100080
2 *Institute of Oceanography, Chinese Academy of Sciences, Qingdao*　266071

Abstract　The observed data of precipitation at 160 observational stations of China, the ERA-40 reanalysis data and the Empirical Orthogonal Function (EOF) and the entropy spectral analysis methods are applied to analyze the interannual variations of summer (June–August) rainfall in China and water vapor transport fluxes over East Asia. The results show that there is an obvious oscillation with a period of two or three years, i. e., the quasi-biennial oscillation, in the interannual variations of summer monsoon rainfall in China, especially in the eastern and southern parts of China including South China, the Yangtze River valley and the Huaihe River valley and North China. And it is also shown that this oscillation is closely associated with the quasi-biennial oscillation in the interannual variations of the water vapor transport fluxes by summer monsoon flow over East Asia. Furthermore, the interannual variations of sea temperature in the surface and subsurface of the tropical western Pacific are analyzed by using the sea surface temperature (SST) data from the NCEP/NCAR reanalysis dataset and the sea temperature data in the subsurface of the western Pacific along 137° E from Japan Meteorological Agency, respectively. And it is revealed that there is also a significant quasi-biennial oscillation in the interannual variations of thermal state of the tropical western Pacific.

---

* 本文原载于：大气科学，第30卷，第4期，545-560，2006年7月出版。

In this paper, the correlative and composite analysis methods are applied to discuss the influence of the quasi-biennial oscillation of thermal state of the tropical western Pacific on summer rainfall in China and water vapor transport over East Asia, and it is shown that the quasi-biennial oscillation in the interannual variations of thermal state of the tropical western Pacific has a great impact on the East Asian summer monsoon and the water vapor transport driven by the monsoon flow. Besides, the influence of the quasi-biennial oscillation in the interannual variations of thermal state of the tropical western Pacific on the quasi-biennual oscillation in the interannual variations of the summer monsoon rainfall in China is simply discussed by using the teleconnection theory of the East Asia/Pacific (EAP) pattern.

From the above-mentioned analyses, the physical mechanism of the quasi-biennial oscillation of summer rainfall in China may be summarized as follows: If the thermal state of the tropical western Pacific is in a warming state during a winter, then the convective activities will be intensified around the Philippines in the following spring and summer, which can cause weak summer monsoon rainfall in the Yangtze River and the Huaihe River valleys through the EAP pattern teleconnection. And due to the intensification of the convective activities around the Philippines, a strong convergence of atmospheric circulation will appear over the tropical western Pacific. This will trigger a strong upwelling in the tropical western Pacific. As a consequence, the thermal state of this region will turn into a cooling one in the following winter. On the other hand, since the tropical western Pacific will in a cooling state during the following winter, the convective activities will weaken around the Philippines in the spring and summer of the third year, which can cause strong summer monsoon rainfall in the Yangtze River and the Huaihe River valleys through the EAP pattern teleconnection. And due to the weakening of the convective activities around the Philippines, a divergence of atmospheric circulation will appear over the tropical western Pacific in the spring and summer of the third year. As a consequence, the thermal state of these ocean regions will again turn into a warming one in the winter of the third year.

1 引言

Reed 等[1]以及 Veryard 和 Ebdon[2]在 20 世纪 60 年代发现了热带平流层下层的纬向平均气流存在着周期为 26 个月的东风和西风互相交替出现的年际变化。通常把热带平流层纬向平均气流这种具有 26 个月周期的年际变化又称为准两年周期振荡（Quasi-biennial oscillation，简称为 QBO）。在此发现之后，许多研究[3~8]表明了在热带海温、季风降水和环流也存在着 2~3a 周期的变化。由于此现象发生在对流层中，为了避免与热带平流层的 QBO 现象相混淆，一般把热带季风环流、降水、海温等具有 2~3 a 周期的年际变化称之为对流层准两年振荡（Tropospheric biennial oscillation，简称为 TBO）。

热带对流层准两年周期振荡（TBO）是亚澳季风区海-气耦合系统变化的基本特征之一。Mooley 和 Parthasarathy[3]以及 Yasunari 和 Suppiah[5]从观测事实的分析指出了在印度尼西亚和印度等热带季风降水的年际变化存在着 TBO。并且，Miao 等[9]、Tian 等[10]、殷宝玉等[11]和 Chang 等[12]的研究也表明了在东亚季风区的季风降水也存在着准两年周期振荡的现象。Lau 等[13]、Barnett[14]等和 Rasmussen 等[15]的研究表明了亚澳季风环流、降水的准两年周期振荡是紧密与 ENSO 的时间尺度相关联的。

然而，关于东亚季风区季风降水的准两年周期的研究迄今还是很不充分，以前的研究所用的观测资料也比较短，并且关于东亚季风区夏季降水的准两年周期振荡产生的机理迄今还不清楚，关于这方面的研究还是比较少。因此，有必要利用更长时间的降水资料和水汽资料进一步分析中国夏季降水和东亚季风区水汽输送的准两年周期振荡（TBO），并利用有关热带西太平洋海洋热状态的观测资料来分析东亚季风区夏季降水的准两年周期振荡的成因。为此，本研究利用 1951~2000 年中国 160 个测站降水资料、CMAP 降水资料（1979~2000 年）[16]和 EAR-40 再分析资料（1958~2000 年）[17]、NCEP/NCAR 再分析资料集的海温资料和日本气象厅沿 137°E 次表面海温剖面资料，分析热带西太平洋海温变化的准两年周期振荡及其对中国夏季风降水和东亚地区水汽输送的准两年周期振荡的影响。

2 中国东部夏季降水的准两年周期振荡

为了研究中国夏季降水的年际变化规律，本研究应用了 EOF（Empirical Orthogonal Function）分析方法来分析中国东部、中部、北部和南部 1951～2000 年夏季降水的年际变化规律。图 1a、b 分别是

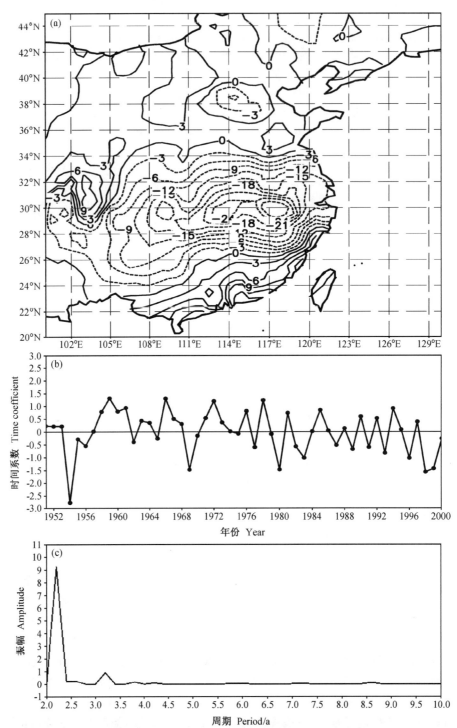

图 1　1951～2000 年中国夏季（6～8 月）降水 EOF 分析第 1 主分量（EOF1）的空间分布（a）和时间系数系列（b）以及时间系数的熵谱分析（c）。EOF1 说明总方差的 16.40%

Fig. 1　The spatial distribution (a) and the corresponding time coefficient series (b) of the first component of EOF analysis (EOF1) of summer (Jun–Aug) rainfall in China from 1951 to 2000, and the entropy spectrum of the time coefficients (c). EOF1 explains 16.40% of the variance

中国夏季降水 EOF 分析第 1 主分量（即 EOF1）的空间分布和时间系数系列（第 1 主分量对方差的贡献为 16.40%）。正如图 1a 所示，中国夏季降水 EOF1 的空间分布在中国东部、中部、北部和南部呈现出一个经向三极子型分布，最强的负信号位于长江中、下游地区和江淮流域，而最大的正信号分别位于华南和华北地区。这表明华北地区夏季风降水异常往往与江淮流域的夏季风降水异常相反，若江淮流域夏季风降水偏多，则华北和华南两地区夏季风降水往往偏少；反之，若江淮流域夏季风降水偏少，则华北和华南地区夏季风降水往往偏多。并且，从图 1b 所示的中国夏季风降水 EOF 分析第 1 主分量时间系数的变化可以看到，从 20 世纪 70 年代中期到 90 年代末中国夏季风降水的准两年周期振荡很明显，而从 50 年代初到 70 年代中期这种周期的振荡并不明显。若应用熵谱分析方法对中国夏季降水 EOF1 的时间系数进行分析，正如图 1c 所示，它呈现出以 2.0 a 周期为主要周期。这清楚说明了中国东部夏季风降水具有准两年周期振荡的特征。这个结果与 Lu 等[18]对韩国和我国东部降水资料的分析结果是一样的。

为了更直观地表明中国夏季降水具有准两年周期振荡的年际变化特征，本节给出江淮流域（包括长江以北和淮河流域的清江、徐州、蚌埠、阜阳、南阳、信阳、东台、南京、合肥、安庆、汉口等 11 个测站）夏季降水距平百分率的年际变化。如图 2 所示，从 20 世纪 70 年代后期到 90 年代末江淮流域夏季降水距平具有 2~3 a 周期变化为主要的年际变化特征。

此外，本文还利用 Xie 和 Arkin[16]的 CMAP 降水资料集，分析东亚地区夏季降水的年际变化规律（图略）。从分析结果可以明显看到，由于 CMAP 降水资料是再分析资料，与用测站资料的分析结果有所不同，在 20 世纪 80 年代东亚夏季降水的准两年振荡并不明显，只是在 90 年代东亚夏季风降水的年际变化具有明显的准两年周期的振荡特征。这与中国东部夏季风降水的年际变化特征相同。

3 东亚夏季风水汽输送的准两年振荡及其对东亚夏季降水准两年振荡的影响

正如前面所述，中国夏季降水的年际变化具有明显的准两年周期的年际变化特征，而中国夏季降水在很大程度上受到东亚夏季风的影响。东亚夏季风从热带西太平洋、南海和孟加拉湾带来大量水汽到东亚地区，从而引起东亚地区的降水。正如黄荣辉等[19]所指出，东亚季风区的降水主要来源于季风气流所引起的水汽平流，因此，东亚夏季风水汽输送的年际变化对东亚夏季降水年际变化有很大影响。为此，本节利用观测资料分析东亚和热带太平洋、南海和孟加拉湾地区水汽输送的年际变化特征。鉴于 NCEP/NCAR 再分析资料在 20 世纪 70 年代以前与实际偏差较大[20]，故本研究利用 EAR-

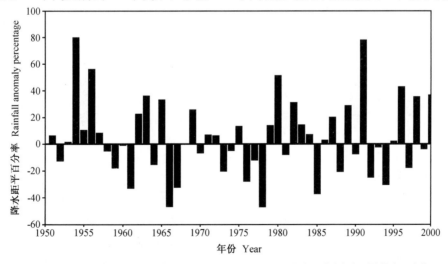

图 2 中国江淮流域 1951~2000 年夏季降水距平百分率的年际变化（取 1961~1990 年各月降水气候平均值为正常值）

Fig. 2 Interannual variations of the summer (Jun - Aug) rainfall anomaly percentage in the Yangtze River and Huaihe River valleys from 1951 to 2000. The climatological mean of monthly rainfalls in this region for 30 years from 1961 to 1990 is taken as the normal in the region

40再分析资料来分析东亚地区和热带西太平洋地区夏季水汽输送的年际变化规律。

假设在100 hPa以上高空没有水汽,水汽输送 $Q=(Q_\lambda,Q_\varphi)$ 能够写成下式:

$$Q_\lambda = \frac{1}{g}\int_{100}^{p_0} q \cdot u \mathrm{d}p, \quad (1)$$

$$Q_\varphi = \frac{1}{g}\int_{100}^{p_0} q \cdot v \mathrm{d}p, \quad (2)$$

其中,Q_λ 和 Q_φ 是纬向和经向的水汽输送,q 是比湿,u 和 v 是风场的纬向和经向分量,p_0 为海平面气压。

本研究利用ERA-40资料集1958～2000年再分析水汽和风场资料,并应用(1)和(2)式计算了从南海、孟加拉湾和热带西太平洋到东亚地区上空夏季(6～8月)水汽的经向和纬向输送通量,并且应用EOF分析方法分析此区域的经向和纬向水汽输送通量的时空变化特征。

3.1 东亚和热带西太平洋经向水汽输送的准两年周期振荡及其对中国夏季降水的影响

黄荣辉等[19]的研究表明了东亚夏季风系统的水汽输送有别于南亚季风系统,东亚夏季风水汽输送的经向分量比纬向分量大,而南亚季风系统则主要是纬向分量,并且表明东亚夏季风降水主要是由东亚夏季偏南气流所驱动的水汽输送(即水汽平流)和风场辐合所造成。因此,分析孟加拉湾、南海和热带西太平洋到东亚地区上空经向水汽输送的时空变化特征及其对中国地区降水的影响是很有必要的。图3a、b是从孟加拉湾、南海和热带西太平洋到东亚地区上空经向水汽输送的EOF分析第1主分量(可以说明总方差24.2%)的空间分布和时间系数。从图3a可以看到,在东亚地区和南海上空有一强的负信号区,而在日本以南热带和副热带西太平洋上空有一个强的正信号区。这表明若从南海到东亚地区上空的经向水汽输送弱,则在日本以南的热带和副热带西太平洋上空水汽的经向输送强;反之,若从南海到东亚上空水汽输送的经向分量强,则在日本以南的热带和副热带西太平洋上空水汽的经向输送弱。这可能与西太平洋副热带高压的位置偏东或偏西有关。从图3b可以看到,从孟加拉湾、南海和热带西太平洋到东亚地区上空水汽输送经向分量的EOF分析第1主分量的时间系数呈现出2～3 a周期的年际振荡。经最大熵谱分析,

这两个振幅相同的年际振荡周期为2.0 a和3.5 a(图3c)。

东亚和热带西太平洋经向水汽输送的准两年周期振荡在EOF分析第2主分量反映更明显。图4a和b分别是从孟加拉湾、南海和热带西太平洋到东亚地区上空水汽输送经向分量的EOF分析第2主分量(可以说明总方差19.0%)的空间分布和时间系数。从图4a可以看到,在中国东南部、韩国、日本以及南海和热带西太平洋上有一强的负信号区,而从中印半岛经我国华中、华北到我国东北有一较强的正信号区。这说明:若从中印半岛经华中、华北到东北经向水汽输送弱,则从南海、热带西太平洋到副热带西太平洋和我国东南沿海的经向水汽输送弱;反之,若从中印半岛经华中、华北到东北经向水汽输送强,则从南海、热带西太平洋到副热带西太平洋和我国东南沿海的经向水汽输送就强。并且,从图4b可以看到,从孟加拉湾、南海和热带西太平洋到东亚地区上空水汽经向输送的EOF分析第2主分量的时间系数呈现出一个明显的准两年周期振荡的年际变化特征。经熵谱分析,这个时间系数序列呈现出以2.0 a为主要周期的年际振荡特征(见图4c)。

上述从南海、孟加拉湾和热带西太平洋到东亚区域的经向水汽输送的EOF分析结果,清楚地表明了南海、孟加拉湾和热带西太平洋到东亚地区的经向水汽输送具有准两年周期的年际变化特征。

为了说明从孟加拉湾、南海和热带西太平洋到东亚地区的夏季风经向水汽输送的年际变化对中国夏季降水的年际变化有很大影响,本研究进一步计算了这两者EOF分析时间系数的相关。计算结果表明:从孟加拉湾、南海和热带西太平洋到东亚地区上空夏季经向水汽输送EOF分析第2主分量的时间系数与中国夏季降水的EOF分析第1主分量时间系数的相关系数,在1958～2000年与1977～2000年期间分别达0.33和0.44,大大超过95%的显著性检验。这两个相关系数可以表明从孟加拉湾、南海和热带西太平洋到东亚地区上空夏季经向水汽输送EOF分析第2主分量时间系数的变化与图1b所示的中国夏季降水EOF分析第1主分量的时间系数的变化有很大相关性。并且,从孟加拉湾、南海和热带西太平洋到东亚地区经向水汽输送EOF分析的第1主分量的时间系数与中国降水EOF

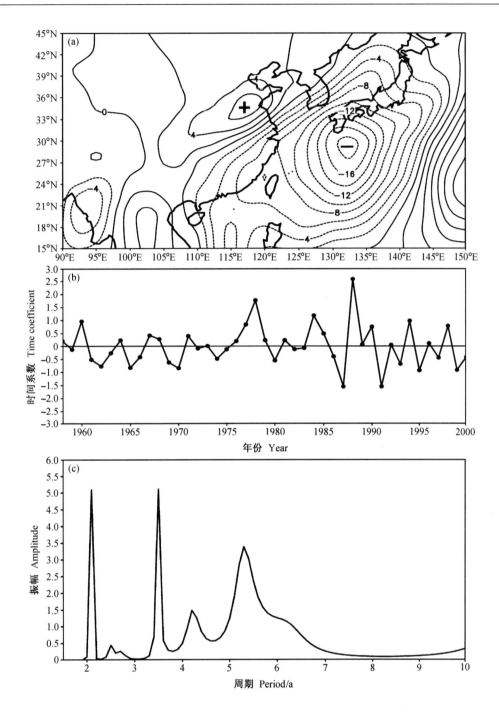

图 3 从孟加拉湾、南海和热带西太平洋到东亚地区上空经向水汽输送通量的 EOF 分析第 1 主分量的空间分布 (a) 和时间系数 (b) 及时间系数的熵谱分析 (c) (资料取自 ERA-40 再分析资料)。EOF1 对方差的贡献为 24.2%

Fig. 3 The spatial distribution (a) and corresponding time coefficient series (b) of the first EOF component of meridional water vapor transports over the region from the Bay of Bengal, the South China Sea (SCS), the tropical western Pacific (TWP) to East Asia during summer (Jun–Aug), and the entropy spectrum of the corresponding time coefficients (c). The moisture and wind fields data are taken from the ERA-40 reanalysis dataset. EOF1 explains 24.2% of the variance

分析第 1 主分量时间系数在 1977~2000 年期间的相关也达到 0.30。这些都表明由于从孟加拉湾、南海和热带西太平洋到东亚地区上空季风气流所引起的经向水汽输送具有准两年周期的振荡,这才造成中

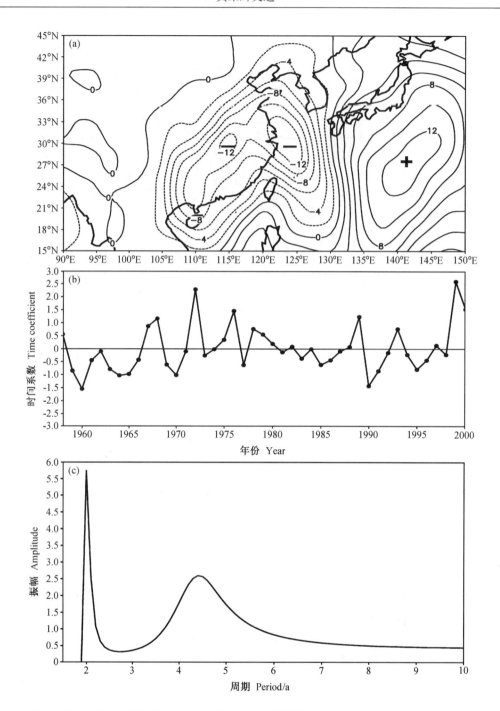

图 4 同图 3，但为 EOF 分析第 2 主分量（EOF2 对方差的贡献为 19.0%）

Fig. 4 Same as Fig. 3, but for the second EOF component (EOF2 explains 19.0% of the variance)

国夏季降水具有准两年振荡的年际变化特征。

3.2 东亚和热带西太平洋夏季纬向水汽输送的准两年周期振荡及其对中国夏季降水年际变化的影响

本研究还对从孟加拉湾、南海和热带西太平洋到东亚地区上空纬向水汽输送进行了 EOF 分析（图 5）。从上述地区夏季纬向水汽输送通量的 EOF 分析第 1 主分量（可以说明总方差 43.6%）的空间分布和时间系数系列可以看到：从孟加拉湾、南海和热带西太平洋到东亚地区上空夏季纬向水汽输送通量 EOF1 的空间分布呈现出一个三极子型分布，一个强的正信号区位于南海和热带西太平洋上空，一个强的负信号区位于华东、韩国和日本上空,另一个较弱的正信号位于东北地区。这表明南海和热带西太平洋上空纬向水汽输送的变化与我国长江淮河流域、韩国和日本上空纬向水汽输送的变化是相反的。若前者强,则后者弱;相反,若前者弱,而后者就强。这与图1a所示的中国夏季降水EOF分析第1主分量的空间分布一致。然而,从孟加拉湾、南海和热带西太平洋东亚上空夏季纬向水汽输送通量EOF分析第1主分量的时间系数2~3 a周期的变化不如4~5 a周期的振荡强,经熵谱分析,它以5 a为主要周期,而2~3 a为第二主周期(图略)。这表明东亚和热带西太平洋上空夏季水汽的纬向输送也具有准两年周期变化的成分。

从孟加拉湾、南海和热带西太平洋到东亚地区

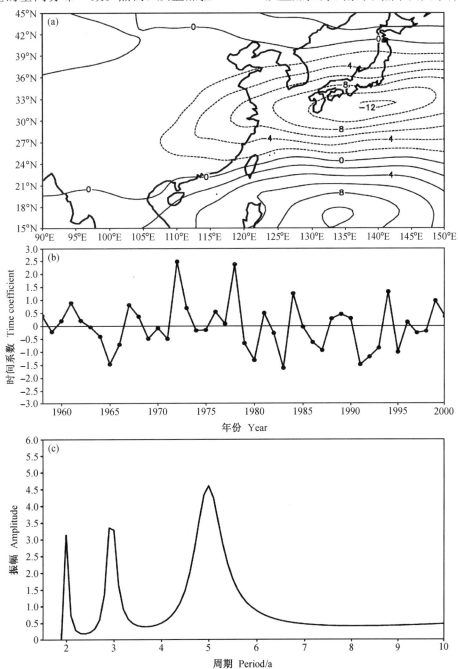

图5 同图3,但为纬向水汽输送(EOF1对方差的贡献为43.6%)

Fig. 5 Same as Fig. 3, but for the zonal water vapor transports (EOF1 explains 43.6% of the variance)

上空，夏季纬向水汽输送 EOF1 的时间系数与中国夏季降水 EOF1 的时间系数的相关在 1958～2000 年与 1977～2000 年期间分别达 0.42 和 0.45，也大大超过 95% 的显著性检验。这表明从孟加拉湾、南海和热带西太平洋到东亚地区的纬向水汽输送所具有准两年周期振荡成分对中国夏季降水的准两年振荡也有一定贡献。本研究还计算了从从孟加拉湾、南海和热带西太平洋到东亚地区经向水汽输送的 EOF2 时间系数与纬向水汽输送的 EOF1 时间系数在 1958～2000 年和 1977～2000 年期间的相关，这两个相关系数分别达到 0.49 和 0.62，大大超过 99% 的显著性检验，这不仅表明了两者相关很大，也说明了从孟加拉湾、南海和热带西太平洋到东亚地区上空夏季经向水汽输送，由于科里奥利力的作用也通过纬向水汽输送来影响中国夏季的降水变化。

从上面分析可以看到：夏季东亚和热带西太平洋上空无论是经向还是纬向的水汽输送通量的年际变化都具有准两年周期变化特征，并且，它们与中国夏季降水有很高的相关系数。这些可以很好地说明中国夏季降水准两年周期振荡的成因。

4 热带西太平洋热状态的年际变化及其对东亚夏季风年际变化的影响

从上面的分析结果可以看到，东亚和热带西太平洋夏季水汽输送（特别是经向水汽输送）具有准两年周期的振荡特征。这种振荡特征是什么原因造成的，目前还不太清楚，对其研究也不多。Lau 和 Shen[13] 提出东亚夏季风准两年周期振荡可能与热带东太平洋 SST 有关。然而，无论是东亚地区，还是南海和热带西太平洋上空的水汽输送或降水直接受热带西太平洋的海洋热状况所影响[21~24]，这些研究从观测事实的分析提出：热带西太平洋是全球海洋温度最高的海域，全球暖海水大部分集中在这里，这个海域又称暖池（warm pool），是大气热量主要供应地之一，并且在暖池上空，由于海表面附近海-气相互作用相当剧烈，又处于 Walker 环流上升支，故它的对流活动强，大尺度气流与水汽的辐合导致了此海域上空大气对流的不断加强和大量的降水。因此，暖池的热状况及其上空的对流活动不仅在维持热带纬圈环流起很大作用，而且在经向对北半球夏季大气环流的变化也有很大作用，特别是对东亚夏季风环流的变异起着十分重要的作用。

图 6 是 Niño.West（0°～14°N，130°E～150°E）区域平均的 SST 距平的年际变化，从图中可以看到热带西太平洋 SST 的变化幅度不大，一般在 1.0℃ 之内。然而，正如 Huang 和 Sun[25] 所指出，热带西太平洋热状态的年际变化还可以进一步从热

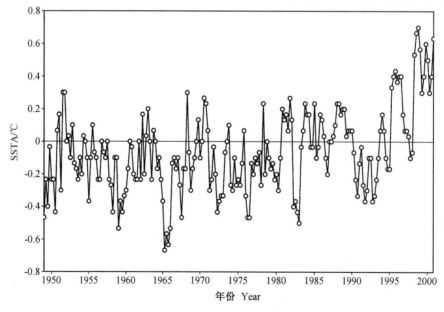

图 6　热带西太平洋 Niño.West（0°～14°N，130°E～150°E）区域平均的 SST 距平的年际变化（单位：℃）

Fig. 6　Interannual variations of the SST anomaly (℃) averaged over the Niño.West region of the tropical western Pacific (0°-14°N, 130°E-150°E)

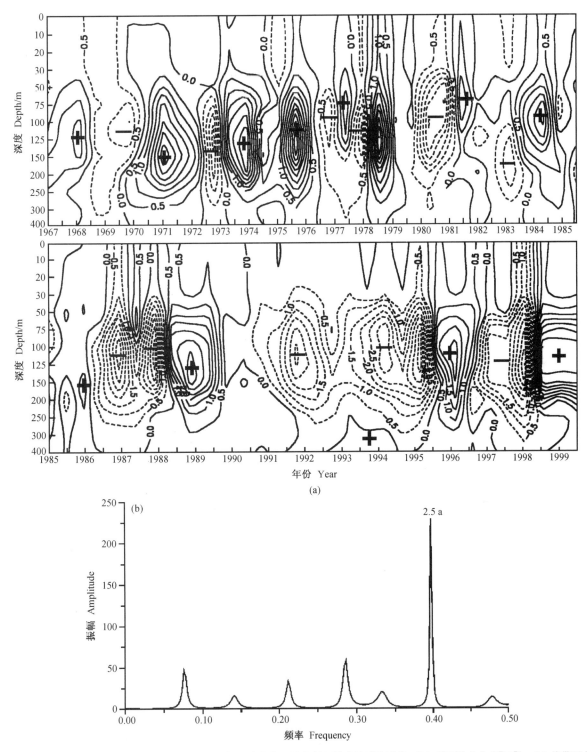

图 7 (a) 热带西太平洋次表层沿 137°E 的 5°N～10°N 平均的海温距平时间-深度剖面图（单位：℃）（资料取自文献[26]）；(b) 热带西太平洋沿 137°E 150 m 深次表层 5°N～10°N 平均的冬季次表层海温距平的熵谱

Fig. 7 (a) The time-depth cross section of the sea temperature (ST) anomalies (℃) averaged over 5°N–10°N along 137°E in the subsurface of the tropical western Pacific (The data are taken from the dataset observed by the Oceanographic Research Vessel "Ryofu-Maru", JMA[26]); (b) the entropy spectrum of the winter sea temperature anomalies averaged over 5°N–10°N along 137°E in the 150 m subsurface of the tropical western Pacific

带西太平洋次表层海温的年际变化看到。为此，本文利用日本气象厅"凌风号"科学考察船所观测的沿137°E表层和次表层的海温资料[26]，分析了热带西太平洋海温的年际变化。从热带西太平洋次表层海温的年际变化方差也可以看到，西太平洋暖池的海温在50~250 m深的次表层变化的均方差较大，最大均方差位于5°N~10°N，150 m深的次表层，均方差可达到4.0℃以上[27]。

图7a是西太平洋暖池次表层沿137°E的5°N~10°N平均的海温距平的时间-深度剖面图。把图7与图6相比较，可以明显看到：暖池次表层的海温距平的变化幅度要比表层海温距平的变化大得多，最大距平可以达到8.0℃左右。因此，利用暖池次表层的海温距平作为热带西太平洋热状态的表征可能更为合适。从图7a可以明显看到，热带西太平洋次表层海温的变化具有准两年周期。图7b是热带西太平洋沿137°E的150 m深次表层5°N~10°N冬季海温距平的熵谱。从此熵谱可以明显看到，热带西太平洋次表层冬季海温以2.5 a周期的振幅最大，这说明热带西太平洋次表层海温具有明显的准两年周期振荡特征。

正如前面所述，Nitta[23]和Huang等[24, 25]的研究指出了热带西太平洋海洋的热状态变化对于东亚夏季风的年际变化有很大的影响，并提出一个影响东亚夏季风变化的东亚/太平洋型（EAP型）遥相关（详见第6节）。黄刚[28]根据Nitta[23]和Huang等[24]所提出的EAP型遥相关型，并利用500 hPa高度场距平定义了一个衡量东亚夏季风年际变化的EAP指数。在此基础上，本文利用夏季（6~8月）850 hPa纬向风也定义了一个衡量东亚夏季风年际变化的季风指数（East Asian summer monsoon index，记为I_{EAM}），

$$I_{EAM} = (\text{Nor} \cdot \Delta U_{850}^A - 2\text{Nor} \cdot \Delta U_{850}^B + \text{Nor} \cdot \Delta U_{850}^C), \quad (3)$$

其中，$\text{Nor} \cdot \Delta U_{850}^A$、$\text{Nor} \cdot \Delta U_{850}^B$和$\text{Nor} \cdot \Delta U_{850}^C$分别是夏季850 hPa面上A（10°N~17.5°N，110°E~140°E）、B（27.5°N~35.0°N，120°E~150°E）、C（45°N~52.5°N，130°E~160°E）区域平均的纬向风标准化距平。图8中实线和虚线分别是本文所定义的1958~2000年东亚季风指数I_{EAM}与Niño.West（0~14°N，130°E~150°E）区域平均的月SST距平以及与Niño3.4区域平均的月SST距平的超前、同时和落后相关系数的变化（样本数在同时相关情况为43个）。I_{EAM}不仅在表征我国东部夏季季风降水的年际变化方面与黄刚[28]所定义的EAP指数具有同样的优点，而且与热带西太平洋SST距平有更高的相关系数。从图8可以看到两者相关系数具有准两年的变化，最大相关系数可达到0.45。这表明热带西太平洋海表温度对于东亚夏季风的准两年周期的变化有很大影响。

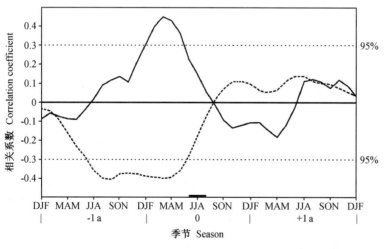

图8 1958~2000年I_{EAM}指数与Niño.West（0~14°N，130°E~150°E）（实线）以及与Niño3.4（虚线）区域平均的月SSTA的落后、同时和超前相关（同时相关的样本数为43）。横坐标0表示同时，"−"与"+"分别表示海温超前和落后

Fig. 8 The lagged correlations between the I_{EAM} defined in this paper and the monthly SST anomalies averaged over the regions of Niño.West (0°-14°N, 130°E-150°E) (solid line) and Niño 3.4 (dashed line) during 1958-2000. "0" denotes the simultaneous correlation, "−" and "+" the SSTA in the month before and after, respectively

5 热带西太平洋热状态的年际变化对东亚夏季水汽输送变化的影响

正如前节所述，热带西太平洋是全球海洋最高的海域，由于大量的水蒸汽从此海域的海洋蒸发到大气，从而给热带西太平洋、南海和东亚地区上空输送大量的水汽，因此，这个海域海温的变化将对东亚和热带西太平洋上空的水汽输送产生重要影响。为了说明这个影响，本研究利用ERA-40再分析资料计算并分析了热带西太平洋不同的热状态所对应的夏季从孟加拉湾、南海和热带西太平洋到东亚地区水汽输送通量的分布情况。图9是1979~2000热带西太平洋处于暖和冷状态夏季东亚和孟加拉湾、南海和热带西太平洋上空水汽通量矢量距平的合成分布图，可以看到从热带西太平洋到东亚地区，水汽输送通量距平矢量呈现出三极子式分布。正如图9a所示，当热带西太平洋处于暖状态时，在热带西太平洋上空有反气旋式水汽通量距平矢量分布；从华南、江南、南海北部到台湾以东的副热带西太平洋上空有一个明显气旋式水汽通量距平矢量的分布；而在我国黄淮流域和华北东部、东北以及日本、朝鲜半岛一带有明显的反气旋式水汽通量距平矢量分布。这说明在热带西太平洋处于暖状态时，有大量的暖湿空气从热带西太平洋向四周辐散，从中印半岛经华南到菲律宾以东地区有强的水汽辐合，造成华南地区的强降水，而在江淮流域有较强的水汽辐散，使得这些地区降水偏少。此外，在我国华北和河套地区也有偏南水汽的辐合，也使得此区域夏季降水偏多。从图9b可以看到，当热带西太平洋处于冷状态时，其水汽输送通量距平的矢量分布正好与上述暖池处于暖状态时的分布情况相反。在这种情况下，在热带西太平洋上空有气旋式水汽通量距平的矢量分布；从南海、华南到台湾以东的副热带西太平洋上空有反气旋式水汽通量距平的矢量分布；从江淮流域、华北东北部和东北地区，经朝鲜半岛到日本一带有明显的气旋式水汽通量距平矢量分布。这说明当热带西太平洋处于冷状态时，热带西太平洋暖湿空气向四周辐散和输送较弱，由于受江南、华南一带副热带高压的作用，大量的水汽从南海、孟加拉湾和副热带西太平洋输送到江淮流域、朝鲜半岛和日本一带，造成水汽在这些地区有较强的辐合，从而在这些地区有强的降水，而在华南和华北夏季降水较弱。

6 热带西太平洋热力变化影响东亚夏季风准两年周期振荡的物理过程

上面的分析已表明了热带西太平洋海表与次表层的海温有准两年周期的振荡，并且它对东亚夏季风的准两年周期变化有重要的影响，这个影响的物理机制是值得进一步探讨的。正如第4节中所述，Nitta[23]、Huang等[24,25]以及黄荣辉和李维京[29]从观测事实、准定常行星波传播理论和数值模拟系统地分析研究了热带西太平洋热状态和菲律宾周围对流活动加强后对北半球大气环流异常，特别是西太平洋副热带高压的影响。

Nitta[23]、Huang等[24]以及黄荣辉和李维京[29]提出热带西太平洋处于暖状态，在菲律宾周围对流强的夏季，从东南亚，经东亚、北太平洋到北美的西海岸明显存在着一个遥相关型。正如图10所示，当热带西太平洋处于暖状态，在菲律宾周围经南海到中印半岛有负距平分布，在我国江淮流域和以北的我国北方与日本本州有一片正距平分布，在这种情况下，西太平洋副热带高压位置偏北；并且，在鄂霍次克海上空有一片负距平，在阿拉斯加和阿留申地区有一片正距平；此外，在北美的北部与美国的西海岸上空有一片负距平，在墨西哥与美国南部上空有一片正距平。这个遥相关型称为东亚-太平洋型（East Asia/Pacific，简称EAP型），它对东亚和北美夏季气候变动（variability）和异常有很大影响。并且，Huang和Sun[25]进一步从观测事实分析、动力理论和数值模拟系统地分析了热带西太平洋的热状态对暖池上空的对流活动和热带西太平洋副热带高压的影响，分析结果表明：热带西太平洋暖池的热状态及其上空的对流活动对西太平洋副热带高压的强度与位置的变化起着十分重要的作用。当热带西太平洋暖池增暖时，从菲律宾周围经南海到中印半岛上空的对流活动将增强，西太平洋副热带高压的位置偏北，我国江淮流域夏季降水偏少；反之，则菲律宾周围的对流活动减弱，副热带高压偏南，江淮流域的降水偏多，黄河流域的降水偏少，易发生干旱。关于热带西太平洋热状态及其上空对流活动对东亚和热带西太平洋夏季风环流变化影响的物理机制已在参考文献[23~25, 28, 29]中详细讨论，本文不再重复。

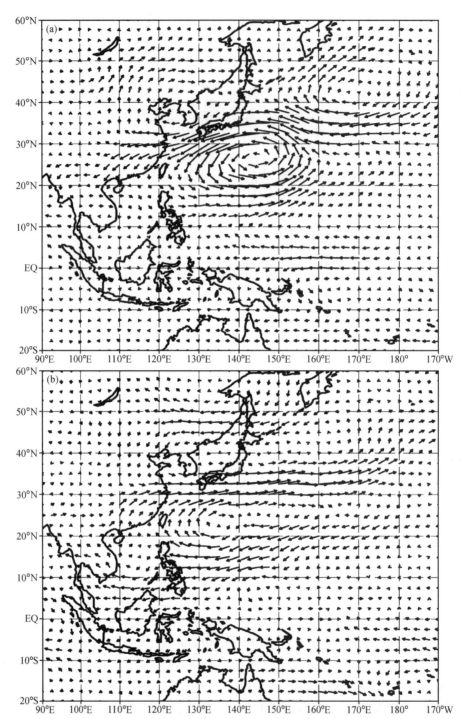

图 9 夏季热带西太平洋不同热状态时,从孟加拉湾、南海和热带西太平洋到东亚地区上空水汽输送通量距平矢量的合成分布(单位:10^3 g·s^{-1}·cm^{-1}):(a)热带西太平洋处于暖状态;(b)热带西太平洋处于冷状态. 资料取之于 ERA-40 再分析资料

Fig. 9 The distributions of composite water vapor transport flux anomalies (10^3 g·s^{-1}·cm^{-1}) over the region from the Bay of Bengal, the SCS and the TWP to East Asia in a warming state (a) and a cooling state (b) of the TWP. The moisture and wind fields data are taken from the ERA-40 reanalysis dataset

7 结论与讨论

本文通过观测资料分析,揭示了中国夏季降水的年际变化具有准两年周期振荡的特征;并利用 ERA-40 再分析资料分析了东亚地区和孟加拉湾、南海和热带西太平洋上空季风气流所引起的水汽输送

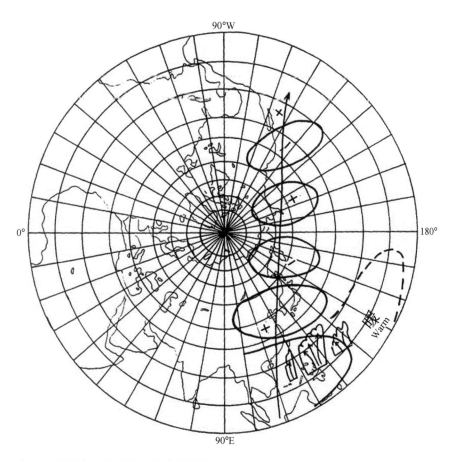

图 10 北半球夏季大气环流异常的东亚/太平洋型遥相关示意图

Fig. 10 Schematic diagram of the East Asia/Pacific pattern teleconnection of the summertime circulation anomalies over the Northern Hemisphere

通量的年际变化,表明此区域季风所驱动的经向水汽输送具有明显准两年周期振荡,而纬向水汽输送也具有准两年周期振荡的成分,从而说明了中国夏季降水的准两年周期振荡(TBO)是由于东亚和热带西太平洋上空夏季风所驱动的水汽输送的准两年周期振荡所引起。本文还利用实测资料分析热带西太平洋次表层海温的准两年周期振荡特征及其对东亚地区夏季风水汽输送和中国夏季降水的准两年周期振荡(TBO)的影响,并从理论上讨论其物理过程。

这些观测事实和理论上的讨论结果(图 11)表明:当某一年冬春季热带西太平洋海温上升,也就是热带西太平洋处于偏暖状态,这就会使得第二年春夏季菲律宾周围的热带西太平洋上空对流活动偏强,由于 EAP 型遥相关波列的影响,将会使得东亚和北半球上空夏季出现似图 11 所示的高度场异常分布,即西太平洋副热带高压偏北,这将导致我国长江、淮河流域、日本和韩国的夏季降水偏少。另一方面,从海洋方面看,由于第二年夏季热带西太平洋面附近上空强对流活动所产生的强辐合将会造成热带西太平洋的海水上翻(upwelling)加强(将在另文讨论),从而导致秋冬季此海域海温开始下降,并使此海域的冬季及以后海表和次表层海温偏低。这样,由于第三年春夏季热带西太平洋海温处于冷状态,这就会使得春夏季菲律宾周围对流活动就减弱,夏季出现与图 10 所示相反的高度场异常分布,即西太平洋副热带高压偏南,从而引起我国长江、淮河流域、日本和韩国的夏季降水偏多。另一方面从海洋方面看,由于第三年春夏季热带西太平洋上空对流活动弱,西太平洋副热带高压偏南,这样造成热带西太平洋海表上空对流层下层气流辐合很弱,并出现反气旋距平环流分布,从而引起热带西太平洋的海水上翻减弱,这导致秋冬季此海域海温开始上升,并使第三年春夏季此海域的海表和次表层海温又变成偏高。这样,热带西太平洋海-气耦合系统的变化经历了一个循环,这个循环周期大约为 2 a 左右。

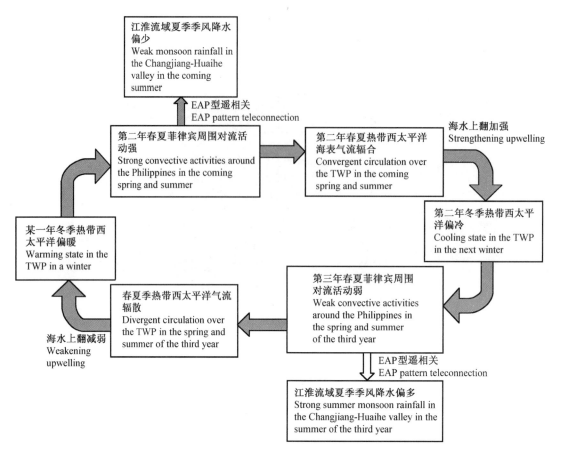

图 11 东亚夏季风准两年周期振荡（TBO）与热带西太平洋海-气耦合系统关联概念图
Fig. 11 Conceptive diagram of the quasi-biennial oscillation of East Asian summer monsoon related to the air–sea coupling system in the tropical western Pacific

从本文分析结果看，中国东部夏季降水的准两年周期振荡主要受热带西太平洋热力状态所影响。当然，南海和孟加拉湾的热力状态对中国夏季降水的准两年周期振荡也有一定影响。限于本文所研究的内容，不可能再阐述这两海域的热力状态的变化及其对中国夏季降水的准两年周期振荡的影响。因此，我们将进一步开展热带东印度洋和南海热力状态变化对中国夏季降水的准两年周期振荡的影响。此外，中国夏季降水的准两年周期振荡在 20 世纪 70 年代中期前并不太明显，而在 70 年代中后期以后比较明显，这种年代际变化成因也有待于进一步探讨。

参考文献（References）

[1] Reed R, Cambell W J, Rasmusson L A, et al. Evidence of a downward propagating annual wind reversal in the equatorial stratosphere. *J. Geophys Res.*, 1961, **66**: 813～818

[2] Veryard R G, Ebdon R A. Fluctuations in tropical stratospheric winds. *Meteor. Mag.*, 1961, **90**: 125～143

[3] Mooley D A, Parthasarathy B. Fluctuations in all-India summer monsoon rainfall during 1871–1978. *Climate Change*, 1984, **6**: 287～301

[4] Meehl G. The annual cycle and interannual variability in the tropical Pacific and Indian Ocean region. *Mon. Wea. Rev.*, 1987, **115**: 27～50

[5] Yasunari T, Suppiah R. Some problems on the interannual variability of Indonesian monsoon rainfall. *Tropical Rainfall Measurements*, edited by J. S. Theon and N. Fugono, Deepak, Hampton, Va., 1988. 113～122

[6] Kiladis G N, Van Loon H. The Southern Oscillation. Part VII: Meteorological anomalies over the Indian and Pacific sectors associated with the extremes of the oscillation. *Mon. Wea. Rev.*, 1988, **116**: 120～136

[7] Yasunari T. Impact of Indian monsoon on the coupled atmosphere/ocean system in the tropical Pacific. *Meteor. Atmos. Phys.*, 1990, **44**: 29～41

[8] Ropelewski C F, Halpert M S, Wang X. Observed tropo-

spherical biennial variability and its relationship to the Southern Oscillation. *J. Climate*, 1992, **5**: 594~614

[9] Miao J H, Lau K-M. Interannual variability of East Asian monsoon rainfall. *Quart. J. Appl. Meteor.*, 1990, **1**: 377~382

[10] Tian S F, Yasunari T. Time and space structure of interannual variations in summer rainfall over China. *J. Meteor. Soc. Japan*, 1992, **70**: 585~596

[11] 殷宝玉, 王连英, 黄荣辉. 东亚夏季风降水的准两年振荡及其可能的物理机制. 见: 黄荣辉等编. 灾害性气候的过程及诊断论文集. 北京: 气象出版社, 1996. 196~205
Yin Baoyu, Wang Lianying, Huang Ronghui. Quasi-biennial oscillation of summer monsoon rainfall in East Asia and its physical mechanism. *Collected Papers of the Project "Disastrous Climate Prediction and Its Impact on Agriculture and Water Resources"*. Huang Ronghui et al., Eds., Beijing: China Meteorological Press, 1996. 196~205

[12] Chang C P, Zhang Y S, Li T. Interannual and interdecadal variations of the East Asian summer monsoon and tropical Pacific SSTs. I, II. *J. Climate*, 2000, **13**: 4310~4340

[13] Lau K-M, Shen P J. Annual cycle, quasi-biennial oscillation and Southern Oscillation in global precipitation. *J. Geophys. Res.*, 1988, **93**: 10975~10988

[14] Barnett T P, Dumenil L, Schlese U, et al. The effect of Eurasian snow cover on regional and global climate variations. *J. Atmos. Sci.*, 1989, **46**: 661~685

[15] Rasmussen E M, Wang X, Ropelewski C F. The biennial component of ENSO variability. *J. Mar. Syst.*, 1990, **1**: 71~90

[16] Xie P, Arkin P A. Analyses of global monthly precipitation using gauge observations, satellite estimates and numerical model prediction. *J. Climate*, 1997, **9**: 804~858

[17] Uppala S. ECWMF Reanalysis, 1957~2001, ERA-40. *ERA-40 Project Report Series*, 2002, **3**: 1~10

[18] Lu R Y, Chung Y S, Huang R H. Interannual variations of the precipitation in Korea and the comparison with those in China and Japan. *J. Korea Environ. Sci. Soc.*, 1995, **4**: 345~356

[19] 黄荣辉, 张振洲, 黄刚, 等. 东亚季风区的水汽输送特征及其与印度季风区的差别. 大气科学, 1998, **22**: 460~469
Huang Ronghui, Zhang Zhenzhou, Huang Gang, et al. Characteristics of the water vapor transport in East Asian monsoon regions and its difference from that in South Asian monsoon region in summer. *Chinese Journal of Atmospheric Sciences (Scientia Atmospherica Sinica)* (in Chinese), 1998, **22**: 460~469

[20] Inoue T, Matsumoto J. A comparison of summer sea level pressure over East Eurasia between NCEP/NCAR reanalysis and ERA-40 for the period 1960-99. *J. Meteor. Soc. Japan*, 2004, **82**: 951~958

[21] Cornejo-Garrido A G, Stone P H. On the heat balance of the Walker circulation. *J. Atmos. Sci.*, 1977, **34**: 1155~1162

[22] Hartmann D, Hendon H, Houze R A. Some implications of the mesoscale circulations in tropical cloud clusters for large-scale dynamics and climate. *J. Atmos. Sci.*, 1984, **41**: 113~121

[23] Nitta Ts. Convective activities in the tropical western Pacific and their impact on the Northern Hemisphere summer circulation. *J. Meteor. Soc. Japan*, 1987, **64**: 373~390

[24] Huang Ronghui, Li Weijing. Influence of the heat source anomaly over the western tropical Pacific on the subtropical high over East Asia. Proc. International Conference on the General Circulation of East Asia, April 10-15, 1987, Chengdu, 40~51

[25] Huang Ronghui, Sun Fengying. Impact of the tropical western Pacific on the East Asian summer monsoon. *J. Meteor. Soc. Japan*, 1992, **70** (1B): 243~256

[26] Japan Meteorological Agency. Monthly Report on Climate System 1978~2000

[27] 叶笃正, 黄荣辉, 等. 长江黄河流域旱涝规律和成因研究. 济南: 山东科技出版社, 1992. 387pp
Ye Duzheng, Huang Ronghui, et al. *Studies on Regularity and Cause of Droughts and Floods in the Yangtze River Valley and the Yellow River Valley*. Ji'nan: Shandong Science and Technology Press, 1992. 387pp

[28] 黄刚. 东亚夏季风环流异常指数与夏季气候变化关系的研究. 应用气象学报, 1999, **10**: 61~69
Huang Gang. Study on the East Asian monsoon circulation index and its relation to summer climate variability in East Asia. *J. Applied Meteor.* (in Chinese), 1999, **10**: 61~69

[29] 黄荣辉, 李维京. 夏季热带西太平洋上空的热源异常对东亚上空副热带高压的影响及其物理机制. 大气科学, 1988, 特刊: 95~107
Huang Ronghui, Li Weijing. Influence of the heat source anomaly over the tropical western Pacific on the subtropical high over East Asia and physical mechanism. *Chinese Journal of Atmospheric Sciences (Scientia Atmospherica Sinica)* (in Chinese), 1988, Special Issue: 95~107

Characteristics and Variations of the East Asian Monsoon System and Its Impacts on Climate Disasters in China*

Huang Ronghui*(黄荣辉), Chen Jilong(陈际龙), and Huang Gang(黄刚)

*Center for Monsoon System Research, Institute of Atmospheric Physics,
Chinese Academy of Sciences, Beijing 100080*

ABSTRACT

Recent advances in studies of the structural characteristics and temporal-spatial variations of the East Asian monsoon (EAM) system and the impact of this system on severe climate disasters in China are reviewed. Previous studies have improved our understanding of the basic characteristics of horizontal and vertical structures and the annual cycle of the EAM system and the water vapor transports in the EAM region. Many studies have shown that the EAM system is a relatively independent subsystem of the Asian-Australian monsoon system, and that there exists an obvious quasi-biennial oscillation with a meridional tripole pattern distribution in the interannual variations of the EAM system. Further analyses of the basic physical processes, both internal and external, that influence the variability of the EAM system indicate that the EAM system may be viewed as an atmosphere-ocean-land coupled system, referred to the EAM climate system in this paper. Further, the paper discusses how the interaction and relationships among various components of this system can be described through the East Asia Pacific (EAP) teleconnection pattern and the teleconnection pattern of meridional upper-tropospheric wind anomalies along the westerly jet over East Asia. Such reasoning suggests that the occurrence of severe floods in the Yangtze and Huaihe River valleys and prolonged droughts in North China are linked, respectively, to the background interannual and interdecadal variability of the EAM climate system. Besides, outstanding scientific issues related to the EAM system and its impact on climate disasters in China are also discussed.

1. Introduction

China is located in the East Asian monsoon region, thus, the climate of China is influenced mainly by the East Asian monsoon (EAM) (e.g., Zhu, 1934; Tu and Huang, 1944). The significant interannual and interdecadal variabilities of the EAM are related to frequent climate disasters (e.g., Huang and Zhou, 2002; Huang et al., 2006a). Especially since the 1980s, severe climate disasters over large areas have caused major damage to agricultural and industrial production in China. Each year, the economic losses due to droughts and floods can reach over 200 billon yuan (i.e., about US$24 billion), accounting for 3%–6% of China's GDP in the early 1990s (e.g., Huang et al., 1999a; Huang and Zhou, 2002). In the summer of 1998, for example, particularly severe floods occurred in the Yangtze River basin and the Songhua and Nen River valley, causing losses as high as 260 billion yuan (i.e., about US$ 31 billion) (e.g., Huang et al., 1998a). In addition, persistent droughts in North China since the late 1970s brought not only huge losses to agriculture and industry, but also seriously affected the region's water resources and ecology resulting in a large increase in the frequency sand-dust storms. To reduce the losses due to climate disasters, research is necessary to increase understanding and the ability to predict climate disasters in China. As a result, such research was one among the first batch of projects supported by the National Key Program for Developing Basic Sciences (e.g., Huang et al., 1999a; Huang, 2001a, 2004) and was a major project supported by the National Natu-

---

* 本文原载于: Advances in Atmospheric Sciences, Vol.24, No.6, 993-1023, 2007.

ral Science Foundation of China.

The occurrence of climate disasters such as droughts and floods in China, especially the severe flooding disaster in the Yangtze River basin in the summer of 1998 and the persistent droughts in North China since the late 1970s, is closely associated with the variability and anomalies associated with the EAM system. Just as monsoons generically may be viewed as atmosphere-ocean-land coupled systems (Webster et al., 1998), the variability and anomaly of the EAM system are affected not only by internal dynamical and thermodynamical processes in the atmosphere, but also by the interactions among various components of a coupled atmosphere-ocean-land system. To study the recurrence and causes of severe climate disasters in China, the variability in various components of the EAM climate system has been analyzed in detail in recent years. These components include the circulation of the EAM, western Pacific subtropical high, atmospheric disturbances in mid-and high latitudes, thermal states of the Pacific warm pool, convective activity around the Philippines, ENSO cycle in the tropical Pacific, dynamical and thermal effects over the Tibetan Plateau, land surface process in the arid and semi-arid areas of Northwest China, and snow cover on the Tibetan Plateau (e.g., Huang et al., 1998a; Huang et al., 2001; Huang et al., 2004a). In addition to descriptive aspects of these features, the nature of the internal and external physical processes that influence the variabilities in the EAM system have been addressed recently.

This paper attempts to summarize the advances in recent studies on the characteristics and variabilities of the EAM system and its impacts on climate disasters in China. The review focuses on the progress achieved by research in China during the recent decades.

2. Characteristic of the EAM system

The Asian-Australian monsoon system is an important circulation system in the global climate system. Many studies have shown that the Asian and Australian monsoons play an important role in global climate variability (e.g., Tao and Chen, 1987; Ding, 1994; Huang and Fu, 1996; Webster et al., 1998; Huang et al., 2003; Huang et al., 2004a). Since there is a close association between the South Asian monsoon (SAM), the East Asian monsoon (EAM) and the North Australian monsoon (NAM), some scholars consider them three subsystems of the Asian-Australian monsoon system (e.g., Webster et al., 1998). However, there are some differences among these monsoon subsystems. For example, the East Asian summer monsoon (EASM) is not only a part of the tropical monsoon, it also has properties of disturbances in middle latitudes (e.g., Tao and Chen, 1987). But, the South Asian summer monsoon (SASM) and the North Australian summer monsoon (NASM) are only tropical. Therefore, the EASM differs from the SASM and the NASM in aspects described in the following sub sections.

2.1 *A relatively independent component of the Asian-Australian monsoon system*

According to Krishnamurti and Ramanathan (1982) study, the principal components of the SASM include: the Mascarene high, Somali cross-equatorial low-level flow, monsoon trough over North India in the lower troposphere, South Asian high, and the north to south cross-equatorial flow in the upper troposphere. Tao's investigation (e.g., Tao and Chen, 1985) showed the main components of EASM monsoon include: the Indian SW monsoon flow, Australian cold anticyclone, cross-equatorial flow east to 100°E, monsoon trough (or ITCZ) over the South China Sea (SCS), tropical easterly flow around the west Pacific the western Pacific subtropical high, mei-yu (or Baiu in Japan, or Changma in Korea) frontal zone, and disturbances in mid-latitudes. Hence, the EASM is a relatively independent monsoon circulation system.

Recently, Chen and Huang (2006) analyzed the climatological characteristics of the wind structure and seasonal evolution of subsystems of the Asian-Australian monsoon system, specifically the SASM over the area (0°–25°N, 60°–100°E) and the EASM over the area (0°–45°N, 100°–140°E) in boreal summer, and the NASM over the area (0°–15°S, 110°–150°E) in austral summer. Results (Figs. 1a and 1c) show that both the SASM and the NASM are solely tropical strong zonal flow with vertical easterly shear, i.e., low-level westerlies and high-level easterlies. The vertical easterly shear in the SASM region is stronger than that in the NASM region. The vertical structure of zonal flow in the EASM region is complex. It includes vertical easterly shear in the region to the south of 25°N, such as over the South China Sea (SCS) and the tropical western Pacific, and vertical westerly shear in the subtropical monsoon region north of 25°N, such as over mainland China, Korea and Japan (Fig. 1b). Thus, the EASM is composed of both tropical and subtropical summer monsoons with a significant meridional flow and vertical northerly shear, i.e., low-level southerlies and high-level northerlies (Fig. 2). Moreover, compared to the meridional components of wind in the SASM and NASM regions, the low-level southerlies in the EASM region are stronger than those in the SASM and NASM regions.

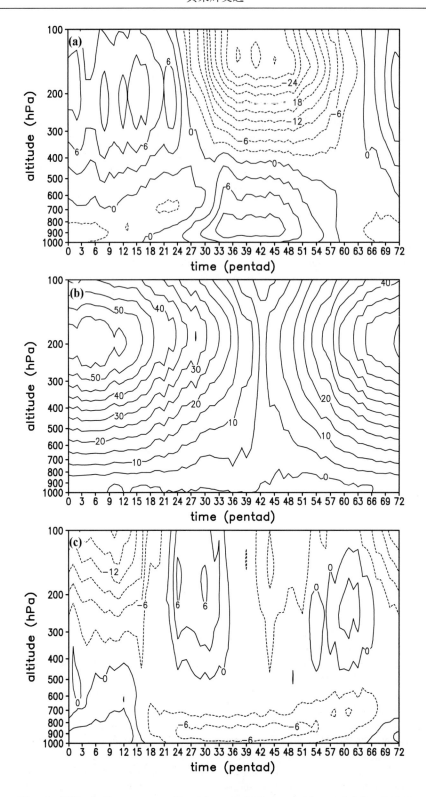

Fig. 1. Altitude-time cross section of zonal wind averaged over (a) South Asia (0°–25°N, 60°–100°E), (b) East Asia (20°–45°N, 100°–140°E) and (c) North Australia (0°–15°N, 110°–150°E). Units: m s^{-1}. Solid and dashed lines indicate westerly and easterly winds, respectively. Shown are 1979–2003 means derived from NCEP/NCAR reanalysis data (e.g., Kalnay et al., 1996).

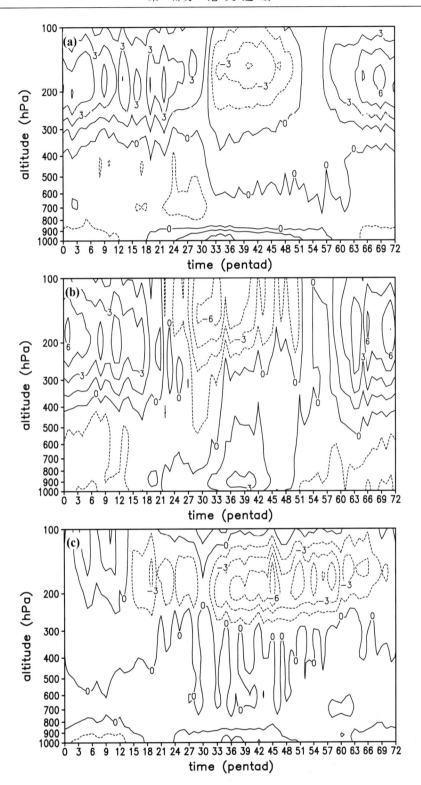

Fig. 2. As in Fig. 1 except for the meridional wind.

2.2 Annual cycle between summer and winter monsoons in the EAM region

It is clear from either precipitation and surface winds or tropospheric air temperatures and circulation patterns that the SAM and the NAM are phenomena having an annual cycle (e.g., Li and Yanai, 1996; Tomas and Webster, 1997; Goswami et al., 2006). The annual cycle is more pronounced in the EAM region.

Tao and Chen (1987) pointed out that the earliest onset of the Asian Summer Monsoon (ASM) is found over the SCS. It subsequently moves northward to South China, the Yangtze River and the Huaihe River valleys, Korea and Japan, and reaches North China and Northeast China in early or mid-July. The annual cycle of the EAM is readily apparent from seasonal variations of the monsoon rainband over East Asia (e.g., Huang et al., 2003). Figure 3 is the latitude-time cross section of 5-day precipitation along 115°E (average for 110°–120°E) averaged over the 40 years from 1961 to 2000. Figure 3 clearly shows that the monsoon rainband is located over South China in spring, then moves northward to the south of the Yangtze River during the period from May to the first 10 days of June, and then abruptly moves northward to the Yangtze River and Huaihe River valleys of China, Japan and South Korea. This is the beginning of the mei-yu season in the Yangtze River and Huaihe River valleys and start of the Baiu in Japan. Thereafter, as seen from Fig. 3, the monsoon rainband moves northward to North China and North Korea in early July. As it does so, the mei-yu season ends in the Yangtze River and Huaihe River valleys (i.e., Jianghuai valley in Chinese) and the rainy season begins in North China and Northeast China. The northward movement of the summer monsoon rainband over East Asia agrees well with the onset of EAM in various regions of East Asia (Tao and Chen, 1987).

The northward movement of the rainband is closely associated with the northward shift of the western Pacific subtropical high (e.g., Huang and Sun, 1992; Ding, 1992; Huang et al., 2003; Lu, 2004; Huang et al., 2005). Yeh et al. (1959) were the first to point out the abrupt change in the planetary–scale the circulation over East Asia occurring during early and mid-June that accompanies the onset of the EASM. Later, Krishnamurti and Ramanathan (1982) and McBride (1987) noted the onset of the SASM and the NASM circulations also coincided with abrupt changes of planetary-scale circulations. Additionally, as shown by Huang and Sun (1992), abrupt changes occurred during early or mid-June in the circulation over East Asia are closely associated with convective activity around the Philippines.

The southward retreat of the EASM is very rapid. Generally, the EASM moves rapidly southward to South China during the two weeks following mid-August. Subsequently, the EAWM begins with strong northerly winds prevailing over East Asia. In October strong northeasterly winds reach the area over the SCS and winds become easterly over the Indo-China Peninsula. Thus, the annual cycle of the EASM and EAWM is characterized mainly by migration in the meridional direction, while migration of the SAM and NAM annual cycle between winter and summer is primarily in the zonal direction.

Differences in the annual cycle of wind fields among these three monsoon systems can be seen also from the altitude-time cross sections of zonal (Fig. 1) and meridional winds (Fig. 2) averaged over East Asia, South Asia, and North Australia. From early June over South Asia (Fig. 1a) strong westerly winds prevail in the lower troposphere below 500 hPa and strong easterly winds are found in the upper troposphere above 500 hPa. However, from early October over this region the westerly wind become easterly in the lower troposphere below 700 hPa, while the easterly wind become westerly in the upper troposphere above 500 hPa. Hence, in the SAM region the annual cycle between summer and winter monsoons is clearly evident in the zonal wind fields. The same phenomenon also appears in the NAM region (Fig. 1c), but the reverse of zonal wind occurs in early November as the summer approaches in the Southern Hemisphere. Comparison of Fig. 1b with Figs. 1a and 1c shows the seasonal reverse of zonal winds in the troposphere over East Asia is not significant. Rather, as shown in Fig. 2b, there is an obvious seasonal reverse of meridional winds in the lower and upper troposphere over East Asia in early June and mid-September, respectively.

From the above analysis, it can be seen that in EAM region the annual migration between winter and summer monsoons is primarily in the meridional direction and appears mainly in the meridional component of wind field. It therefore differs markedly from the annual cycle of wind fields in the SAM and NAM regions.

2.3 The characteristics of water vapor transports in the EASM region

Huang et al. (1998b) showed that the characteristics of water vapor transports in the EAM region are considerably different from those characterizing the SASM area. In the SASM region the zonal transport of water vapor dominates, while the meridional transport of water vapor is very large in the EASM region. Moreover, the convergence of water vapor over the EASM region, which is closely associated with monsoon rainfall there, is to the combined effects of moisture ad-

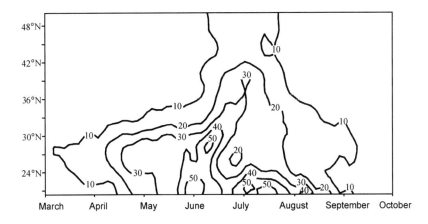

Fig. 3. Latitude-time cross section of 5-day precipitation along 115°E (average over 110°–120°E) for the period 1961–2000. Units: mm.

vection and convergence of the wind fields. Over the SASM region, water vapor convergence results mainly from wind-field convergence. The authors also point out that the anomalous summer monsoon rainfall in East Asia is influenced primarily by three branches of water vapor transport, namely, from the Bay of Bengal, South China Sea and tropical western Pacific. Zhou and Yu (2005) have shown the relationship between summer monsoon rainfall anomaly patterns in China and the water vapor transports. Additionally, they noted that the anomalously strong summer rainfall pattern occurring in the mid- and lower reaches of the Yangtze River valley is closely associated with the convergence of the northward transport of water vapor from the Bay of Bengal and the southward transport of water vapor from mid-latitudes. However, if the anomalous rainfall pattern shifts northward to the Huaihe River valley, it may be supported by the convergence of the water vapor transports from the South China Sea and mid-latitudes.

Recently, Chen and Huang (2007) analyzed the characteristics of water vapor transports over the EASM, the SASM and NASM regions using the ERA-40 reanalysis data for 1979–2002. The result is in good agreement with those by Huang et al. (1998b). The convergence of water vapor transport over the SASM region is mainly due to the convergence of the wind field in the lower troposphere over this region because of the large vertical velocity forced by large vertical shear of the zonal wind, as shown in Fig. 1a. A similar result can be found over the NASM region in the Southern Hemisphere summer (not shown). However, over the EASM region, water vapor transport is due not only to converging wind fields, but also to the moisture advection associated with the southerly monsoon flow in the lower troposphere. This is because there is a smaller vertical velocity corresponding to lesser vertical wind shear (Fig. 1b) than that in the SAM region.

From the above results, it may be concluded that the EAM system is a relatively independent component of the Asian-Australian monsoon system, although the EASM is also influenced by the SASM.

3. Characteristics of temporal and spatial variabilities of the EAM system and their impact on droughts and floods in China

Since the EASM is influenced not only by the SASM and the western Pacific subtropical high (e.g., Tao and Chen, 1987; Huang and Sun, 1992), as well as by mid- and high latitude circulation systems over both Northern Hemispheres (e.g., Tao and Chen, 1987; Gong and Ho, 2003) and Southern Hemisphere (e.g., Nan and Li, 2003; Fan and Wang, 2004; Xue et al., 2004), the interannual and interdecadal variations of EASM system are significant and very complex. In comparison to summertime surface air temperatures, the interannual and interdecadal variabilities of summer monsoon rainfall are more obvious in East Asia and have a large impact on climate disasters in China (e.g., Huang et al., 1999a; Huang and Zhou, 2002). Therefore, the interannual and interdecadal variabilities of summer monsoon rainfall and water vapor transports in East Asia are emphasized in this section.

3.1 Interannual variations of onset and northward advance of the EASM and their impact on droughts and floods in China

The interannual variability of summer monsoon rainfall in East Asia is influenced not only by the strength of the EASM, but also by the date of its onset. According to the studies by Tao and Chen (1987), and

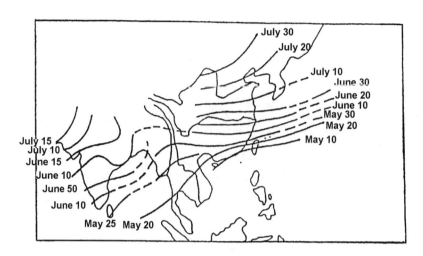

Fig. 4. Climatological-mean onset dates of the EASM (from Tao and Chen, 1987).

He and Luo (1999), the earliest onset of the ASM is found over the SCS and the Indo-China Peninsula, as shown in Fig. 4. Recently, Ding and He (2006) proposed that the earliest onset of the ASM is over the tropical eastern Indian Ocean. Since the onset of the SCSM has a direct impact on the northward advance of the ASM over East Asia, the study of the SCSM onset is emphasized in the review. The appearance of strong convective activity and the southwesterly flow over the SCS signals the onset of the ASM. Generally, the summer monsoon over the SCS is referred to the South China Sea summer monsoon (SCSM) in China. In order to investigate the interannual variability of onset date and progress of the SCSM, it is necessary to define an index for measuring the SCSM onset. However, there are many definitions of the SCSM onset (e.g., Wang et al., 2004) to choose. In comparison with other definitions, the one proposed by Liang and Wu (2002) appears to be more reasonable and used in many studies (e.g., Huang et al., 2005, 2006a).

Huang et al. (2005), and Huang et al. (2006a) analyzed the characteristics of interannual variations of the SCSM onset and its subsequent development. Their results showed that the interannual variability of SCSM onset date is very large and closely associated with the thermal state of the tropical western Pacific in spring. When the tropical western Pacific is warming during spring, the western Pacific subtropical high shifts eastward and twin anomalous cyclones develop early over the Bay of Bengal and Sumatra preceding the onset of the SCSM. In this case, the cyclonic circulation located over the Bay of Bengal can intensify early and develop into a strong trough. As a consequence, the westerly flow and convective activity can intensify over Sumatra, the Indo-China Peninsula and the SCS in mid-May, leading to an early onset of the SCSM (Fig. 5a). On the other hand, when the tropical western Pacific is cooling in spring, the western Pacific subtropical high shifts westward, and the twin anomalous anticyclones are located over the equatorial eastern Indian Ocean and Sumatra from late April to mid-May. Thus, the westerly flow and convective activity does not increase in intensity as early over the Indo-China Peninsula and the SCS. Only when the western Pacific subtropical high moves eastward, the weak trough over the Bay of Bengal can intensify. As a result, the strong southwesterly wind and convective activity over the Indo-China Peninsula and the SCS are delayed generally until late May, thus, the late onset of the SCSM is caused (Fig. 5b).

Following the SCSM onset, the monsoon moves northward over East Asia. Huang and Sun (1992), Huang et al. (2004a), and Huang et al. (2005) also investigated the interannual variations of the northward advance of the EASM. Their results showed the northward advances of the EASM after the onset over the SCS are greatly influenced by the thermal state of the tropical western Pacific in summer. They pointed out that there are close relationships among the thermal states of the tropical western Pacific, the convective activity around the Philippines, the western Pacific subtropical high, and the summer monsoon rainfall in East Asia. As shown in Fig. 5a, when the SST in the tropical western Pacific is above normal in summer, i.e., warm sea water accumulates in the West Pacific warm pool, and a cold tongue extends westward from the Peruvian coast along the equatorial Pacific, convection intensifies from the Indo-China Peninsula to east of the Philippines, and the western Pacific subtropical high shifts anomalously northward. In this case, the summer monsoon rainfall may be below normal in East Asia, especially in the Yangtze River and

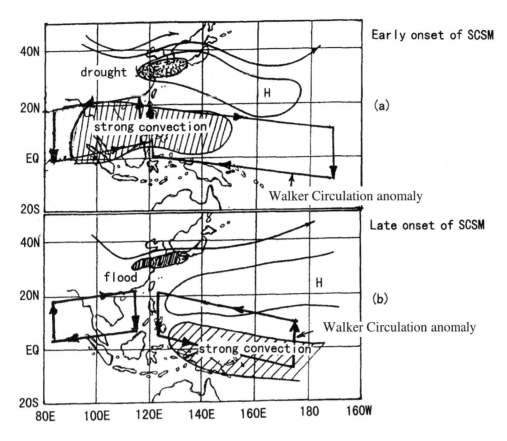

Fig. 5. Schematic map of the relationships among the thermal states of the tropical western Pacific (TWP) (i.e., 0°–14°N, 130°–150°E) in spring, the convective activity around the Philippines, the western Pacific subtropical high, the onset of SCSM and the summer monsoon rainfall in East Asia. (a) warming TWP; (b) cooling TWP.

Huaihe River basins of China, South Korea, and Japan. On the other hand, when the SST in the tropical western Pacific is below normal, i.e., the warm sea water extends eastward from the West Pacific warm pool along the equatorial western Pacific in summer, the convective activity is weak around the Philippines, and the western Pacific subtropical high may shift southward. In this case, the summer monsoon rainfall may be above normal in the Yangtze River and Huaihe River valleys of China, South Korea, and Southwest Japan. Therefore, following a spring with late onset of the SCSM, severe floods may occur in the Yangtze River and Huaihe River valleys and droughts in North China in summer.

Therefore, the thermal state of the tropical western Pacific, especially the anomaly of oceanic heat content (OHC) in the area known as NINO West (i.e., 0°–14°N, 130°–150°E) in spring (March–May) can be considered as a physical factor affecting the SCSM onset and summertime droughts and floods in the Yangtze River and the Huaihe River valleys, South Korea, and Japan.

3.2 The EAP index and interannual variability in the strength of the EASM

Monsoon indices are criteria for measuring the strength of monsoons and are necessary for studying the interannual variability of the ASM. Thus, two types of indices of Asian monsoon strength are used. One is defined from the thermodynamic elements, such as precipitation or OLR (e.g., Tao and Chen, 1987; Murakami and Matsumoto, 1994). The other is defined from the dynamic elements, such as the difference of zonal wind between lower and upper troposphere (e.g., Webster and Yang, 1992; Zeng et al., 1994) or the difference of sea-level pressure between the Eurasian continent and North Pacific (e.g., Guo, 1983). The former is easily influenced by local thermodynamic conditions, but the latter is only suitable for studying the SAM and the NAM regions where there are the large differences in the zonal wind between lower and upper troposphere, as described in section 2. Since differences in zonal wind are small in the EASM region, the definition based on zonal winds is not suitable for use in there. Rather, the study by

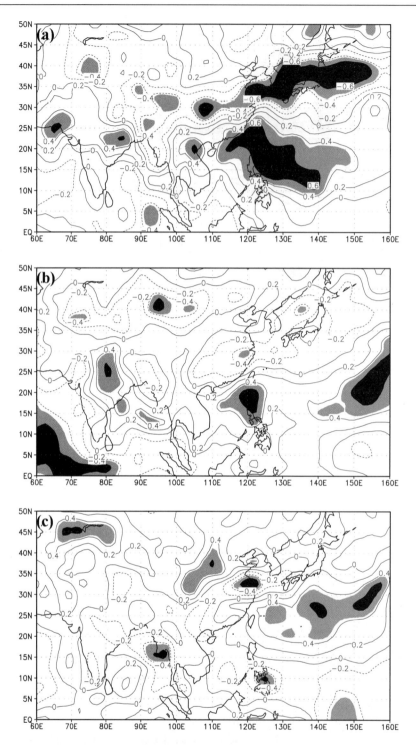

Fig. 6. Distributions of correlation coefficients between summer rainfall anomalies in East Asia with (a) the EAP index, (b) the WY index (e.g., Webster and Yang, 1992), and (c) the SM index (e.g., Guo, 1983). Areas of confidence level over 95% are shaded. Rainfall data is from the Xie-Arkin precipitation data for 1979–1998 (e.g., Xie and Arkin, 1997).

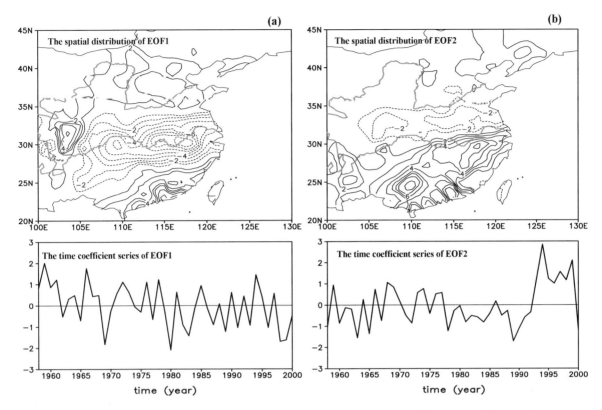

Fig. 7. Spatial distribution and the corresponding time coefficient series of the (a) first and (b) second component of EOF analysis (i.e., EOF1 and EOF2) of summer (JJA) rainfall in China from 1958 to 2000. EOF1 and EOF2 explain 15.6% and 12.7% of the variance, respectively.

Huang (2004) showed that the interannual variability of EASM can be described well by using the East/Pacific teleconnection pattern (EAP) index, which is based upon summer 500 hPa height anomalies corresponding to the EAP teleconnection of summer circulation anomalies (e.g., Nitta, 1987; Huang and Li, 1987, 1988).

Figures 6a–c show correlations between summer rainfall in East Asia and the EAP index, the Webster-Yang (WY) zonal-wind index, and the sea-level pressure based index (SM). Comparison of Fig. 6a with Figs. 6b and 6c clearly shows that the EAP index corresponds well and more so that the WY and SM indices with summer monsoon rainfall in East Asia. Also, from Fig. 6a, it can be seen that, if the EAP index is negative (positive), summer monsoon rainfall tends to be above (below) normal in the Yangtze River and the Huaihe River valleys. For example, in the summers of 1954, 1957, 1980, 1987, 1993, and 1998, the EAP index was largely negative, and severe floods occurred in these regions. Moreover, Fig. 6a shows a meridional tripole pattern in the distribution of correlation coefficients in East and Northeast Asia, which corresponds to interannual variations of summer monsoon rainfall which generally appear as a meridional tripole pattern in rainfall distribution.

3.3 *The quasi-biennial oscillation and tripole pattern distribution of the EASM anomalies over East Asia and their impacts on droughts and floods in China*

The quasi-biennial oscillation of circulation in the tropical troposphere, i.e., the TBO, is a fundamental characteristic of interannual variations in air-sea coupling in the SAM and NAM regions (e.g., Mooley and Parthasarathy, 1984; Yasunari and Suppiah, 1988). Also, Miao and Lau (1990), Lu et al. (1995), Yin et al. (1996), and Chang et al. (2000) proposed that the tropospheric biennial oscillation (TBO) can be found in interannual variations of summer monsoon rainfall in East Asia. Recently, Huang et al. (2006b) showed that there is also an obvious oscillation with a period of two–three years, i.e., the TBO, in the corresponding time-coefficient series of the EOF1 of summer rainfall in China (Fig. 7a), especially from the mid-1970s to the late 1990s. And, as shown by the spatial distribution of EOF1 in Fig. 7a, a strong negative signal occurs in the Yangtze River and the Huaihe River valleys. Moreover, it is also shown that this oscillation is closely associated with the quasi-biennial oscillation in the interannual variations of water vapor transport fluxes by the summer monsoon flow over East Asia (Fig. 8b). Additionally, the spatial distribution of EOF1 of summer rainfall in China (Fig. 7a) and the spatial distribution of the EOF1 of water vapor transports over East and Northeast Asia (Fig. 8a) are similar in exhibiting a meridional tripole pattern.

This meridional tripole pattern is seen clearly in the circulation anomalies at 500 hPa or 700 hPa. Figures 9a and 9b are the composite 500 hPa height anomalies over East Asia for summers with high and low EAP index, respectively. These figures show that with either high or low EAP index summers, which correspond respectively to drought and flood conditions in the Yangtze River and Huaihe River valleys, the summer monsoon circulation anomalies also exhibit the meridional tripole pattern distribution over East and Northeast Asia.

The above results show that the interannual variability of EASM clearly exhibits a quasi-biennial oscillation with a meridional tripole pattern distribution over East Asia and the tropical western Pacific. This may be a significant characteristic of the interannual variability of EASM. The frequency of drought and flood disasters also exhibit a quasi-biennial oscillation in the Yangtze River and the Huaihe River valleys, and the spatial distributions of these disasters often appear in the form of a meridional tripole pattern over East Asia.

3.4 Interannual variability of the East Asian winter monsoon (EAWM) and its relation to the EASM

East Asia is also a region of a strong winter monsoon. The East Asian winter monsoon (EAWM) features strong northwesterlies over North China, Northeast China, Korea and Japan and strong northeasterlies along the coast of East China (e.g., Staff members of Academia Sinica, 1957; Chen et al., 1991; Ding, 1994). A strong winter monsoon can bring not only disasters, such as wintertime low temperatures and severe snow storms in Northwest and Northeast China, North Korea, and North Japan, and springtime severe dust-storms in North and Northwest China, but also can cause strong convection over the maritime continent of Borneo and Indonesia. (e.g., Chang et al., 1979; Lau and Chang, 1987). Also, strong and frequent cold waves caused by the strong EAWM can trigger the occurrence of El Niño (e.g., Li, 1988).

Chen and Graf (1998), and Chen et al. (2000) systematically investigated the interannual variability of EAWM and its relation to the EASM with a new definition of the EAWM index. This index is defined by using the normalized meridional wind anomalies at 10 m over the East China Sea (25°–40°N, 120°–140°E) and the South China Sea (10°–25°N, 110°–130°E), averaged over November to March period following winter. Wu and Wang (2002) proposed an intensity index of the EAWM defined as the sum of zonal sea-surface pressure differences (110°E minus 160°E) over 20°–70°N with a 2.5° × 2.5° resolution in latitude and longitude. Results of the above mentioned studies show there is a significant variability in the interannual variations of EAWM.

Recently, Huang et al. (2007) investigated the interannual variations of the EAWM and its anomalies in the winters of 2005 and 2006 using the intensity index of EAWM defined by Wu and Wang (2002). They pointed out that, as shown in Fig. 10, the interannual variability of EAWM is significant and there is a clear difference between the EAWM intensity in the winter of 2005 (December 2005–February 2006) and the winter of 2006 (December 2006–February 2007). This was reflected in the different climate anomalies in the Northern Hemisphere, especially in East Asia, during these two winters.

The EAWM intensity can influence the following EASM intensity. The study by Chen et al. (2000) showed that following a strong (weak) EAWM, a drought (flood) summer may occur in the Yangtze River and the Huaihe River valleys. For example, in the summer of 1998, the severe flood shown in Fig. 17 in section 5.1 occurred in the Yangtze River valley following the weak EAWM in the winter of 1997.

3.5 Interdecadal variability of the EASM

3.5.1 A dipole pattern distribution in the interdecadal variability of EASM over East Asia

Huang et al. (1999b), and Huang (2001b) analyzed the interdecadal variations of summer monsoon rainfall in East Asia, and Chen et al. (2004) systematically discussed the characteristics of the climate change in China during the last 80 years. These results showed that the interdecadal fluctuations in summer (June–August) monsoon rainfall are more obvious than those in surface air temperature in East Asia. Huang et al. (1999b) pointed out that summer monsoon rainfall began to decrease in North China from the mid-1960s, especially from the late 1970s to the late 1990s. This resulted in prolonged severe droughts in this region, as shown in Fig. 18 of section 5. They also pointed out that the interdecadal variations of summer monsoon rainfall in North China are similar to those in the Sahelian region of Africa. Ren et al. (2004) also discussed the close linkage between the interdecadal variations of summer precipitation in North China and those in the Sahelian region. The opposite phenomenon, namely summer rainfall clearly increasing from the late 1970s

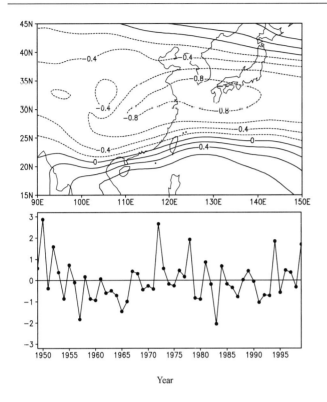

Fig. 8. (a) Spatial distribution and (b) corresponding time coefficient series of the first component of zonal water vapour transports in summer. Moisture and wind fields are taken from NCEP/NCAR reanalysis dataset from 1949 to 1999 (e.g., Kalnay et al., 1996). EOF1 explains 27.0% of the variance.

appears in the Yangtze River and the Huaihe River valleys and Northwest China. This interdecadal oscillation can be seen also in the dipole pattern in the spatial distribution of the EOF2 of summer rainfall in China and corresponding EOF2 time-coefficient series (Fig. 7b). This interdecadal variability of summer monsoon rainfall also appears in the frequency of heavy rainfall. As shown by Bao and Huang (2006), more heavy rainfall occurred in the 1980s than the 1970s in the middle and lower reaches of the Yangtze River and Huaihe River valleys and was followed by a further increase in the 1990s. In conjunction with this, however, a decrease of heavy rainfall occurred in the eastern part of North China since the late 1970s and has continued to the present.

3.5.2 Possible impact of the African summer monsoon on the EASM on an interdecadal timescale

The decrease of summer monsoon rainfall in North China from 1965, especially from the late 1970s, is closely associated with the weakening of the EASM. As pointed out by Huang and Zhou (2004), and Huang et al. (2004b), southerly winds over North China clearly weakened. Recently, Huang et al. (2006c) analyzed the interdecadal variations of drought and flood disasters in China and their association with the EAM system. As shown in Fig. 11, an interdecadal meridional tripole circulation anomaly distribution, similar to the EAP pattern teleconnection, appeared over the West Pacific in the late 1970s. This was influenced by the interdecadal El Niño-like SST anomaly pattern which became apparent in the tropical central and eastern Pacific from the late 1970s and has continued to the present. Comparison of Fig. 11b with Fig. 11a shows the anticyclonic anomaly located over the Mongolian Plateau during 1961–1976 clearly shifted southward to North China, and a cyclonic anomaly appeared over the Yangtze River and the Huaihe River valleys from the late 1970s. Thus, the circulation anomalies exhibited a meridional dipole pattern distribution over East Asia during this period. This led to the weakening of the EASM in North China and the southward and westward shift of the western Pacific subtropical high to south of the Yangtze River. On the other hand, the results also showed that the interdecadal variability of the Walker circulation from the late 1970s to the present resulted in intensification of the descending branch over North Africa, which led to intensification of the anticyclonic anomaly circulation over the Sahelian region and eastern North Africa (Fig. 11b). Due to propagation of quasi-stationary planetary waves, the intensification of the anticyclonic anomaly over the Sahelian region led to the appearance of an anticyclonic anomaly over South China and the Indo-China Peninsula beginning in the mid-late 1970s. Also, as shown in Figs. 11a and 11b, the distribution of interdecadal anomalies in the circulation in the lower troposphere over mid- and high latitudes appeared in a Eurasian (EU) pattern-like teleconnection. Additionally, anticyclonic anomalies over North China appeared after 1976. These led to weakening of the southerly monsoon flow in North China and water vapor convergence in the Yangtze River and the Huaihe River valleys. The net result was an interdecadal variation of droughts and floods in China beginning in the late 1970s.

3.6 Interdecadal variability of the EAWM

The EAWM also has a significant interdecadal variability. As shown in Fig. 10, the EAWM was stronger from the mid-1970s to the late 1980s but tended to be weaker from the late 1980s to late 1990s. This latter period was associated with continuously warm winters in China and an increase of spring rainfall in North and Northwest China (e.g., Huang and Wang, 2006;

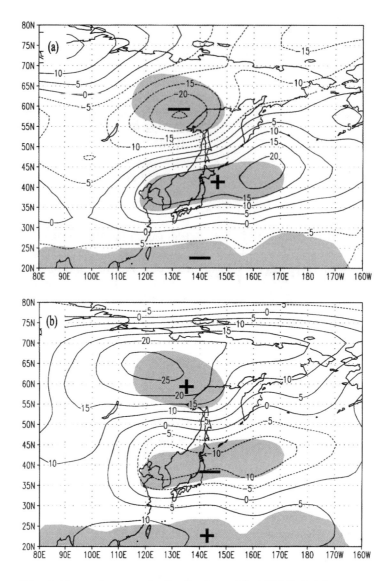

Fig. 9. Composite anomaly distributions of the 500 hPa height anomalies over East Asia and the tropical western Pacific for the summers with (a) high and (b) low EAP index. Units: gpm. Areas of confidence level over 95% are shaded. Height fields are taken from NCEP/NCAR reanalysis (e.g., Kalnay et al., 1996).

Kang et al., 2006).

The interdecadal variations of the EASM and EAWM have a significant impacts not only on drought and flood disasters in China, but also on the marine environment offshore of the Chinese mainland, including the Bohai Sea, Yellow Sea, East China Sea, and South China Sea. According to the study by Cai et al. (2006), since the winter and summer monsoon flows became weak offshore of China from 1976 to the present, winter and summer sea-surface wind stresses, especially the meridional component, have weakened, and SSTs have noticeably increased. These events can provide a favorable marine environment for the frequent occurrence of red tide in oceanic regions offshore of China.

4. The EAM climate system and its variabilities

As shown by Webster et al. (1998), the monsoon is not just an atmospheric circulation system, but rather an atmosphere-ocean-land coupled system. The interannual and interdecadal variabilities of the EASM and EAWM are influenced by many atmospheric circulation systems, such as the SAM, western Pacific

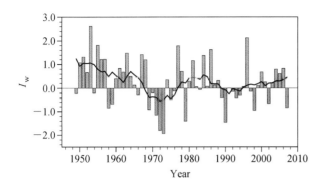

Fig. 10. The interannual variations of EAWM index based upon the definition of EAWM index by Wu and Wang (2002), based upon the NCEP/NCAR reanalysis data (e.g., Kalnay et al., 1996).

subtropical high, atmospheric disturbances in mid-and high latitudes. Additionally, the variability of the EASM and EAWM are affected by features such as the thermal state of the West Pacific warm pool, convection around the Philippines, ENSO cycle in the tropical Pacific, dynamical and thermal influences of the Tibetan Plateau, including cold season snow cover, and land surface process in the arid and semi-arid areas of Northwest China. The components of an atmosphere-ocean-land coupled system associated with the EAM are shown schematically in Fig. 12. This coupled system can be referred to as the East Asian monsoon climate system (e.g., Huang et al., 2004a), which then may be considered a subsystem of the Asian-Australian monsoon coupled system (e.g., Webster et al., 1998).

In order to understand the causes of variabilities in the EAM climate system, it is necessary to assess the interactive processes between the atmosphere and oceans and between the atmosphere and land surfaces. However, because of the complexity of these interactions, only the effects of atmosphere-ocean and atmosphere-land coupling are discussed.

4.1 Thermal effect of the tropical western Pacific on the EAM variability

Oceans have a significant thermal-effect on the Asian monsoon systems. Yang and Lau (1998) studied the influences of SST and ground wetness (GW) on Asian monsoons using a GCM and showed that ocean basin-scale SST anomalies have a stronger impact on the interannual variability of the Asian monsoon systems than GW anomalies. Lu et al. (2002a) discussed the associations between the western North Pacific (WNP) monsoon and the South China Sea monsoon (SCSM). They found that weak (strong) convective activity, which has a large impact on the SCSM onset, is related to El Niño (La Niña) pattern SST anomalies in the preceding winter and in spring.

Studies by many scholars (e.g., Nitta, 1987; Huang and Li, 1987; Kurihara, 1989; Huang and Sun, 1992) showed that the thermal state of the tropical western Pacific and convective activity around the Philippines play important roles in the interannual variability of EASM, as shown in Fig. 5. Nitta (1987), Huang and Li (1987), and Huang and Sun (1992, 1994) made systematic investigations of the thermal influence of the tropical western Pacific and convection around the Philippines on the interannual variability of the EASM circulation based upon observational data and dynamical theory. They proposed the P-J oscillation or the EAP pattern teleconnection of summer circulation anomalies over the Northern Hemisphere. As described in section 3, their results showed that the thermal state of the tropical western Pacific and convective activity around the Philippines have an obvious effect on the meridional shifts of the western Pacific subtropical high in summer. Lu (2001), and Lu and Dong (2001) also showed convective activity over the tropical western Pacific has a significant impact on the zonal shifts of the western Pacific subtropical high. Moreover, Huang et al. (2005) pointed out that the onset date of SCSM is closely associated with the thermal states of the tropical western Pacific and convective activity around the Philippines in spring.

Recently, Huang et al. (2005), and Huang et al. (2006b) investigated the interannual variability in the thermal state of subsurface waters of the tropical west Pacific. It can be seen from Fig. 13 that there is a significant quasi-biennial oscillation in the interannual variations of thermal structure. As shown by Huang et al. (2005), the oscillation has a large impact on the interannual variability of the northward advance of EASM and the water vapor transports driven by the monsoon flow. As shown in Fig. 14a, the composite distribution of water vapor transport anomalies for the summers when the tropical west Pacific is warming is opposite to that for the summers with cooling (Fig. 14b). Also, the anomalies in the water vapor transports shown in Figs. 14a and 14b exhibit a meridional tripole pattern. Thus, the influence of the thermal state of the tropical western Pacific on water vapor transport over East Asia explain well the TBO contribution to the summer monsoon rainfall in China or East Asia (e.g., Huang et al., 2006b).

4.2 ENSO cycle and its impact on the EAM system

It is well known that the ENSO cycle is one of the most striking phenomena of the tropical Pacific and

（图略）

Fig. 11. Interdecadal variations of the summer (JJA) circulation anomalies at 700 hPa over East Asia and tropical western Pacific: (a) 1966–1976; (b) 1977–2000. Normals based upon the climatological monthly mean circulation for 1961–1990. Wind fields obtained from the ERA-40 reanalysis data (e.g., Uppala et al., 2005).

has a great influence on the Asian-Australian monsoon system. Weak SASMs tend to occur in El Niño years (e.g., Webster et al., 1998). Huang and Wu's study (1989) was the first to show that summer monsoon rainfall anomalies in East Asia depend on the stage of ENSO cycle, and these anomalies are closely related to the position of the western Pacific subtropical high. Recently, Huang and Zhou (2002) pointed out from composite analyses of summer monsoon rainfall anomalies for different stages of the ENSO cycle during the period of 1951–2000 that droughts in North China tend to occur in the developing stage (Fig. 15a). During the decaying stage of El Niño events, floods tend to occur in the Yangtze River valley of China, especially in the regions south of the Yangtze River (Fig. 15b). Zhang et al. (1996), Huang et al. (2001), and Zhang (2001) also pointed out that southerly wind anomalies can appear in the lower troposphere over the SCS and the southeastern coast of China during the mature and decaying stages of ENSO. In this case, since an anticyclonic anomaly tends to appear over the Philippine Sea, the corresponding intensification of southerly

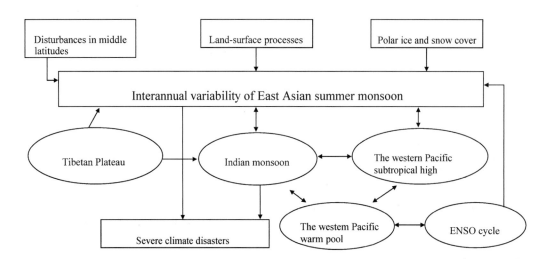

Fig. 12. Schematic map of various components of the EAM climate system.

winds is favorable for the transport of water vapor from the Bay of Bengal and the tropical western Pacific to South China. Thus, during the mature phase of El Niño events, rainfall generally is stronger in South China.

The ENSO cycle also has a significant influence on the annual cycle between the EAWM and the EASM. Chen et al. (2002), and Huang et al. (2004b) proposed that the interannual variations of the EAWM are closely associated with the ENSO cycle. Chen (2002) analyzed the composite distribution of the meridional wind anomalies at 850 hPa for winters preceding El Niño events. He showed that that there are anomalous northerly winds from the coastal area of China to the SCS and, thus, the EAWM is strong. He also pointed out that there is an anomalous cyclonic circulation over the West Pacific and anomalous northeasterly winds over the Yangtze River and Huaihe River valleys and the southeastern coast of China These features indicate a weak western Pacific subtropical high and a weak EASM in a summer when an El Niño is developing. Moreover, following the development stage, an El Niño event generally reaches its maturity, and anomalous southerly winds prevail in the southeastern coast of China and SCS. This points generally to a weak EAWM. In the following summer, the El Niño event may decay, and there is an anomalous anticyclonic circulation over the West Pacific about a strong western Pacific subtropical high. In this case, anomalous southwesterly winds occur over the region from South China to the Yangtze River valley and, therefore, indicate that a strong EASM may appear in when an El Niño event is in its decaying phase.

Since El Niño events can cause severe climate anomalies in many regions of the world, especially in the Asian-Australian monsoon region (e.g., Webster et al., 1998), many meteorologists and oceanographers focus upon studies of the recurrent nature and associated physical mechanisms of the ENSO cycle (e.g., Bjerknes, 1969; Philander, 1981; McCreary, 1983; McCreary and Anderson, 1984; Yamagata and Matsumoto, 1989; Anderson and McCreary, 1985; Cane and Zebiak, 1985; Schopf and Suarez, 1988; Chao and Zhang, 1988; McCreary and Anderson, 1991). Especially because of the interaction between the Asian-Australian monsoon system and ENSO, the relevant physical processes of ENSO cycles are very complex. The tropical western Pacific provides the necessary thermal conditions for ENSO (e.g. Huang and Wu, 1992; Li and Mu, 1999; Li and Mu, 2000; Chao et al., 2002, 2003), and the atmospheric circulation and zonal wind anomalies over this region can provide the necessary dynamical conditions. Li (1988, 1990) pointed out the triggering effect of the anomalous wind fields associated with EAWM on El Niño events. Huang and Fu (1996), Huang et al. (1998c), Huang et al. (2001) analyzed the atmospheric circulation and zonal wind anomalies in the lower troposphere over the tropical western Pacific and their roles in the development and decay processes of El Niño events in the 1980s and 1990s. Figures 16a–d show the distributions of the seasonal-mean circulation anomaly fields at 850 hPa over the tropical western Pacific before the development stage of El Niño evolution. From these figures it can be seen that before the development, there are cyclonic circulation anomalies in the lower troposphere over the tropical western Pacific that were responsible for the westerly wind anomalies over the tropical western Pacific around Indonesia. The westerly wind anomalies in turn were favorable for the formation of the eastward-

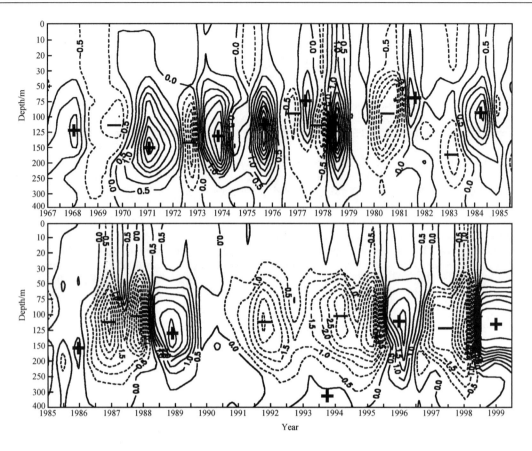

Fig. 13. Time-depth cross section of sea temperature anomalies averaged form 5°N to 10°N along 137°E. Units: °C. Solid and dashed lines indicate positive and negative anomalies, respectively. Data obtained from the Oceanographic Research Vessel "Ryofu-Maru", JMA.

propagating warm Kelvin wave along the equatorial Pacific that gives rise to the El Niño events. Zhang et al. (1996), and Huang et al. (2001) also analyzed the distributions of the seasonal mean circulation anomalies at 850 hPa over the tropical western Pacific for the mature phase of the El Niño events (not shown). The results showed that in the mature phase of El Niño there are also obvious anticyclonic circulation anomalies in the lower troposphere over the tropical western Pacific. These resulted in the easterly wind anomalies over the region from Papua-New Guinea to Sumatra Island along Indonesia conducive to development of an eastward-propagating cold Kelvin wave along the equatorial Pacific that results in the decay and ultimate demise of El Niño events.

Huang et al. (2001), and Huang et al. (2004b) further discussed theoretically the dynamical influence of the westerly wind anomalies over the tropical Pacific on the development and decay of the 1997/98 El Niño event using a simple tropical air-sea coupled model and observed anomalous near sea-surface wind stress of the tropical Pacific during 1997 and 1998. The results identified the triggering effect of zonal wind anomalies over the tropical western Pacific on the equatorial oceanic Kelvin and Rossby waves in the tropical Pacific.

From the above studies it can be seen that the thermal and dynamical influences within and over the tropical western Pacific play an important role in ENSO cycles. Additionally, the coastal boundary of the West Pacific has an important effect on El Niño by reflecting equatorial oceanic Rossby waves, as shown in the theory of the delayed oscillator proposed by Cane and Zebiak (1985), and Schopf and Suarez (1988).

4.3 Thermal effects of the Tibetan Plateau on the EAM System

The Tibetan Plateau has important thermal and dynamical effects on the interannual variability of the EASM. Ye and Gao (1979) first described the thermal effect of the Tibetan Plateau on the ASM. Later, many investigators noted that the heating anomaly over the Tibetan Plateau has a large impact on the ASM anomalies (e.g., Nitta, 1983; Luo and Yanai, 1984; Huang, 1984, 1985). Wu and Zhang (1997) explained how the heating over the Tibetan Plateau acts

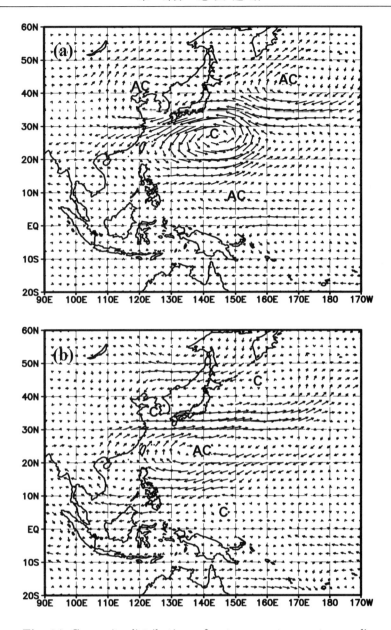

Fig. 14. Composite distributions of water vapor transport anomalies over East Asia and Northeast Asia for summers with (a) warming and (b) cooling of the tropical western Pacific in summer. Units: 10^3 g s^{-1} cm^{-1}. Moisture and wind fields obtained from the ERA-40 reanalysis dataset. (e.g., Uppala et al., 2005)

as an air pump and plays a triggering role in ASM onset. Recently, Zhang et al. (2002) pointed out heating over the Tibetan Plateau has an important effect on the east-west oscillation of the South Asian high, which has a significant influence on the EAM system. Wei and Luo (1996) noted that snow cover significantly influences the heating over the Tibetan Plateau, and there is a large positive correlation between snow cover in the Tibetan Plateau and summer monsoon rainfall in the upper and middle reaches of the Yangtze River.

Recently, Wei et al. (2002, 2003) analyzed the interannual and interdecadal variations of the number of days and depth of snow cover on the Tibetan Plateau based upon the observed daily snow cover at 72 observational stations located in the Tibetan Plateau during 1960–1999. They discovered that there are obvious variations in the number of days and depth of snow cover in the Tibetan Plateau on interannual and inter-

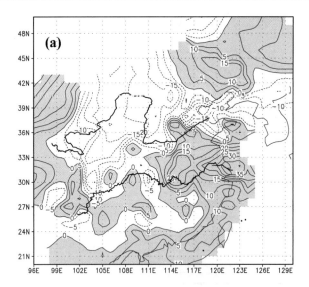

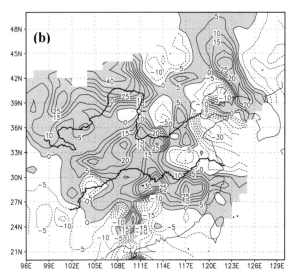

Fig. 15. Composite distributions of summer (June–August) rainfall anomalies (in percentage) in China for (a) the summers in the developing stage and (b) the summers in the decaying stage of El Niño events during the period from 1951 to 2000. Solid and dashed contours indicate positive and negative rainfall anomalies, respectively, and positive values are shaded.

decadal time scales. Moreover, their results showed that there are large positive correlations in the middle and upper reaches of the Yangtze River and negative correlations in South China and Northeast China between the depth of snow cover in the previous winter and spring and the following summer rainfall. This may explain why, if the snowfall in the Tibetan Plateau is heavy in the previous winter and spring of a given year, the following summer rainfall is heavy in the upper and middle reaches of the Yangtze River.

For example, in the winter of 1997 and the spring of 1998, the particularly heavy snowfall on the Tibetan Plateau was followed by the severe flood disaster in the Yangtze River valley in the summer of 1998. Also, comparison of the days and depth of snow cover in the Tibetan Plateau during the period from the late 1970s to the late 1990s with those during the period from the early 1960s to the mid-1970s shows clearly that the days with snow cover have increased and the snow depth has become deeper during the former period. The interdecadal variation of snow cover on the Tibetan Plateau has an important impact on the summer monsoon rainfall in the middle and upper reaches of the Yangtze River (e.g., Wei et al., 2002, 2003; Huang et al., 2004a).

Huang and Zhou (2004) investigated the variability of the northerly wind circulation over the west side of the Tibetan Plateau and its impact on the EASM circulation. Their result showed that after 1965, especially from 1977, northerly winds have become weak, the southerly component of the EASM has become weaker, and the water vapor transported into North China was greatly weakened. This resulted in the decrease of summer rainfall and the severe droughts in North China.

4.4 Thermal effect of sensible heating in the arid and semi-arid regions of Northwest China on the EAM System

Since monsoon circulations result fundamentally from the land-sea thermal contrast, the variability of the EAM system is influenced not only by the thermal states of the tropical western Pacific, but also by the thermal states of the Eurasian continent. Zhou and Huang (2003, 2006) analyzed the interannual and interdecadal variations of the differences between land surface and the surface air temperatures ($T_s - T_a$) and sensible heating in spring in the arid and semi-arid regions of Northwest China and Central Asia and their impact on summer monsoon rainfall in China. Their results show that the strongest sensible heating in the Eurasian continent is located in Northwest China and Central Asia. Thus, this region may be seen as a "Warm lying surface". They also pointed out that the $T_s - T_a$ and sensible heating in this region have an obvious interdecadal variability. Before the late 1970s, the sensible heat anomalies in spring were negative, but the anomalies became largely positive in the late 1970s. Moreover, from correlation analyses it is evident that when $T_s - T_a$ or sensible heating anomalies become positive in Northwest China in spring, rainfall can be heavy in the lower reaches of the Yangtze River and light in North China during the following summer. Therefore, the interdecadal variation of the spring sen-

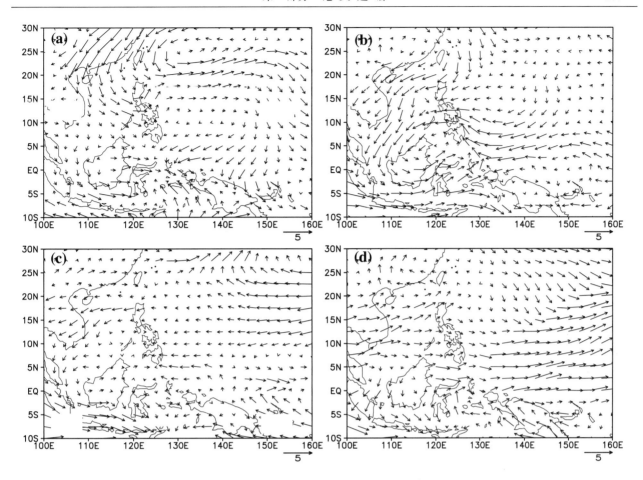

Fig. 16. Distributions of the circulation anomalies at 850 hPa over the tropical western Pacific before the development stage of the El Niño events during the period 1980–1998. Units: m s^{-1}. (a) spring of 1982, (b) winter the 1985, (c) spring of 1991, (d) winter of 1996. Analyses based upon ERA-40 reanalysis data. (e.g., Uppala et al., 2005)

sible heating in Northwest China and Central Asia can be one of the contributors to the interdecadal variability of the EASM that occurred since the late 1970s.

Additionally, the interdecadal cooling at the upper troposphere over the Mongolian Plateau, North China and Northwest China from the late 1970s have an impact on the weakening trend of the summer monsoon over East Asia (e.g., Yu et al., 2004), but the physical mechanism of this impact needs further study.

5. Climate background of the occurrence of climate disasters in China

Due to the severity of climate disasters in China, the causes of floods in the Yangtze River valley and the persistent droughts in North China from the late 1970s have been studied from the perspective of the interannual and interdecadal variabilities of the EAM climate system.

5.1 Climate background of the severe floods in the Yangtze River and Huaihe River valleys

The causes of severe floods in the Yangtze River valley, especially the particularly severe flood in the summer of 1998, as shown in Fig. 17, have been systematically studied from the perspective of interannual variabilities in the anomalies associated with the atmosphere, ocean and land surface components of the EAM climate system (Fig. 12) (e.g., Huang et al., 1998a; e.g., Huang and Zhou, 2002: Huang et al., 2003). Furthermore, the influence of the variability of the EAM climate system on severe flooding disasters in China has been analyzed by assessing the processes associated with the severe flood event in the Yangtze River valley during the summer of 1998 (e.g., Huang et al., 2003, 2004a). The configuration of anomalies associated with various component of the EAM climate system causing the severe floods in the Yangtze River

(图略)

Fig. 17. Schematic diagram of various component anomalies of the EAM climate system associated with the occurrence of severe floods in the Yangtze River and Huaihe River valleys. The distribution of precipitation anomaly percentages shown is for the summer of 1998. Precipitation anomaly percentages over 80% are shaded.

and Huaihe River valleys are summarized in Fig. 17. As shown in Fig. 17, when the tropical western Pacific is cooling, the convection around the Philippines is weak and the western Pacific subtropical high shifts anomalously westward and southward. These conditions are favorable for lengthening the period over. which the summer monsoon rain belt remains over the Yangtze River valley. Also, if the number of days and depth of snow cover on the Tibetan Plateau is anomalously large during the winter and spring, conditions are favorable in the following summer for maintenance of the summer monsoon rain belt, and hence heavy rainfall in the upper and middle reaches of the Yangtze River. In addition, when a low-level trough is located over Inner-Mongolia and Northeast China, cold air can be continuously transported from the trough region to the Yangtze River valley, which is necessary for the maintenance of the mei-yu front over the Yangtze River and Huaihe River valleys. The strong northeastward propagating 30–60 day oscillation from the Bay of Bengal to the Yangtze River valley is also an important factor for severe flooding in the Yangtze and Huaihe River valleys, because it can bring large amounts of water vapor from the Bay of Bengal and the SCS into these valleys. In the summer of 1998, as shown in Fig. 17, large anomalies appeared in all the components of the EAM climate system resulting in the particularly severe flooding and huge economic losses to China.

5.2 Climate background of the persistent droughts in North China

Similarly, the occurring causes of the persistent droughts in North China from the late 1970s have been systematically investigated in terms of the interdecadal variability and anomalies of the various components of the EAM climate system (e.g., Huang et al., 1999b; Huang et al., 2001; Huang and Zhou, 2002; Huang, 2004; Huang et al., 2004a; Huang, 2006; Huang et al., 2006c). The seasonal droughts in summer are emphasized in this paper. Their results clearly show that the interdecadal anomalies of circulation over the tropical western Pacific, which have a large impact on the EASM and the summer monsoon rainfall in China, especially in North China, were influenced by the interdecadal variability of the tropical eastern Pacific SSTs.

(图略)

Fig. 18. Schematic diagram of the climate background causing the persistent droughts in North China. The summer precipitation anomaly percentages in China are the difference between summer rainfall anomalies averaged over 1977–2000 and 1966–1976 at 160 observational stations in China. Areas of positive values are shaded.

Due to the obvious warming of the tropical eastern and central Pacific from the late 1970s, the Walker circulation became weak over the tropical Pacific, which caused weakening of the trade winds over the tropical western Pacific and the EASM. The weakening of the EASM led to the persistent droughts in North China from the late 1970s.

On the other hand, the warming of the tropical eastern and central Pacific from the late 1970s also caused the Walker circulation anomaly over the tropical Atlantic and North Africa (e.g., Huang et al., 2006c). Anomalous ascent and descent were clearly located over the tropical eastern Pacific and North Africa, respectively. The intensification of the anomalous descent over the Sahelian region of North Africa led to a strong anticyclonic anomaly over this region from the late 1970s. Moreover, their results showed that there appears to be a teleconnection pattern of circulation anomalies from the Sahelian region of North Africa to South China near the Arabian Peninsula. This results from propagation of an atmospheric Rossby wave-train and the fact that circulation anomalies over South China can be associated with those over North China in the context of the EAP pattern teleconnection. Because this anticyclonic anomaly over North China suppresses the northward progression of southerly winds to North China, only light monsoon rainfall occurs in this region. As shown by Rodwell and Hoskins (1996), diabatic descent of the Asian summer monsoon can have a significant effect on the circulation anomalies over the desert regions of North Africa, again through propagation of a Rossby wave-train. However, as shown by the study of Huang et al. (2006c), variability in the circulation over the desert regions of North Africa can influence the ASM variability on the interdecadal time-scale.

In addition the interdecadal variation of land surface processes have an impact on ASM variability on the interdecadal time scale. The results analyzed by Wei et al. (2002, 2003) have indicated that the num-

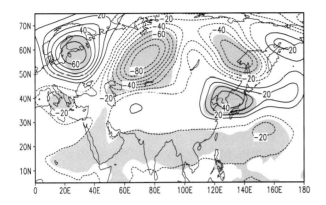

Fig. 19. Distribution of 500-hPa height anomalies regressed by the time coefficients of EOF1. Solid and dashed contours indicate positive and negative height anomalies, respectively. Contour interval is 20 m. Areas of confidence level over 95% are shaded. Analysis based on ERA-40 reanalysis data (e.g., Uppala et al., 2005)

ber of days and thickness of winter and spring snow cover on the Tibetan Plateau increased from the late 1970s. And the $T_s - T_a$ and sensible heating in spring became large in the arid and semi-arid areas of Northwest China from the late 1970s (e.g., Zhou and Huang, 2003, 2006). These factors are conducive to increasing the summer monsoon rainfall in the Yangtze River and the Huaihe River valleys and the decreasing summer rainfall in North China.

From the above-mentioned studies, the climate background of the persistent droughts in North China associated with the interdecadal variations of various components of the EAM climate system may be summarized as shown in Fig. 18.

6. Dynamic processes in the EAM climate system

As shown in section 5, the occurrence of climate disasters in China are closely associated not only with the variability and anomalies of ocean and land, but also with internal dynamical processes in the EAM climate system. Because of these dynamical processes, there are close relationships amongst the variabilities of various components of this system. Therefore, it is necessary to assess how the EAM climate system influences the occurrence of climate disasters in China.

6.1 *The East Asia/Pacific (EAP) pattern teleconnection*

As described in sections 2 and 3, there is an obvious tripole pattern either in the distributions of summer monsoon rainfall anomalies in China and summer water vapor transports over East Asia or in the distributions of summer monsoon circulation anomalies over East Asia. As a result, the distributions of droughts and floods in China are closely associated with this meridional tripole pattern. Huang et al. (2004a), and Huang et al. (2006b) used the EAP pattern teleconnection proposed by Nitta (1987), and Huang and Li (1987, 1988) to interpret the physical mechanism of the meridional tripole pattern. From analyses of observational data, dynamic theory and numerical simulations, they noted that the distribution of atmospheric circulation anomalies with the meridional tripole pattern over East Asia and the West Pacific in summer can be due to the thermal anomalies of the tropical western Pacific or anomalies in convective activity around the Philippines via propagation of a Rossby wave-train. Of course, as shown in Fig. 19, this meridional tripole pattern of circulation anomalies is also associated with the EU pattern teleconnection proposed by Wallace and Gutzler (1981).

Lu and Huang (1996a,b, 1998) investigated the variability of a blocking high over Northeast Asia and its impact on summer monsoon rainfall in the Yangtze River and Huaihe River valleys. They pointed out that there is a good relationship between variations in the blocking high, especially over the Sea of Okhotzk, and summer monsoon rainfall variability in the Yangtze River and Huaihe River valleys. Also, they proposed that this relationship is related to the EAP teleconnection of circulation anomalies over the Northern Hemisphere. When the tropical western Pacific is cooling in summer, the convective activity is weakened around the Philippines and the western Pacific subtropical high shifts southward and westward. In establishing the EAP pattern teleconnection via Rossby wave-train propagation, an anticyclonic anomalous circulation appears over Northeast Asia, and the blocking high is maintained over Northeast Asia for a lengthy period of time. In this case, the mei-yu front is also maintained in the Yangtze River and the Huaihe River valleys for a longer period of time, resulting in severe flooding in these regions.

6.2 *The North Africa/East Asia teleconnection of meridional circulation anomalies in the upper troposphere*

Yang et al. (2002) analyzed the association of Asian-Pacific-American winter climate with the East Asian jet stream (EAJS) on interannual time scales. They proposed that the EAJS is coupled to a teleconnection pattern extending from the Asian continent to North America with the strongest signals over East Asia and the West Pacific in the boreal winter. Moreover, Lu et al. (2002b), and Lin and Lu (2005) analyzed the variability of meridional circulation anomalies for the boreal summers of 1986–2000 using the HadAM3 data from the Hadley Center. Their results showed that there is an obvious teleconnection pattern in the meridional circulation anomalies in the upper troposphere over the regions from North Africa to East Asia. They also pointed out that this teleconnection may be due to the eastward propagation of a Rossby wave-train along the westerly jet stream at 200 hPa (e.g., Lu and Kim, 2004). Recently, Tao and Wei (2006) demonstrated from the analysis of isentropic potential vorticity that the northward advance (southward retreat) of the western Pacific subtropical high may reflect formation of a ridge (trough) along the eastern coast of China associated with the propagation of a Rossby wave-train along the Asian jet in the upper troposphere. From an analysis of the relationship between the summer monsoon rainfall anomalies in East Asia and the circulation anomalies in the upper troposphere over the Eurasian continent, Hsu and Lin (2007) showed that the meridional tripole structure of summertime monsoon rainfall anomalies over East Asia is related to the propagation of a Rossby wave-train along the Asian jet in the upper troposphere, in addition to the EAP pattern teleconnection over East Asia and the western North Pacific. Thus, the above-mentioned meridional tripole structure of circulation anomalies over East Asia may be also associated with the teleconnection pattern along the Asian jet in the upper troposphere.

6.3 Impact of quasi-stationary planetary wave activity variability on the EAWM

Huang and Gambo (1982, 1983a,b), and Huang (1984) investigated the three-dimensional propagation of quasi-stationary planetary waves responding to forcing by topography and stationary heat sources in the troposphere in boreal winter. The study utilized a 34-level model in addition to theoretical consideration based upon the refractive index square and E-P wave fluxes. They pointed out that there are two wave guides in the three-dimensional propagation of quasi-stationary planetary waves. One is the so-called polar waveguide, by which quasi-stationary planetary waves can propagate from the troposphere to the stratosphere at high latitudes. The other is the so-called low-latitude waveguide in which waves can propagate from the lower troposphere over mid- and low latitudes to the upper troposphere over low latitudes. Based on these studies of Chen et al. (2002, 2003, 2005), Chen and Huang (2005) recently examined the interannual variation of propagating wave guides for quasi-stationary waves with E-P wave fluxes using NCEP/NCAR and ERA-40 reanalysis data and an AGCM simulation. Their results showed that these two wave guides for quasi-stationary planetary waves evidently have an interannual oscillation. When the polar wave guide is strong (weak) in winter, the low-latitude wave guide may be weak (strong). Moreover, it was also found that due to wave-flow interaction, the interannual oscillation of these two wave guides has a significant influence on the Arctic Oscillation (AO), which is closely related to the EAWM (e.g., Gong et al., 2001; Gong and Ho, 2003) through NAM proposed by Thompson and Wallace (1998, 2000). When the equatorward propagation of quasi-stationary planetary waves from the lower troposphere over middle latitudes toward the upper troposphere over low latitudes is strong during a winter, the upward propagation of quasi-stationary planetary waves from the troposphere into the stratosphere becomes weak in winter. This can lead to the weakening of the East Asian westerly jet stream because of the convergence of wave activity fluxes for quasi-stationary planetary waves over this region. Thus, the East Asian trough, the Siberian high and the Aleutian low can become weak during winter, which can decrease the northwesterly winds over East Asia and, therefore warming in winter over East Asia.

Recently, Huang and Wang (2006) also investigated the interdecadal variation of quasi-stationary planetary waves and its impact on the EAWM. Their results showed that an obvious interdecadal variability of the EAWM has occurred in East Asia and the EAWM became weak beginning in the late 1980s to late 1990s. As shown by Gong et al. (2001), the weakening of the EAWM may be affected by the AO. Moreover, Huang and Wang (2006) pointed out that there was an obvious interdecadal variation of planetary wave activity in the late 1980s. As shown in Figs. 20b and 20c, the low-latitude wave guide for quasi-stationary planetary waves intensified beginning in 1987, which resulted in intensification of E-P flux convergence of quasi-stationary planetary waves in the upper troposphere over low latitudes. Their result also showed that the intensification of quasi-stationary wave propagation from mid- and high latitudes to low latitudes led to deceleration of zonal-mean zonal winds in the upper troposphere over the latitudes around 35°N. This caused the intensification of the AO, the weakening of EAWM and the prolonged warm winters in China from 1987. Thus, the interdecadal variation of planetary wave activity from the late 1980s to the late 1990s has had an important impact on the interdecadal variation of EAWM.

The interdecadal variation of the AO also has a large impact on the spring climate over the Eurasian continent, especially over East Asia. Yu and Zhou (2004) pointed out that the unique cooling over the

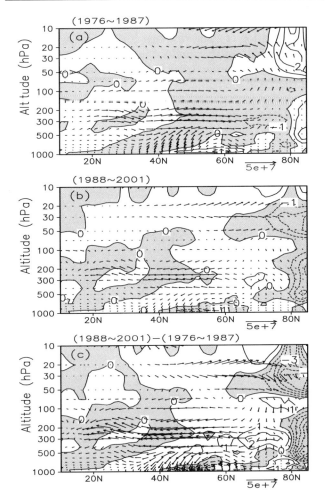

Fig. 20. Composite anomaly distributions of the E-P fluxes ($\times \rho^{-1}$) and their divergence (Units: m s^{-1} d^{-1}) for the quasi-stationary planetary waves 1–3 averaged for (a) the Northern Hemisphere winters of 1976–1987, for (b) the winters of 1988–2001, and (c) difference between them. Normals are the climatological mean E-P fluxes for 30 winters from 1971 to 2000. Solid and dashed contours indicate positive and negative anomalies of the divergence of E-P fluxes respectively. Areas of negative anomalies are shaded. Analyses are derived from the ERA-40 reanalysis data (e.g., Uppala et al., 2005)

subtropical Eurasian continent in spring over the last half century may be related to the interdecadal variation of the NAO. Li et al. (2005) also pointed out that the March–April cooling shift on the lee side of the Tibetan Plateau may not be a local phenomenon. Instead, it is associated with an eastward extension of the cooling signal originating from North Africa, which is related to the NAO of the previous winter. Moreover, Xin et al. (2006) revealed that precipitation in late spring in South China experienced a significant decrease from the late 1970s, and pointed out that this decrease may be associated with the spring cooling in the upper troposphere over Central China, which is strongly related to the NAO during the previous winter.

7. Conclusions and remarks for future studies

From the above review, it can be seen that there have been significant advances in the describing the basic characteristics and properties of the EAM system and the spatial and the temporal variabilities which are closely associated with the occurrence of climate disasters in China. The basic physical processes underlying these variabilities, both internal and external, have also been studied. The advances are summarized as follows:

(1) The EAM system is a relatively independent monsoon subsystem of the Asian-Australian monsoon system. Both the horizontal and vertical structure of wind fields and water vapor transports and the annual cycle of this system differ from the SAM and the NAM systems.

(2) The EAM system has significant interannual and interdecadal variabilities. The interannual variability of this system exhibits an obvious quasi-biennial oscillation, i.e., the TBO, with a meridional tripole pattern in spatial distribution, which can cause a tripole pattern in the spatial distribution of drought and flood disasters in East Asia.

(3) The EAM system variabilities are closely associated with coupling of atmosphere, ocean, and land processes and, hence, the EAM can be referred to as the EAM climate system. The EAM climate system then is a system that includes various components of atmosphere, ocean and land surface processes that influence variability of the EAM.

(4) The occurrence of climate disasters in China is closely associated with the variability and anomalies of various components of the EAM climate system. The climate backgrounds of occurrences of severe floods in the Yangtze River and the Huaihe River valleys and the prolonged droughts in North China from the late 1970s have been tied preliminarily to the interannual and interdecadal variations of the EAM climate system, respectively, and these results have been now applied to the seasonal and annual predictions of climate anomalies in China.

(5) The EAP teleconnection pattern of summertime circulation anomalies and the teleconnection of meridional wind anomalies along the westerly jet stream in the upper troposphere over East Asia can explain well the physical mechanism for the meridional tripole pattern in spatial distributions of the summer

monsoon rainfall, water vapor transports and monsoon circulation anomalies over East Asia.

Nevertheless, many problems on the basic physical-processes of the EAM climate system variability and their impacts on climate disasters in China still remain unclear.

(1) The EAM system is largely independent in its horizontal and vertical wind field structures relative to the SAM and NAM systems. These differences are significant in the annual cycle, monsoon onset, active phase and break periods. However, these monsoon systems are interactively to each other. The associations among the onset, active, break, and annual cycle processes of these monsoon systems requires further study.

(2) Many investigations have shown that the interdecadal variability of the EAM system has a significant impact on its interannual variability, which in turn influences its intraseasonal variations. However, the physical processes involved in the interactions amongst different time-scale features of this system are not yet clear and, thus, important issues for future investigation.

(3) There are complex dynamic and thermal processes in the interannual variability of the EAM climate system. These processes strongly influence the EAM variabilities. The dynamic processes of the quasi-biennial oscillation, i.e., the TBO, with the meridional tripole pattern distribution and the thermal effect of tropical heating on the TBO have been emphasized in recent studies. However, as shown by Yang et al. (2004), extratropical processes also have an important effect on the EAM. Therefore, the effect of extratropical process in the EAM climate system on the EAM variabilities should be considered in future study.

(4) The EAM system has a large interdecadal variability. However, the physical processes of this variability, for example, the dipole pattern distribution of this system, are not yet clear, and studies on this problem to date are not sufficient. These issues here need further study with focus on interactions between various components of the EAM climate system on an interdecadal time-scale.

The above review shows that the EAM system variabilities on interannual and interdecadal time scales and their physical mechanisms remain important research issues for future study. If the physical processes of these variabilities are not revealed further, improvements in seasonal and annual predication of climate disasters in China may be difficult. Therefore, understanding of the internal and external dynamic processes affecting the interannual and interdecadal variabilities of the EAM system remains a main objective of future investigation. We believe that through implementation of some National Research Programs, it is possible to further understand the physical mechanisms of the interannual and interdecadal variabilities of the EAM system.

Acknowledgements. This paper was supported by the "National Key Programme for Developing Basic Sciences" under Grant No. G2006CB403600, and Knowledge Innovation for the 3rd Period, Chinese Academy of Sciences under Grant No. KZCX2-YW-220, and the National Natural Science Foundation of China under Grant Nos. 40730952, 40575026, 40775051 respectively.

REFERENCES

Anderson, D. L. T., and J. P. McCreary, 1985: Slowly propagating disturbances in a coupled ocean-atmosphere model. *J. Atmos. Sci.*, **42**, 615–629.

Bao, M., and R. H. Huang, 2006: Characteristics of the interdecadal variations of heavy rain over China in the last 40 years. *Chinese J. Atmos. Sci.*, **30**, 1057–1067. (in Chinese)

Bjerknes, J., 1969: Atmospheric teleconnection from the equatorial Pacific. *Mon. Wea. Rev.*, **97**, 163–172.

Cai, R. S., J. L. Chen, and R. H. Huang, 2006: The response of marine environment in the offshore area of China and its adjacent ocean to recent global climate change. *Chinese J. Atmos. Sci.*, **30**, 1019–1033. (in Chinese)

Cane, M. A., and S. E. Zebiak, 1985: A theory of El Niño and the Southern Oscillation. *Science*, **228**, 1085–1087.

Chang, C. P., J. Erickson, and L. M. Lau, 1979: Northeasterly cold surges and near-equatorial disturbances over the winter- MONEX area during 1974. Part I: Synoptic aspects. *Mon. Wea. Rev.*, **107**, 812–829.

Chang, C. P., Y. S. Zhang, and T. Li, 2000: Interannual and interdecadal variations of the East Asian summer monsoon and tropical Pacific SSTs. I.II. *J. Climate*, **13**, 4310–4340.

Chao, J. P., and R. H. Zhang, 1988: The air-sea interaction waves in the tropics and their instability. *Acta Meteorologica Sinica*, **2**, 275–287. (in Chinese)

Chao, J. P., S. Y. Yuan, Q. C. Chao, and J. W. Tian, 2002: A data analyses on the evolution of the El Ni/La Nicycle. *Adv. Atmos. Sci.*, **20**, 837–844.

Chao, J. P., S. Y. Yuan, Q. C. Chao, and J. W. Tian, 2003: The origin of warm sea water in the surface layer of the West Pacific warm pool-The analysis of the 1997/1998 El Nievent. *Chinese J. Atmos. Sci.*, **27**, 145–151. (in Chinese)

Chen, J. L., and R. H. Huang, 2006: The comparison of climatological characteristics among Asian and Australian monsoon subsystem. Part I. The wind structure of summer monsoon. *Chinese J. Atmos. Sci.*, **30**, 1091–1102. (in Chinese)

Chen, J. L., and R. H. Huang, 2007: The comparison of

climatological characteristics among Asian and Australian monsoon subsystems. Part II. Water vapor transport by summer monsoon. *Chinese J. Atmos. Sci.*, **31**, 766–778. (in Chinese)

Chen, L. X., Q. G. Zhu, and H. B. Luo, 1991: *East Asian Monsoon*. China Meteorological Press, Beijing, 362pp. (in Chinese)

Chen, L.X., X. J. Zhou, W. L. Li, Y. F. Luo, and W. Q. Zhu, 2004: Characteristics of the Climate Change and its formation mechanism in China during the last 80 years. *Acta Meteorologica Sinica*, **62**, 634–646. (in Chinese)

Chen, W., 2002: The impact of El Niño and La Niña events on the East Asian winter and summer monsoon cycle. *Chinese J. Atmos. Sci.*, **26**, 595–610. (in Chinese)

Chen, W., and Hans-F. Graf, 1998: The interannual variability of East Asian winter monsoon and its relationship to global circulation. *Max-Planck-Institute fur Meteorologie, Report* No. 250, 1–35.

Chen, W., Hans-F. Graf, and R. H. Huang, 2000: Interannual variability of East Asian winter monsoon and its relation to the summer monsoon. *Adv. Atmos. Sci.*, **17**, 48–60.

Chen, W., Hans-F. Graf, and M. Takahashi, 2002: Observed interannual oscillation of planetary wave forcing in the Northern Hemisphere winter. *Geophys. Res. Lett.*, **29**(22), 2073, doi:1029/2002 GL016062.

Chen, W., M. Takahashi, and Hans-F. Graf, 2003: Interannual variations of stationary planetary wave activity in the Northern winter troposphere and stratosphere and their relations to NAM and SST. *J. Geophys. Res.*, **108**(D24), 4797, doi:10.1029/2003 HD003834.

Chen, W., and R. H. Huang, 2005: The three-dimensional propagation of quasi-stationary waves in the Northern Hemisphere winter and its interannual variations. *Chinese J. Atmos. Sci.*, **29**, 139–146. (in Chinese)

Chen, W., S. Yang, and R. H. Huang, 2005: Relationship between stationary planetary wave activity and the East Asian winter monsoon. *J. Geophys. Res.*, **110**, D14110, doi:10.1029/2004JD5669.

Ding, Y. H., 1992: Summer monsoon rainfall in China. *J. Meteor. Soc. Japan*, **70**, 373–396.

Ding, Y. H., 1994: *Monsoon over China*. Kluwer Academic Publishers, 420pp.

Ding, Y. H., and C. He, 2006: The summer monsoon onset over the tropical eastern Indian Ocean: The earliest onset process of the Asian summer monsoon. *Adv. Atmos. Sci.*, **23**, 940–950.

Fan, K., and H. J. Wang, 2004: Antarctic oscillation and the dust weather frequency in North China. *Geophys. Res. Lett.*, **31**, 410201, doi:10, 1029/2004GL019465.

Gong, D. Y., and C. Ho, 2003: Arctic Oscillation Signals in the East Asian summer monsoon. *J. Geophys. Res.*, **108**, 4066, doi:10. 1029/2003 JD 002193.

Gong, D. Y. Wang, S. W., and J. H. Zhu, 2001: East Asian winter monsoon and Arctic Oscillation. *Geophys. Res. Lett.*, **28**, 2073–2076.

Goswami, B. N., G. X. Wu, and T. Yasunari, 2006: The annual cycle, intraseasonal oscillations, and roadblock to seasonal predictability of the Asian summer monsoon. *J. Climate*, **19**, 5078–5099.

Guo, Q. Y., 1983: The strength index of East Asian summer monsoon and its variation. *Chinese J. Geograph. Soc.*, **38**, 207–217. (in Chinese)

He, J. H., and J. J. Luo, 1999: Features of the South China Sea monsoon onset and Asian summer monsoon establishment sequence along with its individual mechanism. *The Recent Advances in Asian Monsoon Research*, He et al., Eds., China Meteorological Press, 74–81. (in Chinese)

Hsu, H. H., and S. M. Lin, 2007: Asymmetry of the tripole rainfall pattern during the East Asian summer. *J. Climate*, **20**, 4443–4458.

Huang, G., 2004: An index measuring the interannual variation of the East Asian summer monsoon-The EAP index. *Adv. Atmos. Sci.*, **21**, 41–52.

Huang, G., and L. T. Zhou, 2004: The variability of the wind system circulating round the west side of the Tibetan Plateau and its relation to the East Asian summer monsoon and summer rainfall in North China. *Climatic and Environmental Research*, **9**, 316–330. (in Chinese)

Huang, R. H., 1984: The characteristics of the forced planetary wave propagations in the summer Northern Hemisphere. *Adv. Atmos. Sci.*, **1**, 85–94.

Huang, R. H., 1985: Numerical simulation of the three-dimensional teleconnections in the summer circulation over the Northern Hemisphere. *Adv. Atmos. Sci.*, **2**, 81–92.

Huang, R. H., 2001a: Progresses in the studies on the formation mechanism and prediction theory of severe climatic disasters in China. *Basic Research in China*, **4**, 4–8. (in Chinese)

Huang, R. H., 2001b: Decadal variability of the summer monsoon rainfall in East Asia and its association with the SST anomalies in the tropical Pacific. *CLIVAR Exchange*, **2**, 7–8.

Huang, R. H., 2004: Review of the studies on the formation mechanism of severe climatic disasters in China. *Basic Research in China*, **4**, 6–13. (in Chinese)

Huang, R. H., 2006: Progresses in research on the formation mechanism and prediction theory of severe climatic disasters in China. *Advances in Earth Sciences.*, **21**, 564–575. (in Chinese)

Huang, R. H., and K. Gambo, 1982: The response of a hemispheric multi-level model atmosphere to forcing by topography and stationary heat sources. Parts I, II. *J. Meteor. Soc. Japan*, **60**, 78–108.

Huang, R. H., and K. Gambo, 1983a: The response of a hemispheric multi-level model atmosphere to forcing by topography and stationary heat sources in summer. *J. Meteor. Soc. Japan*, **61**, 495–509.

Huang, R. H., and K. Gambo, 1983b: On other wave guide in stationary planetary wave propagations in the winter Northern Hemisphere. *Science in China*, **26**, 940–950.

Huang, R. H., and W. J. Li, 1987: Influence of the heat source anomaly over the tropical western Pacific on the subtropical high over East Asia. *Proc. Internal Conf. on the General Circulation of East Asia*, Chengdu, April 10–15, 40–51.

Huang, R. H., and W. J. Li, 1988: Influence of the heat source anomaly over the tropical western Pacific on the subtropical high over East Asia and its physical mechanism. *Chinese J. Atmos. Sci.*, **14**, Special Issue, 95–107. (in Chinese)

Huang, R. H., and Y. F. Wu, 1989: The influence of ENSO on the summer climate change in China and its mechanisms. *Adv. Atmos. Sci.*, **6**, 21–32.

Huang, R. H., and F. Y. Sun, 1992: Impact of the tropical western Pacific on the East Asian summer monsoon. *J. Meteor. Soc. Japan*, **70**(1B), 243–256.

Huang, R. H., and Y. F. Wu, 1992: Research on ENSO cycle dynamics. *Proc. Oceanic Circulation Conf.*, Zeng et al., Eds., China Oceanological Press, Beijing, 41–51.

Huang, R. H., and F. Y. Sun, 1994: Impact of the thermal state and convective activities over the western Pacific warm pool on summer climate anomalies in East Asia. *Chinese J. Atmos. Sci.*, **18**, 262–272. (in Chinese)

Huang, R. H., and Y. F. Fu, 1996: The interaction between the East Asian monsoon and ENSO cycle. *Climatic and Environmental Research*, **1**, 38–54. (in Chinese)

Huang, R. H., and L. T. Zhou, 2002: Research on the characteristics, formation mechanism and prediction of severe climate disasters in China. *Journal of Natural Disasters*, **11**, 1–9. (in Chinese)

Huang, R. H., and L. Wang, 2006: Interdecadal variations of Asian winter monsoon and its association with the planetary wave activity. *Proc. Symposium on Asian Monsoon*, Kuala Lumpur, Malaysia, 4–7 April, 126.

Huang, R. H., Y. H. Xu, and P. F. Wang, 1998a: The features of the particularly severe flood over the Changjiang (Yangtze River) basin during the summer of 1998 and exploration of its cause. *Climatic and Environmental Research*, **3**, 300–313. (in Chinese)

Huang, R. H., Z. Z. Zhang, G. Huang, and B. H. Ren, 1998b: Characteristics of the water vapor transport in East Asian monsoon region and its difference from that in South Asian monsoon region in summer. *Chinese J. Atmos. Sci.*, **22**, 460–469. (in Chinese)

Huang, R. H, X. Y. Zang, R. H. Zhang, and J. L. Chen, 1998c: The westerly anomalies over the tropical Pacific and their dynamical effect on the ENSO cycles during 1980–1994. *Adv. Atmos. Sci.*, **15**, 135–151.

Huang, R. H., R. H. Zhang, and B. L. Yan, 1999a: Progresses in the studies on the interannual variability of East Asian climate system and problems to be studied further. *Basic Research in China*, **2**, 66–73. (in Chinese)

Huang, R. H., Y. H. Xu, and L. T. Zhou, 1999b: The interdecadal variation of summer precipitations in China and the drought trend in North China. *Plateau Meteorology*, **18**, 465–476. (in Chinese)

Huang, R. H., R. H. Zhang, and B. L. Yan, 2001: Dynamical effect of the zonal wind anomalies over the tropical western Pacific on ENSO cycles. *Science in China (Series D)*, **44**, 1089–1098.

Huang, R. H., L. T. Zhou, and W. Chen, 2003: The progress of recent studies on the variabilities of the East Asian monsoon and their causes. *Adv. Atmos. Sci.*, **20**, 55–69.

Huang, R. H., G. Huang, and Z. G. Wei, 2004a: Climate variations of the summer monsoon over China. *East Asian Monsoon*, C. P. Chang Ed., World Scientific Publishing CO. Pte. Ltd., 213–270.

Huang, R. H, W. Chen, B. L. Yan, and R. H. Zhang, 2004b: Recent advances in studies of the interaction between the East Asian winter and summer monsoon and ENSO cycle. *Adv. Atmos. Sci.*, **21**, 407–424.

Huang, R. H., L. Gu, Y. H. Xu, Q. L. Zhang, S. S. Wu, and J. Cao, 2005: Characteristics of the interannual variations of onset and northward advance of the East Asian summer monsoon and their associations with thermal states of the tropical western Pacific. *Chinese J. Atmos. Sci.*, **29**, 20–36. (in Chinese)

Huang, R. H., L. Gu, L. T. Zhou, and S. S. Wu, 2006a: Impact of the thermal state of the tropical western Pacific on onset date and process of the South China Sea summer monsoon. *Adv. Atmos. Sci.*, **23**, 909–924.

Huang, R. H., J. L. Chen, G. Huang, and Q. L. Zhang, 2006b: The quasi-biennial oscillation of summer monsoon rainfall in China and its cause. *Chinese J. Atmos. Sci.*, **30**, 545–560. (in Chinese)

Huang, R. H., R. S. Cai, J. L. Chen, and L. T. Zhou, 2006c: Interdecadal variations of droughts and floodings in China and their association with the East Asian climate system. *Chinese J. Atmos. Sci.*, **30**, 730–743. (in Chinese)

Huang, R. H., K. Wei, J. L. Chen, and W. Chen, 2007: Research on the East Asian winter monsoon anomalies in the winters of 2005 and 2006 and their relation to the quasi-stationary planetary wave activities over the Northern Hemisphere. *Chinese J. Atmos. Sci.*, **31**, 1033–1048. (in Chinese)

Kalnay, E. M., and Coauthors, 1996: The NCEP/NCAR 40-year reanalysis project. *Bull. Amer. Meteor. Soc.*, **77**, 437–471.

Kang, L., W. Chen, and K. Wei, 2006: The interdecadal variation of winter temperature in China and its relation to the anomalies in atmospheric general circulation. *Climatic and Environmental Research*, **11**, 330–339. (in Chinese)

Krishnamurti, T. N., and Y. Ramanathan, 1982: Sensitivity of monsoon onset to differential heating. *J. Atmos. Sci.*, **39**, 1290–1306.

Kurihara, K., 1989: A climatological study on the relationship between the Japanese summer weather and the subtropical high in the western northern Pacific. *Geophys. Mag.*, **43**, 45–104.

Lau, K. M., and C. P. Chang, 1987: Planetary scale aspects of the winter monsoon and atmospheric teleconnections. *Monsoon Meteorology*, C. P. Chang and T. W. Krishnamurti, Eds., Oxford University Press, 161–201.

Li, C. Y., 1988: The frequent actions of the East Asian trough and the occurrences of El Niño. *Science in China (Series B)*, 667–674. (in Chinese)

Li, C. Y., 1990: Interaction between anomalous winter monsoon in East Asia and El Niño events. *Adv. Atmos. Sci.*, **7**, 36–46.

Li, C. Y., and M. Q. Mu, 1999: The occurrence of El Niño event and the sea temperature anomalies in the subsurface of the West Pacific warm pool. *Chinese J. Atmos. Sci.*, **23**, 513–521. (in Chinese)

Li, C. Y., and M. Q. Mu, 2000: ENSO—A cycle of the subsurface sea temperature anomaly in the tropical Pacific driven by the anomalous zonal wind over the equatorial western Pacific. *Advances in Earth Science*, **17**, 631–638. (in Chinese)

Li, C., and M. Yanai, 1996: The onset and interannual variability of the Asian summer monsoon in relation to land-sea thermal contrast. *J. Climate*, **9**, 358–375.

Li, J., R. Yu, T. Zhou, and B. Wang, 2005: Why is there an early spring cooling shift downstream of the Tibetan Plateau. *J. Climate*, **18**, 4660–4668.

Liang, J. Y., and S. S. Wu, 2002: A study of southwest monsoon onset date over the South China Sea and its impact factors. *Chinese J. Atmos. Sci.*, **26**, 844–855. (in Chinese)

Lin, Z. D., and R. Y. Lu, 2005: Interannual meridional displacement of the East Asian upper-tropospheric jet stream in summer. *Adv. Atmos. Sci.*, **22**, 199–211.

Lu, R. Y., 2001: Interannual variability of the summertime North Pacific subtropical high and its relation to atmospheric convection over the warm pool. *J. Meteor. Soc. Japan*, **79**, 771–783.

Lu, R. Y., 2004: Association among the components of the East Asian summer monsoon system in the meridional direction. *J. Meteor. Soc. Japan*, **82**, 155–165.

Lu, R. Y., and R. H. Huang, 1996a: The transformed meridional circulation equation and its application to the diagnostic analysis of the blocking high formation. *Chinese J. Atmos. Sci.*, **20**, 138–148. (in Chinese)

Lu, R. Y., and R. H. Huang, 1996b: Energetics examination of the blocking episodes in the Northern Hemisphere. *Chinese J. Atmos. Sci.*, **20**, 270–278. (in Chinese)

Lu, R. Y., and R. H. Huang, 1998: Influence of East Asia/Pacific teleconnection pattern on the interannual variations of the blocking highs over the Northeastern Asia in summer. *Chinese J. Atmos. Sci.*, **22**, 727–735. (in Chinese)

Lu, R. Y., and B. W. Dong, 2001: Westward extension of North Pacific subtropical high in summer. *J. Meteor. Soc. Japan*, **79**, 1229–1241.

Lu, R. Y., and B. J. Kim, 2004: The climatological Rossby wave source over the STCZs in the summer Northern Hemisphere. *J. Meteor. Soc. Japan*, **82**, 657–669.

Lu, R. Y., Y. S. Chung, and R. H. Huang, 1995: Interannual variations of the precipitation in Korea and the comparison with those in China and Japan. *J. Korean Environmental Science. Soc.*, **4**, 345–365.

Lu, R. Y., C. Ryu, and B. W. Dong, 2002a: Association between the western North Pacific monsoon and the South China Sea monsoon. *Adv. Atmos. Sci.*, **19**, 12–24.

Lu, R. Y., J. H. Oh, and B. J. Kim, 2002b: A teleconnection pattern in upper-level meridional wind over the North African and Eurasian continent in summer. *Tellus*, **54A**, 44–55.

Luo, H. B., and M. Yanai, 1984: The large-scale circulation and heat sources over the Tibetan Plateau and surrounding areas during the early summer of 1979. *Mon. Wea. Rev.*, **108**, 1849–1853.

McBride, J. J., 1987: The Australian summer monsoon. *Monsoon Meteorology*, C. P. Chang and T. N. Krishnamurti, Eds., Oxford University Press, 203–232.

McCreary, J. P., 1983: A model of tropical ocean-atmosphere interaction. *Mon. Wea. Rev.*, **111**, 370–378.

McCreary, J. P., and D. L. T. Anderson, 1984: A simple model of El Niño and the Southern Oscillation. *Mon. Wea. Rev.*, **112**, 934–946.

McCreary, J. P., and D. L. T. Anderson, 1991: An overview of coupled ocean-atmosphere models of El Niño and the Southern Oscillation. *J. Geophys. Res.*, **96**, 3125–3150.

Miao, J. H., and K. M. Lau, 1990: Interannual variability of the East Asian monsoon rainfall. *Journal of Applied Meteorological Science*, **1**, 377–382. (in Chinese)

Mooley, D. A., and B. Parthasarathy, 1984: Fluctuations in all-India summer monsoon rainfall during 1871–1978. *Climate Change*, **6**, 287–301.

Murakami, T., and J. Matsumoto, 1994: Summer monsoon over the Asian continent and western North Pacific. *J. Meteor. Soc. Japan*, **72**, 745–791.

Nan, S., and J. Li, 2003: The relationship between the summer precipitation in the Yangtze River valley and the boreal spring Southern Hemisphere annual mode. *Geophys. Res. Lett.*, **30**, 2266, doi:10.1029/2003GL018381.

Nitta, T., 1983: Observational study of heat sources over the eastern Tibetan Plateau during the summer monsoon. *J. Meteor. Soc. Japan*, **61**, 590–605.

Nitta, T., 1987: Convective activities in the tropical western Pacific and their impact on the Northern Hemisphere summer circulation. *J. Meteor. Soc. Japan*, **64**, 373-390.

Philander, S. G. H., 1981: The response of equatorial ocean to a relaxation of trade winds. *J. Phys. Oceanogr.*, **11**, 176–189.

Ren, B. H., R. Y. Lu, and Z. N. Xiao, 2004: A possible linkage in the interdecadal variability of rainfall over

North China and the Sahel. *Adv. Atmos. Sci.*, **21**, 699–707.

Rodwell, M. J., and B. J. Hoskins, 1996: Monsoon and dynamics of deserts. *Quart. J. Roy. Metor. Soc.*, **122**, 1385–1404.

Schopf, P. S., and M. J. Suarez, 1988: Vacillations in a coupled ocean-atmosphere model. *J. Atmos. Sci.*, **45**, 549–566.

Staff members of Academia Sinica, 1957: On the general circulation over Eastern Asia (I). *Tellus*, **9**, 432–446.

Tao, S. Y., and L. X. Chen, 1985: The East Asian summer monsoon. *Proceedings of International Conference on Monsoon in the Far East*, Tokyo, 5–8 Nov., 1–11.

Tao, S. Y., and L. X. Chen, 1987: A review of recent research on the East Asian summer monsoon in China. *Monsoon Meteorology*, C. P. Cheng and T. N. Krishnamurti, Eds., Oxford University Press, 60–92.

Tao, S. Y., and J. Wei, 2006: The westward, northward advance of the subtropical high over the West Pacific in summer. *Journal of Applied Meteorological Science*, **17**, 513–524. (in Chinese)

Thompson, D. W. J., and J. M. Wallace, 1998: The Arctic Oscillation signature in the wintertime geopotential height and temperature fields. *Geophys. Res. Lett.*, **25**, 1297–1300.

Thompson, D. W. J., and J. M. Wallace, 2000: Annual modes in the extratropical circulation. Part 1: Month to-month variability. *J. Climate*, **13**, 1000–1016.

Tomas, R., and P. J. Webster, 1997: On the location of the intertropical convergence zone and near equatorial convection—The role of inertial instability. *Quart. J. Roy. Meteor. Soc.*, **123**, 1445–1482.

Tu, C. W., and S. S. Huang, 1944: The advance and retreat of the summer monsoon. *Meteorological Magazine*, **18**, 1–20. (in Chinese)

Uppala, S. M., and Coauthors, 2005: The ERA-40 reanalysis. *Quart. J. Roy. Meteor. Soc.*, **131**, 2961–3012.

Wang, B., H. Lin, Y. S. Zhang, and M, M. Lu, 2004: Definition of South China Sea monsoon onset and commencement of the East Asia summer monsoon. *J. Climate*, **17**, 699–710.

Wallace, J. M., and D. S. Gutzler, 1981: Teleconnection in the geopotential height field during the Northern Hemisphere winter. *Mon. Wea. Rev.*, **109**, 748–812.

Webster, P. J., and S. Yang, 1992: Monsoon and ENSO: Selectively interactive system. *Quart. J. Roy. Meteor. Soc.*, **118**, 877–926.

Webster, P. J., V. O. Magana, T. B. Palmer, J. Shukla, R. A. Tomas, M. Yanai, and T. Yasunari, 1998: Monsoons: processes, predictability, and the prospects for prediction. *J. Geophys. Res.*, **103**, 14451–14510

Wei, Z. G., and S. W. Luo, 1996: Impact of the snow cover in western China on the precipitation in China during the rainy season. *Collected Papers of the Project "Disastrous Climate Prediction and Its Impact on Agriculture and Water Resources", II. Processes and Diagnosis of Disastrous Climate*, Huang et al., Eds., China Meteorological Press, Beijing, 40–45 (In Chinese).

Wei, Z. G., R. H. Huang, and W. Chen, and W. J. Dong, 2002: Spatial distributions of interdecadal variations of the snow at the Tibetan Plateau. *Chinese J. Atmos. Sci.*, **26**, 496–508. (in Chinese)

Wei, Z. G., R. H. Huang, and W. J. Dong, 2003: Interannual and interdecadal variations of air temperature and precipitation over the Tibetan Plateau. *Chinese J. Atmos. Sci.*, **27**, 157–170. (in Chinese)

Wu, B. Y., and J. Wang, 2002: Winter Arctic Oscillation, Siberian high and East Asian winter monsoon. *Geophys. Res. Lett.*, **29**(19), 1897, doi:10.1029/2002GL015373.

Wu, G. X., and Y. S. Zhang, 1997: Tibetan Plateau forcing and the timing of the monsoon onset over South Asia and the South China Sea. *Mon. Wea. Rev.*, **126**, 917–927.

Xie, P. A., and P. A. Arkin, 1997: Analyses of global monthly precipitation using gauge observations, satellite estimates and numerical model prediction. *J. Climate*, **9**, 804–858.

Xin, X., R. C. Yu, T. J. Zhou, and B. Wang, 2006: Drought in late spring of South China in recent decades. *J. Climate*, **19**, 3197–3206.

Xue, F., H. J. Wang, and J. H. He, 2004: Interannual variability of Mascarene high and Australian high and their influence on East Asian summer monsoon. *J. Meteor. Soc. Japan*, **82**, 1173–1186.

Yang, S., and K. M. Lau, 1998: Influences of sea surface temperature and ground wetness on Asian monsoon. *J. Climate*, **11**, 3230–3246.

Yang, S., K. M. Lau, and K. M. Kim, 2002: Variations of the East Asian jet stream and Asian-Pacific-American climate anomalies. *J. Climate*, **15**, 306–325.

Yang, S., K. M. Lau, and S. H. Yoo, et al., 2004: Upstream subtropical signals proceeding the Asian summer monsoon circulation. *J. Climate*, **17**, 4213–4229.

Yamagata, T., and Y. Matsumoto, 1989: A simple ocean-atmosphere coupled model for the origin of a warm El Niño/Southern Oscillation event. *Philos. Trans. Roy. Soc. London*, (**A**)**329**, 225–236.

Yasunari, T., and R. Suppiah, 1988: Some problems on the interannual variability of Indonesian monsoon rainfall. *Tropical Rainfall Measurement*, Theon and Fugono, Eds., Daepek, Hampton, Va., 113–132.

Yeh, T. C., D. Z. Ye, S. Y. Tao, and M. C. Li, 1959: The abrupt change of circulation over the Northern Hemisphere during June and October. *The Atmosphere and the Sea in Motion*, 249–267.

Ye, D. Z., and Y. X. Gao, 1979: *Tibetan Plateau Meteorology*. Science Press, Beijing, 279pp. (in Chinese)

Yin, B. Y., L. Y. Wang, and R. H. Huang, 1996: Quasi-biennial oscillation of summer monsoon rainfall in East Asia and its physical mechanism. *Collected papers of the Project "Disastrous Climate Prediction and Its Impact on Agriculture and Water Resources", II. Processes and Diagnosis of Disastrous Climate*, Huang et al. Eds., China Meteorological Press, Beijing, 196–205. (in Chinese)

Yu, R. C., and T. J. Zhou, 2004: Impacts of winter-NAO on March cooling trends over subtropical Eurasian continent in the recent half century. *Geophys. Res. Lett.*, **31**, L12204, doi:10.1029/2004GL01984.

Yu, R. C., B. Wang, and T. J. Zhou, 2004: Tropospheric cooling and summer monsoon weakening trend over East Asia. *Geophys. Res. Lett.*, **31**, L22212, doi:10.1029/2004GL021270.

Zeng, Q. C., B. L. Zhang, and Y. L. Liang, 1994: East Asian summer monsoon—A case study. *Proc. Indian National Science Academy*, **60**(1), 81–96.

Zhang, Q., G. X. Wu, and Y. F. Qian, 2002: The bimodality of the 100 hPa South Asian high and its association on the climate anomaly over Asia in summer. *J. Meteor. Soc. Japan*, **80**, 733–744.

Zhang, R. H., 2001: Relations of water vapor transport from Indian monsoon with those over East Asia and the summer rainfall in China. *Adv. Atmos. Sci.*, **18**, 1005–1017.

Zhang, R. H., A. Sumi, and M. Kimoto, 1996: Impact of El Niño the East Asian monsoon: A diagnostic study of the 86/87 and 91/92 events. *J. Meteor. Soc. Japan*, **74**, 49–62.

Zhou, L. T., and R. H. Huang, 2003: Research on the characteristics of interdecadal variability of summer climate in China and its possible cause. *Climatic and Environmental Research*, **8**, 274–290. (in Chinese)

Zhou, L. T., and R. H. Huang, 2006: Characterisitcs of the interdecadal variability of difference between surface temperature and surface air temperature in spring in the arid and semi-arid region of Northwest China and its impact on summer precipitation in North China. *Climatic and Environmental Research*, **11**, 1–13. (in Chinese)

Zhou, T. J., and R. C. Yu, 2005: Atmospheric water vapor transport associated with typical anomalous summer rainfall patters in China. *J. Geophys. Res.*, **110**, D08104, doi:10.1029/2004JD005413.

Zhu, K. E., 1934: Southeast monsoon and rainfall in China. *Jounal of Geographical Society*, **1**, 1–27. (in Chinese)

我国东部夏季降水异常主模态的年代际变化及其与东亚水汽输送的关系

黄荣辉　陈际龙　刘　永

(中国科学院大气物理研究所季风系统研究中心，北京 100190)

摘　要　本文利用 1958~2000 年 ERA-40 再分析每日资料和我国 516 台站降水资料以及 EOF 方法，分析了我国东部季风区夏季降水异常主模态的年代际变化特征及其与东亚上空水汽输送通量时空变化的关系。分析结果表明了我国东部季风区夏季降水的时空变化存在两种主模态：第 1 主模态不仅显示出明显的准两年周期振荡的年际变化特征且也有明显的年代际变化，在空间上具有经向三极子型分布；第 2 主模态显示出明显的年代际变化特征，且在空间上具有经向偶极子型分布。这表明了这两主模态有明显的年代际变化，在 1958~1977 年期间我国东部夏季降水异常分布为从南到北"＋－＋"经向三极子型分布，而在 1978~1992 年期间降水异常出现了与 1958~1977 年相反的分布，为从南到北"－＋－"经向三极子型分布。然而，在 1993~1998 年期间，由于第 2 主模态的作用增大，我国东部夏季降水异常为从南到北经向三极子型与"＋－"偶极子型模态的结合，这使华南夏季降水明显增加。并且，分析结果还表明：这两主模态的年代际变化与东亚上空夏季水汽输送通量的时空变化密切相关，它不仅与东亚和西北太平洋上空似如东亚/太平洋型（EAP 型）遥相关波列分布的夏季水汽输送通量异常年代际变化有关，而且与欧亚上空中高纬度西风带似如欧亚型（EU 型）遥相关波列的夏季水汽输送通量异常年代际变化密切相关。

Interdecadal Variation of the Leading Modes of Summertime Precipitation Anomalies over Eastern China and Its Association with Water Vapor Transport over East Asia

HUANG Ronghui, CHEN Jilong, and LIU Yong

Center for Monsoon System Research, Institute of Atmospheric Physics, Chinese Academy of Sciences, Beijing 100190

Abstract　Interdecadal variation of the leading modes of summertime precipitation anomalies in the monsoon regions of eastern China and its association with the spatio-temporal variations of summertime water vapor transport fluxes over East Asia are analyzed by using the daily data of the ERA-40 reanalysis and precipitation data at 516 observational stations of China for 1958–2000 and the EOF analysis method. The analysis results show that there are two leading modes in the spatio-temporal variations of summertime precipitation anomalies over the monsoon region of

---

* 本文原载于：大气科学，第 35 卷，第 4 期，589-606，2011 年 7 月出版。

eastern China: The first leading mode exhibits not only a characteristic of obvious interannual variation with a quasi-biennial oscillation, but also a feature of interdecadal variability, and its spatial distribution is of a meridional tripole pattern. And the second leading mode exhibits a characteristic of obvious interdecadal variability, and its spatial distribution is of a meridional dipole pattern. This shows that these two leading modes have a significant interdecadal variability. During the period of 1958-1977, the distribution of summertime precipitation anomalies in eastern China exhibited a "+−+" meridional tripole pattern from the south to the north, and the distribution of precipitation anomalies for 1978-1992 showed a "−+−" meridional tripole pattern in the region, which was opposite to that for 1958-1977, but during the period of 1993-1998, since the role of the second leading mode in summertime precipitation anomalies in eastern China was intensified, the distribution of summertime precipitation in this region showed a combination of "+−+" meridional tripole pattern and "+−" meridional pattern, which caused the increase of summertime precipitation in South China. Moreover, the analysis results also show that the interdecadal variation of these two leading modes is closely associated with the spatio-temporal variations of summertime water vapor transport fluxes over East Asia, which is associated not only with the interdecadal variation of the EAP pattern teleconnection-like wave-train distribution of summertime water vapor transport flux anomalies over East Asia and the western North Pacific, but also with the interdecadal variation of the EU pattern teleconnection-like wave-train distribution of summertime waver vapor transport flux anomalies in the westerly zone over middle and high latitudes of Eurasia.

1 引言

水分循环是涉及到气候系统各子系统的一个重要过程。而在水分循环中，降水和水汽输送起着重要作用，海洋向陆地输送大量水汽，它是大气中水汽的主要源泉。从海洋输送来的水汽在陆地经凝结变成降水，又形成陆地的水分循环。在我国东部季风区，亚洲季风从孟加拉湾、南海和热带西太平洋给此区域带来大量水汽，使此地区有充沛的降水。在70多年前，我国著名气候学家竺可桢（1934）首先提出东亚夏季风对中国降水的影响。之后，涂长望和黄仕松（1944）研究了东亚夏季风的进退对中国降水雨带北进和南撤的影响。他们的研究开辟了关于东亚夏季风变化及其对我国降水影响的研究之路。继他们之后，Tao and Chen（1987）提出了东亚夏季风系统的概念，并且从环流结构论述了东亚夏季风系统是与一个既与南亚季风系统有联系又有相对独立的季风环流系统。

关于东亚夏季风系统降水的年际变化已有不少研究。许多研究指出了东亚夏季风降水存在着准两年周期振荡（缪锦海和 Lau，1990；Tian and Yasunari，1992；Chang et al. 2000），并且，Huang et al.（2004）、黄荣辉等（2006）指出了中国东部夏季降水的准两年周期振荡是与东亚上空夏季风水汽输送通量的准两年周期振荡密切相关，并是热带西太平洋热力变化准两年周期所引起的。此外，Ding（2007）也指出了准两年周期振荡也是亚洲季风年际变化的主模态。近年来，随着全球气候变化研究热潮的兴起，对于东亚夏季风年代际变化研究逐渐深入。黄荣辉等（1999）、周连童和黄荣辉（2003）从东亚夏季风降水的变化提出了在20世纪70年代中后期东亚夏季气候发生了一次明显的年代际变化，并指出这次年代际变化主要特征是：夏季到达华北地区偏南风明显变弱且降水明显偏少，而长江、淮河流域降水明显增强；并且，黄荣辉等（1999）、Huang et al.（2004）指出发生在20世纪70年代中后期及之后的东亚夏季气候年代际变化可能是由于热带东太平洋发生了年代际的"El Niño 现象"所引起，而张庆云等（2007）以及邓伟涛等（2009）指出，中国夏季降水的年代际变化与北太平洋中纬度海温年代际变化（PDO）有关。最近，Huang et al.（2007）利用中国160站降水资料进行分析，分析结果表明了中国东部季风区夏季降水存在着两个主模态，即从南到北经向三极子型和偶极子型分布模态，并表明了这两模态不仅有年际变化，而且有明显的年代际变化。Kwon et al.（2007）的研究表明了在20世纪90年代中期东亚夏季风又发生了一次明显的年代际变化，这次变化是中国南部夏季降水明显增加为特征。Ding et al (2008) 的研究也表明了中国东部夏季降水分别于 1978 年和 1992 年前后发生了两次显著的年代际变化,这两次年代际变化特征是东亚夏季风降水雨带明显南移。邓伟涛等 (2009) 也指出了中国夏季降水型在 20 世纪中后期和 90 年代初发生了明显的年代际变化。

亚洲季风区夏季降水的变化是与此区域上空水汽输送的变化紧密相关。黄荣辉等 (1998)、陈际龙和黄荣辉 (2007) 还研究了东亚季风区与南亚季风区夏季水汽输送特征的差异,指出了东亚季风区夏季水汽输送特征明显不同于南亚季风区的水汽输送特征,东亚季风区夏季经向水汽输送分量很大,而南亚季风区夏季水汽输送以纬向输送为主,且东亚季风区夏季降水主要由水汽平流和季风环流的辐合所引起,而南亚季风区夏季降水主要依赖于季风环流的辐合。最近,关于亚洲季风夏季水汽输送的时空变化特征已引起一些学者的兴趣。Zhang (2001) 指出,印度夏季风系统的水汽输送与东亚夏季风系统的水汽输送呈现反相关。Zhou et al. (2010) 利用 NCEP/NCAR 再分析资料的分析也表明了印度季风区水汽输送 5~7 月份最强,主要是纬向输送,而东亚季风区水汽输送 6~7 月份最强,经向输送分量非常大,特别从南海向北的水汽输送很重要,并且还指出了亚洲季风区夏季水汽输送场 EOF1 的时间系数从 1951~2005 年呈下降趋势。因此,研究我国东部季风区降水的时空变化特征及其与东亚夏季水汽输送的关联,这对于认识东亚季风区夏季旱涝灾害和水分循环的变异特征具有重要科学意义。

鉴于许多研究指出了东亚季风区夏季降水在 20 世纪 70 年代中后期和 90 年代初存在明显的年代际变化(黄荣辉等,1999;黄荣辉等,2006;陈际龙和黄荣辉,2007;张庆云等,2007;Kwon et al., 2007;Ding et al., 2008;邓伟涛等,2009),那么,这些变化与 Huang et al. (2007) 所指出的我国东部季风区降水异常主模态的年代际变化有何关联?并且,由于东亚季风区夏季降水的时空变化是与季风区水汽输送的时空变化密切相关,因此,有必要从更详细的降水观测资料和再分析资料来分析我国东部季风区夏季降水与东亚地区上空水汽输送通量在年代际时间尺度上的变化特征及其它们之间的关联,从而来说明东亚季风区夏季水汽输送的年代际变化对我国东部夏季降水年代际变化的影响。为此,本文从我国 756 个降水观测站中挑选出 516 站降水资料,并利用 ERA-40 每日 4 个时次的比湿和风场再分析资料以及有关 EOF 分析方法,来分析研究我国东部降水异常主模态的年代际变化特征以及与东亚夏季水汽输送变化之间的关系,并且,通过东亚季风区及欧亚上空中高纬度西风带夏季水汽输送通量异常波列分布的年代际变化进一步讨论夏季水汽输送通量异常的变化对我国东部季风区夏季降水异常主模态年代际变化的影响。

2 我国东部夏季降水时空变化的主模态

Huang et al. (2007) 从我国 160 个降水观测站资料分析了我国东部夏季降水的时空变化特征,指出了我国东部季风区夏季降水存在着两个主模态,即经向三极子型和偶极子型模态。但是,他们当时主要分析了降水的年际变化,对模态的年代际变化没有详细分析。为此,本节利用更详细降水资料,即从全国 756 降水观测站资料集挑出 516 站降水资料,并利用 EOF 分析方法来分析我国东部季风区夏季降水时空变化的主模态。

2.1 我国东部季风区夏季降水时空变化的主模态

2.1.1 第 1 主模态

图 1a 和图 1b 分别是我国东部 1958~2000 年夏季降水 EOF1 的空间分布和相应的时间系数序列。从图 1a 可以看到,我国东部季风区夏季降水 EOF1 的空间分布呈现从南到北"+-+"经向三极子型分布,并且,从图 1b 可以明显看到,我国东部季风区夏季降水 EOF1 的时间系数从 20 世纪 70 年代中后期起呈现 2~3 年周期的年际变化特征,即准两年周期振荡。

从图 1b 还可以看到,我国东部季风区夏季降水的 EOF1 时间系数还有明显的年代际变化。此外,从图 1c 所示的我国东部季风区夏季降水时空变化第 1 主模态时间系数的小波分析可以看到:它除了 2~3 年周期变化外,还有一个 8 年左右的年代际变化,在 20 世纪 90 年代初,这个周期的变化特别明显。从 20 世纪 50 年代到 70 年代中后期(大约 1958~1977 年),EOF1 的时间系数为正,结合图 1a 可以看到,这时期华南及华北和东北南部

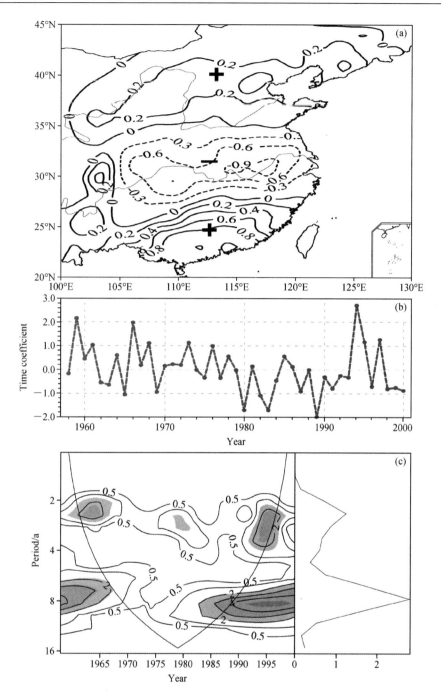

图 1 我国东部 1958～2000 年夏季降水 EOF 分析第 1 主分量 [EOF1（对方差的贡献为 15.3%）] 的空间分布（a，实、虚线表示正、负信号）和相应的时间系数序列（b）以及时间系数的小波分析（c）。降水资料取自中国气象局 756 测站降水资料集（下同）

Fig. 1 (a) Spatial distribution and (b) corresponding time coefficient series of the first component of EOF analysis (EOF1) of summertime precipitation in eastern China for 1958–2000 and (c) the analysis of wavelet spectrum of the time series. The solid and dashed lines in (a) indicate positive and negative signals, respectively, and the EOF1 explains 15.3% of the variance. The precipitation data is from the precipitation dataset at 756 stations of China by China Meteorological Administration (CMA)

地区夏季降水偏多，而长江、淮河流域夏季降水偏少，这表明了这时期我国东部夏季降水异常第一主模态为从南到北"＋－＋"经向三极子型分布；从 70 年代中后期到 90 年代初（大约在 1978～1992 年），EOF1 的时间系数为负，结合图 1a 可以看到，这时期华南和华北地区夏季降水偏少，而长江流域和东北地区夏季降水偏多,这表明了这时期我国东部夏季降水异常第一主模态与1958～1977年时期夏季降水异常主模态的空间分布相反,为从南到北"－＋－"经向三极子型分布;从90年代初到90年代末(大约在1993～1998年),EOF1的时间系数从负又变成正,结合图1a可以看到,华南及华北和东北南部地区夏季降水又偏多,而长江、淮河流域应偏少,这与1958～1977年时期夏季降水异常模态的分布相同,为从南到北经向"＋－＋"经向三极子型分布。然而,这时期我国东部夏季降水的实况却是华北和东北南部降水有所增加,而华南地区夏季降水明显偏多,洪涝灾害频繁发生,而长江、淮河流域夏季降水却没有减少,这可能是我国东部季风区夏季降水时空变化另一模态在起作用。

2.1.2 第2主模态

图2a和图2b分别是1958～2000年我国东部夏季降水EOF2的空间分布和相应的时间系数序列。与图1a所示的我国东部季风区夏季降水EOF1的空间分布相比较,在图2a所示的我国东部季风区夏季降水EOF2的空间分布显示出与图1a不同的分布,它出现了一个从南到北"－＋"经向偶极子型分布。并且,与图1b所示的我国东部季风区夏季降水EOF1时间系列相比较,在图2b所示的我国东部季风区夏季降水EOF2的时间序列不仅有3～4年周期的年际变化,而且更显示出年代际变化。此外,从图2c所示的我国东部季风区夏季降水时空变化第二主模态时间系数的小波分析可以看到:它除了有大约3～4年周期变化外,还有一个6年左右的年际变化,这个周期的变化在20世纪70年代初特别明显。此外,还有一个10年以上的年代际变化(受边界影响意义不大),特别从20世纪90年代初到90年代末(大约1993～1998年),EOF2的时间系数序列从正变成较大的负,结合图2a可以看到,此模态反映了从20世纪90年代初到90年代末华北和东北南部夏季降水应偏少,而长江流域和华南地区夏季降水应偏多,这表明这时期我国东部夏季降水第二模态为从南到北"＋－"经向偶极子型分布。然而,在此时期华北和东北南部夏季降水也有所增多,这是由于第一模态所起的作用。

从上分析可以看到,我国东部季风区夏季降水的时空变化有两个主模态,即在空间分布上从南到北为经向三极子型和经向偶极子型,这两个模态对

方差的贡献为28.5%。并且,无论是经向三极子型模态或是经向偶极子型模态,它们不仅有年际变化,而且有很明显的年代际变化。

2.2 我国东部夏季降水主模态的年代际变化

关于我国东部季风区夏季降水的年际变化特征及其与水汽输送的关系已在黄荣辉等(2006)的文章中讨论了,本文不再重复。因此,本节从图1a、b和图2a、b进一步列表更清楚地分析和讨论上述两个主模态的年代际变化。如表1所示,在1958～1977年时期,我国东部季风区夏季降水异常第1主模态空间分布是"＋－＋"经向三极子型分布,即华北和华南地区夏季降水偏多,而长江、淮河流域夏季降水偏少;在1978～1992年时期,我国东部夏季降水异常第1主模态空间分布型变成与1958～1977年时期相反的分布,即"－＋－"经向三极子分布,也就是华北和华南地区夏季降水偏少,而长江、淮河流域夏季降水偏多。在1993～1998年时期,我国东部夏季降水异常型从第一模态看,应是"＋－＋"经向三极子型,即华北和东北南部夏季降水应偏多,江南和华南地区降水应偏多,而长江和淮河流域降水应偏少。

并且,如表2所示,在1958～1977年时期,我国东部夏季降水第2主模态空间分布从南到北呈现"－＋"经向偶极子型分布,即华北和东北南部地区夏季降水偏多,长江、淮河流域和华南地区夏季

表1 我国东部季风区夏季降水第1模态的年代际变化
Table 1 The interdecadal variation of the EOF1 of summer rainfall in the monsoon region of eastern China

地区	年代		
	1958～1977年	1978～1992年	1993～1998年
华北和东北南部	＋	－	＋
长江和淮河流域	－	＋	－
华南	＋	－	＋

表2 我国东部季风区夏季降水第2模态的年代际变化
Table 2 The interdecadal variation of the EOF2 of summer rainfall in the monsoon region of eastern China

地区	年代		
	1958～1977年	1978～1992年	1993～1998年
华北和东北南部	＋	＋	－
长江和淮河流域	－	－	＋
华南	－	－	＋

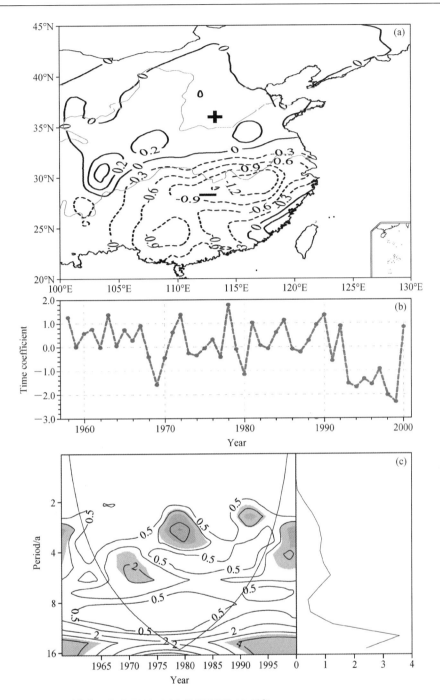

图 2 同图 1,但为 EOF2(对方差的贡献为 13.2%)

Fig. 2 As in Fig. 1 except for EOF2, the EOF2 explains 13.2 % of the variance

降水偏少;在 1978~1992 年时期,我国东部夏季降水第 2 模态空间分布与 1958~1977 年时期的分布型相同;在 1993~1998 年时期,我国东部夏季降水第 2 模态空间分布出现了与 1992 年之前不同的空间分布型,即出现了从南到北"十一"经向偶极子型分布,即华北和东北南部地区夏季降水偏少,而长江、淮河流域和华南地区夏季降水偏多,并且,它的时间系数明显增大。正是由于在此时期我国东部夏季降水时空变化第 1 和第 2 模态(即 EOF1 和 EOF2)在华南地区均偏多产生叠加,致使华南地区在此时期夏季降水严重偏多,洪涝灾害频繁发生,而在长江流域及华北和东北南部地区两模态却产生相反作用,致使长江流域降水没有偏少,而华北和东北南部夏季降水并没有明显偏多。

我国东部季风区夏季降水异常空间分布的年代际变化是与上述两个主模态空间分布的年代际变化密切相关。

2.3 我国东部季风区夏季降水异常的年代际变化特征及其与夏季降水主模态的关系

为了更清楚地看到我国东部季风区夏季降水异常的年代际变化与夏季降水主模态空间分布型的年代际变化的关系,本节首先从我国756站降水资料所挑出516站夏季降水资料来分析我国夏季降水异常的年代际分布。图3a~d分别是我国1958~1977年、1978~1992年、1993~1998年和1999~2009年期间平均的夏季降水距平百分率的分布。从图3a可以看到,在1958~1977年期间,我国东部华北和华南地区夏季降水偏多,出现正距平,洪涝灾害频繁发生,而长江流域和江淮地区夏季降水偏少,出现负距平,这正是上面所述的我国夏季降水异常从南到北的经向"＋－＋"三极子型分布。并且,从图3b可以看到,在1978~1992年期间,我国东部季风区夏季降水距平分布出现了与1958~1977年期间相反的分布,在华北和东北南部及华南地区夏季降水偏少,出现负距平,发生了持续性干旱灾害,而长江流域、汉水和四川盆地夏季降水偏多,出现了正距平,在此时期长江流域出现了多次洪涝灾害,这正是上面所述的我国东部夏季降水异常从南到北的经向"－＋－"三极子型分布。这些与邓伟涛等(2009)的分析结果一致。而从图3c可以看到,在1993~1998年期间,我国东部夏季从南到北出现了降水偏多的现象,不仅在华南地区降

(下接305页)

(图略)

图3 我国各时期平均的夏季(6~8月)降水距平百分率(%)分布图:(a) 1958~1977年;(b) 1978~1992年;(c) 1993~1998年;(d) 1999~2009年。实、虚线:正、负距平;阴影:正距平

Fig. 3 Distributions of summertime (Jun - Aug) precipitation anomaly percentages averaged for (a) 1958 - 1977, (b) 1978 - 1992, (c) 1993 - 1998, and (d) 1999 - 2009 in China. The solid and dashed lines indicate positive and negative anomalies, respectively, and positive anomalies are shaded

水出现了较大的正距平，洪涝灾害频繁发生，而且在长江、淮河流域夏季降水也出现了正距平，华北和东北西部地区夏季降水也有弱的正距平，在此期间华北地区从70年代中后期到90年代初期间所发生的持续干旱有所缓和，这是上面所述的我国东部夏季降水异常从南到北的经向"＋－＋"三极子型与经向"＋－"偶极子型分布的叠加。此外，从图4d可以看到，在1999～2009年期间，我国东部夏季降水在东北和华北地区偏少，出现明显负距平，而从华南地区到淮河流域（除长江沿岸地区）夏季降水偏多，从而形成了"南涝北旱"的降水异常分布型，即出现了从南到北"＋－"经向偶极子型分布。邓伟涛等（2009）指出我国东部夏季降水在20世纪90年代初从经向"＋－＋"三极子分布变成从北到南"－＋"偶极子型分布。本文分析结果表明，从20世纪90年代初偶极子型模态开始起作用，到90年代末我国东部夏季降水异常才从经向三极子型变成偶极子型分布。

在上一节关于我国东部季风区夏季降水异常的年代际变化分析的基础上，本节再详细讨论我国东部季风区夏季降水异常的年代际变化与我国东部夏季降水主模态的关系。图4是沿我国东部115°E（110°E～120°E的平均）夏季（6～8月）降水距平百分率的时间—纬度剖面图。从图4可以清楚看到，我国东部夏季降水异常分布有很明显的年代际变化。如图4所示，在1958～1977年期间，我国东部夏季降水异常分布是：华北和东北南部地区降水明显偏强，有较大的正距平，而在长江流域和江淮流域偏少，有明显的负距平，华南地区也是降水偏强，有正距平，即出现了与表1相同的从南到北"｜－＋"经向三极子型分布；并且，在1978～1992年期间，我国东部夏季降水异常的分布发生很明显的年代际变化，这时期降水异常分布正好与1958～1977年期间的分布相反，即在我国华北和东北南部地区降水偏少，有明显的负距平，出现持续性干旱，长江流域和江淮流域降水偏多，有明显的正距平，洪涝灾害频繁发生，而华南地区有很明显的负距平，降水明显偏少，出现了与表1相同的从南到北"－＋－"经向三极子型分布。这反映这时期东亚夏季偏南季风减弱。但是，这时期东北中部和北部降水却从偏少变成偏多，同样，我国西北干旱/半干旱区夏季降水在这时期也增加（Huang et al.，2004），这反映中纬度西风带水汽输送强盛。关于发生在20世纪70年代中后期东亚夏季风和气候异常的年代际变化及其成因，黄荣辉等（1999）、Huang et al.（2004，2007）已做了详细讨论，并指出这可能是由于热带太平洋发生的"年代际El Niño现象"引起东亚夏季风减弱所致。最近，Ding et al.（2008）从亚洲季风的年代际变化也指出了亚洲夏季风在1978年前后发生了明显减弱的年代际变化。

此外，从图4还可以看到，我国东部夏季降水

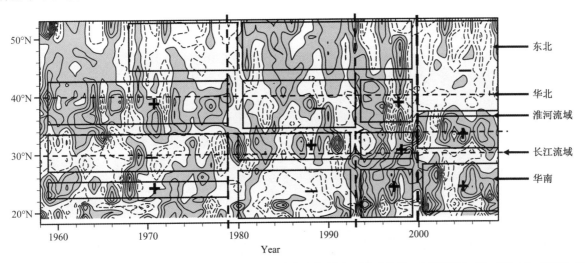

图4 我国东部沿115°E（110°E～120°E的平均）夏季（6～8月）降水距平百分率（％）的纬度—时间剖面图。实、虚线：正、负距平；阴影：正距平

Fig. 4 Latitude–time cross section of summertime (Jun–Aug) precipitation anomaly percentage along 115°E (average for 110°E–120°E) in eastern China. The solid and dashed lines indicate positive and negative anomalies, respectively, and positive anomalies are shaded

异常在 1993 年又发生了明显的年代际变化，在 1993～1998 年期间，我国东部季风区夏季降水距平的分布是：华南地区明显正距平，降水明显增多，而华北和东北南部降水也有正距平，表明此区域降水比 1978～1992 年期间增多，干旱有所缓和，在长江流域和江淮流域也有正距平，表明这两流域降水并没有比 1978～1992 年期间减少。这表明了在这时期我国东部降水异常出现了从南到北一致增多的趋势。正如前面所分析，这是由于在图 2a、b 所示的我国东部夏季季风降水第 2 模态在此时期的影响增大的原因。因此，我国东部季风区在 1993～1998 年期间夏季降水异常是第一模态的从南到北"+-+"经向三极子型分布与第 2 模态的经向"+-"偶极子型分布的叠加。Kwon et al. (2007) 指出了东亚夏季风在 20 世纪 90 年代中期发生了一次明显的年代际变化，Ding et al. (2008) 的研究也表明了亚洲夏季风在 1992 年前后又发生了一次年代际变化。本文分析结果表明，这次中国东部季风降水的年代际变化是第 1 主模态和第 2 主模态共同作用的结果。

同样，从图 4 还可以看到，在 1999 年中国东部夏季降水又发生了一次明显的年代际变化。这次年代际变化表现为我国华北和东北降水明显减少，而淮河流域和华南地区夏季降水继续偏多，从而形成了"南涝（除沿长江附近地区）北旱"的降水异常分布型。这表明了我国东部夏季季风降水异常分布型完全从三极子型变成偶极子型。关于此次降水和环流的年代次变化特征已在 Huang et al. (2010) 详细讨论，本文不再重复。

从上述关于我国东部夏季降水异常分布的年代际变化及其与我国东部夏季降水主模态的关系分析可以看到：在 20 世纪 70 年代中后期之后我国东部夏季降水异常主模态的年代际变化只是与 70 年代中后期之前的降水异常模态符号相反，但仍是三极子型分布。这可能是由于在这两个时期我国夏季降水第 2 模态的时间系数还比较小，使得第 1 模态起主要作用。因此，在这两时期我国东部夏季降水异常的空间分布类似于第 1 模态的空间分布。然而，发生在 20 世纪 90 年代初期夏季降水异常模态的年代际变化却是夏季降水异常模态转换的开始，在 1993～1998 年时期，我国东部夏季降水异常第 2 模态的时间系数明显变大，因此，在此时期我国东部夏季降水异常的空间分布既不同于第 1 模态的空间分布，又不同于第 2 模态的空间分布，而是这两主模态空间分布的叠加，即经向三极子型与经向偶极子型的叠加。到了 90 年代末，我国东部夏季降水异常模态，从经向三极子型完全变成经向偶极子型（除长江流域），这可能由于在 1999 年之后我国东部夏季降水第 2 主模态起主要作用所致。

3 东亚夏季水汽输送通量的时空变化特征及其与中国东部夏季降水异常模态的关系

由于我国东部夏季降水的时空变化是直接与东亚季风区夏季水汽输送的时空变化密切相关，因此，在下面我们将分析东亚夏季水汽输送通量的时空变化特征。

3.1 水汽输送通量

假设 100 hPa 以上大气没有水汽，这样，单位气柱整层大气水汽输送通量 $\boldsymbol{Q}=(\boldsymbol{Q}_\lambda,\boldsymbol{Q}_\varphi)$ 可由下式来计算，即

$$\boldsymbol{Q} = \frac{1}{g}\int_{100}^{p_s}\boldsymbol{V}\cdot q\mathrm{d}p = \frac{1}{g}\int_{100}^{p_s}(u,v)\cdot q\mathrm{d}p. \quad (1)$$

由（1）式分别可得纬向和经向水汽输送通量 Q_λ 和 Q_φ 的计算公式如下：

$$Q_\lambda = \frac{1}{g}\int_{100}^{p_s} u\cdot q\mathrm{d}p, \quad (2)$$

$$Q_\varphi = \frac{1}{g}\int_{100}^{p_s} v\cdot q\mathrm{d}p, \quad (3)$$

上式中 q 为某单位气柱的比湿，u 和 v 分别为东西风和南北风分量，p_s 是地表气压，考虑到我国西部地形复杂，故 p_s 取为随 λ、φ 而变化的地表气压。

依据与利用我国测站降水资料所分析的我国降水年际和年代际变化相比较的结果，似乎利用 ERA-40 再分析资料所计算的水汽输送通量更合理（陈际龙和黄荣辉，2007），因此，在下面的计算就采用 ERA-40 再分析资料。在本研究中首先利用 ERA-40 的 1958～2000 年 6～8 月每日 4 个时次的比湿（q）和风场（u、v）以及地表气压 p_s 的再分析资料，并应用（2）和（3）式计算出我国东部季风区及其周围地区夏季每日 4 个时次的纬向和经向水汽输送通量，并由 4 个时次的水汽输送通量计算出的日平均水汽输送通量，之后，由每日纬向和经向水

汽输送通量再计算出 1958～2000 年夏季各月（6～8月）的水汽输送通量。这样，就可利用所计算的 1958～2000 年夏季各月的水汽输送通量来分析我国东部夏季水汽输送通量的时空变化特征及其与降水变化的关系。

3.2 东亚夏季纬向水汽输送通量时空变化的主模态

3.2.1 第1主模态

本节首先利用从 ERA-40 每日 4 时次再分析资料所计算的 1958～2000 年夏季东亚地区夏季纬向水汽输送通量来分析东亚夏季水汽输送通量的时空变化特征。图5是东亚地区上空 1958～2000 年夏季纬向水汽输送通量的 EOF 分析第1模态（EOF1可以说明方差43.1%）的空间分布和相应的时间系数序列。从图5a 可以明显看到，在我国东部季风区、南海、东海和黄渤海以及日本和朝鲜半岛上空夏季纬向水汽通量的 EOF1 空间分布明显存在着一个从南到北"＋－＋"的经向三极子型的分布。并且，从图5b 可以看到：东亚地区夏季纬向水汽输送通量 EOF1 的时间系数从 20 世纪 70 年代中后期起（大约从 1977 年）呈现 2～3 年周期的年际变化特征，即准两年周期振荡。此外，从图5c 所示的东亚地区夏季纬向水气输送通量时空变化第1主模态

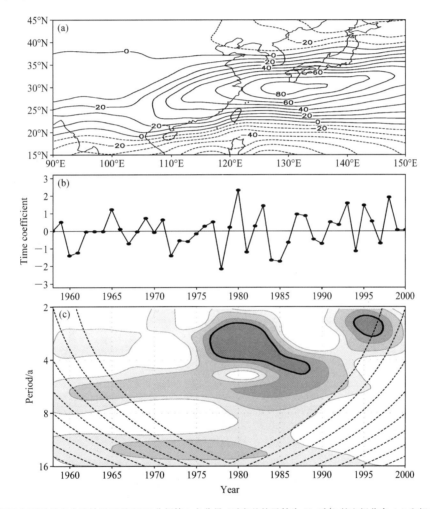

图5 东亚1958～2000年夏季纬向水汽输送通量EOF分析第1主分量（对方差的贡献为43.1%）的空间分布（a）和相应的时间系数序列（b）以及时间系数的小波分析（c）。(a) 实、虚线表示正（向东输送）、负（向西输送）信号；湿度、风场和 p_s 资料取自 ERA-40 再分析资料 (Uppala et al., 2005)

Fig. 5 (a) Spatial distribution and (b) corresponding time coefficient series of the first component of EOF analysis (EOF1) of summertime zonal water vapor transport fluxes over East Asia for 1958 - 2000 and (c) the analysis of wavelet spectrum of the time series. The solid and dashed lines in (a) indicate positive (eastward transport) and negative (westward transport) signals, respectively, and the EOF1 explains 43.1% of the variance. The data of moisture, wind field, and p_s are from the ERA-40 reanalysis (e.g., Uppala et al., 2005)

时间系数的小波分析可以看到：它除了 2～3 年周期变化外（在 20 世纪 70 年代中后期起变得明显），还有一个 10 年以上的年代际变化，但从 20 世纪 80 年代起这个周期的变化变得不明显。

3.2.2 第 2 主模态

由于在以前的研究中，没有很好分析夏季纬向水汽输送通量的 EOF2 特征，为此，本节将较详细分析和讨论东亚地区上空夏季纬向水汽输送通量 EOF2 的空间分布和相应的时间系数序列（见图 6）（EOF2 可以说明方差 17.7%）。与图 5a 所示的 EOF1 空间分布相比较，在图 6a 所示的东亚地区夏季纬向水汽通量 EOF2 的空间分布出现了与图 5(a) 明显不同的分布，在我国华南和南海、我国东部、华北、黄渤海和东海以及朝鲜半岛和日本等地区显示出明显的从南到北"＋－"经向偶极子型分布；并且，在图 6b 所示的东亚地区上空夏季纬向水汽通量 EOF2 的时间系数序列不仅具有 3～4 年周期的年际振荡特征，而且显示出较弱的年代际变化。此外，从图 6c 所示的东亚地区夏季纬向水气输送通量时空变化第 2 主模态时间系数的小波分析可以看到：它除了 2～3 年周期变化外，还有一个 8 年左右的年代际变化，但从 20 世纪 80 年代起这个周期的变化变得不明显。此外，还有一个 16 年左右的年代际变化。从 20 世纪 70 年代中后期到 90 年代初，EOF2 的时间系数比 50 年代中后期到 70 年代中后期有一个变正的趋势，结合图 6a，这表明：在华北、朝鲜半岛和日本地区在 20 世纪 70 年代中后期到 90 年代初向东水汽输送通量偏弱，而华南、江南和热带西太平洋上空在此时期夏季向东水汽输送通量偏强。然而，从 90 年代初到 90 年代末，即 EOF2 时间系数在此时期又发生年代际变负的趋势，结合图 6a，这表明华北、朝鲜半岛和日本

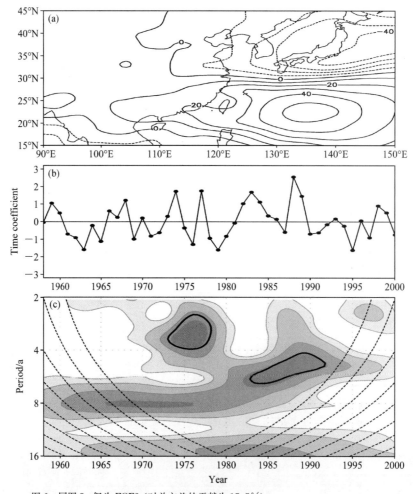

图 6 同图 5，但为 EOF2（对总方差的贡献为 17.7%）

Fig. 6 As in Fig. 5 except for EOF2, the EOF2 explains 17.7 % of the variance

上空向东水汽输送通量增强，而江南和华南地区上空向东的水汽输送通量偏弱，或向西的水汽输送通量偏强。

从上分析可以看到，东亚夏季纬向水汽输送通量时空变化有两个主模态，即在空间分布上经向三极子型与经向偶极子型，这两个模态对总方差的贡献为60.8%。

3.3 东亚夏季经向水汽输送通量时空变化的主模态

3.3.1 第1主模态

图7是利用ERA-40每日4时次再分析资料所计算的东亚上空1958～2000年夏季经向水汽输送通量的EOF分析第1模态（EOF1，可以说明方差23.0%）的空间分布和相应的时间系数序列。从图7a还可以看到，我国东部和华南地区上空与日本及以南的西北太平洋上空夏季经向水汽输送通量存在反相变化。这说明：若我国东部和华南地区上空夏季向北水汽输送增强，则在日本及以南地区上空夏季向北水汽输送偏弱；反之，若我国东部和华南地区上空夏季向北水汽输送偏弱，则在日本及以南地区上空向北水汽输送偏强。这可能是由于夏季西太平洋反气旋环流的变化所造成。并且，从图7b可以看到：东亚地区夏季经向水汽输送通量EOF1的时间系数不仅呈现明显的2～3年周期的年际变化特征，即准两年周期振荡，而且呈现明显的年代际变化。此外，从图7c所示的东亚地区夏季经向水气输送通量时空变化第1主模态时间系数的小波分析可以看到：它除了2～3年周期变化外，从20世纪80年代，8年左右的年代际变化以及10年以上的年代际变化变得更明显。在1958～1971年期间，EOF1的时间系数为正；在1972～1992年期间，EOF1时间系数有变负的趋势；在1993～1998年期间，EOF1时间系数又有变正的趋势。把图7b所

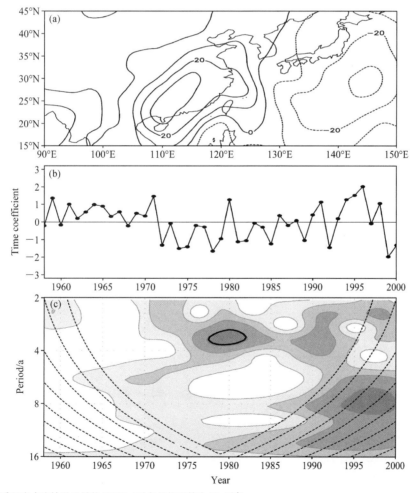

图7 同图5，但为夏季经向水汽输送通量的EOF1（对方差的贡献为23.0%）

Fig. 7 As in Fig. 5 except for the EOF1 of summertime meridional water vapor transport fluxes, and EOF1 explains 23.0 % of the variance

示的时间系数的年代际变化再结合图 7a 所示的 EOF2 空间分布可以看到：20 世纪从 50 年代末到 70 年代初我国东部和华南地区夏季向北水汽输送偏强；从 70 年代初到 90 年代初期间，上述地区夏季向北水汽输送偏弱，这造成了华北地区在此时期夏季降水偏少，出现持续性干旱；从 90 年代初到 90 年代末，上述地区夏季向北水汽输送又偏强，这使得华北地区夏季降水有所增加。

3.3.2 第 2 主模态

图 8 是东亚地区上空夏季经向水汽输送通量 EOF2 的空间分布和相应的时间系数序列。与图 7a 所示的 EOF1 空间分布相比较，在图 8a 所示的东亚地区夏季经向水汽输送 EOF2 的空间分布显示出从我国中部经我国东部、黄渤海和东海、朝鲜半岛和日本上空呈现一个纬向"－＋－"的波列结构；并且从图 8b 可以看到，我国东部及周围地区夏季经向水汽输送 EOF2 的时间系数呈现出 3～4 年周期的年际变化特征。此外，从图 8c 所示的东亚地区夏季经向水气输送通量时空变化的第 2 主模态时间系数的小波分析可以看到它除了 3～4 年周期的年际变化外，还有一个 6～7 年周期的年际变化，这个周期变化在 20 世纪 80～90 年代变得明显。

从上分析可以看到，东亚夏季经向水汽输送通量时空变化也有两个主模态（对方差的贡献 38.7%），即西南—东北向的三极子型分布型和纬向三极子分布型。这两个模态在我国东部都是出现南北一致分布的变化。

若把上述由 ERA-40 每日 4 个时次再分析资料的比湿和风场所分析东亚地区夏季纬向和经向水汽输送通量的 EOF1 和 EOF2 与由 ERA-40 再分析资料的月平均比湿和风场所分析的东亚地区夏季纬向和经向水汽通量的 EOF1 和 EOF2（Huang et al.

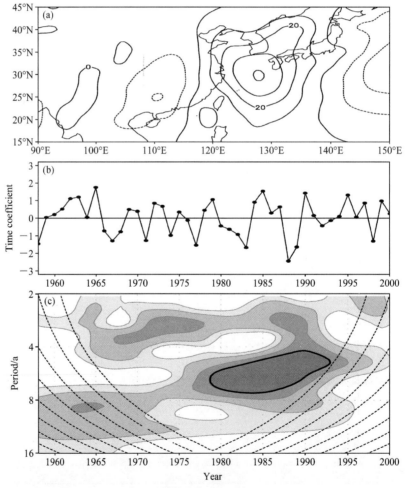

图 8　同图 7，但为 EOF2（对方差的贡献为 15.7%）
Fig. 8　As in Fig. 7 except for EOF2, the EOF2 explains 15.7% of the variance

2004；黄荣辉等，2006）做比较，则可以看到，无论它们的空间分布模态和相应的时间系数大致相同，但它们还是存在一些差别。但从分析结果看，似乎用 ERA-40 每日 4 个时次的比湿和风场再分析资料计算的水汽输送通量所分析的 EOF1 和 EOF2 的结果更合理。

4 我国东部夏季降水异常模态的年代际变化与东亚水汽输送通量时空变化的关系

关于我国东部季风区夏季降水准两年周期的年际变化及其与东亚夏季水汽输送通量准两年周期的年际变化的关系及其机理，这已在黄荣辉等（2006）的文章中详细讨论了，并且 Chen et al.（2008）也讨论了东亚夏季风水汽输送的异常模态及其与我国夏季降水型之间的关系。故本文不再重复。本文着重分析在年代际时间尺度上我国东部夏季降水异常主模态与东亚上空夏季水汽输送通量时空变化的关系。

4.1 我国东部季风区夏季降水异常的年代际变化与东亚上空水汽输送变化的关系

从图 5a 和图 6a 可以看到，东亚上空夏季纬向水汽输送有两个主模态，即经向三极子模态和经向偶极子模态。由于第 1 主模态，如图 5b 所示，主要反映东亚上空夏季纬向水汽输送的年际变化，而第 2 主模态，如图 6b 所示，主要反映了东亚上空夏季纬向水汽输送较弱的年代际变化特征。并且，从图 7a 和图 8a 可以看到，东亚上空夏季经向水汽输送通量也有两个主模态，这两个模态的空间分布在我国东部表现为从南到北一致的变化，而从图 7b 可以看到东亚夏季经向水汽输送通量的第一主模态存在明显的年代际变化。结合图 7b 与图 7a 可以看到：在 1958～1971 年期间，我国东部季风区上空夏季从南向北的经向水汽输送通量偏强；1972～1992 年期间，我国东部季风区上空夏季从南到北的经向水汽输送通量偏弱；1993～1998 年期间，我国东部季风区夏季从南到北的经向水汽输送通量又偏强。然而，从 1999 年之后，我国东部季风区夏季从南到北的经向水汽输送通量又偏弱。

把图 1a、图 2a 与图 5a、图 6a 做比较，可以看到：图 1a 和图 2a 所示的我国东部夏季降水异常两个主模态的空间分布与图 5a 和图 6a 所示的东亚上空夏季纬向水汽输送通量时空变化两个主模态的空间分布有相似之处，即经向三极子型和经向偶极子型分布。然而，图 3 和图 4 所示的我国东部夏季降水异常的年代际变化却主要受图 7a 和图 7b 所示的东亚上空经向水汽输送通量第 1 主模态年代际变化的调控。

4.2 我国东部季风区夏季降水异常主模态的年代际变化与欧亚上空水汽输送异常波列的年代际变化之间的关系

从黄荣辉和陈际龙（2010）的推导可以看到，某个区域的降水主要依赖于此区域水汽输送的辐合，若某区域水汽输送通量距平矢量为气旋性分布，则此区域有水汽辐合的异常，降水将增多；反之，若水汽输送通量距平矢量为反气旋性分布，则此区域有水汽辐散的异常，降水将减少。因此，我国东部夏季降水异常模态的年代际变化可以从东亚及周围地区夏季水汽输送通量距平矢量分布的年代际变化更清楚看到，为此，我们又进一步分析欧亚地区上空夏季水汽输送通量距平矢量分布的年代际变化。

图 9a-c 分别是 1958～1977 年、1978～1992 年和 1993～1999 年期间平均的东亚及周围地区夏季水汽通量距平矢量的分布。从图 9a 可以看到，在 1958～1977 年期间，从我国华南和南海经我国长江和淮河流域及日本和韩国到我国东北和日本海上空夏季水汽输送通量距平矢量呈现一个似如 Nitta（1987）、Huang and Li（1987）以及黄荣辉和李维京（1988）所提出的东亚/太平洋型（EAP 型）波列分布。这个分布是：在从南海到菲律宾周围上空有一反气旋式分布，在我国华南和长江流域到日本南部的副热带西太平洋上空有一气旋式分布，在我国从华北东部经黄淮地区到日本海上空有一反气旋式分布，在日本海和库页岛上空有一气旋式分布；并且，在中纬度西风带从西北欧经蒙古高原到我国东部季风区和日本上空夏季水汽输送距平矢量也呈现一个西北－东南向似如 Wallace and Gutzler（1981）所提出的欧亚型（EU 型）遥相关波列分布，这个分布是：在西北欧上空有一反气旋式分布，在乌拉尔山上空有一气旋式分布，在贝加尔湖和蒙古高原上空有一反气旋式分布，在华北地区上空有一气旋式分布，在黄淮地区、黄海和日本上空有一反气旋式分布。这引起了在华北东部和黄淮地

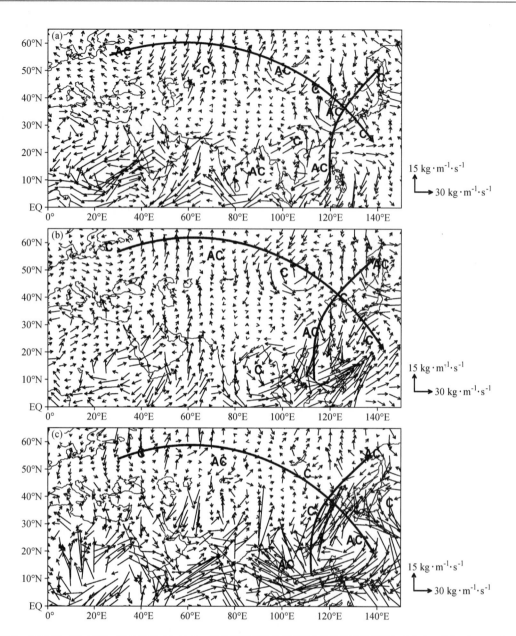

图 9 欧亚上空 (a) 1958～1977 年、(b) 1978～1992 年和 (c) 1993～1998 期间平均的夏季水汽输送通量距平矢量分布（单位：kg·m^{-1}·s^{-1}）。湿度、风场和 p_s 取自 ERA-40 再分析资料（Uppala et al., 2005）

Fig. 9 Anomaly distributions of summertime water vapor transport flux vectors averaged for (a) 1958–1977, (b) 1978–1992, and (c) 1993–1998 (units: kg·m^{-1}·s^{-1}). The data of moisture, wind field, and p_s are from the ERA-40 reanalysis (e. g., Uppala et al., 2005)

区上空有较强的向北水汽输送通量距平，即向北水汽输送偏强。正是由于这两个水汽输送通量距平矢量波列的交叉和相互作用，我国东部夏季降水异常才出现了从南到北"＋－＋"的经向三极子型分布。

然而，在1958～1977年期间欧亚地区上空夏季水汽输送通量距平矢量的波列分布到了1978年之后产生了相反变化。从图9b与图9a的比较可以发现，在图9b所示的1978～1992年期间无论在欧亚中高纬西风带地区或是东亚季风区上空夏季水汽输送通量距平矢量的波列分布都与图9a所示的1958～1977年期间的水汽输送通量距平矢量的波列分布相反。正如图9b所示，在从南海到菲律宾周围及东北侧上空有一气旋分布，在我国华南和长江流域有一反气旋式分布，在我国黄海和朝鲜半岛有一弱的气旋式分布，日本南部西太平洋上空有一强的气旋式分布,在库页岛和鄂霍次克海上空有一反气旋式分布;并且,在中高纬度西风带的西北欧上空有一气旋式分布,在乌拉尔山周围有一气旋式分布,在贝加尔湖和蒙古高原有一较强的气旋式分布,在朝鲜半岛有一气旋式分布。这引起了从黄淮地区经黄海和东海到东南沿海上空有向南的水汽输送距平,表明此时期在我国东部季风区夏季向北水汽输送偏弱。因此,在1978~1992年期间,我国东部从南到北夏季降水异常存在"-+-"的经向三极子型分布。

如图9c所示,在1993~1998年期间,欧亚上空夏季水汽输送通量距平矢量分布存在着与图9a和图9b均有一些不同的分布。这时期在欧亚大陆中高纬度西风带地区水汽输送通量距平虽然其分布型有点类似于图9a所示的分布型,但通量距平值显著变小,且EU型波列分布不很清楚,特别在贝加尔湖和蒙古高原上空的反气旋式分布范围收缩,这表明来自大西洋沿中高纬度西风带的水汽输送变弱。并且,在东亚地区的水汽输送通量距平矢量从南海到菲律宾东北侧有一强的反气旋式分布,从我国东部经朝鲜半岛到日本上空有一气旋式分布,在库页岛和鄂霍次克海上空有一反气旋式分布,这类似于EAP型的波列分布。这引起了在我国东部沿海及黄海和东海上空有强向北水汽输送距平,表明在此期间有强的向北水汽输送。因此,在1993~1998年期间,我国东部夏季降水异常分布呈现出经向三极子型和经向偶极子型的结合。

从上分析可以看到,我国东部季风区夏季降水异常主模态的年代际变化不仅与从菲律宾周围往北传播的似如EAP型遥相关的水汽输送通量距平波列分布的年代际变化有关,而且也与中高纬度西风带传播的似如EU型遥相关的水汽输送通量异常波列变化有关。

5 结论和讨论

本文利用1958~2000年ERA-40每日4个时次再分析资料和我国516台站降水资料以及EOF方法分析了我国东部季风区夏季降水异常主模态的年代际变化特征及其与东亚上空水汽输送通量时空变化的关系。分析结果表明了我国东部季风区夏季降水的时空变化存在着两个主模态:第1主模态不仅显示出明显的准两年周期振荡的年际变化特征,而且也有明显的年代际变化,在空间上具有经向三极子型分布;第2主模态显示出明显的年代际变化特征,且在空间上具有经向偶极子型分布。这表明了我国东部季风区夏季降水异常主模态有明显的年代际变化,在1958~1977年期间我国东部夏季降水异常分布为从南到北"+-+"经向三极子型分布,而在1978~1992年期间降水异常出现了与1958~1977年相反的分布,为从南到北"-+-"经向三极子型分布。然而,在1993~1998年期间,由于第2主模态的作用增大,我国东部夏季降水异常为从南到北经向"+-+"三极子型与"+-"偶极子型的叠加。并且,分析结果还表明:我国东部夏季降水异常主模态的年代际变化与东亚上空夏季水汽输送通量的时空变化密切相关,这说明了我国东部夏季降水异常的年代际变化不仅与夏季从南海和菲律宾周围经我国东部和日本南到鄂霍次克海上空似如EAP型遥相关波列的水汽输送通量距平分布的年代际变化有关,而且与夏季欧亚中高纬度西风带似如欧亚型(EU型)遥相关波列水汽输送通量异常分布的年代际变化有关。

本文分析结果一方面说明了利用测站资料作EOF分析与利用再分析资料作EOF分析,其结果会有差别;另一方面也表明了中国夏季降水的年代际转型与东亚和西太平洋地区低层环流的转型并不同步。这可能是由于降水不仅仅依赖于由季风环流所驱动的水汽平流,而且还依赖于环流场的辐合(黄荣辉和陈际龙,2010),而这两者的年代际变化并不一致。从东亚地区夏季纬向水汽输送的EOF2和经向水汽输送的EOF1时间系数变化来看,在20世纪80年代末有一个明显的变化,东亚和西太平洋上空向北水汽输送从弱又变为偏强。张人禾等(2008)从利用低层大气环流所定义的2个东亚夏季风指数都表明东亚夏季风在20世纪80年代末发生了一次年代际转型。然而,本文利用中国测站资料来研究夏季降水的年代际转型却发生在20世纪90年代初。

从上面分析可以看到,关于东亚地区夏季风降水异常年代际变化的成因是比较复杂的,它不仅与影响东亚地区夏季水汽输送的太平洋中纬度海温年代际变化(PDO)有关(张庆云等,2007;邓伟涛等,2009),而且与欧亚上空中高纬度西风带水汽输送的大西洋海温年代际变化和中高纬度环流有关。并且，夏季蒙古高原上空的环流和热带西太平洋上空环流的年代际变化对于我国东部季风区夏季降水异常的主模态年际变化起着十分重要的作用，因此，关于这两区域上空环流年代际变化的成因应该进一步深入研究。

参考文献（References）

Chang C P, Zhang Y S, Li T. 2000. Interannual and interdecadal variations of the East Asian summer monsoon and tropical Pacific SSTs. I, II [J]. J. Climate, 13: 4310 – 4340.

陈际龙, 黄荣辉. 2007. 亚澳季风各子系统气候学特征的异同研究. II. 夏季风水汽输送 [J]. 大气科学, 31: 766 – 778. Chen Jilong, Huang Ronghui. 2007. The comparison of climatological characteristics among Asian and Australian monsoon subsystem. Part II: Water vapor transport by summer monsoon [J]. Chinese J. Atmos. Sci. (in Chinese), 31: 766 – 778.

Chen Jilong, Chan Johny C L, Zhou Wen, et al. 2008. Anomalous modes of moisture transport by East Asian summer monsoon and associated rainfall patterns in China [C]. Extented Abstracts of Third WCRP International Conference on Reanalysis, 2008. http: wcrp. ipst. jussieu. fr/workshop/Reanalysis 2008/Documents/G3 – 332 ea. pdf.

邓伟涛, 孙照渤, 曾刚, 等. 2009. 中国东部夏季降水型的年代际变化及其与北太平洋海温的关系 [J]. 大气科学, 21: 835 – 845. Deng Weitao, Sun Zhaobo, Zeng Gang, et al. 2009. Interdecadal variation of summer precipitation pattern over eastern China and its relationship with the North Pacific SST [J]. Chinese J. Atmos. Sci. (in Chinese), 33: 835 – 845.

Ding Yihui. 2007. The variability of the Asian summer monsoon [J]. J. Metoer. Soc. Japan, 85B: 21 – 54.

Ding Yihui, Wang Zunya, Sun Ying. 2008. Interdecadal variation of the summer precipitation in East China and its association with decreasing Asian summer monsoon. Part I: Observed evidences [J]. Int. J. Climatol., 28: 1139 – 1161.

Huang Gang, Liu Yong, Huang Ronghui. 2011. The interannual variability of summer rainfall in the arid and semiarid regions of northern China and its association with the Northern Hemisphere circumglobal teleconnection [J]. Adv. Atmos. Sci., 28: 257 – 268.

Huang Ronghui, Li Weijing. 1987. Influence of the heat source anomaly over the tropical western Pacific on the subtropical high over East Asia [C]. Proceedings of International Conference on the General Circulation of East Asia. Chengdu, 10 – 15, April, 1987, 40 – 51.

黄荣辉, 李维京. 1988. 夏季热带西太平洋上空的热源异常对东亚上空副热带高压的影响及其物理机制 [J]. 大气科学, 12 (特刊): 107 – 106. Huang Ronghui, Li Weijing. 1988. Influence of heat source anomaly over the western tropical Pacific on the subtropical high over East Asia and its physical mechanism [J]. Chinese J. Atmos. Sci. (Scientia Atmospherica Sinica) (in Chinese), 29 (Special Issue): 20 – 36.

黄荣辉, 张振洲, 黄刚, 等. 1998. 夏季东亚季风区水汽输送特征及其与南亚季风区水汽输送的差别 [J]. 大气科学, 22: 460 – 469. Huang Ronghui, Zhang Zhenzhou, Huang Gang, et al. 1998. Characteristics of the water vapor transport in East Asian monsoon region and its difference from that in South Asian monsoon region in summer [J]. Chinese J. Atmos. Sci. (Scientia Atmospherica Sinica) (in Chinese), 22: 460 – 469.

黄荣辉, 徐予红, 周连童. 1999. 我国夏季降水的年代际变化及华北干旱化趋势 [J]. 高原气象, 18: 465 – 476. Huang Ronghui, Xu Yuhong, Zhou Liantong. 1999. The interdecadal variation of summer precipitation in China and the drought trend in North China [J]. Plateau Meteor. (in Chinese), 18: 465 – 476.

Huang Ronghui, Huang Gang, Wei Zhigang. 2004. Climate variations of the summer monsoon over China [M]//Chang C P. East Asian Monsoon. World Scientific Publishing Co. Pte. Ltd., 213 – 270.

黄荣辉, 陈际龙, 黄刚, 等. 2006. 中国东部夏季降水的准两年周期振荡及其成因 [J]. 大气科学, 30: 545 – 560. Huang Ronghui, Chen Jilong, Huang Gang, et al. 2006. The quasi-biennial oscillation of summer monsoon rainfall in China and its cause [J]. Chinese J. Atmos. Sci. (in Chinese), 30: 545 – 560.

黄荣辉, 蔡榕硕, 陈际龙, 等. 2006. 我国旱涝气候灾害的年代际变化及其与东亚气候系统变化的关系 [J]. 大气科学, 30: 730 – 743. Huang Ronghui, Cai Rongshuo, Chen Jilong, et al. 2006. Interdecadal variations of the drought and flooding disasters in China and their association with the East Asian climate system [J]. Chinese J. Atmos. Sci. (in Chinese), 30: 730 – 743.

Huang Ronghui, Chen Jilong, Huang Gang. 2007. Characteristics and variations of the East Asian monsoon system and its impacts on climate disasters in China [J]. Adv. Atmos. Sci., 24: 993 – 1023.

黄荣辉, 陈际龙. 2010. 我国东、西部夏季水汽输送特征及其差异 [J]. 大气科学, 34 (6): 1035 – 1045. Huang Ronghui, Chen Jilong. 2009. Characteristics of the summertime water vapor transports over the eastern part of China and those over the western part of China and their difference [J]. Chinese J. Atmos. Sci. (in Chinese), 34 (6): 1035 – 1045.

Kwon Min-Ho, Jhun Jong-Ghap, Ha Kyung-Ja. 2007. Decadal change in east Asian summer monsoon circulation in the mid-1990s [J]. Geophys. Res. Lett., 34, L21706, doi: 10. 1029/2007GL031977.

缪锦海, Lau K M. 1990. 东亚季风降水的年际变化 [J]. 应用气象学报, 1: 377 – 382. Miao J H, Lau K M. 1990. Interannual variability of the East Asian monsoon rainfall [J]. Quart. J. Appl. Meteor. (in Chinese), 1: 377 – 382.

Nitta Ts. 1987. Convective activities in the tropical western Pacific and their impact on the Northern Hemisphere summer circulation

[J]. J. Meteor. Soc. Japan, 64: 373-400.

Tao Shiyan, Chen Longsun. 1987. A review of recent research on the East Asian summer monsoon in China [M] // Chang C P, Krishnamurti T N. Monsoon Meteorology. Oxford University Press, 60-92.

Tian S F, Yasunari T. 1992. Time and space structure of interannual variations in summer rainfall over China [J]. J. Meteor. Soc. Japan, 70: 585-596.

涂长望，黄仕松. 1944. 夏季风进退 [J]. 气象杂志, 18: 1-20. Tu Changwang, Huang Shisong. 1944. The advance and retreat of the summer monsoon [J]. Meteor. Mag. (in Chinese), 18: 1-20.

Uppala S M, Kallberg P W, Simmons A J, et al. 2005. The ERA-40 reanalysis [J]. Quart. J. Roy. Meteor. Soc., 131: 2961-3012.

Wallace J M, Gutzler D S. 1981. Teleconnection in the geopotential height field during the Northern Hemisphere winter [J]. Mon. Wea. Rev., 100: 748-812.

张庆云，吕俊梅，杨莲梅，等. 2007. 夏季中国降水型的年代际变化与大气内部动力过程及外强迫因子关系 [J]. 大气科学, 31: 1290-1300. Zhang Qingyun, Lü Junmei, Yang Lianmei, et al. 2007. The interdecadal variation of precipitation pattern over China during summer and its relationship with the atmospheric internal dynamic processes and extra-forcing factors [J]. Chinese J. Atmos. Sci. (in Chinese), 31: 1290-1300.

Zhang Renhe. 2001. Relations of water vapor transport from Indian monsoon with that over East Asia and the summer rainfall in China [J]. Adv. Atmos. Sci., 18: 1005-1017.

张人禾，武炳义，赵平，等. 2008. 中国东部夏季气候20世纪80年代后期的年代际转型及其可能成因 [J]. 气象学报, 66: 697-706. Zhang Renhe, Wu Bingyi, Zhao Ping, et al. 2008. The decadal shift of the summer climate in the late 1980s over East China and its possible causes [J]. Acta Meteorologica Sinica (in Chinese), 66: 697-706.

周连童，黄荣辉. 2003. 关于我国夏季气候年代际变化特征及其可能成因的研究 [J]. 气候与环境研究, 8: 274-290. Zhou Liantong, Huang Ronghui. 2003. Research on the characteristics of interdecadal variability of summer climate in China and its possible cause [J]. Climatic and Environmental Research (in Chinese), 8: 274-290.

Zhou Xiaoxia, Ding Yihui, Wang Panxing. 2010. Moisrure transport in the Asian summer monsoon region and its relationship with summer precipitation in China [J]. Acta Meteorologica Sinica, 24: 31-42.

竺可桢. 1934. 东南季风与中国之雨量 [J]. 地理学报, 1: 1-27. Zhu Kezhen. 1934. Southeast monsoon and rainfall in China [J]. Acta Geographica Sinica (in Chinese), 1: 1-27.

Characteristics, Processes, and Causes of the Spatio-temporal Variations of the East Asian Monsoon System*

Huang Ronghui*(黄荣辉), Chen Jilong(陈际龙),
Wang Lin(王林) and Lin Zhongda(林中达)

*Center for Monsoon System Research, Institute of Atmospheric Physics,
Chinese Academy of Sciences, Beijing 100190*

ABSTRACT

Recent advances in the study of the characteristics, processes, and causes of spatio-temporal variabilities of the East Asian monsoon (EAM) system are reviewed in this paper. The understanding of the EAM system has improved in many aspects: the basic characteristics of horizontal and vertical structures, the annual cycle of the East Asian summer monsoon (EASM) system and the East Asian winter monsoon (EAWM) system, the characteristics of the spatio-temporal variabilities of the EASM system and the EAWM system, and especially the multiple modes of the EAM system and their spatio-temporal variabilities. Some new results have also been achieved in understanding the atmosphere–ocean interaction and atmosphere–land interaction processes that affect the variability of the EAM system. Based on recent studies, the EAM system can be seen as more than a circulation system, it can be viewed as an atmosphere–ocean–land coupled system, namely, the EAM climate system. In addition, further progress has been made in diagnosing the internal physical mechanisms of EAM climate system variability, especially regarding the characteristics and properties of the East Asia-Pacific (EAP) teleconnection over East Asia and the North Pacific, the "Silk Road" teleconnection along the westerly jet stream in the upper troposphere over the Asian continent, and the dynamical effects of quasi-stationary planetary wave activity on EAM system variability. At the end of the paper, some scientific problems regarding understanding the EAM system variability are proposed for further study.

1. Introduction

The East Asian monsoon (EAM) system features strong southerly wind in summer, [i.e., the East Asian summer monsoon (EASM)], and strong northerly wind in winter, [i.e., the East Asian winter monsoon (EAWM)]. Generally, the southerly wind transports water vapor over eastern China, Korea, and Japan in summer, but in winter the dry northwesterly wind prevails over North China, Central China, Northeast China, Korea, and Japan, and northeasterly wind prevails over East China, South China, and the Indo-China Peninsula. The strong summer monsoon flow transports a large amount of water vapor into East Asia, causing severe climate disasters such as floods in South China, and droughts in the Yangtze-Huaihe River valley, and extremely hot summers in eastern China, Korea, and Japan (Huang et al., 1998a, 2008; Huang and Zhou, 2002). In contrast, the strong winter monsoon flow can bring a large amount of cold and dry air into East Asia and can cause climate disasters such as low temperatures and severe snow storms in Northwest China, Northeast China, North China, Korea, and Japan in winter and severe dust storms in Northwest China and North China in spring (Chen et al., 2000; Huang et al., 2007b; Gu et al., 2008; Wang and Chen, 2010b). In addition, the strong EAWM can trigger strong convective activities over the Maritime Continent around Borneo and Indonesia (Chang et al., 1979; Lau and Chang, 1987; Wang and Chen,

---

* 本文原载于: Advances in Atmospheric Sciences, Vol.29, No.5, 910-942, 2012.

2010a). Moreover, many studies have shown that the EAM system plays an important role in global climate variability, especially in climate variability over East Asia (e.g., Tao and Chen, 1987; Chen et al., 1991; Ding, 1994, Chang et al., 2000a, b; Huang et al., 2003, 2004a, 2007a).

Climate variability in China is mainly influenced by the EAM system (Zhu, 1934; Tu and Huang, 1944; Tao and Chen, 1987). The interannual and interdecadal variabilities of EAM system are quite significant. Therefore, climatic disasters such as droughts and floods in summer and severe cold surges, freezing rain, and low temperature in winter frequently occur in China (Huang and Zhou, 2002; Huang, 2006; Gu et al., 2008; Huang et al., 2011c, 2012b; Barriopedro et al., 2012). Especially since the 1980s, severe climatic disasters have caused huge damage to agricultural and industrial production across China. The economic losses due to droughts and floods can reach more than 200 billon CNY (∼US\$30 billion) each year (Huang et al., 1999; Huang and Zhou, 2002). For example, the particularly severe floods that occurred in the Yangtze River basin and the Songhuajiang River and Nenjiang River valleys in the summer of 1998 caused economic losses as high as 260 billion CNY (∼US\$38 billion; Huang et al., 1998b). From the winter of 2009 to the summer of 2010, China experienced many severe climatic disasters: the severe low temperatures and snowstorms in Northeast China, Northwest China, and North China from November 2009 to January 2010 (Wang and Chen, 2010b), the particularly severe drought in Southwest China from the autumn of 2009 to the spring of 2010 (Barriopedro et al., 2012; Huang et al., 2012b), severe floods in South China and the Yangtze River basin during May–July, particularly severe floods in the middle and eastern Northeast China from late July to early August and in late August, and severe hot summer in Northwest China, South China, the Yangtze River basin and North China. The economic losses caused by these severe climatic disasters far exceeded those in 1998. Moreover, the persistent droughts in North China and southern Northeast China since the late 1990s have not only caused huge losses in agriculture and industry but also have seriously affected the water resources and ecological environment in these regions. For example, a ∼20% reduction was observed in the nationwide hydroelectrical production due to the depleted water reservoirs during the 2009–2010 Southwest China droughts (Barriopedro et al., 2012).

These climatic disasters are closely associated with the spatio-temporal variabilities of the EAM system (Huang, 2006). EAM system variability is influenced not only by the internal dynamical and thermodynamical processes of the atmosphere but also by the interactions among various spheres of the atmosphere–ocean–land coupled system. Webster et al. (1998) proposed that the Asian–Australian monsoon could be viewed as an atmosphere–ocean–land coupled system. Similarly, the EAM system can also be regarded as a atmosphere–ocean–land coupled system. We named it the EAM climate system to distinguish it from the general circulation system proposed by Tao and Chen (1987) (see also Huang et al., 2004a, 2007a). The EAM climate system includes EAM circulation, the western Pacific subtropical high, the mid- and high-latitude disturbances in the atmosphere, thermal states of the western Pacific warm pool and associated convective activity around the Philippines, the thermal state of the tropical Indian Ocean, the ENSO cycle in the tropical Pacific in the Ocean, the dynamical and thermal effects over the Tibetan Plateau, the land surface processes in the arid and semiarid areas of Northwest China, and snow cover over the Eurasian continent and the Tibetan Plateau. The characteristics of the spatio-temporal variabilities of this system and their impacts on climatic disasters in China have been analyzed further in recent years. The internal and external physical processes that influence these variabilities have also been discussed in more detail recently.

In this study, we summarized the advances in the studies on the characteristics, causes, and processes of the spatio-temporal variability of the EAM system, especially of the spatio-temporal variations of the EASM and EAWM systems, the impacts and processes of the atmosphere–ocean–land coupled system on EAM system variability and the internal dynamic and thermodynamic mechanisms of EAM system variability. The review mainly focuses on research progress in China during the past five years.

2. Climatological characteristics of the EASM system

Because there are close associations among the South Asian monsoon (SAM), the East Asian monsoon (EAM), and the North Australian monsoon (NAM), some scholars have considered them as three subsystems of the Asian–Australian monsoon system (e.g., Webster et al., 1998). However, the characteristics of EAM system are different from those of SAM and NAM systems. According to the study by Tao and Chen (1987), the EASM system has both tropical and subtropical properties because it is influenced by the western Pacific subtropical high and the disturbances over middle latitudes in addition to tropical systems. In contrast, the South Asian summer monsoon (SASM) is only a tropical monsoon. Therefore, the characteristics of the spatio-temporal variation of the EASM system may be different from those of the South Asian summer monsoon (SASM) system.

2.1 *Characteristics of circulation structure in the EASM system*

Tao and Chen (1985) showed that the main components of the EASM system include the Indian SW monsoon flow, the Australian cold anticyclone, the cross-equatorial flow along the east to 100°E, the monsoon trough (or ITCZ) over the South China Sea (SCS) and the tropical western Pacific, the western Pacific subtropical high and the tropical easterly flow, the mei-yu (or baiu in Japan, or changma in Korea) frontal zones, and the disturbances over mid-latitudes. Their results suggest that the EASM system is a relatively independent monsoon circulation system, although it is linked to the SASM.

Recently, Chen and Huang (2006) analyzed the climatological characteristics of wind structure of the EASM system over the area (0°–45°N, 100°–140°E) and the SASM system over the area (0°–25°N, 60°–100°E) in boreal summer, respectively. Their results show that the vertical structure of zonal flow in the EASM region includes a vertical easterly shear in the region south of 25°N, such as the SCS and the tropical western Pacific, and the vertical westerly shear in the subtropical region to the north of 25°N, such as the mainland of China, Korea, and Japan (Fig. 1a). In contrast, the SASM system is purely a tropical monsoon, with strong zonal flow and vertical easterly shear, i.e., low-level westerly wind and high-level easterly wind (Fig. 1b).

Moreover, as shown in Fig. 2a, the EASM system is composed of tropical and subtropical summer monsoons with significant meridional flow and verti-

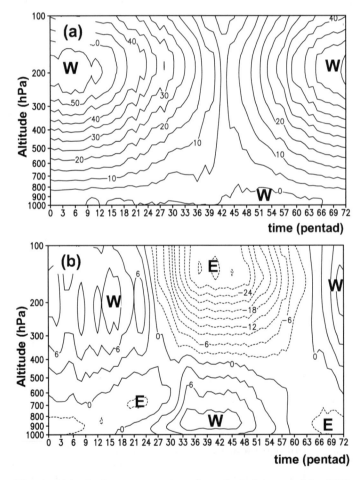

Fig. 1. Altitude-time cross section of zonal wind averaged for 1979–2003 over (a) East Asia (20°–45°N, 100°–140°E) and (b) South Asia (0°–25°N, 60°–100°E) based on NCEP/NCAR reanalysis data (Kalnay et al., 1996). Unit: m s^{-1}. The solid and dashed lines in panels (a) and (b) indicate the westerly and easterly winds, respectively.

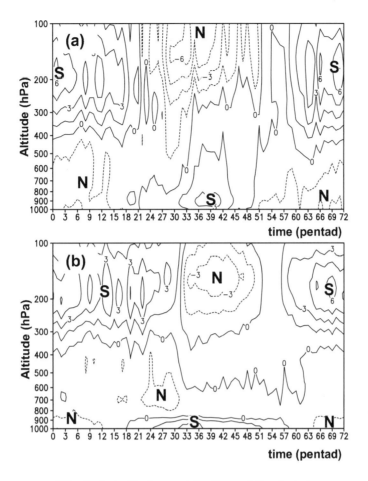

Fig. 2. As in Fig. 1, except for the meridional wind.

cal northerly shear, i.e., low-level southerly wind and high-level northerly wind. Compared with the meridional component of wind field in the SASM system (Fig. 2b), the low-level southerly wind and the upper-level northerly wind in the EASM region (Fig. 2a) are stronger than those in the SASM system.

The differences between the annual wind cycle in the EAM system and in the SAM system can also be seen from the altitude–time cross sections of zonal and meridional winds over East Asia and South Asia (Figs. 1 and 2). The strong westerly wind prevails in the lower troposphere below 500 hPa, and the strong easterly wind prevails in the upper troposphere (above 500 hPa) from early June over South Asia (Fig. 1b). However, from early October, the westerly wind becomes easterly in the lower troposphere below 700 hPa, while the easterly wind becomes westerly in the upper troposphere above 500 hPa over this region. Therefore, in the SAM region, the annual wind cycle between summer and winter monsoons is obvious according to the zonal wind field. However, a comparison of Fig. 1a with Fig. 1b shows that the seasonal reversal of zonal wind in the troposphere over East Asia is not as significant as that in the SAM region. As shown in Fig. 2a, the seasonal reversal of meridional wind obviously occurs in both the lower troposphere and the upper troposphere over East Asia in early June and mid-September, respectively.

From these analyses, the annual cycle between summer and winter monsoons mainly appears in the meridional component of wind field in the EAM region, different from that in the SAM system.

2.2 Characteristics of water vapor transports in the EASM region

Recently, Chen and Huang (2007) analyzed water vapor transports over the EASM and the SASM regions using ERA-40 reanalysis data for the period 1979–2002. Their results showed that the characteristic of water vapor transports in the EASM region is greatly different from that in the SASM region. The climatological EASM rainfall is mainly influenced by

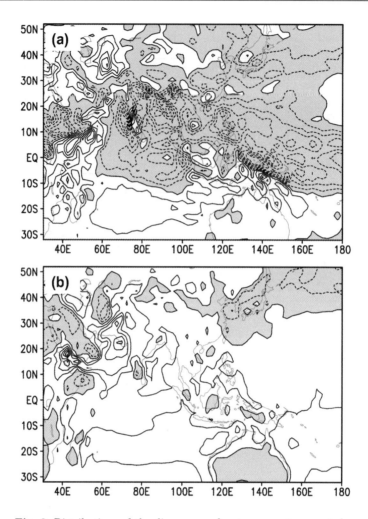

Fig. 3. Distributions of the divergence of water vapor transport due to (a) the divergence of wind field and (b) the moisture advection in the Asian monsoon region during boreal summer based on ERA-40 reanalysis data (Uppala et al., 2005). Units: mm d^{-1}. The solid and dashed lines in panels (a) and (b) indicate divergence and convergence, respectively. Areas of convergence are shaded.

three branches of water vapor transport coming from the Bay of Bengal, the South China Sea, and the tropical western Pacific. Their study showed that in the SASM region, the zonal transport of water vapor is dominant and the meridional transport is relatively smaller, but the meridional transport of water vapor is dominant in the EASM region. As shown in Figs. 3a and 3b, the convergence of water vapor transports that are closely associated with monsoon rainfall are caused by moisture advection and the convergence of wind fields in the EASM region, but those are mainly due to the convergence of wind fields in the SASM region. This result is in good agreement with the analysis by Huang et al. (1998a).

Huang and Chen (2010) analyzed the differences between the summertime water vapor transports in the EASM region and those in the arid and semiarid regions of Northwest China using the daily ERA-40 reanalysis data. The results show some obvious differences between the summertime water vapor transport in the EASM region and that in arid and semiarid regions. Because a large amount of water vapor is transported by the Asian summer monsoon flow from the Bay of Bengal, the South China Sea, and the tropical western Pacific into the EASM region, the meridional water vapor transport fluxes are larger than the zonal water vapor transport fluxes in South China and the Yangtze River valley. In contrast, the summertime zonal water vapor transport fluxes are larger than the meridional water vapor transport fluxes in the arid and semiarid region of Northwest China due to the strong mid-latitude westerlies. The water vapor divergence in these regions mainly depends on moisture advection. Moreover, either zonal or meridional water vapor transport fluxes in the arid and semiarid regions are approximately 10% of those in the EASM region, which causes very small amounts of summertime rainfall in Northwest China, resulting in drought conditions.

2.3 Characteristics of rainfall cloud system in the EASM system

Recently, Du et al. (2011) reported that the rainfall cloud system in the EASM region is different from that in the SASM region, due to the differences of the circulation structure and water vapor transport between the EASM region and the SASM regions. They analyzed the characteristics of spatio-temporal distributions of convective rainfall and stratiform rainfall from Tropical Rainfall Measuring Mission (TRMM) precipitation data for 12 years. Their results show that the spatial distributions of convective rainfall and stratiform rainfall mainly exhibit a variation with latitude in the EASM region (Fig. 4). In the subtropical monsoon region to the north of 25°N, the ratio of stratiform rainfall with respect to the total rainfall is ∼60% in summer. This ratio increases in proportion to latitude north of 25°N, and it varies with regions. In the low latitudes south of 25°N, in contrast, summertime rainfall is mainly due to convective cloud systems, contributing ∼50% of the total annual rainfall, and this ratio is constant in this region. Thus, summertime rainfall cloud systems in the EASM system are mainly due to stratiform cloud systems to the north of 25°N and to convective cloud systems south of 25°N. In contrast, summertime rainfall cloud systems in the SASM system are mainly due to convective cloud systems. Fu et al. (2003), and Liu and Fu (2010) analyzed the seasonal characteristics of convective rainfall and stratiform rainfall in Asian monsoon region and the climatological characteristics of convective rainfall and stratiform rainfall in summer over South China, using the TRMM data for the period 1998–2007. Their results also show that stratiform rainfall amount is comparative to convective rainfall amount in South China. These studies can explain that the summertime rainfall cloud systems in the EASM system are a mixing of stratiform cloud systems with convective cloud systems. This mixture may make it difficult to numerically model cumulus parameterization in the EASM system (e.g., Cheng et al., 1998).

These results make it clear that the EASM system is different from the SASM system and is a relatively independent monsoon system, although it is also in-

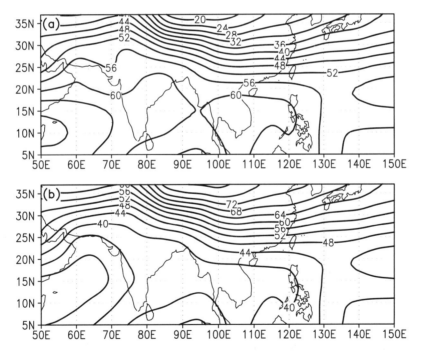

Fig. 4. Distributions of the average summertime rainfall (percentage) for 1988–2009 due to (a) convective rainfall and (b) stratiform rainfall based on TRMM data.

fluenced by the SASM system.

3. Characteristics of the spatio-temporal variabilities of the EASM system

EASM system variability is influenced not only by the SASM and the western Pacific subtropical high (e.g., Tao and Chen, 1987; Huang and Sun, 1992) but also by mid- and high-latitude disturbances (e.g., Tao and Chen, 1987); therefore, the spatio-temporal variations of the EASM system are significant and very complex. These variations have an important effect on the spatio-temporal variations of climate disasters in China (e.g., Huang et al., 2007a, 2008, 2011c). In this section, we focus on the spatio-temporal variabilities of EASM system, especially summer monsoon rainfall in East Asia.

3.1 Interannual variations of onset and northward advances of the EASM system

The early or late onset of the ASM has an important impact on the interannual variability of summer monsoon rainfall in East Asia (Huang et al., 2005, 2006c). To investigate the interannual variability of onset date and process of the SCSM, we defined an index for measuring SCSM onset. Generally, the appearance of strong convective activity and the southwesterly flow over the SCS signals the onset of the SCSM. However, many definitions of the SCSM onset have been devised (e.g., Wang et al., 2004). For example, Tao and Chen (1987) suggested that the earliest onset of Asian summer monsoon (ASM) occurs over the SCS and the Indo-China Peninsula; it is generally called the South China Sea monsoon (SCSM) in China. Ding and He (2006) proposed that the earliest onset of the ASM occurs over the tropical eastern Indian Ocean. Compared with other definitions of SCSM onset, we also preferred the definition proposed by Liang and Wu (2002) (e.g., Huang et al., 2005, 2006c). Huang et al. (2005, 2006c) pointed out that the SCSM onset is closely associated with the thermal state of the tropical western Pacific and the convective activity around the Philippines. To explain the impact of convective activity over the tropical western Pacific on the SCSM onset, the relationship between the SCSM onset date and convective activity around the Philippines (i..e,, $10°-20°N$, $110°-140°E$) in spring was analyzed by Huang et al. (2006c) using the observation data of high cloud amount (HCA) obtained by geostationary meteorological satellite (GMS) satellite. An out-of-phase relationship between the SCSM onset date and the HCA around the Philippines in spring is observed (Fig. 5). The correlation coefficient between them reaches -0.76, which exceeds the 99% confidence level. Therefore, the convective activity around the Philippines has a great influence on the onset of SCSM. In a spring with strong convective activity around the Philippines, which generally occurs during a warming

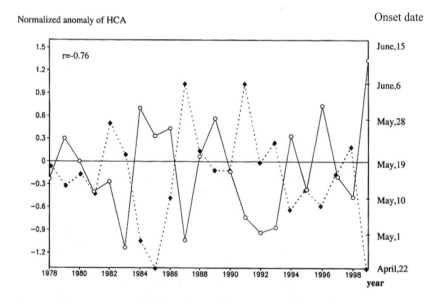

Fig. 5. Interannual variations of normalized high cloud amount (HCA) averaged for the area around the Philippines ($10°-20°N$, $110°-140°E$) in spring (solid line) and the SCSM onset date (dashed line) based on observed HCA data from Monthly Report on Climate System, JMA.

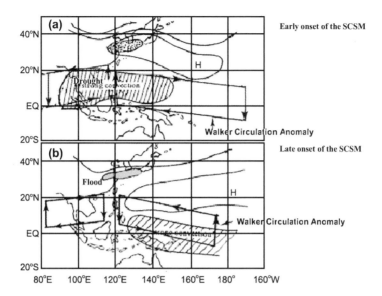

Fig. 6. Schematic map of the relationships among the thermal states of the tropical western Pacific (TWP, 0°–14°N, 130°–150°E) in spring, the convective activity around the Philippines, the western Pacific subtropical high, the onset of SCSM and the summer monsoon rainfall in East Asia in (a) the warming and (b) the cooling state of the TWP.

period in the tropical western Pacific, the SCSM onset is early. On the other hand, the SCSM onset is late in a spring with weak convective activity around the Philippines, which is generally found in a cooling period of the tropical western Pacific. The interannual variability of SCSM onset date is very large and is closely associated with the thermal states of the tropical western Pacific, especially the convective activity around the Philippines in spring.

Huang et al. (2006c) also analyzed the influence of the thermal state of the tropical western Pacific on the SCSM onset process, Their results show that when the tropical western Pacific is in a warming state in spring, the twin anomalous cyclones can appear early over the Bay of Bengal and Sumatra before the onset of the SCSM, due to the eastward shift of the western Pacific subtropical high. Thus, the southwesterly flow and strong convective activities are intensified over Sumatra, the Indo-China Peninsula, and the SCS in mid-May. This leads to an early onset of the SCSM (Fig. 6a). On the other hand, when the tropical western Pacific is in a cooling state in spring, the twin anomalous anticyclones are located over the equatorial eastern Indian Ocean and Sumatra from late April to mid-May due to the westward shifts of the western Pacific subtropical high. Thus, the southwesterly flow and convective activities cannot be intensified early over the Indo-China Peninsula and the SCS. Only when the western Pacific subtropical high moves eastward, the weak trough located over the Bay of Bengal can be intensified and becomes a strong trough. As a result, the strong southwesterly wind and convective activities generally intensify over the Indo-China Peninsula and the SCS in late May. This leads to late onset of the SCSM (Fig. 6b). In addition, some studies suggest that the onset of SCSM is also associated with the atmospheric circulation anomalies over middle and high latitudes, and with 30–60-day low frequency oscillation over the tropics (He et al., 2003; Wen et al., 2006).

Following the onset of the SCSM, the monsoon moves northward over East Asia. Huang and Sun (1992) discussed well the dependence of the northward advance of the EASM system on the thermal state of the tropical western Pacific. Huang et al. (2005) further investigated the interannual variations of the intraseasonal variability, i.e., the northward advances and southward retreat of EASM system. Their results show that the northward advances of the EASM system after its onset over the SCS are also influenced by the thermal state of the tropical western Pacific in summer, and that there are close relationships among the thermal states of the tropical western Pacific, the convective activity around the Philippines, the western Pacific subtropical high, and the summer monsoon rainfall in East Asia. As shown in Fig. 6a, when the thermal state in the tropical western Pacific is warming (i.e., the warm sea water is accumulated in the

West Pacific warm pool in summer), convective activity is intensified from the Indo-China Peninsula to the east of the Philippines. In this case, the western Pacific subtropical high shifts unusually northward, and the summer monsoon rainfall may be below normal in the Yangtze River and Huaihe River basins of China, South Korea, and Japan. Drought may occur in these regions. On the other hand, when the thermal state in the tropical western Pacific is in a cooling state, the convective activity is weak around the Philippines in summer, in this case, the western Pacific subtropical high may shift southward (Fig. 6b). The summer monsoon rainfall may be above normal in the Yangtze-Huaihe River valley of China, South Korea, and Southwest Japan. Therefore, following a spring with late onset of the SCSM, severe floods may occur in the Yangtze-River valley, and droughts will be caused in North China in summer, such as in the summers of 1998 and 2010.

These results suggest that the thermal state of the tropical western Pacific, especially the oceanic heating content (OHC) anomaly in the Niño west region (i.e., 0°–14°N, 130°–150°E) in spring (March–May) and summer can be considered as a physical factor affecting the SCSM onset and summertime droughts and floods in the Yangtze-Huaihe River valley, South Korea and Japan. A conceptual map of seasonal prediction for summertime droughts and floods in eastern China is shown in Fig. 6.

3.2 Characteristics of multi-modes in the spatio-temporal variabilities of the EASM system

The multi-modal characteristics of the spatio-temporal variabilities of the EASM system are shown not only in monsoon rainfall but also in the water vapor transport in this system. Huang et al. (2006b, 2007a) analyzed the spatio-temporal variabilities of the summertime (June–August) precipitation in eastern China using the observed precipitation data at 160 stations in China. They proposed that the spatial distribution of the EASM system variability is characterized by multiple modes. In the following section, the characteristics of multi-modes of the spatio-temporal variabilities of the EASM system are discussed.

Recently, Huang et al. (2011c) further analyzed the leading modes of summertime precipitation anomalies in eastern China using precipitation data of 516 observational stations of China using the empirical orthogonal function (EOF) method. The result also shows that two leading modes characterize the spatio-temporal variations of the summertime precipitation anomalies in eastern China. The first mode exhibits a meridional tripole pattern in spatial distribution (Fig. 7a) with typical quasi-biennial period (Fig. 7b). In addition, a clear interdecadal change can be seen around the 1980s. As shown in Fig. 8b, the second mode exhibits obvious interdecadal variability, characteristic of the meridional dipole pattern in spatial distribution shown in Fig. 8a. Therefore, the spatio-temporal variabilities of EASM system are characterized by multiple modes.

The interannual variability of EASM system with the meridional tripole pattern in spatial distribution has a significant impact on the distributions of summertime droughts and floods in eastern China. For example, the summertime monsoon rainfall amounts in the summers of 1980, 1983, 1987, and 1998 were above normal in the Yangtze-Huaihe River valley and below normal in South China and North China. During these summers, severe floods were recorded in the Yangtze-Huaihe River valley, while droughts were recorded in South China and North China. In contrast, the summertime monsoon rainfall amounts in the summers of 1976 and 1994 were below normal in the Yangtze-Huaihe River valley and above normal in South China and North China. Thus, severe droughts were recorded in the Yangtze-Huaihe River valley, while floods occurred in South China in these two summers.

These studies show that quasi-biennial oscillation appears not only in the interannual variability of the SASM system (e.g., Yasunari and Suppiah, 1988) but also in the interannual variability of the EASM system. Therefore, as explained by Ding (2007), the quasi-biennial oscillation may be a leading mode of the Asian summer monsoon, including the EASM and SASM systems.

The multi-modal characteristics of the spatio-temporal variability of the EASM system are shown not only in the spatio-temporal variability of monsoon rainfall but also in water vapor transport associated with the EASM system. Recently, Huang et al. (2011c) analyzed further the spatio-temporal variability of water vapor transport fluxes in the EASM system using the moisture and wind fields of ERA-40 reanalysis (6-h intervals, figures not shown). According to their analysis, the meridional tripole pattern and meridional dipole pattern are also the first two leading modes in the zonal water vapor transport fluxes in the EASM region. Obvious interannual variability can also be seen with quasi-biennial rhythm and interdecadal change in the corresponding time series of the EOF1 of zonal water vapor transports in the EASM region. From the late 1950s to the early 1970s, the northward water vapor transport was strong, but it became weaker from the early 1970s to the early 1990s, and it again became strong from the early 1990s to the late 1990s. However, from the analysis by Zhou et al. (2010) using

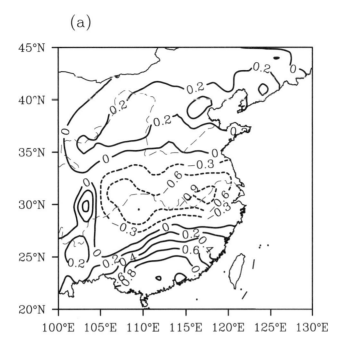

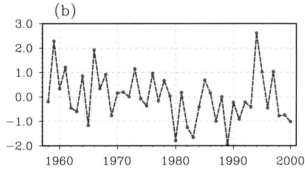

Fig. 7. (a). Spatial distribution and corresponding time coefficient series of the first EOF mode of summertime (JJA) rainfall in eastern China for 1958–2000 based on observed precipitation data at 756 China stations. The solid and dashed lines in (a) indicate positive and negative values, respectively. EOF1 explains 15.3% of the total variance.

NCEP/NCAR reanalysis data, it can be seen that the corresponding time coefficients of EOF1 of summertime water vapor transport field exhibited a decreasing trend from 1951 to 2005. Huang et al. (2011c) also showed that the multiple modes of the spatio-temporal variability of monsoon rainfall in eastern China are closely associated with those of water vapor transport over East Asia.

3.3 Interdecadal variability of the EASM system

The EASM system has not only obvious interannual variability but also significant interdecadal variability, especially in monsoon rainfall over East Asia.

Huang et al. (1999) analyzed the interdecadal variations of summer monsoon rainfall and reported that the summer monsoon rainfall in North China obviously decreased from the late 1970s to the early 1990s, causing prolonged severe droughts in this region. They also described an opposite phenomenon that appeared in the Yangtze River and the Huaihe River valleys and Northwest China. In these regions, summer rainfall obviously increased from the late 1970s. Chen et al. (2004) systematically discussed the characteristics of the climate change in China during the last 80 years. Their results also show that the interdecadal fluctuations of summer (June–August) monsoon rainfall are more obvious than surface air temperature in East

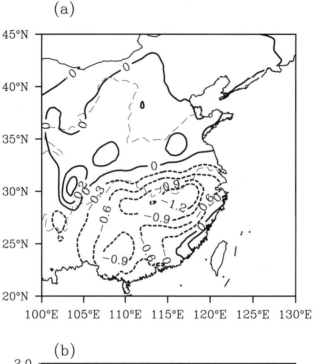
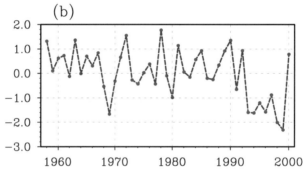

Fig. 8. As in Fig. 7, except for the EOF2, which explains 13.2% of the total variance.

Asia.

Recently, many studies have shown that the EASM rainfall underwent significant interdecadal variations not only in the late 1970s but also in the early 1990s (e.g., Kwon et al., 2007; Ding et al., 2008; Deng et al., 2009; Wu et al., 2010; Huang et al., 2011c). This interdecadal oscillation may be also presented in the corresponding time coefficient series of the EOF1 and EOF2 of summer rainfall in eastern China (Figs. 7b, 8b). The oscillation exhibits a meridional tripole pattern or a meridional dipole pattern in the spatial distribution (Figs. 7a and 8a). As described by Huang et al. (2011c), these two leading modes of monsoon rainfall variability in the EASM region exhibit significant interdecadal variability. This variability can be clearly seen from the interdecadal variation of summer monsoon rainfall anomalies in eastern China during the periods 1958–1997, 1978–1992, 1993–1999, and 1999–2009 (figures not shown), and latitude-time cross section of summer (June–August) rainfall anomalies in eastern China (averaged over 110°–120°E, Fig. 9). As shown in Fig. 9, the distributions of summertime monsoon rainfall anomalies in eastern China exhibited a "+, −, +" meridional tripole pattern from the south to the north during 1958–1977 period. During 1978–1992, opposite anomaly distributions to those during 1958–1977 appeared in this region. But during 1993–1998, because the role of the second leading mode of the summertime monsoon rainfall variability over eastern China was intensified (Fig. 9), the anomaly distributions of summertime rainfall appeared in a meridional tripole pattern combination of "+, −, +" and a meridional dipole pattern "+, −" from the south to the north, which caused the obvious increase of summer monsoon rainfall in South China.

The results analyzed by Huang et al. (2011c) also

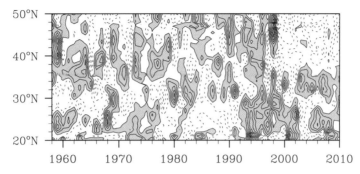

Fig. 9. Latitude-time cross section of summertime (JJA) rainfall anomalies (percentage) along 115°E (average for 110°–120°E) in eastern China based on precipitation data at 756 observation stations. The solid and dashed lines indicate positive and negative anomalies, respectively. Positive anomalies are shaded.

show that the interdecadal variability of leading modes of summertime monsoon rainfall variability in eastern China is closely associated with the interdecadal variability of water vapor transport fluxes by summer monsoon flow over East Asia. Furthermore, their results show that the interdecadal variability of water vapor transport fluxes over East Asia is associated with not only the interdecadal variation of the East Asia/Pacific (EAP) teleconnection of summer circulation anomalies over East Asia (e.g., Nitta, 1987; Huang and Li, 1987, 1988) but also the interdecadal variation of the Eurasian (EU) teleconnection of summer circulation anomalies over middle and high latitudes (e.g., Wallace and Gutzler, 1981). Li et al. (2011) also pointed out that there is a long-term change in summer water vapor transport over South China during recent decades.

However, according to the analysis of Huang et al. (2011c), a interdecadal variation of summer water vapor transport over East Asia, which is closely associated with the summer monsoon circulation, occurred in the late 1980s. This result agrees with those obtained from monsoon circulation by Wu et al. (2009) and Zhang et al. (2008). Thus, there is small difference between the interdecadal variability of summer monsoon rainfall in eastern China (e.g., Kwon et al., 2007; Ding et al., 2008; Deng et al., 2009; Huang et al., 2011c) and the interdecadal variability of summer monsoon circulation over East Asia (e.g., Zhang et al. 2008; Wu et al., 2009). Summer rainfall variability depends not only on the moisture advection by the monsoon flow but also on the convergence of the monsoon circulation.

The studies mentioned here mainly focused on the interdecadal variability of summer climate over East Asia occurred in about 1978 and 1992, respectively (e.g., Huang et al., 1999, 2004a, 2006a, 2011c; Zhang et al., 2007; Kwon et al., 2007; Ding et al., 2008; Deng et al., 2009). Recently, Huang et al. (2011b) described a climate jump in the late 1990s in the summer monsoon rainfall in China, especially in North China, Northeast China, and Northwest China. This climate jump was characterized by the decrease of summer rainfall in North China, Northeast China and the eastern part of Northwest China and the increase in the Huaihe River valley (Fig. 9). Thus, the rainfall anomaly distributions caused by this climate jump appeared a meridional dipole pattern, with droughts in the northern part but floods in the southern part of China. The interdecadal variation of the summer rainfall anomalies in eastern China is closely associated with the zonal mean flow in the upper troposphere over East Asia (e.g., Huang et al., 2012a). As shown in Fig. 10, a transition of zonal mean flow anomalies at 200 hPa over East Asia from the meridional tripole pattern to the meridional dipole pattern over the region from 10°N to 55°N of East Asia occurred in the late 1990s. This trend agrees with the interdecadal variability of the JJAS mean rainfall in the late 1990s. Moreover, Huang et al., (2011b) studied the interdecadal variability of July–September (JJAS) mean rainfall in northern China using the daily observational rainfall data for 1961–2008 in China. Their study show that the summer monsoon rainfall became weak in northern China including the southern part of Northwest China, North China, and the Northeast China. This result further demonstrates the climate jump in the late 1990s.

The analysis by Huang et al. (2011b) of the upper-level circumglobal teleconnection in boreal summer also revealed the cause of the interdecadal variation of the JJAS rainfall in northern China during in the late 1990s. To study the relationship between them, two indices (CGTI-1 and CGTI-2) were defined following Ding and Wang (2005). These two indices were defined as the normalized geopotential height anomaly at 200

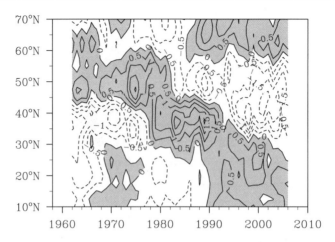

Fig. 10. Latitude-time cross section of the 9-year running mean summer time (JJA) zonal wind anomalies at 200 hPa averaged for 100°–140°E over East Asia based on NCEP/NCAR reanalysis data. Units: m s^{-1}. Solid and dashed lines indicate positive and negative anomalies, respectively. The positive anomalies are dashed. The climatological mean is based on the period 1971–2008.

hPa averaged over the area (35°–40°N, 60°–70°E) and the difference of the normalized geopotential height at 200 hPa between the area (40°–50°N, 20°–10°W) and the area (60°–65°N, 15°–25°E) in June–September. The correlation coefficient between the CGTI1 index and the normalized JJAS mean rainfall anomaly averaged for North China was 0.50 (Fig. 11a), which exceeded the 99% confidence level. The correlation coefficient between the CGTI-2 index and the JJAS mean rainfall averaged for Northwest China reached 0.53 (Fig. 11b), also exceeding the 99% confidence level. From Figs. 11a and b, it can be seen that both the CGTI-1 index and the CGTI-2 index as well as the JJAS mean rainfall in both North China and Northwest China, were mostly below normal after 1999, different from those before 1999. Therefore, from the interdecadal variation of the upper-level circumglobal teleconnection, it can also be seen that an interdecadal variation of the EASM system occurred in the late 1990s.

4. Characteristics of the spatio-temporal variabilities of the EAWM system

East Asia is a region of strong winter monsoon. The EAWM system includes the following: (1) the Siberian High, (2) the Aleutian Low, (3) strong northwesterly wind over North China, Northeast China, Korea and Japan, strong northeasterly wind along the coast of Southeast China, South China, and the Indo-China Peninsula, (4) the East Asian trough at 500 hPa over Northeast China, and (5) the East Asian jet with its maximum in the upper troposphere over Southeast Japan (e.g., Chen et al., 1991; Ding, 1994). The East Asian jet is associated with intense baroclinicity, large vertical wind shear and strong cold advection (e.g., Ding, 1994; Chen et al., 2003; Wang et al., 2010). The spatio-temporal variabilities of the EAMW system are also very complex.

4.1 Interannual variability of the EAWM system

The strength of the EAWM system has been a major concern in most previous studies on the interannual variability of the EAWM system. As reviewed by Wang and Chen (2010a), many indices have been defined to describe the EAWM strength (e.g., Chen and Graf, 1998; Wu and Wang, 2002; Jhun and Lee, 2004; Li and Yang, 2010; Wang and Chen, 2010a). For example, Chen and Graf (1998) and Chen et al. (2000) defined the EAWM index as the normalized meridional wind anomalies at 10 m averaged over the East China Sea (25°–40°E, 120°–140°E) and the South China Sea (10°–25°N, 110°–130°E) from November to March of the following year. Wu and Wang (2002) defined the EAWM index as the sum of zonal sea-level pressure differences (110°E minus 160°E) over 20°–70°N with a 2.5°×2.5° interval in latitude and longitude. Their results show significant interannual variations in the EAWM strength.

Recently, Huang et al. (2007b) investigated the characteristics of the interannual variability of the EAWM strength using the index defined by Wu and Wang (2002). They reported that the strength of

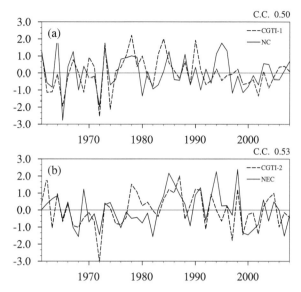

Fig. 11. The time series of (a) the CGT-I index and (b) CGTI-2 index and the normalized JJAS mean rainfall anomalies averaged for (a) North China and (b) Northwest China. The definitions of the CGTI-1 index and the CGTI-2 index follow Ding and Wang (2005).

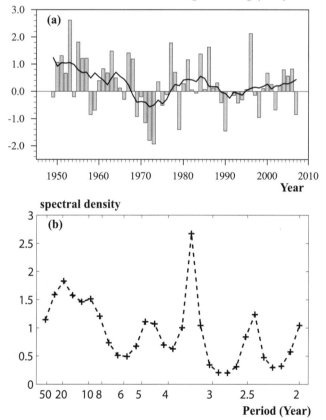

Fig. 12. (a) Interannual variations of the EAWM index and (b) the dominant period analyzed by using the entrophy spectrum analysis based on NCEP/NCAR reanalysis data. The definition of the EAWM index follows Wu and Wang (2002).

the EAWM has a significant interannual variability with an oscillation of approximately 4 years (Fig. 12). Their results also show an obvious difference between the EAWM intensity in the winter of 2005 (December 2005–February 2006) and 2006 (December 2006–February 2007), which caused the different winter climate anomalies in the Northern Hemisphere, especially in East Asia. A similar phenomena also occurred in January 2008 and the winter of 2009. Prolonged cooling and severe snowstorms struck most of China in January 2008, causing 129 deaths and the economic losses of 150 billion CNY (∼US$21 billion) (e.g., Gu et al., 2008; Wang et al., 2009b). In the winter of 2009, frequent severe cold weather afflicted many areas of mid-latitude Northern Hemisphere, including northern China, with record-breaking blizzards, snowstorms, and low temperatures, which caused extensive traffic damage and great losses to agriculture and fisheries (e.g., Wang and Chen, 2010b).

4.2 Characteristic of multiple modes in the spatio-temporal variabilities of the EAWM system

In recent years, the variability of EAWM system has been recognized as complex and as having multi-modal characteristics (e.g., Kang et al., 2006, 2009; Wang et al., 2009a, 2010; Wang and Chen, 2010a). Based on the observations from 160 stations in China, Kang et al. (2006, 2009) first identified two leading modes of wintertime surface air temperature in China on both interdecadal and interannual time scales. The first leading mode reflects a variability of surface air temperature in whole China, which is related to the strength of the EAWM system. The second leading mode describes a surface air-temperature oscillation between the northern and southern parts of China. Moreover, Wu et al. (2006) also suggested that there are also two distinct modes in the spatio-temporal variations of the EAWM system, and that these two leading distinct modes mainly reflect the variability of meridional and zonal winds in winter, respectively. In addition, from the EOF analysis of the wintertime surface air temperature in East Asia, Wang et al. (2010) also obtained a northern mode and a southern mode for the wintertime surface air temperature variability in East Asia.

Like the second leading modes of the spatio-temporal variations of the EAWM system, the pathway of the EAWM exhibits significant interannual variation, associated with the tilt of the East Asian trough (EAT) axis in winter (Wang et al., 2009a). When the tilt of the EAT axis is small, the southern pathway is strong. In this case, the major cold air flows along the coast of China and penetrates into

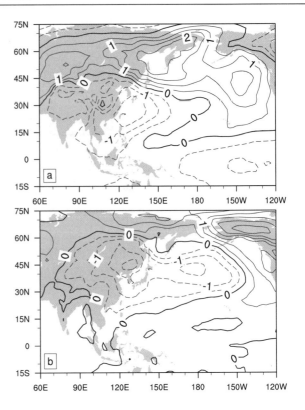

Fig. 13. Distributions of the 850-hPa air temperature anomalies for five strong winters with (a) strong southern EAWM pathway and (b) strong eastern EAWM pathway based on ERA-40 reanalysis data. The solid and dashed contours indicate positive and negative anomaly, respectively. Contour intervals are 0.5°C.

the Southern Hemisphere, which leads to significant cooling in the SCS and Southeast Asia and warming in the northern part of East Asia (Fig. 13a). Thus, the tropical rain belt in Southeast Asia can be pushed southward by stronger northerly wind, and more precipitation can be observed in the East Asian continent (Wang et al., 2009a). In contrast, when the tilt of the EAT axis is large, the eastern pathway is strong. In this case, more cold air flows into the North Pacific with a weakened southern branch of airflow, and the climate anomalies are generally reversed (Fig. 13b). Therefore, the variability of the EAWM pathway has a modulation effect on the regional climate in East Asia and Southeast Asia. Further analysis implies that the North Pacific SST can influence the interannual variability of the EAWM pathway through the atmosphere–ocean interaction. When warm SST anomalies are observed over the North Pacific from the preceding summer to autumn, the EAWM tends to take the southern pathway in the following winter, and when cold SST anomalies are observed, the

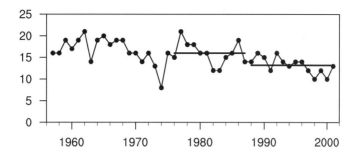

Fig. 14. Number of general cold surges that occurred in China for the extended winter, NDJFMA, of 1957–2001, recorded in the CMA cold surge almanac.

EAWM tends toward the eastern pathway in the following winter (Wang et al., 2009a).

4.3 Interdecadal variability of the EAWM system

The EAWM system is characterized by obvious interdecadal variability in addition to interannual variability (e.g., Jhun and Lee, 2004; Huang and Wang, 2006; Wang et al., 2009b; Wang and Chen, 2010a). Observational studies have revealed that the strength of the EAWM system was significantly weakened after 1987 (e.g., Huang and Wang, 2006; Kang et al., 2006; Wang et al., 2009b; Wang and Chen, 2010a). Compared with the period 1976–1987, both the Siberian High/Aleutian Low (Wang et al., 2009a) and the sea surface wind stress along the coasts of East Asia (Cai et al., 2006) weakened for the period of 1988–2001, which accompanied fewer cold waves (Wang et al., 2009b) and frequent warm winters (Wang and Ding, 2006; Kang et al., 2006) in China. Wang et al. (2009b) analyzed the numbers of cold waves in China for the winters during 1957–2001 according to the data recorded in the NCC-CMA cold-wave almanac and revealed that an average of 15.4 general cold waves occurred each winter. There were 16 cold waves per year for the strong EAWM period 1976–1987, but there were only 13.2 cold waves per year for the weak EAWM period 1988–2001 (Fig. 14).

Although the decreasing trend for the number of cold waves over China for 1988–2001 is clear (e.g., Wang and Ding, 2006; Wang et al., 2009b), the frequency of severe cold winters tended to increase from the early 21st century (e.g., Ding and Ma, 2007; Huang et al., 2008; Ma et al., 2008; Hong and Li, 2009; Lu and Chang, 2009; Wen et al., 2009; Wang and Chen, 2010b). It suggests that the continuous warm winters after 1987 in China have ended before 2008 (e.g., Ma et al., 2008; Wang et al., 2009b). Thus, it is possible that the EAWM system may undergo another interdecadal variation from recent years (Wei et al., 2011).

4.4 Impact of the EAWM system variability on wintertime rainfall in China and the marine environment over the offshore area of China

The studies mentioned in the previous section show that the EAWM system variability has an important impact on surface air temperature over China. The EAWM system variability also has a significant impact on precipitation. Wang and Feng (2011) identified two major modes of wintertime (December–February) precipitation over China through EOF method. The first mode (EOF1) reflects the strength of precipitation over southeastern China with a clear 2–4-year period. Its interdecadal variations suggest that the wintertime precipitation over southeastern China was below normal before the mid-1980s and above normal thereafter, and underwent a slight decreasing trend in recent years. The second mode (EOF2) delineates an out-of-phase relationship between South China and the middle and lower reaches of the Yangtze River as well as the northern part of Xinjiang, with a clear 2–4-year period, too. They further suggested that EOF1 is closely related to ENSO and the strength of the EAWM. The variation of EOF2 is closely associated with a barotropic wave train across the Eurasian continent originating from the North Atlantic.

Zhou (2011) investigated the impact of the EAWM system variability on late winter (January–March, JFM) rainfall over southeastern China using observational data for 1951–2003. His study showed a significant correlation between the EAWM variability and JFM rainfall over southeastern China. In a winter (JFM) with the weak EAWM, the southwesterly wind anomalies at 700 hPa dominate over the SCS, which can transport more water vapor from the Bay of Bengal and the SCS into southeastern China. In this case, the westerly jet is weakened over East Asia and displaces southward, contributing to the intensification of ascending motion over southeastern China, and

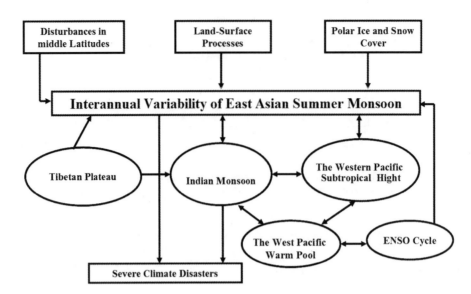

Fig. 15. Schematic map of components of the EAM climate system.

causing the increase of wintertime (JFM) rainfall in southeastern China. Gu et al. (2008) investigated the impact of the EAWM variability on wintertime snow storms in China, especially on severe blizzards, freezing rain, and low temperatures in January 2008. Their results show that the East Asian trough stayed in a stable state for long time in January 2008, leading to the continuous southward-intrusion of cold air along the Mongolian Plateau to Central China, Southwest China, and South China. Meanwhile, the western Pacific subtropical high shifted anomalously northward and westward, which led to the northward transport of a large amount of wet air along the west edge of the subtropical high from the Bay of Bengal and the SCS, and the southwest flow with wet air converged with cold air from the Mongolian Plateau in the Yangtze River valley. Thus, prolonged heavy snowfall and low temperature occurred in Central China, South China, the eastern part of Southwest China, and the Yangtze River valley in January 2008.

The EAWM system variability has a significant impact not only on severe cold waves and heavy snowfall in China but also on the marine environment in the offshore area of China including the Bohai Sea, the Yellow Sea, the East China Sea, and the SCS and its adjacent ocean. Cai et al. (2006, 2011) suggest that the sea surface wind stresses, especially the meridional sea surface wind stresses, have been weakened over the offshore area of China and its adjacent ocean from the 1980s, due to the weakened EASM and the EAWM systems. The weakened wind stress has caused obvious warming in the underling ocean, providing a favorable marine environment for the frequent occurrence of red tides in the offshore area of China.

5. Influence of atmosphere–ocean–land interactions on the EAM system variability

As shown by Webster et al. (1998), the monsoon is not only an atmospheric circulation system but also an atmosphere–ocean–land coupled system. The interannual and interdecadal variabilities of EAM system are influenced not only by many circulation system such as the SASM system, the western Pacific subtropical high, the disturbances in middle and high latitudes (e.g., Tao and Chen, 1987) but also by other factors such as thermal states of the West Pacific warm pool and convective activity around the Philippines, ENSO cycles in the tropical Pacific, the thermal state of the tropical Indian Ocean, the dynamical and thermal effects over the Tibetan Plateau, and the land surface process in the arid and semiarid area of Northwest China and the snow cover on the Tibetan Plateau and the Eurasian continent and so on. As shown in Fig. 15, these factors affecting the EAM system variability are various components of an atmosphere–ocean–land coupled system, and these components are interactions. Therefore, this coupled system may be called the EAM climate system (e.g., Huang et al., 2004a, 2006a, 2007a; Huang, 2009).

To understand the causes of the EAM climate system variability, it is necessary to analyze the interactive processes between the atmosphere and oceans and between the atmosphere and land surface in the EAM climate system. However, because the interactions among these components of the EAM climate system are complex, only the impacts of atmosphere–ocean and atmosphere–land coupling components of this climate system on the EAM system variability are simply discussed in the following sections.

5.1 *Thermal effect of the tropical western Pacific on the EAM system variability*

The tropical western Pacific is the region with the highest SSTs in the global sea surface. Due to the warm state of this region, the atmosphere–ocean interaction is very strong, and the ascending branch of the Walker circulation extends over the region. This leads to strong convergence of air and moisture and strong convective activity and heavy rainfall there (Cornejo-Garrido and Stone, 1977; Hartmann et al., 1984). Thus, the tropical western Pacific has an important thermal effect on EAM system variability.

Many important studies (e.g., Nitta, 1987; Huang and Li, 1987, 1988; Kurihara, 1989; Huang and Sun, 1992) show that the thermal states of tropical western Pacific and convective activity around the Philippines play important roles in the interannual variability of the EASM system. Huang et al. (2005) investigated the relationship between the intraseasonal variations of the western Pacific subtropical high and the thermal state of the tropical western Pacific in summer; they showed a close relationship between the anomalous northward shift of the western Pacific subtropical high and the intensified convective activities around the Philippines. When the tropical western Pacific is in a warming state in summer, and convective activities are strong around the Philippines, there are abrupt northward shifts of the western Pacific subtropical high from South China to the Yangtze-Huaihe River valley in early or mid-June and from the Yangtze-Huaihe River valley to North China and Northeast China in late June or early July. Thus, the summer monsoon rainfall may be weak in the Yangtze-Huaihe River valley. In contrast, when the tropical western Pacific is in a cooling state in summer and convective activities are weak around the Philippines, no abrupt northward shifts of the western Pacific subtropical high occur from South China to the Yangtze-Huaihe River valley in early or mid-June and from the Yangtze-Huaihe River valley to North China and Northeast China in late June or early July. Thus, the mei-yu front is maintained for long time, and the summer monsoon rainfall may be strong in the Yangtze-Huaihe River valley. The schematic map is shown in Fig. 6.

Huang et al. (2004b, 2006b) investigated the interannual variability of thermal state of the tropical western Pacific and its impact on interannual variability with the quasi-biennial oscillation of the EASM system. There is a dominant quasi-biennial oscillation in the interannual variations of thermal state of the tropical western Pacific, and it reveals that the oscillation has a great impact on the interannual variability with the quasi-biennial oscillation of the EASM system (Huang et al., 2006b). When the tropical western

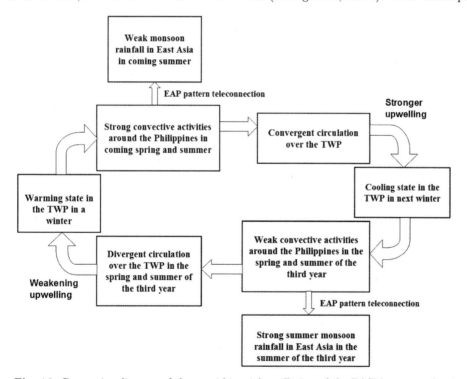

Fig. 16. Conceptive diagram of the quasi-biennial oscillation of the EASM system related to the atmosphere–ocean coupling system over the tropical western Pacific.

Pacific is in a warming state in a winter and coming spring, convective activity generally is stronger around the Philippines in coming spring and summer (Fig. 16b). This can cause the northward shift of the western Pacific subtropical high due to the EAP pattern teleconnection (e.g., Nitta, 1987; Huang and Li, 1987, 1988), which leads to weak monsoon rainfall in the Yangtze River valley of China, Japan, and South Korea. At the same time, the convergence due to the strong convective activity around the Philippines can cause strong upwelling in the tropical western Pacific, which brings a cooling state into the tropical western Pacific in the next winter. Because a cooling state appears in the tropical western Pacific in the following winter, weak convective activity occurs around the Philippines in the spring and summer of the third year. This causes the southward shift of the western Pacific subtropical high due to the EAP pattern teleconnection, which leads to strong monsoon rainfall in the Yangtze River valley of China in the summer of the third year. At the same time, the divergence due to the weak convective activity around the Philippines causes weak upwelling in the tropical western Pacific in the spring and summer of the third year, which brings another warming state into the tropical western Pacific. The period of the process is approximately 2–3 years.

The thermal state of the tropical western Pacific has a significant effect not only on the interannual variability of the EASM system but also on the intraseasonal variability of this system, as described in section 3 (e.g., Huang et al., 2005, 2006b).

5.2 Impact of ENSO cycles in the tropical Pacific on the EAM system variability

It is well known that ENSO cycles are one of the most important phenomena of atmosphere–ocean interaction in the tropical Pacific and has a great influence on EAM system variability. Huang and Wu (1989) first showed that summer monsoon rainfall anomalies over East Asia depend on the stages of ENSO cycle in the tropical Pacific. This dependence is also reflected in the runoff from (at least) spring (e.g., Chen et al., 2009). From the composite analyses of summer monsoon rainfall anomalies for different stages of ENSO cycles during the period 1951–2000, Huang and Zhou (2002) concluded that droughts in North China tend to occur in the developing stage of El Niño events, but floods tend to occur in the Yangtze River valley of China during the decaying stage of El Niño events (Fig. 17). For example, severe floods occurred in the Yangtze River valley in the summers of 1998 and 2010, These two summers belong to the decaying state of El Niño events that occurred in 1997 and 2009. The modeling study by Sun and Yang (2005) also showed different interannual anomalous atmospheric circulation patterns over East Asia during different stages of an El Niño event.

Huang and Wu (1989) were the first to describe the important impact of El Niño decaying phase on EASM system variability, generally called the delayed impact of El Niño. Later, Zhang et al. (1996) and Zhang and Huang (1998) investigated the process and mechanism of the delayed impact of El Niño events on EASM system variability and pointed out that this delayed effect may be due to the appearance of an anomalous anticyclonic circulation in the lower troposphere over the tropical western Pacific after the mature phase of El Niño events. The southwesterly anomalies on the western side of the anticyclonic anomaly circulation intensify the southwesterly flow, causing the southwest monsoon over Southeast China to strengthen. Wang et al. (2003) further studied the formation mechanism of the anomalous anticyclonic circulation over the tropical western Pacific. Their study showed that in a winter during the mature stage of El Niño, the weakening convective activities over the tropical western Pacific caused by the weakening Walker circulation and negative anomalies of SST in the tropical western Pacific can trigger a cooling Rossby wave train in the tropical western Pacific and can form an anomalous anticyclonic circulation in the lower troposphere over the tropical western Pacific. This anomalous circulation can be maintained into following spring.

ENSO cycles also significantly influence the transitions between the EAWM and EASM. Chen (2002) and Huang et al. (2004b) proposed that the EAM system variability is associated with ENSO cycles. Chen (2002) analyzed the composite distributions of meridional wind anomalies at 850 hPa and rainfall anomalies for various stages of ENSO cycles. These composite distributions show that in the winter before the developing stage of El Niño, anomalous northerly winds occur along the coastal areas of China (strong EAWM), and an anomalous cyclonic circulation occurs over the tropical western Pacific, which contribute to the development of El Niño. In the following summer, the western Pacific subtropical high is weak, leading to northeasterly anomalies over the Yangtze-Huaihe River valley and the southeastern coast of China. Thus, a weak EASM occurs in the summer when El Niño is developing. When El Niño reaches its mature phase in the following winter, an anticyclonic anomaly circulation begins to appear over the tropical western Pacific (e.g., Zhang et al., 1996; Zhang and Huang, 1998; Wang et al., 2003), and anomalous southwesterly winds prevail in the southeastern coast of China and the SCS. This suggests that a weak EAWM may appear in the winter

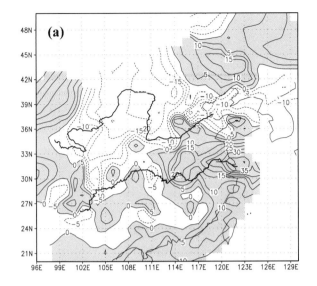

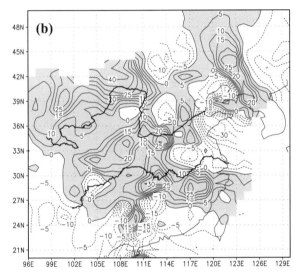

Fig. 17. Composite distributions of summer (JJA) rainfall anomalies (in percentages) over China (a) for the summers in the developing stage and (b) for the summers in the decaying stage of El Niño events occurred during the period from 1951 to 2000. The solid and dashed contours indicate positive and negative anomalies, respectively. Positive values are shaded.

when an El Niño event is in its mature phase. When El Niño begins to decay, the anomalous anticyclonic circulation is intensified over the tropical western Pacific, strengthening the western Pacific subtropical high. In this case, anomalous southwesterly winds are distributed over the region from South China to the Yangtze River valley, which shows that a strong EASM may appear over the Yangtze-Huaihe River valley in the summer when El Niño is in its decaying phase. The recent study by Wang and Wu (2012) further suggests that the role of ENSO is particularly evident during the transitions from weak EAWM to weak EASM.

Recently, several studies on the leading modes of the tropical Pacific SST variability have been published (e.g., Ashok et al., 2007; Weng et al., 2007; Zhang and Huang, 2008; Huang and Huang, 2009). They reveal two leading modes in the tropical Pacific SST anomalies. These two leading modes are characterized by positive SST anomalies in the tropical eastern Pacific and the tropical central Pacific, i.e, the eastern Pacific warming pattern and the central Pacific warming pattern (i.e., El Niño Modoki), respectively. Huang and Huang (2009) analyzed the impact of El Niño Modoki on EAM system variability, especially on summertime rainfall over eastern China. Their results show that summertime rainfall anomaly distributions during the developing and decaying stages of an El Niño Modoki are very different from those shown in Fig. 17. In the summer, with the developing stage of El Niño Modoki, summer monsoon rainfall may be below normal in the Huaihe River valley and above normal in South China (Fig. 18b). In contrast, in the summer with the decaying stage of El Niño Modoki, summer rainfall may be above normal in the Huaihe River valley and below normal in North China and the region south of the Yangtze River (Fig. 18b). In addition, Feng et al. (2010) described the two types of Pacific warming that have distinct impacts on the wintertime rainfall over southeast China and Southeast Asia. In El Niño winters, wet conditions occur over South China, and dry conditions occur over the Philippines, Borneo, Celebes, and Sulawesi. In contrast, for El Niño Modoki winters, the negative rainfall anomalies around the Philippines are weaker and are located more northward compared to their El Niño counterpart.

5.3 Impact of PDO on the EAM system variability

Studies have shown significant decadal oscillations in the North Pacific SST, i.e., the Pacific Decadal Oscillation (PDO); this oscillation has an important impact on the atmospheric circulation over the North Pacific (e.g., Trenberth and Hurrell, 1994; Bond and Harrison, 2000). The interannual and interdecadal variabilities of EASM system mentioned previously are closely associated with not only atmospheric but also oceanic anomalous patterns. Specifically, the PDO has a significant impact on EAM system variability on an interdecadal time scale. As mentioned above, the mature phase of El Niño is usually accompanied by a weaker EAWM system and the mature phase of La Niña is usually accompanied by a stronger EAWM

(图略)

Fig. 18. Composite distributions of summer rainfall anomalies (percentage) in China for (a) the summers of 1994, 2002, and 2006 (i.e., the developing stage of El Niño Modoki) and (b) the summer of 1995, 2003, and 2007 (i.e. the decaying stage of El Niño Modoki). The solid and dashed contours indicate positive and negative anomalies, respectively. Positive values are shaded.

system. However, according to Wang et al. (2008), the impact of the ENSO cycle on the EAWM system variability is modulated by the PDO. When the PDO is in its high phase, there is no significant relationship between the ENSO and EAWM systems on the interannual time scale because the ENSO–EAWM teleconnection is not significant. When the PDO is in its low phase, however, the ENSO cycle has strong impact on the EAWM system, causing significantly low temperatures over East Asia. The PDO was in its high phase after the mid-1970s; therefore, the impact of ENSO cycles on the EAWM system became weak during the following decades. However, in the winters before the mid-1970s, the ENSO cycle had a significant influence on the North Pacific Oscillation (NPO) and on EAWM system variability (Wang et al., 2007).

The PDO also has an important influence on EASM system variability. Previous studies have elucidated the close relationship between PDO and interdecadal variability of climate in China (e.g., Zhu and Yang, 2003; Yang et al., 2005; Xu et al., 2005; Yang and Zhu, 2008). Recently, Zhang et al. (2007) analyzed the interdecadal variations of summertime rainfall in eastern China and their association with the

(Continued on page 337)

temporal evolution of the PDO. Their results show that the interdecadal variations of summertime rainfall pattern in eastern China and the EASM circulation are closely related to the PDO in the North Pacific. Moreover, Deng et al. (2009) concluded that the first interdecadal variation of summertime rainfall pattern over eastern China occurred in the mid-and late 1970s may have been influenced by the transition of the PDO from a negative to a positive phase, and that the second interdecadal variation of the rainfall pattern that occurred in the late 1980s and the early 1990s may have been association with the warming of the western North Pacific to the south of Japan, especially the warming around the Philippines.

5.4 Impact of the North Indian Ocean on the EASM system variability

The tropical Indian Ocean also has an important thermal effect on EASM system variability. The studies by Annamalai et al. (2005) and Yang et al. (2007) suggest that the SST anomalies in the Indian Ocean have an important effect on EASM system variability and climate anomalies over East Asia. Xie et al. (2009) proposed that the Indian Ocean has a capacitor-like effect on the climate variability over the Indo-Western Pacific during the summer following El Niño.

Recently, many studies have focused the thermal role of the tropical northern Indian Ocean in the delayed impact of El Niño on the EASM system. For example, Li et al. (2008), Huang and Hu (2008), and Huang et al. (2010) systematically studied the thermal effect of the tropical northern Indian Ocean on the anomalous anticyclonic circulation in the low troposphere over the western North Pacific in summer, which causes the delayed impact of El Niño on the EASM system. Huang and Hu (2008) further reported that the interannual variation of the western North Pacific anomalous anticyclone is closely associated with SST anomalies in the tropical and northern Indian Ocean in summer, but that it does not have an obvious correlation with the SST anomalies in the western South Indian Ocean. The low-level anomalous anticyclonic circulation over the Northwest Pacific caused by the summertime warming SSTs in the tropical northern Indian Ocean was well simulated by Huang and Hu (2008) using the ECHAM 5.0 climate model. Using five AGCMs, Li et al. (2008) demonstrated that Indian Ocean warming can trigger an anticyclonic anomaly circulation in the lower troposphere over the subtropical western Pacific, intensifying the southwesterly flow to East China; it can also trigger a Gill-type response with the intensified South Asian high in the upper troposphere. The two circulation systems are favorable to the enhancement of the EASM system.

Moreover, Huang et al. (2010) also investigated the interdecadal variation of the relationship between the SST in the tropical northern Indian Ocean and the Northwest Pacific low-level anticyclonic anomaly circulation in boreal summer using both observation data and AGCM simulations. Their results show that the low-level anticyclonic anomaly circulation over the Northwest Pacific is positively correlated with the summertime SST in the tropical northern Indian Ocean after the mid-1970s, but the correlation between them was weak for the period 1958–1976. Their numerical simulations with a 21-member ensemble AGCM (ECHAM5.0) also showed an interdecadal variation in the relationship between the low-level anticyclonic anomaly circulation over the Northwest Pacific and the summertime SST in the tropical Indian Ocean after the mid-1970s.

These studies show that a basin-scale warming trend occurred in the tropical Indian Ocean from the mid-1970s, which caused the intensification of the low-level anticyclonic anomaly circulation and the strong southwesterly flow over East Asia from the mid-1970s. However, according to the results of Huang et al. (2004a, 2006a, 2010), the EASM system has become substantially weaker since the mid-1970s, especially from the late 1990s, although it was stronger during the period 1993–1998. Thus, these results cannot confirm a causal relationship between the tropical Indian Ocean warming and the interdecadal weakening of the EASM system from the mid-1970s (e.g., Li et al., 2008; Huang et al., 2010).

The tropical Indian Ocean has not only a significant impact on the low-level anomaly circulation over the Northwest Pacific but also an important thermal effect on the South Asian High. The variations of the South Asian High (SAH) are closely associated with precipitation and circulation over Asia (e.g., Tao and Zhu, 1964). According to the result of Zhang et al. (2002), in the summer when the SAH shifts eastward, the summer monsoon rainfall may be above normal in Southern Japan, the Korea Peninsula and the Yangtze River valley of China. On the other hand, in the summer when the SAH shifts westward, the summer monsoon rainfall may be below normal in these regions. Zhang et al. (2005) pointed out that the SAH is also linked to global circulation and precipitation in summer. A strengthened SAH is generally accompanied by strengthened and westward western Pacific subtropical high, weakened mid-Pacific trough, and intensified Mexican high, Moreover, increasing precipitation may appear in South Asia, Central America, Australia, and Central Africa, and decreasing precipitation may be caused over the Pacific and the Mediterranean Sea. Additionally, a strengthened SAH can lead to the intensification of the subtropical high over the extratropical North Pacific (Zhao et al., 2009). Therefore, SAH variability is also a factor affecting the EASM system.

From their numerical simulations with five AGCMs, Li et al. (2008) stated that the tropical Indian Ocean warming can trigger the intensification of the SAH in the upper troposphere over South Asia. Recently, the simulation results of Huang et al. (2011a), using the ECHAM 5.0 AGCM also showed that when the tropical Indian Ocean is in a warming state, the SAH is strengthened and its center shifts southward over South Asia. Furthermore, they proposed a possible mechanism of the connection between the thermal state of the tropical Indian Ocean and the SAH variability: The warming SST in the tropical Indian Ocean causes the increase of the equivalent potential temperature in the atmospheric boundary layer and can alter the temperature profile of the moist atmosphere over the tropical Indian Ocean. This induces significant positive geopotential height anomalies over South Asia, and the SAH is thus intensified.

5.5 The Impact of land surface processes on the EASM system variability

Because monsoon circulations result fundamentally from the ocean–land thermal contrast, EAM system variability is influenced not only by the thermal state of the tropical Pacific and tropical Indian Ocean but also by the thermal state of the Eurasian continent.

The spring Eurasian snow cover greatly influences the thermal state of the Eurasian continent; thus, it has an important impact on the following summer climate over East Asia. Recently, Wu et al. (2009), and Zhang et al. (2008) analyzed the cause of the significant decadal shift of the summer climate over eastern China that occurred in the late 1980s, which was proposed from the interdecadal variability of the summer monsoon index defined by Wang et al. (2001). This decadal shift of the summer climate over eastern China in the late 1980s, proposed by Wu et al. (2009) and Zhang et al. (2008), may be similar to the decadal shift of the summer rainfall to South China in the early 1990s reported in previous studies (e.g., Ding et al., 2008; Deng et al., 2009; Huang et al., 2011c). As mentioned in subsection 3.3, Wu et al. (2009) and Zhang et al. (2008) revealed that the decadal climate shift of the summer monsoon rainfall belt to South China is closely associated with the decadal variability of the spring snow cover over the Eurasian continent. Their studies also showed a strong negative correlation between these two aspects. They also investigated the physical processes of this correlation; their results show that the snow cover variability in spring Eurasian continent can excite a Rossby wave train over high latitudes from spring to summer, leading to an anomalous high over North China and a weak low over South China. Under these conditions, more monsoon rainfall can be caused in South China.

The snow cover over the Tibetan Plateau is a part of the snow cover over the Eurasian continent. Ye and Gao (1979) first reported that the Tibetan Plateau has important thermal and dynamic effects on the interannual variability of the EASM system. Since then, many investigators have also emphasized the thermal effect of the Tibetan Plateau on the EASM system variability (e.g., Huang, 1984, 1985; Wu and Zhang, 1998).

In the past 10 years, some studies have focused the thermal effect of snow cover over the Tibetan Plateau on the EASM system variability. Wei et al. (2002, 2003) analyzed the interannual and interdecadal variations of the days and depth of snow cover over the Tibetan Plateau using the observation data of daily snow cover at 72 observation stations in the Tibetan Plateau for the period 1960–1999. They discovered obvious interannual and interdecadal variations of the days and depth of snow cover on the Tibetan Plateau. Moreover, these variations of snow cover over the Tibetan Plateau also have an important impact on the summer monsoon rainfall in the middle and upper reaches of the Yangtze River valley (e.g., Wei et al., 2002, 2003; Huang et al., 2004a).

5.6 Thermal effect of the sensible heating in the arid and semiarid regions of Northwest China on the EASM system variability

EAM system variability is greatly influenced by the thermal states of the Eurasian continent, especially the thermal states in the arid and semiarid regions of Northeast China and Central Asia, because the sensible heating in these regions is larger than that in the EASM region in spring and summer (Fig. 19). Zhou and Huang (2006, 2010) and Zhou (2010) analyzed the interannual and interdecadal variations of the difference between the surface temperature and the surface air temperature, i.e, $T_s - T_a$, and sensible heating in spring in the arid and semiarid regions of Northwest China and Central Asia and their impact on summer monsoon rainfall in China. Their results show that the strongest sensible heating over the Eurasian continent is located at Northwest China and Central Asia in spring, thus, this region may be seen as a "warm lying surface" in the Eurasian continent. The $T_s - T_a$ and sensible heating over this region exhibit obvious interdecadal variability. Before the late 1970s, the sensible hearting anomalies in spring were negative, but

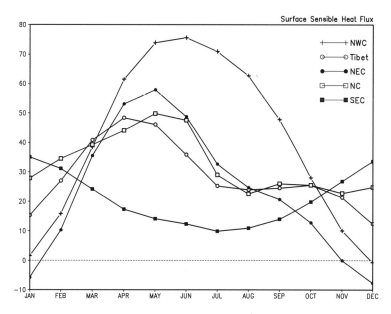

Fig. 19. Climatological mean monthly sensible heating flux in Northwest China (NWC), Tibetian Plateau (Tibet), Northeast China (NEC), North China (NC) and Southeast China (SEC) based on ERA-40 reanalysis data. Units: W m^{-2}.

the sensible heating anomalies became largely positive from the late 1970s onward. Zhou and Huang (2008, 2010) discussed a possible physical process of the thermal effect of the sensible heating anomalies in the arid and semiarid regions of Northwest China on EASM system variability. Their analyses showed that since the sensible heating became strong during the period 1978–2000, the cyclonic circulation anomaly intensified over Northwest China. This was closely associated with the anomalous ascending motion over Northwest China, which could contribute to anomalous descending motion over North China. Moreover, their results also showed that during the period 1978–2000, the air temperature at 300 hPa decreased obviously, but it increase near the surface over Northwest China in the summer. This caused the enhancement of vertical convective instability, which contributed to the strengthening of ascending motion and the increase of rainfall in Northwest China after 1978. Moreover, influenced by the anomalous ascending motion over Northwest China, an opposite phenomenon occurred over North China, anomalous descending motion occurred over North China and the vertical convective instability became weak over North China. This contributed to the decrease of summer rainfall in North China after the late 1970s. Therefore, the intensification of spring sensible heating over Northwest China and Central Asia may have been one of the causes of the interdecadal variability of the EASM system occurred in the late 1970s.

6. The internal dynamic processes in the EAM climate system

As shown in section 5, EAM system variability is closely associated not only with the variabilities of various component of the EAM climate system, including atmosphere–ocean–land interactions but also with the internal dynamical and thermo-dynamical processes in the EAM climate system. Because of the dynamical and thermo-dynamical processes in this system, there are close relationships among the variabilities of its various components. Therefore, it is necessary to discuss the internal dynamical and thermo-dynamical processes in the EAM climate system.

6.1 *The property of the East Asia/Pacific (EAP) teleconnection and its role in the EASM system variability*

The interannual and interdecadal variabilities of the EASM system are dominated by the meridional teleconnection. The meridional teleconnection is generally referred to as the Pacific-Japan (PJ) pattern (Nitta, 1987) or the EAP pattern (Huang and Li, 1987, 1988). Studies on the role of the EAP pattern teleconnection in the processes of the EASM system variability, its characteristics and properties have advanced our understanding in recent years.

During recent years, many studies on the role of the EAP pattern teleconnection in the meridional tripole pattern distribution of the EASM system variability have used reanalysis data and observational data (e.g., Huang et al., 2006b, 2007b, 2011c). As described in section 3, an obvious meridional tripole pattern occurs either in the anomaly distributions of summer monsoon rainfall in eastern China or in the anomaly distributions of the summertime water vapor transport fluxes over East Asia on interannual and interdecadal time scales (e.g., Huang et al., 2006b, 2011c). As a result, the distributions of droughts and floods in eastern China also exhibit a characteristic of the meridional tripole pattern. Huang et al. (2006b, 2007b) used the EAP pattern teleconnection proposed by Nitta (1987), and Huang and Li (1987, 1988) to interpret the physical mechanism of the meridional tripole pattern of the spatio-temporal variabilities of the EASM system. Huang et al. (2006b, 2007b, 2011c) explained that the distributions of summertime atmospheric circulation anomalies with the meridional tripole pattern over East Asia and the western North Pacific are caused by the heating anomaly due to thermal anomaly of the tropical western Pacific or convective activity anomaly around the Philippines through the EAP pattern teleconnection. Of course, this meridional tripole pattern of circulation anomalies over East Asia is also associated with thermal anomalies over the North Atlantic through the EU pattern teleconnection over middle and high latitudes, as proposed by Wallace and Gutzler (1981).

An important advance in recent research on the internal dynamical processes is the further understanding of EAP-pattern teleconnection. In the middle 1980s, the EAP-pattern teleconnection was considered a northward-propagating Rossby wave train excited by the anomalous heating due to convective activity around the Philippines (Nitta, 1987; Huang and Li, 1987, 1988). Therefore, the EAP pattern teleconnection can be seen as a thermal mode of summertime circulation variability over East Asia and the Northwest Pacific. However, Kosaka and Nakamura (2006) proposed that this meridional teleconnection could efficiently gain kinetic energy and available potential energy from the basic flow over East Asia and the western North Pacific. Thus, the meridional teleconnection can be also considered as a dynamical mode of summertime circulation variability over these regions. Lu et al. (2006) reported two leading modes in the atmospheric circulation anomalies over East Asia and the western North Pacific. The first mode is associated with the circulation variation over the tropical region, which is mainly attributed to external SST forcing. The second mode is a meridional teleconnection mode, which mainly results from internal atmospheric variability. Therefore, the EAP pattern teleconnection may be considered as a combination of the thermal mode and the dynamical mode of summertime circulation variability over East Asia and the western North Pacific.

Moreover, Kosaka and Nakamura (2006) also presented the three-dimensional structure of the EAP pattern teleconnection and described its meridional and vertical coupled characteristics. Lu and Lin (2009) proposed that this vertical and meridional coupled teleconnection can be depicted by zonal shift of the western Pacific subtropical high in the subtropical lower troposphere and the meridional displacement of the East Asian jet stream (EAJS) in the upper troposphere over Asia. Actually, the meridional displacement of the EAJS is the leading mode of interannual variation of upper-tropospheric zonal wind anomalies over East Asia and the western North Pacific.

On the other hand, Lu and Lin (2009) suggested that the precipitation anomaly in the EASM system plays a crucial role in maintaining this meridional teleconnection. The climatological mean of the subtropical precipitation over East Asia is ~ 7.0 mm d^{-1} and the interannual standard deviation is 1.0 mm d^{-1}, which are comparable to their counterparts over the tropical western Pacific. Thus, as a strong heating source, the strong precipitation in the EASM system can significantly feedback to the meridional teleconnection. This explains further that the meridional teleconnection is more properly viewed as a thermal-dynamical mode.

6.2 The "Silk Road pattern" teleconnection along the Asian jet in the upper troposphere and its role in EASM system variability

Yang et al. (2002) analyzed the association of Asian–Pacific–American winter climate with the East Asian jet stream (EAJS) on an interannual time scale. They proposed that the EAJS is coupled to a teleconnection pattern extending from the Asian continent to North America with the strongest signals over East Asia and the West Pacific in boreal winter. Moreover, Lu et al. (2002), Lu and Kim (2004), and Lin and Lu (2005) analyzed the variability of meridional circulation anomalies in the upper troposphere over the Northern Hemisphere for the boreal summers of 1986–2000 using HadAM3 data. Their results showed that there is an obvious teleconnection pattern in the meridional circulation anomalies along the Asian jet in the upper troposphere over the region from West Asia to East Asia. At the same time, Enomoto et al. (2003) further analyzed the formation of the high ridge near Japan (i.e., the Bonin high) resulting from the propagation of stationary Rossby waves along the Asian jet in the upper troposphere. This teleconnection is called the "Silk Road pattern" in their study. Lu and Kim (2004) also explained that this teleconnection may be due to the eastward propagation of the Rossby wave-train along the westerly jet stream at 200 hPa.

Recently, from the analysis of isentropic potential vorticity, Tao and Wei (2006) demonstrated that the northward advance or southward retreat of the western Pacific subtropical high may be associated with the propagation of the Rossby wave train along the Asian jet in the upper troposphere because it may form a high ridge or a low trough along the eastern coast of China. From the analysis of the relationship between the summer monsoon rainfall anomalies in East Asia and the circulation anomalies in the upper troposphere over the Eurasian continent, Hsu and Lin (2007) also concluded that the meridional triple structure of summertime monsoon rainfall anomalies over East Asia is related to the propagation of the Rossby wave-train along the Asian jet at the upper troposphere in addition to the EAP pattern teleconnection over East Asia and the western North Pacific.

According to the study by Lu (2004), the EAP pattern teleconnection exhibits an intraseasonal difference between early summer and late summer in addition to its interannual variability. Generally, it is weak in June but strong in July and August. Lu (2004) attributed this intraseasonal difference to two reasons. First, June is a transition month from spring with a strong vertical westerly shear over the western North Pacific to late summer (July and August) with a strong vertical easterly shear; therefore, the vertical shear over the western North Pacific is near zero in June. The near-zero vertical shear is unfavorable for the coupling of external mode and internal mode excited by the precipitation anomaly in the western North Pacific, so it weakens the meridional teleconnection (e.g., Lu, 2004; Lin and Lu, 2008). Second, the EAJS abruptly jumps northward from 40°N in mid-July to 45°N in late July (e.g., Lin and Lu, 2008). When the jet axis is located more northward, the vertical shear is favorable for the coupling of external mode and internal mode. Both the lower- and the upper-level responses are significantly stronger, so the meridional teleconnection becomes stronger (e.g., Kosaka and Nakamura, 2010; Ye and Lu, 2011). This result suggests that the EAP pattern teleconnection can be modulated by the "Silk Road pattern" teleconnection.

From the studies mentioned here, it can be concluded that the meridional tripole structure of summertime circulation anomalies over East Asia are also associated not only with the EAP teleconnection but also with the "Silk Road pattern" teleconnection along the Asian jet in the upper troposphere. Furthermore, the "Silk Road pattern" teleconnection can modulate the EAP teleconnection.

6.3 Impact of quasi-stationary planetary wave activity on the EAWM system variability

As discussed in subsections 6.1 and 6.2, both the EAP teleconnection and the "Silk Road pattern" teleconnection are closely associated with the propagations of quasi-stationary planetary waves over the Northern Hemisphere in summer. Thus, quasi-stationary planetary wave activity has a significant impact on EASM system variability. Similarly, quasi-stationary planetary wave activity also has an important effect on EAWM system variability.

Early in the 1980s, Huang and Gambo (1982a,b) investigated the three-dimensional propagation of quasi-stationary planetary waves responding to forcing by topography and stationary heat sources in the troposphere in boreal winter with a 34-level model in addition E-P flux of waves. Two wave guides were discernible in the three-dimensional propagations of quasi-stationary planetary waves in the Northern Hemisphere winter. That is, in addition to the polar wave guide by which the quasi-stationary planetary waves propagate from the troposphere to the stratosphere over high latitudes (e.g., Dickinson, 1968), there is a second wave guide, namely, the low-latitude wave guide. The waves can propagate from the lower troposphere over middle and high latitudes to the upper troposphere over low latitudes along this low-latitude wave guide. Based on these studies, Chen et al. (2002, 2003, 2005) and Chen and Huang (2005) systematically studied the interannual variations of propagating wave guides for quasi-stationary waves with E-P fluxes using the NCEP/NCAR and ERA-40 reanalysis data and AGCM simulation data. Their results suggest an out-of-phase oscillation between these two wave guides for quasi-stationary planetary waves. When the polar wave guide is strong in a winter, the low-latitude wave guide is weak; when the polar wave guide is weak, the low-latitude wave guide is strong. Moreover, the anomalous propagation of quasi-stationary planetary waves characterized as the convergence/divergence of the wave E-P fluxes can induce a dipole mode in the anomaly distribution of zonal mean zonal wind anomalies. This shows that, due to the wave-flow interaction, the interannual oscillation of these two wave guides of quasi-stationary planetary waves has a significant influence on Arctic Oscillation (AO), which is closely related to EAWM system variability (e.g., Gong et al., 2001; Gong and Ho, 2003) through the Northern Annular Mode (NAM, Thompson and Wallace, 1998, 2000).

Recently, some studies have shown that the stratospheric circulation anomalies can influence the EAWM system variability. Along with the anomalous cold event that occurred over South China in January 2008, a downward propagation of stratospheric zonal wind anomalies was observed in the polar region of the Northern Hemisphere (Gu et al., 2008; Yi et al., 2009). In December 2009, the anomalous cold events over East Asia were also accompanied by significant downward propagations of stratospheric signals (Wang and Chen, 2010b). Therefore, the preceding upward propagation of anomalous planetary waves from the troposphere into the stratosphere and the succeeding downward propagation of anomalous zonal flow and low temperatures in the stratosphere and their impact on the troposphere may have key roles in the mechanisms of these cold events.

The activity of planetary waves also undergoes obvious interdecadal variation, which is well related to EAWM system variability (e.g., Huang and Wang, 2006; Wang et al., 2009b). Compared to the period 1976–1987, the southward propagation of quasi-stationary planetary waves after 1988 was enhanced along the low-latitude wave guide in the troposphere, and the upward propagation of waves into the stratosphere was reduced along the polar wave guide (Fig. 20). This can cause a weakened subtropical jet around 35°N due to the convergence of the wave E-P fluxes. The East Asian jet stream was then weakened, which led to the weakening of the EAWM system from 1988. In addition, the amplitude of quasi-stationary planetary waves was significantly decreased around 45°N, which was related to the reduced upward propagation of waves from the lower troposphere after 1988. The decreased amplitude of planetary waves weakened both the Siberian High and the Aleutian Low and led to a decrease in the pressure gradient between them; then, the EAWM system was weakened. Further analyses indicate that the planetary wave zonal wavenumber 2 played the dominant role in this process of the interdecadal variability (Wang et al., 2009b).

7. Conclusions and remarks for future studies

In recent years, many significant advances have taken place in understanding the characteristics and causes of the spatio-temporal variabilities of the EAM system associated with climate disasters in China. The basic physical processes, both internal and external, that influence these variabilities have also been studied further.

However, it should be pointed out that many problems regarding the basic physical processes of the EAM climate system variability and their impacts on climate disasters in China remain unclear. These problems include the association among the onset, activity, and decline of the EASM and the EAWM systems over East Asia as well as cycle processes between them, the association of the EAM system with the SAM and the NAM systems, the interactions and their physical processes among different time-scale variabilities of various components of the EAM system, the effect of extratropical process on the EAM variabilities, the physical mechanism of the second leading mode (i.e., the meridional dipole pattern distribution) of the EASM system variability, the evolution trend of the EAM system under the background of the global warming, and so on. These important issues must be investigated further. Specifically, the EAM climate system vari-

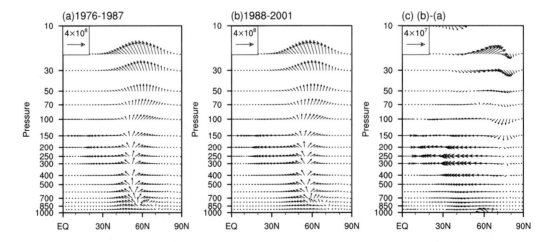

Fig. 20. Composite distribution of the E-P fluxes ($\times \rho^{-1}$) for the quasi-stationary planetary waves 1–3 averaged for the boreal winters of (a) 1976–1987, and (b) 1988–2001, and (c) their difference based on ERA-40 reanalysis data.

abilities on interannual and interdecadal time scales and their physical mechanisms should be emphasized in future studies. If the physical processes of these variabilities can be revealed further, the seasonal and annual prediction of climate disasters in China, such as droughts, floods, cold winter and hot summer, can be improved. Therefore, future studies must aim to understand the dynamical and thermodynamical processes affecting the spatio-temporal variabilities of the EAM climate system. Through the implementation of National Research Programs, it will be possible to more completely understand the physical mechanisms of the spatio-temporal variabilities of the EAM system.

Acknowledgements. The comments from two anonymous reviewers and the editor are appreciated. We thank Dr. DU Zhencai for his help with the references. This paper was supported jointly by the National Basic Research Program of China 973 Projects (Grant No. 2010CB950403), the National Special Scientific Research Project for Public Interest (Meteorology) (Grant No. GYHY201006021), the Chinese Academy of Sciences (Grant No. KZCX2-EW-QN204), and the National Natural Science Foundation of China (Grant No. 40975046).

REFERENCES

Annamalai, H., P. Liu, and S. P. Xie, 2005: Southwest Indian Ocean SST variability: Its local effect and remote influence on Asian monsoons. *J. Climate*, **18**, 4150–4167.

Ashok, K., S. K. Behera, S. A. Rao, H. Y. Weng, and T. Yamagata, 2007: El Niño Modoki and its possible teleconnection. *J. Geophys. Res.*, **112**, doi: 10.1029/2006JC003798.

Barriopedro, D., C. M. Gouveia, R. M. Trigo, and L. Wang, 2012: The 2009–2010 drought in China: Possible causes and impacts on vegetation. *J. Hydrometeorology*, doi: 10.1175/JHM-D-11-074.1. (in press)

Bond, N. A., and D. E. Harrison, 2000: The Pacific decadal oscillation, air-sea interaction and central north Pacific winter atmospheric regimes. *Geophys. Res. Lett.*, **27**, 731–734.

Cai, R. S., J. L. Chen, and R. H. Huang, 2006: The response of marine environment in the offshore area of China and its adjacent ocean to recent global climate change. *Chinese J. Atmos. Sci.*, **30**, 1019–1033. (in Chinese)

Cai, R. S., J. L. Chen, and H. J. Tan, 2011: Variations of the sea surface temperature in the offshore area of China and its relationship with the East Asian monsoon under the global warming. *Climatic and Environmental Research*, **16**, 94–104. (in Chinese)

Chang, C. P., J. E. Erickson, and K. M. Lau, 1979: Northeasterly Cold Surges and near-Equatorial Disturbances over the Winter Monex Area during December 1974 .1. Synoptic Aspects. *Mon. Wea. Rev.*, **107**, 812–829.

Chang, C. P., Y. S. Zhang, and T. Li, 2000a: Interannual and interdecadal variations of the East Asian summer monsoon and tropical Pacific SSTs. Part I: Roles of the subtropical ridge. *J. Climate*, **13**, 4310–4325.

Chang, C. P., Y. S. Zhang, and T. Li, 2000b: Interannual and interdecadal variations of the East Asian summer monsoon and tropical Pacific SSTs. Part II: Meridional structure of the monsoon. *J. Climate*, **13**, 4326–4340.

Chen, J. L., and R. H. Huang, 2006: The comparison of climotological characteristics among Asian and Australian monsoon subsystem. Part I. The wind structure of summer monsoon. *Chinese J. Atmos. Sci.*, **30**, 1091–1102. (in Chinese)

Chen, J. L., and R. H. Huang, 2007: The comparison of climotological characteristics among Asian and Australian monsoon subsystem. Part II. Water vapor transport by summer monsoon. *Chinese J. Atmos. Sci.*, **31**, 766–778. (in Chinese)

Chen, L. X., Q. G. Zhu, H. B. Luo, and J. H. He, 1991: *East Asian Monsoon*. China Meteorological Press, Beijing, 362pp. (in Chinese)

Chen, L. X., X. J. Zhou, W. L. Li, Y. F. Luo, and W. Q. Zhu, 2004: Characteristics of the climate change and its formation mechanism in China during the last 80 years. *Acta Meteorolologica Sinica*, **62**, 634–646. (in Chinese)

Chen, W., 2002: Impacts of El Niño and La Niña on the cycle of the East Asian winter and summer monsoon. *Chinese J. Atmos. Sci.*, **26**, 595–610. (in Chinese)

Chen, W., and H.-F. Graf, 1998: The interannual variability of East Asian winter monsoon and its relation to global circulation. Report No. 250.

Chen, W., and R. H. Huang, 2005: The three-dimensional propagation of quasi-stationary planetary waves in the Northern Hemisphere winter and its interannual variations. *Chinese J. Atmos. Sci.*, **29**, 137–146. (in Chinese)

Chen, W., H. F. Graf, and R. H. Huang, 2000: The interannual variability of East Asian winter monsoon and its relation to the summer monsoon. *Adv. Atmos. Sci.*, **17**, 48–60.

Chen, W., H. F. Graf, and M. Takahashi, 2002: Observed interannual oscillations of planetary wave forcing in the Northern Hemisphere winter. *Geophys. Res. Lett.*, **29**, doi: 10.1029/2002gl016062.

Chen, W., M. Takahashi, and H. F. Graf, 2003: Interannual variations of stationary planetary wave activity in the northern winter troposphere and stratosphere and their relations to NAM and SST. *J. Geophys. Res.*, **108**, doi: 10.1029/2003jd003834.

Chen, W., S. Yang, and R. H. Huang, 2005: Relationship between stationary planetary wave activity and the East Asian winter monsoon. *J. Geophys. Res.*, **110**, doi: 10.1029/2004JD005669.

Chen, W., L. Wang, Y. Xue, and S. Sun, 2009: Variabilities of the spring river runoff system in East China and their relations to precipitation and sea surface temperature. *Int. J. Climatol.*, **29**, 1381–1394.

Cheng, A., W. Chen, and R. Huang, 1998: The Sensitivity of Numerical Simulation of the East Asian Monsoon to Different Cumulus Parameterization Schemes. *Adv. Atmos. Sci.*, **15**, 204–220.

Cornejo-Garrido, A. G., and P. H. Stone, 1977: On the Heat Balance of the Walker Circulation. *J. Atmos. Sci.*, **34**, 1155–1162.

Deng, W. T., Z. B. Sun, G. Zeng, and D. H. Ni, 2009: Interdecadal variation of summer precipitation pattern over eastern China and its relationship with the North Pacific SST. *Chinese J. Atmos. Sci.*, **33**, 835–846. (in Chinese)

Dickinson, R. E., 1968: Planetary Rossby Waves Propagating Vertically through Weak Westerly Wind Wave Guides. *J. Atmos. Sci.*, **25**, 984–1002.

Ding, Q. H., and B. Wang, 2005: Circumglobal teleconnection in the Northern Hemisphere summer. *J. Climate*, **18**, 3483–3505.

Ding, Y. H., 1994: Monsoons over China. Dordrecht/Voston/London: Springer/Kluwer Academic Publishers, 432pp.

Ding, Y. H., 2007: The variability of the Asian summer monsoon. *J. Meteor. Soc. Japan*, **85**, 21–54.

Ding, Y. H., and C. He, 2006: The summer monsoon onset over the tropical eastern Indian Ocean: The earliest onset process of the Asian summer monsoon. *Adv. Atmos. Sci.*, **23**, 940–950.

Ding, Y. H., and X. Q. Ma, 2007: Analysis of isentropic potential vorticity for a strong cold wave in 2004/2005 winter. *Acta Meteorologica Sinica*, **65**, 695–707. (in Chinese)

Ding, Y. H., Z. Y. Wang, and Y. Sun, 2008: Inter-decadal variation of the summer precipitation in East China and its association with decreasing Asian summer monsoon. Part I: Observed evidences. *Int. J. Climatol.*, **28**, 1139–1161.

Du, Z. C., R. H. Huang, G. Huang, and J. L. Chen, 2011: The characteristics of spatial and temporal distributions of convective rainfall and stratiform rainfall in the Asian monsoon region and their possible mechanisms. *Chinese J. Atmos. Sci.*, **35**, 993–1008. (in Chinese)

Enomoto, T., B. J. Hoskins, and Y. Matsuda, 2003: The formation mechanism of the Bonin high in August. *Quart. J. Roy. Meteor. Soc.*, **129**, 157–178.

Feng, J., L. Wang, W. Chen, S. Fong, and K. Leong, 2010: Different impacts of two types of Pacific Ocean warming on the Southeast Asian rainfall during boreal winter. *J. Geophys. Res.*, **115**, D24122, doi: 10.1029/2010JD014761.

Fu, Y. F., Y. H. Lin, G. S. Liu, and Q. Wang, 2003: Seasonal characteristics of precipitation in 1998 over East Asia as derived from TRMM PR. *Adv. Atmos. Sci.*, **20**, 511–529.

Gong, D. Y., S. W. Wang, and J. H. Zhu, 2001: East Asian winter monsoon and Arctic Oscillation. *Geophys. Res. Lett.*, **28**, 2073–2076.

Gong, D. Y., and C. H. Ho, 2003: Arctic oscillation signals in the East Asian summer monsoon. *J. Geophys. Res.*, **108**, doi: 10.1029/2002JD002193.

Gu, L., K. Wei, and R. H. Huang, 2008: Severe disaster of blizzard, freezing rain and low temperature in January 2008 in China and its association with the anomalies of East Asian monsoon system. *Climatic and Environmental Research*, **13**, 405–418. (in Chinese)

Hartmann, D. L., H. H. Hendon, and R. A. Houze, 1984: Some implications of the mesoscale circulations in tropical cloud clusters for large-scale dynamics and climate. *J. Atmos. Sci.*, **41**, 113–121.

He, H. Y., C. H. Sui, M. Q. Jian, Z. P. Wen, and G. D. Lan, 2003: The evolution of tropospheric temperature field and its relationship with the onset of Asian summer monsoon. *J. Meteor. Soc. Japan*, **81**, 1201–1223.

Hong, C. C., and T. Li, 2009: The extreme cold anomaly over Southeast Asia in February 2008: Roles of ISO and ENSO. *J. Climate*, **22**, 3786–3801.

Hsu, H. H., and S. M. Lin, 2007: Asymmetry of the tripole rainfall pattern during the east Asian summer. *J. Climate*, **20**, 4443–4458.

Huang, G., and K. M. Hu, 2008: Impact of North Indian Ocean SSTA on Northwest Pacific lower layer anomalous anticyclone in summer. *Journal of Nanjing Institute of Meteorology*, **31**, 749–757. (in Chinese)

Huang, G., K. M. Hu, and S. P. Xie, 2010: Strengthening of tropical Indian Ocean teleconnection to the Northwest Pacific since the mid-1970s: An atmospheric GCM study. *J. Climate*, **23**, 5294–5304.

Huang, G., X. Qu, and K. M. Hu, 2011a: The impact of the tropical Indian Ocean on South Asian high in boreal summer. *Adv. Atmos. Sci.*, **28**, 421–432.

Huang, G., Y. Liu, and R. H. Huang, 2011b: The interannual variability of summer rainfall in the arid and semiarid regions of northern China and its association with the Northern Hemisphere circumglobal teleconnection. *Adv. Atmos. Sci.*, **28**, 257–268.

Huang, P., and R. Huang, 2009: Relationship between the modes of winter tropical Pacific SSTAs and the intraseasonal variations of the following summer rainfall anomalies in China. *Atmospheric and Oceanic Science Letters*, **2**, 295–300.

Huang, R. H., 1984: The characteristics of the forced stationary planetary wave propagations in summer Northern Hemisphere. *Adv. Atmos. Sci.*, **1**, 84–94.

Huang, R. H., 1985: The numerical simulation of the three-dimensional teleconnections in the summer circulation over the Northern Hemisphere. *Adv. Atmos. Sci.*, **2**, 81–92.

Huang, R. H., 2006: Progresses in research on the formation mechanism and prediction theory of severe climatic disasters in China. *Advances in Earth Science*, **21**, 564–575. (in Chinese)

Huang, R. H., 2009: Recent progresses in studies of variations and anomalies of East Asian monsoon (EAM) climate system and formation mechanism of severe climate disasters in China. *Bulletin of the Chinese Academy of Sciences*, **23**, 222–225. (in Chinese)

Huang, R. H., and K. Gambo, 1982a: The response of a hemispheric multilevel model atmosphere to forcing by topography and stationary heat-sources. 1. Forcing by topography. *J. Meteor. Soc. Japan*, **60**, 78–92.

Huang, R. H., and K. Gambo, 1982b: The response of a hemispheric multilevel model atmosphere to forcing by topography and stationary heat-sources. 2. Forcing by stationary heat-sources and forcing by topography and stationary heat-sources. *J. Meteor. Soc. Japan*, **60**, 93–108.

Huang, R. H., and W. J. Li, 1987: Influence of the heat source anomaly over the tropical western Pacific on the subtropical high over East Asia. *Proc. Int. Conf. on the General Circulation of East Asia*, Chengdu, 40–51.

Huang, R. H., and W. J. Li, 1988: Influence of heat source anomaly over the western tropical Pacific on the subtropical high over East Asia and its physical mechanism. *Chinese J. Atmos. Sci.*, **12**, 107–116. (in Chinese)

Huang, R. H., and Y. F. Wu, 1989: The influence of ENSO on the summer climate change in China and its mechanism. *Adv. Atmos. Sci.*, **6**, 21–32.

Huang, R. H., and F. Y. Sun, 1992: Impacts of the tropical western Pacific on the East Asian summer monsoon. *J. Meteor. Soc. Japan*, **70**, 243–256.

Huang, R. H., and L. T. Zhou, 2002: Research on the characteristics, formation mechanism and prediction of severe climatic disasters in China. *Journal of Natural Disasters*, **1**, 1–9. (in Chinese)

Huang, R. H., and L. Wang, 2006: Interdecadal variation of Asian winter monsoon and its association with the planetary wave activity. *Proc. Symposium on Asian Monsoon*, Kuala Lumpur, Malaysia, 126.

Huang, R. H., and J. L. Chen, 2010: Characteristics of the summertime water vapor transports over the eastern part of China and those over the western part of China and their difference. *Chinese J. Atmos. Sci.*, **34**, 1035–1045. (in Chinese)

Huang, R. H., Z. Z. Zhang, G. Huang, and B. H. Ren, 1998a: Characteristics of the water vapor transport in East Asian monsoon region and its difference from that in South Asian monsoon region in summer *Chinese J. Atmos. Sci.*, **22**, 460–469. (in Chinese)

Huang, R. H., Y. H. Xu, P. F. Wang, and L. T. Zhou, 1998b: The features of the catastrophic flood over the Changjiang River Basin during the summer of 1998 and cause exploration. *Climatic and Environmental Research*, **3**, 300–313. (in Chinese)

Huang, R. H., Y. H. Xu, and L. T. Zhou, 1999: The interdecadal variation of summer precipitations in China and the drought trend in North China. *Plateau Meteorology*, **18**, 465–476. (in Chinese)

Huang, R. H., L. T. Zhou, and W. Chen, 2003: The progresses of recent studies on the variabilities of the East Asian monsoon and their causes. *Adv. Atmos. Sci.*, **20**, 55–69.

Huang, R. H., G. Huang, and Z. G. Wei, 2004a: Climate variations of the summer monsoon over China. *East Asian Monsoon*, Chang, Ed., World Scientific Publishing Co. Pte. Ltd., Singapore, 213–270.

Huang, R. H., W. Chen, B. L. Yang, and R. H. Zhang, 2004b: Recent advances in studies of the interaction between the East Asian winter and summer monsoons and ENSO cycle. *Adv. Atmos. Sci.*, **21**, 407–424.

Huang, R. H., L. Gu, Y. H. Xu, Q. L. Zhang, S. S. Wu, and J. Cao, 2005: Characteristics of the interannual variations of onset and advance of the East Asian summer monsoon and their associations with thermal states of the tropical western Pacific. *Chinese J. Atmos. Sci.*, **29**, 20–36. (in Chinese)

Huang, R. H., R. S. Cai, J. L. Chen, and L. T. Zhou, 2006a: Interdecaldal variations of drought and flooding disasters in China and their association with the East Asian climate system. *Chinese J. Atmos. Sci.*, **30**, 730–743. (in Chinese)

Huang, R. H., J. L. Chen, G. Huang, and Q. L. Zhang, 2006b: The quasi-biennial oscillation of summer monsoon rainfall in China and its cause. *Chinese J. Atmos. Sci.*, **30**, 545–560. (in Chinese)

Huang, R. H., L. Gu, L. T. Zhou, and S. S. Wu, 2006c: Impact of the thermal state of the tropical western Pacific on onset date and process of the South China Sea summer monsoon. *Adv. Atmos. Sci.*, **23**, 909–924, doi: 10.1007/s00376-006-0909-1.

Huang, R. H., J. L. Chen, and G. Huang, 2007a: Characteristics and variations of the East Asian monsoon system and its impacts on climate disasters in China. *Adv. Atmos. Sci.*, **24**, 993–1023, doi: 10.1007/s00376-007-0993-x.

Huang, R. H., K. Wei, J. L. Chen, and W. Chen, 2007b: The East Asian winter monsoon anomalies in the winters of 2005 and 2006 and their relations to the quasi-stationary planetary wave activity in the northern hemisphere. *Chinese J. Atmos. Sci.*, **31**, 1033–1048. (in Chinese)

Huang, R. H., L. Gu, J. L. Chen, and G. Huang, 2008: Recent progresses in studies of the temporal-spatial variations of the East Asian monsoon system and their impacts on climate anomalies in China. *Chinese J. Atmos. Sci.*, **32**, 691–719. (in Chinese)

Huang, R. H., J. L. Chen, and Y. Liu, 2011c: Interdecadal variation of the leading modes of summertime precipitation anomalies over Eastern China and its association with water vapor transport over East Asia. *Chinese J. Atmos. Sci.*, **35**, 589–606. (in Chinese)

Huang, R. H., Y. Liu, and T. Feng, 2012a: Characteristics and causes of the interecadal jump of summertime monsoon rainfall and circulation in eastern China occurred in the late 1990s. *Chinese Sci. Bull.*. (in press)

Huang, R. H., Y. Liu, L. Wang, and L. Wang, 2012b: Analyses of the causes of severe drought occurred in Southwest China from the fall of 2009 to the spring to 2010. *Chinese J. Atmos. Sci.*, **36**, 443–457. (in

Chinese)
Iguchi, T., T. Kozu, R. Meneghini, J. Awaka, and K. Okamoto, 2000: Rain-profiling algorithm for the TRMM precipitation radar. *J. Appl. Meteor.*, **39**, 2038–2052.
Jhun, J. G., and E. J. Lee, 2004: A new East Asian winter monsoon index and associated characteristics of the winter monsoon. *J. Climate*, **17**, 711–726.
Kalnay, E., and Coauthors, 1996: The NCEP/NCAR 40-year reanalysis project. *Bull. Amer. Meteor. Soc.*, **77**, 437–471.
Kang, L. H., W. Chen, and K. Wei, 2006: The Interdecadal Variation of Winter Temperature in China and Its Relation to the Anomalies in Atmospheric General Circulation. *Climatic and Environmental Research*, **11**, 330–339. (in Chinese)
Kang, L. H., W. Chen, L. Wang, and L. J. Chen, 2009: Interannual variations of winter temperature in China and their relationship with the atmospheric circulation and sea surface temperature. *Climatic and Environmental Research*, **14**, 45–53. (in Chinese)
Kosaka, Y., and H. Nakamura, 2006: Structure and dynamics of the summertime Pacific-Japan teleconnection pattern. *Quart. J. Roy. Meteor. Soc.*, **132**, 2009–2030.
Kosaka, Y., and H. Nakamura, 2010: Mechanisms of Meridional Teleconnection Observed between a Summer Monsoon System and a Subtropical Anticyclone. Part I: The Pacific-Japan Pattern. *J. Climate*, **23**, 5085–5108.
Kurihara, K., 1989: A climatological study on the relationship between the Japanese summer weather and the subtropical high in the western northern Pacific. *Geophys. Mag.*, **43**, 45–104.
Kwon, M., J. G. Jhun, and K. J. Ha, 2007: Decadal change in east Asian summer monsoon circulation in the mid-1990s. *Geophys. Res. Lett.*, **34**, L21706, doi: 21710.21029/22007GL031977.
Lau, K. M., and C. P. Chang, 1987: Planetary scale aspects of the winter monsoon and atmospheric teleconnections. *Monsoon Meteorology*, Chang and Krishnamurti, Eds., Oxford University Press, Oxford, 161–201.
Li, S. L., J. Lu, G. Huang, and K. M. Hu, 2008: Tropical Indian Ocean basin warming and East Asian summer monsoon: A multiple AGCM study. *J. Climate*, **21**, 6080–6088.
Li, X. Z., Z. P. Wen, and W. Zhou, 2011: Long-term Change in Summer Water Vapor Transport over South China in Recent Decades. *J. Meteor. Soc. Japan*, **89A**, 271–282.
Li, Y. Q., and S. Yang, 2010: A dynamical index for the East Asian winter monsoon. *J. Climate*, **23**, 4255–4262.
Liang, J. Y., and S. S. Wu, 2002: A study of southwest monsoon onset date over the South China Sea and its impact factors. *Chinese J. Atmos. Sci.*, **26**, 829–844. (in Chinese)
Lin, Z. D., and R. Y. Lu, 2005: Interannual meridional displacement of the east Asian upper-tropospheric jet stream in summer. *Adv. Atmos. Sci.*, **22**, 199–211.
Lin, Z. D., and R. Y. Lu, 2008: Abrupt northward jump of the East Asian upper-tropospheric jet stream in mid-summer. *J. Meteor. Soc. Japan*, **86**, 857–866.
Liu, P., and Y. F. Fu, 2010: Climatic characteristics of summer convective and stratiform precipitation in Southern China based on measurements by TRMM precipitation radar. *Chinese J. Atmos. Sci.*, **34**, 802–814. (in Chinese)
Lu, M. M., and C. P. Chang, 2009: Unusual late-season cold surges during the 2005 Asian winter monsoon: Roles of Atlantic blocking and the Central Asian anticyclone. *J. Climate*, **22**, 5205–5217.
Lu, R. Y., 2004: Associations among the components of the east Asian summer monsoon system in the meridional direction. *J. Meteor. Soc. Japan*, **82**, 155–165.
Lu, R. Y., and B. J. Kim, 2004: The climatological Rossby wave source over the STCZs in the summer northern hemisphere. *J. Meteor. Soc. Japan*, **82**, 657–669.
Lu, R. Y., and Z. D. Lin, 2009: Role of subtropical precipitation anomalies in maintaining the summertime meridional teleconnection over the western North Pacific and East Asia. *J. Climate*, **22**, 2058–2072.
Lu, R. Y., J. H. Oh, and B. J. Kim, 2002: A teleconnection pattern in upper-level meridional wind over the North African and Eurasian continent in summer. *Tellus (A)*, **54**, 44–55.
Lu, R. Y., Y. Li, and B. W. Dong, 2006: External and internal summer atmospheric variability in the western North Pacific and East Asia. *J. Meteor. Soc. Japan*, **84**, 447–462.
Ma, X. Q., Y. H. Ding, H. Xu, and J. H. He, 2008: The relation between strong cold waves and low-frequency waves during the winter of 2004/2005. *Chinese J. Atmos. Sci.*, **32**, 380–394. (in Chinese)
Nitta, T., 1987: Convective activities in the tropical western Pacific and their impact on the Northern Hemisphere summer circulation. *J. Meteor. Soc. Japan*, **65**, 373–390.
Sun, X. G., and X. Q. Yang, 2005: Numerical modeling of interannual anomalous atmospheric circulation patterns over East Asia during different stages of an El Niño event. *Chinese Journal of Geophysics*, **48**, 501–510. (in Chinese)
Tao, S. Y., and F. K. Zhu, 1964: The 100-mb flow patterns in southern Asia in summer and its relation to the advance and retreat of the West-Pacific subtropical anticyclone over the far east. *Acta Meteorologica Sinica*, **34**, 385–396. (in Chinese)
Tao, S. Y., and L. X. Chen, 1985: The East Asian summer monsoon. *Proc. Int. Conf. on Monsoon in the Far East,* Tokyo, 1–11.
Tao, S. Y., and L. X. Chen, 1987: A review of recent research on the East Asia summer monsoon in China. *Monsoon Meteorology,* Chang and Krishnamurti, Eds., Oxford University Press, Oxford, 60–92.
Tao, S. Y., and J. Wei, 2006: The westward, northward

advance of the subtropical high over the West Pacific in summer. *J. Appl. Meteor. Sci.*, **17**, 513–525.

Thompson, D. W. J., and J. M. Wallace, 1998: The Arctic Oscillation signature in the wintertime geopotential height and temperature fields. *Geophys. Res. Lett.*, **25**, 1297–1300.

Thompson, D. W. J., and J. M. Wallace, 2000: Annular modes in the extratropical circulation. Part I: Month-to-month variability. *J. Climate*, **13**, 1000–1016.

Trenberth, K. E., and J. W. Hurrell, 1994: Decadal atmosphere-ocean variations in the Pacific. *Climate Dyn.*, **9**, 303–319.

Tu, C. W., and S. S. Huang, 1944: The advance and retreat of the summer monsoon. *Acta Meteorologica Sinica*, **18**, 82–92. (in Chinese)

Uppala, S. M., and Coauthors, 2005: The ERA-40 re-analysis. *Quart. J. Roy. Meteor. Soc.*, **131**, 2961–3012.

Wallace, J. M., and D. S. Gutzler, 1981: Teleconnections in the Geopotential Height Field during the Northern Hemisphere Winter. *Mon. Wea. Rev.*, **109**, 784–812.

Wang, B., R. G. Wu, and K. M. Lau, 2001: Interannual variability of the Asian summer monsoon: Contrasts between the Indian and the western North Pacific-East Asian monsoons, *J. Climate*, **14**, 4073–4090.

Wang, B., R. G. Wu, and T. Li, 2003: Atmosphere-warm ocean interaction and its impacts on Asian-Australian monsoon variation. *J. Climate*, **16**, 1195–1211.

Wang, B., LinHo, Y. S. Zhang, and M. M. Lu, 2004: Definition of South China Sea monsoon onset and commencement of the East Asia summer monsoon. *J. Climate*, **17**, 699–710.

Wang, B., Z. W. Wu, C. P. Chang, J. Liu, J. P. Li, and T. J. Zhou, 2010: Another Look at Interannual-to-Interdecadal Variations of the East Asian Winter Monsoon: The Northern and Southern Temperature Modes. *J. Climate*, **23**, 1495–1512.

Wang, L., W. Chen, and R. H. Huang, 2007: Changes in the variability of North Pacific Oscillation around 1975/1976 and its relationship with East Asian winter climate. *J. Geophys. Res.*, **112**, doi: 10.1029/2006JD008054.

Wang, L., W. Chen, and R. H. Huang, 2008: Interdecadal modulation of PDO on the impact of ENSO on the east Asian winter monsoon. *Geophys. Res. Lett.*, **35**, doi: 10.1029/2008GL035287.

Wang, L., W. Chen, W. Zhou, and R. H. Huang, 2009a: Interannual variations of East Asian trough axis at 500 hPa and its association with the East Asian winter monsoon pathway. *J. Climate*, **22**, 600–614.

Wang, L., R. H. Huang, L. Gu, W. Chen, and L. H. Kang, 2009b: Interdecadal variations of the East Asian winter monsoon and their association with quasi-stationary planetary wave activity. *J. Climate*, **22**, 4860–4872.

Wang, L., and W. Chen, 2010a: How well do existing indices measure the strength of the East Asian winter monsoon? *Adv. Atmos. Sci.*, **27**, 855–870.

Wang, L., and W. Chen, 2010b: Downward Arctic Oscillation signal associated with moderate weak stratospheric polar vortex and the cold December 2009. *Geophys. Res. Lett.*, **37**, doi: 10.1029/2010gl042659.

Wang, L., and J. Feng, 2011: Two major modes of the wintertime precipitation over China. *Chinese J. Atmos. Sci.*, **35**, 1105–1116. (in Chinese)

Wang, L., and R. Wu, 2012: The in-phase transition from the East Asian winter to summer monsoon: Role of the Indian Ocean. *J. Geophys. Res.*, **117**, D11112, doi: 10.1029/2012JD017509.

Wang, Z. Y., and Y. H. Ding, 2006: Climate change of the cold wave frequency of China in the last 53 years and the possible reasons. *Chinese J. Atmos. Sci.*, **30**, 1068–1076. (in Chinese)

Webster, P. J., V. O. Magaña, T. N. Palmer, J. Shukla, R. A. Tomas, M. Yanai, and T. Yasunari, 1998: Monsoons: Processes, predictability, and the prospects for prediction. *J. Geophys. Res.*, **103**, 14451–14510.

Wei, K., W. Chen, and W. Zhou, 2011: Changes in the East Asian Cold Season since 2000. *Adv. Atmos. Sci.*, **28**, 69–79, doi: 10.1007/s00376-010-9232-y.

Wei, Z. G., R. H. Huang, W. Chen, and W. J. Dong, 2002: Spatial distributions and interdecadal variations of the snow at the Tibetan Plateau weather stations. *Chinese J. Atmos. Sci.*, **26**, 496–508. (in Chinese)

Wei, Z. G., R. H. Huang, and W. J. Dong, 2003: Interannual and interdecadal variations of air temperature and precipitation over the Tibetan Plateau. *Chinese J. Atmos. Sci.*, **27**, 157–170. (in Chinese)

Wen, M., S. Yang, A. Kumar, and P. Q. Zhang, 2009: An analysis of the large-scale climate anomalies associated with the snowstorms affecting China in January 2008. *Mon. Wea. Rev.*, **137**, 1111–1131.

Wen, Z. P., R. H. Huang, H. Y. He, and G. D. Lan, 2006: The influences of anomalous atmospheric circulation over mid-high latitudes and the activities of 30-60d low frequency convection over low latitudes on the onset of the South China Sea summer monsoon. *Chinese J. Atmos. Sci.*, **30**, 952–964. (in Chinese)

Weng, H. Y., K. Ashok, S. K. Behera, S. A. Rao, and T. Yamagata, 2007: Impacts of recent El Niño Modoki on dry/wet conditions in the Pacific rim during boreal summer. *Climate Dyn.*, **29**, 113–129.

Wu, B. Y., and J. Wang, 2002: Winter Arctic Oscillation, Siberian High and East Asian winter monsoon. *Geophys. Res. Lett.*, **29**, doi: 10.1029/2002GL015373.

Wu, B. Y., R. H. Zhang, and R. D'Arrigo, 2006: Distinct modes of the East Asian winter monsoon. *Mon. Wea. Rev.*, **134**, 2165–2179.

Wu, B. Y., K. Yang, and R. H. Zhang, 2009: Eurasian snow cover variability and its association with summer rainfall in China. *Adv. Atmos. Sci.*, **26**, 31–44, doi: 10.1007/s00376-009-0031-2.

Wu, G. X., and Y. S. Zhang, 1998: Tibetan Plateau forcing and the timing of the monsoon onset over South Asia and the South China Sea. *Mon. Wea. Rev.*, **126**, 913–927.

Wu, R. G., Z. P. Wen, S. Yang, and Y. Q. Li, 2010: An

interdecadal change in Southern China summer rainfall around 1992/93. *J. Climate*, **23**, 2389–2403.

Xie, S. P., K. M. Hu, J. Hafner, H. Tokinaga, Y. Du, G. Huang, and T. Sampe, 2009: Indian Ocean capacitor effect on Indo-Western Pacific climate during the summer following El Niño. *J. Climate*, **22**, 730–747.

Xu, G. Y., X. Q. Yang, and X. G. Sun, 2005: Interdecadal and interannual variation characteristics of rainfall in North China and its relation with the northern hemisphere atmospheric circulations. *Chinese Journal of Geophysics*, **48**, 511–518. (in Chinese)

Yang, J. L., Q. Y. Liu, S. P. Xie, Z. Y. Liu, and L. X. Wu, 2007: Impact of the Indian Ocean SST basin mode on the Asian summer monsoon. *Geophys. Res. Lett.*, **34**, doi: 10.1029/2006GL028571.

Yang, S., K. M. Lau, and K. M. Kim, 2002: Variations of the East Asian jet stream and Asian-Pacific-American winter climate anomalies. *J. Climate*, **15**, 306–325.

Yang, X. Q., and Y. M. Zhu, 2008: Interdecadal climate variability in China associated with the Pacific Decadal Oscillation. Chapter 3, *Regional Climate Studies in China*, Fu et al., Eds., Springer, 97–118.

Yang, X. Q., Q. Xie, Y. M. Zhu, X. G. Sun, and Y. J. Guo, 2005: Decadal-to-interdecadal variability of precipitation in North China and associated atmospheric and oceanic anomaly patterns. *Chinese Journal of Geophysics*, **48**, 789–797. (in Chinese)

Yasunari, T., and R. Suppiah, 1988: Some problems on the interannual variability of Indonesian monsoon rainfall. *Tropical Rainfall Measurements,* Theon and Fugono, Eds., Deepak, Hampton, Va, 113–112.

Ye, D. Z., and Y. X. Gao, 1979: *Tibetan Plateau Meteorology*. Science Press, Beijing, 279pp. (in Chinese)

Ye, H., and R. Y. Lu, 2011: Subseasonal variation in ENSO-related East Asian rainfall anomalies during summer and its role in weakening the relationship between the ENSO and summer rainfall in Eastern China since the late 1970s. *J. Climate*, **24**, 2271–2284.

Yi, M. J., Y. J. Chen, R. J. Zhou, and S. M. Deng, 2009: Analysis on isentropic potential vorticity for the snow calamity in South China and the stratospheric polar vortex in 2008. *Plateau Meteorology*, **28**, 880–888. (in Chinese)

Zhang, P. Q., S. Yang, and V. E. Kousky, 2005: South Asian high and Asian-Pacific-American climate teleconnection. *Adv. Atmos. Sci.*, **22**, 915–923.

Zhang, Q., G. X. Wu, and Y. F. Qian, 2002: The bimodality of the 100 hPa South Asia High and its relationship to the climate anomaly over East Asia in summer. *J. Meteor. Soc. Japan*, **80**, 733–744.

Zhang, Q. Y., J. M. Lü, and L. M. Yang, 2007: The interdecadal variation of precipitation pattern over China during summer and its relationship with the atmospheric internal dynamic processes and extra-forcing factors. *Chinese J. Atmos. Sci.*, **31**, 1290–1300. (in Chinese)

Zhang, R. H., and R. H. Huang, 1998: Dynamical roles of zonal wind stresses over the tropical Pacific on the occurring and vanishing of El Niño. Part I: Diagnostic and theoretical analyses. *Chinese J. Atmos. Sci.*, **22**, 587–599. (in Chinese)

Zhang, R. H., A. Sumi, and M. Kimoto, 1996: Impact of El Niño on the East Asian monsoon: A diagnostic study of the '86/87 and '91/92 events. *J. Meteor. Soc. Japan*, **74**, 49–62.

Zhang, R. H., B. Y. Wu, P. Zhao, and J. P. Han, 2008: The decadal shift of the summer climate in the late 1980s over Eastern China and its possible causes. *Acta Meteorologica Sinica*, **22**, 435–445.

Zhang, Z. H., and G. Huang, 2008: Different types of El Niño events and their relationships with China summer climate anomaly. *Journal of Nanjing Institute of Meteorology*, **31**, 782–789. (in Chinese)

Zhao, P., X. D. Zhang, Y. F. Li, and J. M. Chen, 2009: Remotely modulated tropical-North Pacific ocean-atmosphere interactions by the South Asian high. *Atmospheric Research*, **94**, 45–60.

Zhou, L. T., 2010: Characteristics of temporal and spatial variations of sensible heat flux in the arid and semi-arid region of Eurasia. *Transactions of Atmospheric Sciences*, **33**, 299–306. (in Chinese)

Zhou, L. T., 2011: Impact of East Asian winter monsoon on rainfall over southeastern China and its dynamical process. *Int. J. Climatol.*, **31**, 677-686.

Zhou, L. T., and R. H. Huang, 2006: Characterisitcs of the interdecadal variability of difference between surface temperature and surface air temperature in spring in the arid and semi-arid region of Northwest China and its impact on summer precipitation in North China. *Climatic and Environmental Research*, **11**, 1–13. (in Chinese)

Zhou, L. T., and R. H. Huang, 2008: Interdecadal variability of sensible heat in arid and semi-arid regions of Northwest China and its relation to summer precipitation in China. *Chinese J. Atmos. Sci.*, **32**, 1276–1288. (in Chinese)

Zhou, L. T., and R. H. Huang, 2010: Interdecadal variability of summer rainfall in Northwest China and its possible causes. *Int. J. Climatol.*, **30**, 549–557.

Zhou, X. X., Y. H. Ding, and P. X. Wang, 2010: Moisture Transport in the Asian Summer Monsoon Region and Its Relationship with Summer Precipitation in China. *Acta Meteorologica Sinica*, **24**, 31–42.

Zhu, K. Z., 1934: The enigma of southeast monsoon in China. *Acta Geographica Sinica*, **1**, 1–28. (in Chinese)

Zhu, Y. M., and X. Q. Yang, 2003: Relationships between Pacific Decadal Oscillation (PDO) and climate variabilities in China. *Acta Meteorologica Sinica*, **61**, 641–654. (in Chinese)

关于中国西北干旱区陆-气相互作用及其对气候影响研究的最近进展

黄荣辉[1]　周德刚[1]　陈　文[1]　周连童[1]　韦志刚[2]
张　强[3]　高晓清[2]　卫国安[2]　侯旭宏[2]

(1 中国科学院大气物理研究所季风系统研究中心，北京 100190；2 中国科学院寒区旱区环境与工程研究所，兰州 730000；3 中国气象局兰州干旱气象研究所，兰州 730020)

摘　要　本文综述了中国西北干旱区陆—气相互作用及其对气候影响研究的最近进展。文中不仅回顾了"中国西北干旱区陆—气相互作用观测试验"经过连续 12 年的观测和多次加强期观测所取得的干旱区陆面过程参数的分析以及边界层和陆—气相互作用特征等的分析和研究，而且综述了应用这些参数来优化有关陆面过程模式的参数化方案和改进有关陆面过程模式的研究；并且，本文还综述了关于中国西北干旱区感热输送特征以及西北干旱区陆—气相互作用对中国东部气候的影响及其机理，并揭示了中国西北干旱区春、夏季具有高感热输送特征，此高感热对中国东部夏季气候变异有重要影响。此外，本文还指出今后在此方面应进一步观测和深入研究的科学问题。

Recent Progress in Studies of Air–Land Interaction over the Arid Area of Northwest China and Its Impact on Climate

HUANG Ronghui[1], ZHOU Degang[1], CHEN Wen[1], ZHOU Liantong[1], WEI Zhigang[2],
ZHANG Qiang[3], GAO Xiaoqing[2], WEI Guoan[2], and HOU Xuhong[2]

1 *Center for Monsoon System Research, Institute of Atmospheric Physics, Chinese Academy of Sciences, Beijing* 100190
2 *Cold and Arid Regions Environmental and Engineering Research Institute, Chinese Academy of Sciences, Lanzhou* 730000
3 *Institute of Arid Meteorology, China Meteorological Administration, Lanzhou* 730020

Abstract　We review the recent progress in studies of air–land interaction over the arid area of Northwest China and its impact on climate. The paper examines the analyses of the data observed continuously during the Field Experiment on Air–Land Interaction in the Arid Area of Northwest China (NWC-ALIEX) for 12 years and during many Intense Observation Periods of this project. The analyses of the land-surface parameters and studies of the characteristics of the boundary layer and air–land interaction in the arid area from continuous observations were reviewed as well as applications of these parameters to optimize the parameterization scheme in land surface models and to improve related land surface models. Moreover, we examined the characteristics of sensible heat transfer in the arid area of Northwest

China and the impact of the air–land interaction in this area on climate variability in eastern China and its physical mechanism. We found a feature of high sensible heat transfer in spring and summer in the arid area of Northwest China, which has a strong impact on climate variability in eastern China. Finally, the issues that need further study and evaluation are also pointed out in this paper.

1 引言

从20世纪70年代中期开始，大气科学家对气候的认识有一个飞跃，认识到气候不仅仅是大气中动力、热力过程所形成，而且是地球系统中大气圈、水圈、冰雪圈、岩石圈和生物圈相互作用的结果，甚至与人类活动也有一定关系。而能量与水分循环是联系这个气候系统中各圈层的两个重要过程，它与地球气候系统各成员的变化紧密关联在一起的。

由于水分和能量循环对于气候变化与异常起着十分重要作用，因此，在世界气候研究计划（WCRP）之下特别制定了全球能量和水分循环试验计划（Global Energy and Water Cycle Experiment, 简称GEWEX）。GEWEX计划是全球最大的气候和环境科学试验，它在全球有代表性的下垫面开展了陆—气相互作用以及能量和水分循环观测试验（世界气象组织，2006）。

从20世纪80年代中期起，在国际上进行了许多大型陆—气相互作用观测试验，如：（1）湿润区水分收支和蒸发通量的水文—大气试验(Hydrologic Atmospheric Experiment for the Study of Water Budget and Evaporation Flux at the Climatic Scale, 简称为HAPEX/MOBILMY)，它主要研究湿润地区陆面过程的特征，在大、中、小三种尺度方面，关于湿润区的水汽和蒸发通量与气候变异之间的关系的研究取得进展（André et al., 1986）；（2）在20世纪90年代初，在欧洲进行了欧洲干旱化区域野外观测试验（European Field Experiment in a Desertification-threatened Area，简称EFEDA），此计划的完成不仅对欧洲半干旱地区陆面特征有了新的认识，而且在干旱化过程及机理方面取得进展（Bolle et al., 1993）；（3）在90年代，在加拿大进行了为期4年的北半球生态系统—大气研究计划（The Boreal Ecosystem-Atmosphere Study，简称BOREAS），此计划对加拿大北部森林与大气之间的相互影响及其对气候变化的影响进行了深入研究（Sellers et al., 1995, 1997）。通过这些大型的野外观测试验获取了大量有关不同下垫面陆-气相互作用的观测资料，同时也发展了一系列陆-气相互作用模式，如Biosphere–Atmosphere Transfer Scheme（简称BATS）和Simple Biosphere Model（简称SiB）等著名的陆面过程模式。

鉴于我国西北干旱区陆—气相互作用对东亚季风系统有严重影响（Huang et al., 2002），并且目前干旱区陆面过程有关参数还远远满足不了此区域能量和水分循环特征分析和气候数值模式的需要，为此，在"国家重大基础研究发展规划"首批启动项目"我国重大气候灾害的形成机理与预测理论研究"的资助下，继西北干旱区黑河地区地—气相互作用试验（胡隐樵，1994）之后，我们在我国西北干旱区进行陆—气相互作用的野外观测试验，并开展了有关分析研究。在此项目的资助下，在敦煌的双敦子戈壁下垫面、临泽的巴丹吉林沙漠下垫面，以及在五道梁的高寒地区分别建立了干旱区陆—气相互作用观测试验站（图1），并于2000～2003年在这三个观测试验站同时进行了"中国西北干旱区陆—气相互作用观测试验"(The Field Experiment on Air–land Interaction in the Arid Area of Northwest China, 简称为NWC-ALIEX)（黄荣辉，2006）。近几年来，在国家自然科学基金委员会重点基金项目"我国西北干旱区陆—气相互作用特征及其对气候影响机理研究"、中国科学院知识创新工程项目"干旱/半干旱地带陆—气相互作用及模型设计机理和数值模拟"、国家重大基础研究发展规划另一项目"全球变暖背景下东亚能量和水分循环变异及其对我国极端气候的影响"等的资助下，以及中国科学院资源与环境科学技术局"野外观测试验站运转费"的支持下，于2004～2012年相继在敦煌戈壁下垫面又进行了较长期的干旱区陆—气相互作用观测试验，把NWC-ALIEX计划延续至今。

通过此观测试验10多年的连续观测，积累了大量有关西北干旱区陆—气相互作用和陆面过程参数的原始观测资料，经过多年的分析研究，得到

图 1 中国西北干旱区陆-气相互作用试验观测站分布图

Fig. 1 Distribution of observational stations of the Field Experiment on Air-Land Interaction in the Arid Area of Northwest China (NWC-ALIEX)

了许多有关干旱区陆面过程参数、陆-气相互作用特征及其对气候影响方面很有意义的科学成果。由于有关此试验早期一些观测成果,张强等(2005)已作了全面总结,并且有关此试验的科学意义、试验方案和数据整编以及早期的观测、分析和数值模拟的科学成果已有著作进行了系统总结(黄荣辉等,2011),因而,此文重点是综述近几年来利用此试验 10 多年来的连续观测和近几年的加强期观测资料所得的部分科学成果,其中部分结果是至今还没有公开发表过的研究成果。

2 敦煌干旱区的陆面过程参数及陆-气相互作用特征

由于西北干旱区人烟稀少,缺乏饮用水,不利于野外观测,因此在广阔的沙漠和戈壁上缺乏气象参数。这不仅给分析中国西北气候特征带来困难,而且也造成干旱区气候数值模式中陆面过程参数的不确定性。为此,最近王超等(2010a)以及 Zhou and Huang(2011)对 NWC-ALIEX 连续 12 年的观测试验资料以及对加强期观测试验资料的质量问题进行了详细分析。分析结果表明:此观测试验的资料质量是比较好的,正常数据占总数据的 91.2%。非正常数据只占总数据的 8.8%(王超等,2010a)。因此,可利用此野外观测试验所取得资料来分析敦煌典型干旱区一些重要的陆面过程参数并揭示西北干旱区一些重要陆面过程特征。

2.1 敦煌干旱区地表附近动量总体输送系数和热力总体交换系数

Huang et al.(2005)和张强等(2005)利用 NWC-ALIEX 早期 1 年的观测数据,提出敦煌干旱地表动量粗糙度长度为 1.9 ± 0.7 mm,而感热粗糙度长度为 0.43 ± 0.32 mm,地表附近总体动量和热量输送系数分别为 $(1.63\pm1.68)\times10^{-3}$ 和 $(1.49\pm1.08)\times10^{-3}$(组合法)。最近,周德刚等(2012)利用 NWC-ALIEX 的多年观测试验资料,剔除人工物对观测的干扰后确定了敦煌干旱区动量粗糙度 z_{0m} 大约为 0.61 ± 0.02 mm,而热力粗糙度 z_{0h} 在白天平均约在 0.05 mm 左右。因此,用多年观测试验资料所确定的敦煌干旱地表动量粗糙度要比用 1 年的观测资料计算值要小一些,但它们量级都是一致的,也比黑河试验观测所得结果($1.7\sim4.5$ mm)也要小一些(左洪超和胡隐樵,1992);而热量粗糙度用多年观测试验资料所确定的值要比用 1 年的观测资料所计算的值(0.43 ± 0.32 mm)要小一个量级。并且,从超声探测仪观测的感热等资料和地—气温差资料计算的敦煌干旱区热力总体交换系数 C_h 大约为 $(2.3\pm0.2)\times10^{-3}$；而动量总体输送系数 C_D 大约为 $(3.0\pm0.2)\times10^{-3}$。由于受超声仪器的更换、资料处理方法的差异以及观测本身误差的影响，用后期多年观测试验资料所计算的动量输送系数要比应用 1 年的观测资料所计算的值要小一些，而热量输送要大一些，但两者量级相当，并与黑河观测试验所得结果也较接近。

这些参数对于优化陆面过程模式中有关参数化方案起到重要作用，从而对有关陆面过程的改进起到重要作用。

2.2 敦煌干旱区到达地表总辐射、净辐射和地表反照率的变化特征

为了说明 NWC-ALIEX 所测得的辐射具有代表性，首先把此观测试验在敦煌干旱区所得的 12 年辐射各分量与西北干旱区及北半球各地的辐射各分量求相关，结果表明：敦煌干旱区辐射各分量与北半球同纬度地区（特别是西北干旱区）的辐射各分量有很好的正相关，这说明敦煌测得的辐射各分量对于西北干旱区有较好的代表性。

（1）不同地表状况辐射各分量的日积分值

Huang et al.（2005）和张强等（2005）利用 NWC-ALIEX 在敦煌干旱区 2002 年 5~6 月加强期观测试验资料研究了敦煌戈壁和绿洲地表的热量和辐射平衡特征，不仅指出了敦煌戈壁区 5~6 月份在晴天时有很大的总辐射，最大可达 1000 W m^{-2}，这是其它区域较少观测到的，而且指出了敦煌戈壁地表和绿洲农田辐射平衡中各辐射分量的日积分值有很大不同。最近，王超等（2012）又利用 NWC-ALIEX 在敦煌稀疏植被 2009 年 7~9 月加强期观测资料分析了敦煌稀疏植被地表辐射平衡中各辐射分量的日积分值。表 1 比较了在敦煌不同地表辐射平衡中各辐射分量的日积分值。从表 1 可以看到：敦煌戈壁地表的总辐射、大气长波辐射和地表向上长波辐射的日积分值大于绿洲农田地表的总辐射、大气长波辐射和地表向上长波辐射，而戈壁地表的反辐射和净辐射日积分值要小于绿洲农田的地表反辐射和净辐射；并且从表 1 所列的敦煌稀疏植被辐射各分量的日积分值也可以看到，敦煌在 7~9 月总辐射和净辐射日积分值比 5~6 月份的总辐射和净辐射值小，这可能是由于夏至之后，日照时间变短，但稀疏下垫面大气的长波辐射、地表向上的长波辐射的日积分值比戈壁下垫面的大气长波辐射和地表向上长波辐射的日积分值小，但比绿洲下垫面的相应辐射分量的日积分值大。因此，上述结果不仅表明了西北干旱区辐射平衡中各分量因地表植被状况不同而有很大不同，而且也表明了植被是有利于地表温度的保持，地表植被愈少，地表向上的长波辐射愈大。

（2）戈壁区总辐射通量和净辐射通量的季节变化

从敦煌戈壁区 2001~2007 年的 7 年辐射各分量观测资料的分析结果表明：敦煌干旱区各年总辐射通量在夏季最大，约 760 W m^{-2}，春季略小于夏季，秋季次之，冬季最小（图 2a）；并且，净辐射也是在夏季最大，约 370 W m^{-2}，春季略小于夏季，秋季次之，冬季最小（图 2b）。

把敦煌干旱区所观测的太阳总辐射和反射辐射与 NCEP/NCAR 再分析的敦煌附近地表总辐射和反射辐射通量资料（Kalnay et al., 1996）相比较，如图 3a 和图 3b 所示，NCEP/NCAR 再分析的总辐射和反射辐射通量要比敦煌干旱区观测试验实际测得的总辐射和反射辐射通量偏大，而实际观测到的总辐射和反射辐射通量的波动要比 NCEP/NCAR 再分析总辐射和反射辐射通量的波动大。

（3）戈壁区地表反照率及其日变化和季节变化特征

Huang et al.（2005）和张强（2005）利用 NWC-ALIEX 观测试验早期 1 年的观测资料，提出敦煌戈壁地表反照率大约为 0.255 ± 0.021。最近，从 NWC-ALIEX 的多年辐射资料的分析结果提出敦煌

表 1 夏季敦煌三种地表的辐射各分量的日积分值

Table 1　Diurnal integrated value of various components of summertime radiation at three kinds of Dunhuang's surface

	总辐射 (MJ m^{-2} d^{-1})	大气向下长波辐射 (MJ m^{-2} d^{-1})	地表向上长波辐射 (MJ m^{-2} d^{-1})	地表反辐射 (MJ m^{-2} d^{-1})	净辐射 (MJ m^{-2} d^{-1})
戈壁（2000 年 5~6 月）	32.43	29.38	44.06	7.42	10.33
稀疏植被（2009 年 7~9 月）	25.28	29.31	39.92	6.13	8.57
绿洲农田（2000 年 5~6 月）	31.17	27.51	36.15	8.08	14.54

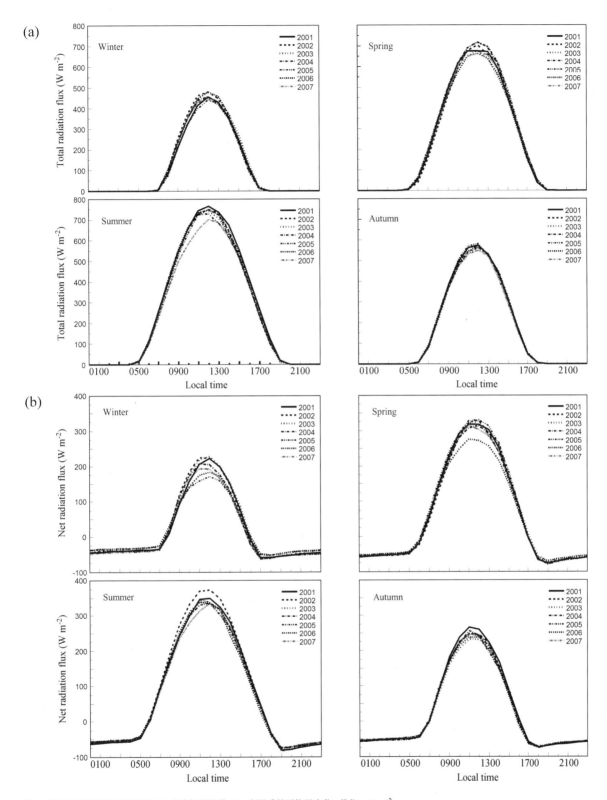

图 2 敦煌戈壁地表总辐射通量 (a) 和净辐射通量 (b) 在四季的平均日变化. 单位：W m^{-2}

Fig. 2 Average diurnal variations of four-seasonal (a) total radiation fluxes and (b) net radiation fluxes at the surface of Dunhuang Gobi area. Units: W m^{-2}

戈壁区地表反照率在夏季的白天大约为 0.255±0.002，这与应用 1 年的观测试验资料所得的结果很一致，但比黑河试验在戈壁地表测得的地表反照率（0.228）（邹基玲等，1992）要大一些。并且，利

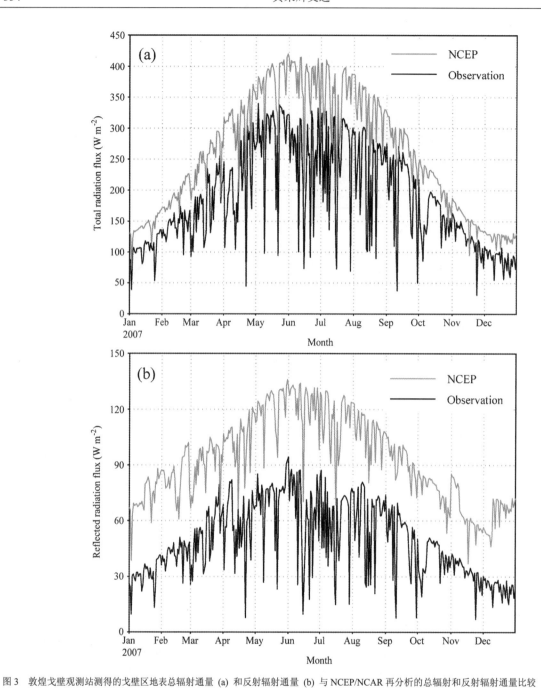

图 3　敦煌戈壁观测站测得的戈壁区地表总辐射通量 (a) 和反射辐射通量 (b) 与 NCEP/NCAR 再分析的总辐射和反射辐射通量比较

Fig. 3　Comparisons between (a) total radiation flux and (b) reflected radiation flux at the surface of Gobi area observed from Dunhuang Gobi observation station and those from NCEP/NCAR reanalysis (e.g., Kalnay, et al., 1996). Units: W m^{-2}

用 NWC-ALIEX 的多年辐射资料所得的敦煌干旱区地表反照率的日变化呈 "U" 字型。如图 4 所示，它一般在当地时间 11 时到 13 时最小，而在上午 11 时之前及下午 13 时之后呈现增大趋势；此外，分析还表明，它有一定的季节变化，一般在冬季最大（由于有雪的缘故），而夏季最小，春、秋比夏季稍大一些。

最近，王超等（2012）利用 NWC-ALIEX 2009 年夏季加强期观测资料分析了敦煌稀疏植被下垫面的反照率，提出了敦煌稀疏植被下垫面的反照率为 0.24，比戈壁下垫面的反照率小，这个结果是合理的。

2.3　敦煌戈壁区的边界层特征及边界层结构

敦煌干旱区由于水汽含量很低，水汽凝结高度高，因此，一般对流边界层都比较高，NWC-ALIEX 的观测结果充分证明了这点。

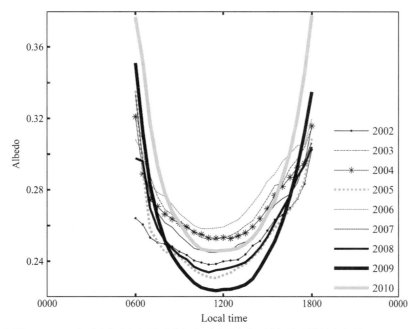

图 4 敦煌戈壁观测站观测的 2002~2010 年夏季典型晴天下戈壁地表反照率的日变化。图中时间是敦煌当地时间

Fig. 4 Diurnal variation of albedo at the surface of Gobi area in typical clear summer day observed from Dunhuang Gobi station during the period 2002–2010. The time in Fig. 4 is the local time of Dunhuang station

(1) 对流边界层特征

张强等（2004）以及张强和王胜（2009）利用 NWC-ALIEX 早期观测资料，指出敦煌戈壁区具有高边界层特征。最近，韦志刚等（2010）和惠小英等（2011）从 NWC-ALIEX 2008 年加强期观测所测得资料的分析结果，也揭示了敦煌戈壁区边界层具有高对流边界层及明显的日变化特征。如图 5a 所示，在北京时 10 时（地方时约 08 时）敦煌戈壁区对流边界层已有所发展，对流边界层从北京时 12 时到 14 时发展很快，新发展的对流与残留层的对流相混合，使对流边界层迅速增厚，在北京时 16 时到 23 时（地方时 14 时～21 时），对流边界层高度可达最高高度，它可达距地 4200 m 高度。这表明了在敦煌戈壁区边界层具有高对流边界层特征；并且，从图 5a 还可看到，敦煌戈壁区稳定边界层从北京时 19 时（地方时约 17 时）左右开始发展，高度可达距地面 1300 m 高度，稳定边界层一般在北京时 12 时到 14 时消退。这些表明利用最近敦煌戈壁区边界层观测资料的分析结果与以前利用敦煌气象站加强探空资料分析的结果（张强等，2004）较为一致。

最近，Wei et al.（2009）还利用平凉黄土塬半干旱区陆—气相互作用的观测试验资料分析了西北半干旱区的对流边界层厚度，结果表明：黄土塬的对流边界层厚度可达 1000 m 高度，而稳定边界层厚度达 650 m 高度。这说明黄土高原的对流边界层厚度比敦煌戈壁区和青藏高原对流边界层高度（可达 2700 m）都低。

上述结果表明了夏季敦煌戈壁区的对流边界层厚度确实较厚，它比青藏高原和黄土高原的边界层的高度都高，具有高对流边界层特征。这可能是由于敦煌戈壁区十分干旱，年降水量不足 50 mm，而年蒸发能力达 3400 mm，地表感热输送很强，而水汽含量很低，水汽不易在较低高度凝结所致。

(2) 边界层结构

韦志刚等（2010）从 NWC-ALIEX 的 2008 年 8 月加强期观测资料的分析结果揭示了敦煌戈壁区对流边界层结构。如图 5b 所示，敦煌戈壁区也具有五层结构，即近地层、混合层、逆温层、中性层、次逆温层。

2.4 敦煌戈壁区地—气温差的平均日变化、季节变化和典型天气下的日变化特征

(1) 平均的日变化

王超等（2010b）利用敦煌戈壁观测试验站 2008 年 12 月 1 日至 2009 年 12 月 31 日的观测资料，分析敦煌戈壁区地—气温差的变化，结果表明：敦煌

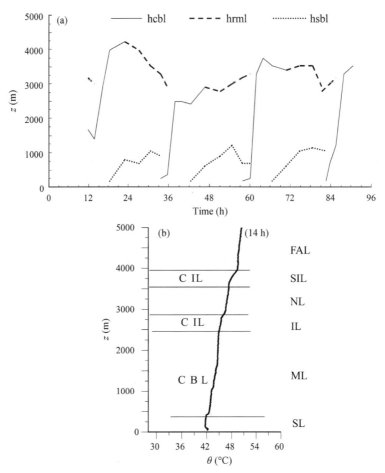

图 5 敦煌戈壁观测站测得 2008 年 8 月 15～18 日戈壁区边界层高度的变化（a）和 8 月 11 日 14 时测得的边界层结构（b）。图中时间为北京时；hcbl、hrml 和 hsbl 分别表示对流边界层高度、残留混合层高度和稳定边界层高度；FAL、SIL、NL、IL、ML 和 SL 分别表示自由大气层、次逆温层、中性层、逆温层、混合层和近地层

Fig. 5 (a) The variation of boundary layer with time during August 11–15, 2008 and (b) the structure of boundary layer of Gobi area at 1400 BT (Beijing time) in August 11, 2008 observed from Dunhuang Gobi station. The time in Fig. 5 is Beijing time; hcbl, hrml, and hsbl indicate the heights of convective boundary layer, residual mixed layer, and stable boundary layer, respectively; FAL, SIL, NL, IL, ML, and SL indicate free atmosphere layer, second inversion layer, neutral layer, inversion layer, mixed layer, and surface layer, respectively

戈壁区地—气温差不仅有很大的日变化，而且有很明显的季节变化；并且，分析结果还表明：敦煌戈壁地表地—气温差变化幅度较地温和气温小，它的分布在 −1°C 附近有一个非常明显的峰值，观测期间出现的最小地—气温差为 −13.8°C，它出现在夜间，而最大地—气温差为 35°C，它出现在白天，平均地—气温差为 4.1°C。此外，如图 6a 所示，敦煌戈壁地表地—气温差的日变化规律性强，白天为正，夜间为负，以 12 时（当地时间）大致对称，夜间变化幅度小，白天温差变化幅度大。

（2）季节变化

如图 6b 所示，不同季节敦煌戈壁区地—气温差的日变化趋势大致相同，但变化幅度、温差大小和温差正负转变的时间有所不同。此区域地气温差在 12 月份最小，6 月份最大。若按季节划分，则夏季最大，为 25.8°C，冬季最小，为 10.2°C。

（3）不同天气条件下的地—气温差

不同天气条件下敦煌戈壁区地—气温差的日变化有所不同。如图 6c 所示，虽然在晴天（2009 年 4 月 3 日）、阴天（4 月 11 日）和沙尘暴（4 月 23 日）天气下，敦煌戈壁区地—气温差的变化趋势是一致的，但变化幅度、峰值和温差正负转换的时间都有所不同。在晴天，地—气温差值最大，出现在当地中午 12 时（北京时 14 时）；在阴天，地—气温差值次之，出现在当地时间午后 14 时（北京时 16 时）比晴天天气条件下地—气温差最大值出

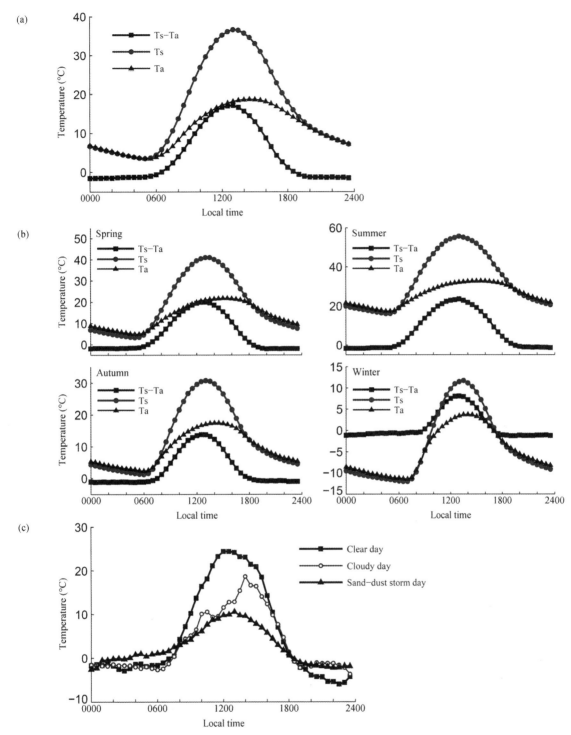

图 6 敦煌戈壁观测站所观测的戈壁区地—气温差（T_s-T_a）的平均的日变化（a），季节变化（b）和不同天气条件下的日变化（c），单位：℃。图c中晴天是2009年4月3日，阴天是4月11日，沙尘暴天气是4月23日

Fig. 6 (a) Diurnal and (b) seasonal variations of difference between surface temperature and surface air temperature (T_s-T_a), (c) diurnal variation of (T_s-T_a) under different weather conditions in the Gobi area observed from Dunhuang Gobi observational station. Units: ℃. In Fig. 6c, the sunny day was on April 3, 2009, the cloudy day was on April 11, 2009, and the day of sand-dust storm was on April 23, 2009

现的时间晚 2 h；在沙尘暴天气时，地—气温差最小，出现在当地时间午后 13 时（北京时 15 时），比晴天地—气温差最大值出现时间稍晚，而比阴天地—气温差最大值出现时间偏早。

2.5 敦煌干旱区表层和次表层土壤温度

张强等（2005）从 NWC-ALIEX 早期 1 年的观测资料分析了敦煌戈壁干燥土壤热力参数，特别是推算出戈壁干燥土壤的热容量、热传导率和热扩散率。NWC-ALIEX 连续多年对敦煌戈壁区近地面 4 层（1 m、2 m、8 m、18 m）的大气风温湿分量、4 层土壤湿度（5 cm、10 cm、20 cm、80 cm）以及 7 层土壤温度（0 cm、5 cm、10 cm、20 cm、40 cm、80 cm、180 cm）进行了观测，取得了大量观测资料。最近，我们利用此资料，分析了敦煌戈壁干燥土壤表层、80 cm 和 180 cm 土壤温度的年际变化趋势。如图 7 所示，夏季敦煌戈壁区表层、次表层温度有微弱的逐渐下降趋势，而深层土壤温度有微弱的升高趋势。由于从敦煌干旱区土壤温度的变化可以推算敦煌戈壁干燥土壤热容量的变化，因此上述结果表明了敦煌戈壁土壤表层和次表层热容量有微弱的下降趋势，但深层土壤热容量都有微弱的上升趋势。

此外，我们还利用在敦煌戈壁区连续多年所测得干燥土壤 5 cm、10 cm、20 cm、80 cm 深度四层土壤含水量资料分析了敦煌戈壁干燥土壤表层和次表层土壤体积含水量（图略），结果表明了敦煌戈壁干燥土壤含水量与降水量有很好的正相关，但由于敦煌干旱区年降水量很少，因此，它并没有明显的年际变化。

上述表明了通过分析 NWC-ALIEX 所得的多年观测资料，提出了许多敦煌干旱区陆面过程的重要参数，这些参数不仅对于优化陆面过程模式中的有关参数化方案及改进干旱区陆面过程模式有直接的应用，而且在关于中国西北干旱区陆—气相互作用对气候影响的研究也是非常重要的。

3 观测的陆面过程参数在陆面过程模式中参数化方案的优化及模拟结果改进的应用及作用

Henderson-Sellers et al. (1993) 比较了国际上流行的 25 个陆面过程模式，表明了不同陆面过程模式对地表附近感热和潜热的模拟结果相差甚大。

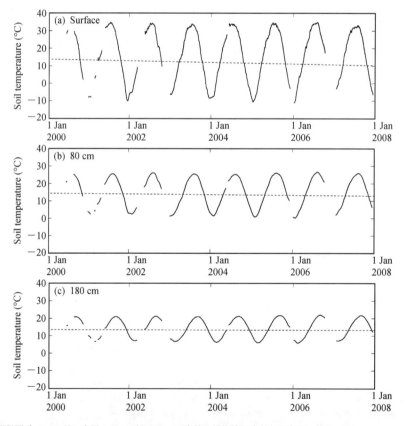

图 7 敦煌戈壁观测站所测得夏季（6~8 月）表层、80 cm 和 180 cm 深度的干燥土壤温度的年际变化。单位：°C

Fig. 7 Interannual variability of the summertime (June–August) dry-soil temperature at surface, 80-cm and 180-cm depths observed from Dunhuang Gobi observational station. Units: °C

其中部分原因是模式中一些陆面要素的参数化方案不恰当所造成的。为此，我们近几年来利用 NWC-ALIEX 所测得的干旱区陆面过程参数对有关陆面过程模式中的陆面参数化方案进行优化，从而改进了这些陆面过程模式对陆面过程的模拟结果。

3.1 观测的陆面过程参数在干旱区陆面过程模式参数化方案的优化及其模拟结果改进的作用

上述总结了根据 NWC-ALIEX 观测试验结果，提出了许多干旱区重要陆面过程参数，这些参数对于陆面模式中参数的优化有着重要作用。Chen et al.（2009）利用 NWC-ALIEX 连续 8 年的观测数据以及近年来关于陆面过程参数化方案的最新研究结果，对 BATS 陆面过程模式（Dikinson et al., 1993）中反照率、粗糙度长度、土壤体积热容量和土壤热传导率 4 个参数进行优化。并且，他们还利用改进过的陆面过程模式按照不同的参数优化方案组合形式设计了典型干旱区地气交换过程的控制试验和重要陆面过程参数影响的敏感性试验方案，对敦煌 2000 年 5 月～2004 年 7 月的陆面过程进行了离线 (off-line) 数值模拟分析。在此基础上，通过对比各数值试验的模拟结果，细致地分析了不同参数在不同时段对地—气相互作用中感热通量、潜热通量和地表温度模拟的影响（朱德琴，2006）。

经参数优化的 BATS 陆面过程模式可以显著地提高模式的模拟能力。如图 8a 所示，对地表温度而言，参数优化后的模式能更准确地模拟地表温度的日变化特征。这主要是由于在陆面模式中改进了反照率参数，从而改进了模式对净短波辐射和净长波辐射的模拟，改进后的模式在净短波辐射和净长波辐射的模拟上均比原模式的模拟结果有了一定程度的改善，可以比较好地模拟出它们的日变化和季节变化特征。并且，经参数优化后的模式可以比较好地模拟深层土壤温度的季节性变化特征，全年各季节土壤温度的模拟对反照率和粗糙度长度都比较敏感，尤其是夏半年更明显，而在冬半年它对

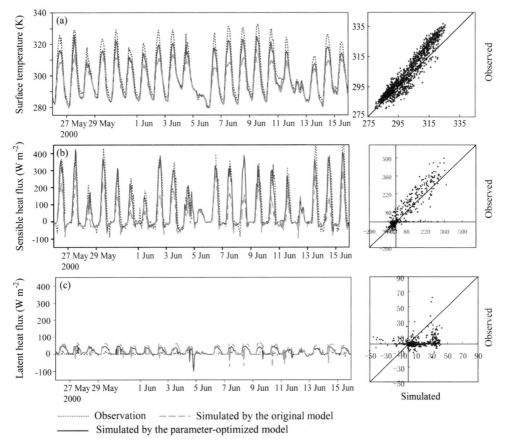

图 8 利用 NWC-ALIEX 观测试验得到的干旱区陆面参数对 BATS 陆面过程模式中有关参数进行优化后所模拟的（a）地表温度（单位：K）以及地表附近（b）感热输送通量（单位：W m^{-2}）和（c）潜热输送通量（单位：W m^{-2}）的日变化（朱德琴，2006）

Fig. 8 Diurnal variations of (a) surface temperature (unit: K), (b) sensible heat transfer flux (unit: W m^{-2}), and (c) latent heat transfer flux (unit: W m^{-2}) simulated by using the parameter-optimized BATS land surface model, where the related parameters were observed from the NWC-ALIEX (Zhu, 2006)

于土壤体积热容量和热传导率更为敏感,在夏季对此两参数的影响比较小。此外,如图8b和图8c所示,与加强观测期的感热通量观测值对比,参数优化后的陆面过程模式能比较好地模拟感热通量的日变化特征,并对潜热通量的模拟也有一定的改进。

3.2 观测的陆面过程参数在国际常用陆面过程模式的参数化方案优化及其模拟结果改进的作用

为了更好验证上述从观测试验所得的参数在陆面过程参数化方案的优化及其模拟改进的作用,房云龙等(2010)选取目前国际上较为常用的四个陆面过程模式——BATS(Dikinson et al., 1993)、SSiB (Xue et al., 1991)、NCAR-LSM (National Center for Atmospheric Research, Land Surface Model) (Bonan, 1996) 以及 CoLM(Common Land Model)(Dai et al., 2001, 2003),并根据 NWC-ALIEX 的观测资料对这四个模式中的地表反照率和粗糙度长度两个参数的参数化方案进行统一改进和优化;在此基础上对敦煌干旱区陆面过程进行单站离线 (off-line) 数值试验,并对比各模式的模拟性能。结果显示:经反照率和粗糙度长度参数化方案优化之后,各模式对地表温度的日变化和感热通量的季节变化等的模拟能力均有较大程度的提高(图9、10)。

并且,如表2所示,各模式对感热通量模拟的均方根误差和平均偏差都有显著的减小,这说明改进这四个模式中两个参数就可以更好地模拟干旱区地表特征;此外,优化参数后,模式之间模拟性能的差异也有所减小。这个结果说明了对于干旱区陆面过程模式中参数化方案的正确选取与物理过程的方案选取可能同等重要。

表2 BATS、SSiB、NCAR-LSM 和 COLM 陆面过程模式经参数优化后与未经优化对感热通量模拟结果的比较
Table 2 Comparisons between sensible heat fluxes simulated by using the parameter-optimized surface land models of BATS, SSiB, NCAR-LSM, and CoLM and those simulated by using original models. Units: W m^{-2}

模式	均方根误差(W m^{-2})		平均偏差(W m^{-2})	
	未优化	优化后	未优化	优化后
BATS	7.95	5.55	6.41	4.34
SSiB	7.46	4.61	6.77	4.05
NCAR-LSM	15.99	8.64	13.44	7.36
CoLM	5.05	2.29	4.08	1.80

3.3 经参数化方案优化后陆面过程模式对气候模拟的改进

朱德琴(2006)利用参数优化后 BATS 陆面过程模式和 RegCM 区域气候模式(Dikinson et al., 1989; Giorgi, 1990)对我国夏季降水进行了数值模拟。如图11所示,改进后的陆面过程模式不仅可以较好地模拟干旱区陆面过程,而且能较好地模拟周边的气候变化,特别是改进了对我国东部夏季季风降水的数值模拟。把图11a和图11b分别与图11c相比较可以明显看到,经陆面参数化方案优化以后区域气候模式对我国夏季降水有很大改进,所模拟的我国夏季降水与实况较接近。

上述结果说明了通过利用NWC-ALIEX所测得的多年资料对陆面过程参数进行优化,经参数优化后的陆面过程模式不仅大大改进了干旱区的地表温度和感热输送的模拟结果,而且应用到区域气候模式中,可以改进我国东部夏季季风降水的数值模拟结果。这表明了我国干旱区陆面过程参数的优化和陆面过程模式的改进对于我国气候数值模拟是相当重要的。

4 中国西北干旱区感热输送特征的研究

中国西北地区由于山脉的阻挡,水汽输送到这里很少,云量很少,故在春、夏季的净短波辐射大,这就导致此区域在春、夏季地—气温差很大,从而造成感热输送很大。然而,中国西北干旱区,一方面由于对地表感热输送的直接观测的站点很少,即便有少数观测站,但观测的时间也较短;另一方面,由于气象台站只有常规气象要素的观测,并没有感热通量的直接观测。为了估算我国西北地区的地表感热通量,Zhou and Huang (2010a) 利用 NWC-ALIEX 在干旱下垫面观测试验所取得的地表能量通量湍流输送,并评价了利用该输送参数来计算一般气象台站感热通量的可行性,从而利用气象台站四个时次(02时、08时、14时、20时)观测资料估算了西北干旱区感热输送(周连童, 2009a)。

4.1 感热通量在敦煌地区地表能量平衡中的作用

干旱区地表能量平衡方程如下:
$$R_n = H + L + G, \quad (1)$$

式中,R_n 为净辐射,H 是地表附近的感热通量,L 是潜热通量,G 是地表土壤热通量。(1)式表明:干旱地表所得太阳的净辐射与地表向大气输送的

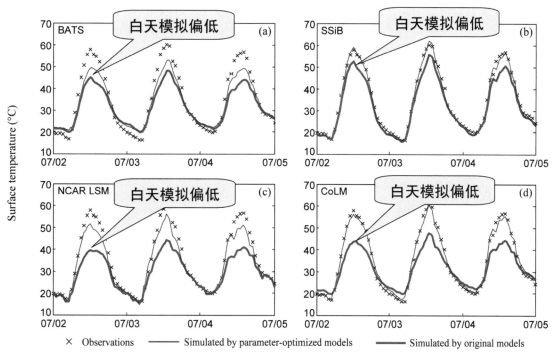

图 9 BATS (a)、SSiB (b)、NCAR-LSM (c) 和 CoLM (d) 陆面模式经参数化方案优化与未优化对地表温度日变化模拟结果的比较。单位：°C

Fig. 9 Comparisons between the diurnal variations of surface temperature simulated by using the parameter-optimized surface land models of (a) BATS, (b) SSiB, (c) NCAR-LSM, and (d) CoLM and those simulated by original models. Units: °C

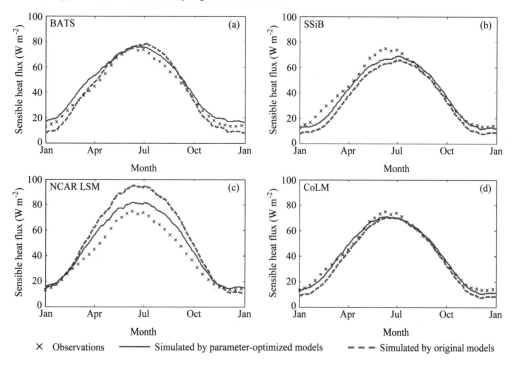

图 10 同图 9，但为感热通量的季节变化。单位：W m^{-2}

Fig. 10 As in Fig. 9, except for the seasonal variations of sensible heat flux. Units: W m^{-2}

感热、潜热以及土壤向下或向上的热通量相平衡。通过 NWC-ALIEX 的加强期观测可以测得敦煌干旱区的感热通量、潜热通量和地表土壤热通量；并且，由于敦煌是典型干旱区，潜热输送通量很小，因此还可利用地表能量平衡方程以及土壤热通量和净辐射的观测数据计算出敦煌戈壁区地表的感热通

(图略)

图 11 观测的我国 1996 年夏季（6~8 月）降水距平的分布（a）及与经参数化方案优化后 BATS 陆面过程模式和 RegCM 区域气候模式相耦合的模拟结果（b）和与原模式模拟结果（c）的比较。单位：mm。（朱德琴，2006）

Fig. 11 (a) Observed anomaly distribution of 1996 summertime (Jun–Aug) precipitation in China, and its comparisons with (b) the anomaly distribution simulated by using the RegCM regional climate model coupled with the parameter-optimized BATS land surface model and (c) the one simulated by original BATS model. Units: mm. (Zhu, 2006)

量，或者利用感热输送通量和净辐射的观测数据计算出地表土壤热通量。通过与加强观测期的观测值对比，表明了利用（1）式所计算的感热通量或地表土壤热通量与观测值对应较好（图 12a、12b）。

图 12a 为典型夏季敦煌戈壁地表能量平衡的日变化。从图 12a 可见，地表能量平衡主要以感热通量为主，而潜热通量很小，感热通量、潜热通量、地表热通量日积分值分别占净辐射日积分值的 95.7%、3.1%和 1.1%，感热通量是潜热通量的 30 倍左右，这也说明了敦煌地区地表呈现出极端干旱的特征。胡隐樵等（1994）根据黑河流域戈壁下垫面观测的结果也指出，在黑河流域干旱区潜热通量比感热通量约小一个量级，这表明了 NWC-ALIEX 观测试验结果与黑河流域试验的结果也比较吻合。并且，从图 12a 还可以看到，敦煌戈壁观测站夏季 12 d 平均的感热通量的极值大约在 300 W m^{-2}，而个别观测日的极值可以接近 500 W m^{-2}，感热通量极值出现的时间大概在当地时间 13~14 时，比净辐射极值出现的时间晚 1~2 h，夜间的感热通量比较平稳，基本维持在 -20 W m^{-2}。此外，从图 12a 还可以看到：地表热通量在 04~14 时为正，表示在这段时间内热量由大气向土壤传导，此时土壤为升温过程；其他时间，地表热通量为负，表示热量由土壤向大气传导，土壤为降温过程。地表热通量的日积分值为 0.086 MJ m^{-2} d^{-1}，这说明在夏季有净的热量进入土壤，所以从季节变化的角度来看，整个夏季（或者夏半年）是中国西北干旱区土壤的升温时期。

图 12b 为典型冬季地表能量平衡的日变化。从图 12b 可以明显看到，与夏季不同的是冬季净辐射峰值出现的时间比夏季略提前，为当地时间 11:30，数值略大于 150 W m^{-2}，感热通量和地表热通量的峰值分别约为 100 W m^{-2} 和 120 W m^{-2}，比夏季小得多。感热通量在 09 时之后开始为正，14 时达到最大值，最低值出现在日出之前，为 -30 W m^{-2} ~ -20 W m^{-2}。由于冬季的日出时间较晚，因此，直到 06:30 地表热通量才开始大于 0，并一直维持到 14 时左右，这段时间为地表温度的升温时段，升温时间比夏季大约少了 2 h。在冬季敦煌地区净辐射、感热通量和地表热通量的日积分值分别为 0.84、1.21 和 -0.35 MJ m^{-2} d^{-1}。这说明在冬季西北干旱区地表热通量的积分值为负，这也就是说，在冬季西北干旱区地表得到太阳的净辐射的能量不足以平衡感热通量，需要地表热通量来补充净辐射不足的能量，这就造成土壤处于降温状态。因此，冬季西北干旱区土壤为热源。

4.2 中国西北干旱区春、夏季地表感热输出特征

周连童（2009a, 2010）以及 Zhou et al.（2010）从 NCEP/NCAR 和 ERA-40 的感热输送再分析资料，并结合 NWC-ALIEX 所观测的结果，分析了欧亚大陆春、夏季的感热分布。如图 13a 和图 13b 所示：中国西北干旱和半干旱区是欧亚大陆上最高的感热中心之一，中国西北干旱和半干旱区除了南部和西部外，大部分区域感热输送通量在夏季可达 120 W m^{-2} 以上。并且，他们还由 ERA-40 再分析的感热资料计算了我国各区域 1958~2002 年气候平均感热输送及其季节变化（图 14）。从图 14 可以明显看到，中国西北干旱区气候平均的春、夏季感热输送在中国各区域感热输送中都是位居第一，在夏季高达 75 W m^{-2}，它比东北、华北、青藏高原的感热都大，特别是它比我国东部季风湿润区的感热输送要大得多。

4.3 中国西北干旱区感热时空变化特征及与中亚干旱区感热变化的关联

最近，周连童（2009a）比较了 NCEP/NCAR 和 ERA-40 感热再分析资料，并与利用实测资料计算得到的感热做比较，相对而言，敦煌干旱区 ERA-40 的感热再分析资料与应用实际观测资料所计算的敦煌干旱区的感热变化趋势更接近。因而，周连童（2010）利用 1958~2002 年 ERA-40 再分析的春、夏感热输送资料做了 EOF 分析，结果显示：无论春季或夏季，欧亚大陆干旱/半干旱区的感热输送都有三个主模态。如图 15a 所示，第一模态表示中国西北干旱区与中亚地区春季感热变化在空间分

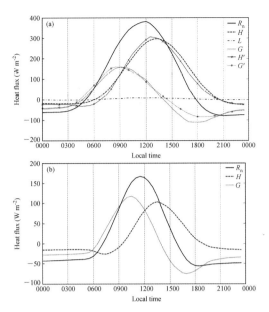

图12 夏季（a）和冬季（b）敦煌戈壁地表能量平衡的日变化。单位：W m^{-2}。R_n：净辐射；H：感热输送通量观测值；L：潜热输送通量观测值；G：地表土壤热通量观测值；H'：感热输送通量计算值；G'：地表土壤热通量计算值

Fig. 12 Diurnal variation of surface energy balance at Dunhuang Gobi site in boreal (a) summer and (b) winter. Units: W m^{-2}. R_n: net radiation; H: observed sensible heat flux; L: observed latent heat flux; G: observed surface soil heat flux; H': calculated sensible heat flux; G': calculated surface soil heat flux

布上的一致性，并从图15b可看到，从20世纪70年代中期开始，欧亚大陆干旱/半干旱区的春季感热输送明显增强。第二、三模态（图略）显示出中国西北干旱和半干旱区中亚地区感热变化在空间布上的一致性，并从图15b可看到，从20世纪70年代中期开始，欧亚大陆干旱/半干旱区的春季感热输送明显增强。第二、三模态（图略）显示出中国西北干旱和半干旱区中亚地区感热变化在空间分布的不一致性，且存在着年际、年代际变化特征。并且，夏季中亚和中国西北干旱区感热EOF分析的第1主模态在空间分布和时间系数的变化也与春季感热EOF分析的空间分布较一致，即西北干旱区与中亚夏季感热变化具有一致性，但时间系数两者呈相反变化趋势，即夏季无论中亚或中国西北干旱区感热输送呈减少趋势。因此，Zhou et al.（2010）指出：在中国西北干旱区春季感热的年代际变化与夏季感热的年代际变化趋势是不同的。此外，小波分析结果也显示出欧亚大陆春、夏季感热输送存在着明显的年际、年代际变化，其年代际变化信号要强于年际变化的信号。

这些结果揭示了在春、夏季中国西北干旱区具有高感热输送特征，它是欧亚大陆地表附近最大感热输送中心之一，并与中亚地区感热变化有重要关联。

5 中国西北干旱区陆—气相互作用对中国东部气候变异的影响及其机理

上述分析表明了中国西北干旱区是欧亚大陆春、夏季最大感热中心，这表明此地区也是陆—气相互作用相当强的区域。由于中国西北干旱区十分广阔，因此，此区域的陆—气相互作用过程的变化不仅会引起此区域气候的变化，而且会引起东亚季风区（特别是中国东部）气候的变化（周连童和黄荣辉，2006，2008）。因而，本节分析中国西北干旱区地表热状况和陆—气相互作用的变化对中国东部夏季降水的年代际跃变的影响。

5.1 中国西北干旱区春季感热对中国东部华北和长江流域夏季降水的影响

周连童和黄荣辉（2006）、Zhou（2009）分析了西北干旱区49站1961~2000年地温、气温和地—气温差的时空演变特征以及春、夏季感热的年代际变化，指出：中国西北干旱区春、夏季地—气温差和春季感热在1976年前后发生了明显增强的

（图略）

图13 1958~2002年气候平均的欧亚大陆春季（a）和夏季（b）感热输送通量分布。单位：W m^{-2}。阴影区表示大于80 W m^{-2}，资料取自ERA-40再分析的感热资料（Uppala et al., 2005）

Fig. 13 Distributions of climatological mean sensible heat flux for boreal (a) springs and (b) summers of 1958–2002 over the Eurasian continent. Units: W m^{-2}. The areas of sensible heat flux over 80 W m^{-2} are shaded, and data of sensible heat flux are from the EAR-40 reanalysis (e.g., Uppala et al., 2005)

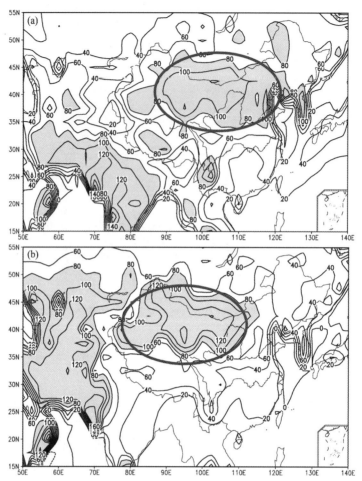

图 13 1958~2002 年气候平均的欧亚大陆春季（a）和夏季（b）感热输送通量分布。单位：W m^{-2}。阴影区表示大于 80 W m^{-2}，资料取自 ERA-40 再分析的感热资料（Uppala et al., 2005）

Fig. 13 Distributions of climatological mean sensible heat flux for boreal (a) springs and (b) summers of 1958–2002 over the Eurasian continent. Units: W m^{-2}. The areas of sensible heat flux over 80 W m^{-2} are shaded, and data of sensible heat flux are from the EAR-40 reanalysis (e.g., Uppala et al., 2005)

区感热输送呈减少趋势。因此，Zhou et al.（2010）指出：在中国西北干旱区春季感热的年代际变化与夏季感热的年代际变化趋势是不同的。此外，小波分析结果也显示出欧亚大陆春、夏季感热输送存在着明显的年际、年代际变化，其年代际变化信号要强于年际变化的信号。

这些结果揭示了在春、夏季中国西北干旱区具有高感热输送特征，它是欧亚大陆地表附近最大感热输送中心之一，并与中亚地区感热变化有重要关联。

5 中国西北干旱区陆—气相互作用对中国东部气候变异的影响及其机理

上述分析表明了中国西北干旱区是欧亚大陆春、夏季最大感热中心，这表明此地区也是陆—气相互作用相当强的区域。由于中国西北干旱区十分广阔，因此，此区域的陆—气相互作用过程的变化不仅会引起此区域气候的变化，而且会引起东亚季风区（特别是中国东部）气候的变化（周连童和黄荣辉，2006，2008）。因而，本节分析中国西北干旱区地表热状况和陆—气相互作用的变化对中国东部夏季降水的年代际跃变的影响。

5.1 中国西北干旱区春季感热对中国东部华北和长江流域夏季降水的影响

周连童和黄荣辉（2006）、Zhou（2009）分析了西北干旱区 49 站 1961~2000 年地温、气温和地—气温差的时空演变特征以及春、夏季感热的年代际变化，指出：中国西北干旱区春、夏季地—气温差和春季感热在 1976 年前后发生了明显增强的

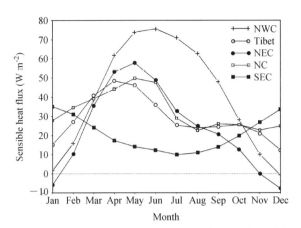

图 14 中国不同区域 1958~2002 年气候平均的感热通量随月份的变化。单位：W m^{-2}。资料取自 ERA-40 再分析资料（Uppala et al., 2005）

Fig. 14 Monthly variation of climatological mean sensible heat flux for 1958–2002 in various regions of China. Units: W m^{-2}. Data are from the ERA-40 reanalysis (e.g., Uppala et al., 2005)

年代际变化特征；并且，他们还指出了此区域春季地—气温差和感热的增强对于中国东部的夏季降水具有重要影响。周连童和黄荣辉（2006，2008）以及 Zhou and Huang（2010b）根据实测的气象观测值所计算的感热与全国夏季降水进行相关分析，指出：西北干旱区春季感热与长江流域和东北地区夏季降水有正相关，而与华北和西南地区夏季降水有负相关（图略）。并且，他们分析了西北和华北地区实测的夏季降水的年际和年代际变化，指出了西北与华北地区夏季降水的年代际变化的异同。如图 16 所示，西北干旱区夏季降水与华北地区的夏季降水有相反的年代际变化，中国华北地区从 20 世纪 70 年代中后期起夏季降水具有明显的减少趋势（除 90 年代中期），而中国西北干旱区夏季降水从 70 年代中后期起却有明显的增加趋势。

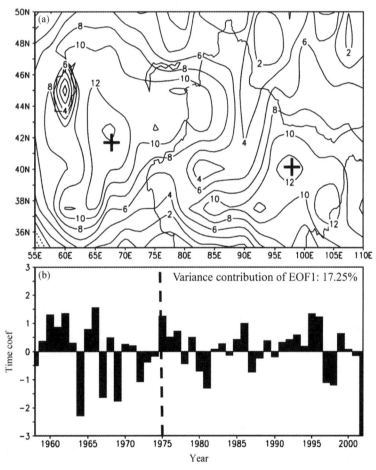

图 15 中亚和中国西北干旱区春季感热输送 EOF 分析第 1 主分量的空间分布（a）和时间系数（b）。图 15a 中实、虚线分别表示正、负信号，EOF1 占总方差的 17.25%

Fig. 15 (a) The spatial distribution and (b) corresponding time coefficients of the first component of EOF analysis (EOF1) of spring sensible heat flux in Central Asia and the arid area of Northwest China. The solid and dashed lines in Fig.15a indicate positive and negative signals, respectively, and EOF1 explains 17.25% of the variance

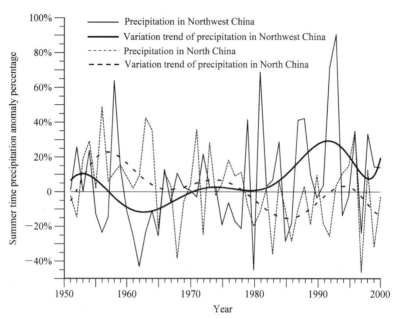

图 16 中国西北干旱区夏季（6～8 月）降水距平百分率的年际变化（细实线）和变化趋势（粗实线）及其华北地区夏季降水距平的年际变化（细虚线）和变化趋势（粗虚线）

Fig. 16 Interannual variation of summertime (Jun–Aug) precipitation anomaly percentage in the arid area of Northwest China (thin solid line) and its variation trend (heavy solid line) and interannual variation of summertime precipitation anomaly in North China (thin dashed line) and its variation trend (heavy dashed line)

5.2 中国西北干旱区陆—气相互作用变化对中国东部季风区夏季降水的影响过程

周连童和黄荣辉（2008）以及周连童（2009b）指出了中国西北干旱区春季地—气温差从 20 世纪 70 年中后期开始的年代际增强及其对华北地区持续性干旱的影响过程。他们的研究结果表明：从 20 世纪 70 年代中后期起西北干旱区春季地—气温差显著增强，致使该地区春季感热增强和上升气流异常增强，并且这种上升气流异常可维持到夏季，造成西北地区上空对流层低层在春、夏季出现气旋性环流异常型，这有利于西北地区降水偏多。由于受西北干旱区上升气流异常的影响，夏季在华北地区上空出现下沉气流的异常且出现年代际的反气旋性环流异常型，从而造成华北地区夏季降水偏少。并且，Zhou and Huang（2010b）利用位涡理论揭示西北干旱区春季感热对华北夏季降水影响的动力过程，指出：由于在夏季西北地区上空对流层高层温度从 20 世纪 70 年代中后期起出现显著的降低，而西北干旱区上空对流层低层温度明显升高，这种温度差异造成该地区垂直对流不稳定性增强，而在华北地区上空出现垂直对流不稳定性减弱，因而引起从 20 世纪 70 年代中后期开始西北地区降水增多，而华北地区降水减少。他们从上述结果提出了中国西北干旱区春季感热增强对华北夏季降水影响的概念图（见图 17）。

5.3 中国西北干旱区与东部季风区夏季降水相关联的机理

黄荣辉等（2012）从欧亚上空整层水汽输送通量距平的分布，说明了西北干旱区和中国东部季风区水汽输送异常是通过沿欧亚上空副热带急流传播的"Silk Road"型遥相关波列而相关联的。Chen and Huang（2012）还利用行星波传播理论从动力学上论述了中国西部干旱区和高原区与中国中、东部夏季降水异常的相关联是通过沿欧亚地区上空副热带急流传播的"Silk Road"型遥相关波列（Lu et al., 2002; Enomoto et al., 2003; Enomoto, 2004）。正如图 18 所示，欧亚大陆对流层上层环流（特别是副热带急流）的变化不仅影响我国西北干旱区夏季降水的年际变化，而且通过"Silk Road"型遥相关波列影响我国东部季风区的夏季降水的年际变化，从而进一步揭示了我国西北干旱区与东部季风区夏季降水异常相关联的机理。

5.4 中国西北干旱区植被退化对区域气候的影响

Li and Xue（2010）利用耦合了陆面过程模型 SSiB（Xue et al., 1991）的 NCEP GCM (Kalnay et al., 1990; Kanamitsu et al., 2002) 全球大气环流数值模式（NCEP GCM/SSiB）模拟了西北地区及青藏高原植被退化对周围夏季气候的可能影响。在模拟试验

(图略)

图 17 中国西北干旱区春季感热增强对华北夏季降水影响的示意图

Fig. 17 Schematic map of the impact of intensification of spring sensible heat in the arid area of Northwest China on summertime precipitation in North China

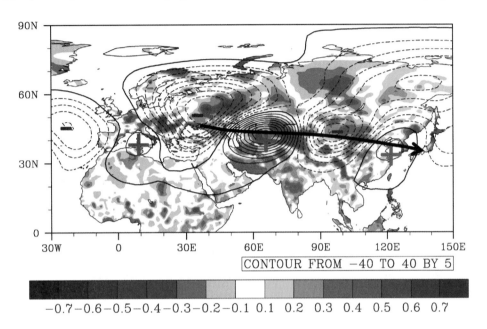

图 18 欧亚上空 7 月 200 hPa 经向风 EOF1 时间系数与欧亚大陆降水相关系数（阴影区）分布图。图中实、虚线表示欧亚大陆上空 200 hPa 高度场对 EOF1 时间系数的回归分布，欧亚大陆降水数据来自 NOAA 的全球陆地降水资料（PREC/L），风场数据来自 ERA-40 再分析资料 (e.g., Uppala et al., 2005).

Fig. 18 Distributions of the correlation coefficients between the first component of EOF analysis of the 200-hPa wind fields and precipitation anomalies in the Eurasian continent in July (shading areas) and the regression of 200-hPa geopotential height field over the Eurasian continent to its corresponding time coefficients of EOF1 (solid and dashed contours). The data of precipitations in the Eurasian continent are from NOAA's Precipitation Reconstruction over Land (PREC/L), and the data of wind fields are from the ERA-40 reanalysis (e.g., Uppala et al., 2005)

中利用了两个完全不同的陆面植被覆盖并输入大气环流模型中的 SSiB 模型，通过对比来得到一些较为明显的气候信号。其中的一个试验是植被覆盖情况来源于反演的卫星资料（Case S1），另一个试验是西北地区大部分地区为裸土(Case S2)。数值模拟结果表明：中国西北地区植被的退化（从植被覆盖到裸土）将减少地表吸收的辐射，并引起较弱的地表热力作用，这使得中国西北干旱区大部分区域上空对流层中层有反气旋异常环流，会导致此区域大部分地区降水减少。然而，中国西北干旱区植被退化会使高原东北侧上空对流层上层产生反气旋异常，在高原的东北部上空对流层中层会产生气旋环流异常，从而引起了高原东北侧上空产生垂直上升运动的异常，这些环流的变化导致了青藏高原东北部的降水增多。这个数值模拟表明：当西北干旱区植被退化，会导致此区域大部分地区降水减少，但也会引起高原东北侧局部地区降水反而增多。因此，某区域一旦植被破坏导致干旱，总体来说，如Charney (1975) 所指出该区域会变得愈来愈干旱，但还存在区域差异，甚至某一局部区域降水反而增多。

上述分析结果表明了中国西北干旱区陆—气相互作用（特别是春季感热）对东部季风区夏季降水变异有很大影响；并且，还表明了这两区域夏季降水异常之间有很大关联，这个关联不仅通过垂直纬向环流，而且通过"Silk Road"遥相关波列。此外，区域气候数值模式的模拟结果也表明了西北干旱区植被减少所引起陆—气相互作用的变化会使中国西北大部分干旱地区变得更干旱，但它也会造成青藏高原东北部上空反气旋环流和下沉气流的减弱以及中层气旋环流异常，因而会使得青藏高原东北部的降水增多。

6 总结及需进一步研究的问题

本文的重点是综述进行了12年之久的"中国西北干旱区陆—气相互作用观测试验（NWC-ALIEX）"最近几年所取得的观测和研究进展，特别是综述了近几年关于敦煌干旱区一些重要陆面过程参数的提出，干旱区陆—气相互作用特征的分析以及这些参数在陆面过程模式中参数化方案的优化和陆面过程数值模式改进应用的研究进展。并且，本文还系统地综述了中国西北干旱区春、夏季的高感热输送特征以及高感热对中国东部夏季降水影响及其过程和机理。这些研究结果也是近年来国际上关于全球气候变化方面研究所关注的内容。但必须指出，这些研究成果还是初步的，许多问题还有待于利用更长时间的观测试验资料以及更多典型下垫面的长时间观测资料的分析以及利用更好的陆面过程模式的数值模拟进一步做深入研究。特别是以下几个问题今后还需进一步研究：

（1）本文所综述的西北干旱区陆面过程参数主要是针对戈壁典型干旱下垫面，这也是目前国内外许多有关陆—气相互作用观测试验所缺少的。然而，应该看到，西北干旱区还有大片的稀疏植被和绿洲区域，目前还未能在这些区域对陆面过程坚持较长时间观测以及很好地分析总结其观测结果，这可能是今后应加强观测试验研究的一个问题。

（2）利用卫星对地表状况的遥感资料并与观测试验所观测的陆面参数相结合来反演区域陆面过程参数，这可以把单站测得的陆面过程参数推广到面上或区域尺度。但由于费用较大，目前还未能接收和反演更多有关干旱区陆面过程的卫星遥感资料，因此，今后应大力发展利用卫星遥感资料给出时间尺度较长并有区域代表性的我国西北干旱区陆面过程参数的动态变化情况。

（3）西北干旱区沙尘暴频繁发生，这也是我国西北干旱区主要的天气气候特征之一。一方面沙尘暴发生与我国西北干旱区多沙漠有关，这里干旱而少雨造成了陆面干燥而松软，因此，只要气旋锋面或中尺度天气系统经过就容易引起此区域沙尘暴的发生；另一方面，沙尘暴发生或过境时由于大量沙尘会散射太阳辐射，故会给此区域的陆面过程或陆—气相互作用带来很剧烈的影响。因此，西北干旱区沙尘暴与陆面过程是相互作用的。NWC-ALIEX 所设的观测站中敦煌和临泽观测站都是沙尘暴高发区，在这12年中曾观测到多次沙尘暴发生或沙尘暴过境时的陆面过程变化情况。然而，很多沙尘暴发生和过境对陆面过程的影响及其"阳伞效应"还有待于进一步分析总结。

（4）水分能以气、液、固三种形态参与气候系统中各圈层相互作用的物理过程、生物过程，甚至化学过程，因此，水分循环过程是导致气候系统变化的重要过程。然而，干旱区由于降水稀少，土壤和空气十分干燥，再加上春、夏季高温，蒸发能力很大，而实际蒸发却很小；并且，在干旱区能量循环由于缺乏水分，主要靠地表向大气的感热输送，潜热输送十分小。因此，在干旱区水分和能量循环不同于半干旱区和半湿润区，更不同于我国东部季风湿润区。目前，在水分十分缺乏的典型干旱区，陆面过程中地表蒸发和土壤水分输送的计算在理论上还存在不少困惑，因此，在本文综述中涉及西北干旱区水分循环很少。关于这一方面迫切需要进一步深入观测和研究。

参考文献（References）

André J C, Goutorbe J P, Perrier A. 1986. HAPEX-MOBLIHY: A hydrologic atmospheric experiment for the study of water budget and evaporation flux at the climate scale [J]. Bull. Amer. Meteor. Soc., 67 (2): 138–144.

Bolle H J, André J C, Arrue J L, et al. 1993. EFEDA: European field experiment in a desertification-threatened area [J]. Ann. Geophys., 11 (2–3): 173–189.

Bonan G B. 1996. A Land Surface Model (LSM version 1.0) for Ecological, Hydrological, and Atmospheric Studies: Technical Description and User's Guide [R]. NCAR Tech. Note NCAR/TN-417+STR, National Center for Atmospheric Research, Boulder, CO, 150pp.

Charney J G. 1975. Dynamics of desert and drought in the Sahel [J]. Quart. J. Roy. Meteor. Soc., 101 (428): 193–202.

Chen Guosen, Huang Ronghui. 2012. Excitation mechanisms of the teleconnection patterns affecting the July precipitation in Northwest China [J]. J. Climate, 25: 7834–7851.

Chen Wen, Zhu Deqin, Liu Huizhi, et al. 2009. Land–air interaction over arid/semi-arid areas in China and its impact on the East Asian summer monsoon. Part I: Calibration of the land surface model (BATS) using multicriteria methods [J]. Advances in Atmospheric Sciences, 26 (6): 1088–1098.

Dai Yongjiu, Zeng Xubin, Dickinson R E. 2001. Common Land Model: Technical Documentation and User's Guide [M/OL]. ftp://159.226.119.9/model/CoLM/clmdoc.pdf. [2012/10/25]

Dai Yongjiu, Zeng Xubin, Dickinson R E, et al. 2003. The common land model [J]. Bull. Amer. Meteor. Soc., 84 (8): 1013–1023.

Dikinson R E, Errico R M, Giorgi F, et al. 1989. A regional climate model for the western United States [J]. Climatic Change, 15 (3): 383–422.

Dikinson R E, Henderson-Sellers A, Kennedy P J. 1993. Biosphere–atmosphere Transfer Scheme (BATS) Verson 1e as Coupled to the NCAR Community Climate Model [R]. NCAR Tech. Note NCAR/TN-387+STR, 72pp.

Enomoto T. 2004. Interannual variability of the Bonin high as associated with the propagation of Rossby waves along the Asian Jet [J]. J. Meteor. Soc. Japan, 82 (4): 1019–1034.

Enomoto T, Hoskins B J, Matsuda Y. 2003. The formation mechanism of the Bonin high in August [J]. Quart. J. Roy. Meteor. Soc., 129: 157–178.

房云龙, 孙菽芬, 李倩, 等. 2010. 干旱区陆面过程模型参数优化和地气相互作用特征的模拟研究 [J]. 大气科学, 34 (2): 290–306. Fang Yunlong, Sun Sufeng, Li Qian, et al. 2010. The optimization of parameters of land surface model in arid region and the simulation of land–atmosphere interaction [J]. Chinese Journal of Atmospheric Sciences (in Chinese), 34 (2): 290–306.

Giorgi F. 1990. Simulation of regional climate using a limited area model nested in a general circulation model [J]. J. Climate, 3: 941–963.

Henderson-Sellers A, Yang Z L, Dickinson R E. 1993. The project for intercomparison of land-surface parameterization scheme [J]. Bull. Amer. Meteor. Soc., 74: 1335–1349.

胡隐樵, 高由禧, 王介民, 等. 1994. 黑河实验 (HEIFE) 的一些研究成果 [J]. 高原气象, 13: 225–236. Hu Yinqiao, Gao Youxi, Wang Jiemin, et al. 1994. Some achievements in scientific research during HEIFE [J]. Plateau Meteorology (in Chinese), 13: 225–236.

黄荣辉. 2006. 我国在气候灾害的形成机理和预测理论研究 [J]. 地球科学进展, 21: 564–575. Huang Ronghui. 2006. Progresses in research on the formation mechanism and prediction theory of severe climatic disasters in China [J]. Advances in Earth Sciences (in Chinese), 21: 564–575.

黄荣辉, 陈文, 张强, 等. 2011. 中国西北干旱区陆—气相互作用及其对东亚气候变化的影响 [M]. 北京: 气象出版社, 356pp. Huang Ronghui, Chen Wen, Zhang Qiang, et al. 2011. Land–Atmosphere Interaction over Arid Region of Northwest China and Its Impact on East Asian Climate Variability (in Chinese) [M]. Beijing: China Meteorological Press, 356pp.

Huang Ronghui, Liu Yong, Feng Tao. 2012. Characteristics and internal dynamic causes of the interdecadal jump of summer monsoon rainfall and circulation in eastern China occurred in the end of the 1990s [J]. Chinese Science Bulletin, (accepted)

Huang Ronghui, Wei Guoan, Zhang Qiang, et al. 2002. The Field Experiment on Air–Land Interaction in the Arid Area of Northwest China (NWC-ALIEX) and the preliminary scientific achievements of this experiment [C]// Proceedings of the International Workshop on the Air-Land Interaction in Arid and Semi-Arid Areas and Its Impact on Climate. 17–21 August, 2002, Dunhuang, Gansu Province, China.

Huang Ronghui, Wei Guoan, Zhang Qiang, et al. 2005. The preliminary scientific achievements of the Field Experiment on Air–Land Interaction in the Arid Area of Northwest China (NWC-ALAIEX) [C]// Proceedings of the 4th CTWF International Workshop on the Land Surface Models and Their Applications. 15–18 Nov., 2005, Zhuhai. China.

惠小英, 高晓清, 韦志刚, 等. 2011. 利用探空气球升速判定敦煌夏季白天边界层高度的分析 [J]. 高原气象, 30: 614–619. Hui Xiaoying, Gao Xiaoqing, Wei Zhigang, et al. 2011. Analysis on the determination of boundary layer height in daytime of Dunhuang summer using ascent rate of sounding balloon [J]. Plateau Meteorology (in Chinese), 30: 614–619.

Kalnay E, Kanamitsu M, Baker W E. 1990. Global numerical weather prediction at the National Meteorological Center [J]. Bull. Amer. Meteor. Soc., 71: 1410–1428.

Kalnay E, Kanamitsu M, Kistler R, et al. 1996. The NCEP/NCAR 40-year reanalysis project [J]. Bull. Amer. Meteor. Soc., 77: 437–471.

Kanamitsu M, Kumar A, Juang H M H, et al. 2002. NCEP dynamical seasonal forecast system 2000 [J]. Bull. Amer. Meteor. Soc., 83: 1019–1037

Li Qian, Xue Yongkang. 2010. Simulated impacts of land cover change on summer climate in the Tibetan Plateau [J]. Environ. Res. Lett., 5: 015102, doi:10.1088/1748-9326/5/1/015102.

Lu Riyu, Oh J H, Kim B J. 2002. A teleconnection pattern in upper-level meridional wind over the North African and Eurasian continent in summer [J]. Tellus, 54A: 44–55.

Sellers P, Hall F, Ranson K J, et al. 1995. The Boreal Ecosystem-Atmosphere Study (BORES): An overview and early results from the 1994 field year [J]. Bull. Amer. Meteor. Soc., 76: 1549–1577.

Sellers P J, Hall F G, Kelly R D, et al. 1997. BOREAS in 1997: Experiment overview, scientific results, and future directions [J]. J. Geophys. Res., 102(D24): 28731–28769. doi:10.1029/97JD03300.

世界气象组织 (WMO). 2006. 世界气候研究计划 2005–2015 年战略框架: 地球系统的协调观测和预报 (COPES) [M]. 李建平, 刘屹岷, 周天军, 等, 译. 北京: 气象出版社, 109pp. World Meteorological Organization (WMO). 2006. The World Climate Research Programme (WCRP) Strategic Framework 2005–2015: Coordinated Observation and Prediction of the Earth System (COPES) (in Chinese) [M]. Li Jianping, Liu Yimin, Zhou

Tianjun, et al., translated. Beijing: China Meteorological Press, 109pp.

Uppala S M, Kallberg P W, Simmons A J, et al. 2005. The ERA-40 reanalysis [J]. Quart. J. Roy. Meteor. Soc., 131: 2061–3012.

王超, 韦志刚, 李振朝. 2010a. 敦煌戈壁塔站资料的质量控制 [J]. 干旱气象, 28 (2): 121–127. Wang Chao, Wei Zhigang, Li Zhenchao. 2010a. A quality control routine for Dunhuang Gobi meteorology tower data [J]. Journal of Arid Meteorology (in Chinese), 28 (2): 121–127.

王超, 韦志刚, 李振朝, 等. 2010b. 敦煌戈壁地区地气温差变化特征分析 [J]. 干旱区环境与资源, 25 (11): 72–78. Wang Chao, Wei Zhigang, Li Zhenchao, et al. 2010b. The variation characteristics of differences between surface and air temperature in Dunhuang Gobi [J]. Journal of Arid Land Resources and Environment (in Chinese), 25 (11): 72–78

王超, 韦志刚, 高晓清, 等. 2012. 夏季敦煌稀疏植被下垫面物质和能量交换的观测研究 [J]. 高原气象, 31: 622–628. Wang Chao, Wei Zhigang, Gao Xiaoqing, et al. 2012. An observation study of surface air exchanges and energy budget at a sparse vegetation site of Dunhuang in summer [J]. Plateau Meteorology (in Chinese), 13: 622–628.

韦志刚, 陈文, 黄荣辉. 2010. 敦煌夏末大气垂直结构和边界层高度特征 [J]. 大气科学, 34: 905–913. Wei Zhigang, Chen Wen, Huang Ronghui. 2010. Vertical atmospheric structure and boundary layer height in the summer clear days over Dunhuang [J]. Chinese Journal of Atmospheric Sciences (in Chinese), 34: 905–913.

Wei Zhigang, Wen Jun, Li Zhenchao. 2009. Vertical atmospheric structure of the late summer clear days over the east Gansu Loess Plateau in China [J]. Advances in Atmospheric Sciences, 26: 381–389.

Xue Y K, Sellers P J, Kinter J L, et al. 1991. A simplified biosphere model for global climate studies [J]. J. Climate, 4: 345–364.

张强, 王胜. 2009. 西北干旱区夏季大气边界层结构及其陆面过程特征 [J]. 气象学报, 66: 599–608. Zhang Qiang, Wang Sheng. 2009. A study on atmospheric boundary layer structure on a clear day in the arid region in Northwest China [J]. Acta Meteorologica Sinica (in Chinese), 66: 599–608.

张强, 卫国安, 侯平. 2004. 初夏敦煌荒漠戈壁大气边界结构特征的一次观测研究 [J]. 高原气象, 23: 587–597. Zhang Qiang, Wei Guoan, Hou Ping. 2004. Observation studies of atmosphere boundary layer characteristic over Dunhuang Gobi in early summer [J]. Plateau Meteorology (in Chinese), 23: 587–597.

张强, 黄荣辉, 王胜, 等. 2005. 西北干旱区陆—气相互作用试验(NWC-ALIEX) 及其研究进展 [J]. 地球科学进展, 20: 427–441. Zhang Qiang, Huang Ronghui, Wang Sheng, et al. 2005. NWC-ALIEX and its research advances [J]. Advances in Earth Sciences (in Chinese), 20: 427–441.

周德刚, 黄刚, 马耀明. 2012. 中国西北干旱区戈壁下垫面夏季的热力输送 [J]. 大气科学学报, 35 (5): 541–549. Zhou Degang, Huang Gang, Ma Yaoming. 2012. Summer heat transfer over a Gobi underlying surface in the arid region of Northwest China [J]. Transactions of Atmospheric Sciences (in Chinese), 35 (5): 541–549.

Zhou Degang, Huang Ronghui. 2011. Characterization of turbulent flux transfer over a Gobi surface with quality-controlled observations [J]. Science China Earth Sciences, 54: 753–763.

周连童. 2009a. 比较NCEP/NCAR 和ERA-40 再分析资料与观测资料计算得到的感热资料的差异 [J]. 气候与环境研究, 14: 9–20. Zhou Liantong. 2009a. A comparison of NCEP/NCAR, ERA-40 reanalysis and observational data of sensible heat in Northwest China [J]. Climatic and Environmental Research (in Chinese), 14: 9–20.

周连童. 2009b. 引起华北地区夏季出现持续干旱的环流异常型 [J]. 气候与环境研究, 14: 120–130. Zhou Liantong. 2009b. Circulation anomalies pattern causing the persistent drought in North China [J]. Climatic and Environmental Research (in Chinese), 14: 120–130.

Zhou Liantong. 2009. Difference in the interdecadal variability of spring and summer sensible heat fluxes over Northwest China [J]. Atmospheric and Oceanic Science Letters, 2: 119–123.

周连童. 2010. 欧亚大陆干旱半干旱区感热通量的时空变化特征[J]. 大气科学学报, 33: 299–306. Zhou Liantong. 2010. Characteristics of temporal and spatial variations of sensible heat flux in the arid and semi-arid region of Eurasia [J]. Transactions of Atmospheric Sciences (in Chinese), 33: 299–306.

周连童, 黄荣辉. 2006. 中国西北干旱、半干旱区春季地气温差的年代际变化特征及其对华北夏季降水年代际变化的影响 [J]. 气候与环境研究, 11: 1–13. Zhou Liantong, Huang Ronghui. 2006. Characteristics of interdecadal variability of the difference between surface temperature and surface air temperature in spring in arid and semi-arid region of Northwest China and its impact on summer precipitation in North China [J]. Climatic and Environmental Research (in Chinese), 11: 1–13.

周连童, 黄荣辉. 2008. 中国西北干旱、半干旱区感热的年代际变化特征及其与中国夏季降水的关系 [J]. 大气科学, 32: 1276–1288. Zhou Liantong, Huang Ronghui. 2008. Interdecadal variability of sensible heat in arid and semi-arid regions of Northwest China and its relation to summer precipitation in China [J]. Chinese Journal of Atmospheric Sciences (in Chinese), 32: 1276–1288.

Zhou Liantong, Huang Ronghui. 2010a. An assessment of the quality of surface sensible heat flux derived from reanalysis data through comparison with station observations in Northwest China [J]. Advances in Atmospheric Sciences, 27: 500–512, doi: 10.1007/s00376-009-9081-8.

Zhou Liantong, Huang Ronghui. 2010b. Interdecadal variability of summer rainfall in Northwest China and its possible causes [J]. Int. J. Climatol., 30: 549–557, doi:10.1002/joc.1923.

Zhou Liantong, Wu Renguang, Huang Ronghui. 2010. Variability of surface sensible heat flux over Northwest China [J]. Atmospheric and Oceanic Science Letters, 3: 75–80.

朱德琴. 2006. 我国干旱/半干旱地区陆面过程及其对区域气候影响的数值模拟研究 [D]. 中国科学院大气物理研究所博士学位论文, 133pp. Zhu Deqin. 2006. Land surface process over arid/semi-arid areas in China and the simulations study of its impacts on regional climate [D]. Ph. D dissertation (in Chinese), Institute of Atmospheric Physics, Chinese Academy of Sciences, 133pp.

邹基玲, 候旭宏, 季国良. 1992. 黑河地区夏末太阳辐射特征的初步分析 [J]. 高原气象, 11: 381–388. Zou Jiling, Hou Xuhong, Ji Guoliang. 1992. Preliminary study of surface solar radiation properties in "HEIFE" area in late summer [J]. Plateau Meteorology (in Chinese), 11: 381–388.

左洪超, 胡隐樵. 1992. 黑河试验区沙漠和戈壁的总体输送系数 [J]. 高原气象, 11: 371–380. Zuo Hongchao, Hu Yinqiao. 1992. The bulk transfer coefficient over desert and Gobi in Heihe region [J]. Plateau Meteorology (in Chinese), 11: 371–380.

20世纪90年代末中国东部夏季降水和环流的年代际变化特征及其内动力成因[*]

黄荣辉[①②] 刘永[②] 冯涛[②]

(① 中国科学院大气物理研究所大气科学和地球流体力学数值模拟国家重点实验室, 北京 100029;
② 中国科学院大气物理研究所季风系统研究中心, 北京 100190)

摘要 观测资料分析表明, 中国东部夏季降水在20世纪90年代末发生了年代际突变, 在1999~2010年期间降水异常从以往的经向三极子型分布变成了经向偶极子分布, 形成了"南涝北旱"(除长江沿岸地区)的特征; 中国东部这次降水的年代际突变与东亚上空对流层环流及散度、垂直运动以及整层水汽输送散度的经向偶极子型年代际异常分布相对应. 并且, 本文还从大气内动力和热力过程讨论了1999~2010年期间东亚地区上空夏季对流层中、上层纬向气流和经向气流异常对中国东部夏季降水年代际突变的影响, 其结果表明, 由于在此时期东亚上空副热带急流北移减弱, 使得东亚上空纬向气流异常形成经向偶极子型. 这一方面使得东亚对流层上层沿副热带急流传播的"丝绸之路(Silk Road)"型、沿东亚经向传播的东亚/太平洋(EAP)型和沿极锋急流传播的欧亚(EU)型遥相关波列发生异常, 从而引起中国北方为下沉运动异常, 而南方为上升运动异常; 另一方面造成了中国东部对流层中层北方有冷平流异常, 而南方有暖平流异常, 这也引起了中国北方有下沉运动异常, 而南方有上升运动异常, 因而在1999~2010年期间夏季中国形成南涝北旱的降水异常.

　　我国地处东亚季风区, 夏季气候受东亚夏季风系统的严重影响. 由于东亚夏季风具有很明显的年代际变化特征, 因此, 我国夏季气候(特别是夏季风降水)表现出显著的年代际变化特征[1~8]. Yamamoto等人[9]指出20世纪50年代北半球气温发生了一次年代际气候突变, 这次气温突变特征是东亚地区气温有明显变化; 严中伟等人[10]指出了20世纪60年代中期北半球夏季降水发生了一次明显的代年际变化, 这次突变的特征是我国华北地区夏季降水开始减少; 黄荣辉等人[3]指出在70年代中后期东亚地区夏季气候发生了一次明显的年代际突变, 并指出这次突变的特征与1965年前后的气候年代际突变有明显的不同, 其中发生在1965年前后的东亚气候年代际突变主要特征是中国华北地区夏季降水开始减少, 而发生在70年代中后期的东亚气候年代际突变的主要特征不仅是华北地区夏季降水明显减少, 而且长江流域夏季降水明显增加. 最近, Kwon等人[11]和Ding等人[12]指出东亚地区气候在90年代初又发生了一次年代际突变, 这次突变的特征是中国华南地区夏季降水明显增多; 并且, Zhu等人[13]以及黄荣辉等人[8]的研究也初步表明了在90年代末东亚地区夏季气候可能又发生了一次明显的年代际变化, 这次变化的主

---

[*] 本文原载于: 科学通报, 第58卷, 第8期, 617-628, 2013年3月出版.

要特征是我国东北、华北地区以及西北东部夏季降水明显减少,而淮河流域的降水明显增多.

有关这几次年代际变化的可能原因,特别是针对 20 世纪 70 年代末和 90 年代初期这两次气候突变的成因,气象学家做了大量的研究工作,发现东亚夏季气候的年代际变化并不是孤立的,而是和全球其它地区气候突变具有一致性[2-7]. 如黄荣辉等人[3]指出了从 20 世纪 70 年代中后期起由于赤道东太平洋明显增温,并出现了类似 El Niño 型的海表温度异常分布,这使得东亚夏季风从 70 年代中后期起明显减弱;在这之后,不少研究从东亚上空环流的年代际变异方面也指出东亚夏季风在 70 年代中后期出现明显的减弱,并与西北太平洋副热带高压在 70 年代中后期强度增强并向西向南偏移有密切关系[4,5,14,15],与之对应,中国华北地区降水持续性减弱;并且,也有学者指出发生在 70 年代中后期东亚夏季气候的年代际突变是与东亚上空南亚高压增强[16,17]、东亚上空对流层变冷[18,19]以及副热带高空西风急流增强并向南偏移[11]也有很好关系. 最近,有些学者对发生在 90 年代初东亚夏季降水的年代际突变的成因做了一些研究. 特别是 Kwon 等人[11]指出中国华南地区夏季降水在 90 年代初的年代际增多是与东亚高空副热带西风急流出现了年代际减弱有关,而 Ding 等人[20]和 Wu 等人[21]认为这次东亚气候的年代际突变是由于西北太平洋地区对流层低层环流出现反气旋异常所致.

黄荣辉等人[8]的研究表明了中国东部夏季降水年代际突变明显表现在夏季降水异常主模态的年代际变化,在 1993~1998 年期间,中国东部夏季降水的年代际异常中偶极子模态(南北振荡型)的作用在增大,即经向三极子模态与经向偶极子模态并列;并且,他们的研究也初步表明了在 90 年代末中国东部夏季降水异常的主模态有从经向三极子模态变成经向偶极子模态的趋势. 为了更详细和更系统地分析在 20 世纪 90 年代末所发生的中国东部夏季降水年代际突变特征及其成因,本文利用更长的中国东部夏季降水资料以及 NCEP/NCAR 再分析资料来分析和诊断中国东部夏季降水在 1999~2010 年期间的年代际变化特征及其与东亚地区上空大尺度季风环流的年代际变化的联系;进一步从动力学上分析研究了发生在 90 年代末这次中国东部夏季降水变化的大气内动力成因,特别是从热力学方程诊断东亚上空对流层中层定常、瞬变气流的变化通过温度平流来影响垂直运动来引起降水变化,从而来说明 20 世纪 90 年代末我国东部夏季降水突变之成因.

1 资料和方法

本文使用的降水资料是中国气象局国家气象信息中心提供的 756 站逐日降水资料,考虑到站点观测的时间长度和连续性,本研究选取其中 481 个站点的夏季降水资料,时间跨度为 1958 年 1 月~2010 年 12 月. 本文使用大气环境场资料为美国国家环境预报中心和大气研究中心(NCEP/NCAR)再分析资料[22],主要包括水平风场、垂直运动场、温度场和大气比湿等变量. 所使用的分析方法是常规的合成分析方法及 Student's t 显著性检验方法. 气候均值突变检测方法采用滑动 T 检验以及 Lepage 检验[23,24]两种方法. 需要说明的是下文中所提及的距平均表示与气候态均值之差,这里我们采用 1971~2000 年气候平均作为气候态均值.

2 20 世纪 90 年代末中国东部夏季降水的年代际突变特征

我国东部夏季气候突变主要表现在降水[3],因此,首先从夏季降水出发来分析发生在 90 年代末我国东部夏季气候的突变.

2.1 20 世纪 90 年代末中国东部各区域夏季降水的年代际突变特征

为了分析 20 世纪 90 年代末我国东部夏季降水年代际突变的区域性特征,本研究利用 1958~2010 年我国东部(100°~135°E) 481 测站的降水资料分析了我国东部各区域夏季降水的年际变化. 图 1(a)~(d)分别是中国东北(40°~50°N)、华北地区(35°~40°N)、江淮流域(30°~35°N)及华南地区(20°~30°N)夏季降水距平百分率的年际变化. 从图 1(a)和(b)可以看到,在 1999~2010 年期间,东北、华北地区夏季降水比 20 世纪 90 年代的降水有明显的减少;并且如图 1(c)所示,江淮流域夏季降水在这期间有明显的增多;此外,从图 1(d)可以看到,在此期间,华南地区夏季降水仍继续偏多,但比 1993~1998 年期的降水有所减少.

为了更清楚地表示中国东部夏季降水在 20 世纪 90 年代末所发生的年代际突变特征,本文还给出了中国东部地区 100°~120°E 纬向平均的降水异常百分率

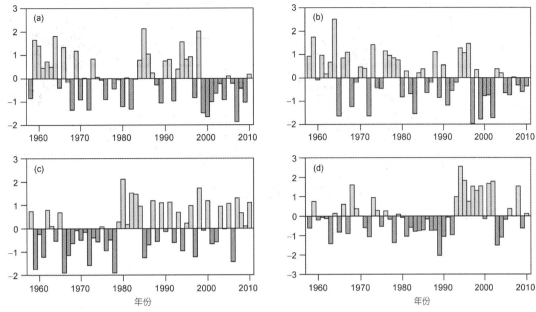

图1 中国东北地区(a)、华北地区(b)、江淮流域(c))和华南地区(d)区域平均的夏季降水距平百分率(%)的年际变化

9 a 滑动平均的纬度-时间分布图(图2). 如图2所示, 中国东部地区夏季降水年代际突变在20世纪90年代以前表现出以长江流域为中心的经向三极子型的分布, 表明该时期中国东部夏季降水异常主要还是受经向三极子型主模态所控制; 而在20世纪90年代初到90年代末, 全国呈现一致性降水正异常, 表明该时期中国东部夏季降水异常不仅受经向三极子型模态所控制, 而且经向偶极子型模态已在起作用. 然而, 从20世纪90年代末到21世纪初, 中国降水异常由经向三极型分布转变为经向偶极型分布, 即呈现出南正北负的特征, 这与Zhu等人[13]的结果一致.

2.2 对20世纪90年代末中国东部夏季降水年代际突变的显著性检验

为了更好地说明我国东部夏季降水在90年代末所发生的年代际变化, 本文还利用T检验和Lepage对气候突变的检验方法[23](见Liu等人[24]), 检验了我国东部夏季降水在1999年出现显著性年代际突变的站点分布情况和对应的降水异常的空间分布特征. 从图3所示的通过95%信度检验站点分布可以看出: 在20世纪90年代末到21世纪初期间, 中国夏季降水的年代际变化主要发生在华北和东北地区, 这两个地区大部分站点的夏季降水在1999~2010年期间明显减少, 它们的异常值都通过滑动T和Lepage显著性检验.

上述分析结果清楚地表明了中国东部夏季降水在20世纪90年代末发生了一次年代际突变, 这次突变的特征是, 在1999~2010年期间, 中国东部季风区夏季降水异常从经向三极子型分布变成经向偶极子型分布, 中国东北和华北地区夏季降水明显偏少, 而淮河流域夏季降水明显增多, 华南地区夏季降水继

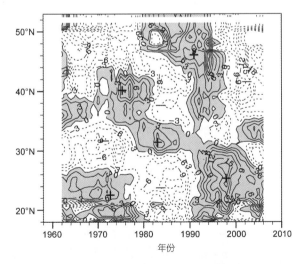

图2 中国东部100°~120°E纬向平均的夏季降水异常百分率的9 a 滑动平均的纬度-时间分布

实、虚线表示正、负距平, 阴影为正距平

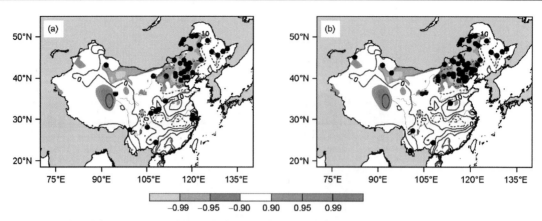

图 3 中国东部 1999~2010 年期间平均的夏季降水距平百分率分布(等值线)和发生突变的站点分布

实线、虚线分别表示正、负距平. 阴影区为通过95%的Student's t 检验区域. (a)和(b)中等值线和阴影表征内容相同. 黑色圆点为通过95%显著性信度的发生突变站点, 其中(a)为 Lepage 检验的结果, (b)为滑动 T 检验结果

续偏多, 即出现了南涝北旱(除长江沿岸地区)的经向偶极子型分布.

3　20世纪90年代末中国东部夏季降水异常模态的年代际变化特征

为了更好地说明中国东部季风区夏季降水在 20 世纪 90 年代末发生了年代际突变, 本文进一步利用 1958~2010 年中国东部 481 测站夏季降水资料来分析中国东部夏季降水异常在 1999~2010 年期间主模态的变化特征.

3.1　经向三极子分布特征

图 4(a)和(c)分别是中国东部 1958~2010 年夏季降水 EOF1 的空间分布和相应的时间系数序列. 从图 4(a)可以看到, 中国东部季风区夏季降水 EOF1 的空间分布从南到北呈现"+, −, +"经向三极子型分布; 并且, 从图 4(c)可以清楚看到, 中国东部季风区夏季降水 EOF1 的时间系数从 20 世纪 70 年代中后期起呈现 2~3 a 周期的年际变化特征, 即准两年周期振荡, 而且还可以看出, 中国东部季风区夏季降水的 EOF1 时间系数表现出明显的年代际变化. 从 20 世纪 50 年代到 70 年代中后期(大约 1958~1977 年), EOF1 的时间系数为正, 结合图 4(a)可以看到, 这时期华南及华北和东北南部地区夏季降水偏多, 而长江、淮河流域夏季降水偏少, 这表明了该时期我国东部夏季降水异常第一主模态从南到北为"+, −, +"经向三极子型分布; 而从 70 年代中后期到 90 年代初(大约在 1978~1992

年), EOF1 的时间系数为负, 结合图 4(a)可以看到, 这时期华南和华北地区夏季降水偏少, 而长江流域和东北地区夏季降水偏多, 这表明了这时期中国东部夏季降水异常第一主模态与 1958~1977 年时期夏季降水异常主模态呈现相反的空间分布, 即从南到北"−, +, −"经向三极子型分布; 从 90 年代初到 90 年代末(大约在 1993~1998 年), EOF1 的时间系数从负又变成正, 结合图 4(a)可以看到, 中国华南、华北和东北地区夏季降水应偏多, 而长江、淮河流域偏少, 这与 1958~1977 年时期夏季降水异常模态的分布相同, 为从南到北的经向"+, −, +"经向三极子型分布. 然而, 这时期中国东部夏季降水的实况却是华北和东北南部降水有所增加, 而华南地区夏季降水明显偏多, 洪涝灾害频繁发生, 而长江、淮河流域夏季降水却没有减少, 这可能是我国东部季风区夏季降水时空变化的第二模态在起作用(见图 4(b)). 从图 4(c) 可以看到, 从 1999 年之后 EOF1 的时间系数变小, 而 EOF2 的时间系数变大(见 3.2 小节), 这表明我国东部夏季降水的主模态在 20 世纪 90 年代末发生一次明显的年代际变化.

3.2　经向偶极子分布特征

图 4(b)和(d)分别是 1958~2010 年中国东部夏季降水 EOF2 的空间分布和相应的时间系数序列. 图 4(b)所示的中国东部季风区夏季降水 EOF2 的空间分布显示出与图 4(a)不同的分布, 它呈现从南到北"+, −"经向偶极子型分布. 并且, 与 4(c)相比较, 在图 4(d)所示的中国东部季风区夏季降水 EOF2 的时间序列

不仅有 3~4 a 周期的年际变化,而且具有更显著的年代际变化,特别从 20 世纪 90 年代末到 21 世纪初,EOF2 的时间系数序列从负变成较大的正. 这表明了从 20 世纪 90 年代末到 21 世纪初中国东部夏季降水异常主模态明显从经向三极型分布占有优变成经向偶极子型分布占优,结合图 4(b),这反映了中国东部夏季降水异常从 20 世纪 90 年代末起变成"南正北负",即南涝北旱的分布.

从上分析可以看到,中国东部季风区夏季降水的时空变化有两个主模态,即在空间分布上存在着从南到北为经向三极子型和经向偶极子型,它们不仅有年际变化,而且有很明显的年代际变化,从 20 世纪 90 年代末起到 21 世纪初,中国东部夏季降水异常主模态从经向三极子型分布变成经向偶极子型分布,这与中国东部夏季降水异常分布相对应.

4 1999~2010 年期间东亚夏季大气环流的年代际变化及其与中国东部夏季降水变化的关系

发生在 20 世纪 90 年代末中国东部夏季降水的年代际突变是与东亚地区上空的大气环流变化相对应. 为此,本研究利用 NCEP/NCAR 再分析资料分析了 1999~2010 年 700 和 200 hPa 东亚上空的风场和散度场、500 hPa 的垂直速度以及整层水汽输送的年代际异常特征.

4.1 东亚高低层大尺度环流异常

图 5(a)和(b)分别是东亚地区上空 1999~2010 年期间平均的夏季 700 和 200 hPa 距平风场和散度距平分布. 从图 5(a)可以看到:1999~2010 年期间在中国华南地区上空 700 hPa 有气旋性环流异常,而在蒙古高原上空及其周围的东北、华北和西北地区上空 700 hPa 有较强的反气旋环流异常. 这表明,在 1999~2010 年期间东亚上空对流层低层的距平环流为经向偶极子型分布;并且,从图 5(a)所示的东亚上空 1999~2010 年期间 700 hPa 风场的散度距平也可以看到在华南地区上空的对流层下层有气流的辐合异常,而在蒙古高原及其周围的东北、华北和西北地区有气流的辐散异常,这同样是经向偶极子型分布. 从图 5(b)可以看到,在 1999~2000 年期间在中国内蒙、东北和华北中西部上空 200 hPa 有强的东北风异常,而在黄海、东海和日本上空有偏南风异常. 这造成了在华南和江南地区上空对流层上层有强的气流辐散异常,而在内蒙、东北和华北地区中西部上空对流层上层有强的气流辐合异常,这同样是经向偶极子型分布,并且呈

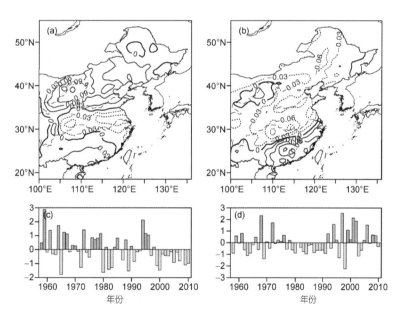

图 4 中国东部 1958~2010 年夏季降水 EOF 分析第 1 和第 2 主分量(EOF1, EOF2)的空间分布((a), (b))和相应的时间系数序列((c), (d))

(a)和(b)中实、虚线分别表示正、负信号. EOF1 和 EOF2 对方差的贡献分别为 12.4%, 9.0%

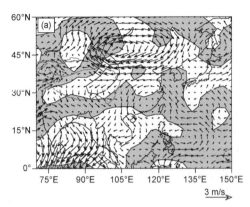 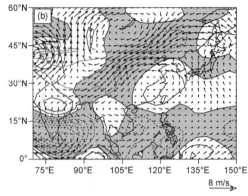

图 5 东亚地区上空 1999~2010 年平均的夏季 700 hPa (a)和 200 hPa (b)距平风场(单位: m s^{-1})和散度距平(单位: 10^{-7} s^{-1})分布
实、虚线分别表示散度的正、负距平, 阴影区表示辐合异常

现与对流层下层散度距平的经向偶极子型相反的分布特征. 正是由于东亚上空高低层环流异常的配合, 从而利于淮河流域和华南夏季降水的增多以及东北和华北地区及西北东部夏季降水的减少.

上述分析结果清楚表明, 1999~2010 年期间夏季东亚上空对流层上、下层环流和散度异常出现相反的经向偶极子型分布, 这有利于我国南方夏季降水增多, 北方降水减少, 从而造成了我国东部夏季降水异常出现年代际的经向偶极子型分布.

4.2 垂直运动异常

由于在 1999~2010 年期间夏季东亚上空对流层上、下层环流都发生了明显的年代际变化, 这种变化势必导致此地区垂直运动的年代际变化. 图 6 是东亚上空 1999~2010 年平均的夏季 500 hPa 垂直运动异常的分布. 从图 6 可以看到, 在中国南方上空有上升运动的异常, 而在北方有下沉运动的异常, 即出现了经向偶极子型分布. 东亚上空垂直运动距平的经向偶极子型分布造成了在 1999~2010 年期间夏季中国降水呈现南多北少的偶极子型分布, 使华南地区和淮河流域夏季多次发生严重洪涝灾害, 东北和华北出现了持续干旱现象.

从上分析结果可以看到, 中国东部夏季降水在 20 世纪 90 年代末发生的年代际突变与东亚上空夏季大气环流的年代际变化具有很好的对应性.

5 中国东部夏季降水在 20 世纪 90 年代末年代际突变的大气内动力成因分析

东亚夏季气候的年代际突变不仅与太平洋和印度

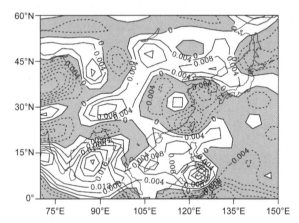

图 6 同图 5, 但是 500 hPa 垂直运动
单位: 10^{-5} hPa s^{-1}. 阴影区为上升运动异常

洋的热力强迫的年代际变化有关, 而且与东亚上空大气环流内部结构变化有密切关系. 黄荣辉和严邦良[25,26]指出, 即使是同等外强迫, 但在不同基流背景下对外源强迫响应的大气环流和行星波动异常分布也有很大的不同. 对应东亚夏季降水每次年代际变化, 东亚上空副热带急流在强度和位置都出现了明显的年代际变化[11]. 因此, 夏季东亚上空对流层气流的年代际变化对于我国东部夏季气候的变化可能有重要影响.

5.1 东亚上空对流层中、上层纬向气流的年代际变化

为了研究东亚上空对流层上、中层气流的年代际异常对中国东部夏季降水年代际变化的内动力作用, 本节首先利用再分析资料分析了东亚上空夏季 100°~140°E 平均的 200 hPa 纬向风异常的 9 a 滑动平均(图

7(a)). 如图 7(a)所示, 对应于中国东部夏季降水的年代际突变, 东亚上空副热带急流的强度和位置都出现了明显的年代际变化, 在 20 世纪 70 年代中后期前东亚上空夏季纬向气流的异常分布从南到北是"+, −, +"经向三极子型分布; 而从 20 世纪 70 年代中后期到 90 年代初期东亚上空纬向气流的异常变成了"−, +, −"的经向三极子型, 出现了与 70 年代中后期之前的异常型相反的分布; 并且, 从 20 世纪 90 年代初期开始到 2010 年, 东亚上空夏季纬向气流的异常从南到北呈现出"+, −"偶极子型, 显示出与 90 年代初以前的异常型截然不同的分布, 这表明东亚上空对流层上层的副热带急流减弱和北移, 使得东亚对流层上层的纬向气流异常在 20 世纪 90 年代末从经向三极子型分布变成经向偶极子型分布. 此外, 从图 7(a)还可以看到: 从 20 世纪 60 年代初到 21 世纪初, 东亚和东北亚上空夏季西风气流的负距平不断从高纬度向中纬度传播, 同时, 在东亚中纬度地区上空的西风气流的正距平也随之不断从东亚中纬度地区上空向副热带和低纬地区上空传播.

图 7(b)是东亚上空 500 hPa 纬向风异常 9 a 滑动平均的纬度-时间剖面图, 从图 7(b)可看到东亚上空夏季 500 hPa 纬向风异常的年代际变化与图 7(a)所示的 200 hPa 纬向风变化基本一致.

将图 7 与图 3 相比较, 可以看到东亚地区上空 200 和 500 hPa 夏季纬向风异常的经向分布型的年代际变化与中国东部夏季降水异常的经向分布型的年代际变化是一致, 在 20 世纪 90 年代末都表现为从经向三极子型分布变成经向偶极子型分布. 这表明了在 1999~2010 年期间东亚上空对流层上、中层纬向气流的年代际变化影响了对流层下层东亚夏季风环流的变化, 从而使中国东部夏季降水有明显的年代际变化.

5.2 1999~2010 年期间夏季欧亚上空气流异常对中国东部夏季水汽输送异常的动力作用

黄荣辉等人[8]指出中国东部夏季降水的年代际变化不仅受东亚/太平型(EAP 型)遥相关波列的影响, 而且还受到欧亚上空中高纬度西风带欧亚型(EU 型)的影响. 并且, Lu 等人[27]提出夏季在欧亚大陆上空对流层上层的经向风异常存在着一遥相关波列, 之后, 日本学者 Enomoto 等人[28]以及 Enomoto[29]把此遥相关波列称为"丝绸之路"("Silk Road")型遥相关. 最近 Chen 和 Huang[30]指出: 此遥相关型不仅影响着中亚地区和我国西北地区的夏季降水异常, 而且还通过与 EAP 型遥相关波列的相互作用影响中国东部夏季降水异常. Kosaka 等人[31]的研究表明, 东亚夏季梅雨降水的年际变化受到 3 支遥相关波列的影响, 即沿亚州上空副热带急流传播的"Silk Road"遥相关波列, 沿东亚上空经向传播的 PJ 振荡以及沿极锋急流传播的欧亚(EU型)遥相关波列. 因此, 在 1999~2010 年期间这 3 支遥相关波列的异常势必影响到东亚夏季梅雨降水的异常.

上面分析结果表明了东亚高低空环流在 1999~2010 年期间都发生了明显的年代际变化, 这势必影响到此地区整层水汽输送的年代际变化. 为了更好地分析这 3 支遥相关型波列的变化对中国东部夏季降水异常的影响, 本文还利用 NCEP/NCAR 再分析的水汽资料分析了欧亚上空水汽输送通量距平分布的年代际变化. 图 8(a)~(d)分别是 1958~1977 年、1978~

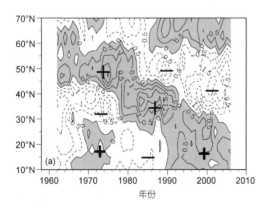 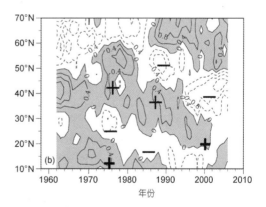

图 7 东亚地区 100°~140°E 平均的 200 hPa (a)和 500 hPa (b)纬向风异常 9 a 滑动平均的纬度-时间剖面
实、虚线分别表示西风和东风异常(单位: m s^{-1}). 阴影区表示西风异常

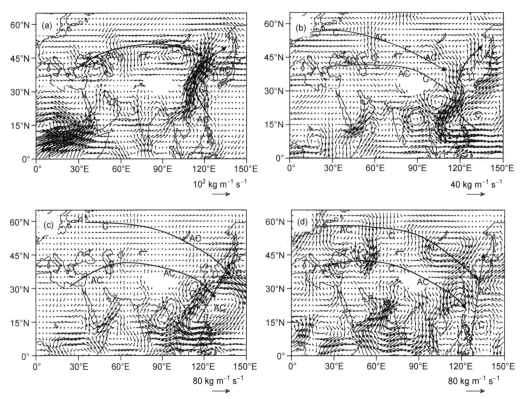

图 8 欧亚上空 1958~1977 年(a), 1978~1992 年(b), 1993~1998 年(c)和 1999~2010 年(d)期间平均夏季 1000~300 hPa 整层水汽输送通量距平矢量分布

1992 年、1999~2010 年期间平均的欧亚上空夏季 1000~300 hPa 整层水汽输送通量距平矢量分布.

从图 8(a)可以看到, 从我国华南和南海经长江和淮河流域到我国东北和日本海的东亚上空夏季水汽输送通量距平矢量分布出现"反气旋-气旋-反气旋"异常分布, 它们似如 EAP 型[32~34]遥相关波列; 并且, 从里海周围经蒙古高原到我国东部和日本海上空夏季水汽输送通量距平矢量分布出现"气旋-反气旋-气旋-反气旋"异常分布, 它们似如"Silk Road"型遥相关波列分布, 此波列比较偏北, 这可能是由于在这期间欧亚上空副热带急流偏北所致(见图 7(a)); 同时在欧亚高纬度地区上空夏季水汽输送通量距平分布上存在着 EU 型遥相关波列[35], 但此波列偏北且很弱. 此外, 从图 8(a)还可以看到, 在中国东部有很强的向北的水汽输送异常, 这表明此时期东亚夏季风偏强.

如图 8(b)所示, 在 1978~1992 年期间欧亚上空夏季水汽输送通量距平矢量分布出现了与图 8(a)相反的分布. 从南海和菲律宾经长江流域和我国东北和朝鲜半岛到鄂霍次克海上空水汽输送通量距平矢量分布出现"气旋-反气旋"异常分布, 它们呈现一个与图 8(a)相反的 EAP 遥相关波列; 并且, 从里海周围经中亚和我国西北及内蒙到我国华北上空水汽输送距平矢量分布出现"反气旋-气旋-反气旋-气旋-反气旋"异常分布, 它们呈现出一个与图 8(a)相反的"Silk Road"型遥关波列分布, 随着欧亚上空副热带急流的向南加强(见图 7(a)), 此水汽输送异常遥相关波列也向南移动; 同时在欧亚中高纬度地区上空水汽输送距平分布上从西北欧经乌拉尔地区和蒙古高原到我国东北呈现出"气旋-反气旋-气旋-反气旋-气旋"似如 EU 型遥相关波列的水汽输送异常分布. 此外, 从图 8(b)还可以看到, 在中国东部有很强的向南水汽输送异常, 这表明此时期东亚夏季风偏弱.

正如图 8(c)所示, 在 1993~1998 年期间欧亚上空夏季水汽输送通量距平矢量分布出现了与图 8(a)和(b)有一些不同的分布. 在欧亚中高纬度上空水汽输送通量距平虽然其分布有些类似于图 8(a)所示的分布型, 但 EU 型遥相关波列不清楚, 这表明来自大西洋沿西风带的水汽输送变弱; 并且, 从中国南海经中国东部和西北太平洋到日本海和日本上空水汽输送异常有"反气旋-气旋"异常分布,这样在中国东部及黄、东海上空有强的向北水汽输送距平,这引起了在1993~1998年期间中国东部夏季降水偏多;同时欧亚上空沿副热带急流方向的水汽输送距平从里海周围经中亚和我国西北到我国东部呈现"反气旋-气旋-反气旋-气旋"类似"Silk Road"型遥相关波列的异常分布,随着欧亚上空副热带急流的向南加强(见图7(a)),此水汽输送异常遥相关波列也略向南移动. 此外,从图8(c)还可以看到,在我国东部有较强的向北水汽输送异常,这表明此时期东亚夏季风又变得偏强.

上述从NCEP/NCAR再分析资料所计算的欧亚上空整层水汽输送通量的年代际异常分布型与用ERA-40再分析的结果有所不同(见黄荣辉等人[8]中图9),特别是对于1958~1977年时间段,两套资料计算的水汽输送距平在东亚地区东部差异较大,NCEP资料结果表现为较强的南风异常输送,而ERA资料结果则表现为从南至北三极型"反气旋-气旋-反气旋"的异常分布,这种差异可能与两套资料的模式依赖性有关.

然而,正如图8(d)所示,1999~2010年期间在欧亚上空的中高纬度地区上空水汽输送异常显示出与1993~1998年不同的波列分布,在此地区上空出现了一个明显的EU型水汽输送异常分布型,即在西北欧上空有反气旋水汽输送异常,在乌拉尔地区为气旋性水汽输送的异常分布,在蒙古高原、内蒙、东北和华北上空为一较强的反气旋性的水汽输送异常分布,在日本海和日本上空为一气旋性异常分布,相比于图8(b)所示的1993~1998年期间的水汽输送异常型的分布,1999~2010年期间的水汽输送异常EU型遥相关波列分布明显;并且,从经向方向看,在中国南方有一气旋性水汽输送异常分布,而在北方有反气旋性水汽输送异常分布,这表明在1999~2010年期间从中国华南到蒙古高原地区整层水汽输送异常也出现了经向偶极子型分布;同时从图8(d)可以看到,从里海周围经中亚和我国西北到我国华中和华东地区上空水汽输送通量异常出现"反气旋-气旋-反气旋-气旋"异常分布,它们似如"Silk Road"型遥相关波列的分布,此波列分布虽与1993~1998年期间的波列相同,但我国南方的气旋型异常分布偏东,并且蒙古高原上空反气旋型水汽输送加强,这使得我国东部出现向南的水汽输送异常. 此外,从图8(d)还可以看到,从中国东北经华北到达西南地区为向南的水汽输送异常,这表明在1999~2010年期间东亚夏季风偏弱,西南季风从孟加拉湾经中印半岛和中国西南到华北和东北地区的水汽偏弱,因此,在上述地区夏季发生不同程度的持续干旱,而在华南和淮河流域夏季降水明显增多,并引起了这些地区有些年的夏季发生了严重洪涝灾害.

从上述整层水汽输送的年代际异常分析结果可以看到: 在20世纪90年代末所发生的中国东部夏季季风降水的年代际突变不仅与从菲律宾周围往北传播的似如EAP型遥相关波列分布的年代际变化有关,而且也与欧亚中高纬度沿极锋急流传播的似如EU型遥相关波列变化以及沿副热带急流方向传播的"Silk Road"遥相关波列变化有关.

6 1999~2010年期间夏季东亚上空对流层中层气流异常对中国东部夏季降水异常的热力作用

最近,Sampe和Xie[36]指出: 来自高原南侧的暖湿气流在青藏高原东侧的东亚副热带上空对流层中层产生暖平流,从而触发此地区上升运动的异常,并导致对流不稳定;并且,来自南海和热带西太平洋低层气流携带大量的水汽使得对流不稳定得以持续;此外,还由于沿中纬度副热带急流传播的瞬变天气尺度扰动加强了东亚上空上升运动和对流不稳定性. 这些因子使得东亚初夏梅雨得到长时期维持,从而引起东亚夏季降水的异常.

6.1 1999~2010年期间东亚夏季对流层中层温度平流的异常

在 p 坐标下热力学方程可写成下式:

$$\frac{\partial T}{\partial t} = \frac{Q}{C_p} - \vec{V} \cdot \nabla T - \left(\frac{P}{P_0}\right)^{R/C_p} \cdot \omega \frac{\partial \theta}{\partial p}, \quad (1)$$

其中 T 为温度, θ 为位温, $\vec{V}$ 为水平风矢量, ω 为 p 坐标下的垂直速度, R 为气体常数, C_p 为定压比热, Q 为绝热加热, P_0 为海面气压,并取 $P_0=1000$ hPa.

由于本文所研究的对象是年代际时间尺度的气候变化,文中所计算的水平温度平流,即 $(-\vec{V} \cdot \nabla T)$ 采用月时间尺度,因此,在(1)式中 $\frac{\partial T}{\partial t}$ 很小;并且,本文所研究的范围是中纬度和副热带地区,在中纬度地区, 非绝然加热 Q 较小, 而在副热带地区 Q 虽然较大的, 但它总是与对流活动密切相关, 即与垂直运动 ω 有关. 因此, 从方程(1)可得到垂直运动与水平温度平流相平衡的近似关系, 即

$$\left(\frac{P}{P_0}\right)^k \omega \frac{\partial \theta}{\partial p} \propto -\vec{V} \cdot \nabla T, \quad (2)$$

上式 $k=R/C_p$. 式(2)表明了在中纬度和副热带地区垂直运动与水平温度平流相平衡, 即在中纬度和副热带的某地区上空的等压面上有暖平流, 则将引起该地区有上升运动; 反之, 某地区上空有冷平流, 则将引起该地区有下沉运动.

根据 Sampe 和 Xie[36]的研究, 把水平温度平流分成定常和瞬变两部分, 即

$$\overline{\vec{V} \cdot \nabla T} = \overline{\vec{V}} \cdot \nabla \overline{T} + \overline{\vec{V}' \cdot \nabla T'}, \quad (3)$$

式(3)右边第 1 项为定常气流引起的水平温度输送, 第 2 项为瞬变气流引起的水平温度输送. 为了研究 1999~2010 年期间夏季对流层中层气流异常通过暖平流异常对中国东部垂直运动异常的热力作用. 本节利用 NCEP/NCAR 再分析的月平均风场和温度资料计算了每年夏季(6~8 月)东亚地区 500 hPa 温度平流, 即 $(-\overline{\vec{V}} \cdot \nabla \overline{T})$ 距平(见图9).

从图9可以看到, 从 20 世纪 90 年代初, 东亚上空对流层中层温度平流异常从经向三极子型转变成经向偶极子型; 并且, 与图 7(a)和(b)相比较可以得出, 随着东亚上空对流层上、中层纬向西风异常从中纬度向低纬度传播, 其上空对流层中层水平暖平流异常也向低纬度传播. 这样, 造成了如图10(a)所示的1999~2010 年期间夏季平均的 500 hPa 水平温度平流即

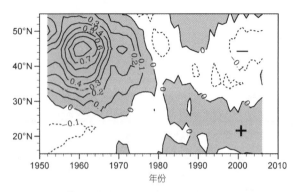

图 9 东亚地区 100°~140°E 纬向平均的 500 hPa 月平均气流引起的夏季(6~8 月)温度平流距平的 9 a 滑动平均纬度-时间剖面

单位: K d^{-1}. 图中实(虚)线分别表示正(负)距平. 阴影区表示正距平异常

$(-\overline{\vec{V}} \cdot \nabla \overline{T})$ 异常在中国北方产生了负异常, 即冷平流异常, 而在从黄淮地区到华南的南方地区出现了正距平, 即暖平流异常. 同样地, 本文也计算瞬变扰动 (2~9 d)对对流层中层温度平流的贡献, 即 $(-\overline{\vec{V}' \cdot \nabla T'})$, 如图 10(b)所示, 在东亚地区东部(100°E 以东区域), 其结果与图 10(a)所示的年代际的定常气流引起温度平流异常的分布相同, 但量值比定常气流引起的温度平流小, 大约是定常气流引起的温度平流值的 1/4.

6.2 1999~2010 年期间东亚地区夏季对流层中层温度平流异常与垂直运动异常的关系

Sampe 和 Xie[36]的研究表明了在东亚梅雨期对流层中层温度平流和垂直运动有很好的对应关系, 这表明了东亚对流层中层暖平流对于梅雨锋形成和维

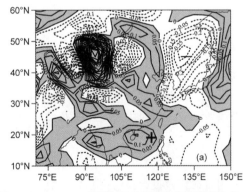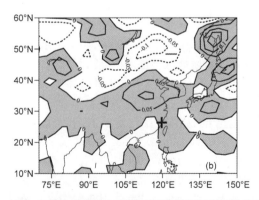

图 10 1999~2010 年期间夏季平均的东亚地区上空 500 hPa 定常(a)和瞬变(b)气流引起的温度平流异常分布

单位: K d^{-1}. 阴影区表示暖平流异常

持起到重要作用. 为此, 依据上面所计算的温度平流异常来推断垂直运动异常.

为了比较东亚地区夏季对流层中层温度平流异常与垂直运动异常的关系, 把图 10 与图 6 相比较, 可以看到, 两者异常特征比较吻合. 图 10 中冷平流异常与图 6 中我国东北、华北和西北东部的下沉气流异常相对应, 而暖平流异常与图 6 中淮河流域和华南北部的异常的上升运动相对应. 这表明由于 1999~2010 年期间夏季东亚地区上空对流层中、上层纬向气流异常从经向三极子型变成经向偶极子型, 使得对流层中层定常扰动和瞬变扰动引起的温度平流也从经向三极子型变成经向偶极子型, 从而造成了垂直运动异常变成经向偶极子型分布, 因而引起中国东部地区在此时期夏季降水异常变成经向偶极子型分布, 即南涝北旱型.

7 结论和讨论

本文利用我国详细的测站降水资料以及 NCEP/NCAR 再分析资料分析了中国东部夏季降水在 20 世纪 90 年代末的年代际突变特征, 表明了在 1999~2010 年期间中国东部夏季降水异常从经向三极子型分布变成经向偶极子型分布, 即中国东北和华北地区及西北地区东部夏季降水显著减少, 并通过了滑动 T 检验和 *Lepage* 检验, 而淮河流域以南地区(除长江沿岸地区)夏季降水偏多. 并且, 本文还分析和讨论了与之对应的东亚地区上空大气环流的异常, 表明此时期东亚地区上空对流层高、低层环流及散度、垂直运动都呈现出与中国东部夏季降水年代际变化相对应的经向偶极子型分布.

本文还利用水汽输送通量的异常分布通量从大气的内动力分析诊断了 1999~2010 年期间东亚地区上空对流层中上层纬向气流异常的偶极子型(特别是副热带西风急流减弱、北移)对沿欧亚上空副热带急流传播的 "Silk Road" 型和沿东亚上空经向传播的 EAP 型以及沿极锋急流传播的 EU 型遥相关波列的影响, 从而从定常波活动的年代际异常讨论了对流层中上层气流异常对东亚上空在此时期垂直运动异常的动力作用; 并且从热力学方程分析诊断了东亚上空对流层中层气流的异常对东亚上空温度平流的影响, 从而从温度平流异常讨论了对流层中层气流对东亚此时期垂直运动和降水异常的热力作用.

本文只从大气内动力和热力方面讨论了东亚地区对流层中上层气流异常对中国东部夏季降水在 1999~2010 年期间年代际变化的作用. 然而, 对于 20 世纪 90 年代末到 21 世纪初中国东部夏季的年代际变化, 热带太平洋和印度洋的热力变化也有很重要影响. 为此, 本研究还做了数值试验, 受篇幅所限, 不再详述.

参考文献

1 Chen L, Dong M, Shao Y. The characteristics of interannual variations on the East Asian monsoon. J Meteorol Soc Jpn, 1992, 70: 397–421
2 Ding Y H. Summer monsoon rainfalls in China. J Meteorol Soc Jpn, 1992, 70: 373–396
3 黄荣辉, 徐予红, 周连童. 我国夏季降水的年代际变化及华北干旱化趋势. 高原气象, 1999, 18: 465–477
4 Huang R H, Zhou L T, Chen W. The progresses of recent studies on the variabilities of the East Asian monsoon and their causes. Adv Atmos Sci, 2003, 20: 55–69
5 Huang R H, Chen W, Zhang R H. Recent advances in studies of the interaction between the East Asian winter and summer monsoons and ENSO cycle. Adv Atmos Sci, 2004, 21: 407–424
6 Ding Y H, Chan J C L. The East Asian summer monsoon: An overview. Meteorol Atmos Phys, 2005, 89: 117–142
7 Huang R H, Chen J L, Huang G. Characteristics and variations of the East Asian monsoon system and its impacts on climate disasters in China. Adv Atmos Sci, 2007, 24: 993–1023
8 黄荣辉, 陈际龙, 刘永. 我国东部夏季降水主模态的年代际变化及其与东亚水汽输送的关系. 大气科学, 2011, 35: 589–606
9 Yamamoto R, Iwashima T, Songa N K, et al. An analysis of climatic jump. J Meteorol Soc Jpn, 1986, 64: 273–281
10 严中伟, 季劲钧, 叶笃正. 60 年代北半球夏季气候跃变: I. 降水和温度变化. 中国科学 B 辑, 1990, 20: 97–103
11 Kwon M H, Jhun J G, Ha K J. Decadal change in East Asian summer monsoon circulation in the mid-1990s. Geophys Res Lett, 2007, 34: L21706
12 Ding Y H, Wang Z Y, Sun Y. Interdecadal variation of the summer precipitation in East China and its association with decreasing Asian summer monsoon. Part I: Observed evidences. Int J Climatol, 2008, 28: 1139–1161

13 Zhu Y, Wang W, Zhou W, et al. Recent changes in the summer precipitation pattern in East Asia and the background circulation. Clim Dyn, 2010, 36: 1463–1473

14 Gong D Y, Ho C H. Shift in the summer rainfall over the Yangtze River valley in the late 1970s. Geophys Res Lett, 2002, 29: 1436

15 Xie S P, Hu K M, Hafner J, et al. Indian Ocean capacitor effect on Indo-western Pacific climate during the summer following El Niño. J Clim, 2009, 22: 730–747

16 Li S L, Lu J, Huang G, et al. Tropical Indian Ocean basin warming and East Asian summer monsoon: A multiple AGCM study. J Clim, 2008, 21: 6080–6088

17 Huang G, Liu Y, Huang R H. The interannual variability of summer rainfall in the arid and semiarid regions of Northern China and its association with the northern hemisphere circumglobal teleconnection. Adv Atmos Sci, 2011, 28: 257–268

18 Yu R C, Wang B, Zhou T J. Tropospheric cooling and summer monsoon weaking trend over East Asia. Geophys Res Lett, 2004, 31: L22212

19 Yu R C, Zhou T J. Seasonality and three-dimensional structure of interdecadal change in the East Asian monsoon. J Clim, 2007, 20: 5344–5355

20 Ding Y H, Sun Y, Wang Z Y, et al. Inter-decadal variation of the summer precipitation in China and its association with decreasing Asian summer monsoon. Part II: Possible causes. Int J Climatol, 2009, 29: 1926–1944

21 Wu R G, Wen Z P, Yang S, et al. An interdecadal change in southern China summer rainfall around 1992/93. J Clim, 2010, 23: 2389–2403

22 Kalnay E, Kanamitsu M, Kistler R, et al. The NCEP/NCAR 40-year reanalysis project. Bull Amer Meteorol Soc, 1996, 77: 437–471

23 Lepage Y. A combaination of Wilecoxon's and Ansar-Bradley's statistics. Biometrika, 1971, 58: 213–217

24 Liu Y, Huang G, Huang R H. Inter-decadal variability of summer rainfall in Eastern China detected by the Lepage test. Theor Appl Climatol, 2011, 106: 481–488

25 Huang R H, Yang B L. Influence of the basic flow in the tropics on the stationary planetary waves at middle and high latitudes during the Northern Hemisphere winter. Acta Meteorol Sin, 1989, 3: 437–447

26 黄荣辉, 严邦良, 岸保勘三郎. 基本气流在 ENSO 对北半球冬季大气环流影响中的作用. 大气科学, 1991, 15: 44–53

27 Lu R Y, Oh J H, Kim B J. A teleconnection pattern in upper-level meridional wind over the North African and Eurasian continent in summer. Tellus, 2002, 54A: 44–55

28 Enomoto T, Hoskins B J, Matsuda Y. The formation mechanism of Bonin high in August. J Roy Meteorol Soc, 2003, 129: 157–178

29 Enomoto T. Interannual variability of the Bonin high as associated with the propagation of Rossby waves along the Asian jet. J Meteorol Soc Jpn, 2004, 82: 1019–1024

30 Chen G S, Huang R H. Excitation mechanism of the teleconnection patterns affecting the July precipitation in Northwest China. J Clim, 2012, 25: 7834–7851

31 Kosaka Y, Xie S P, Nakamura H. Dynamics of interannual variability in summer precipitation over East Asia. J Clim, 2011, 24: 5435–5453

32 Nitta T. Convective activities in the tropical western Pacific and their impact on the Northern Hemisphere summer circulation. J Meteorol Soc Jpn, 1987, 64:373–400

33 Huang R H, Li W J. Influence of the Heat source anomaly over the tropical western Pacific on the Subtropical high over East Asia. Proceedings of International Conference on the General Circulation of East Asia. 1987, 40–51

34 黄荣辉, 李维京. 夏季热带西太平洋上空的热源异常对东亚上空副热带高压的影响及其物理机制. 大气科学, 1988, 12(特刊): 20–36

35 Wallace J M, Gutzler D S. Teleconnection in the geopotential height field during the Northern Hemisphere winter. Mon Weather Rev, 1981, 100: 748–812

36 Sampe T, Xie S P. Large-scale dynamics of the Meiyu-Baiu rainband: Environmental forcing by the westerly jet. J Clim, 2010, 23: 113–134

20世纪90年代末东亚冬季风年代际变化特征及其内动力成因*

黄荣辉[1]　刘　永[1]　皇甫静亮[1]　冯　涛[2]

(1 中国科学院大气物理研究所季风系统研究中心，北京 100190；
2 南京大学大气科学学院，南京 210093)

摘　要　为纪念陶诗言先生对东亚冬季风研究的杰出贡献，本文利用我国测站、NCEP/NCAR 和 ERA-40/ERA-Interim 再分析资料分析了我国冬季气温和东亚冬季风在 20 世纪 90 年代末所发生的年代际跃变特征及其内动力成因。分析结果表明：从 20 世纪 90 年代末之后，我国冬季气温和东亚冬季风发生了明显的年代际跃变。从 1999 年之后，随着东亚冬季风从偏弱变偏强，我国冬季气温变化从全国一致变化型变成南北振荡型（即北冷南暖型），并由于从 1999 年之后我国北方冬季气温从偏高变成偏低，故冬季低温雪暴冰冻灾害频繁发生，同时，我国冬季气温和东亚冬季风年际变化在此时期从以往 3～4 a 周期年际变化变成 2～8 a 周期；并且，结果还表明了东亚冬季风此次年代际变化是由于西伯利亚高压和阿留申低压的加强所致。本文还从北极涛动（AO）和北半球准定常行星波活动的动力理论进一步讨论了此次东亚冬季风年代际跃变的内动力成因及其机理，结果表明：从 20 世纪 90 年代末之后，北半球冬季准定常行星波在高纬地区沿极地波导传播到平流层加强，而沿低纬波导传播到副热带对流层上层减弱，这造成了行星波 E-P 通量在高纬度地区对流层和平流层辐合加强，而在副热带地区对流层中、上层辐散加强，因而导致了北半球高纬度地区从对流层到平流层纬向平均纬向流和欧亚上空极锋急流减弱，而副热带急流加强，这造成了 AO 减弱和东亚冬季风加强。

Characteristics and Internal Dynamical Causes of the Interdecadal Variability of East Asian Winter Monsoon near the Late 1990s

HUANG Ronghui[1], LIU Yong[1], HUANGFU Jingliang[1], and FENG Tao[2]

1 *Center for Monsoon System Research, Institute of Atmospheric Physics, Chinese Academy of Sciences, Beijing* 100190
2 *School of Atmospheric Sciences, Nanjing University, Nanjing* 210093

Abstract　In memory of the excellent contributions made by academician Tao Shiyan to the study on East Asian winter monsoon (EAWM), the characteristics and internal dynamical causes of winter surface air temperature in China and the EAWM occurring near the late 1990s are analyzed in this paper by using observational data in China and reanalysis data of the NCEP/NCAR and ERA-40/ERA-Interim. The analyzed results show a significant jump of interdecadal variability of winter surface-air temperature in China and the EAWM occurrence in the late 1990s. With the strengthening of the

---

* 本文原载于：大气科学，第 38 卷，第 4 期，627-644，2014 年 7 月出版.

EAWM, this variability of winter surface air-temperature in China has undergone a change from a "similar pattern in the whole China" to a "south-north oscillation pattern" (i.e., cold in the north but warm in the south) since 1999. Because the winter surface-air temperature in northern China shifted into a colder state during 1999–2012, wintertime disasters of low temperature, snowstorms, and blizzards have frequently occurred in this region. In addition, the dominant period of the interannual variability of winter surface-air temperature and the EAWM turned into 2–8 years from previous 3–4 years. Moreover, the results show that this interdecadal jump of the EAWM is attributed to the strengthening of the Siberian high and the Aleutian low. The internal dynamical causes and physical mechanism of this interdecadal variability of the EAWM are discussed further from the dynamical theories of Arctic Oscillation (AO) and quasi-stationary planetary wave activity. Since the late 1990s, the propagation of quasi-stationary planetary waves into the stratosphere over high latitudes of the Northern Hemisphere along the polar wave-guide was enhanced, while the propagation into the upper troposphere over the subtropics along the low-latitude wave-guide weakened, which caused the strengthening of the convergence of Eliassen-Palm (E-P) fluxes of quasi-stationary planetary waves in the troposphere and stratosphere over high latitudes and strengthening of the divergence of E-P fluxes in the middle and upper troposphere over the subtropics of the Northern Hemisphere. This led to the weakening of winter time zonal-mean flow from the troposphere to the stratosphere over high latitudes of the Northern Hemisphere and the polar front jet and strengthening of the wintertime subtropical jet during 1999–2012, which caused the weakening of the wintertime AO and strengthening of the EAWM.

1 引言

东亚冬季风（以下简称为 EAWM）的变化与异常对我国冬季气候灾害有严重影响，特别是寒潮。寒潮是影响我国冬季寒害、雪灾、早霜和晚霜等灾害性气候的重要成因。如 2008 年 1 月，由于东亚冬季风的异常，在我国南方发生了严重的低温雨雪冰冻灾害，造成了 1500 多亿元的经济损失，2012 年冬季我国北方又发生了低温雪灾冰冻严重灾害。因此，EAWM 年际和年代际变异是我国大气科学重要的研究课题。

早在 20 世纪 50 年代，陶诗言先生对东亚冬季风活动，特别是对于东亚寒潮活动路径及其与寒潮爆发有关的东亚冬季风环流的变化过程做出了系统而开创性的研究（陶诗言，1952，1956，1957，1959；陶诗言等，1965；陶诗言和张庆云，1998）。陶诗言先生是 EAWM 研究的开拓者，他首先提出了东亚寒潮爆发有三条路径，他按路径把寒潮分三种型：即西北型寒潮、超极地型寒潮和沿贝加尔湖以东自北向南入侵东亚的寒潮；并且，他还提出 EAWM 环流系统的特征和结构。在临近陶诗言先生仙逝一周年之际，我们深切缅怀陶诗言先生。他在亚洲季风、东亚寒潮、中小尺度天气系统和暴雨等领域做出开拓性的系统研究，为中国天气预报提供了理论依据和方法，是中国现代天气预报理论和方法的开拓者和奠基人之一。他严谨治学、平易近人、虚怀若谷、淡泊名利、实事求是、勤奋一生，彰显了一代气象大师的崇高品格和治学风范。斯人已逝，文章不朽。在临近陶先生逝世一周年之际，特撰写此文以纪念陶诗言先生对我国气象学发展的伟大贡献。

继陶诗言先生研究之后，我国和国外许多学者在 EAWM 的变异及其机理做出不少研究。Chen and Graf（1998），Chen et al.（2002），Wu and Wang（2002），Jhun and Lee（2004），Li and Yang（2010），Wang and Chen（2010）和 Wang and Chen（2013a）等各自定义了 EAWM 指数并研究了 EAWM 的年际变化。黄荣辉等（2007）利用 Wu and Wang（2002）所定义的 EAWM 指数详细分析了 EAWM 的年际变化，他们研究表明了 EAWM 有显著的 3~4 年周期年际变化，并指出 2005 年和 2006 年 EAWM 有很明显的差别。同样，近年来国内许多学者研究表明了 EAWM 有显著的年代际变化（Huang and Wang, 2006; Wang et al., 2009; Wang and Chen, 2010; Wang and Chen, 2013b）。特别是王遵娅和丁一汇（2006）指出从 1988 年之后，东亚寒潮发生频次减少；并且，Huang and Wang（2006）以及 Wang et al.（2009）提出在 1988 年之后 EAWM 经历了一次明显变弱的年代际变化，中国东部和北部经历了连续多年的暖冬。

最近，黄荣辉等（2013）指出：中国东部夏季降水和东亚夏季风在上世纪 90 年代末又发生了一个明显的年代际跃变，在 1999~2010 年期间，中国东部夏季降水从以往的经向三极子型分布变成了经向偶极子型分布，形成了"南涝北旱"（除长江沿岸地区）的特征；并且，中国东部这次降水的年代际跃变是与东亚上空夏季风环流、水汽输送的年代际变化相关联。从 1999 年之后，不仅东亚夏季风发生了明显的年代际跃变，而且 EAWM 也经历了一次明显的年代际跃变。从 1999 年之后，我国北方和东部的持续暖冬结束了，出现了冷冬和暖冬相间的变化，我国北方从气温偏高变成整体偏冷的现象，特别从 2008 年起我国北方和东部经常发生低温雨雪冰冻灾害，造成了严重经济损失。为此，本文利用我国的测站资料以及再分析资料来分析和诊断中国冬季气温和 EAWM 强度的年代际变化特征以及它们之间的联系，并进一步从海平面气压（SLP）以及北极涛动（AO）和北半球冬季准定常行星波动力学来分析和讨论在上世纪 90 年代末发生的中国气温和 EAWM 强度年代际跃变的内动力成因。

2 资料和方法

本文使用的气温资料是中国气象局气象信息中心提供的 756 站气温资料，考虑到站点的观测时间长度和连续性，本研究选取其中 553 个站点。并且，本文所用的 SLP 和高度场资料分别是取自美国 NCEP/NCAR（Kalnay et al., 1996）和欧洲中期天气预报中心的 ERA-40（Uppala et al., 2005）的再分析资料，并使用了 ERA-Interim 再分析资料。文中用到的北极涛动指数（AO）来自于美国气候预测中心（Climate Prediction Center, CPC）。本文所使用的分析方法是 EOF 分析、小波分析以及合成分析方法，并使用了 Student t 检验和 Lepage 检验（Lepage, 1971；Liu et al., 2011）。

本研究中采用 Wu and Wang（2002）定义的 EAWM 指数来表征东亚冬季风异常的变化 [如公式（1）所示]，它是根据西伯利亚高压和阿留申低压强度之差来定义。根据 Wang and Chen（2013a）的研究，该指数不仅计算简单，而且是一个与我国冬季气温相关很好的指数。

$$I_{EAWM} = \frac{(M_t - \overline{M})}{\sigma_M}, t: 1956, 1959, \ldots, 2012 \quad (1)$$

$$M_t = \sum_{i=1}^{20}(p_{s,110°E} - p_{s,160°E}), i=1, 2, \ldots, 20$$

式中，$p_{s,110°E}$ 和 $p_{s,160°E}$ 分别为 110°E 和 160°E 冬季（12月~次年2月）平均海平面气压，M_t 为 1958~2012 年某一年冬季的 110°E 和 160°E 海平面气压差沿 20°N~70°N 的 50 个纬距之和，$\overline{M}$ 表示 1958~2012 年冬季 M_t 的数学期望值，σ_M 为 M_t 的方差，i 表示从 20°N 到 70°N 间隔为 2.5° 的 20 个纬度点。若 I_{EAWM} 的正值愈大，则表明西伯利亚高压与阿留申低压之差值愈大，即 EAWM 愈强。

东亚冬季风异常与北半球行星波活动有着紧密的联系。例如，Huang and Wang（2006）以及 Wang et al.（2009）研究发现发生在 1980 年代中后期的 EAWM 年代际跃变与北半球准定常行星波活动有着紧密的关系。由于 Eliassen-Palm 通量（E-P 通量）平行于行星波经向传播的群速度，可以用来表征行星波的传播特征，本研究采用 Edmon et al.（1980）定义的准地转二维 E-P 通量，表达式如下：

$$\begin{cases} F(\phi) = -\rho a \cos\phi \overline{u'v'}, \\ F(p) = \rho a \cos\phi \dfrac{\overline{\theta'v'}}{\overline{\theta}_p} \end{cases} \quad (2)$$

式中，ρ 是空气密度，a 是地球半径。E-P 通量的散度为 $\nabla \cdot \boldsymbol{F}$，其表达式为

$$\nabla \cdot \boldsymbol{F} = \frac{1}{a\cos\phi}\frac{\partial}{\partial \phi}[F(\phi)\cos\phi] + \frac{\partial}{\partial p}[F(p)]. \quad (3)$$

依据 Edmon et al.（1980）给出的球面等压坐标下的波—流相互作用方程式，即

$$\frac{\partial \overline{u}}{\partial t} - fv^* = \frac{1}{\rho a \cos\varphi}\nabla \cdot \boldsymbol{F}, \quad (4)$$

其中，$\overline{u}$ 为平均纬向风，f 是科里奥利参数，v^* 为平均剩余经向气流，可以看出 E-P 通量的散度可以用来诊断行星波对平均流的强迫作用。

3 20 世纪 90 年代末发生的中国冬季气温的年代际跃变特征

为了分析从 20 世纪到新世纪初中国冬季（12月~次年2月）气温的年代际跃变特征，本文利用 1960~2012 年我国 553 观测站冬季气温的月平均资料并进行 EOF 分析。从 EOF 分析可得我国冬季气温有两种主要模态：全国一致型变化模态、南北振荡型变化模态。

3.1 全国一致型变化模态

图 1a 和图 1b 分别是中国冬季气温 EOF1 分析第 1 主分量（即 EOF1）的空间分布和时间系数序列。从图 1a 可以看到，中国冬季气温 EOF1 的空间分布呈现出全国一致型变化分布（除青藏高原的西部外），它占据总方差的 56.6%。这表明我国冬季气温变化的第 1 主模态为全国一致型的变化；并且，从

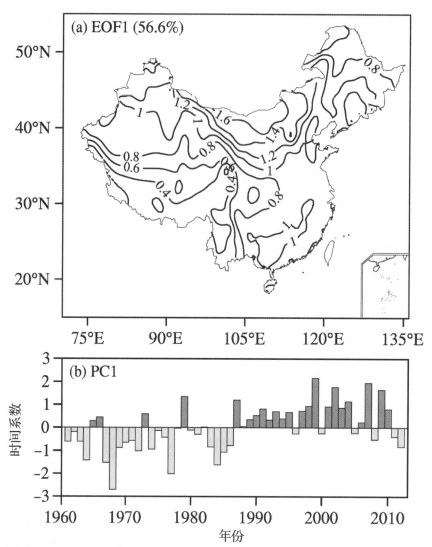

图 1 中国 1960～2012 年冬季（12 月至次年 2 月）气温 EOF 分析第 1 主分量（EOF1）的空间分布（a）和相应的时间系数序列（b）。图（a）中实、虚线分别表示正、负信号，EOF1 能说明总方差的 56.6%

Fig. 1 (a) The spatial distribution and (b) corresponding time-coefficient series of the first principle component (EOF1) of wintertime (December–February of next year) surface air temperature in China during 1960–2012. The solid and dashed lines in Fig.1a denote positive and negative values, respectively, and the EOF1 explains 56.6% of the total variance

图 1b 可以看到，在 1960～1987 年期间，我国冬季气温 EOF1 时间系数普遍为负，结合图 1a，这表明此时期全国冬季气温普遍偏低，这时期我国冬季寒潮爆发频次偏多（Huang et al. 2012）；而在 1988～1998 年期间，我国冬季气温 EOF1 时间系数普遍为正，结合图 1a，这表明此时期全国冬季气温普遍偏高，这时期我国冬季寒潮爆发频次偏少（Huang et al., 2012），全国气温偏暖。

3.2 南北振荡型变化模态

图 2a 和图 2b 分别是中国冬季气温 EOF 分析第 2 主分量（即 EOF2）的空间分布和时间系数序列。从图 2a 可以看到，中国冬季气温 EOF2 的空间分布呈现出南北振荡型变化特征，即我国东北和华北与我国南方和西南区域气温变化呈现出相反的变化特征。当我国东北、西北和华北地区冬季偏冷，则我国华东、华中、西南和华南气温偏高，如 2012 年冬季我国东北、华北和西北气温偏低，发生了严重低温和雪灾，而我国华南、华中和西南地区气温偏高；反之，当我国东北、西北和华北地区冬季偏暖，则我国华东、华中、西南和华南气温偏低，如 2008 年 1 月我国华中、华南、西南气候偏低发生了严重低温雨雪冰冻灾害，而我国东北、华北气温偏高。并且，从图 2b 可以看到，南北振荡型也有明显的年代际振荡特征。在 1964～1987 年期间，我

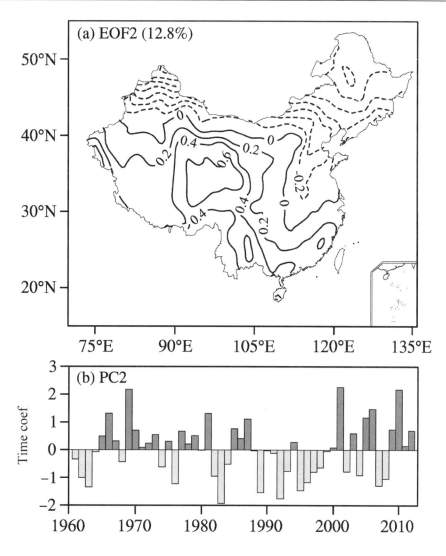

图 2 同图 1，但为 EOF2，EOF2 能说明总方差的 12.8%
Fig. 2 Same as Fig. 1 but for EOF2, which explains 12.8% of the total variance

国冬季气温 EOF2 的时间系数为正，结合图 2a，这表明了此时期我国东北、华北和西北气温偏低，而华南、华东、西南和华东气温偏高；在 1988～1998 年期间，我国气温 EOF2 的时间系数为负，结合图 2a，这表明了此时期我国东北、华北和西北气温偏高，出现暖冬，而华南、华东、西南和华东气温偏低；在 1999～2012 年期间，我国冬季气温 EOF2 的时间系数又从负变成正负相间，即在 1999～2012 年期间，我国北方冬季气温出现冷暖相间的现象，特别从 2009 年以后变成正，结合图 2a，这表明了此时期我国东北、华北和西北气温从偏高变成偏低。

上面分析结果表明：中国冬季气温的时空变化有两个主模态，即在空间分布上有全国一致变化分布型和南北振荡变化分布型。这与康丽华等（2006）和 Wang et al.（2010）在研究中国冬季气温年际变化所得结论相同。

3.3 20 世纪 90 年代末与 80 年代中后期发生的中国冬季气温年代际跃变对气温年际变化影响的差别

从上分析可以看到，中国冬季气温在 1988 年前后和 1999 年前后发生了明显年代际跃变。这两次中国冬季气温的年代际跃变的特征有明显不同，发生在 1988 年前后的年代际跃变的特征是中国北方（包括东北、华北和西北）出现持续暖冬现象；而发生在 1999 年前后的年代际跃变的特征是中国北方先出现冷暖相间现象，特别从 2008 年之后出现持续偏冷现象，而我国西南、华中和华南出现偏暖现象。

为了更好地比较中国冬季气温这两次年代际跃变对中国气温年际变化影响的差别,本研究应用小波分析方法对中国冬季气温 EOF1 和 EOF2 的时间系数进行小波分析 (见图 3a 和图 3b)。从图 3a 所示的 EOF1 时间系数小波分析结果可以看出,中国冬季气温第一模态在 1980 年代初之前呈现为显著的 3~4 a 周期,但在 1980 年代中期至 1990 年代末期年际变化不明显,此时期对应中国冬季气温第二模态占主导 (图 3b),也表现为显著的 3~4 a 周期。这与黄荣辉等 (2007) 利用熵谱分析方法所得 EAWM 年际变化周期一致。同时我们可以看出,自 1999 年以来,第一模态表现为以准两年周期变化为主,第二模态表现为以 2 a 和 4 a 左右为峰值的变化周期,这表明两个模态的周期变化均与 1999 年之前的年际变化周期有明显不同。

3.4 20 世纪 90 年代末与 80 年代中后期发生的中国冬季气温年代际跃变站点分布的差别

为了更进一步比较中国冬季气温这两次年代际跃变特征的差别,本研究分别应用 Lepage 和滑动 Student t (MTT) 检验方法来分析中国冬季气温跃变站点分布的差别 (见图 4)。从图 4 可以明显看到,在 1988 年和 1999 年前后中国冬季气温出现明显年代际跃变测站的站点都较多。并且,虽然发生在 1988 年前后中国冬季气温的年代际跃变的站点数要多于发生在 1999 年前后年代际跃变的站点数,但如图 5a 所示,中国冬季气温在 1988 年前后发生年代际跃变的站点主要分布在华北、东北以及黄淮和江淮流域,而在 1999 年前后所发生的年代际跃

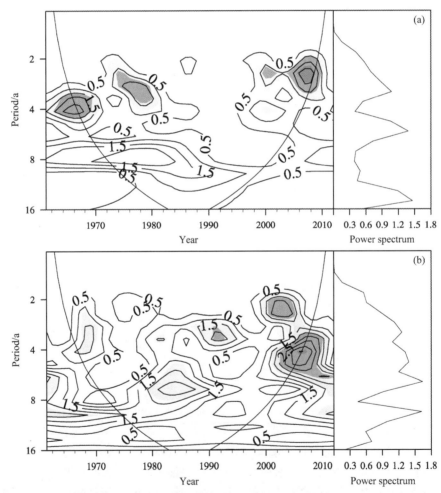

图 3 中国冬季 (12 月至次年 2 月) 气温 (a) EOF1 和 (b) EOF2 时间系数的小波分析。左图等值线为功率谱,右图曲线为全球小波谱,阴影为通过 95% 的 Chi-square 显著性检验的区域

Fig. 3 The wavelet analyses of the corresponding time coefficients of (a) EOF1 and (b) EOF2 of wintertime surface air temperature in China. The contour on the left denotes the power spectrum, and the curve on the right denotes the global wavelet spectrum; shading depicts power spectrum significant beyond 95% level based on the Chi-square test

变的站点不仅位于中国东北、华北、西北东部，而且还位于华东、华中、西南和华南广大地区。这表明中国更多地区冬季气温在 1999 年前后发生了明显年代际跃变。

上述分析结果表明：中国冬季气温在 20 世纪 90 年代末所发生的年代际跃变不仅表现在中国北方冬季气温的下降，而且冬季气温的年际变化从之前的 3～4 a 周期变成 2～8 a 周期。并且，这次年代际跃变发生在中国更广泛地区。

4 20 世纪 90 年代末发生的 EAWM 年代际跃变特征

中国冬季气温的年代际跃变是与 EAWM 的强度年代际跃变密切相关。为此，本节首先要分析 EAWM 强度的年际和年代际变化。

4.1 20 世纪 90 年代末发生的 EAWM 的年代际跃变特征及其与西伯利亚高压和阿留申低压强度变异的关系

利用公式（1）并分别利用 NCEP/NCAR 和 ERA-40 再分析资料（为了与 NCEP/NCAR 资料时间长度一致，本文在 2002 年之后应用 ERA-interim 再分析资料与 ERA-40 相连接）的海平面气压资料计算了 1958～2012 年的 EAWM 指数（见图 6a 和图 6b）。把图 6a 与图 6b 相比较，可以看到，用 NCEP/NCAR 再分析资料与 ERA-40 再分析资料所计算的 EAWM 指数除 20 世纪 50 和 60 年代有一定

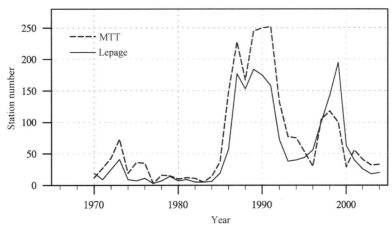

图 4　通过 Lepage（实线）和 MTT（虚线）检验（95%信度）的中国局部冬季气温年代际跃变的站点数

Fig. 4　The station numbers of wintertime surface air temperature anomalies beyond the 95% confidence level based on Lepage (solid line) and MTT (dashed line) tests, respectively

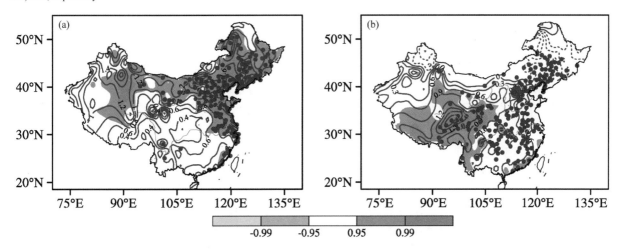

图 5　中国局部冬季气温年代际平均的距平分布(单位：°C，等值线)及通过 Lepage 检验的气温年代际跃变的站点（红点）分布：(a) 1988～1998 年；(b) 1999～2012 年

Fig. 5　The distributions of interdecadal mean wintertime surface air temperature anomalies in China averaged for (a) 1988–1998 and (b) 1999–2012. The solid and dashed lines depict positive and negative anomalies, respectively; green shading and red dots denote areas and stations with the surface air temperature anomalies beyond the 95% confidence level based on Lepage test, respectively

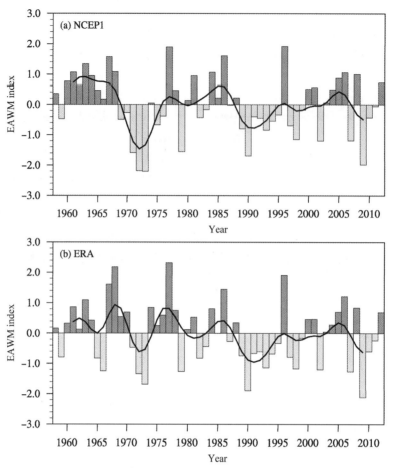

图 6 利用（a）NCEP/NCAR 和（b）ERA-40（从 2002 年之后用 ERA-interim）再分析资料所计算的 1958/1959～2012/2013 年 EAWM 指数的年际变化。曲线表示 9 年滑动平均

Fig. 6 The interannual variations of EAWM (East Asian winter monsoon) indexes during 1958/1959–2012/2013 calculated by using (a) NCEP/NCAR and (b) ERA-40 reanalysis data (the ERA-interim reanalysis data is used after 2002), respectively. The curve indicates 9-year running mean

差别外，从 70 年代中期到 2012 年两者计算结果比较一致。为此，在本文利用 NCEP/NCAR 再分析资料来研究 1976 年 EAWM 的年际和年代际变化（见图 6）。从图 6a 与图 6b 可以清楚看到，在 1988 年前后和 1999 年前后 EAWM 发生了明显的年代际变化，从 1988 年之后 EAWM 从强变弱，而从 1999 年，EAWM 又从弱变成强弱相间。

EAWM 的变化是与西伯利亚高压和阿留申低压的变化密切相关（Wu and Wang, 2002）。发生在 20 世纪 90 年代末和 80 年代中后期 EAWM 年代际跃变可以更清楚从图 7a–c 所示各时期平均的海平面气压（SLP）距平分布看到。把图 7b 与图 7a 比较可以看到：从 1988 年之后，西伯利亚高压变弱，出现负距平，而阿留申低压也变弱，也出现正距平，根据公式（1），EAWM 指数由正变负；并且，由于东西气压差变小，故偏北风变弱，导致

了 EAWM 变弱。而把图 7c 与图 7b 比较可以清楚看到：从 1999 年之后，西伯利亚由弱变强，出现正距平，而阿留申低压也由弱变强，出现负距平，根据公式（1），EAWM 指数由负变强；并且，由于东西气压差加大，故偏北风加强且导致了 EAWM 变强。

然而，把图 7c 与图 7a 相比较可以明显看到，1999 年之后 EAWM 虽然加强，但它的强度不如 1976～1987 年时期的 EAWM 强度。

4.2 20世纪90年代末与80年代中后期发生的EAWM年代际跃变对EAWM年际变化影响的差别

为了更好地比较发生在 20 世纪 90 年代末与 80 年代中后期 EAWM 年代际跃变对 EAWM 年际变化影响的差别，本研究还应用小波分析方法对 EAWM 指数（即 I_{EAWM}）进行分析（见图 8a 与图 8b）。从图 8a 和图 8b 都可以看到，无论利用 NCEP/NCAR

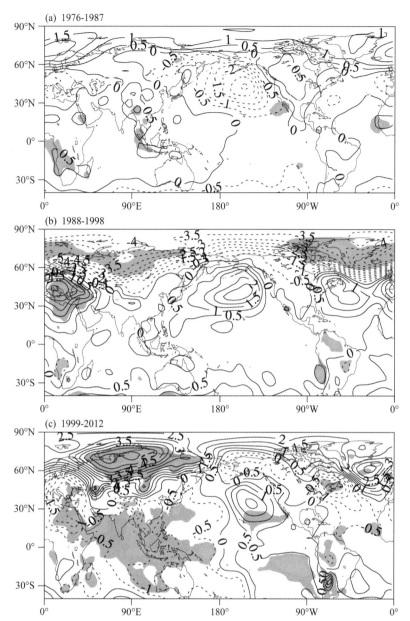

图 7 北半球各时期平均的冬季（12 月至次年 2 月）海平面气压距平分布（单位：hPa）：(a) 1976～1987 年；(b) 1988～1998 年；(c) 1999～2012 年。实、虚线分别表示正、负距平，阴影表示超过 95%的显著性检验，资料取自 NCEP/NCAR 再分析资料

Fig. 7 Distributions of interdecadal mean wintertime (December–Feberuary of next year) sea level pressure anomalies (hPa) over the Northern Hemisphere averaged for (a) 1976–1987, (b) 1988–1998, and (c) 1999–2012. The solid and dashed lines denote positive and negative anomalies, respectively. The anomalies beyond the 95% confidence level are shaded. The data are from the NCEP/NCAR reanalysis data

的 SLP 再分析资料所计算的 I_{EAWM} 或利用 ERA-40 的 SLP 再分析资料所计算的 I_{EAWM} 在 1999 之前呈现出 3～4 a 周期的年际变化特征，特别是在 1970 年代和 1990 年代，这与 Huang et al.（2012）利用熵谱分析方法所得的 EAWM 年际变化周期一致。而在 1999 之后，EAWM 却呈现出显著的准两年周期的年际变化，这与 1999 之前 EAWM 年际变化的周期有很大差别。

4.3 20 世纪 90 年代末与 80 年代中期发生的 EAWM 欧亚大陆冬季气温年代际跃变的差别

为了更好地揭示发生在 20 世纪 90 年代末 EAWM 的年代际跃变与发生在 80 年代中后期年代际跃变的差别，并鉴于上述从 70 年代之后利用两种再分析资料所计算的 EAWM 指数有一定的一致性，本研究就利用 NCEP/NCAR 地表气温分析了欧亚大陆及西太平洋上空 1976～1987 年、1988～1998

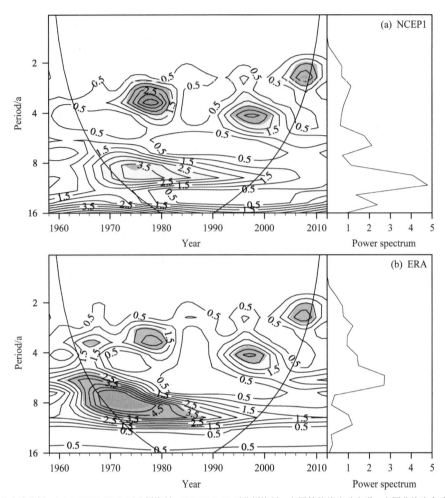

图 8 EAWM 指数的小波分析：(a) NCEP/NCAR 再分析资料；(b) ERA-40 再分析资料。左图等值线为功率谱，右图曲线为全球小波谱，阴影为通过 95% 的 Chi-square 显著性检验的区域

Fig. 8 The wavelet analyses of the EAWM indexes calculated by using (a) NCEP/NCAR and (b) ERA-40 reanalysis data. The contour on the left denotes the power spectrum, and the curve on the right denotes the global wavelet spectrum; shading depicts power spectrum significant beyond 95% level based on the Chi-square test

年，1999~2012 年期间平均的冬季气温距平分布（见图 9a–c）。把图 9b 与图 9a 相比较可以看到：1976~1987 年期间，整个欧亚大陆地表附近的气温偏低，而在 1988~1998 年期间，北冰洋地区外，欧亚大陆地表附近的气温偏高。这表明了随着 1988 年之后东亚冬季风变弱，欧亚大陆地表附近的气温普遍升高，出现持续暖冬现象。并且，把图 9c 与图 9b 作比较可以看到：在 1999~2012 年期间，除在北冰洋地区气温由负距平变成正距平外，在中高纬度的欧亚地区地表附近气温变成负距平，而在南亚、东南亚和我国南方气温仍为正距平，这表明了随着 EAWM 变强，欧亚大陆中高纬度地冬季地表气温又变低。此外，若把图 9c 与图 9a 和图 9b 比较可以明显看到：在 1999~2012 年期间，中国冬季气温距平是北负南正，即我国北方偏冷，南方偏暖，

而在 1976~1987 年和 1988~1998 年期间中国冬季气温是全国一致偏冷和全国一致偏暖。因此，发生在 1990 年代末的 EAWM 年代际跃变使我国冬季气温从全国一致变化型转变成南北振荡型（即北冷南暖型）的变化。

以上分析结果表明了发生在 20 世纪 90 年代末的 EAWM 跃变与发生在 80 年代中后期的跃变不同，这次跃变使我国冬季气温从全国一致变化型转变成南北振荡型的变化。

5 东亚冬季风年代际变化与北半球冬季准定常行星波活动年代际变化的关系

发生在 20 世纪 90 年代末与 80 年代中后期 EAWM 的年代际跃变是与北半球冬季准定常行星

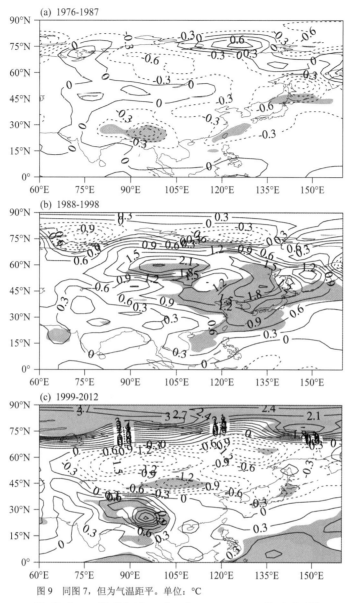

图 9 同图 7，但为气温距平。单位：°C

Fig. 9 Same as Fig.7, but for surface air temperature anomalies (units: °C)

波活动的异常有密切相关。以前的研究表明：北半球冬季准定常行星波在三维大气中传播存在两支波导，即极地波导和低纬波导（Huang and Gambo, 1982, 1983）；它们不仅有年际变化（Chen et al., 2003），而且有很明显的年代际变化（Huang and Wang, 2006; Wang et al., 2009）。并且，Chen et al.（2005）以及陈文和黄荣辉（2005）研究表明了北半球冬季准定常行星波活动的变化通过北半球环状模（NAM）严重地影响 EAWM 的强弱。黄荣辉等（2007）研究了 2005 年和 2006 年 EAWM 异常的差别及其与这两年北半球冬季准定常行星波活动变异的关系；并且，Huang and Wang（2006）以及 Wang et al.（2009）研究了发生在 1980 年代中后期的 EAWM 年代际跃变特征及其与北半球准定常行星波活动的关系。为此，有必要分析和研究发生在 1990 年代末的 EAWM 年代际跃变与北半球准定常行星波活动的关系。

5.1 1988～1998 年与 1976～1987 年期间北半球冬季准定常行星波活动特征及其它们之间的差别

我们利用（2）和（3）式计算出 1976～1987 年、1988～1998 年、1999～2012 年各时期每年冬季准定常行星波的 E-P 通量及其散度。图 10a–c 分别是所计算的 1976～1987 年期间、1988～1998 年期间平均的冬季北半球 1～3 波合成的准定常行星波及其散度分布以及它们之差（后者减前者）。从图 10a 可以看到：在 1976～1987 年期间，北半球冬季准定常行星波在 60°N 附近的上空通过极地波导上传到平流层偏强，而在对流层通过低纬波导向低纬度对流层上层传播偏弱。并且，如图 10a 所示，由于极地波导偏强，而低纬波导偏弱，这引起了北半球 50°～70°N 地区上空的对流层和平流层的 E-P 通量散度为负，即 E-P 通量辐合强，而在 35°N 附近上空对流层上层 E-P 通量散度为正，即 E-P 通量辐散。

同时，从图 10b 可看到：在 1988～1998 年期间，北半球冬季准定常行星波在 60°N 附近上空通过极地波导上到平流层比 1976～1987 年期间的冬季明显偏弱，而在对流层通过低纬波导向低纬度对流层上层传播显然比 1976～1987 年冬季的传播偏强。由于 1988～1998 年期冬季极地波导偏弱，而低纬波导偏强，因而引起了此时期北半球冬季 50°～70°N 地区上空的对流层和平流层的 E-P 通量辐合比 1976～1987 年期间冬季弱，而在 35°N 附近上空对流层 E-P 通量的辐散比 1976～1987 年期间偏弱。这些差别可以从图 10c 所示的这两时期的 E-P 通量散度之差更清楚看到，如图 10c 所示，从 40°～60°N 的对流层和 50°～70°N 的平流层 E-P 通量的散度之差都为正值。这表明 1988～1998 年期间北半球冬季准定常行星波 E-P 通量辐合比 1976～1987 年期间的辐合变弱（即辐散加强），而在 30°～40°N 对流层上层 E-P 通量辐散也变弱（即辐合加强）。

5.2 1999～2012 年与 1988～1998 年期间北半球冬季准定常行星波活动的差别

图 11a–c 分别是所计算的 1988～1998 年期间和 1999～2012 年期间平均的北半球冬季 1～3 波合成的准定常行星波及其散度以及之差。从图 11b 与图 11a 可看到：在 1999～2012 年期间，冬季准定常行星波在 40°～60°N 上空通过极地波导上传到平流层比 1988～1998 年期间的冬季明显偏强，而在对流层通过低纬波导向低纬度对流层上层传播显然比 1988～1998 年冬季的传播偏弱；并且，由于在 1999～2012 年期间冬季极地波导变得偏强，而低纬波导变得偏弱，因而在 1999～2012 年期间北半球冬季 50°～70°N 地区上空的对流层和平流层的 E-P 通量辐合比 1988～1998 年期冬季偏强，而在 30°～40°N 附近上空对流层 E-P 通量的辐散偏强。这从图 11c 所示的这两时期的 E-P 通量散度之差可以更明显看到，如图 11c 所示，从 40°～60°N 的对流层和 50°～70°N 的平流层 E-P 通量的散度之差都是负值，这表明 1999～2012 年期间北半球冬季准定常行星波 E-P 通量辐合比 1988～1998 年期间的辐合增强。而在 30°～40°N 对流层上层 E-P 通量辐散也加强。

上述结果表明了 1999～2012 年期间，北半球冬季准定常行星波沿极地波导往平流层传播加强，而沿低纬波导往副热带对流层上层传播减弱，这引起了高纬度地区上空准定常行星波 E-P 通量辐合加强，而副热带上空 E-P 通量辐散加强。

6 北半球冬季准定常行星波活动的年代际变化对北极涛动(AO)和 EAWM 年代际变化的影响

上述分析结果表明了从 20 世纪 90 年代末开始的北半球冬季准定常行星波的传播发生了明显的年代际变化，出现了明显不同于 1988～1998 年期间的传播特征。依据（4）式，半球冬季准定常行星波 E-P 通量的辐散或辐合的变化对于纬向平均气流的变化有着重要影响。若北半球冬季准定常行星波 E-P 通量的散度为负（辐合），即 $\nabla \cdot \boldsymbol{F} < 0$，则北半球纬向平均西风将减弱；反之，若北半球冬季准定常行星波 E-P 通量的散度为正（辐散），即 $\nabla \cdot \boldsymbol{F} > 0$，则北半球纬向平均西风将加强。因此，在上节所述北半球冬季准定常行星波传播及其 E-P 通量散度分布的年代际变化将直接对北半球冬季纬向平均气流造成重要影响。为此，本节首先利用 NCEP/NCAR 再分析资料分析北半球冬季纬向平均纬向风的年代际变化。

6.1 北半球冬季准定常行星波活动对纬向平均纬向流年代际跃变的影响

图 12a–c 分别是利用 NCEP/NCAR 风场再分析资料所计算的 1988～1998 年，1999～2012 年时期平均的纬向平均纬向风距平分布。从图 12a 可以看到：在 1976～1987 年期间冬季，北半球 40°N 以北地区对流层和平流层下层纬向平均西风为负距平，在平流层中层纬向平均西风为正距平，而在 40°N 以南地区对流层纬向平均西风为正距平。并且，从图 12b 可以看到：到了 1988～1998 年期间冬季，北半球 40°N 以北地区对流层和平流层纬向平均西风为正距平，特别在 60°N 平流层有大的西风距平。

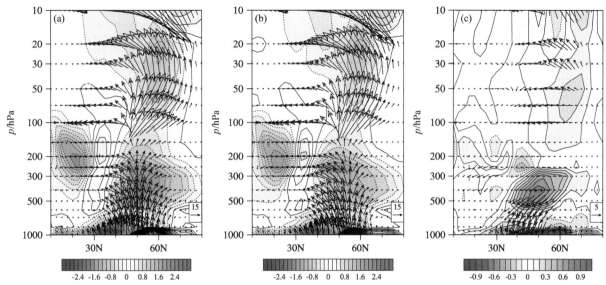

图 10 （a）北半球 1976～1987 年、（b）1988～1998 年平均的冬季准定常行星波 1～3 波合成的 E-P 通量（×ρ^{-1}）（单位：$m^3 s^{-2}$）及其散度（单位：$m^2 s^{-1} d^{-1}$）分布以及（c）它们之差（1988～1998 年冬季减去 1976～1987 年冬季）。E-P 通量散度中红色表示正值（辐散），蓝色表示负值（辐合），资料取自 NCEP/NCAR 再分析资料

Fig. 10 Composite distributions of E-P fluxes (×ρ^{-1}) (Uint: $m^3 s^{-2}$) of quasi-stationary planetary waves for wave numbers 1–3 and their divergences (Units: $m^2 s^{-1} d^{-1}$) over the Northern Hemisphere averaged for the winters of (a) 1976–1987 and (b) 1988–1998, and (c) the differences between them (1988–1998 minus 1976–1987). The red and blue areas indicate positive (divergence) and negative (convergence) values, respectively. The data are from the NCEP/NCAR reanalysis data

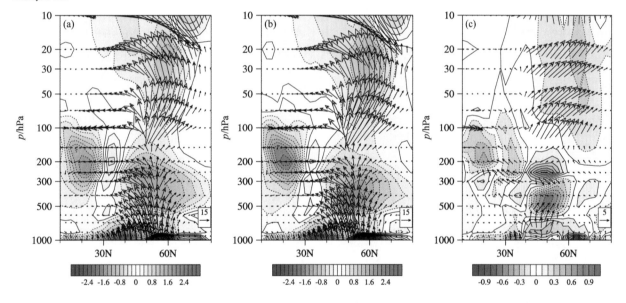

图 11 同图 10，但为（a）1988～1998 年和（b）1999～2012 年

Fig. 11 As Fig.10, but for (a) 1988–1998 and (b) 1999–2012

把图 12b 与图 12a 相比较，可以看到：此时期 40°N 以北高纬度地区西风增强，这正是由于此时期极地波导减弱导致的高纬地区行星波 E-P 通量的辐散加强而引起；而在 40°N 以南的对流层纬向平均西风为负距平，这表明此时期冬季副热带地区西风气流偏弱，这正是由于此时期低纬波导变强所导致的北半球副热带地区行星波 E-P 通量的辐合加强而引起。然而，从图 12c 可以看到：到了 1999～2012 年期间冬季，北半球 45°N 以北地区对流层和平流层纬向平均西风为负距平，特别在 60°N 平流层为显著的西风负距平，这表明此时期 45°N 以北高纬度地区西风偏弱，这正是由于此时期极地波导加强

导致的北半球高纬地区行星波 E-P 通量的辐合加强而引起；而在 30°～45°N 对流层纬向平均西风为正距平，这表明此时期冬季副热带地区西风气流又加强，这正是由于此时期低纬波导变弱导致的在北半球副热带地区行星波 E-P 通量辐散加强而引起。

上述结果可以从图 13a 和图 13b 所示的 1988～1998 年和 1999～2012 年期间平均的冬季欧亚上空 200 hPa 纬向风距平分布得到进一步证实。如图 13a 所示，在 1988～1998 年期间冬季欧亚大陆高纬度地区上空 200 hPa 有西风距平，这表明此地区西风偏强，而副热带地区上空 200 hPa 有东风距平，这表明此地区西风偏弱。然而，到了 1999～2012 年期间，如图 13b 所示，在欧亚大陆高纬度地区上空 200 hPa 出现东风距平，而副热带地区上空 200 hPa 有西风距平。这表明此时期高纬度地区上空西风偏弱，而副热带地区上空西风偏强。

无论从上述纬向平均纬向风距平分布或是从欧亚大陆上空 200 hPa 纬向风距平分布都可以看到：北半球冬季纬向风在 20 世纪 90 年代末发生了明显的年代际跃变，在高纬度地区纬向风变弱，即极峰急流变弱，而副热带地区纬向风变强，即副热带急流变强。这与发生在 1980 年代中后期的纬向风的年代际跃变特征有明显的不同。

6.2 北半球冬季准定常行星波活动对 AO 年代际变化的影响

上述结果表明了北半球冬季准定常行星波活动的变化通过波—流相互作用将直接影响着北半球冬季纬向气流，而冬季纬向气流的变化将通过影响北半球环状模（NAM）进而影响 AO（Thompson and Wallace，2000）。

Chen et al.（2005）从北半球冬季准定常行星波 E-P 通量的散度定义了一个行星波活动指数，他们所定义的指数 I_{pwa} 是

$$I_{pwa} = \text{Nor.}(\nabla \cdot \boldsymbol{F}_A - \nabla \cdot \boldsymbol{F}_B), \quad (5)$$

式中，$\nabla \cdot \boldsymbol{F}_A$ 和 $\nabla \cdot \boldsymbol{F}_B$ 分别为（500 hPa，50°N）和（300 hPa，40°N）区域行星波 E-P 通量的散度，Nor. 表示利用与（1）式相同的算法对 $(\nabla \cdot \boldsymbol{F}_A - \nabla \cdot \boldsymbol{F}_B)$ 进行标准化运算。

利用 1950～2012 年冬季准定常行星波 E-P 通量的散度，从（5）式便可以计算出各年冬季的行星波活动指数 I_{pwa}（见图 14）。从图 14 可看到：北半球冬季准定常行星波指数 I_{pwa} 与 AO 指数有很好的正相关，它们之间的相关系数达到 0.67，超过了

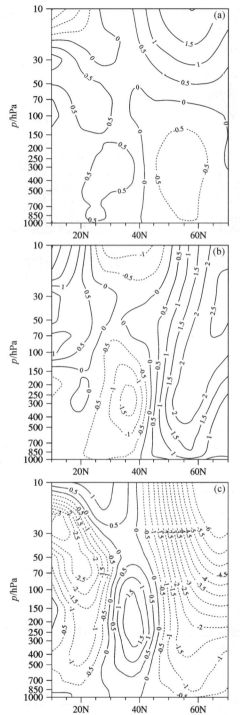

图 12 （a）1976～1987 年期间、(b) 1988～1998 年期间和（c）1999～2012 年期间平均的北半球冬季纬向平均纬向风距平随高度和纬度的分布。单位：m s^{-1}。1971～2000 年北半球各层气候平均的纬向风速取为正常值。图中实、虚线分别表示正、负距平。风场资料取自 NCEP/NCAR 再分析资料

Fig. 12 Composite distributions of zonal mean zonal wind anomalies (unit: m s^{-1}) with height and latitude for the winters of (a) 1976-1987, (b)1988-1998, and (c) 1999-2012. The climatological mean zonal winds from 1971 to 2000 are defined as the normals. The solid and dashed lines denote positive and negative anomalies, respectively. The data are from the NCEP/NCAR reanalysis data

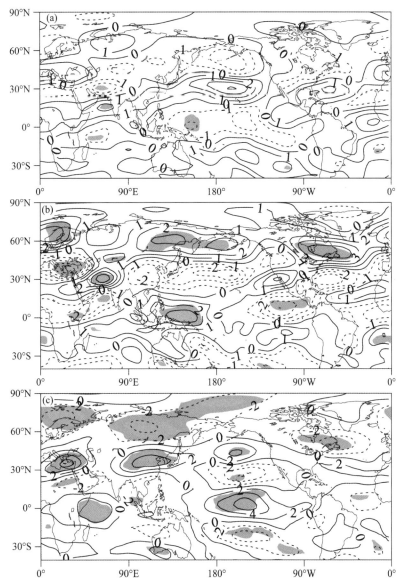

图 13 同图 12，但为 200 hPa 面上纬向风距平（单位：m s^{-1}）。阴影区表示超过 95%显著性检验

Fig. 13 As in Fig.12, except for the zonal wind anomalies at 200 hPa (unit: m s^{-1}). The areas over 95% confidence level are shaded

99%的信度。这就说明了北半球冬季准定常行星波活动的变化通过波—流相互作用影响了高纬度和副热带上空纬向流的变化，并通过北半球环状模（NAM）的变化影响到 AO 的变化。并且，如图 14 所示，1976~1987 年期间冬季行星波活动处于低指数，AO 指数为负；而在 1988~1998 年期间冬季，行星波活动指数变成高指数，此时期 AO 指数也变为正；到了 1999~2012 年期间冬季，行星波活动指数又变成低指数，此时期 AO 指数也随之由正变负。

从上分析结果可以看到：从 20 世纪 90 年代末以后，由于北半球冬季准定常行星波传播发生了年代际跃变，使得行星波活动指数由正变负，导致 AO 指数也由正变负。这与在 1988~1998 年期间冬季北半球冬季准定常行星波活动指数由负变正且导致 AO 也由负变正截然不同。

6.3 北半球冬季 AO 年代际变化对 EAWM 年代际跃变的影响

上述结果表明了在 1999~2012 年期间冬季 AO 为负位相，而在 1988~1998 年期间冬季 AO 正位相。Gong et al.（2001）以及 Wu and Wang（2002）的研究表明，AO 对于 EAWM 有很明显的影响，他们指出：若某一年冬季 AO 指数为负，则该年 EAWM 偏强；反之，若某一年冬季 AO 为正，则该年 EAWM 偏弱。

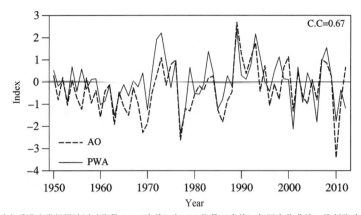

图 14　1950～2012 年北半球冬季准定常行星波活动指数 I_{pwa}（实线）与 AO 指数（虚线）年际变化曲线。资料取自 NCEP/NCAR 再分析资料
Fig. 14　Interannual variability of I_{pwa} (solid line) and AO index (dashed line). The data are from the NCEP/NCAR reanalysis data

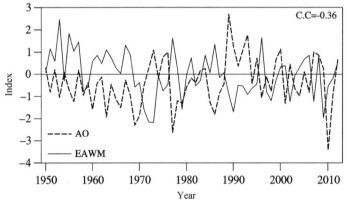

图 15　1950～2012 年东亚冬季风指数 I_{EAWM}（实线）与 AO 指数（虚线）的年际变化
Fig. 15　Interannual variability of the EAWM index (solid line) and AO index (dashed line). The data are from the NCEP/NCAR reanalysis data

图 15 是 1950～2012 年北半球冬季 AO 指数与 EAWM 指数（I_{EAWM}）的年际变化曲线。从图 15 可以看到，EAWM 指数与冬季 AO 指数有很好的负相关，它们之间相关系数达到−0.36，达到 99%的信度。并且，从图 15 还可看到：1988～1998 年期间冬季 AO 指数由负变正，而 I_{EAWM} 由正变负，这表明此时期，由于 AO 处于正位相，它引起了 EAWM 偏弱；并且，在 1999～2012 年期间冬季 AO 指数由正变负，而 I_{EAWM} 由负变正，这表明此时期，由于 AO 处于微弱的正位相，它引起了 EAWM 稍偏强。若把 1999～2012 年期间与 1976～1987 年期间冬季 AO 指数相比，则 1999～2012 年期间冬季 AO 指数远不如 1976～1987 年期间冬季 AO 指数的负值。因此，1999～2012 年期间的 EAWM 远不如 1976～1987 年期间冬季风强。

7　结论和讨论

为纪念陶诗言先生对东亚冬季风研究的杰出贡献，本文利用我国 553 站观测气温、NCEP/NCAR 和 ERA 再分析资料分析了 20 世纪 90 年代末我国冬季气温和 EAWM 的年代际跃变特征。分析结果表明：从 20 世纪 90 年代末之后，我国冬季气温发生了明显的年代际跃变，我国北方冬季气温从偏高变成偏低，低温雨雪冰冻灾害频繁发生，这与 EAWM 由偏弱变成偏强的年代际变化有关，即与西伯利亚高压和阿留申低压加强有关；并表明这次跃变不仅使中国冬季气温从全国一致变化型变成南北振荡型（即北冷南暖型），而且使我国冬季气温和 EAWM 的年际变化也发生了显著的年代际变化。

并且，本文还从冬季北极涛动（AO）和北半球准定常行星波活动的年代际变化来讨论这次年代际跃变的机理。分析结果表明：在 1999～2012 年期间，北半球冬季准定常行星波在高纬度地区沿极地波导传播到平流层加强，而沿低纬波导往低纬度对流层上层传播减弱，它造成了行星波 E-P 通量在高纬度地区上空辐合加强，而在副热带地区上空

E-P 通量辐散加强,从而引起了北半球高纬度地区从对流层到平流层纬向平均纬向流和欧亚上空极锋急流减弱,而副热带急流加强,这导致了 AO 减弱,因而利于西伯利亚高压和阿留申低压的加强,即 EAWM 加强。这些变化与发生在 1988～1998 年期间的北半球冬季准定常行星波活动的年代际变化特征有明显不同。

本文只是从大气内部动力成因来讨论了在 20 世纪 90 年代末发生的 EAWM 年代际跃变的机理。海洋和陆面过程等大气外强迫因子的年代际变化对于这次年代际跃变也起了重要作用,这将以后进一步再讨论。

参考文献（References）

Chen Wen, Graf H F. 1998. The interannual variability of East Asian winter monsoon and its relation to global circulation [R]. Max-Planck-Institute of Meteorology Report., No. 250.

陈文, 黄荣辉. 2005. 北半球冬季准定常行星波的三维传播及其年际变化. 大气科学, 29: 137–146. Chen Wen, Huang Ronghui. 2005. The three-dimensional propagation of quasi-stationary planetary waves in the Northern Hemisphere winter and its interannual variations [J]. Chinese J. Atmos. Sci. (in Chinese), 29: 137–146.

Chen Wen, Graf H F, Huang R H. 2002. The interannual variability of East Asian winter monsoon and its relation to the summer monsoon [J]. Adv. Atmos. Sci., 17: 48–60.

Chen W, Takuhashi M, Graf H F. 2003. Interannual variations of stationary planetary wave activity in the northern winter troposphere and stratosphere and their relations to NAM and SST [J]. J. Geophys. Res., 18 (D24), 4797, doi:10.1029/2003JD003834.

Chen Wen, Yang Song, Huang Ronghui. 2005. Relationship between stationary planetary wave activity and the East Asian winter monsoon [J]. J. Geophys. Res., 110, D1410, doi:10.1029/2004 JD005669.

Edmon M J Jr, Hoskins B J, McIntyre M E. 1980. Eliassen-Palm cross sections for the troposphere [J]. J. Atmos. Sci., 37: 2600–2617.

Gong Daoyi, Wang Shaowu, Zhu Jinhong. 2001. East Asian winter monsoon and Arctic Oscillation [J]. Geophys. Res. Lett., 28: 2073–2076.

Huang Ronghui, Gambo K. 1982. The response of a hemispheric multi-level model atmosphere to forcing by topography and stationary heat sources. Part I. Forcing by topography, and Part II: Forcing by stationary heat sources and forcing by topography and stationary heat sources [J]. J. Meteor. Soc. Japan, 60: 78–108.

Huang Ronghui, Gambo K. 1983. On other wave guide in stationary planetary wave propagations in the winter Northern Hemisphere [J]. Science in China, 26: 940–950.

Huang Ronghui, Wang Lin. 2006. Interdecadal variation of Asian winter monsoon and its association with the planetary wave activity [C] // Proc. International Symposium on Asian Monsoon. Kuala Lumpur, Malaysia, 126.

黄荣辉, 刘永, 冯涛. 2013. 20 世纪 90 年代末中国东部夏季降水和环流的年代际变化特征及其内动力成因 [J]. 科学通报, 58: 617–628. Huang Ronghui, Liu Yong, Feng Tao. 2013. Interdecadal change of summer precipitation over eastern China around the late-1990s and associated circulation anomalies, internal dynamical causes [J]. Chinese Sci. Bull., 58: 1339–1349.

黄荣辉, 魏科, 陈际龙, 等. 2007. 东亚 2005 年和 2006 年冬季风异常及其与准定常行星波活动关系的分析研究 [J]. 大气科学, 31: 1033–1048. Huang Ronghui, Wei Ke, Chen Jilong, et al. 2007. The East Asian winter monsoon anomalies in the winter of 2005 and 2006 and their relations to the quasi-stationary planetary wave activity in the Northern Hemisphere [J]. Chinese J. Atmos. Sci. (in Chinese), 31: 1033–1048.

Huang Ronghui, Chen Jilong, Wang Lin, et al. 2012. Characteristics, processes, and causes of the spatio-temporal variabilities of the East Asian monsoon system [J]. Adv. Atmos. Sci., 29: 910–942.

Jhun J G, Lee E J. 2004. A new East Asian winter monsoon index and associated characteristics of the winter monsoon [J]. J. Climate, 17: 711–726.

Kalnay E, Kanamitsu M, Kistler R, et al. 1996. The NCEP/NCAR 40-year reanalysis project [J]. Bull. Amer. Meteor. Soc., 77: 437–471.

康丽华, 陈文, 魏科. 2006. 我国冬季气温年代际变化及其与大气环流异常变化的关系 [J]. 气候与环境研究, 11: 330–339. Kang Lihua, Chen Wen, Wei Ke. 2006. The interdecadal variation of winter temperature in China and its relation to the anomalies in atmospheric general circulation [J]. Climate Environ. Res. (in Chinese), 11: 330–339.

Lepage Y. 1971. A combination of Wilecoxon's and Ansar-Baradley's statistics [J]. Biometrika, 58: 213–217.

Li Y Q, Yang S. 2010. A dynamical index for the East Asian winter monsoon [J]. J. Climate, 23: 4255–4262.

Liu Y, Huang G, Huang R H. 2011. Inter-decadal variability of summer rainfall in eastern China detected by the Lepage test [J]. Theor. Appl. Climatol., 106: 481–488.

陶诗言. 1952. 冬季从印缅移过来的高空低槽 [J]. 气象学报, 23: 171–192. Tao Shiyan. 1952. The low trough in the upper level moved from the region of Indo-Burma [J]. Acta Meteor. Sin. (in Chinese), 23: 171–192.

陶诗言. 1956. 冬季中国上空平直西风环流条件下的西风波动 [J]. 气象学报, 27: 345–360. Tao Shiyan. 1956. The upper air cold trough over China during high index circulation over Far East [J]. Acta Meteor. Sin. (in Chinese), 27: 345–360.

陶诗言. 1957. 阻塞形势破坏时期的东亚一次寒潮过程 [J]. 气象学报, 28: 63–74. Tao Shiyan. 1957. A synoptic and aerological study on a cold wave in the Far East during the period of the break down of the blocking situation over Euroasia and Atlantic [J]. Acta Meteor. Sin. (in Chinese), 28: 63–74.

陶诗言. 1959. 十年来我国对东亚寒潮的研究 [J]. 气象学报, 30: 226–230. Tao Shiyan. 1959. Study on East Asian cold waves in China during recent 10-years (1949–1959) [J]. Acta Meteor. Sin. (in Chinese), 30: 226–230.

陶诗言, 张庆云. 1998. 亚洲冬季风对 ENSO 现象的响应 [J]. 大气科学, 22: 399–407. Tao Shiyan, Zhang Qingyun. 1998. Response of the Asian winter and summer monsoon to ENSO events [J]. Chinese J. Atmos. Sci.

(in Chinese), 22: 399–407.

陶诗言, 李毓芳, 温玉璞. 1965. 东亚对流层上部和平流层中下部大气环流的初步研究 [J]. 气象学报, 37: 155–165. Tao Shiyan, Li Yufang, Wen Yupu. 1965. A preliminary study on the general circulation of East Asia in the upper troposphere and stratosphere [J]. Acta Meteor. Sin. (in Chinese), 37: 155–165.

Thompson D W J, Wallace J M. 2002. Annular modes in the extratropical circulation. Part I: Month-to month variability [J]. J. Climate, 13: 1000–1016.

Uppala S M, Kållberg P W, Simmons A J, et al. 2005. The ERA-40 re-analysis [J]. Quart. J. Roy. Meteor. Soc., 131: 2961–3012.

Wang B, Wu Z W, Chang C P, et al. 2010. Another look at interannual-to-interdecadal variations of the East Asian winter monsoon: The northern and southern temperature modes [J]. J. Climate, 23: 1495–1512.

Wang Lin, Chen Wen. 2010. How well do existing indices measure the strength of the East Asian winter monsoon? [J]. Adv. Atmos. Sci., 27: 855–870.

Wang Lin, Chen Wen. 2013a. An intensity index for the East Asian winter monsoon [J]. J. Climate, 24, doi:10.1175/JCLI-D-13-00086.1.

Wang Lin, Chen Wen. 2013b. The East Asian winter monsoon: Re-amplification in the mid-2000s [J]. Chinese Sci. Bull., 59 (4):430–436, doi:10.1007/s11434-013-0029-0.

Wang Lin, Huang Ronghui, Gu Lei. 2009. Interdecadal variations of the East Asian winter monsoon and their association with quasi-stationary planetary wave activity [J]. J. Climate, 22: 4860–4872.

王遵娅, 丁一汇. 2006. 近 53 年中国寒潮的变化特征及其可能原因 [J]. 大气科学, 30: 1068–1076. Wang Zunya, Ding Yihui. 2006. Climate change of the cold wave frequency of China in the last 53 years and the possible reasons [J]. Chinese J. Atmos. Sci. (in Chinese), 30: 1068–1076.

Wu Bingyi, Wang Jia. 2002. Winter Arctic oscillation, Siberian high and East Asian winter monsoon [J]. Geophys. Res. Letter, 29 (19), 1897, doi:10.1029/2002 GL015373.

Differences and Links between the East Asian and South Asian Summer Monsoon Systems: Characteristics and Variability[*]

Ronghui Huang[1,2,3], Yong Liu[*1], Zhencai Du[1], Jilong Chen[1], and Jingliang Huangfu[1]

[1]*Center for Monsoon System Research, Institute of Atmospheric Physics, Chinese Academy of Sciences, Beijing 100190, China;* [2]*State Key Laboratory of Numerical Modeling for Atmospheric Sciences and Geophysical Fluid Dynamics, Institute of Atmospheric Physics, Chinese Academy of Science, Beijing 100029, China;* [3]*Institute of Earth Sciences, University of Chinese Academy of Sciences, Beijing 100049, China*

ABSTRACT

This paper analyzes the differences in the characteristics and spatio–temporal variabilities of summertime rainfall and water vapor transport between the East Asian summer monsoon (EASM) and South Asian summer monsoon (SASM) systems. The results show obvious differences in summertime rainfall characteristics between these two monsoon systems. The summertime rainfall cloud systems of the EASM show a mixed stratiform and cumulus cloud system, while cumulus cloud dominates the SASM. These differences may be caused by differences in the vertical shear of zonal and meridional circulations and the convergence of water vapor transport fluxes. Moreover, the leading modes of the two systems' summertime rainfall anomalies also differ in terms of their spatiotemporal features on the interannual and interdecadal timescales. Nevertheless, several close links with respect to the spatiotemporal variabilities of summertime rainfall and water vapor transport exist between the two monsoon systems. The first modes of summertime rainfall in the SASM and EASM regions reveal a significant negative correlation on the interannual and the interdecadal timescales. This close relationship may be linked by a meridional teleconnection in the regressed summertime rainfall anomalies from India to North China through the southeastern part over the Tibetan Plateau, which we refer to as the South Asia/East Asia teleconnection pattern of Asian summer monsoon rainfall. The authors wish to dedicate this paper to Prof. Duzheng YE, and commemorate his 100th anniversary and his great contributions to the development of atmospheric dynamics.

1. Introduction

In the 100th anniversary of Prof. Duzheng YE (i.e. Tu-Cheng YEH), as one of his former students, I (Ronghui HUANG) and my coauthors wish to dedicate this study to commemorate this outstanding meteorologist and his substantial contributions to the development of the atmospheric sciences and the foundations of the modern era of this discipline in China. Accordingly, we begin with a brief summary of Prof. YE's research experience and achievements.

Prof. YE was without doubt one of the world's most outstanding meteorologists and, domestically, one of the major founders of modern-day atmospheric sciences in China. Over a more than 70-year research career he made many important contributions to the development of atmospheric sciences, including proposing the theory of energy dispersion of Rossby waves, establishing a theory of atmospheric general circulation over East Asia, initiating the study of Tibetan Plateau meteorology, proposing a scale theory regarding the adaptation process of atmospheric motion, and pioneering the new concept of adaptation to global warming (Yeh, 2008). To develop the atmospheric sciences in his home country, Prof. Ye left Chicago University for China in 1950. Subsequently, he invested considerable energy into improving and organizing the research activities of this discipline within the Chinese research community. Together with his colleagues (e.g. Profs. Shiyan TAO and Zhenchao GU), Prof. YE dedicated a great deal of effort into developing the study of atmospheric sciences in China, especially those related to East Asian general circulation. From the 1950s to the 1960s, Prof. YE systematically studied the characteristics and variabilities of East

---

[*] 本文原载于: Advances in Atmospheric Sciences, Vol.34, 1204-1218, 2017.

Asian general circulation, through observational and theoretical analyses, in cooperation with Profs. Shiyan TAO and Baozhen ZHU and others (staff members of the Section of Synoptic and Dynamic Meteorology, Institute of Geophysics and Meteorology, Academia Sinica, 1957, 1958a, 1958b; Ye and Zhu, 1958). The studies conducted by Profs. YE, TAO and colleagues showed that the seasonal variation of the general circulation over East Asia from winter to summer is very distinct and fairly abrupt (Ye et al., 1959). Moreover, they also pointed out that this abrupt change in planetary-scale circulation brings about the onset of the East Asian summer monsoon (EASM). This abrupt change in monsoon circulation was further demonstrated in the 1980s by Krishnamurti and Ramanathan (1982) and McBride (1987), in studies on the Indian summer monsoon and Australian monsoon.

Enlightened by the theory of atmospheric general circulation over East Asia proposed by Prof. YE, scientists have been able to carry out studies on the characteristics, causes and mechanisms of the spatiotemporal variabilities of the EASM system. It is known, for instance, that the EASM system has notable interannual and interdecadal variabilities and, owing to that, climatic disasters such as droughts and floods occur frequently in summer in China (Huang and Zhou, 2002; Huang et al., 2004, 2006a, 2007, 2008, 2011a; Huang, 2006). Indeed, since the 1980s, severe and frequent climatic disasters over large areas have caused vast amounts of damage to agricultural and industrial production in China (Huang et al., 1998a, 1999; Huang and Zhou, 2002; Wang and Gu, 2016). Therefore, it is of great importance to study the variability and possible causes of the EASM, as well as its links to nearby or remote climate systems.

The Asian monsoon region is one of several monsoon climate regions worldwide and, generally, it is considered to comprise two subcomponents (Webster et al., 1998; Ding et al., 2013)—namely, the East Asian and South Asian summer monsoons (hereafter referred to as the EASM and SASM, respectively). It has been documented by many studies that the characteristics of the EASM and SASM system are different (Tao and Chen, 1987; Huang et al., 1998b, 2012; Wang et al., 2001). For instance, the EASM has both tropical and subtropical characteristics because, not only it is influenced by the western Pacific subtropical high and the disturbances over the middle latitudes, but also by tropical circulation systems, such as Maritime Continent convection; whereas, the SASM only has tropical properties, such as the Mascarene high, Somali jet, and so on. However, the EASM and SASM systems also show links. For example, strong SASM flow delivers a large amount of water vapor into the EASM region from the Bay of Bengal, which can cause strong rainfall and severe floods in this region. Previous studies have also indicated the existence of a close relationship between the summer precipitation of the SASM and EASM (Guo and Wang, 1988; Guo, 1992; Kripalani and Singh, 1993; Kripalani and Kulkarni, 2001; Ding and Wang, 2005; Lin et al., 2016; Wu, 2002, 2017), although Wu (2017) pointed out that this relationship has become unstable and weakened since the late 1970s.

Because of the complexity of the EASM and SASM systems, the differences and links between them—in particular, with respect to the dominant modes of summertime rainfall and water vapor transport—are still not clearly understood. Accordingly, using long-term observational and reanalysis data, the present study aims to systematically investigate these issues. Specifically, we examine: (1) the differences in the characteristics of rainfall cloud systems and their association with the vertical shear of zonal and meridional circulations and the convergence of water vapor transport; (2) the differences in the spatiotemporal variabilities of summertime rainfall and water vapor transport between the EASM and SASM systems; and (3) the links between the two systems in terms of their spatiotemporal variabilities of summertime rainfall and water vapor transport.

The remainder of the paper is structured as follows: Section 2 details the datasets and methods used in the study. Section 3 describes the climatological characteristics of the rainfall and general circulation of the EASM and SASM, and highlights the differences between them. Sections 4 and 5 analyze the features and differences of the interannual and interdecadal variabilities of summer rainfall in the two monsoon regions. The links in terms of summer rainfall variability between the EASM and SASM are explained in section 6 and, finally, a conclusion and further discussion are provided in section 7.

2. Data and methodology

The data used in the present work are as follows: (1) NCEP–NCAR reanalysis data, covering the period from 1961 to 2014 (Kalnay et al., 1996); (2) Tropical Rainfall Measuring Mission (TRMM) Precipitation Radar Rainfall L3 monthly data, version 7, with a $0.5° \times 0.5°$ mesh (TRMM_3A25), covering the period from 1998 to 2015 (TRMM, 2011); (3) Precipitation data from the Climate Research Unit (CRU) TS3.23 precipitation dataset, with a $0.5° \times 0.5°$ horizontal resolution in the Asian region from 1961 to 2014 (Harris et al., 2014); (4) and precipitation data from 597 observational stations in China for the period 1961 to 2014, archived and updated by the China Meteorological Administration (http://data.cma.cn/data/cdcdetail/dataCode/SURF_CLI_CHN_MUL_DAY_V3.0.html).

The present work employs empirical orthogonal function (EOF) analysis to investigate the spatiotemporal features of summer (June–July–August) precipitation in the EASM and SASM regions, wherein the first two leading modes of summer rainfall in the two monsoon regions are adopted, respectively. Based on the method of North et al. (1982), the first two modes of summer rainfall in each monsoon region are completely separated. The present work also uses correlation analysis and regression analysis and the two-tailed Student's t-test to detect the statistical significance of a signal, and the effective degrees of freedom is calculated according to the method of Davis (1976). In addition, to discuss the interdecadal variability of summer rainfall in the two monsoon regions, we adopt a 9-yr running mean method to obtain the interdecadal component of the summer precipitation.

The atmospheric water vapor transport and its convergence are estimated based on the following expressions. The vertically integrated water vapor transport flux vector, $Q = (Q_\lambda, Q_\varphi)$, and its zonal and meridional components, Q_λ and Q_φ, are described as follows:

$$Q_\lambda = \frac{1}{g} \int_{300}^{p_0} (qu)dp ; \qquad (1)$$

$$Q_\varphi = \frac{1}{g} \int_{300}^{p_0} (qv)dp . \qquad (2)$$

Here, g is the gravitational acceleration; q is the specific humidity; and u and v represent the zonal and meridional winds, respectively. For the sake of simplicity and to avoid the problem of data limitation (the top level in the specific humidity field only reaches up to 300 hPa), thus, the integration is from the bottom level ($p_0 = 1000$ hPa) and to the top level (300 hPa).

In a spherical coordinate system, the divergence of water vapor transport fluxes, $\nabla \cdot Q$, can be calculated using the following formula:

$$\nabla \cdot Q = \frac{1}{a\cos\varphi}\left(\frac{\partial Q_\lambda}{\partial \lambda} + \frac{\partial Q_\varphi \cos\varphi}{\partial \varphi}\right), \qquad (3)$$

where (λ, φ) represents the longitude and latitude, respectively, and a is the Earth's radius. Furthermore, if topography is not considered, $\nabla \cdot Q$ can be divided into two parts (thermal and dynamic components), as follows:

$$\begin{aligned}\nabla \cdot Q &= \frac{1}{g}\int_{300}^{p_0} \nabla\cdot(Vq)dp \\ &= \frac{1}{g}\int_{300}^{p_0} V\cdot\nabla q\, dp + \frac{1}{g}\int_{300}^{p_0} q(\nabla\cdot V)dp.\end{aligned} \qquad (4)$$

The first of the two terms on the right-hand side of Eq. (4) is the quantity due to moisture advection (i.e. the thermal component), and the second is the quantity due to divergence of the wind field (i.e. the dynamic component). Based on the first part of Eq. (4), moist advection can cause local convergence of the water vapor flux, but dry advection leads to local divergence of the water vapor flux. As for the second part, local wind circulation convergence benefits local convergence of the water vapor flux, and vice versa.

3. Rainfall and circulations of the EASM and SASM regions: climatological characteristics and differences

As mentioned, the EASM system is a relatively independent subtropical monsoon circulation system and has different horizontal circulations from the SASM system (Tao and Chen, 1987). However, importantly, differences between the two monsoon systems may also exist in terms of the vertical structure of the zonal and meridional circulations, which may further lead to differences in their rainfall cloud systems and water vapor transport. Therefore, the climatological characteristics of the two regions' rainfall cloud systems, vertical structure of horizontal circulations, and water vapor transport, and their differences, are explored in this section.

3.1. Rainfall cloud systems

Using the TRMM_3A25 precipitation data, the distribution of the ratio of rainfall due to convective and stratiform cloud systems to total rainfall over the Asian summer monsoon region is analyzed. As shown in Fig. 1, the rainfall ratio related to convective cloud systems in the EASM region decreases with latitude, with the value reaching 50% in the low latitudes to the south of 25°N (Fig. 1a). Meanwhile, the rainfall ratio related to stratiform cloud systems exhibits the opposite feature, increasing from low to high latitudes, with the value increasing to 65% to the north of 25°N (Fig. 1b). Specifically, the rainfall over South China due to convective and stratiform cloud systems is comparable, which agrees with the findings of Fu et al. (2003) and Liu and Fu (2010). These features suggest that summertime rainfall cloud systems over the EASM region are a mix of stratiform and convective cloud systems—something that may lead to difficulty in parameterizing cumulus cloud when numerically modeling the EASM system (Cheng et al., 1998).

As for the SASM region, the ratio of the rainfall induced by convective/stratiform cloud systems to total rainfall is uniformly spread over continental India. Furthermore, the rainfall induced by convective cloud systems contributes more than 60% of the total rainfall (Fig. 1a), while the ratio of the rainfall induced by stratiform cloud systems to total rainfall is about 35% over India (Fig. 1b). This indicates that the summertime rainfall cloud systems over the SASM region are dominated by convective cloud, which is different to the situation over the EASM region.

3.2. Zonal and meridional circulation

As mentioned, the vertical structure of the horizontal circulations over the EASM and SASM regions are different, which may be responsible for their differences in monsoon rainfall cloud systems. Thus, the differences in the vertical shear of zonal and meridional circulations between these two monsoon regions are investigated in this part of the study. Figure 2 displays the vertical distribution of the climatological mean zonal and meridional wind averaged over the domains (20°–45°N, 100°–140°E) and (0°–25°N, 60°–100°E), representing the EASM and SASM regions, respectively. As shown in Fig. 2a, during boreal summer, the climatological mean zonal circulation over the EASM region is characterized by westerly wind throughout the troposphere in the vertical direction, showing obvious vertical westerly shear. This may be unfavorable for the intensification of convective cloud systems, but favorable for the formation of stratiform cloud systems. Thus, there is a mix of stratiform and convective cloud systems in the EASM region. However, the SASM system belongs purely to the tropical monsoon, with notable westerly wind prevailing in the lower troposphere below 500 hPa and easterly wind in the mid/upper troposphere (Fig. 2b), showing strong vertical easterly shear. But, the wind speed is weaker than that in the EASM region. The low-level west-

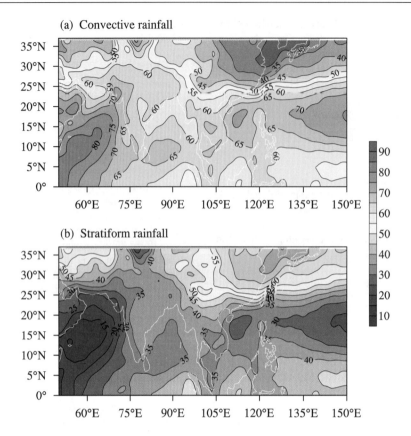

Fig. 1. Climatological-mean distributions of summertime rainfall (%) due to (a) convective rainfall and (b) stratiform rainfall, for 17 summers during 1998–2014, based on TRMM data.

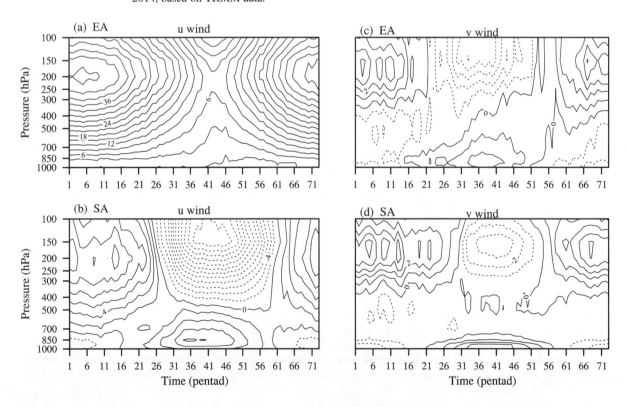

Fig. 2. Time–height cross section of climatological (a, b) zonal and (c, d) meridional winds, averaged for 26 summers during 1979–2014 over (a, c) East Asia (20°–45°N, 100°–140°E) and (b, d) South Asia (0°–25°N, 60°–100°E). Solid and dashed lines indicate westerly/southerly and easterly/northerly winds, respectively. Units: m s^{-1}.

erly wind and vertical easterly shear may be favorable for the intensification of convective activity in the SASM region, resulting in the monsoon rainfall mainly triggered by convective cloud systems in this region (Halverson et al., 2002).

As for the meridional circulation, significant meridional flow with low-level southerly and high-level northerly prevails in the EASM region (Fig. 2c), featuring strong vertical northerly shear and suggesting it is composed of tropical and subtropical summer monsoon characteristics. In contrast, the meridional circulation in the SASM region is characterized by low-level southerly flow (below 800 hPa, approximately) and strong upper-level northerly flow (Fig. 2d). Furthermore, it is clear that southerly wind prevails over larger ranges in the EASM region compared with the SASM region—a situation that is the reverse for the northerly flow. This shows that the vertical northerly shear of meridional circulation over the SASM region is stronger than that over the EASM region.

As indicated above, there are obvious differences in the vertical shear of horizontal circulation between the EASM and SASM systems, which may lead to their differences in monsoon rainfall cloud systems. This suggests that different ways can be used to measure the strength and variability of the two monsoon systems and, as such, different monsoon indices have been defined for the two monsoon systems. For example, Huang (2004) and Zhao et al. (2015) defined East Asia–Pacific (EAP) indices that measure the strength of the EASM system using the 500-hPa geopotential height and the zonal wind at 200 hPa, respectively. Meanwhile, Wang and Fan (1999) and Webster and Yang (1992) defined the strength of the SASM system using the zonal wind in the lower troposphere and the difference in the zonal wind between the upper and lower troposphere, respectively.

3.3. *Water vapor transport*

Water vapor transport and its convergence are of great importance to local rainfall variation. As analyzed above, the horizontal and vertical features of the circulation over the EASM and SASM regions show significant differences, and this may lead to differences in the characteristics of water vapor transport and its convergence over the two monsoon regions. Therefore, the water vapor transport, its convergence, and associated thermal and dynamical components, are analyzed in this subsection.

Figure 3a displays the climatological mean distribution of summertime water vapor transport fluxes over the EASM and SASM regions during 1979–2014, calculated based on Eqs. (1) and (2). Clearly, the distribution of water vapor transport flux is significantly different in the two monsoon regions. There are three sources for water vapor transport in the

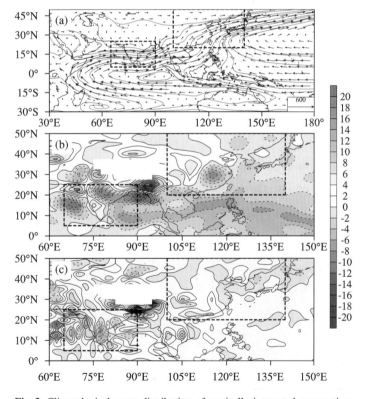

Fig. 3. Climatological mean distribution of vertically integrated summertime water vapor transport fluxes (a), and its divergence due to (b) wind divergence and (c) moisture advection, averaged for 26 summers during 1979–2014. The EASM (right-hand box) and SASM (left-hand box) regions are plotted. The unit for (a) is kg m^{-1} s^{-1}, and that for (b) and (c) is kg m^{-2} s^{-1}.

EASM region: from the Bay of Bengal, the South China Sea, and the tropical western Pacific. Owing to the abundant water vapor delivered by the summer monsoon flow from these source regions, the meridional water vapor transport fluxes are significant over South China and the Yangtze River valley. However, with respect to the SASM region, the zonal transport fluxes of water vapor are notably larger than the meridional transport fluxes.

Figures 3b and c show the climatological thermal and dynamical contributions to the divergence of the water vapor transport flux, i.e. by moisture advection and wind divergence, respectively. It is clear that, in the EASM region, the southern part is dominated by the dynamical contribution of wind divergence, while the northern part is dominated by moisture advection. As for the SASM region, the contribution of the wind divergence component is relatively smaller than that of moisture advection, especially over the northern India. By contrast, both the dynamical and thermal contributions are relatively smaller in the EASM region than in the SASM region.

From the above analyses, we can conclude that there are remarkable differences in the rainfall cloud systems, vertical structure of horizontal circulation, and water vapor transport, of the EASM and SASM regions, and that these are responsible for their differences in monsoon rainfall characteristics during boreal summer.

4. Interannual variability of summertime rainfall in the EASM and SASM regions and their differences

As indicated in the previous section, the climatological features of the rainfall cloud systems, circulation, and water vapor transport in the EASM and SASM regions are different from one another, making it conceivable that the spatiotemporal variability of the summer rainfall in the two monsoon regions may also be different. Therefore, using the observational station and gridded data and the EOF method, this section analyzes the spatiotemporal variability of the summer rainfall in the EASM and SASM regions.

4.1. *Characteristics of the interannual variability of summertime rainfall in the EASM region*

Figures 4 and 5 depict the spatial distributions and corresponding time-coefficient series and their wavelet analyses for the two leading modes of summertime rainfall anomalies in eastern China, respectively. As shown in Fig. 4a, the first leading mode characterizes a meridional tripole pattern in eastern China, and exhibits an obvious interannual variability with a period of 2–3 years, i.e. quasi-biennial oscillation, before the early 1990s, which may be influenced by the interannual variability of the thermal states of the western Pacific warm pool (Huang et al., 2006b, 2006c). Meanwhile, the interdecadal variability of the first leading mode is also remarkable from the late 1970s (Figs. 4b and c). As for the second leading mode, it features a meridional dipole pattern in eastern China (Fig. 5a); plus, it also exhibits an obvious interannual variability with a period of 2–3 years, especially from the mid-1960s to the late-1970s and from the early-1990s to the late-1990s, while its variability shifts and has a period of 3–8 years from the late-1990s onwards (Figs. 5b and c). Therefore, the interannual variability of summertime rainfall in the EASM region is unstable and characterized by multiple modes with an interdecadal change from the late-1990s. Accordingly, climatic disasters have occurred frequently in China (Huang et al., 2008, 2011b).

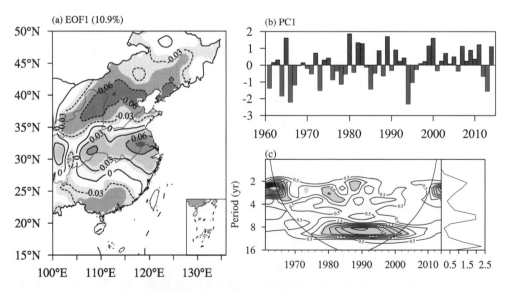

Fig. 4. Spatial distribution of the first leading mode of summer rainfall anomalies in eastern China (a), and the (b) corresponding time-coefficient series and (c) wavelet analysis for 1961–2014. Solid and dashed lines in (a) indicate positive and negative signals, respectively. EOF1 explains 10.9% of the total variance. Shading in (c) depicts the power significant beyond the 95% confidence level.

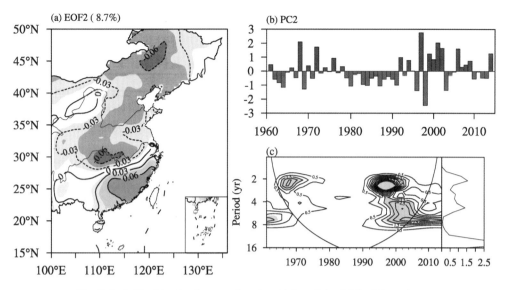

Fig. 5. As in Fig.4 but for the second mode, which explains 8.7% of the variance.

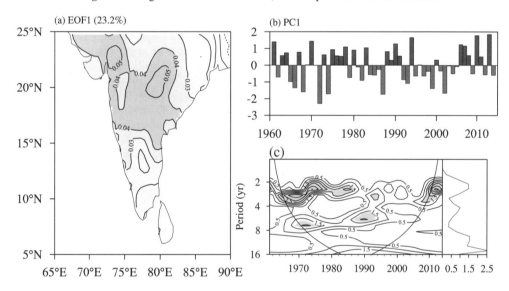

Fig. 6. As in Fig. 4 but for summer rainfall in the SASM (Indian) region. EOF1 explains 23.2% of the variance.

4.2. Characteristics of the interannual variability of summertime rainfall in the SASM region

For comparison, the spatiotemporal variability of summertime rainfall in the SASM region is also analyzed. Figure 6 depicts the spatial distribution, corresponding time-coefficient series and wavelet analyses of the first mode of summertime rainfall in the SASM region for 1961–2014. The first mode of summertime rainfall anomalies in the SASM region explains about 23.2% of the total variance and exhibits a uniform pattern in its spatial distribution. Additionally, it reveals a notable interannual variability with quasi-biennial oscillation (Figs. 6b and c), which agrees with the findings of Yasunari and Suppiah (1988). As shown in Fig. 7a, the second mode of summertime rainfall in the SASM region accounts for about 15.0% of the total variance and displays a meridional dipole pattern from the south to the north of India.

The interannual variability of the second mode is also significant, with quasi-biennial oscillation during the periods from the mid-1970s to the mid-1980s and from the mid-1990s to the late-2000s, and 4–5 years during the period from the mid-1980s to the mid-1990s (Figs. 7b and c).

4.3. Differences in the characteristics of the interannual variability of summertime rainfall in the EASM and SASM regions

From the above analyses, a quasi-biennial oscillation of summertime rainfall appears in both the EASM and SASM region. This is consistent with Ding (2007), who pointed out that quasi-biennial oscillation may be a leading mode of the Asian summer monsoon system, including the EASM and SASM systems. However, comparing Fig. 3a with Fig. 6a, an obvious difference is apparent in terms of their spatial dis-

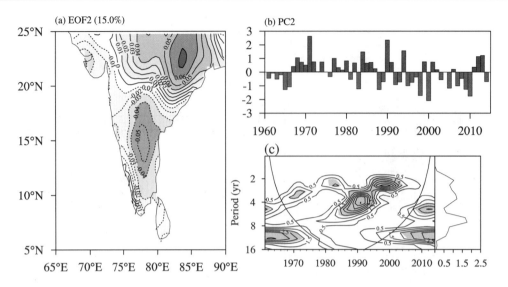

Fig. 7. As in Fig. 6 but for the second mode, which explains 15.0% of the variance.

tributions. The spatial distribution of the first mode of summertime rainfall anomalies in the EASM region exhibits a meridional tripole pattern, whereas it exhibits a uniform pattern in the SASM region.

5. Interdecadal variability of summertime rainfall in the EASM and SASM regions and their differences

From section 4 we can state that, in addition to significant interannual variability, the summertime rainfall in the EASM and SASM regions reveal notable interdecadal variabilities. Huang et al. (1999) pointed out that, during the late-1970s to early-1990s, summer rainfall in North China decreased sharply, causing prolonged and severe droughts in the region, but the opposite situation occurred in the Yangtze River and Huaihe River valley regions. Recently, several studies have revealed that EASM rainfall also experienced a significant interdecadal change in the early-1990s (Kwon et al., 2007; Ding et al., 2008, 2009; Wu et al., 2010; Zhang, 2015), and in the late-1990s (Liu et al., 2011; Huang et al., 2011a,b, 2013). To compare the interdecadal variability of summertime rainfall between the EASM and SASM regions, 9-yr running mean summertime precipitation data for the two regions, along with the EOF method, are used.

5.1. Characteristics of the interdecadal variability of summertime rainfall in the EASM region

As shown in Fig. 8a, the first mode of the 9-yr running mean summertime rainfall in the EASM region, which explains 26.1% of the total variance, exhibits a notable meridional dipole structure, i.e. a south–north oscillation pattern. Furthermore, the corresponding time coefficients are positive during the period from the mid-1960s to the early-1990s (Fig. 8b) and, in combination with Fig. 8a, the summertime rainfall anomalies during this period feature a negative–positive meridional dipole pattern from the south to the north of eastern China, with negative anomalies in South China, the Yangtze River and Huaihe River valleys, and positive anomalies in North China. However, the opposite structure appears in this region from the early-1990s; the distribution of summertime rainfall anomalies exhibits a positive–negative meridional dipole pattern from the south to the north of eastern China.

As for the second mode of the interdecadal variability of the summer rainfall in the EASM region, it explains 19.2% of the variance and exhibits a meridional tripole pattern in its spatial distribution (Fig. 8c). The time coefficients shown in Fig. 8d are negative from the mid-1960s to the late-1970s and, corresponding to the spatial feature in Fig. 8c, it reveals the summertime rainfall anomalies during this period to be positive in North and South China and negative in the Yangtze River and Huaihe River valleys. Comparing Figs. 8d and b, we can see that the time coefficients of EOF2 are much larger than those of EOF1 prior to the 1980s. Thus, the distribution of summertime rainfall anomalies presents a positive–negative–positive meridional tripole pattern during this period. Also, from the late-1970s to early-1990s, the time coefficients of EOF2 change from negative to positive and, in combination with its spatial distribution shown in Fig. 8c, it can be seen that the summertime rainfall decreases in North and South China and increases in the Yangtze River and Huaihe River valleys, i.e. the distribution of summertime rainfall anomalies presents a negative–positive–negative meridional tripole pattern from the south to the north of eastern China, which is opposite to the situation in the previous period. As for the period from the early- to the late-1990s, although the time coefficients of EOF2 (Fig. 8d) are still positive, those of EOF1 become negative and larger than those of EOF2. In combination with Figs. 8a and c, the distribution of summertime rainfall anomalies presents an above-normal

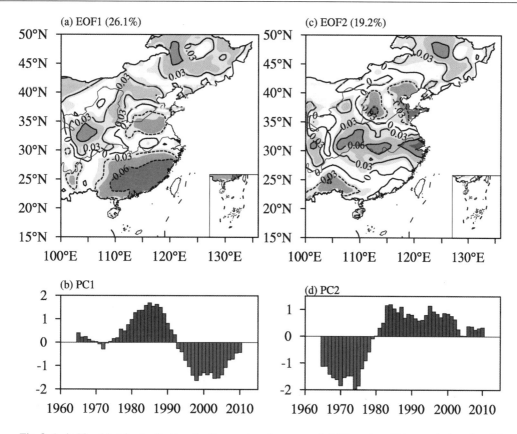

Fig. 8. As in Fig. 4 but for the first two leading modes of summer rainfall in eastern China on the interdecadal timescale: (a, c) spatial distribution; (b, d) corresponding time-coefficient series. EOF1 and EOF 2 explain 26.1% and 19.2% of the total variance, respectively.

pattern over the whole region, except for the northern area of North China. Also, since the negative time coefficients of EOF1 become larger from the late-1990s, the distribution of summertime rainfall anomalies presents a positive–negative meridional dipole pattern from the south to the north of eastern China—namely, a "south-flood–north-drought" pattern.

Moreover, to display the characteristics of the interdecadal variability even more clearly, we explore the features of the 9-yr running mean summertime rainfall anomalous percentages zonally averaged over the domain 100°–120°E. As shown in Fig. 9a, during the period from the mid-1960s to the late-1970s, the distribution of rainfall anomalies exhibits a positive–negative–positive meridional tripole pattern from the south to the north in eastern China. Furthermore, an opposite anomaly distribution, with a negative–positive–negative meridional tripole pattern, appears in eastern China during the period from the late-1970s to the early-1990s. As for the period from the early- to late-1990s, the rainfall anomalies in eastern China show a uniformly above-normal pattern, except in the Huaihe River valley. Meanwhile, from the late-1990s, the distribution of summertime rainfall anomalies presents a positive–negative meridional dipole pattern from the south to the north in eastern China. These features are coherent with the interdecadal variability described by the leading modes of the 9-yr running mean summertime rainfall in the EASM region. In addition, the positive and negative anomalies of summertime rainfall in eastern China indicate an anomalous signal moving from the north to the south on the interdecadal timescale (Fig. 9a).

5.2. *Characteristics of the interdecadal variability of summertime rainfall in the SASM region*

For comparison, the interdecadal variability of summertime rainfall in the SASM region is analyzed using the CRU precipitation data for 1961–2014. As shown in Fig. 10a, the first mode of summertime rainfall in the SASM region explains 28.8% of the total variance and exhibits a meridional dipole pattern with large anomalies distributed to the flanks of 20°N. The related time coefficients reveal an obvious interdecadal change, with below-normal values before the early-1990s and above-normal values thereafter (Fig. 10b). In combination with Fig. 10a, it is apparent that the summertime rainfall anomalies during this period are negative over southern India and positive over northern India, presenting a negative–positive meridional dipole pattern from the south to the north of the country. However, from the early-1990s, the distribution of summertime rainfall anomalies is the opposite, with a positive–negative meridional dipole pattern from the south to the north of India.

Meanwhile, the second mode of 9-yr running mean sum-

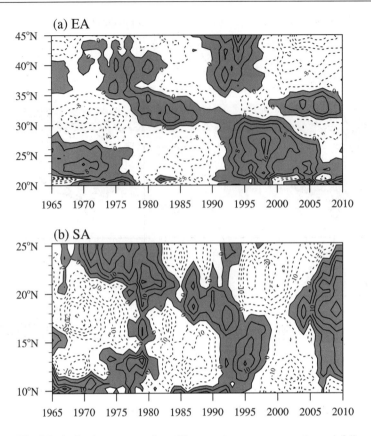

Fig. 9. Latitude–time cross section of 9-yr running mean summertime rainfall anomalies (%) averaged for 100°–120°E in eastern China (a) and for 75°–85°E in the SASM (Indian) region (b). Solid and dashed lines indicate positive and negative anomalies, respectively, and positive anomalies are shaded.

mertime rainfall in the SASM region accounts for 21.8% of the total variance and shows a uniform spatial pattern (Fig. 10c), but with large anomalies to the north of 20°N, which represents a slight southward shift compared with the northern anomalous center in Fig. 10a. The time coefficients are positive prior to the early-1970s and, comparing Figs. 10d and b, the time coefficients of EOF1 are larger than those of EOF2 during this period. Thus, the summertime rainfall anomalies present a negative–positive meridional dipole pattern in India during this period (Fig. 9b). However, from the early-1970s to early-1980s, the time coefficients shown in Fig. 10d change from positive to negative and are larger than those of EOF1 (Fig. 10b); plus, in combination with Fig. 10c, the summertime rainfall anomalies are positive over the whole of India. As for the period from the early-1980s to early-1990s, although the time coefficients of EOF2 are still negative, they are smaller than those of EOF1, and thus the summertime rainfall anomalies also present a negative–positive meridional dipole pattern from the south to the north of India (Fig. 9b). The time coefficients of EOF2 change from negative to positive from the early- to late-1990s. Meanwhile, the time coefficients of EOF1 become positive, and thus the distribution of summertime rainfall anomalies presents a positive–negative meridional dipole pattern from the south to the north of India. As for the period from the late-1990s to the mid-2000s, since the time coefficients of EOF2 are larger than those of EOF1, the distribution of summertime rainfall anomalies thus presents a positive–negative meridional dipole pattern from the south to the north of India.

The 9-yr running mean summertime rainfall anomalous percentages zonally averaged over the domain 75°–85°E are also examined. As shown in Fig. 9b, from the mid-1960s to the early-1990s, the distribution of summertime monsoon rainfall anomalies in the SASM region exhibits a negative–positive meridional dipole pattern from the south to the north of India, except for a positive pattern over the whole of India during the period from the mid-1970s to the early-1980s. From the early- to late-1990s, an opposite anomalous distribution, with a positive–negative meridional dipole pattern from the south to the north of India, appears in this region. Furthermore, during the period from the late-1990s to the mid-2000s, there are negative anomalies over the whole of India. Meanwhile, from the mid-2000s, the summertime rainfall anomalies shift to a positive–negative meridional dipole pattern from the south to the north of India. Besides, the positive and negative anomalies of summertime rainfall in the

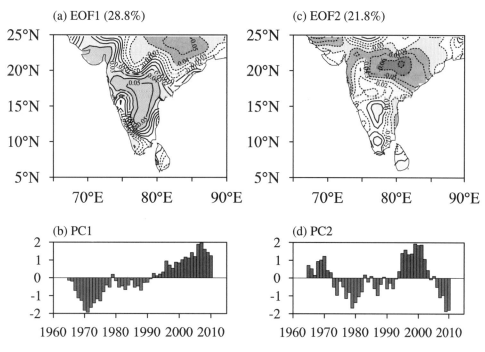

Fig. 10. As in Fig. 8 but for summer rainfall in the SASM (Indian) region. EOF1 and EOF2 explain 28.8% and 21.8% of the total variance, respectively.

SASM region also show a trend of movement from the north to the south.

5.3. Differences in the characteristics of the interdecadal variability of summertime rainfall in the EASM and SASM regions

As indicated above, on the interdecadal timescale, the summertime rainfall in both the EASM and SASM regions also show significant differences. The first mode of interdecadal variability of summertime rainfall in both the EASM and SASM regions presents a meridional dipole pattern from the south to the north in its spatial distribution, and both experience an interdecadal variation in the early-1990s. Also, the second mode of the interdecadal variability of summertime rainfall in the EASM region exhibits a tripole structure in spatial pattern and experiences an interdecadal variation in the late-1970s. Meanwhile, the second mode of the interdecadal variability of summertime rainfall in the SASM region shows a uniform spatial pattern and has three interdecadal variations—in the mid-1970s, the early-1990s, and the mid-2000s. Therefore, the interdecadal variability of summertime rainfall in the SASM region may be more complex than that in the EASM region since the early-1990s.

6. Links between the spatiotemporal variabilities of summertime rainfall in the EASM and SASM systems

Although the spatiotemporal variabilities of the EASM system are different from those of the SASM system, there are nevertheless some close links between them. To investigate these links, monsoon indices that measure the strength and variability of the EASM and SASM systems are used to analyze their relationship. Specifically, we use the EAP index to measure the strength of the EASM system, as defined by Huang (2004), and the Webster and Yang (1992) (WY) and Wang and Fan (1999) (WF) indices to measure the strength of the SASM system.

6.1. Interannual and interdecadal variabilities

Figure 11 shows the distributions of the linear regression of summertime rainfall anomalies in the EASM and SASM regions with respect to the EAP index, WY index and WF index, separately. Clearly, the regressed rainfall anomalies against the EAP index show a meridional tripole pattern, with negative anomalies in the Yangtze River valley and positive anomalies in North China and the southeastern coast of China (Fig. 11a). This resembles the first mode of summer rainfall in the EASM region and suggests the EAP index describes the interannual variability of the EASM well. The rainfall anomalies over India, regressed against the WY and WF indices, show strong similarities, especially in northern India (Figs. 11b and c); and combining the first mode of summer rainfall in the SASM region, the WF index may be more suitable to measure the variability of the EASM.

There is high correlation between the EAP and WF indices; their correlation coefficient is around 0.32, reaching the 98% confidence level. This high correlation indicates a close link between the first mode of summertime rainfall in the EASM and SASM regions. This link can also be seen from the correlation between the time coefficient series of the

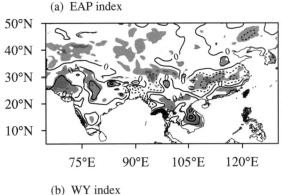
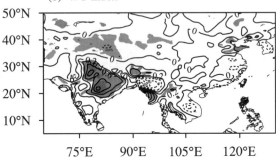
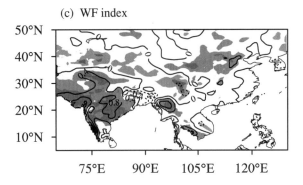

Fig. 11. Linear regression of summertime rainfall anomalies in Eurasia against (a) the EAP index, (b) the WY index and (c) the WF index. Shaded areas denote anomalies exceeding the 95% confidence level.

first mode of summertime rainfall in the EASM and SASM regions. Their correlation coefficient reaches −0.23, which is significant beyond 90% confidence level. This means that, if the monsoon rainfall anomaly in India in summer is positive, then it is also positive in South China and North China, while it is negative in the Yangtze River valley. Furthermore, this relationship is unstable, becoming stronger from the early-1990s; the correlation coefficient is only −0.19 during 1961–93, but reaches −0.33 during 1994–2014, at the greater than 90% confidence level. There is also a close link between the interdecadal variability of summertime rainfall in the EASM and SASM regions. The correlation between the first mode of summer rainfall in the EASM and SASM regions reaches −0.58, with statistical significance above the 90% confidence level, suggesting a closer relationship on the interdecadal timescale.

Previous studies have also pointed out the positive correlation between the summertime monsoon rainfall in India and that in North China (Kripalani and Kulkarni, 2001; Wu, 2002, 2017), and that it mainly exists in the quasi-biennial oscillation mode of summertime rainfall in these two monsoon regions (Lin et al., 2016). These studies, however, were based on the correlation between averaged summer monsoon rainfall anomalies over India and North China, which is different to our study. The relationship in the present study is between the first mode of summertime rainfall, with a meridional tripole pattern from the south to the north of eastern China, and the first mode of summertime rainfall, with a uniform pattern in India.

Moreover, it is clear that the regressed rainfall anomalies against the three indices show close similarities over the Eurasian continent. There is a significant meridional teleconnection pattern in the anomalous rainfall from South Asia to North China, showing a positive–negative–positive tripole structure from the south to the north of the Asian summer monsoon region. We refer to this teleconnection as the South Asia/East Asia teleconnection of Asian summer monsoon rainfall anomalies. This agrees with previous findings of a meridional tripole teleconnection existing not only in the EASM, which is related to the Pacific–Japan oscillation, or EAP teleconnection pattern (Nitta, 1987; Huang and Li, 1988; Huang and Sun, 1992), but also in the Asian summer monsoon rainfall anomalies from South Asia to North China (Liu and Ding, 2008; Ding et al., 2013).

6.2. *Spatiotemporal variabilities of summertime water vapor transport*

The close links between the first modes of interannual and interdecadal variability of summertime rainfall in the EASM and the SASM systems may be closely related to the links between the spatiotemporal variabilities of summertime water vapor transport. It is clear from the climatological mean distribution of water vapor transport fluxes over the Asian summer monsoon shown (Fig. 4) that a large amount of water vapor is transported into the EASM region from the Bay of Bengal in summer. Thus, the spatiotemporal variabilities of summertime water vapor transport over the EASM region should be closely associated with those over the SASM region. The correlations between the time-coefficient series of the first two leading modes of summertime water vapor transport fluxes over the EASM and SASM regions are investigated (figure not shown). The correlation coefficients between corresponding pairs are 0.35 and 0.38, respectively, which all reach the 99% confidence level. These correlations can explain the close link between the spatiotemporal variabilities of summertime water vapor transport in the EASM and SASM regions.

7. Conclusions and discussion

Using long-term observational and reanalysis data, the differences and links between the climatological features and interannual/interdecadal variabilities of summer rainfall in the EASM and SASM regions are analyzed. The main results can be summarized as follows:

The climatological characteristics of summer rainfall in the EASM and SASM regions show obvious differences in terms of the rainfall cloud systems. In the EASM region, there is a mix of stratiform and cumulus rainfall, whereas the SASM is dominated by cumulus cloud systems. This difference may be due to the regions' differences in vertical wind shear in the zonal and meridional winds and the convergence of water vapor transport. The contribution of moisture advection and wind convergence to the convergence of the water vapor transport flux are comparable in the EASM region. However, the contribution of wind convergence is superior to that of moisture advection in the SASM region.

The first two leading modes of the spatiotemporal variabilities of summertime rainfall in the EASM and SASM regions exhibit remarkable interannual variability, with quasi-biennial oscillation and interdecadal variability, and obviously different spatial distributions. In the EASM region, the first and second modes of the interannual variability of summertime rainfall present a meridional tripole pattern and meridional dipole pattern, respectively. Meanwhile, those in the SASM region present a uniform pattern and meridional dipole pattern, respectively. Furthermore, on the interdecadal timescale, the first and second modes of summertime rainfall in the two monsoon regions are similar in pattern to the second and first modes of their interannual variability in their spatial distributions, respectively.

The variability of the summer rainfall in the EASM and SASM regions shows close links on both the interannual and interdecadal timescale. The first mode of interannual variability of summer rainfall in the two monsoon regions has a high negative correlation, and this relationship becomes stronger from the early-1990s. Furthermore, on the interdecadal timescale, the first modes of the summer rainfall in the two monsoon regions also have a high negative correlation. This high correlation of summer rainfall in the EASM and SASM regions may be linked by a meridional tripole teleconnection pattern from South Asia to North China, through the southeast part of the Tibetan Plateau, in the regressed summertime rainfall anomalies over the Asian summer monsoon regions, which agrees with previous studies (Liu and Ding, 2008; Ding et al., 2013). We refer to this pattern as the South Asia/East Asia teleconnection of Asian summer monsoon rainfall.

Overall, it is clear that there are some close links between the EASM and SASM systems, but that the summertime rainfall cloud systems and the spatiotemporal variabilities of summertime rainfall in the EASM region are different from those in the SASM region. Therefore, the EASM system may be a relatively independent monsoon subsystem linked closely to the SASM system in the Asian–Australian monsoon system. Also, it should be pointed out that many problems remain regarding our understanding of the basic physical processes involved in the differences and links between the EASM and SASM systems, and these need to be urgently studied in future work. In particular, the thermal effect of the western Pacific warm pool and the tropical Indian Ocean on the spatiotemporal variabilities of the EASM and SASM systems and their links should be systematically investigated.

Acknowledgements. This study was supported jointly by the National Key Research and Development Program (Grant No. 2016YFA0600603), the National Basic Research of China (Grant No. 2013CB430201) and the National Natural Science Foundation of China (Grant Nos. 41605058, 41375065, 41461164005, 41230527, and 41375082).

REFERENCES

Cheng, A. N., W. Chen, and R. H. Huang, 1998: The sensitivity of numerical simulation of the East Asian monsoon to different cumulus parameterization schemes. *Adv. Atmos. Sci.*, **15**, 204–220, doi: 10.1007/s00376-998-0040-6.

Davis, R. E., 1976: Predictability of sea surface temperature and sea level pressure anomalies over the North Pacific Ocean. *J. Phys. Oceanogr.*, **6**, 249–266, doi: 10.1175/1520-0485(1976)006<0249:POSSTA>2.0.CO;2.

Deng, W. T., Z. B. Sun, G. Zeng, and D. H. Ni, 2009: Interdecadal variation of summer precipitation pattern over eastern China and its relationship with the North Pacific SST. *Chinese Journal of Atmospheric Sciences*, **33**, 835–846, doi: 10.3878/j.issn.1006-9895.2009.04.16. (in Chinese)

Ding, Q. H., and B. Wang, 2005: Circumglobal teleconnection in the Northern Hemisphere summer. *J. Climate*, **18**, 3483–3505, doi: 10.1175/JCLI3473.1.

Ding, Y. H., 2007: The variability of the Asian summer monsoon. *J. Meteor. Soc. Japan*, **85B**, 21–54, doi: 10.2151/jmsj.85B.21.

Ding, Y. H., Z. Y. Wang, and Y. Sun, 2008: Inter-decadal variation of the summer precipitation in East China and its association with decreasing Asian summer monsoon. Part I: Observed evidences. *International Journal of Climatology*, **28**, 1139–1161, doi: 10.1002/joc.1615.

Ding, Y. H., and Coauthors, 2013: Interdecadal and interannual variabilities of the Asian summer monsoon and its projection of future change. *Chinese Journal of Atmospheric Sciences*, **37**, 253–280, doi: 10.3878/j.issn.1006-9895.2012.12302. (in Chinese)

Fu, Y. F., Y. H. Lin, G. S. Liu, and Q. Wang, 2003: Seasonal characteristics of precipitation in 1998 over East Asia as derived from TRMM PR. *Adv. Atmos. Sci.*, **20**, 511–529, doi: 10.1007/BF02915495.

Guo, Q. Y., 1992: Teleconnection between the floods/droughts in North China and Indian summer monsoon rainfall. *Acta Geographica Sinica*, **47**, 394–402, doi: 10.11821/xb199205002. (in Chinese)

Guo, Q. Y., and J. Q. Wang, 1988: A comparative study on summer monsoon in China and India. *Journal of Tropical Meteorology*, **4**, 53–60. (in Chinese)

Halverson, J. B., T. Rickenbach, B. Roy, H. Pierce, and E. Williams, 2002: Environmental characteristics of convective systems during TRMM-LBA. *Mon. Wea. Rev.*, **130**, 1493–1509, doi: 10.1175/1520-0493(2002)130<1493:ECOCSD>2.0.CO;2.

Harris, L., P. D. Jones, T. J. Osborn, and D. H. Lister, 2014: Updated high-resolution grids of monthly climatic observations-the CRU: TS3.10 Dataset. *International Journal of Climatology*, **34**, 623–642, doi: 10.1002/joc.3711.

Huang, G., 2004: An index measuring the interannual variation of the East Asian summer monsoon-The EAP index. *Adv. Atmos. Sci.*, **21**, 41–52, doi: 10.1007/BF02915679.

Huang, G., Y. Liu, and R. H. Huang, 2011a: The interannual variability of summer rainfall in the arid and semiarid regions of northern China and its association with the Northern Hemisphere circumglobal teleconnection. *Adv. Atmos. Sci.*, **28**(2), 257–268, doi: 10.1007/s00376-010-9225-x.

Huang, R. H., 2006: Progresses in research on the formation mechanism and prediction theory of severe climatic disasters in China. *Advances in Earth Science*, **21**, 564–575, doi: 10.11867/j.issn.1001-8166.2006.06.0564.

Huang, R. H., and W. J. Li, 1988: Influence of the heat source anomaly over the tropical western Pacific on the subtropical high over East Asia and its physical mechanism. *Chinese Journal of Atmospheric Sciences*, **12**, 107–116, doi: 10.3878/j.issn.1006-9895.1988.t1.08. (in Chinese)

Huang, R. H., and F. Y. Sun, 1992: Impacts of the tropical western Pacific on the East Asian summer monsoon. *J. Meteor. Soc. Japan*, **70**, 243–256, doi: 10.2151/jmsj1965.70.1B_243.

Huang, R. H., and L. T. Zhou, 2002: Research on the characteristics, formation mechanism and prediction of severe climatic disasters in China. *Journal of Natural Disasters*, **11**, 1–9, doi: 10.3969/j.issn.1004-4574.2002.01.001. (in Chinese).

Huang, R. H., Y. H. Xu, P. F. Wang, and L. T. Zhou, 1998a: The features of the catastrophic flood over the Changjiang River basin during the summer of 1998 and cause exploration. *Climatic and Environmental Research*, **3**, 300–313, doi: 10.3878/j.issn.1006-9585.1998.04.02. (in Chinese)

Huang, R. H., Z. Z. Zhang, G. Huang, and B. H. Ren, 1998b: Characteristics of the water vapor transport in East Asian monsoon region and its difference from that in South Asian monsoon region in summer. *Scientia Atmospherica Sinica*, **22**, 460–469, doi: 10.3878/j.issn.1006-9895.1998.04.08. (in Chinese)

Huang, R. H., Y. H. Xu, and L. T. Zhou, 1999: The interdecadal variation of summer precipitations in China and the drought trend in North China. *Plateau Meteorology*, **18**, 465–476, doi: 10.3321/j.issn:1000-0534.1999.04.001. (in Chinese).

Huang, R. H., G. Huang, and Z. G. Wei, 2004: Climate variations of the summer monsoon over China. *East Asian Monsoon*, C. P. Chang, Ed., World Scientific Publishing Co. Pte. Ltd., 213–270.

Huang, R. H., R. S. Cai, J. L. Chen, and L. T. Zhou, 2006a: Interdecadal variations of drought and flooding Disasters in China and their association with the East Asian climate system. *Chinese Journal of Atmospheric Sciences*, **30**, 730–743, doi: 10.3878/j.issn.1006-9895.2006.05.02. (in Chinese)

Huang, R. H., J. L. Chen, G. Huang, and Q. L. Zhang, 2006b: The quasi-biennial oscillation of summer monsoon rainfall in China and its cause. *Chinese Journal of Atmospheric Sciences*, **30**, 545–560, doi: 10.3878/j.issn.1006-9895.2006.04.01. (in Chinese)

Huang, R. H., L. Gu, L. T. Zhou, and S. S. Wu, 2006c: Impact of the thermal state of the tropical western Pacific on onset date and process of the South China Sea summer monsoon. *Adv. Atmos. Sci.*, **23**, 909–924, doi: 10.1007/s00376-006-0909-1.

Huang, R. H., J. L. Chen, and G. Huang, 2007: Characteristics and variations of the East Asian monsoon system and its impacts on climate disasters in China. *Adv. Atmos. Sci.*, **24**, 993–1023, doi: 10.1007/s00376-007-0993-x.

Huang, R. H., L. Gu, J. L. Chen, and G. Huang, 2008: Recent progresses in studies of the temporal-spatial variations of the East Asian monsoon system and their impacts on climate anomalies in China. *Chinese Journal of Atmospheric Sciences*, **32**, 691–719, doi: 10.3878/j.issn.1006-9895.2008.04.02. (in Chinese)

Huang, R. H., J. L. Chen, and Y. Liu, 2011b: Interdecadal variation of the leading modes of summertime precipitation anomalies over eastern China and its association with water vapor transport over East Asia. *Chinese Journal of Atmospheric Sciences*, **35**, 589–606, doi: 10.3878/j.issn.1006-9895.2011.04.01. (in Chinese)

Huang, R. H., J. L. Chen, L. Wang, and Z. D. Lin, 2012: Characteristics, processes, and causes of the spatio-temporal variabilities of the East Asian monsoon system. *Adv. Atmos. Sci.*, **29**, 910–942, doi: 10.1007/s00376-012-2015-x.

Huang, R. H., Y. Liu, and T. Feng, 2013: Interdecadal change of summer precipitation over Eastern China around the late-1990s and associated circulation anomalies, internal dynamical causes. *Chinese Science Bulletin*, **58**, 1339–1349, doi: 10.1007/s11434-012-5545-9.

Kalnay, E., and Coauthors, 1996: The NCEP/NCAR 40-year reanalysis project. *Bull. Amer. Meteor. Soc.*, **77**, 437–471, doi: 10.1175/1520-0477(1996)077<0437:TNYRP>2.0.CO;2.

Kripalani, R. H., and S. V. Singh, 1993: Large scale aspects of India-China summer monsoon rainfall. *Adv. Atmos. Sci.*, **10**, 71–84, doi: 10.1007/BF02656955.

Kripalani, R. H., and A. Kulkarni, 2001: Monsoon rainfall variations and teleconnections over South and East Asia. *International Journal of Climatology*, **21**, 603–616, doi: 10.1002/joc.625.

Krishnamurti, T. N., and Y. Ramanathan, 1982: Sensitivity of the monsoon onset to differential heating. *J. Atmos. Sci.*, **39**, 1290–1306, doi: 10.1175/1520-0469(1982)039<1290:SOTMOT>2.0.CO;2.

Kwon, M. H., G. J. Jhun, and K. J. Ha, 2007: Decadal change in East Asian summer monsoon circulation in the mid-1990s. *Geophys. Res. Lett.*, **34**, L21706, doi: 10.1029/2007GL031977.

Lin, D. W., C. Bueh, and Z. W. Xie, 2016: Relationship between summer rainfall over North China and India and its genesis analysis. *Chinese Journal of Atmospheric Sciences*, **40**, 201–214, doi: 10.3878/j.issn.1006-9895.1503.14339. (in Chinese)

Liu, P., and Y. F. Fu, 2010: Climatic characteristics of summer convective and stratiform precipitation in southern China based on measurements by TRMM precipitation radar. *Chinese Journal of Atmospheric Sciences*, **34**, 802–814, doi: 10.3878/j.issn.1006-9895.2010.04.12. (in Chinese)

Liu, Y., G. Huang, and R. H. Huang, 2011: Inter-decadal variability of summer rainfall in Eastern China detected by the Lepage test. *Theor. Appl. Climatol.*, **106**(3–4), 481–488, doi: 10.1007/s00704-011-0442-8.

Liu, Y. Y., and Y. H. Ding, 2008: Analysis and numerical simulation of the teleconnection between Indian summer monsoon and precipitation in North China. *Acta Meteorologica Sinica*, **66**, 789–799, doi: 10.11676/qxxb2008.072. (in Chinese)

McBride, T. L., 1987: The Australian summer monsoon. *Monsoon Meteorology*, C. P. Chang and T. N. Krishnamurti, Eds., Oxford University Press, 203–232.

Nitta, T., 1987: Convective activities in the tropical western Pacific and their impact on the Northern Hemisphere summer circulation. *J. Meteor. Soc. Japan*, **64**, 373–390, doi:

10.2151/jmsj1965.65.3_373.

North, G. R., T. L. Bell, R. F. Cahalan, and F. J. Moeng, 1982: Sampling errors in the estimation of empirical orthogonal functions. *Mon. Wea. Rev.*, **110**(7), 699–706, doi: 10.1175/1520-0493(1982)110<0699:SEITEO>2.0.CO;2.

Staff Members of the Section of Synoptic and Dynamic Meteorology, Institute of Geophysics and Meteorology, Academia Sinica, 1957: On the general circulation over Eastern Asia (I). *Tellus*, **9**, 432–446, doi: http://dx.doi.org/10.1111/j.2153-3490.1957.tb01903.x.

Staff Members of the Section of Synoptic and Dynamic Meteorology, Institute of Geophysics and Meteorology, Academia Sinica, 1958a: On the general circulation over Eastern Asia (II). *Tellus*, **10**, 58–75, doi: 10.1111/j.2153-3490.1958.tb01985.x.

Staff Members of the Section of Synoptic and Dynamic Meteorology, Institute of Geophysics and Meteorology, Academia Sinica, 1958b: On the general circulation over Eastern Asia (III). *Tellus*, **10**, 299–312, doi: 10.1111/j.2153-3490.1958.tb02018.x.

Tao, S. Y., and L. X. Chen, 1987: A review of recent research on the East Asian summer monsoon in China. *Monsoon Meteorology*, C. P. Chang and T. N. Krishnamurti, Eds., Oxford University Press, 60–92.

Tropical Rainfall Measuring Mission (TRMM), 2011: TRMM_3A25: TRMM Precipitation Radar Rainfall L3 1 month (5 x 5) and (0.5 x 0.5) degree V7. Greenbelt, MD, Goddard Earth Sciences Data and Information Services Center (GES DISC), Accessed https://disc.gsfc.nasa.gov/datacollection/TRMM_3A25_7.html.

Wang, B., and Z. Fan, 1999: Choice of South Asian summer monsoon indices. *Bull. Amer. Meteor. Soc.*, **89**, 629–638, doi: 10.1175/1520-0477(1999)080<0629:COSASM>2.0.CO;2.

Wang, B., R. G. Wu, and K.-M. Lau, 2001: Interannual variability of the Asian summer monsoon: Contrasts between the Indian and the western North Pacific-East Asian monsoons. *J. Climate*, **14**, 4073–4090, doi: 10.1175/1520-0442(2001)014<4073:IVOTAS>2.0.CO;2.

Wang, L., and W. Gu, 2016: The Eastern China flood of June 2015 and its causes. *Science Bulletin*, **61**, 178–184, doi: 10.1007/s11434-015-0967-9.

Webster, P. J., and S. Yang, 1992: Monsoon and ENSO: Selectively interactive systems. *Quart. J. Roy. Meteor. Soc.*, **118**, 877–926, doi: 10.1002/qj.49711850705.

Webster, P. J., V. O. Magaña, T. N. Palmer, J. Shukla, R. A. Tomas, M. Yanai, and T. Yasunari, 1998: Monsoons: Processes, predictability, and the prospects for prediction. *J. Geophys. Res.*, **103**, 14 451–14 510, doi: 10.1029/97JC02719.

Wu, R. G., 2002: A mid-latitude Asian circulation anomaly pattern in boreal summer and its connection with the Indian and East Asian summer monsoons. *International Journal of Climatology*, **22**, 1879–1895, doi: 10.1002/joc.845.

Wu, R. G., 2017: Relationship between Indian and East Asian summer rainfall variations. *Adv. Atmos. Sci.*, **34**, 4–15, doi: 10.1007/s00376-016-6216-6.

Wu, R. G., Z. P. Wen, S. Yang, and Y. Q. Li, 2010: An Interdecadal Change in Southern China Summer Rainfall around 1992/93. *J. Climate*, **23**, 2389–2403, doi: 10.1175/2009JCLI3336.1.

Yasunari, T., and R. Suppiah, 1988: Some problems on the interannual variability of Indonesian monsoon rainfall. *Tropical Rainfall Measurements*, J. C. Theon and N. Fugono, Eds., Deepak, Hampton, 113–122.

Yeh, T. C., and P. C. Chu, 1958: *Some Fundamental Problems of the General Circulation of the Atmosphere*. Science Press, Beijing, 159 pp. (in Chinese)

Yeh, T. C., S. Y. Tao, and M. T. Li, 1959: The abrupt change of circulation over the Northern Hemisphere during June and October. *The Atmosphere and the Sea in Motion*, B. Bolin, Ed., The Rockefeller Institute Press and Oxford University Press, 249–267.

Yeh, T. C., 2008: *Selected Papers of Ye Duzheng*. Anhui Educational Publishing House, 498pp. (in Chinese)

Zhang, R. H., 2015: Changes in East Asian summer monsoon and summer rainfall over eastern China during recent decades. *Science Bulletin*, **60**, 1222–1224, doi: 10.1007/s11434-015-0824-x.

Zhao, G. J., G. Huang, R. G. Wu, W. C. Tao, H. N. Gong, X. Qu, and K. M. Hu, 2015: A new upper-level circulation index for the East Asian summer monsoon variability. *J. Climate*, **28**, 9977–9996, doi: 10.1175/JCLI-D-15-0272.1.

四、西北太平洋台风气候学研究

西北太平洋热带气旋移动路径的年际变化及其机理研究*

黄荣辉　陈光华

(中国科学院大气物理研究所季风系统研究中心，北京，100080)

摘　要

利用JTWC的热带气旋资料、NCEP/NCAR再分析的风场资料以及Scripps海洋研究所的海温资料分析了西北太平洋热带气旋(TC)移动路径的年际变化及其机理。结果表明，西北太平洋TC移动路径有明显的年际变化并与西太平洋暖池热状态有很密切的关系。当西太平洋暖池处于暖状态，西北太平洋上空TC移动路径偏西，影响中国的台风个数偏多；相反，当西太平洋暖池处于冷状态，西北太平洋的TC移动路径偏东，影响日本的台风个数偏多，而影响中国的台风个数可能偏少。本研究以西太平洋暖池处于冷状态的2004年与西太平洋暖池处于暖状态的2006年的西北太平洋TC移动路径的差别进一步论证了这一分析结果并从动力理论方面分析了在西太平洋暖池不同热状态下，季风槽对赤道西传天气尺度的Rossby重力混合波转变成热带低压型波动(TD型波动)的影响，以此揭示西太平洋暖池的热状态对西北太平洋TC生成位置与移动路径年际变化的影响机理。分析结果表明，当西北太平洋暖池处于暖状态时，季风槽偏西，使得热带太平洋上空对流层低层Rossby重力混合波转变成TD型波动的位置也偏西，从而造成TC生成平均位置偏西，并易于出现西行路径；相反，当西太平洋暖池处于冷状态时，季风槽偏东，这造成了对流层低层Rossby重力混合波转变成TD型波动的区域，以及TC生成的平均位置都偏东，从而导致TC移动路径以东北转向为主。

1　引　言

今年是谢义炳院士的90华诞。谢义炳院士是世界知名的气象学家，他不仅是中国现代大气环流理论和天气动力学研究的开拓者之一，而且也是中国现代大气科学教育的开拓者之一。他一生不仅为国家培养了众多的大气科学专家和气象工作者，而且在全球大气环流理论、湿斜压天气动力学、以及寒潮、锋面、台风与暴雨等研究领域做出了系统而创新性研究，特别在对流层中、上层冷性涡旋的发展理论方面做出了开创性的工作。他为中国大气科学的发展提出了许多有前瞻性、战略性的重大研究方向。他倡议并亲自组织和主持了第一、二届全国大气科学前沿学科研讨会，病重期间还亲自为第二届全国大气科学前沿学科研讨会的论文集"现代大气科学前沿与展望"[1]写祝词，他的这个倡议迄今仍然得到大气科学有关院校、研究机构和业务单位的支持，他所提出的中国大气科学重大研究方向还在继续发展，在他仙逝之后又召开了第三、四届全国大气科学前沿学科研讨会。

我作为聆听过谢义炳院士亲自授课的北大地球物理系的一名学子，并长期得到他关心、指导和启发的学生，我和我的学生从20世纪80年代中期起，致力于西太平洋暖池的热力变化及其对季风和台风活动年际变化的影响机理研究。近年来我们开展了西太平洋暖池热力作用下季风变异对台风活动年际变化的影响及其机理的研究。

中国处于西太平洋的西岸，而全球约三分之一

---

* 本文原载于：气象学报，第65卷，第5期，683-694，2007年10月出版.

(年均约30个)的TC在西太平洋形成,由于受东风带和副热带高压的影响,在西太平洋生成的热带气旋中一大部分移向中国、日本、菲律宾、越南和韩国等地并登陆,给这些国家造成巨大的经济损失和人员伤亡。中国是世界上少数遭受台风灾害最严重的国家之一,平均每年大概有7—8个台风登陆中国,最多12个左右。一般,每年影响中国(但不一定在中国登陆)的台风有10个以上。谢义炳先生早在20世纪50—60年代就对台风移动路径、赤道辐合带(ITCZ)对于台风发生、发展的动力作用等方面做了重要研究[2-4]。特别是在有的研究强调台风生成、发展的热力条件时[5],他毅然在"赤道辐合带上扰动不稳定性的简单理论分析"一文中提出了赤道辐合带上扰动切变不稳定理论,指出了台风生成、发展的动力条件。随后,中国许多学者对台风的生成、发展和移动路径作了不少研究,如陈联寿、丁一汇[6]、丁一汇等[7]和陈联寿等[8]。近年来李英等[9-10]、徐亚梅、伍荣生[11]和Chan[12-13]等做了系统的研究。同样,在国际上关于TC也有大量研究,Elsberry[14]、王斌等[15]对20世纪80—90年代国际上有关台风的观测事实、动力理论及其数值模拟和台风路径的预报等研究进展作了系统的回顾;并且王斌等[15]对于国际上热带气旋动力学的研究进展也做了深入的综述。

然而,应该看到以前的许多研究主要强调TC的天气学方面,如台风的强度、结构、移动路径和登陆后的天气以及台风生成和发展的机理等。相对而言,TC的气候学研究起步较晚,并且,在TC的气候学研究中较多的研究集中讨论TC的季节内变化,如Liebmann等[16],祝从文等[17]指出西太平洋台风活跃与30—60 d振荡(MJO)西风位相有密切关系,Harr和Elsberry[18],王慧等[19]指出西太平洋季风活跃对TC的活跃有很大影响。最近,围绕着全球变暖背景下台风的强度和生成个数是否增加这个问题,国际上兴起了台风气候学的研究热潮,特别是关于西北太平洋台风活动的年际变化和年代际变化正在吸引着许多学者的关注[20-22]。

以往关于西北太平洋TC年际变化的研究主要集中于考察TC活动与赤道中东太平洋海表温度所引起的ENSO事件以及热带西太平洋暖池热状况的联系(Chan[12-13,23],陈光华和黄荣辉[24-25],吴迪生[26])。这些研究指出ENSO事件与西北太平洋TC生成总数之间没有明显的关系,但在El Niño年(暖池偏冷年)的夏秋季,西北太平洋东南海域TC活动频繁,而西北太平洋的西北海域活动减弱;并且,在El Niño年TC平均生命史要比La Nina(暖池偏暖年)年的长。但是,关于西北太平洋TC移动路径的年际变化研究较少,并且对TC平均生成位置年际变化的机理也还需要进一步研究。西北太平洋TC移动路径的年际变化对于登陆中国台风个数的年际变化有重要影响。为此,有必要开展西北太平洋海域TC移动路径的年际变化规律和特征的研究,以便提高登陆中国台风个数的跨季度预测水平,减轻台风给中国造成的经济损失和人员伤亡。此外,由于无论ENSO循环或亚洲季风都与西太平洋暖池的热力状态密切相关[27-30],西太平洋暖池的热力状态势必影响西北太平洋TC的生成和移动路径。因此,本研究将利用再分析资料和海洋有关的观测资料分析西北太平洋TC生成区域和移动路径的年际变化,以及它们与西太平洋暖池热力状态的关系,并从动力理论来分析和讨论西太平洋暖池的热力状态通过热带波动来影响西北太平洋TC活动年际变化的机理。

2 数据和资料来源

本文首先利用美国联合台风预警中心(Joint Typhoon Warning Center, JTWC)1959—2006年TC生成位置、强度和移动路径的资料。为简单起见,本文强调具有一定强度TC活动的年际变化,故下文中的TC指的是热带风暴强度以上的气旋。

本文所用太平洋次表层海洋数据资料取自美国Scripps海洋研究所联合海洋环境数据分析中心(JEDA)的$2.5°×2.5°$月平均资料;日本气象厅海洋调查船"启风丸(Keifu Maru)"所观测的137°E的西太平洋海温资料。大气环流资料取自NCEP/NCAR再分析的逐日数据资料。

3 西北太平洋TC移动路径的年际变化及其与西太平洋暖池热力状态的关系

西太平洋暖池是全球海温最高的海域,全球有95%海表温度超过28.5℃的暖海水集中在此区域,故又称暖池(The Warm Pool)。该区域也是Walker环流的上升支,因此,其海域海-气相互作用十分剧烈,强烈的暖湿气流辐合导致该区域出现强烈的上升气流和对流活动。许多研究都表明此海域的热力状态对于热带西太平洋和东亚、东南亚地区季风的年际变化有很大影响[27-30];并且,陈光华和黄荣辉[24-25]的研究表明此海域的热状态对于热带西太平洋热带气旋生成位置有很大影响。因此,此海域的热状态对于西北太平洋 TC 的移动路径也会有很大影响。

3.1 西太平洋暖池热力的年际变化

若把 1959—2003 年西北太平洋热带气旋和台风路径都画在一张图上,可以看到 TC 的路径是杂乱无章,似乎没有规律。然而,我们把这些 TC 的移动路径依西太平洋暖池次表层的热状态划分,则可以得到很有趣的现象。众所周知,某海域海-气相互作用是直接与该海域海洋的热容量(Oceanic Heat Content)密切相关,而海洋热容量异常一般取决于该海域表层和次表层海温的异常。Cornejo-Garrido 和 Stone[31]的研究表明了西太平洋暖池有很大的热容量,正是由于如此巨大的热容量才造成这一海域上空有强的上升气流,因此,此海域是全球海洋中海-气相互作用最强的区域。

由于西太平洋暖池在次表层的 100—150 m 深的海温有很明显的年际变化[27-30]。因此,本研究取(0°—16°N、125°—165°E)海域的 120 m 深的海温作为衡量西太平洋暖池热状态的标准。从西太平洋暖池(0°—16°N,125°—165°E)海域 120 m 深 7—10月平均的海温距平(图 1)可以看到,此海域次表层的海温有很大的年际变化。定义此区域平均的海温距平大于+1.5 ℃的年份为暖年,小于-1.5 ℃为冷年,则可得到 1970,1975,1978,1998,1999 和 2000 年为西太平洋暖池的暖年,而 1972,1982,1987,1991,1993 和 1997 年为西太平洋暖池的冷年。并且,陈光华和黄荣辉[24-25]的研究结果表明了西北太平洋 TC 生成与 NINO3.4 的 SST 和 SOI 指数相关并不好,未能达到 95% 的信度检验,只有与上述西太平洋暖池区域次表层 120 m 深的 ST 有很好相关,它们相关超过 95% 的信度检验。

3.2 西北太平洋 TC 的移动路径与西太平洋暖池热力状态的关系

图 2 是本研究所定义的西太平洋暖池处于暖年(1970,1975,1978,1998,1999 和 2000 年)以及处于冷年(1972,1982,1987,1991,1993 和 1997 年)

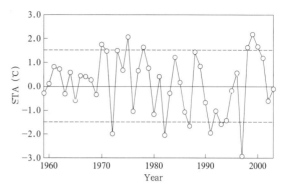

图 1 1959—2003 年西太平洋暖池(0°—16°N,125°—165°E)海域次表层 120 m 深 7—10 月平均海温距平的年际变化
(1971—2000 年各月海温的气候平均值取为正常值)

Fig. 1 Interannual variations of the subsurface sea temperature anomaly(relative to the climatological normals of July, August, September and October from 1971 to 2000;℃) at 120 m depth averaged over the period from July to October and the West Pacific warm pool area of 0°—16°N, 125°—165°E during 1959—2003

7—10 月西北太平洋 TC 移动路径的合成分布。从图 2a 可以看到,当西太平洋暖池处于暖年的夏季和初秋,西北太平洋上空的 TC 移动路径偏西、偏北,其中一大部分 TC 进入 140°E 以西海域,特别是进入中国东南沿海,并常在中国的台湾、福建、浙江等地登陆。然而,从图 2b 可以看到与图 2a 所示的有较大差别的西北太平洋 TC 移动路径,即当西太平洋暖池次表层海温处于冷年的夏季和初秋,西北太平洋上空的 TC 移动路径偏东,其中一大部分 TC 在 130°—140°E 附近就转向日本及日本以东地区移动,并且另有一部分台风向中国华南地区和越南方向移动,这会导致在中国广东省和越南北部登陆的 TC 较多,而在中国东南沿海省份登陆的 TC 较少。

为了更进一步分析西北太平洋 TC 移动路径与西太平洋暖池热力状态的关系,本研究分析了西太平洋暖池次表层热状态处于暖年与处于冷年西北太平洋各区域年平均 TC 活动个数之差(图 3)。从图 3 可以看到,西太平洋暖池的暖年与冷年西北太平洋上空各格点上年平均 TC 活动个数之差的正值位于菲律宾附近、中国东部、南部的沿海地区以及东海和南海一带,而负值位于菲律宾以东的热带西太平洋、日本及以东地区。显然,当西太平洋暖池处于暖

(图略)

图 2 西太平洋暖池处于暖年(a)和冷年(b)7—10月西北太平洋上空的TC移动路径的合成分布

Fig. 2 Composite distributions of the moving track of tropical cyclones (TCs) from July to October over the Northwest Pacific for the warm (a) and cold (b) state years of the West Pacific warm pool

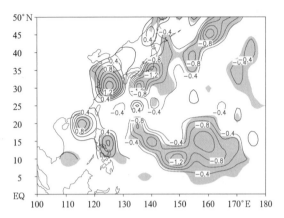

图 3 西太平洋暖池为暖年与冷年的西北太平洋上空各网格点上年平均TC活动个数之差

(阴影部分表示显著性检验超过95%区域)

Fig. 3 Distribution of the difference of the annual mean number of TC activities from July to October at various grid-points over the Northwest Pacific between the warm and cold state years of the West Pacific warm pool

(The areas over the 95% significance level are shaded)

年时,正的差值位于西北太平洋的偏西位置,而负值位于偏东位置。这表明:当西太平洋暖池处于暖年时,则西北太平洋TC的活动路径偏西,而西太平洋暖池处于冷年时,则西北太平洋TC的移动路径偏东。因此,在西太平洋暖池偏暖状态时,影响菲律宾、中国东南沿海和东部的TC可能偏多;而当西太平洋暖池处于偏冷状态下,影响中国东南沿海和东部的TC可能偏少,而影响菲律宾以东的热带西太平洋以及日本南部诸岛和日本东部的TC偏多。

4 热带西太平洋TC生成源地对西北太平洋TC移动路径的影响

陈光华和黄荣辉[24-25]的研究表明西北太平洋TC的生成源地依赖于西太平洋暖池的热力状态。当西太平洋暖池处于暖年(图4a),由于西太平洋副热带高压偏北,季风槽位于热带西太平洋偏西、偏北侧,这使得TC容易在热带西太平洋的偏西侧生成;相反,当西太平洋暖池处于冷年(图4b),由于西太平洋副热带高压偏南,季风槽位于热带西太平洋偏东、偏南侧,这使得TC容易在热带西太平洋的偏东、偏南侧生成。

为了研究热带西太平洋TC生成源地对它们在西北太平洋上空的移动路径的影响(图4a和4b),本研究把热带西太平洋以15°N,150°E为界依西北、西南、东南和东北象限分别划分成1,2,3和4区。从图4a可以看到:当西太平洋暖池处于暖年,则TC在第1,2区生成多,而在西太平洋暖池处于冷年,则TC在第2,3区生成多;并且从图4a和4b还可以看到,无论西太平洋暖池处于暖年或冷年,TC在第4区的生成个数少。为此,在下面只讨论TC在1,3和2区生成对它们在西北太平洋上空移动路径的影响。

从热带西太平洋1区生成的TC在西北太平洋上空的移动路径分布(图5a)可以明显看到,1区生成的TC较多在西北太平洋上空往西北方向移动,其中较多的TC进入130°E以西的西北太平洋,这导致了较多的TC登陆或影响中国。因此,在1区生

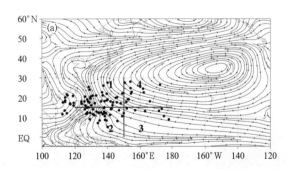

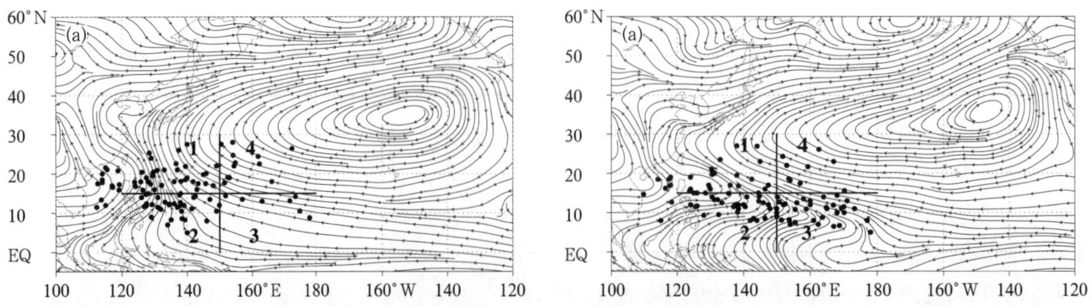

图 4 西太平洋暖池处于暖(a)和冷年(b)时,西北太平洋上空7—10月850 hPa流场和TC生成源地的合成分布

Fig. 4 Composite distributions of 850 hPa stream field and origins (thick black dots) of tropical cyclones in July—October for the warm (a) and cold state (b) years of the West Pacific warm pool

(图略)

图 5 在热带太平洋 1(a)，2(b) 和 3(c) 区域上空生成的 TC 在西北太平洋上空的移动路径分布

Fig. 5 Moving tracks of TCs formed in 1(a), 2(b), and 3(c) regions (see Fig. 4a) of the tropical western Pacific

成的 TC 在西北太平洋移动路径偏西北方向，这种情况下登陆中国的台风个数较多。图 5c 表示在西太平洋暖池处于冷年下 3 区生成的 TC 在西北太平洋上空的移动路径。从图 5c 也可以明显看到，3 区生成的 TC 大部分在西北太平洋上空往西北移动到 140°E 附近就转向东北移动，很少 TC 能移入 130°E 以西海域的上空。因此，在这种情况下，登陆中国的台风个数就偏少，而影响日本及以东地区的台风增多。此外，从 2 区生成的 TC 在西北太平洋上空的移动路径(图 5b)可以看到，在 2 区生成的 TC 在西北太平洋上空大部分往偏西方向移动，经菲律宾进入中国南海、华南地区和越南北部，但也有一部分在西北太平洋上空往西北方向移动，并在 130°E 附近转向东北方向移动。由于在西太平洋暖池处于暖年或冷年下，有一部分 TC 都可以在 2 区生成，因此，单纯由西太平洋暖池热力状态来预报 TC 在西北太平洋上空的移动路径，对于在 2 区生成的 TC 就会产生一些偏差。然而，不管西太平洋处于暖或冷状态，2 区生成的 TC 大部分在西北太平洋往西方向移动，并经菲律宾进入中国南海和华南地区以及越南北部，这对于预报登陆中国华南地区的 TC 路径是有一定帮助的。

5 2004 年与 2006 年西北太平洋 TC 的移动路径的差别

图 6a 和 6b 分别是由日本气象厅"启风丸"(Keifu Muru)海洋调查船所测得 2004 年 7 月和 2006 年 7 月沿 137°E 西太平洋海温距平的纬度-深度剖面[32]。从图 6a 可以看到，在 0°—12°N 的热带西太平洋海洋次表层的海温距平为较大的负值，在 5°N、100—150 m 深处的海温约比气候平均值偏低了 7 ℃，因此，可以说 2004 年夏季西太平洋暖池处于冷年。而从图 6b 所示的 2006 年 7 月热带西太平洋的海温距平分布可以看到，2006 年 7 月西太平洋暖池呈现出与图 6a 相反的海温距平分布，在 5°—14°N 纬度带次表层的海温从负变成正，此纬度 100—150 m 深的海温距平为 2.0 ℃以上，这说明在 2006 年 7 月 5°—14°N 的热带西太平洋被暖水所控制。因此，可以说 2006 年夏季西太平洋暖池处于暖年。

图 7a 和 7b 分别是 2004 和 2006 年的 5—11 月西北太平洋 TC 的移动路径分布。从图 7a 可以看到：2004 年西北太平洋上空 TC 的移动路径偏东，大部分 TC 首先往西北方向移动，到了 130°E 附近

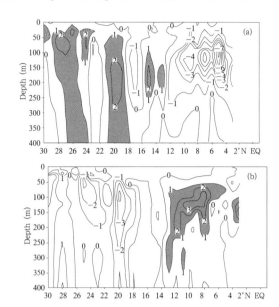

图 6 日本气象厅"启风丸"(Keifu Maru)号海洋调查船测得 2004 年 7 月 (a)和 2006 年 7 月(b)沿 137°E 西太平洋海温距平的纬度-深度剖面[32]
(单位：℃，阴影区表示高于 1.0 ℃海温距平)

Fig. 6 Latitude-depth cross sections of the West Pacific subsurface sea temperature anomaly along 137°E in (a) July 2004 and (b)July 2006, observed by the Oceanographic Research Vessel Keifu Maru of JMA[32]
(The ST anomalies over 1.0 ℃ are shaded；℃)

转向东北移动，而进入 130°E 以西洋面的台风较少。因此，在 2004 年影响日本的台风异常之多，多达 10 个，这给日本造成严重的经济损失和不少的人员伤亡，而登陆中国的台风就较少，这使得 2004 年中国由台风造成的经济损失和人员伤亡也较少。图 7b 显示出与图 7a 差别较大的 2006 年西北太平洋 TC 的移动路径。从图 7b 可以看到：2006 年西北太平洋上空 TC 的移动路径偏西，大部分 TC 往西或西北方向移动，并进入 130°E 以西的洋面，而较少台风在 130°E 以东洋面转向东北移动。因此，在 2006 年影响中国的台风不仅多，而且强，有 8 个台风登陆中国，其中有 6 个是强台风，另外还有 5 个未登陆但对中国有影响的台风和登陆中国但已变成热带低压[26]。这些给中国造成严重的经济损失和重大人员伤亡，特别是"碧利斯"和"桑美"超强台风给中国带来数百亿的巨大经济灾害，近 1000 人死亡和失踪。但在 2006 年，日本因台风造成的损失却不大。

(图略)

图 7 2004 年(a)和 2006 年(b)的 5—11 月西北太平洋 TC 的移动路径分布

Fig. 7 Tracks of TCs over the Northwest Pacific in the May—November of (a) 2004 and (b) 2006

此外，从图 7a 所示的 TC 生成地点也可以看到，2004 年大部分 TC 在 8°—12°N、140°—165°E 区域生成；而 2006 年（图 7b）大部分 TC 在 10°—20°N、130°—150°E 区域生成。因此，2004 年由于西太平洋暖池处于冷年，使得热带西太平洋 TC 生成源地偏东，偏南；相反，2006 年由于西太平洋暖池处于暖年，使得热带西太平洋 TC 生成源地偏西、偏北。可见 2004 与 2006 年西太平洋暖池热状态的差别以及这两年西北太平洋 TC 移动路径的差别可以证实在第 3、4 节分析所得的结论基本上是正确的。

6 热带太平洋上空季风槽对 Rossby 重力混合波向 TD 型波动转变的作用

在西北太平洋暖池处于暖年，热带西太平洋季风槽位置偏西北，故 TC 生成的地点也偏西北；相反，在西北太平洋暖池处于冷年，热带西太平洋季风槽位置偏东南，故 TC 生成的地点也偏东南。为什么热带西太平洋季风槽的位置对于 TC 生成地点有如此大的影响？这一方面是由于季风槽不仅提供了风场的纬向和经向切变，为低层气旋性环流的发展提供有利的动力条件，而且季风槽强烈的辐合所形成的强烈上升运动和对流活动为 TC 发展提供了有利的动力和热力条件；另一方面，季风槽南边纬向气流的强烈辐合也为热带 Rossby 重力混合波转变为 TD 型波动提供了动力条件。Takayabu 和 Nitta[33]指出发源于中东太平洋的 Rossby 重力混合波沿赤道西传，在进入热带西太平洋区域后，将逐渐脱离赤道向西北方向传播，并逐渐转变为热带低压型扰动，被称为 TD 型扰动，而 TD 型扰动可以为 TC 生成提供初始涡旋。但是，季风槽区的环流在两类波动的转化过程中所起的作用还不是很清楚，因此有必要从动力学上进一步分析热带西太平洋季风槽对这两类大气高频波动转换的动力作用，从而进一步来揭示西太平洋暖池不同热状态对 TC 生成的影响。

6.1 季风槽槽区南侧纬向风辐合对 Rossby 重力混合波向 TD 型波动转变的动力作用

根据 Lighthill 的推导[34]，对于一维流体波动的位相函数可写成

$$\theta = kx - \omega_a t \quad (1)$$

其中 k 为波数，ω_a 为波动的绝对频率，它为 k 和 x 的函数，即

$$\omega_a = \omega[k(x), x]$$

由式(1)可知

$$k = \frac{\theta}{x}$$

$$\omega_a = -\frac{\theta}{t} \quad (2)$$

这样，从式(2)便可得到

$$\frac{k}{t} = -\frac{\omega_a}{x} \quad (3)$$

若一个固有频率为 ω_r 的波动在基本流 $\overline{U}(x)$ 上传播，则此波动的绝对频率 ω_a 可写成为

$$\omega_a = \overline{U}(x)k + \omega_r \quad (4)$$

并且，根据 Matsuno[35]于 1966 年的推导，在热带地区波数为 k 的 Rossby 重力混合波的固有频率 ω_r 是

$$\omega_r = \frac{1}{2}kC_0 - \sqrt{(\frac{1}{2}kC_0)^2 + \beta C_0} \quad (5)$$

这样，可得

$$\omega_a = \overline{U}(x)k + \frac{1}{2}kC_0 - \sqrt{(\frac{1}{2}kC_0)^2 + \beta C_0}$$

上式 C_0 为重力波的相速度，即 $C_0 = \sqrt{gH}$，β 是热带地区的科氏参数随纬度的变化。这样，由式(3)和(4)便可得

$$\frac{k}{t} = -\frac{}{x}\omega_a(x,k) =$$
$$-k\frac{\overline{U}(x)}{x} - \overline{U}(x)\frac{k}{x} - \frac{\omega_r}{x} \quad (6)$$

若定义 $\frac{d_g k}{dt}$ 为波数 k 的波动沿它的群速度 C_g 的随体微商

$$\frac{d_g k}{dt} = \frac{k}{t} + C_g \cdot \frac{k}{x} \quad (7)$$

由式(4)可得，$C_g = \frac{\omega_a}{k} = \overline{U}(x) + \frac{\omega_r}{k}$

若把上式和式(6)代入式(7)，则可得

$$\frac{d_g k}{dt} = -k\frac{\overline{U}(x)}{\partial x} - \overline{U}(x)\frac{k}{x} - \frac{\omega_r}{x} +$$
$$\left(\overline{U}(x) + \frac{\omega_r}{k}\right) \cdot \frac{k}{x} =$$
$$-k\frac{\overline{U}(x)}{x} - \overline{U}(x)\frac{k}{x} - \frac{\omega_r}{x} +$$
$$\overline{U}(x) \cdot \frac{k}{x} + \frac{\omega_r}{k} \cdot \frac{k}{x} =$$
$$-k\frac{\overline{U}(x)}{x} \quad (8)$$

Webster 和 Zhang[36]对于热带 Rossby 波也得到同样的方程。可见式(8)不仅适合于热带 Rossby 波,也适合于热带 Rossby 重力混合波。由式(8)可知,若在热带西太平洋上空对流层低层纬向风强烈辐合,则波动的波数将增大,波长将缩短。一般,在季风槽区的槽部南侧,它的东边是东风,而西边是西风,这样就有强的 $\frac{\mathrm{d}\overline{U}(x)}{\mathrm{d}x} < 0$,即纬向风有强烈的辐合,这使得 Rossby 重力混合波的波数 k 逐渐增大,波长缩短。这说明了波长较长的 Rossby 重力混合波在季风槽区南侧的强辐合区可以产生波性质的转变,演变成波长较短的 TD 型波动。此外,这两类波动的转换过程还伴随有位相波速减缓,波动周期增加的特征。正如文献[3]所述,谢义炳先生早在20世纪60年初就指出在热带辐合带上可以发展高频波动。

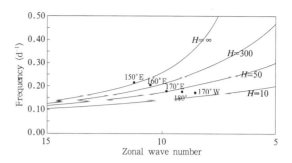

图 8 由式(5)所计算的理论上 Rossby 重力混合波的固有频率 ω_r(实线)与从 3—6 d 带通滤波的观测风场计算而得到的实际 Rossby 重力混合波的固有频率(黑点)

Fig. 8 Comparison between the variations of the eign-frequency (ω) of the mixed Rossby-gravity waves calculated from theoretical Formula (5) (solid lines) and the 3—6 day filtered observed wind field (black dots)

6.2 热带西太平洋上空 Rossby 重力混合波向 TD 型波动的转变

由于热带西传的高频波动的主要周期为 3—6 d。因此,本节对 850 hPa 风场进行 3—6 d 带通滤波,然后利用理论上得到的线性 MRG 频散关系与资料计算所得的结果进行对比,以此考察热带西传高频波动的波数和频率在赤道太平洋上空不同位置的变化特征。

首先利用无基流情况下线性的 MRG 波动频散关系式(5)(其中 $C_0 = \sqrt{gH}$,H 代表大气相当厚度),绘制在不同 H 情况下波数和频率的分布。如果考虑基本纬向气流的情况,则 $\omega_r = \omega_a - \overline{u}(x)k$,其中 ω_r 为固有频率,ω_a 为多普勒频率。选取赤道上不同经度位置的经向风作为参考点,画出点相关滞后分布图,可以得到赤道不同经度上的多普勒频率 ω_a 和波数 k,同时根据基本纬向流的大小,通过以上公式可以计算出固有频率 ω_r。图 8 中点出赤道上150°E、160°E、170°E、180°、170°W 所计算的频率和波数的位置。从图中可以看出,存在于中太平洋地区(170°E、180°、170°W 作为代表)上空的波动基本满足线性 MRG 波所要求的波数频率条件。随着波动向西移动,波数和频率的分布将趋近于大 H 值所对应的线性 MRG 波数和频率。在 150°E 位置时,所对应的 $H \to \infty$,这说明了当赤道波动由中太平洋地区传播到 WNP 后,波动的性质已经由原来的线性 MRG 波向另一类波动类型转变。

7 西太平洋暖池不同热状态下热带太平洋上空大气高频波动之间的转变

由于季风槽槽部南侧对流层低层纬向风的强辐合,使得西传 Rossby 重力混合波能转换成 TD 型波动,并偏离热带地区向西北方向移动。然而,由于西太平洋暖池的热状态影响季风槽的位置,从而影响着纬向风辐合区的位置,这将导致热带 Rossby 重力混合波向 TD 型波动转变区域的不同。在西太平洋暖池处于暖年时,季风槽位于偏西、偏北,故流场的辐合区位于热带西太平洋的偏西侧上空;而在西太平洋暖池处于冷年时,季风槽位于热带西太平洋的偏南、偏东侧上空,故流场的辐合区位于热带西太平洋偏东侧上空。因此,很有必要分析在西太平洋暖池不同热状态下热带太平洋上空大气高频波动之间的转变,从而揭示西太平洋暖池热状态影响西北太平洋 TC 活动变化的机理。

由于 OLR 从 1980 年才有较好的资料,故在以下分析中对 1980—2001 年 7—10 月的 850 hPa 风场资料进行带通滤波(只保留 3—6 d 周期的波动),以此来分析暖池处于暖年和冷年时,热带西太平洋 Rossby 重力混合波转换成 TD 型波动的区域差异。

7.1 西太平洋暖池处于暖年

图 9a 是西太平洋暖池处于暖年下,以(0°N、160°E)经向风为参考点线性回归的热带太平洋上空 850 hPa 高频变化风场和 OLR 距平场的分布。从

图9a可以看到:超前1天回归的850 hPa扰动风场呈现出一个以(0°N、160°E)为中心的气旋式扰动环流,并且,由负的OLR距平区所代表的强对流区并未出现在气旋扰动环流中心,而位于在气旋式扰动风场中心的西南侧,而弱对流区处于中心的西北侧;并且,同时回归的850 hPa扰动风场呈现出原先气旋式扰动环流已向西移动,参考点的扰动风场出现出一致的南风气流,而它的东侧为反气旋式环流所控制,在气旋和反气旋环流之间的南风气流北侧为对流活动活跃区,而南侧为对流活动衰减区;落后1天回归的850 hPa风场和OLR距平场呈现出与超前1天回归场正相反分布特征。此外,从回归的扰动风场分布还可以看到,这种线性回归得到的扰动风场分布正是相对于赤道成反对称分布。因此,这种扰动风场和对流活动的配置正是与热带Rossby重力混合波相关的扰动风场与对流活动的配置相吻合,这也与Tukayabu和Nitta[33]的分析结果一致。此外,这种波动只限于对流层低层(图略)。

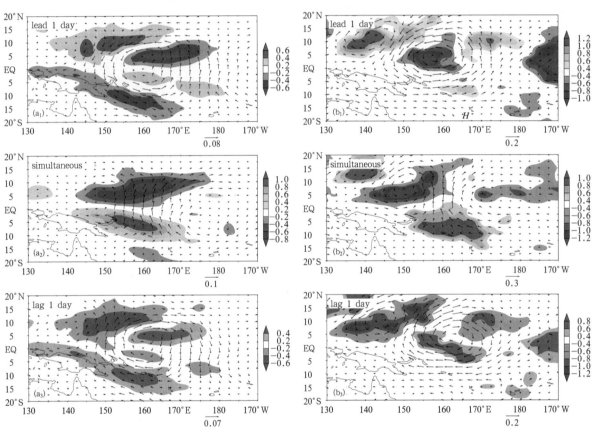

图9 西太平洋暖池处于暖年(a)和处于冷年(b)下以(0°N、160°E)经向风为参考点超前1 d(a_1、b_1),同时(a_2、b_2)和落后1 d(a_3、b_3)线性回归的热带太平洋上空7—10月850 hPa扰动风场和OLR距平场的合成分布

(蓝色阴影区表示强对流区)

Fig. 9 Composite distributions of 850 hPa linearly regressed perturbation wind field and OLR anomalies over the period from July to October for the warm (a) and cold (b) state years of the Western Pacific warm pool under the case of taking the meridional component of wind at point (0°N,160°E) as the reference point

(a_1,b_1. lead 1 day, a_2,b_2. simultaneous, a_3,b_3. lag 1 day; the areas of strong convective activities are shaded with blue color)

从以上分析可以看到,在热带西太平洋暖池处于暖年下,热带西太平洋偏东侧还存在明显热带Rossby重力混合波。这也说明,在赤道160°E的经度上,Rossby重力混合波还未转变为TD型扰动,而发生传播方向的偏离。

7.2 西太平洋暖池处于冷年

图9b是西太平洋暖池处于冷年下,以(0°N、160°E)经向风为参考点线性回归的热带西太平洋上空850 hPa 7—10月高频风场和OLR距平场分布。从图9b可以看到,在西太平洋暖池处于冷年时,热

带西太平洋上空的波动特征完全与图 9a 所示的热带太平洋上空的波动特征不同,无论是超前 1 天,或是同时和落后 1 天的 850 hPa 回归扰动风场都呈现出西南—东北倾斜的波动结构特征,并且负的 OLR 距平区所表示的强对流活动中心位于此气旋扰动风场的中心,这与西太平洋暖池处于暖年热带中太平洋 Rossby 重力混合波的强对流偏离气旋扰动环流中心截然不同;并且从图 9b 可以看到 这种波动往热带西太平洋的西北方向传播,这种波动的特征呈现出波长较短的热带 Rossby 波特征。这种波动正是 Takayabu 和 Nitta[33]于 1993 年提出的 TD 型波动,它在对流层高低层耦合比较紧密(图略)。

上述分析结果清楚表明:当西太平洋暖池处于暖年时,由于热带西太平洋季风槽偏西,季风槽槽部南侧纬向风的强辐合区也偏西,这使得热带西太平洋上空对流层低层的 Rossby 重力混合波转变成 TD 型波动的位置也偏西,因此。在这种情况下,在 0°N、160°E 附近的高频波动还表现为 Rossby 重力混合波的特征;而当西太平洋暖池处于冷年时,由于季风槽偏东,季风槽南侧的强辐合区也偏东,这使得热带太平洋对流层低层的 Rossby 重力混合波转变成 TD 型波动也偏东,因此,在这种情况下,在 0°N、160°E 附近的高频波动已表现为 TD 型波动了。

8 结论和讨论

本文利用美国 JTWC 的 TC 资料,NCEP/NCAR 的风场再分析资料以及 Scripps 海洋研究所的海温资料和 JMA 的 137°E 海温剖面观测资料系统地分析了西北太平洋上空 TC 移动路径的年际变化及其机理。分析结果表明了西北太平洋上空 TC 移动路径有明显的年际变化,并与西太平洋暖池热状态(特别是次表层海温)有很大关系。当西太平洋暖池处于暖年,西北太平洋上空的 TC 移动路径偏西,影响中国的台风个数偏多(图 10a);相反,当西太平洋暖池处于冷年,西北太平洋的 TC 移动路径偏东,影响日本的台风个数偏多,而影响中国的台风个数可能偏少(图 10b)。本研究以西太平洋暖池处于冷年的 2004 年与处于暖年的 2006 年西北太平洋 TC 移动路径的差别进一步论证了这一分析结果。

并且,本研究还进一步从动力理论上分析和讨论了西太平洋暖池的热状态对季风槽及其因此而产生的热带太平洋上空对流层低层的 Rossby 重力混合波转变成 TD 型波动的过程,从而揭示西太平洋暖池的热状态对西北太平洋上空 TC 移动路径年际变化的影响机理。分析表明了西北太平洋 TC 移动路径直接与它们的生成位置有关,而它们的生成又与热带西太平洋季风槽的位置有关。当西太平洋暖池处于暖年(图 10a),季风槽偏西,季风槽南侧纬向风的强辐合区也偏西,因此造成了热带太平洋上空对流层低层 Rossby 重力混合波转变成 TD 型波动的位置也偏西,这就使得 TC 生成偏西,从而导致它们的移动路径也偏西;相反,当西太平洋暖池处于冷年(图 10b),季风槽偏东,季风槽南侧的纬向风的强辐合区也偏东,因此造成了热带西太平洋对流层低层 Rossby 重力混合波转变成 TD 型波动的位置也偏东,这就使得 TC 生成偏东,从而导致它们的移动路径也偏东。

以上结论是针对西太平洋暖池不同热状态而分析的结果。但是,由于无论在西太平洋暖池处于暖年或冷年,热带西太平洋上空的 TC 都有可能在 15°N 以南、150°E 以西的 2 区生成,对于此区域生成的 TC 移动路径大部分在热带西太平洋上空往西向中国南海、华南地区和越南北部移动。因此,此区域生成的台风不能单纯由西太平洋暖池热状态而决定,还是要进一步从季风槽南侧纬向风的强辐合所在地区和西太平洋副热带高压位置变化来分析。

(图略)

图 10 西太平洋暖池热状态、季风槽、副热带高压与西北太平洋 TC 移动路径的关系
(a. 暖池为暖年,b. 暖池为冷年)
Fig. 10 Schematic diagrams of the relationship among the thermal state of West Pacific warm pool, the monsoon trough, the western subtropical high and moving tracks of TCs over the Northwest Pacific for (a) the warm and (b) cold state years of West Pacific warm pool

参考文献

[1] 谢义炳. 会议贺词//国家自然科学基金委员会等编. 现代大气科学前沿与展望. 北京:科学出版社,1995:3
[2] 谢义炳,陈受钧. 东南亚基本气流与台风发生的一些事实统计与分析. 气象学报,1963,33:206-217
[3] 谢义炳,黄寅亮. 赤道辐合带上扰动不稳定性的简单理论分析. 气象学报,1964,34:198-210.
[4] 谢义炳,张镡等. 初论西风带和热带辐合带环流系统的相互作用. 大气科学,1977,2:132-137
[5] Gray W M. Global view of the origin of tropical disturbances and storms. Mon Wea Rew, 1968, 96: 669-700
[6] 陈联寿,丁一汇. 西太平洋台风概论. 北京:科学出版社,1979:491pp
[7] 丁一汇,范惠君,薛秋芳等. 热带辐合区多台风同时发展的初步研究. 大气科学,1977,2(1):89-98
[8] 陈联寿,徐祥德,解以杨等. 台风异常运动及其外区热力不稳定非对称结构的影响效应. 大气科学,1997,21(1):83-90
[9] 李英,陈联寿,张胜军. 登陆我国热带气旋的统计特征. 热带气象学报,2004,20(1):14-22
[10] 李英,陈联寿,王继志. 登陆我国热带气旋长久维持与迅速消亡的大尺度环流特征. 气象学报,2004,61(2):167-179
[11] 徐亚梅,伍荣生. 热带气旋碧丽斯(2000)发生的数值模拟:非对称的发展及转换. 大气科学,2005,29(1):79-89
[12] Chan J C L. Tropical cyclone activity in the northwest Pacific in relation to the stratospheric quasi-biennial oscillation. Mon Wea Rev, 1995, 123: 2567-2571
[13] Chan J C L. Interannual and interdecadal variations of tropical cyclone activity over the western North Pacific. Meteor Atmos Phys, 2005, 89: 143-152.
[14] Elsberry R L. Global perspectives on tropical cyclone. WMO, TD—No. 693, 1995, Ch. 4: 106-197.
[15] 王斌,Elsberry R L,王玉清等. 热带气旋运动的动力学研究进展. 大气科学,1998,22(4):535-547
[16] Liebmann B, Hendon H H, Glick J D. The relationship between tropical cyclones of the western Pacific and Indian Oceans and the Madden-Julian oscillation. J Meteor Soc Japan, 1994, 72: 401-411
[17] 祝从文,Nakazawa T,李建平. 大气季节内振荡对印度洋—西太平洋地区热带低压/气旋生成的影响. 气象学报,2004,62(1):42-50
[18] Harr P A, Elsberry R L. Tropical cyclone track characteristics as a function of large-scale circulation anomalies. Mon Wea Rev, 1991, 119: 1448-1468
[19] 王慧,丁一汇,何金海. 西北太平洋夏季风的变化对台风生成的影响. 气象学报,2006,64(2):345-356
[20] Webster P J, Holland G H, Curry J A, et al. Changes in tropical cyclone number, duration, and intensity in a warming environment. Science, 2005, 309: 1844-1846
[21] Emanuel K A. Increasing destructiveness of tropical cyclones over the past 30 years. Nature, 2005, 436: 686-688
[22] Chan J C L. Comment on "Changes in tropical cyclone number, duration, and intensity in a warming environment". Science, 2006, 311: 1713-1714
[23] Chan J C L. Tropical cyclone activity over the western North Pacific associated with El Nino and La Nina events. J Climate, 2000, 13: 2960-2972
[24] 陈光华,黄荣辉. 西北太平洋热带气旋和台风活动若干气候问题的研究. 地球科学进展,2006,21(6):610-616
[25] 陈光华,黄荣辉. 西北太平洋暖池热状态对热带气旋活动的影响. 热带气象学报,2006,22(6):527-532
[26] 吴迪生,白毅平,张红梅等. 赤道西太平洋暖池次表层水温变化对热带气旋的影响. 热带气象学报,2003,19(3):253-259
[27] Nitta Ts. Convective activities in the tropical western Pacific and their impact on the Northern Hemisphere summer circulation. J Meteor Soc Japan, 1987, 64: 373-390
[28] 黄荣辉,李维京. 夏季热带西太平洋上空的热源异常对东亚上空副热带高压的影响及其物理机制. 大气科学,1988(特刊):95-107
[29] Huang Ronghui, Sun Fengying. Impact of the tropical western Pacific on the East Asian summer monsoon. J Meteor Soc Japan, 1992, 70 (IB): 243-256
[30] Huang Ronghui, Huang Gang, Wei Zhigang. Climate variations of the summer monsoon over China // Chang C P. East Asian Monsoon. World Scientific Publishing CO. Pte Ltd, 2004:213-270
[31] Cornejo-Garrido A G, Stone P H. On the heat balance of the Walker circulation. J Atmos Sci, 1977, 34: 1155-1162
[32] Japan Meteorological Agency. Depth-latitude section of subsurface temperature along 137°E observed by the Research Vessel "Keifu Maru" in July 2004 and July 2006. Monthly Report on Climate System, July 2004 and July 2006, 2
[33] Takayabu Y Ts, Nitta Ts. 3—5 day-period disturbances coupled with convection over the tropical Pacific Ocean. J Meteor Soc Japan, 1993, 71: 221-246
[34] Lighthill J. Wave in Fluides. London: Cambridge Univ Press, 1978: 504pp
[35] Matsuno T. Quasi-geostrophic motion in the equatorial area. J Meteor Soc Japan, 1966,44: 25-42
[36] Webster P J, Chang H R. Equatorial energy accumulation and emanation regions: Impacts of a zonally varying basic state. J Atmos Sci, 1988, 45: 803-828

台风在我国登陆地点的年际变化及其与夏季东亚/太平洋型遥相关的关系*

黄荣辉 王磊

(中国科学院大气物理研究所季风系统研究中心，北京 100190)

摘 要 本文利用 1979~2007 年日本气象厅 JRA-25 风场和高度场再分析资料和美国 JTWC 热带气旋的观测资料分析了 7~9 月份西北太平洋台风和热带气旋 (TC) 在我国登陆地点的年际变化及其与北半球夏季大气环流异常的东亚/太平洋型 (即 EAP 型) 遥相关的关系，特别是分析了 7~9 月份在厦门以北登陆台风和 TC 数量的年际变化与夏季 (6~8 月) EAP 指数的相关。分析结果表明：当夏季 (6~8 月) EAP 指数为高指数时，则 7~9 月份在东亚和西北太平洋上空 500 hPa 高度场异常将出现"−，＋，−"EAP 型遥相关的波列分布，这时西太平洋副热带高压的位置偏北、偏东。在这种情况下，西北太平洋上较多的台风和 TC 的移动路径偏北，这引起了 7~9 月份在我国厦门以北沿海登陆的台风和 TC 数量偏多。反之，当夏季 (6~8 月) EAP 指数为低指数时，在东亚和西北太平洋上空 500 hPa 高度场异常为"＋，−，＋"的 EAP 型遥相关的波列分布，这时西太平洋副热带高压的位置偏南、偏西。在这种情况下，西北太平洋上较多的台风和 TC 移动路径偏南，这引起了 7~9 月份在我国厦门以北沿海登陆的台风和 TC 数量偏少，较多的台风和 TC 在厦门以南的华南沿海登陆。

Interannual Variation of the Landfalling Locations of Typhoons in China and Its Association with the Summer East Asia/Pacific Pattern Teleconnection

HUANG Ronghui and WANG Lei

Center for Monsoon System Research, Institute of Atmospheric Physics, Chinese Academy of Sciences, Beijing 100190

Abstract The interannual variation of the landfalling locations of the western North Pacific typhoons and tropical cyclones (TCs) in China from July to September and its association with the summer East Asia/Pacific pattern (i.e., EAP pattern) teleconnection of the atmospheric circulation anomalies over the Northern Hemisphere are analyzed by using the data of wind and geopotential height fields from JRA-25 analysis of Japan Meteorological Agency during 1979 - 2007 and the observational data of tropical cyclones from the US Joint Typhoon Warning Center (JTWC). Especially the correlation between interannual variation of the numbers of landfalling typhoons and TCs to the north of Xiamen and the summer (June - August) EAP index is analyzed. The results show that in summer (June - August)

---

* 本文原载于：大气科学，第 34 卷，第 5 期，853-864，2010 年 9 月出版。

with a high EAP index, the wave-train distribution of "−, +, −" EAP pattern teleconnection of geopotential high anomalies will appear over East Asia and the western North Pacific, and the position of the western Pacific subtropical high will shift northward and eastward at 500 hPa from July to September. In this case, most of the moving tracks of typhoons and TCs over the western North Pacific will be more northward, which can cause more landfalling typhoons and TCs to the north of Xiamen. On the contrary, in summer (June – August) with a low EAP index, the wave-train distribution of "+, −, +" EAP pattern teleconnection of geopotential height anomalies at 500 hPa will appear over East Asia and the western North Pacific, and the position of the western Pacific subtropical high will shift southward and eastward from July to September. In this case, most of the moving tracks of typhoons and TCs over the western North Pacific will be more southward. Therefore, this will cause less landfalling typhoons and TCs on the coast of Southeast China to the north of Xiamen from July to September, and more typhoons and TCs will make landfall on the coast of South China to the south of Xiamen.

1 引言

西北太平洋（WNP）是全球热带气旋（TC）和台风活动最为频繁的海域，每年全球 TC 和台风约三分之一在 WNP 海域上空生成。由于受西太平洋副热带高压的影响，在 WNP 海域上空生成的 TC 和台风一部分移向中国、日本、韩国、菲律宾和越南等地登陆，给这些国家造成巨大的经济损失和重大人员伤亡。我国是全世界受台风灾害最为严重的国家之一，平均每年约 7～8 个台风登陆我国。据近几年来的统计，我国每年台风灾害约造成 250 亿元以上的经济损失，死亡人数高达数百人，登陆台风多的年份，台风造成的经济损失和人员伤亡更加严重。如 2006 年 7 月 14 日"碧丽斯"台风在福建霞浦登陆后往西移动，给福建、江西和湖南南部、广东北部造成了 600 多人死亡，200 多人失踪；2006 年 8 月 10 日在浙江苍南登陆的"桑美"台风是 50 多年来登陆我国大陆最强的台风，中心风力达到 17 级，中心气压为 920 hPa，给福建沙埕港造成了 980 艘船沉没和 200 多人死亡；2009 年 8 月 7 日"莫拉克"台风在我国台湾花莲登陆之后又在福建霞浦再度登陆，在我国台湾南部造成 3000 mm 以上超记录的降水，不仅给福建北部、浙江南部造成了重大经济损失，而且给我国台湾南部造成近千亿元新台币巨大经济损失以及 461 死亡和 192 人失踪（见2009 年 8 月 30 日灾害防御简讯）。因此，关于台风生成、移动路径、强度和登陆地点的预报一直是我国气象界热门的研究课题之一。

为了更准确地进行台风生成、移动路径、强度和登陆地点的预报，近年来国际上许多科学家开展关于 TC 和台风生成的年际变化的研究。如 Chan (1985) 提出 WNP 海域 TC 和台风生成与 ENSO 循环有密切关系。而 Wang and Chan (2002) 提出热带西太平洋上空 TC 和台风生成个数在 ENSO 事件发生年似乎没有明显的变化，但 ENSO 事件对 TC 和台风移动路径有一定影响，在 ENSO 年 WNP 海域上空 TC 和台风向北转向偏多。陈光华和黄荣辉 (2006)、黄荣辉和陈光华 (2007)、Chen and Huang (2008) 的研究表明了热带西太平洋热力状态严重地影响着西北太平洋上 TC 和台风的生成和移动路径。他们的研究表明：热带西太平洋的热力（特别是海洋热容量）调制着热带西太平洋季风槽位置的变化，而季风槽通过对赤道波动的转换严重影响着西北太平洋上 TC 和台风活动以及登陆我国的地点。

西北太平洋台风和 TC 的移动路径及其在我国的登陆地点受着牵引气流的影响，特别是受西太平洋副热带高压位置的影响（陈联寿和丁一汇，1979；张庆云和彭京备，2003；王磊等，2009b）。而由于东亚/太平洋型遥相关之原因，西太平洋副热带高压严重地受着西太平洋暖池热力状态的影响（Nitta, 1987; Huang and Li, 1987；黄荣辉和李维京，1988; Huang and Sun, 1992）。正如 Huang and Sun (1992) 所指出：当夏季西太平洋暖池处于偏暖状态，菲律宾周围对流活动强，这种情况下，由于东亚/太平洋型（EAP 型）遥相关，西太平洋副热带高压位置偏北；相反，当夏季西太平洋暖池处于偏冷状态，菲律宾周围对流活动弱，这种情况下西太平洋副热带高压位置偏南。因此，由于受到西太平洋副热带高压的影响，西北太平洋台风的移动路径及其在我国的登陆地点可能与夏季 EAP 型遥相关存在着密切的关系。为此，本文利用1979～2007年日本气象厅JRA-25再分析资料和美国JTWC (Joint Typhoon Warning Center) 热带气旋的观测资料来分析西北太平洋TC和台风在我国登陆地点的年际变化及其机理，特别是探讨它与北半球夏季大气环流异常的东亚/太平洋型（EAP型）遥相关的关系。

2 西北太平洋台风和TC在我国登陆地点的年际变化及其与夏季EAP指数的关系

根据王磊等（2009a）的研究，7～9月份登陆我国的台风和TC数量约占登陆我国台风和TC全年总数的80%左右。因此，本研究主要研究7～9月份西北太平洋和TC在我国登陆地点的年际变化。

2.1 7～9月份台风和TC在我国登陆地点的年际变化

黄荣辉和陈光华（2007）的研究表明：西北太平洋上空TC和台风移动路径有明显的年际变化，它与西太平洋暖池的热力状态（特别是次表层海温）有密切的关系。当西太平洋暖池处于暖状态，西北太平洋的TC和台风的移动路径偏西、偏北，影响中国的台风个数偏多；相反，当西太平洋暖池处于冷状态，西北太平洋的TC和台风的移动路径偏东、偏南，并且易于130°E附近向东北转向，从而导致影响日本的TC和台风个数增多，而影响中国的TC和台风个数偏少。

由于受西太平洋暖池热力状态的影响，西北太平洋的台风和TC移动路径有很大年际变化，西北太平洋上空TC和台风的移动路径势必影响它们在我国登陆地点的位置，故它们在我国登陆地点也会有很大的年际变化。为此，本节利用美国JTWC的TC和台风移动路径和登陆地点资料来分析西北太平洋上空TC和台风在我国登陆地点的年际变化。图1是1979～2007年7～9月份登陆我国台风和TC数量的年际变化（实线）。从图1可以看到，7～9月份登陆我国台风和TC数有很大的年际变化，在一般年份，7～9月份有5个台风和TC登陆我国，最多于1994年登陆我国的台风和TC数量达到10个，而最少于1986年登陆我国的台风和TC数量只有2个。

如图2所示，由于厦门的位置在7～9月份处于西太平洋副热带高压和热带西太平洋季风槽西侧之间，因此，它是西北太平洋台风和热带气旋在我国登陆地点偏北和偏南的"分水岭"。为了更好分析西北太平洋上空台风和TC在我国登陆地点的年际变化，故我们把在我国登陆的台风和TC划分为在厦门以北和厦门以南登陆两种。图3a是1979～2007年7～9月份在我国厦门以北登陆的台风和TC数量的年际变化。从图3a可以看到，7～9月份在我国厦门以北登陆的台风和TC的个数有很明显的年际变化，最多于1994年在厦门以北登陆的台风和TC达到5个，而最少于1999年没有台风和TC在厦门以北登陆。并且，从图3a还可以看到，从20世纪80年代至2007年7～9月份在我国厦门以北登陆的台风和TC数量有增多的趋势。图3b是1979～2007年7～9月份在我国厦门以南登陆的

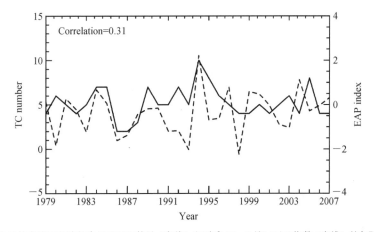

图1 1979～2007年7～9月份在我国登陆的台风和TC数量（实线）和夏季（6～8月）EAP指数（虚线）的年际变化

Fig. 1 Interannual variations of the numbers of the landfalling typhoons and tropical cyclones (TCs) in China (solid line) from Jul to Sep and the summer (Jun-Aug) EAP (East Asia/Pacific) index (dashed line) during 1979-2007

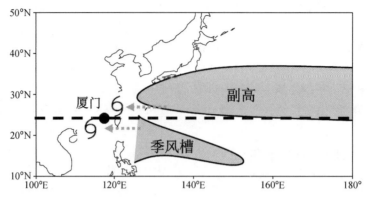

图 2 厦门的地理位置与西太平洋副热带高压和季风槽位置示意图
Fig. 2 Schematic diagram of the geographic position of Xiamen and the positions of the western Pacific subtropical high and the monsoon trough

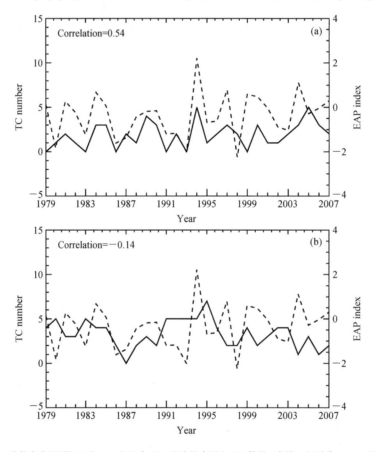

图 3 1979～2007 年 7～9 月份在我国厦门以北（a）和以南（b）登陆的台风和 TC 数量（实线）和夏季（6～8 月）EAP 指数（虚线）的年际变化
Fig. 3 Interannual variations of the numbers of the landfalling typhoons and TCs (solid line) (a) to the north of Xiamen and (b) to the south of Xiamen from Jul to Sep and the summer (Jun–Aug) EAP index (dashed line) during 1979–2007

台风和 TC 数量的年际变化。从图 3b 可以看到，7～9 月份在我国厦门以南登陆的台风和 TC 也有明显的年际变化，最多于 1995 年，在厦门以南登陆的台风和 TC 数量达到 7 个，而最少于 1987 年没有台风和 TC 在厦门以南登陆。并且，从图 3b 还可以看到，从 20 世纪 80 年代 7～9 月份在厦门以南登陆的台风和 TC 数量有减少的趋势，如 1998、2004、2006 年在厦门以南登陆的台风和 TC 数量只有 1 个。这个结果与 Wu and Wang（2004）以及 Wu et al.（2005）的研究结果是一致的，他们指出，从 1965 年以来西北太平洋路径总体向西偏移，影响南海的台风明显减少。

2.2 7～9 月份台风和 TC 在我国登陆地点的年际变化与夏季 EAP 指数的关系

Nitta（1987）从观测事实指出了夏季菲律宾周围与日本周围大气环流异常存在着相反的振荡，即P-J 振荡。与此同时，Huang and Li（1987），黄荣辉和李维京（1988）利用北半球夏季准定常行星波的传播规律，并从观测事实、动力理论和数值模拟研究了北半球夏季大气环流异常的遥相关及 Rossby 波列在北半球大气的传播特征，从而提出东亚/太平洋型遥相关，即 EAP 型遥相关。随后，Huang（2004）利用上述东亚/太平洋型遥相关（或 EAP 型遥相关），并利用东亚夏季风系统异常的空间分布特征定义了一个能够很好地表征东亚夏季风系统变化的指数，即 EAP 指数（I_{EAP}）。这个指数定义如下：

$$I_{EAP} = -0.25Z'_s(60°N, 125°E) +$$
$$0.50Z'_s(40°N, 125°E) - 0.25Z'_s(20°N, 125°E), \quad (1)$$

式中，$Z'_s = Z'\sin45°/\sin\varphi$ 是某一格点夏季（6～8 月）平均的 500 hPa 标准化高度距平，Z' 是该格点夏季平均的 500 hPa 高度距平，φ 是该格点所在的纬度。

利用 NCEP/NCAR 再分析资料，并应用 Huang（2004）所提出 EAP 指数的定义计算了 1979～2007 年每年夏季（6～8 月）EAP 指数（见图 1～3 的虚线）。从图 3a 中可以看到，7～9 月份在厦门以北登陆台风和 TC 的数量（实线）与夏季（6～8 月）EAP 指数有很好的正相关，它们之间相关达到 0.54，超过 99% 的信度。这表明：当夏季（6～8 月）EAP 指数为高指数时（即西太平洋副热带高压偏北），则登陆我国厦门以北台风和 TC 的数量偏多；相反，当夏季（6～8 月）EAP 指数为低指数时（即西太平洋副热带高压偏南），则登陆我国厦门以北台风和 TC 的数量偏少。然而，从图 3b 可以看到，每年 7～9 月份在厦门以南登陆台风和 TC 的数量与夏季（6～8 月）EAP 指数却没有什么相关，它们相关系数只有 -0.14。这表明 7～9 月份登陆厦门以南的台风和 TC 多少并不依赖于西太平洋副热带高压的位置，它可能受热带西太平洋季风槽及其它环流系统的影响。由于 7～9 月份在我国厦门以南登陆台风和 TC 的数量与夏季 EAP 指数并没有什么相关，故 7～9 月份在我国东南沿海登陆台风和 TC 的数量与夏季 EAP 指数之间的相关系数如图 2 所示只有 0.31，只达到 90% 的信度。这表明 7～9 月份在我国东南沿海地区登陆的台风和 TC 的数量与夏季 EAP 指数也存在一定的关系。

3 7～9 月份西北太平洋台风和 TC 移动路径及其在我国登陆地点的年际变化与西太平洋副热带高压位置的关系

台风和 TC 的移动路径受环境气流的引导，特别是受对流层中层气流的引导（Hope and Neumann，1970；Kutzbach，1979；Chan and Gray，1982；McBride，1995）。研究表明，7～9 月份西北太平洋台风和 TC 的移动路径受到西太平洋副热带高压位置的严重影响。西北太平洋副热带高压的位置不仅影响西北太平洋台风和 TC 的形成（Chia and Ropelewiski，2002），而且影响西北太平洋台风和 TC 的移动路径（Ho et al.，2004）。张庆云和彭京备（2003）指出：在西太平洋副热带高压的脊线偏北时，则登陆我国台风的数量偏多；相反，若西太平洋副热带高压的脊线偏南时，则登陆我国台风的数量就偏少。王磊等（2009b）指出：当 7～9 月份西太平洋副热带高压偏西、偏北，则登陆我国厦门以北的台风和 TC 的数量偏多；相反，当西太平洋副热带高压偏东、偏南，则登陆我国厦门以北的台风和 TC 的数量偏少。

夏季西太平洋副热带高压的位置是受热带西太平洋热力状态的严重影响。黄荣辉等许多研究表明了热带西太平洋的热力状态不仅影响南海夏季风的爆发，而且影响着西太平洋副热带高压的位置（Huang and Sun，1992；黄荣辉等，2005；Huang et al.，2006，2007）。他们的研究表明：当热带西太平洋处于暖状态，菲律宾周围对流活动强，则夏季西太平洋副热带高压偏北，在这种情况，EAP 指数为高指数；相反，当热带西太平洋处于冷状态，菲律宾周围对流活动弱，则夏季西太平洋副热带高压偏南，在这种情况，夏季 EAP 指数为低指数。并且，黄荣辉和陈光华（2007）的研究表明：当热带西太平洋处于暖状态（次表层海温偏高），则西北太平洋台风和 TC 移动路径偏西，并导致在我国东南沿海登陆的台风和 TC 数量偏多；相反，当热带西太平洋处于冷状态（次表层海温偏低），则西北太平洋台风和 TC 移动路径偏东，易于 130°E 附近转向东北方向移动，并导致在日本登陆的台风和

TC 数量增多。因此，热带西太平洋热力状态影响着西太平洋副热带高压，而西太平洋副热带高压影响着西北太平洋台风和 TC 的移动路径。为此，本节在上一节研究基础上利用 JRA-25 再分析资料来分析 7～9 月份西北太平洋台风和 TC 的移动路径及其在我国的登陆地点与西太平洋副热带高压的关系。

图 4 是 7～9 月份西北太平洋台风和 TC 在我国厦门以北登陆数量偏多和偏少年份 7～9 月平均的东亚和西太平洋上空 500 hPa 环流合成图。从图 4a 和图 4b 可以看到，西北太平洋台风和 TC 在我国厦门以北沿海登陆数量多的年份与在我国厦门以北沿海登陆数量少的年份 7～9 月份东亚和西北太平洋上空的环流有明显的差别，其主要差别是西太平洋副热带高压的位置。在我国厦门以北沿海登陆的台风和 TC 数量偏多的年份，如图 4a 所示，7～9 月份平均的西太平洋副热带高压的位置偏北、偏东，它的脊

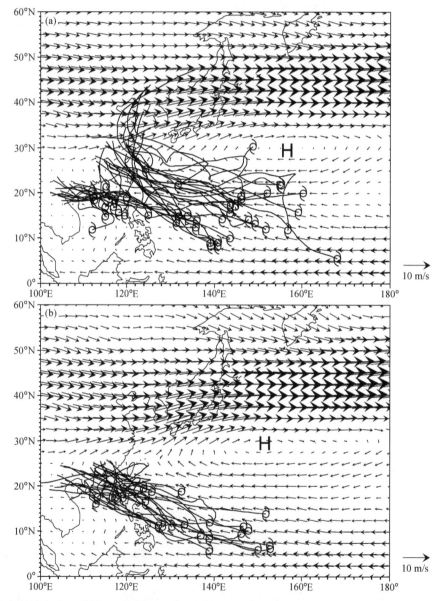

图 4 在我国厦门以北登陆的台风和 TC 数量（a）偏多和（b）偏少年份 7～9 月份平均的东亚和西北太平洋上空 500 hPa 风场合成图（风场资料取自 JRA-25 再分析资料）。细实线：台风和 TC 路径

Fig. 4 Composite distributions of 500-hPa wind fields over East Asia and the western North Pacific averaged for Jul–Sep for the years with (a) more and (b) less landfalling typhoons and TCs to the north of Xiamen from Jul to Sep. The thin solid lines indicate moving tracks of typhoons and TCs, and the data of wind fields are from JRA–25 reanalysis

线大约位于 30°N，西伸到 130°E 左右；相反，在我国厦门以北沿海登陆的台风和 TC 数量偏少的年份，如图 4b 所示，7～9 月份平均的西太平洋副热带高压的位置偏南、偏西，它的脊线大约位于 28°N，西伸到 120°E 左右。因此，如图 4a 所示，正是由于西太平洋副热带高压位置偏北、偏东，它引起了西北太平洋台风和 TC 移动路径偏北，这才导致在厦门以北登陆的台风和 TC 数量偏多；相反，如图 4b 所示，由于西太平洋副热带高压偏南、偏西，它引起了西北太平洋台风和 TC 的移动路径偏南，这才导致在厦门以北登陆的台风和 TC 数量偏少。

为了凸显西北太平洋上空台风和 TC 在我国的登陆地点与西太平洋副热带高压的关系，我们进一步分析 7～9 月份西北太平洋上空台风和 TC 在我国厦门以北登陆数量偏多和偏少年份 7～9 月平均的东亚和西太平洋上空 500 hPa 环流异常合成图（见图 5）。从图 5a 可以看到，在厦门以北登陆台风和 TC 数量偏多的年份，7～9 月份平均的东亚和西北太平洋上空 500 hPa 环流异常是：在我国东南沿海和菲律宾以东的热带西太平洋上空为气旋（C）

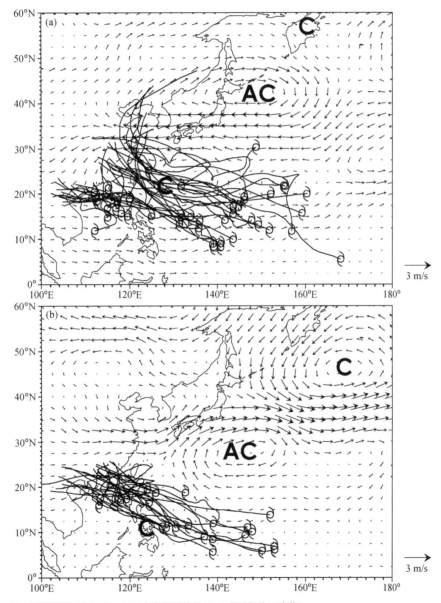

图 5　同图 4，但为 500 hPa 风场距平（取 1979～2007 年气候平均的 7～9 月风场为正常值）

Fig. 5　As in Fig. 4, except for wind field anomalies at 500 hPa. The climatological mean wind field for the period of Jul – Sep of 1979 – 2007 is taken as the normal

性距平环流，而在我国华北、东北和日本北部及以东上空的西北太平洋上空为反气旋（AC）性距平环流，这正是西太平洋副热带高压偏北的情况，并且在勘察加半岛以东的西北太平洋上空为气旋性距平环流。这是"一，+，一"的 EAP 型遥相关的波列分布（Nitta, 1987; Huang and Li, 1987; 黄荣辉和李维京，1988）。在这种情况下，从日本以南和我国江淮流域上空为明显的东风距平环流，因此，如图 5a 中细实线所示，西北太平洋台风和 TC 的移动路径偏北，从而导致台风和 TC 在厦门以北登陆的数量偏多。另一方面，从图 5b 可以看到，在厦门以北登陆台风和 TC 数量偏少的年份，7～9 月份平均的东亚和西北太平洋上空 500 hPa 环流异常正好与在厦门以北登陆台风和 TC 数量偏多的情形相反，即，在菲律宾以南的热带西太平洋上空为气旋性距平环流，而在我国台湾以东和日本南部的西北太平洋为反气旋性距平环流，这正是西太平洋副热量高压偏南的情形，并且在堪察加半岛以南和以东西北太平洋上空为气旋性距平环流。把图 5b 与图 5a 相比较，可以看到，虽然在图 5b 所示的东亚和西北太平洋上空 500 hPa 环流异常也呈"一，+，一"的 EAP 型遥相关的波列分布，但图 5b 所示的整个环流系统异常比图 5a 所示的环流系统异常的位置不仅偏南 15 个纬距，而且偏西了 5 个经度。因此，在这种情况下，如图 5b 所示，在菲律宾以东热带西太平洋上空为明显的偏东风距平环流，而我国华南和长江中、下游地区和东海上空为西风距平环流，这导致西北太平洋台风和 TC 的移动路径偏南，从而引起西北太平洋台风和 TC 在厦门以南的华南沿海登陆偏多，而在厦门以北的东南沿海地区登陆的台风和 TC 数量偏少。

上述分析结果表明，7～9 月份在厦门以北登陆的台风和 TC 的数量与西太平洋副热带高压的位置之间存在着密切关系。

4 7～9 月份西北太平洋台风和 TC 在我国登陆地点的年际变化、西太平洋副热带高压位置和夏季 EAP 遥相关型之间的关系

4.1 7～9 月份西太平洋副热带高压位置与夏季 EAP 型遥相关之间的关系

Huang（2004）的研究表明了夏季（6～8 月）EAP 指数的年际变化能够很好地反映西太平洋副热带高压位置的年际变化。当夏季 EAP 指数为正，则西太平洋副热带高压位置偏北；相反，当夏季 EAP 指数为负，则西太平洋副热带高压的位置偏南。7～9 月份西太平洋副热带高压位置与夏季 EAP 指数之间是否还存在一定关系，这是值得进一步研究的问题。为此，本节利用 JRA-25 再分析资料来分析 7～9 月份平均的西太平洋副热带高压的位置与夏季（6～8 月）EAP 指数之间的关系。

为了分析夏季（6～8 月）EAP 指数与 7～9 月份西太平洋副热带高压位置的关系，我们选取 6～8 月份平均的 EAP 指数等于或大于 0.5，即 $I_{EAP} \geqslant 0.5$，为高指数，而 6～8 月份平均的 EAP 指数小于或等于-0.5，即 $I_{EAP} \leqslant -0.5$，为低指数。这样，夏季高 EAP 指数年有 1981、1984、1985、1989、1994、1997、1999、2000、2004 年；而夏季低 EAP 指数年有 1980、1983、1986、1987、1988、1991、1993、1995、1996、1998、2002、2003 年。研究表明：由于西太平洋副热带高压位置的关系，在高 EAP 指数年时，江淮流域夏季风降水偏少，并往往出现干旱；而在低 EAP 指数年时，江淮流域夏季风降水偏多，并往往出现洪涝（Huang et al, 2007）。

图 6 是分别对应于夏季（6～8 月）EAP 指数为高指数（a）和低指数（b）7～9 月份东亚和西北太平洋上空 500 hPa 位势高度距平合成图。从 6a 可以看到，当夏季 EAP 指数为高指数时，7～9 月份平均的东亚和西北太平洋上空 500 hPa 高度场异常是：在我国东南沿海和菲律宾以东的热带西太平洋上空为负距平区域，而在我国华北、朝鲜半岛和日本以东西北太平洋上空为正距平区域，这正是西太平洋副热带高压偏北的情况，并且，在鄂霍茨克海和堪察加半岛上空为负距平区域。这个分布型与夏季的"一，+，一" EAP 型遥相关的波列分布相似。另一方面，从图 6b 可以看到，当夏季 EAP 指数为低指数时，7～9 月份平均的东亚和西北太平洋上空 500 hPa 高度距平分布正好与高 EAP 指数年时的高度场距平分布相反，即，在华南、南海和菲律宾以东的热带西太平洋上空为正距平区域，而在我国江淮流域、华北、朝鲜半岛、日本及以东的西北太平洋上空为负距平区域,这正是西太平洋副热带高压位置偏南的情况，并且，在鄂霍茨克海和堪察加半岛上空为正距平区域。这个分布型与夏季

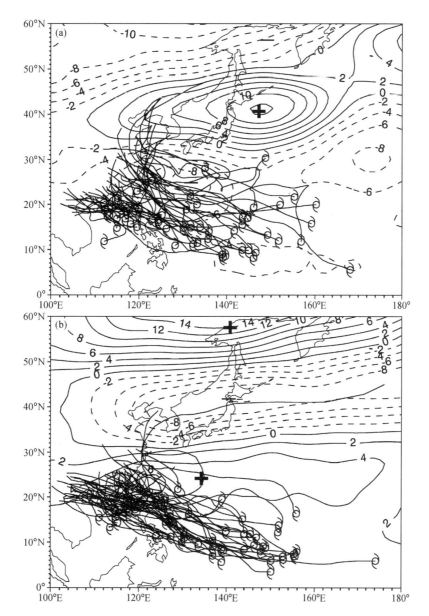

图 6 夏季（6~8月）EAP指数为（a）高指数年和（b）低指数年7~9月份东亚和西北太平洋上空500 hPa高度场距平合成分布图（单位：gpm）。取1979~2007年7~9月份高度场气候平均为正常值，高度场资料取自JRA-25

Fig. 6 Composite distributions of geopotential height anomalies (gpm) at 500 hPa over East Asia and the western North Pacific averaged for Jul–Sep for the years with (a) the high summertime (Jun–Aug) EAP index and (b) the low summertime EAP index. The climatological mean geopotential height field for the period of Jul–Sep of 1979–2007 is taken as the normal, and the data of geopotential height are from the JRA-25 reanalysis

"+，-，+" EAP型遥相关的波列分布相似。

从上述分析可以看到，夏季EAP指数能够很好反映7~9月份东亚和西北太平洋上空500 hPa高度场距平的分布，特别是能够很好反映7~9月份西太平洋副热带高压的南北位置。因此，夏季EAP指数的年际变化不仅能够反映夏季东亚/太平洋遥相关的年际变化，而且能够反映7~9月份西

太平洋副热带高压南北位置的年际变化。

4.2 7~9月份在我国厦门以北沿海登陆台风和TC的数量、西太平洋副热带高压与夏季EAP型遥相关的关系

上两节分析结果表明，不仅7~9月份在我国厦门以北登陆的台风和TC数量与夏季EAP指数有很好的关系，而且7~9月份西太平洋高压的位

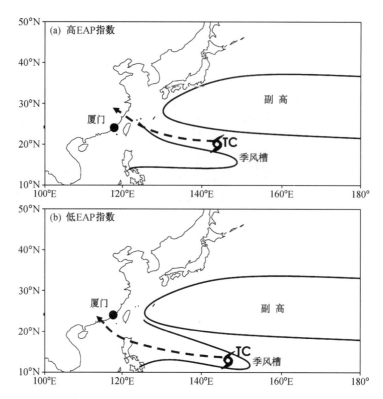

图 7 夏季（6～8月）EAP指数、7～9月份西太平洋副热带高压的位置、7～9月份在厦门以北登陆台风和TC数量之间的关系示意图

Fig. 7 Schematic diagrams of the relationship among the summer (Jun – Aug) EAP index, the position of the western Pacific subtropical high from Jul to Sep, and the numbers of landfalling typhoons and TCs to the north of Xiamen from Jul to Sep

置与夏季EAP指数也有很好的相关。因此，7～9月份西北太平洋的台风和TC在我国厦门以北登陆的数量、西太平洋副热带高压的位置与夏季EAP指数之间也可能有很好的关系。为此，本节在前面分析的基础上进一步分析上述三者之间的关系。

图6a和图6b中细实线分别是在夏季EAP指数为高指数年和低指数年7～9月份西北太平洋台风和TC的移动路径。从图6a细实线可以看到：当夏季（6～8月）EAP指数为高指数时，7～9月份在东亚和西北太平洋上空500 hPa高度场异常出现"–,+,–"的EAP型遥相关的波列分布，这时西太平洋副热带高压的位置偏北。在这种情况下，西北太平洋较多的台风和TC的移动路径偏北，从而这引起7～9月份在我国厦门以北沿海登陆的台风和TC数量偏多。相反，当夏季（6～8月）EAP指数为低指数时，7～9月份在东亚和西北太平洋上空500 hPa高度场异常出现"+,–,+"的EAP型遥相关的波列分布。在这种情况下，如图6b细实线所示，西北太平洋较多的台风和TC的移动路径偏南，这引起7～9月份在我国厦门以北沿

海登陆的台风和TC数量偏少，而较多的台风和TC在厦门以南的华南地区登陆。

上述分析结果表明了7～9月在我国厦门以北登陆的台风和TC数量以及西太平洋副热带高压的南北位置与夏季（6～8月）北半球大气环流异常的东亚/太平洋型遥相关之间有很好的关系。为了更形象地显示它们三者之间的关系，图7给出了夏季（6～8月份）EAP指数、7～9月份西太平洋副热带高压位置、7～9月份在厦门以北登陆的台风和TC数量之间的关系示意图。

5 结论与讨论

本文利用1979～2007年日本气象厅JPA-25风场和高度场再分析资料和美国JTWC热带气旋的观测资料分析了7～9月份西北太平洋台风和TC在我国登陆地点的年际变化及其与东亚和西北太平洋上空500 hPa环流异常的东亚/太平洋型（即EAP型）遥相关的关系，特别是分析7～9月份在厦门以北登陆台风和TC数量的年际变化与夏季（6～8月份）EAP指数的相关。分析结果表明：当

夏季（6~8月）EAP指数为高指数时，7~9月份在东亚和西北太平洋上空500 hPa高度场异常为"—，+，—"的EAP型遥相关的波列分布，这时西太平洋副热带高压的位置偏北，在这种情况下，西北太平洋上较多的台风和TC的移动路径偏北，这引起了7~9月份在我国厦门以北沿海登陆的台风和TC数量偏多。反之，当夏季（6~8月）EAP指数为低指数时，7~9月份在东亚和西北太平洋上空500 hPa高度场异常为"+，—，+"的EAP型遥相关的波列分布，这时西太平洋副热带高压的位置偏南。在这种情况下，西北太平洋上台风和TC的移动路径偏南，这引起了7~9月份在我国厦门以北沿海登陆的台风和TC数量偏少，较多的台风和TC在厦门以南的华南沿海登陆。上述结果可以作为西北太平洋台风登陆我国地点的季节预报的参考。

本文的分析结果表明了夏季EAP指数为高指数时，仍有部分台风和TC在厦门以南的华南沿海地区登陆，这造成了7~9月份在厦门以南登陆的台风和TC数量与夏季EAP指数相关并不好。这部分台风主要在第2区（即15°N以南，150°E以西）生成，这部分台风移动路径偏西，易在菲律宾、我国华南沿海和越南登陆。这表明7~9月份在厦门以南登陆的台风和TC数量并不完全依赖于西太平洋副热带高压的位置，它可能还受到热带西太平洋季风槽以及另外大尺度环流系统的影响，这是需进一步分析研究的问题。

参考文献 (References)

Chan J C L. 1985. Tropical cyclone activity in the northwest Pacific in relation to the El Niño/southern oscillation phenomenon [J]. Mon. Wea. Rev., 113: 599 – 606.

Chang J C L, Gray W M. 1982. Tropical cyclone movement and surrounding flow relationships [J]. Mon. Wea. Rev., 110 (10): 1354 – 1374.

陈光华，黄荣辉. 2006. 西北太平洋暖池热状态对热带气旋活动的影响 [J]. 热带气象学报, 22: 527 – 532. Chen Guanghua, Huang Ronghui. 2006. The effect of warm pool thermal states on tropical cyclones in western North Pacific [J]. J. Tropical Meteor. (in Chinese), 22: 527 – 532.

Chen Guanghua, Huang Ronghui. 2008. Influence of monsoon over the warm pool on interannual variation of tropical cyclone activity over the western North Pacific [J]. Adv. Atmos. Sci., 25: 319 – 328.

陈联寿，丁一汇. 1979. 西北太平洋台风概论 [M]. 北京：科学出版社. 491pp. Chen Lianshou, Ding Yihui. 1979. The Perspective of Typhoon in the Western Pacific (in Chinese) [M]. Beijing: Science Press. 491pp.

Chia H H, Ropelewiski C F. 2002. The interannual variability in the genesis location of tropical cyclones in the northwest Pacific [J]. J. Climate, 15: 2934 – 2944.

Ho C H, Baik J J, Kim J H, et al. 2004. Interdecadal changes in summertime typhoon tracks [J]. J. Climate, 17: 1767 – 1776.

Hope J R, Neumann C J. 1970. An operational technique for relating the movement of existing tropical cyclones to past tracks [J]. Mon. Wea. Rev., 98 (12): 925 – 933.

Huang Gang, 2004. An index measuring the interannual variation of the East Asian summer monsoon—The EAP index [J]. Adv. Atmos. Sci., 21: 41 – 52.

Huang Ronghui, Li Weijing. 1987. Influence of the heat source anomaly over the tropical western Pacific on the subtropical high over East Asia [C]. Proceedings of International Conference on the General Circulation of East Asia, Chengdu, April 10 – 15, 1987, 40 – 51.

黄荣辉，李维京. 1988. 夏季热带西太平洋上空的热源异常对东亚上空副热带高压的影响及其机理 [J]. 大气科学, 12 (特刊): 107 – 116. Huang Ronghui, Li Weijing. 1988. Influence of heat source anomaly over the western tropical Pacific on the subtropical high over East Asia and its physical mechanism [J]. Chinese J. Atmos. Sci. (in Chinese), 12 (Special Issue): 107 – 116.

Huang Ronghui, Sun Fengying. 1992. Impact of the tropical western Pacific on the East Asian summer monsoon [J]. J. Meteor. Soc. Japan, 70 (1B): 243 – 256.

黄荣辉，顾雷，徐予红，等. 2005. 东亚夏季风爆发和北进的年际变化特征及其与热带西太平洋热状态的关系 [J]. 大气科学, 29: 20 – 36. Huang Ronghui, Gu Lei, Xu Yuhong, et al. 2005. Characteristics of the interannual variations of onset and advance of the East Asian summer monsoon and their associations with thermal states of the tropical western Pacific [J]. Chinese J. Atmos. Sci. (in Chinese), 29: 20 – 36.

Huang Ronghui, Gu Lei, Zhou Liantong, et al. 2006. Impact of the thermal state of the tropical western Pacific on onset date and process of the South China Sea summer monsoon [J]. Adv. Atmos. Sci., 23: 909 – 924.

Huang Ronghui, Chen Jilong, Huang Gang. 2007. Characteristics and variations of the East Asian monsoon system and its impacts on climate disasters in China [J]. Adv. Atmos. Sci., 24: 993 – 1023.

黄荣辉，陈光华. 2007. 西北太平洋热带气旋移动路径的年际变化及其机理研究 [J]. 气象学报, 65: 683 – 694. Huang Ronghui, Chen Guanghua, 2007. Research on interannual variations of tracks of tropical cyclones over the Northwest Pacific and their physical mechanism [J]. Acta Meteorologica Sinica (in Chinese), 65: 683 – 694.

Kutzbach G. 1979. The Thermal Theory of Cyclones [M]. Boston: Amer. Meteor. Soc., 255pp.

McBride J L. 1995. Tropical cyclone formation [C]. Global Perspectives on Tropical Cyclones, WMO/TD-No. 693, World Meteorological Organization, Geneva, Switzerland, 63 - 105.

Nitta T. 1987. Convective activities in the tropical western Pacific and their impact on the Northern Hemisphere summer circulation [J]. J. Meteor. Soc. Japan, 64: 373 - 400.

Wang B, Chan J C L. 2002. How strong ENSO events affect tropical storm activity over the western North Pacific [J]. J. Climate. 15: 1643 - 1655.

王磊, 陈光华, 黄荣辉. 2009a. 近 30a 登陆我国的西北太平洋热带气旋活动的时空变化特征 [J]. 南京气象学院学报, 32: 182 - 188. Wang Lei, Chen Guanghua, Huang Ronghui. 2009a. Spatio-temporal distributive characteristics of tropical cyclone activities over the Northwest Pacific in 1979 - 2006 [J]. J. Nanjing Institute of Meteor. (in Chinese), 32: 182 - 188.

王磊, 陈光华, 黄荣辉. 2009b. 影响登陆我国不同区域热带气旋活动的大尺度环流定量分析 [J]. 大气科学, 33: 916 - 922. Wang Lei, Chen Guanghua, Huang Ronghui. 2009. Quantitative analysis on large scale circulation system modulating landfalling tropical cyclone activity in the diverse Chinese regions [J]. Chinese J. Atmos. Sci. (in Chinese), 33: 916 - 922.

Wu L, Wang B. 2004. Assessing impacts of global warming on tropical cyclone tracks [J]. J. Climate, 17: 1686 - 1698.

Wu L, Wang B, Braun S A. 2005. Impact of air - sea interaction on tropical cyclone track and intensity [J]. Mon. Wea. Rev., 133: 3299 - 3314.

张庆云, 彭京备. 2003. 夏季东亚环流年际和年代际变化对登陆中国台风的影响 [J]. 大气科学, 27: 97 - 106. Zhang Qingyun, Peng Jingbei. 2003. The interannual and interdecadal variations of East Asian summer circulation and its impact on the landing typhoon frequency over China during summer [J]. Chinese J. Atmos. Sci. (in Chinese), 27: 97 - 106.

关于西北太平洋季风槽年际和年代际变异及其对热带气旋生成影响和机理的研究

黄荣辉[1,2]　皇甫静亮[1]　武　亮[1]　冯　涛[3]　陈光华[1]

(1. 中国科学院大气物理研究所季风系统研究中心,北京 100190;2. 中国科学院大气物理研究所大气科学和地球流体力学国家重点实验室,北京 100029;3. 南京大学大气科学学院,江苏南京 210023)

摘　要：总结和综述近年来中国科学院大气物理研究所季风系统研究中心,关于西北太平洋季风槽的年际和年代际变异及其对热带气旋和台风(TCs)生成的影响和机理的气候学研究进展,并综述一些有关的国内外研究。给出了夏、秋季西北太平洋季风槽的气候特征以及利于 TCs 生成的四类季风槽环流型,表明了西北太平洋季风槽强度和位置有明显的年际和年代际变异。特别是揭示了西北太平洋季风槽的年际和年代际变异不仅通过影响西北太平洋上空对流层低层气流的涡度和对流层高层的散度、对流层中、下层的水汽以及对流层上下层风场的垂直切变等利于 TCs 生成的大尺度环境因子的分布而影响 TCs 的生成,而且通过对热带对流耦合波动的转化和提供扰动能量而对 TCs 生成起着重要的动力作用。还指出今后有关西北太平洋季风槽和 TCs 活动一些亟需进一步研究的气候学问题。

1　引　言

热带西太平洋是全球海洋温度最高的海域,全球约 90% 的暖海水集中在此,故此海域又称为暖池(warm pool)。它是全球海-气相互作用最剧烈的区域,也是全球大气中潜热释放的一个重要源区[1-2]。由于西太平洋暖池具有很高的海温,其上空是 Walker 环流的上升支,因此,此海域是气流和水汽的强辐合区,这导致此区域有很强的对流活动和强降水。

西北太平洋位于西太平洋暖池的核心海域,它不仅是全球高海温的海域之一,而且其上空也是全球热带气旋和台风(TCs)主要生成区域之一,据统计全球约三分之一的 TCs 在此海域上空生成[3-4]。在西北太平洋上空生成的 TCs 中一大部分向西或西北方向移动,并在中国、菲律宾、越南、日本和韩国登陆,给这些国家造成严重的经济损失和重大人员伤亡。中国是世界上遭受 TCs 灾害最严重的国家之一,据统计,平均每年有 7~8 个 TCs 登陆我国东南沿海和华南地区,最多可达 12 个,每年给我国东南沿海和华南地区造成约 200 多亿元的经济损失和数百人的人员伤亡[5-6]。

西北太平洋上空为什么是 TCs 易于生成的区域?这不仅是由于西北太平洋表层和次表层海温很高,满足 TCs 生成的热力条件[7],而且在夏、秋季西北太平洋上空经常是季风槽所在之处[5,8]。据统计,全球 75% 以上的 TCs 在季风槽中生成[9],在西北太平洋季风槽中生成的 TCs 也占西北太平

---

* 本文原载于：热带气象学报,第 32 卷,第 6 期,767-785,2016 年 12 月出版.

洋上空生成的 TCs 总数的 80% 以上[10-11]。西北太平洋季风槽是西北太平洋对流层低层夏、秋季一个重要的环流系统。当亚洲西南季风爆发以后，从南半球吹来的偏南气流到了北半球，由于科氏力的作用就会向东偏，并与亚洲西南季风叠加，从而在南海和西北太平洋上空形成强的西南气流；并且，此西南季风气流在南海和西北太平洋上空与从热带东、中太平洋对流层低层吹来的偏东气流相遇时，将与东风气流合并沿西太平洋副热带高压的西南部向西北方向吹，这就在南海和西北太平洋上空对流层低层形成季风槽。文献[12-13]对南海季风槽和孟加拉季风槽的形成、结构及其对天气气候的影响做了深入研究。关于西北太平洋季风槽对 TCs 生成的作用，国际上已有一些研究[14-16]。最近，研究表明了西北太平洋季风槽不仅可以为 TCs 生成提供低层气流的辐合和气旋性相对涡度、高层气流的辐散，较小的垂直风切以及充足的水汽等有利的大尺度环境条件，而且可以为 TCs 生成提供初始扰动和动力条件[5, 10-11, 17-19]。

鉴于西北太平洋上 TCs 给我国和其它有关国家和地区造成严重灾害，许多学者对于西北太平洋上 TCs 的生成、结构、发展和移动路径做出了系统的研究[20-22]。然而，相对而言，关于西北太平洋上 TCs 活动的气候学研究较少。最近，围绕全球变暖背景下 TCs 生成个数是否增加这个问题，国际上兴起 TCs 气候学的研究热潮，特别是全球变暖背景下 TCs 活动的年际和年代际变异特征正吸引许多学者的关注[22-24]。因此，西北太平洋上空 TCs 活动的年际和年代际变异特征、成因及其机理的研究也受到我国及国际有关学者的重视。

近年来，鉴于夏、秋季热带气旋和台风等天气灾害给我国造成经济损失的严重性，其活动（生成和移动路径）的年际和年代际变异成因的复杂性以及开展西北太平洋 TCs 生成和移动路径、登陆地点季度预测的迫切性，我们研究小组开展了关于西北太平洋 TCs 气候学的研究，特别是开展了西北太平洋季风槽的年际和年代际变异及其对 TCs 活动（包括生成和移动路径）影响的研究，并取得一些研究结果。本文主要总结和综述近年来我们研究小组关于西北太平洋季风槽年际和年代际变异及其对 TCs 活动的影响过程和机理等方面的研究进展以及与此有关的部分国内外研究。由于之前的一些关于西北太平洋 TCs 活动变异的研究，根据研究对象不同而所用资料的月份也有所不同，为了一致起见，本文所给出研究结果主要是 7—9 月。这主要由于 7—9 月是西北太平洋季风槽较强时期，也是西北太平洋 TCs 生成的强盛期。因此，本文中所给出的许多幅图是新分析的结果，与以前发表的图有所不同。并且，由于篇幅有限，本文所综述的国内外有关研究可能挂一漏百，甚至是很重要的研究。

2 西北太平洋季风槽的气候特征以及利于 TCs 生成的几种环流型

2.1 西北太平洋季风槽的气候特征

正如引言中所述，西北太平洋季风槽是亚洲西南季风、跨赤道气流与西太平洋副热带高压南沿的偏东气流，在热带西北太平洋汇合而形成的一个大尺度气旋性环流系统。在西北太平洋季风槽的作用下，每年 6—11 月在西北太平洋上空会生成很多 TCs，由于 7—9 月是西北太平洋季风槽最强时期，也是西北太平洋 TCs 生成的强盛时期，因此，本研究首先分析 7—9 月西北太平洋季风槽的分布及其变化。图 1 是利用 NCEP/NCAR 再分析的风场资料[25]和美国海洋大气管理局（NOAA）的对外长波辐射（OLR）资料所分析的 1979—2012 年 7—9 月平均的西北太平洋 850 hPa 流线分布和 OLR 分布。从图 1 可以看到，西北太平洋季风槽从南海上空经菲律宾向东南延伸到 150°E 附近的西北太平洋上空，在槽的南边盛行西南季风气流，而在槽的北边盛行东南气流，在季风槽区有较强的对流活动，特别沿季风槽的槽线有明显的强对流活动中心。

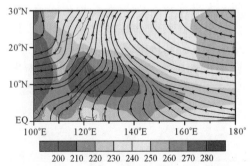

图 1　1979—2012 年 7—9 月平均的西北太平洋上空 850 hPa 流线和 OLR（彩色，单位：W/m^2）分布

风场资料取自 NCEP/NCAR 再分析资料[25]，OLR 资料取自 NOAA 的 OLR 资料集。

由于季风槽一旦建立就能使西北太平洋上空

对流层环流和水汽以及风场的垂直切变产生重要变化,因此分析季风槽与这些大尺度环境因子的关系是很有必要的。图 2a～2d 分别是 1979—2012 年 7—9 月平均的西北太平洋 850 hPa 气流的相对涡度、200 hPa 气流的散度、200 hPa 与 850 hPa 之间风场的垂直切变以及 700～500 hPa 之间平均的相对湿度的分布。这些大尺度环境因子参照 Feng 等[11]所用公式计算而得。从图 2a 可以清楚看到:在 7—9 月南海上空对流层低层有一个较强的正相对涡度中心,而在西北太平洋 120～160 °E,5～20 °N 海域上空对流层低层有东西向带状正相对涡度的分布,这与图 1 所示的季风槽位置相吻合;并且,从图 2b～2d 分别可看到:在南海和西北太平洋 100～160 °E,0～20 °N 海域上空 200 hPa 有较强的正散度分布;在西北太平洋西侧 10～20 °N 到东侧 0～13 °N 海域的上空对流层上层与对流层低层之间风场有较弱的垂直切变;在 100～150 °E,0～15 °N 海域上空对流层中、下层有较大的相对湿度,即充足的水汽。这些都表明西北太平洋季风槽会使西北太平洋上空对流层低层气流产生强的辐合、对流层高层气流强的辐散、对流层上下层之间小的风场垂直切变以及对流层中、下层充足的水汽。这些大尺度环境因子的分布正是利于西北太平洋上空 TCs 的生成。

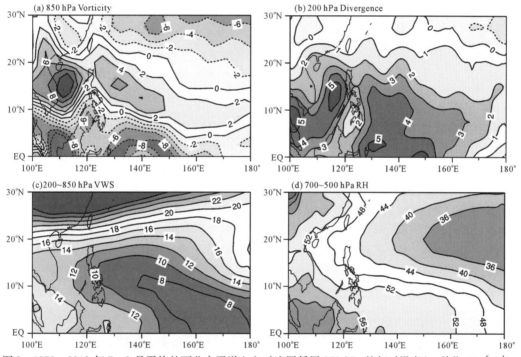

图 2　1979—2012 年 7—9 月平均的西北太平洋上空对流层低层 850 hPa 的相对涡度(a,单位:$10^{-6}s^{-1}$)、对流层上层 200 hPa 的散度(b,单位:$10^{-6}s^{-1}$)、200 hPa 与 850 hPa 之间风场的垂直切变(c,单位:m/s)以及 700～500 hPa 平均的相对湿度(d,单位:%)分布　风场和湿度资料取自 NCEP/NCAR 再分析资料[25]。

正如图 3 所示,夏、秋季南海和西北太平洋上空对流层低层的季风槽是从东南亚吹来的西南季风和从南半球吹来的跨赤道气流与从热带东、中太平洋上空沿西太平洋副热带高压南沿吹来的东风气流相交汇而形成的一个对流层低层气旋性大尺度环流系统。西北太平洋季风槽可以造成 5～20 °N 附近西北太平洋上空对流层低层强的辐合、对流层上层强的辐散、对流层高低层的小风速垂直切变和对流层中、下层充足的水汽。这些正好为西北太平洋上空 TCs 生成提供有利的大尺度环境因子,也就是西北太平洋 TCs 大部分在季风槽中生成之原因。

图 3　西北太平洋季风槽及其周围环流特征示意图

2.2 利于TCs生成的几类季风槽环流型

西北太平洋季风槽在不同年份和日期有不同的环流型。Feng 等[10]系统地分析了有利于西北太平洋TCs生成的西北太平洋季风槽的环流型,由于西北太平洋上空大部分TCs在7—10月生成,因此,他们利用NCEP-DOE再分析资料分析了1991—2010年7—10月西北太平洋上空计425个TCs生成的对流层低层850 hPa的大尺度环流型,提出了西北太平洋上空利于TCs生成有五类大尺度环流型。这些环流型分别是:(1)季风切变型(MS型),在此环流型中计生成140个TCs,占32.9%;(2)季风辐合型(MC型),在此环流型中计生成141个TCs,占33.3%;(3)季风倒槽型(RMT型),在此环流型中计生成64个TCs,占15.1%;(4)季风涡旋型(MG型),在此环流型中计生成18个TCs,占4.2%;(5)热带东风环流型(TE型),在此环流型中计生成37个TCs,占8.7%。还有在其它类型环流中生成25个TCs,占5.9%。其中在与季风槽有关的前三类环流型中生成的TCs占这20年西北太平洋上空生成TCs总数的81.2%,若再计入在季风涡旋型中生成的TCs,可占TCs生成总数的85.6%。图4a～4d分别是上述利于西北太平洋上TCs生成的前4类季风槽的对流层低层850 hPa大尺度环流的合成分布。

一般在西太平洋暖池偏冷时(图4a),西北太平洋季风槽较强,其位置可东伸到西北太平洋东侧上空。由于这时期从南海经西北太平洋西侧和中部上空吹到西北太平洋东侧上空的亚洲偏西的季风气流经常与沿西太平洋副热带高压南沿吹的偏东信风相遇,故在季风槽东部形成很强的气流辐合,即 $\frac{\partial \bar{u}}{\partial x}$ 为较大的负值,因此这时期易生成TCs的大尺度季风槽环流型是MC型。而一般在西太平洋暖池偏暖时(图4b),西北太平洋季风槽偏弱,其位置主要位于南海和西北太平洋的西侧上空,由于这时期从南海吹来的亚洲西南季风与沿西太平洋副热带高压南沿吹来偏东气流经常在西北太平洋上空西侧汇合并向西北方向吹,故季风槽在西北太平洋西侧上空收缩,并形成很强的气流经向切变,故在季风槽中 $\frac{\partial \bar{u}}{\partial y}$ 为较大的负值,在这时期易生成TCs的大尺度季风槽环流型是MS型。在前汛期,特别在南海夏季风爆发之后,位于从南海到西北太平洋西侧上空的锋面往往是西南-东北走向,在锋面周围气流往往形成一种如图4c所示的槽线,是西南-东北走向的气旋性气流,这时期易生成TCs的大尺度季风槽环流型称RMT型。此外,在8—9月期间,由于西北太平洋季风槽较强,往往在南海和菲律宾以东海域上空的对流层低层的季风槽中心产生较强涡旋,在此涡旋周围易生成TCs,这时期易生成TCs的大尺度季风槽环流型称MG型。

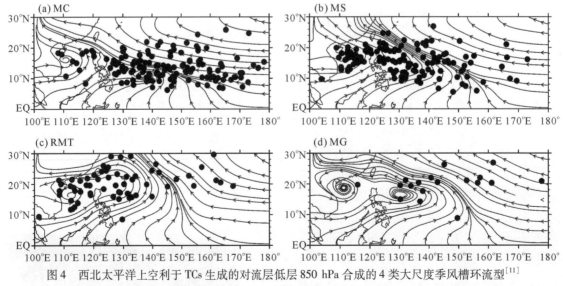

图4 西北太平洋上空利于TCs生成的对流层低层850 hPa合成的4类大尺度季风槽环流型[11]
a. MC型;b. MS型;c. RMT型;d. MG型。·表示TCs生成位置。风场资料取自NCEP-DOE的AMIP-II再分析资料[26],TCs资料来自美国JTWC资料集。

3 西北太平洋季风槽年际变异及其对TCs生成的影响过程

西北太平洋季风槽有明显的年际变异(variability),并对TCs生成的年际变异有直接的影响。这个影响主要通过影响西北太平洋上空对流层低层的相对涡度、对流层高层的辐散、对流层上下层风场的垂直切变以及对流层中、下层的水汽等利于TCs生成的大尺度环境因子。关于西北太平洋季风槽年际变异成因已有一些研究,如黄荣辉等[5,27]提出西太平洋暖池次表层热力的年际变异对季风槽的年际变异有很大影响;冯涛等[28]研究了跨赤道气流变化对季风槽和TCs生成的年际变异的影响。本节主要讨论西北太平洋季风槽环流型、位置和强度的年际变异特征及其对TCs生成的影响。

3.1 季风槽环流型的年际变异对TCs生成的影响

西北太平洋季风槽不仅有几天和季节内变异,而且有明显的年际变异。最近,冯涛等[29]分析和比较了西北太平洋季风槽位置和环流型的年际变异及其对利于TCs生成大尺度环境因子变化的影响,他们利用NCEP-DOE的AMIP-II再分析资料并以2004年与2006年的7—9月季风槽环流型差异为例,来说明季风槽环流型的年际变异对利于TCs生成大尺度环境因子的影响。本文利用NCEP/NCAR再分析资料分析了2004年与2006年7—9月850 hPa西北太平洋季风槽环流型的差别及其对TCs生成位置的影响。

分析结果表明:如图5a1所示,2004年7—9月季风槽位置东伸到西北太平洋上空的偏东侧,并在季风槽东端有强的纬向风辐合,因而在2004年7—9月西北太平洋上空对流层低层大尺度环流型主要为季风辐合型(MC型),这引起了在2004年7—9月有较多强的TCs在西北太平洋偏东侧上空的MC环流型中生成(图5a1)。在此期间,西北太平洋上空计生成14个TCs,其中有9个TCs生成在140°E以东的MC型中,仅8月就有6个TCs生成在MC型中。而2006年7—9月季风槽的位置(图5a2)虽也东伸到西北太平洋的偏东侧上空,但在季风槽的东端呈现强的纬向风经向切变,因而在2006年7—9月西北太平洋上空对流层低层的大尺度环流型主要为季风切变型(MS型),这引起了在2006年7—9月TCs的生成位置比2004年7—9月TCs生成位置略偏西、偏北(图5a2)。在此期间,西北太平洋上计生成14个TCs,其中有8个TCs在140°E以西生成,这当中有5个TCs在MS环流型生成。这些结果与冯涛等[29]的分析结果基本一致。这说明西北太平洋季风槽的环流型对TCs生成有明显的影响。

3.2 季风槽位置的年际变异对TCs生成的影响过程

西北太平洋季风槽位置有很大的年际变异。黄荣辉等[5]及Chen等[17]分析了西北太平洋上空夏、秋季季风槽位置的年际变异及其对TCs生成和移动路径年际变异的影响。他们的分析结果表明:西北太平洋上空夏、秋季的季风槽位置有显著的年际变异,并与西太平洋暖池热状态的年际变异有很密切的关系,它对TCs生成和移动路径有明显的影响。在西太平洋暖池处于暖状态的夏、秋季,西北太平洋上空季风槽位置偏西、偏北,西北太平洋上空TCs生成区域和移动路径也偏西、偏北,从而造成影响我国的台风个数偏多;相反,在西太平洋暖池处于冷状态的夏、秋季,西北太平洋上空季风槽位置偏东、偏南,西北太平洋的TCs生成区域偏东、偏南,其移动路径偏东,且易于在130°E转向,从而造成影响日本和韩国的台风个数偏多,而影响我国的台风个数可能偏少。

文献[10-11,27]的研究表明,西北太平洋季风槽位置对西北太平洋TCs生成的大尺度环境因子有重要影响。文中指出:西北太平洋季风槽位置通过影响对流层下层强的气旋性相对涡度、对流层上层强的辐散、小的对流层上下层风场的垂直切变以及对流层中、下层充足水汽等大尺度环境因子的分布位置而影响TCs的生成位置。最近,本文作者利用NCEP/NCAR再分析资料分析了1990年与1998年7—9月850 hPa西北太平洋季风槽位置的差别及其对TCs生成位置的影响。分析结果表明:1990年7—9月期间季风槽位置东伸到西北太平洋偏东、偏南侧上空(图5b1),它在西北太平洋的偏东、偏南侧上空为TCs生成提供强的低层气旋性相对涡度、强的高层辐散、小的对流层上下层风场的垂直切变和对流层中、下层充足的水汽等利于TCs生成的大尺度环境因子(图略),因此,1990年7—9月期间TCs主要在西北太平洋上空的偏东、偏南侧生成(图5b1);而在1998

年7—9月,西太平洋季风槽位置在西北太平洋上空的偏西、偏北侧(图5b2),它在西北太平洋的偏西、偏北侧上空为TCs生成提供较强的低层气旋性相对涡度、较强的高层辐散、较小的对流层上下层风场的垂直切变和对流层中、下层充足的水汽等利于TCs生成的大尺度环境因子(图略),因此,1998年7—9月期间TCs生成很少,并主要在西北太平洋上空的偏西、偏北侧生成(图5b2)。

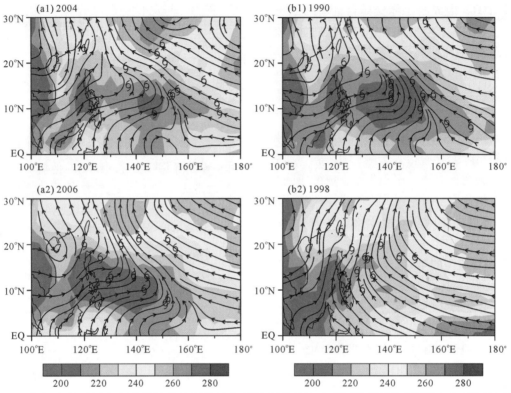

图5 2004年(a1)和2006年(a2)以及1990年(b1)和1998年(b2)7—9月平均的西北太平洋上空850 hPa流线及TCs生成位置(以♀标志) 风场资料取自NCEP/NCAR再分析资料[25],TCs资料取自IBTrACS资料。

为了更好地分析西北太平洋季风槽位置的年际变异通过利于TCs生成的大尺度环境因子的分布而影响TCs生成的位置。黄荣辉等[27]利用合成方法分别分析了1979—2010年7—9月西太平洋暖池处于暖状态(即季风槽位置偏西),以及冷状态时(即季风槽位置偏东)西北太平洋上空对流层低层850 hPa的相对涡度、对流层上层200 hPa的散度、200~850 hPa之间风场的垂直切变以及700~500 hPa平均的水汽分布(图6、7)。

由于西太平洋暖池的表层海温变异很小,一般在1.0 ℃以内,而它的次表层海温变异很大,在50~250 m深次表层海温变化方差很大,最大方差位于5~10 °N,120 m深,夏季可达4.0 ℃,冬季可达3.0 ℃以上[30]。因此,西太平洋暖池热力异常主要取决于次表层海温的变异。最近Lin等[31]指出,海洋次表层海温对超强台风Haiyan的发展起着重要的热力作用;Huang等[32]研究表明了海洋次表层热力状态对TCs强度有重要影响。为此,本研究定义西太平洋暖池在125~165 °E,0~16 °N海域的次表层120 m夏季海温距平$\Delta T_{sub} \geq 1.5$ ℃为偏暖年,而$\Delta T_{sub} \leq -1.5$ ℃为偏冷年。这样在1959—2003年西太平洋暖池偏暖年有:1970、1975、1998、1999、2000年;而偏冷年有:1972、1982、1987、1991、1993、1997年。从图6可以看到:在西太平洋暖池处于偏暖的7—9月期间,由于西太平洋季风槽的位置偏西、偏北,850 hPa强的气旋性相对涡度、200 hPa强的辐散、对流层上下层小的风速垂直切变以及对流层中、下层大的湿度等利于TCs生成的大尺度环境因子分布区域都位于西北太平洋偏西、偏北侧的上空,因此,在西太平洋暖池偏暖(即季风槽位置偏西)的7—9月,TCs生成位置位于西北太平洋上空的偏西、偏北侧;相反,在西太平洋暖池处于偏冷的7—9月(图7),由于西太平洋季风槽的位置偏东、偏南,上述利于 TCs 生成的大尺度环境因子分布区域都位于西北太平洋偏东、偏南侧的上空,因而 TCs 生成位置在西北太平洋上空的偏东、偏南侧。

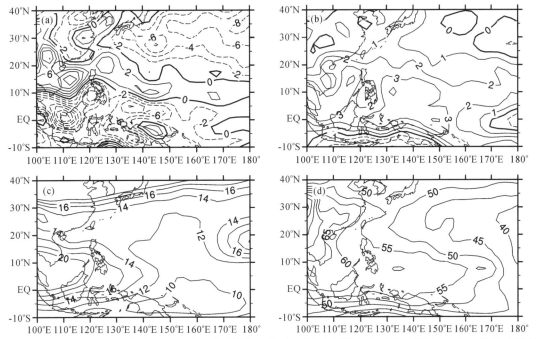

图 6 在西北太平洋暖池偏暖的 7—9 月(季风槽位置偏西)西北太平洋上空对流层低层 850 hPa 的相对涡度(a,单位:$10^{-6}s^{-1}$)、对流层上层 200 hPa 散度(b,单位:$10^{-6}s^{-1}$)、200~850 hPa 之间风速的垂直切变(c,单位:m/s)以及 700~500 hPa 平均相对湿度(d,单位:%)的合成分布 风场和湿度资料取自 NCEP/NCAR 再分析资料[25]。

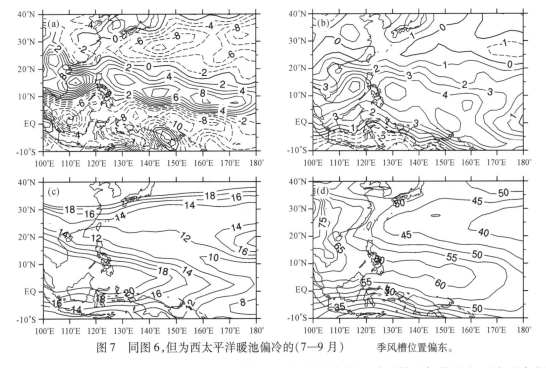

图 7 同图 6,但为西太平洋暖池偏冷的(7—9 月)季风槽位置偏东。

3.3 季风槽强度的年际变异及其对 TCs 生成的影响过程

3.3.1 季风槽(强盛期)强度指数

上述研究表明了季风槽位置和环流型有很大的年际变异。季风槽不仅位置和环流型有很大年际变异,而且它的强度也有明显的年际变异。张翔等[33]研究了西北太平洋季风槽强度的年际变异,并从 6—11 月 850 hPa 季风槽环流的相对涡度定义了一个季风槽强度指数。为了研究西北太平洋上空季风槽处于较强时期它的强度变化，本文提出一个在季风槽强盛时期的西北太平洋季风槽强度指数，即定义西北太平洋季风槽（强盛期）强度指数（简称 MTI）为某年 7—9 月西北太平洋上空 135~165 °E,5~20 °N 区域平均的对流层低层 850 hPa 相对涡度 ζ' 与 1979—2013 年 7—9 月此区域 850 hPa 相对涡度的气候平均值 $\bar{\zeta}$ 之差与此区域 7—9 月平均的 850 hPa 相对涡度的均方差 $\sqrt{\sigma^2}$ 之比，即 MTI 可定义为，

$$\text{MTI} = \frac{\zeta' - \bar{\zeta}}{\sqrt{\sigma^2}} \quad (1)$$

从式(1)可看到，若 MTI>0，则西北太平洋季风槽偏强；相反，若 MTI<0，则西北太平洋季风槽偏弱。图 8 是应用上述定义和利用 NCEP/NCAR 再分析资料所计算的 1979—2013 年西北太平洋季风槽处于较强时期的 7—9 月季风槽强度指数 MTI 的年际变化。从图 8 可以看到：1979、1980、1982、1986、1987、1990、1991、1992、1993、1994、1997、2002、2003、2004、2005、2006、2009 年 MTI 为正，即西北太平洋季风槽偏强；而 1981、1983、1984、1988、1989、1995、1996、1998、1999、2000、2001、2007、2008、2010、2011、2012、2013 年 MTI 为负，即西北太平洋季风槽偏弱。特别是 1982、1990、1991、1997、2002、2009 年 MTI 为较大正值，即西北太平洋季风槽较强，而 1984、1988、1996、1998、1999、2008、2010 年 MTI 为较大负值，即西太

平洋季风槽较弱。因此，本文所定义的 MTI 年际变异与张翔等[33]从 6—11 月 850 hPa 季风槽环流的相对涡度所定义的 MTI 年际变异有所不同。

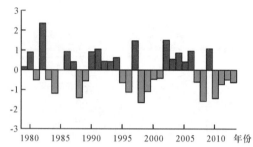

图 8　1979—2013 年 7—9 月北太平洋季风槽指数（MTI）的年际变化　风场资料取自 NCEP/NCAR 再分析资料[25]。

3.3.2　季风槽强度的年际变异对 TCs 生成的影响过程

图 9a 和 9b 分别是 7—9 月西北太平洋季风槽偏强（即 MTI>0）和偏弱（MTI<0）年份相应的 7—9 月平均的季风槽环流和对流活动的合成分布。从图 9a 所示的流线和 OLR 分布可以看到：在季风槽偏强年份的 7—9 月，季风槽从南海上空对流层低层向东伸到西北太平洋东侧 160 °E 附近上空的对流层低层，季风槽位置也偏南，强对流活动分布也东伸到西北太平洋东侧；从图 9b 所示的流线和 OLR 分布可以看到：在季风槽偏弱年份的 7—9 月，季风槽向西北太平洋西北侧收缩，从南海上空对流层低层只东伸到西北太平洋 135 °E 附近上空的对流层低层，强对流活动分布主要位于南海和西北太平洋西侧上空。

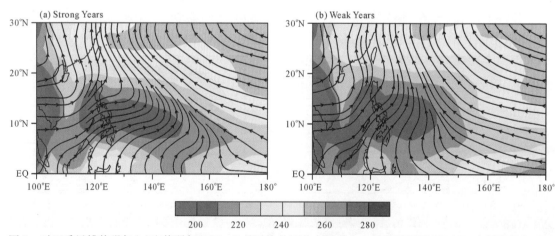

图 9　对于季风槽偏强年(a)和偏弱年(b)7—9 月平均的西北太平洋上空对流层低层 850 hPa 水平流场和 OLR（彩色，单位：W/m²）合成分布　风场资料取自 NCEP/NCAR 再分析资料[25]，OLR 资料取自 NOAA 的 OLR 资料集。

上述分析结果表明：西北太平洋季风槽强度有很大年际变异，一般在强季风槽年份的 7—9 月，季风槽位置东伸到西北太平洋的偏东、偏南侧上空，强对流活动区位置也偏南，并东伸到西北太

平洋东侧上空,依据3.2节的分析结果,利于TCs生成的大尺度环境因子分布区域也会东伸到西北太平洋的偏东、偏南侧上空,故TCs生成也偏于西北太平洋中、东侧上空。而在季风槽弱年份的7—9月,季风槽位置位于西北太平洋上空的偏西、偏北侧,强对流活动区位置也位于西北太平洋上空的偏西、偏北侧,并主要位于南海和西北太平洋西侧上空,依据3.2节的分析结果,利于TCs生成的大尺度环境因子分布也会位于西北太平洋的偏西、偏北侧上空,故TCs生成位置也会偏于西北太平洋的西侧上空。

4 西北太平洋季风槽的年代际变异及其对TCs生成的影响过程

西太平洋季风槽不仅有明显的年际变异,而且有显著的年代际变异。最近,Huangfu等[34]深入研究了夏、秋季西北太平洋季风槽在1990年代末发生的年代际跃变及其对TCs生成的影响。研究表明:从1990年末起夏、秋季西北太平洋季风槽的平均位置比1979—1998年期间的平均位置要偏西、偏北。由于西北太平洋季风槽的年代际变异对TCs生成位置的年代际变异有着重要影响,并且7—9月是西太平洋季风槽处于较强时期和TCs生成强盛期,因此本文利用NCEP/NCAR再分析资料分析了1990年代末前后7—9月西北太平洋季风槽位置的年代际变异及其对TCs生成的影响。

4.1 西北太平洋季风槽的年代际变异特征

图10a和10b分别是1979—1998年与1999—2013年7—9月平均的西北太平洋上空对流层低层850 hPa的流线及OLR的分布。从图10a可以看到:在1979—1998年的7—9月,西北太平洋季风槽东伸到155°E附近的西北太平洋偏东、偏南侧上空,强对流活动区也东伸到西北太平洋的偏东、偏南侧上空;而在1999—2013年的7—9月(图10b),西北太平洋季风槽主要位于西北太平洋偏西、偏北侧上空,它只东伸到148°E附近,强对流活动区域也位于西北太平洋的偏西、偏北上空。由于在1999—2013年的7—9月,西太平洋暖池偏暖,热带太平洋出现了类似于"ENSO-modoki型"的年代际增温[27],这导致了这时期西北太平洋季风槽平均位置位于西太平洋的偏西、偏北侧上空,强对流活动区域也位于西北太平洋的偏西、偏北侧上空。

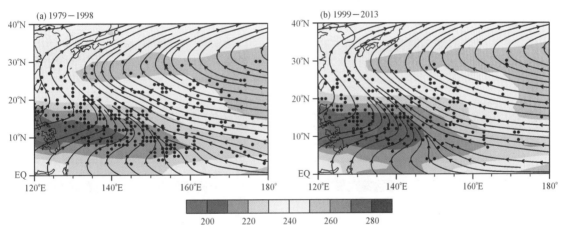

图10 1979—1998年(a)和1999—2013年(b)7—9月平均的西北太平洋上空对流层低层850 hPa上流线和OLR分布
● 表示TCs生成位置。风场资料取自NCEP/NCAR再分析资料[25],热带气旋资料取自IBTrACS资料集。

4.2 季风槽年代际变异对TCs生成位置年代际变异的影响过程

为了更好地分析西北太平洋季风槽位置的年代际变异,将通过影响利于TCs生成的大尺度环境因子的分布而对TCs生成位置的年代际变异产生影响,我们分别分析了1979—1998年与1999—2013年7—9月平均的西北太平洋上空对流层低层850 hPa的相对涡度、对流层上层200 hPa的散度、200 hPa与850 hPa之间风速的垂直切变以及700~500 hPa平均相对湿度等利于TCs生成的大尺度环境因子的分布(图11、12)。从图11a~11d可以看到:在1979—1998年7—9月,由于西北太平洋季风槽位置东伸到西北太平洋的东侧上空(图10a),这时期西太平洋暖池偏冷[27],因此,对流层低层气流强的气旋性相对涡度和对流层上层强的气流辐散、对流层上下层之间小的风速垂直切变以及对流层中、下层充足的水汽等利于TCs生成大尺度环境因子的分布区域东伸到西北太平

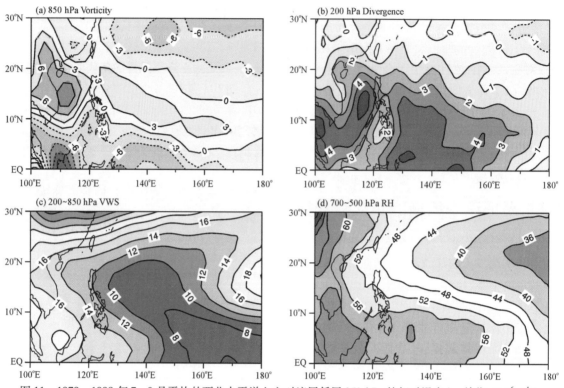

图11　1979—1998年7—9月平均的西北太平洋上空对流层低层850 hPa的相对涡度(a,单位:$10^{-6}s^{-1}$)、对流层上层200 hPa散度(b,单位:$10^{-6}s^{-1}$)、200~850 hPa之间风速的垂直切变(c,单位:m/s)以及700~500 hPa平均相对湿度(d,单位:%)　风场和湿度资料取自NCEP/NCAR再分析资料[25]。

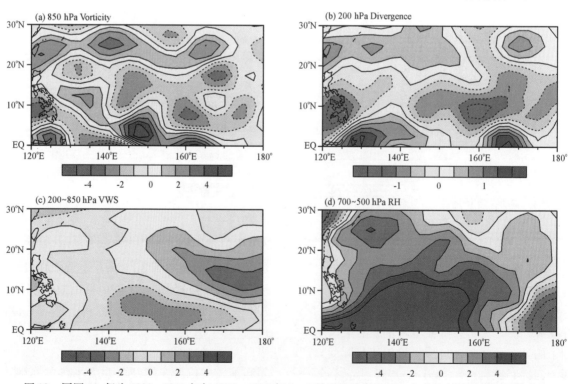

图12　同图11,但为1999—2013年与1979—1998年7—9月平均的利于TCs生成的大尺度环境因子之差

洋中、东侧上空,这使得这时期 TCs 在西北太平洋中、东侧上空生成较多(图10a)。另外,为了更清楚看到西北太平洋季风槽在 20 世纪末发生的年代际变异对利于 TCs 生成的大尺度环境因子变化的影响,分析了 1999—2013 年与 1979—1998 年 7—9 月平均的利于 TCs 生成的大尺度环境因子之差(图12)。从图12a~12d 并结合图11a~11d 可以看到:在 1999—2013 年 7—9 月,由于西北太平洋季风槽的平均位置在西北太平洋的偏西侧上空(图 10b),因而在这时期对流层低层强的气旋性相对涡度和对流层上层强的气流辐散、对流层上下层之间小的风速垂直切变以及对流层中、下层充足的水汽区域的平均位置要比 1979—1998 年期间的平均位置略偏西,它们主要位于南海和西北太平洋西侧上空。这些利于 TCs 生成的大尺度环境因子的分布使得这时期 TCs 在西北太平洋西侧上空生成较多(图10b)。

从上述分析结果可以看到:西北太平洋季风槽有很明显的年代际变异,它通过影响西北太平洋上空对流层低层气流的气旋性相对涡度以及对流层上层气流的辐散、对流层上下层之间风场的垂直切变、对流层中、下层水汽等分布而对 TCs 生成年代际变异产生影响。从 1990 年代末之后,由于热带太平洋热力呈现类似于"ENO-modoki 型"的年代际增温,西北太平洋季风槽平均位置比 1979—1998 年 7—9 月平均位置要略偏西、偏北,这导致了利于 TCs 生成的大尺度环境因子分布也比 1979—1998 年 7—9 月平均的因子分布要略偏西、偏北,因而在西北太平洋上空的偏西、偏北侧上空 TCs 生成较多一些(图10b)。

5 西北太平洋季风槽变异对 TCs 生成变异的动力作用

上述研究都表明了西北太平洋季风槽的位置和强度的年际和年代际变异对于 TCs 生成有重要影响。这不仅是由于季风槽可以为 TCs 生成提供有利的大尺度环境因子,而且还可以通过季风槽纬向风的辐合和切变使得 Rossby-重力混合波(即 MRG 波)转变成热带低压型波动(即 TD 型)扰动[5, 18, 35-37]。由于 TD 型扰动可以认为是 TCs 生成的先兆扰动(precursor disturbance),因此,西北太平洋季风槽对于 TCs 生成具有重要的动力作

用[5, 18, 37]。

5.1 西北太平洋上空的对流耦合波动

自从 Matsuno[38] 从理论上提出赤道波动以后,热带大气和海洋环流动力学得到很大发展。近几年,热带对流活动与赤道波动的相互作用已成为热带大气动力学研究的一个热点问题。由于它是深入研究西北太平洋 TCs 生成动力学的关键,文献[39-41]系统地研究了热带对流耦合波动。最近,Wu 等[42-43] 利用 Cloud Archive User Service(CLAUS)的云顶亮温(T_b)资料、NOAA 的射出长波辐射(OLR)资料以及再分析风场资料,并基于 Wheeler-Kiladis 滤波方法[44]研究了包括 Rossby-重力混合波(MRG 波)、热带低压型波动(TD 型波动),西传赤道 Rossby 波(ER 波)等热带对流耦合波动的特征及其在北太平洋季风槽中的演变过程。

5.2 西北太平洋季风槽对 TCs 生成的动力作用

TCs 生成先兆扰动的 TD 型波动(或 TD 型扰动)的波数和频率大于 MRG 波,而波长短于 MRG 波[42-43]。Takayabu 等[35]指出了 MRG 波进入热带西太平洋后将逐渐转变为向西北方向传播的 TD 型扰动。MRG 波如何转变成 TD 型扰动?关于此问题,文献[5, 18, 37]做了深入研究,他们把 Webster 等[45]所推导的赤道 Rossby 波在沿波传播路径传播时波数变化与纬向风辐合的关系推广到 MRG 波与 TD 型波动的波数域中,即,

$$\frac{d_g k}{dt} = -k \frac{\partial \bar{u}}{\partial x} \quad (1)$$

式中,k 是 MRG 波的纬向波数,$\bar{u}(x)$ 是纬向风的时间平均,d_g/dt 是沿群速度(即波的传播路径)方向的随体微商。

式(1)表明了在纬向风辐合区域传播的赤道 MRG 波的纬向波数将增加,其波长将变短。黄荣辉等[5]、Chen 等[18, 37]指出:由于西北太平洋季风槽的东端是纬向风的强辐合之处,这样,当波长较长而波数较小的 MRG 波西进到季风槽的东端,此波动在季风槽纬向风强辐合的作用下,它将转变成波数较大而波长较短的 TD 型扰动,这将为 TC 生成提供先兆(precursor)扰动。他们还利用同样原理,得到:

$$\frac{d_g m}{dt} = -m \frac{\partial \bar{u}}{\partial y} \quad (2)$$

式中 m 是经向波数。

式(2)表明了在纬向风的经向切变强的区域，赤道 MRG 波的经向波数将增加，其波长将变短。文献[5,18,37]的研究表明：西北太平洋季风槽中是纬向风强切变之处，这样，当 MRG 波西进入季风槽东端时，此波动在季风槽东端纬向风辐合的作用下，即在式(1)中 $\frac{\partial \bar{u}}{\partial x} < 0$，故 $\frac{d_g k}{dt} > 0$，则其纬向波数增大，而波长将变短；而当此波继续传播到季风槽中，在季风槽纬向风的经向切变作用下，即在式(2)中 $\frac{\partial \bar{u}}{\partial y} < 0$，故 $\frac{d_g m}{dt} > 0$，则其经向波数增大。因此，在季风槽纬向风的纬向辐合和经向切变的共同作用下，此波动不仅纬向波数继续变大，而且经向波数将变大，波长继续变短，这样不断地变形、分裂，最终形成 TC。

黄荣辉等[5]、Chen 等[17-18]利用 1980—2001年再分析的 850 hPa 风场和 OLR 资料进行了 3~6 d 的带通滤波，并以 160°E,0°的经向风为参考点分别对西太平洋暖池偏暖年和偏冷年的 7—10 月热带西太平洋上空 850 hPa 高频(3~6 d)变化风场和 OLR 距平场求线性回归，从而分析了与西太平洋暖池次表层热力相关的季风槽位置的不同对 MRG 波转变成 TD 型扰动的影响。最近本文作者利用 1979—2013 年再分析的 850 hPa 风场和 OLR 资料进行了 3~6 d 的带通滤波，并以 160°E,0°的经向风为参考点，分别对季风槽偏强年和偏冷年的 7—9 月热带西太平洋上空 850 hPa 高频(3~6 d)变化风场和 OLR 场进行线性回归，并将回归系数场乘以 0.1 m/s，从而分析了西北太平洋季风槽偏强和偏弱对 MRG 波转化成 TD 型扰动的动力作用。分析结果表明：在西北太平洋季风槽偏强年的 7—9 月，这时西北太平洋上空季风槽位置东伸到西北太平洋东侧上空(图 9a)，如图 13a~13c 所示，强对流活动位于波列的气旋环流中心区，这表明西北太平洋偏东侧上空已存在明显的 TD 型扰动，在 160°E 附近的对流层低层 MRG 波已转变成 TD 型扰动；相反，在西北太平洋季风槽偏弱年的 7—9 月，这时季风槽位置位于西北太平洋西侧上空(图 9b)，如图 14a~14c 所示，强对流活动并不位于波列中气旋环流的中心区，而是位于波列的气旋环流和反气旋环流之间，表明这种情况西北太平洋偏东侧上空没有明显的 TD 型扰动，在

160°E 附近的对流层低层 MRG 波还没转变成 TD 型扰动。Wu 等[43]的研究也表明了在西北太平洋季风槽东伸的年份，西北太平洋东南侧上空 TCs 生成是与季风槽把 MRG 波转变成 TD 型扰动的动力作用有密切关系。

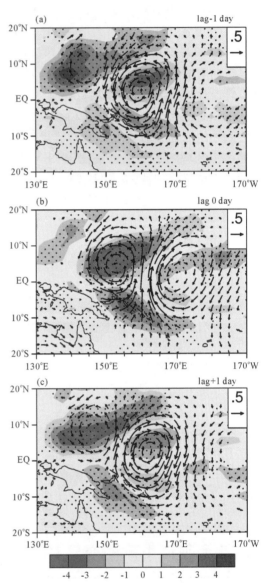

图 13 西北太平洋季风槽偏强年的 7—9 月以 160°E,0°的经向风为参考点的超前 1 d(a)、同时(b)和落后 1d(c) 线性回归的西北太平洋上空 850 hPa 面上 3~6 d 滤波的高频风场(矢量，单位:0.5 m/s)和 OLR 距平场(阴影区，单位:W/m²)的合成分布　风场资料取自 NCEP/NCAR 再分析资料[25]。

上述分析结果表明了西北太平洋季风槽在 MRG 波转变成 TC 先兆扰动的 TD 型扰动起着重要的动力作用。

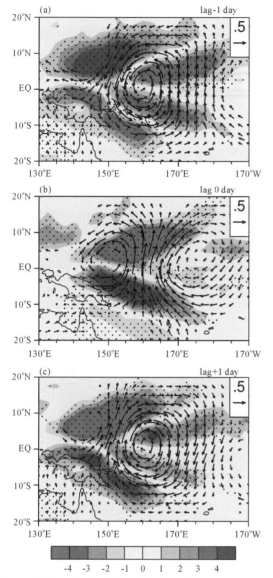

图 14　同图 13，但为西北太平洋季风槽偏弱年的 7—9 月

5.3　从正压能量转换来看西北太平洋季风槽中扰动发展的动力作用

为了更深入地说明西北太平洋季风槽对 TC 生成的动力作用，Wu 等[46]进一步利用正压能量转换方程来诊断季风槽对 TC 生成的动力作用。他们所应用的正压能量方程为，

$$\frac{\partial K'}{\partial t} = -\overline{(u')^2}\frac{\partial \bar{u}}{\partial x} - \overline{u'v'}\frac{\partial \bar{u}}{\partial y} - \overline{u'v'}\frac{\partial \bar{v}}{\partial x} - \overline{v'^2}\frac{\partial \bar{v}}{\partial y}$$

（3）

式中，u'、v' 分别为瞬变扰动的纬向和经向风速，$\bar{u}$、$\bar{v}$ 分别是时间平均的纬向和经向风速，K' 是瞬变扰动的动能。Wu 等[46]利用式（3）对 6—10 月西北太平洋上空对流层低层 850 hPa 的平均气流与瞬变扰动之间的正压能量转化进行了定量计算和分析。最近，本文作者对 7—9 月季风槽强盛期的西北太平洋上空对流层低层 850 hPa 的平均气流与瞬变扰动之间的正压能量转化进行了计算和分析。如图 15a 与 15b 所示，无论在西太平洋暖池偏暖年（即季风槽位置偏西、强度偏弱）或是在偏冷年（即季风槽位置偏东、强度偏强）的 7—9 月，西北太平洋对流层低层沿着季风槽存在着明显的平均气流的动能向扰动动能转化；并且，若把图 15b 与 15a 相比较可看到，该能量转化区随着季风槽的增强东伸也加强东进。因此，季风槽位置和强度的变化引起了大尺度平均气流的动能向较小尺度瞬变扰动动能转化的变化，从而影响 TCs 的生成位置和强度的相应变化。

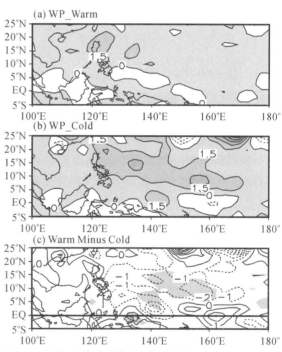

图 15　西太平洋暖池偏暖（季风槽位置偏西，a）和偏冷（季风槽位置偏东，b）年的 7—9 月 850 hPa 面上扰动动能倾向 $\partial K'/\partial t$ 的合成分布　　单位：$10^{-5} m^2/s^3$。

Huang 等[47]根据式（3）从动力理论分析了扰动在季风槽中的演变过程，提出了西北太平洋季风槽在从 MRG 波向 TD 型扰动转变且发展成 TC 过程中的动力作用概念图（图 16）。图 16 表明：当 MRG 波从赤道中、东太平洋向西传播过程中，波在季风槽以东时，其结构呈现 MRG 波；而波传播接近西太平洋季风槽东端的位置时，波的强度开始加强，水平尺度收缩，并呈东北-西南倾斜；当此种波动沿着季风槽向西北行，波的强度不断加强并呈"腰子型"结构，之后分裂成尺度更小的涡旋，其垂直结构呈正压结构、对流中心与涡旋中心重合，此时波更具有 TD 波的性质。

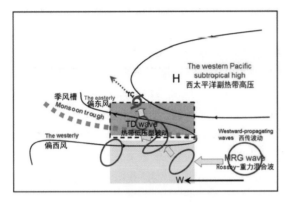

图 16　在西北太平洋季风槽 MRG 波向 TD 型扰动转变的动力过程概念图

6　西北太平洋季风槽对 TCs 生成动力作用的简单数值模拟

上节从动力理论讨论了西北太平洋季风槽对 TCs 生成的动力作用，为了进一步研究季风槽对 TCs 生成的作用，Wu 等[46]利用一组与 Aiyyer 等[48]所用模型相类似的浅水波方程组对季风槽在对流耦合波动转化中的动力作用进行简单的数值试验。

6.1　浅水波方程组及简单的数值试验

Wu 等[46]使用了 β 平面上的浅水波方程，将方程中各项分解为基本态（不随时间变化）和扰动态（随时间变化），同时忽略非线性项影响，从而得到一组基本态方程和一组扰动方程。其中扰动方程组可以写成，

$$\frac{\partial u'}{\partial t} + \bar{u}\frac{\partial u'}{\partial x} + u'\frac{\partial \bar{u}}{\partial x} + \bar{v}\frac{\partial u'}{\partial y} + v'\frac{\partial \bar{u}}{\partial y} - \beta y v' = -g\frac{\partial h'}{\partial x} \quad (4)$$

$$\frac{\partial v'}{\partial t} + \bar{u}\frac{\partial v'}{\partial x} + u'\frac{\partial \bar{v}}{\partial x} + \bar{v}\frac{\partial v'}{\partial y} + v'\frac{\partial \bar{v}}{\partial y} + \beta y u' = -g\frac{\partial h'}{\partial y} \quad (5)$$

$$\frac{\partial h'}{\partial t} + \frac{\partial}{\partial x}\left[(\bar{h}+H)u' + h'\bar{u}\right] + \frac{\partial}{\partial y}\left[(\bar{h}+H)v' + h'\bar{v}\right] = 0 \quad (6)$$

方程(4~6)中，u'、v'、h' 分别是扰动的纬向、经向风速和高度，而 $\bar{u}$、$\bar{v}$、$\bar{h}$ 分别是基本态（即季风槽及其周边的环流）纬向、经向风速及高度，g 是重力加速度，β 是科氏参数的经向梯度，H 是平均流体深度，约为 25 m。

为了比较西北太平洋上空强与弱季风槽大尺度背景场中 MRG 波和 ER 波演变的差别，本数值试验分别以强与弱季风槽年西北太平洋上空 850 hPa 流场的合成分布作为大尺度背景流场（图17），并分别把 MRG 波和 ER 波的经典理论解（图略）作为扰动场，使用上述线性浅水波数值模型分别对在图 17a 和 17b 所示的西太平洋强与弱季风槽基本气流作用下的天气尺度 MRG 波和 ER 波的演变进行简单数值试验。

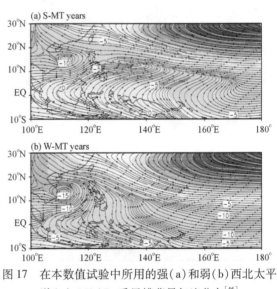

图 17　在本数值试验中所用的强(a)和弱(b)西北太平洋上空 850 hPa 季风槽背景气流分布[46]

6.2　在季风槽中 MRG 波演变的数值试验

6.2.1　在强季风槽中的演变

图 18a 是应用上述线性浅水波数值模型对 MRG 波在西北太平洋上空 850 hPa 强季风槽背景流场（图 17a）作用下的演变所进行数值试验的第 7 d 计算结果。从图 18a 可以看到：在强季风槽的作用下，MRG 波的波长变短，并沿季风槽向西北太平洋上空的西北方向传播，其结构先变成在季风槽的槽线以南呈东北-西南倾斜，而在季风槽的槽线以北呈西北-东南倾斜的"腰子型"结构的扰动，之后逐渐变成尺度较小偏离赤道向西北方向传播的涡旋，即 TD 型波动；并且如图 18a 中阴影区所示，季风槽通过正压能量转换使扰动强度增长和结构的变化。

6.2.2　在弱季风槽中的演变

图 18b 是应用上述线性浅水波数值模型对 MRG 波在西北太平洋上空 850 hPa 弱季风槽背景流场（图 17b）作用下的演变所进行数值试验的第 7 d 计算结果。从图 18b 同样看到：在弱季风槽的

作用下,MRG 波的波长也变短,也沿季风槽向西北太平洋上空的西北方向传播,其结构也是先变成在季风槽线以南呈东北-西南倾斜而在季风槽线以北呈西北-东南倾斜的"腰子型"结构,之后逐渐变成尺度较小、偏离赤道向西北方向传播的涡旋,即 TD 型波动,这些与强季风槽作用下 MRG 波演变特征相同。

6.2.3 在强与弱季风槽中演变的差异

从图 18a 与图 18b 的比较可以看到:在强季风槽的作用下,由于季风槽东伸,故 MRG 波变成 TD 型波列的位置偏东,而在弱季风槽的作用下,由于季风槽位置较偏西,故 MRG 波变成 TD 型波列的位置略偏西;并且,图 18b 中阴影所示的弱季风槽正压能量转换要比图 18a 中阴影所示的强季风槽正压能量转换略弱,因此在弱季风槽作用下,MRG 波变成 TD 型波动强度要比强季风槽中 TD 型波动略弱。

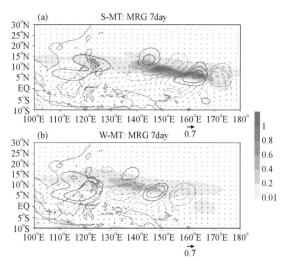

图 18 利用线性浅水模型对 MRG 波在强季风槽(a)和弱季风槽(b)基本态作用下演变所进行数值模拟的第 7 d 风场(箭头)和高度场(等值线,仅标出 ±0.2、0.4、0.8、1.6 和 3.2 m)[46] 阴影是 7 d 平均的正压能量转化项($10^{-5} m^2/s^3$)。

上述数值试验结果表明:季风槽大尺度环流不仅为西北太平洋上空对流层低层扰动增长提供能量,使得扰动强度增强,而且季风槽大尺度气流的纬向辐合和经向切变使得扰动的尺度收缩,扰动的结构发生改变,使 MRG 波向水平尺度较小的 TD 型波动转变,从而为西北太平洋 TC 生成提供先兆扰动。最近,冯涛等[29]的研究结果也证实了上述简单数值试验结果,他们的分析表明:在西太平洋暖池偏冷情况下,由于西北太平洋季风槽东伸,故 MRG 波转化成 TD 型波列位于西北太平洋偏东、偏南侧上空,从而导致了在西北太平洋偏东、偏南侧上空不断出现 TD 型波列,这些波列为 TCs 生成提供先兆扰动;相反,在西北太平洋偏暖情况下,由于西北太平洋季风槽位置偏西,故 MRG 波转化成 TD 型波列往往位于西北太平洋偏西、偏北侧上空,从而导致了在西北太平洋偏西、偏北侧上空经常出现 TD 型波列,为 TCs 生成提供先兆扰动。

6.3 在季风槽中 ER 波演变的数值试验

6.3.1 在强季风槽中的演变

图 19a 是应用上述线性浅水波数值模型对 ER 波在西北太平洋上空 850 hPa 强季风槽背景流场作用下的演变进行数值试验的第 7 d 计算结果。从图 19a 可看到:在强季风槽作用下,ER 波的波长变短,扰动沿季风槽向西北太平洋上空偏西方向传播,其结构呈东北-西南向倾斜,图 19a 中阴影所示的正压能量转换使扰动强度增长和结构变化。

6.3.2 在弱季风槽中的演变

图 19b 是应用上述线性浅水波数值模型对 ER 波在西北太平洋上空 850 hPa 弱季风槽背景流场作用下的演变进行数值试验的第 7 d 计算结果。由图 19b 同样看到:在弱季风槽作用下,ER 波的波长也变短,扰动也沿季风槽向西北太平洋上空偏西方向传播,其结构也呈东北-西南向倾斜。但与图 19a 所示的波列位置和强度相比,在弱季风槽作用下,波列的位置略偏西,并且这种情况波列的扰动强度和正压能量转换都不及强季风槽作用下扰动的强度和正压能量转换。

6.3.3 在季风槽中 MRG 波与 ER 波演变的差异

把图 18a 和 18b 与图 19a 和 19b 相比较可清楚看到:ER 波与 MRG 波在季风槽中的演变存在差异。在季风槽的作用下,向西传播的 MRG 波可以变成尺度较小的偏离赤道向西北太平洋上空西北方向传播的 TD 型波动,从而为 TC 生成提供先兆扰动;而同样在季风槽的作用下,向西传播的 ER 波不能变成尺度较小的偏离赤道向西北太平洋上空的西北方向传播的 TD 型波动,只是变成向西传播的东北-西南倾斜的扰动,其波长比 TD 型波动的波长要长。但应该看到:在季风槽作用下,ER 波所演变的扰动强度和正压能量转换都比

MRG波所变成的TD型扰动强度和正压能量转换大。因此,在季风槽中ER波所演变的扰动虽不能成为TC生成的"先兆扰动",但它可以为TC的生成和发展提供扰动能量。

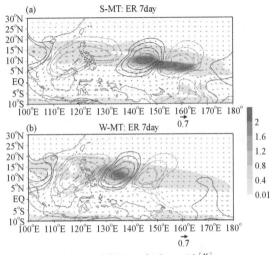

图19 同图18,但为ER波[46]

上述数值试验清楚表明了西太平洋季风槽位置和强度不仅影响ER波的演变,而且对于西北太平洋上空的MRG波的演变起着重要的动力作用,季风槽通过MRG波向TD型波动的转化为西北太平洋上空TC生成先兆扰动的TD型波列提供扰动能量,对TC的生成起着重要的动力作用。

7 总结及今后需进一步研究的科学问题

本文回顾和综述了近年来关于西北太平洋季风槽的年际和年代际变异特征及其对TCs生成的影响过程和机理的研究进展。指出了西北太平洋季风槽是夏、秋季西北太平洋上空对流层低层与亚洲季风、跨赤道气流以及偏东信风气流相关联的一种大尺度环流系统。它可以为西北太平洋上空TCs生成提供强的对流层低层的气旋性相对涡度和强的对流层上层气流的辐散、对流层中、下层充足的水汽以及对流层上下层小的风场垂直切变等有利因子。本文还揭示了受西太平洋暖池和热带太平洋热力的年际和年代际变异的影响,西北太平洋季风槽位置和强度有很明显的年际和年代际变异。如图20a所示,当季风槽向西北太平洋偏东侧伸展时,它的强度偏强,利于TCs生成的大尺度环境因子分布也向西北太平洋东侧上空伸展,故TCs生成位置偏于西北太平洋中、东侧上空;相反,如图20b所示,当季风槽向西北太平洋偏西侧上空收缩时,它的强度偏弱,利于TCs生成的大尺度环境因子分布也向西北太平洋偏西侧上空收缩,故TCs生成位置也偏于西北太平洋西侧上空。本文还提出:由于1990年代末之后热带太平洋发生了类似于"ENSO-modoki型"的年代际增温,热带中、西太平洋海温升高,西北太平洋季风槽平均位置向西北太平洋西侧上空收缩,利于TCs生成的大尺度环境因子分布也偏于西北太平洋西侧上空,造成了从1990年代末之后TCs生成的平均位置偏于西北太平洋西侧上空。本文还通过动力理论分析和简单数值模拟说明了西北太平洋季风槽不仅通过影响利于TCs生成的大尺度环境因子分布来影响TCs生成,而且还通过纬向风的纬向辐合和经向切变使MRG波转变成波长较短的TD型波动,从而为TC生成提供"先兆扰动",并为TD型波动强度增强提供扰动能量。

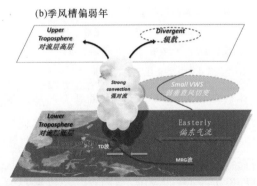

图20 夏、秋季西北太平洋季风槽位置和强度的年际变异对TCs生成影响过程示意图

尽管近年来我们在关于西太平洋季风槽年际和年代际变异及其对西北太平洋上空TCs活动的影响过程和机理方面的研究取得若干重要进展,但还有许多科学问题亟需进一步深入研究。

(1)西北太平洋上越赤道气流、西太平洋副热带高压等大尺度环流以及 MJ 振荡(MJO)、准双周振荡(QBWO)和赤道波动对于西北太平洋季风槽和 TCs 活动的季节内、年际和年代际变异有很大影响。因此,西北太平洋上空的季风槽与不同时空尺度的大气环流、波动、对流和 TCs 的相互作用过程和机理亟需今后进一步深入研究。

(2)西北太平洋季风槽的变异不仅受热带太平洋 ENSO 循环的影响,而且受西太平洋暖池和热带印度洋的海-气相互作用所调控。因此,西太平洋暖池和热带印度洋的海-气相互作用变异调控季风槽的季内、年际和年代际变异的过程及机理也值得进一步研究。

(3)研究表明了西北太平洋季风槽可以把西北太平洋上空对流层低层 MRG 波转化成空间尺度较小的 TD 型波动(TC 生成的先兆扰动)。然而,TD 型扰动空间尺度大,它在什么样的动力和热力作用下尺度收缩、强度加强,从而发展成 TC 的过程应作进一步研究;并且,ER 波在季风槽的动力作用下也可以变成尺度较小的波动,此波动虽不是 TD 型波动,它对 TC 生成到底起什么作用还不清楚,也值得进一步深入研究。

(4)目前,国内外许多关于西北太平洋 TCs 气候学的研究,其中较多关注季风槽对 TCs 生成的作用,而关于季风槽对 TCs 移动路径及登陆地点的影响关注较少。因此,今后在关于西北太平洋 TCs 活动的气候学研究方面应该对于 TCs 移动路径的年际和年代际变异及与季风槽的关联给予关注。

(5)目前,由于我们对 TCs 生成与移动的物理过程了解还不十分清楚,特别是积云对流在模式中描述不准确,并且受气候数值模式中强迫因子、初始条件、边界条件和模式的分辨率的约束,在气候模式的 TCs 模拟中难于生成 TCs,往往要采取"人造台风"方法来模拟季风槽大尺度环流背景下 TCs 活动的变化,这就大大增加了 TCs 活动变化模拟的困难和不确定性。因此,这需要今后长期不断改进 TCs 和台风气候数值模式的研究。

(6)目前,为了减轻 TCs 给我国造成的严重灾害,有关部门亟需对西北太平洋 TCs 活动及登陆我国台风的季度和年际变异进行预测。为此,一方面需要系统地深入分析 ENSO 循环以及西太平洋暖池和热带印度洋的海-气相互作用与西北太平洋上空季风槽变异和 TCs 活动之间的关联;另一方面应该深入对西北太平洋季风槽变异及其对 TCs 活动的季节内、年际和年代际变异的影响过程和机理进行深入的探讨。此外,要对 TCs 活动状况作出客观定量的季节和年际预测,还需要不断改进以往 TCs 气候数值模式,提高模式对 TCs 的模拟和预测能力。

总之,为了满足国民经济建设和社会发展的需求以及减轻台风造成的灾害,当前开展西北太平洋 TCs 活动和登陆我国地点的季度和年际预测是非常必要的。为此,我国今后应特别重视西北太平洋季风槽和 TCs 活动及登陆我国台风的气候学研究,以便为 TCs 活动和登陆地点的短期气候预测提供科学依据,并争取在这方面的理论研究做出国际一流的研究成果。

参 考 文 献

[1] CORNEJO-GARRIDO A G, STONE P H. On the heat balance of the Walker circulation[J]. J Atmos Sci, 1977, 34(8): 1 155-1 162.
[2] HARTMANN F, HENDON H, HOUZE R A. Some implications of the meso-scale circulation in tropical cloud clusters for large-scale dynamics and climate[J]. J Atmos Sci,1984, 41(1): 113-121.
[3] ELSBERRY R. Global Perspective on Tropical Cyclone[M]. WMO TD-NO 693,Ch.4, WMO, Geneva, Switzerland, 1995:106-197.
[4] ELSBERRY R. Monsoon-related tropical cyclones in East Asia[M]// East Asian Monsoon. CHANG C P. World Scientific Publishing Co Pte Ltd, 2004: 463-498.
[5] 黄荣辉,陈光华. 西北太平洋热带气旋移动路径的年际变化及其机理研究[J]. 气象学报,2007, 65(5):683-694.
[6] 黄荣辉,王磊. 台风在我国登陆地点的年际变化及其与夏季东亚/太平洋型遥相关的关系[J]. 大气科学,2010, 34(5):853-864.
[7] GRAY W M. Global view of the origin of tropical disturbances and storms[J]. Mon Wea Rew, 1968, 96(10): 669-700.
[8] WU L, WEN Z P, HUANG R H. Possible linkage between the monsoon trough variability and the tropical cyclone activity over the western North Pacific[J]. Mon Wea Rew, 2012, 140(1): 140-150.

[9] MCBRIDE J L. Tropical cyclone formation. chap3 global perspectives on tropical cyclones[M]//Tech Doc WMO/ TD No693 WMO. Geneva, Switzerland, 1995: 63-105.

[10] 冯涛,黄荣辉,陈光华,等. 近年来关于西北太平洋热带气旋和台风活动的气候学研究进展[J]. 大气科学,2013, 37(2):364-382.

[11] FENG T, CHEN G H, HUANG R H, et al. Large-scale circulation patterns favorable to tropical cyclogenesis over the western North Pacific and associated barotropic energy conversions[J]. Internati J Climat, 2014, 34(1): 216-227.

[12] 潘静,李崇银. 夏季南海季风槽与印度季风槽的气候特征之比较[J]. 大气科学,2006, 30(3):377-390.

[13] 李崇银,潘静. 南海夏季风槽的年际变化和影响研究[J]. 大气科学,2007, 31(6):1 049-1 058.

[14] LANDER M A. An exploratory analysis of the relationship between tropical storm-formation in the western North Pacific and ENSO[J]. Mon Wea Rev, 1994, 122(4): 636-651.

[15] BRIEGEL L M, FRANK W M. Large-scale influences on tropical cyclogenesis in the western North Pacific[J]. Mon Wea Rev, 1997, 125(7): 1 397-1 413.

[16] RITCHIE E A, HOLLAND G J. Large-scale patterns associated with tropical cyclogenesis in the western Pacific[J]. Mon Wea Rev, 1999, 127(9): 2 027-2 043.

[17] CHEN G H, HUANG R H. Influence of monsoon over the warm pool on interannual variation of tropical cyclone activity over the western North Pacific[J]. Adv Atmos Sci, 2008, 25(2): 319-328.

[18] CHEN G H, HUANG R H. Role of equatorial wave transitions in tropical cyclogenesis over the western North Pacific[J]. Atmos Ocea Sci Lett, 2008, 1(1): 64-68.

[19] HUANGFU J L, HUANG R H, CHEN W. Interdecadal increase of tropical cyclone genesis frequency over the western North Pacific in May [J]. Internation J Climatol, 2016, doi:10.1002/joc.4760.

[20] 陈联寿,丁一汇. 西太平洋台风概论[M]. 北京:科学出版社,1979: 491.

[21] 王斌,ELSBERRY R L,王玉清,等. 热带气旋运动的动力学研究进展[J]. 大气科学,1998, 22(4):535-547.

[22] CHAN J C L. Interannual and interdecadal variations of tropical cyclone activity over the western North Pacific[J]. Meteor Atmos Phys, 2005, 89(1): 143-152.

[23] WEBSTER P J, HOLLAND G H, CURRY I A, et al. Changes in tropical cyclone number, duration, and intensity in a warming environment [J]. Science, 2005, 309(5 742): 1 844-1 846.

[24] EMANUEL K A. Increasing destructiveness of tropical cyclones over the past 30 years[J]. Nature, 2005, 436(7 051): 686-688.

[25] KALNAY E, COAUTHORS. The NCEP/NCAR 40-years reanalysis project[J]. Bull Amer Meteor Soc, 1996, 77(3): 437-471.

[26] KANAMITSU M, EBISUZAKI W, WOOLLEN J. et al, NCEP DOE AMIP-II reanalysis (R-2)[J]. Bull Amer Meteo Soc, 2002, 83(11): 1 631-1 643.

[27] 黄荣辉,皇甫静亮,刘永,等. 西太平洋暖池对西北太平洋季风槽和台风活动的影响过程及其机理的最近研究进展[J]. 大气科学, 2016, Doi.10.3878/j.issn1060-9895.112.15251.

[28] 冯涛,黄荣辉,陈光华,等. 热带西太平洋越赤道气流的年际变化对西北太平洋热带气旋生成的影响[J]. 热带气象学报,2014, 30 (1): 11-22.

[29] 冯涛,黄荣辉,杨修群,等. 2004 年与 2006 年 7—9 月西北太平洋上空大尺度环流场及其对热带低压型波列和热带气旋生成影响的差别[J]. 大气科学,2016, 40(1): 157-175.

[30] 叶笃正,黄荣辉. 长江黄河流域旱涝规律和成因研究[M]. 济南:山东科技出版社,1996: 387.

[31] LIN I I, PUN I P, LIEN C C. "Category-6" supertyphoon Haiyan in global warming hiate contribution from subsurface ocean warming[J]. Geophys Res Lett, 2014, 41(23): 8 542-8 553.

[32] HUANG P, LIN I I, CHOU, et al. Change in ocean subsurface environment to suppress tropical cyclone intensification under global warming [J]. Nature Communicat, 2015, 6:7188, Doi:10.1038/ncomms8188.

[33] 张翔,武亮,皇甫静亮,等. 西北太平洋季风槽的季节和年际变化特征[J]. 气候与环境研究,2016. doi:10.3878/j. issn. 1006-9585.

[34] HUANGFU J L, HUANG R H, CHEN W. Influence of tropical western Pacific warm pool thermal state on the interdecadal change of the onset of the South China Sea summer monsoon in the late-1990s[J]. AtmosOce Sci Lett, 2015, 8(1): 95-99.

[35] TAKAYABU Y N, NITTA T. 3~5 day-period disturbances coupled with convection over the tropical Pacific Ocean[J]. J Meteor Soc Japan, 1993, 71(2): 221-245.

[36] DICKINSON M, MOLINARI J. Mixed Rossby-gravity waves and western Pacific tropical cyclogenesis Part 1: Synoptic evolution[J]. J Atmos Sci, 2002, 59(14): 2 183-2 196.

[37] CHEN G H, HUANG R H. Interannual variations in mixed Rossby-gravity waves and their impacts on tropical cyclogenesis over the western North Pacific[J]. J Clim, 2009, 22(3): 535-549.

[38] MATSUNO T. Quasi-geostrophic motions in the equatorial area[J]. J Meteor Soc Japan,1966, 44(1): 25-42.

[39] YANG G Y, HOSKINS B, SLINGO J. Convectively coupled equatorial waves Part I: Horizontal and vertical structure[J]. J Atmos Sci, 2007, 64(10): 3 406-3 423.

[40] YANG G Y, HOSKINS B, SLINGO J. Convectively coupled equatorial waves Part II: Propagaation characteristics[J]. J Atmos Sci, 2007, 64(10): 3 424-3 437.

[41] KILADIS G H, WHEELER M C, HAERTEL P T, et al. Convectively coupled equatorial waves[J]. Rev Geophys, 2009, 47: RG2003, doi: 10.1029/2008 RG000266.

[42] WU L, WEN Z P, LI T. ENSO-phase dependent TD and MRG wave activity in the western North Pacific[J]. Clim Dynam, 2014, 42(5): 1 217-1 227.

[43] WU L, WEN Z P, WU R G. Influence of the monsoon trough on westward-propagating tropical waves over the western North Pacific, Part I: Observations[J]. J Clim, 2015, 28(18):7 108-7 127.

[44] WHEELER M, KILADIS G N. Convectively coupled equatorial waves: Analysis of clouds and temperature in the wavenumber-frequency domain[J]. J Atmos Sci, 1999, 56(3): 374-399.

[45] WEBSTER P J, CHANG H R. Equatorial energy accumulation and emanation regions: Impacts of a zonally varying basic state[J]. J Atmos Sci, 1988, 45(5): 803-828.

[46] WU L, WEN Z P, WU R G. Influence of the monsoon trough on westward-propagating over the western North Pacific, Part II: Energetics and numerical experiments[J]. J Clim, 2015, 28(23): 9 332-9 349.

[47] HUANG R H, WU L, CHEN G H. Interannual variations of the activities of tropical cyclones over the Northwest Pacific and their association with the West Pacific warm pool thermal states[C]//Proceedings of XXVIUGG General Assembly. Australia. 2011.

[48] AIYYER A R, MOLINARI J. Evolution of mixed Rossby-gravity waves in idealized MJO environments[J]. J Atmos Sci, 2003, 60(23): 2 837-2 855, doi:10.1175/1520-0469(2003)060,2837:EOMRWI.2.0.CO.

五、平流层重力波研究

平流层球面大气地转适应过程和惯性重力波的激发[*]

黄荣辉 陈金中

(中国科学院大气物理研究所,北京 1000800)

摘 要 观测表明,在1979年2月平流层爆发性增温期间由于准定常行星波的上传,平流层存在着强烈非地转运动,从而导致很强的散度场和很强的大振幅重力波活动。作者用 Hough 函数构造一个线性正压全球谱模式,以1979年2月22日平流层爆发性增温期间10 hPa 上的实际地转偏差作为初始扰动来模拟平流层在爆发性增温时非地转扰动的地转适应过程,并从散度场的变化来讨论平流层爆发性增温时重力波的激发和传播。模拟结果表明:由于在平流层爆发增温时,伴随着大范围的非地转运动的产生,气压场将很快与流场相互调整(此过程以气压场向流场调整为主),经过2~4 h 气压场与流场达到地转关系,并且在适应过程中散度场产生剧烈变化,这说明在适应过程中所激发的惯性重力波的活动是很剧烈的。

1 引言

非地转初始扰动的演变和发展实际上就是所谓的地转适应过程。长久以来,许多研究者从不同方面讨论了地转适应过程。按照经典大气动力理论,大气运动最根本的原因是由于大气质量分布不均匀,即由气压梯度力所产生,然而 Rossby 提出相反的看法,他指出大气质量的分布是大气运动的结果。Rossby[1,2]分析了一个初始只有流场而无气压梯度相平衡的演变过程,演变结果表明了流场变化不大,而产生了与科里奥利力相平衡的气压梯度,因此,他指出气压场向流场适应。之后,Cahn[3]、Charney[4]指出气压场向风场适应是通过惯性重力波的频散来完成的。这些研究说明了地转适应的存在性。叶笃正[5]首先指出,地转适应中存在着尺度问题,即在大尺度环流中,风场适应于气压场,而在中、小尺度系统中,气压场适应于风场。曾庆存[6~8]从数学上严格地证明这一理论,并且,他指出在高空气压场向风场适应,即运动变化的原因是动力性的。叶笃正和李麦村[9]对地转适应问题的研究作了总结。

20世纪70年代以后地转适应过程理论已日趋完善,并且开展了地转适应的数值模拟。Wiin-Nielson[10]、Janic 和 Wiin-Nielson[11]分别用柱面和一维球面模式讨论了理想化的非地转初始扰动的演变,并讨论了 β 效应对地转适应的影响。Walterscheid 和 Bouecher[12]讨论了由于非地转激发出来波动场的结构和特征。Fritts 和 Luo[13]利用一个二维模式讨论了急流附近在地转适应过程中激发出来的重力波的特征,指出急流附近的

---

[*] 本文原载于: 大气科学, 第26卷, 第3期, 289-306, 2002年5月出版.

非地转运动是重力惯性波的一个重要源区。Matsuda 和 Takayama[14]用一个浅水波全球谱模式研究了全球地转适应过程的特征，讨论了"地转适应"这一概念在整个球面上的适用性，并且在数值模式中，与叶笃正和李麦村[9]一样，他们用散度场来表征重力波的活动。

然而，上面的研究大部分集中讨论对流层大气环流中的地转适应问题与惯性重力波的激发。根据观测，在平流层爆发性增温期间经常可观测到大振幅惯性重力波[15]。陈金中和黄荣辉[16]的研究指出，在平流层发生爆发性增温期间，由于平流层流场和温度场急剧变化，在欧洲和北美上空平流层发生了很大的非地转运动，并引起很大的散度场，从而激发了大振幅的惯性重力波，这表明在平流层爆发期间存在着大范围的地转适应过程。因此，本文用一个正压球面模式来研究平流层爆发性增温期间的实际非地转初始扰动的地转适应过程，并讨论在地转适应过程中散度场的变化，以了解在适应过程中惯性重力波的活动情况。

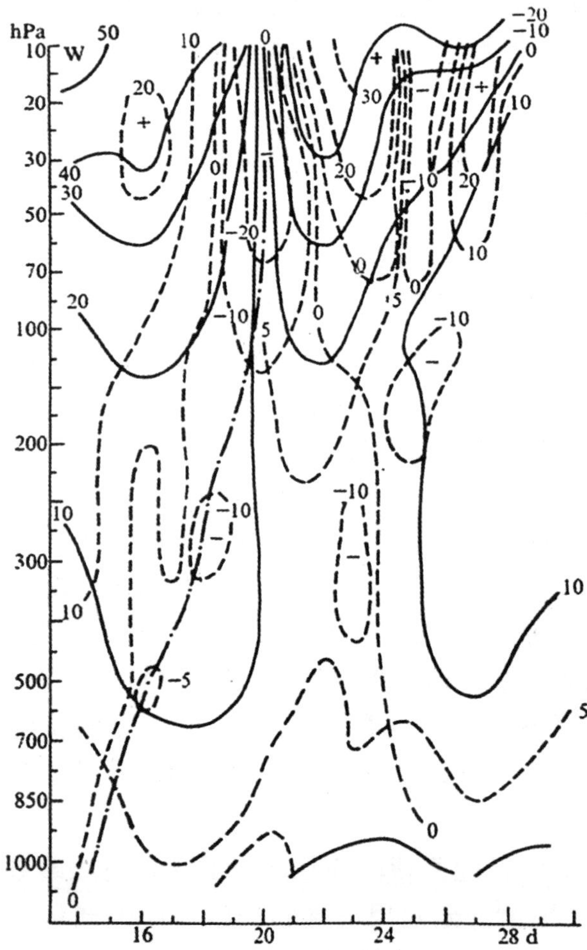

2 平流层爆发性增温时的非地转运动

黄荣辉与邹捍[17]分析和诊断了1979年2月平流层爆发性增温过程中行星波的垂直上传及其与基本气流相互作用。分析结果表明：由于在1979年2月上半月北半球高纬度对流层下层的基本气流由东风变成西风，使得波数2准定常行星波迅速上传至平流层；上传的行星波产生强的E-P通量的辐合，由于上传的行星波与基本气流相互作用，从而使得西风迅速减弱并变成东风（见图1），并形成了两个很强的反气旋环流；这两个很强的反气旋环流不仅导致了平流层爆发性增温，而且也导致了如图2所示的在（60°E，60°N）和（90°W，60°N）为中心强的非地转风偏差（参见文献[16]）。这个非地转偏差造成了流场的非地转不稳定，从而激发了惯性重力波，伴随之将产生地转适应过程。这个地转适应的物理过程将是本文所讨论的。

图1 1979年2月下半月沿75°N纬圈行星波E-P通量散度（虚线，单位：10^{14} m）与平均纬向风（实线，单位：m s^{-1}）的高度-时间剖面图

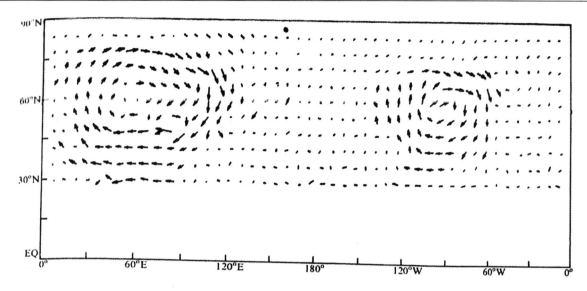

图 2 1979 年 2 月 22 日 10 hPa 的地转偏差的分布

3 球面正压大气数值模式与初值

3.1 数值模式

在平流层，由于气温不随高度变化，层结稳定，垂直运动较小，并且温度场的分布与高度场分布比较一致，因此，可以用一个正压模式来研究平流层大气运动的变化。

球面正压大气的运动方程可写成

$$\frac{\partial u}{\partial t} + u\frac{\partial u}{a\cos\varphi \partial \lambda} + v\frac{\partial u}{a\partial \varphi} - 2\Omega\sin\varphi v - \frac{uv}{a}\tan\varphi = -\frac{\partial \Phi}{a\cos\varphi \partial \lambda}, \quad (1)$$

$$\frac{\partial v}{\partial t} + u\frac{\partial v}{a\cos\varphi \partial \lambda} + v\frac{\partial v}{a\partial \varphi} + 2\Omega\sin\varphi u + \frac{u^2}{2}\tan\varphi = -\frac{\partial \Phi}{a\partial \varphi}, \quad (2)$$

$$\frac{\partial \Phi}{\partial t} + u\frac{\partial \Phi}{a\cos\varphi \partial \lambda} + v\frac{\partial \Phi}{a\partial \varphi} + \Phi\left(\frac{\partial u}{a\cos\varphi \partial \lambda} + \frac{\partial v\cos\varphi}{a\cos\varphi \partial \varphi}\right) = 0, \quad (3)$$

其中 u、v 分别是纬向、经向速度，$\Phi = gh$ 是位势，Ω 是地球自转角速度。

对方程（1）～（3）进行线性化和无量纲化，取下列无量纲量：

$$\tilde{u} = \frac{1}{\sqrt{gh_e}}u', \quad \tilde{v} = \frac{1}{\sqrt{gh_e}}v', \quad \tilde{h} = \frac{1}{h_e}h', \quad \tilde{t} = 2\Omega t;$$

并取 $\tilde{h}_0(\varphi) = \frac{1}{h_e}\bar{h}(\varphi)$ 为无量纲纬向平均高度，$\bar{\tilde{\omega}} = \frac{\bar{u}}{2\Omega a\cos\varphi}$ 为无量纲纬向基本气流。令 $\hat{u} = \bar{h}_0(\varphi)\tilde{u}$，$\hat{v} = \bar{h}_0(\varphi)\tilde{v}$，为简单起见，把 $\hat{u}$、$\hat{v}$、$\hat{h}$ 记为 u、v、h，则方程（1）～（3）可变成

$$\frac{\partial u}{\partial t} - \sin\varphi v + \frac{r}{\cos\varphi}\frac{\partial h}{\partial \lambda} = -\bar{\tilde{\omega}}\frac{\partial u}{\partial \lambda} - \left(\cos\varphi\frac{d\bar{\tilde{\omega}}}{d\varphi} - 2\bar{\tilde{\omega}}\sin\varphi\right)v, \quad (4)$$

$$\frac{\partial v}{\partial t} + \sin\varphi\, u + \frac{r\partial h}{\partial \varphi} = -\bar{\bar{\omega}}\frac{\partial v}{\partial \lambda} - 2\bar{\bar{\omega}}\sin\varphi\, u, \quad \text{.} \tag{5}$$

$$\frac{\partial h}{\partial t} + \frac{r}{\cos\varphi}\left(\frac{\partial u}{\partial \lambda} + \frac{\partial v\cos\varphi}{\partial \varphi}\right) = -\bar{\bar{\omega}}\frac{\partial h}{\partial \lambda} - \frac{r}{\cos\varphi}(\bar{h}_0 - 1)\left(\frac{\partial u}{\partial \lambda} + \frac{\partial v\cos\varphi}{\cos\varphi\, \partial \varphi}\right) - vr\frac{d\bar{h}_0}{d\varphi}, \tag{6}$$

方程组 (4) ~ (6) 中 u、v、h 是无量纲化的小扰动量, $\gamma = \varepsilon^{-1/2}$, $\varepsilon = 4a^2\Omega^2/gh_e$ 为 Lamb 数。方程组 (4) ~ (6) 可写成如下矢量形式:

$$\frac{\partial \boldsymbol{W}}{\partial t} + F\boldsymbol{W} = C\boldsymbol{W}, \tag{7}$$

其中:

$$F = \begin{bmatrix} 0 & -\sin\varphi & \dfrac{is\gamma}{\cos\varphi} \\ \sin\varphi & 0 & \gamma\dfrac{d}{d\varphi} \\ \dfrac{is\gamma}{\cos\varphi} & \dfrac{\gamma}{\cos\varphi}\dfrac{d}{d\varphi}[(\ )\cos\varphi] & 0 \end{bmatrix}, \tag{8}$$

$$C = \begin{bmatrix} \bar{\bar{\omega}}\dfrac{\partial}{\partial \lambda} & -\left(\cos\varphi\dfrac{d\bar{\bar{\omega}}}{d\varphi} - 2\bar{\bar{\omega}}\sin\varphi\right) & 0 \\ -2\bar{\bar{\omega}}\sin\varphi & \bar{\bar{\omega}}\dfrac{\partial}{\partial \lambda} & 0 \\ -\dfrac{\gamma}{\cos\varphi}(\bar{h}_0-1)\dfrac{\partial}{\partial \lambda} & -\dfrac{\gamma}{\cos\varphi}(\bar{h}_0-1)\dfrac{\partial(\ )\cos\varphi}{\cos\varphi\, \partial\varphi} - \gamma\dfrac{d\bar{h}_0}{d\varphi} & -\bar{\bar{\omega}}\dfrac{\partial}{\partial \lambda} \end{bmatrix}, \tag{9}$$

$$\boldsymbol{W} = \begin{bmatrix} u \\ v \\ h \end{bmatrix}. \tag{10}$$

由于本文讨论平流层球面地转适应过程问题,利用 Hough 函数来构造谱模式有一定优越性。Hough 函数是 Laplace 潮汐方程的特征解。Kasahara[18] 的研究表明:虽然 Hough 函数不是有基流时线性正压浅水波方程的特征解,但对于相同数量的模式变量,Hough 函数展开法是一种比球函数更有效的工具,利用 Hough 函数展开法可以方便地把快波(重力波)和慢波(涡漩波)分开;并且,Hough 函数展开法中三个要素 u、v、h 只有一个共同的谱系数,这意味着三个要素之间有着一种内在的动力学约束,物理意义更明显。Kasahara[19] 和董双林等[20] 都专门讨论过 Hough 函数的计算。因此,本研究利用 Hough 函数 $H_l^s e^{is\lambda}$ 展开矢量方程 (7) 变量(见附录),并设解为

$$\boldsymbol{W} = \sum_{s=-\infty}^{+\infty}\sum_{l=1}^{+\infty} w_l^s(t) H_l^s e^{is\lambda}, \tag{11}$$

$$H_l^s = -\begin{bmatrix} \hat{U}_l^s \\ \hat{V}_l^s \\ \hat{Z}_l^s \end{bmatrix}, \tag{12}$$

并有 $H_l^{-s} = [H_l^s]$，其中 $w_l^s(t)$ 只是与 t 有关的谱系数，l 是经圈模的求和指数。Longuet-Higgins[21]系统地研究了潮汐方程的特征频率和特征函数，指出方程包含着两类波动。因此，上述展开式包括了 LWG（西传惯性重力波），LR（西传 Rossby 波），LEG（东传惯性重力波）这几种模态。在 $s=0$ 时，只要重力波模态成对选取，则展开式（11）是实数，即

$$w_l^s(t) = \frac{1}{2\pi} \int_0^{2\pi} \int_{-1}^{1} \boldsymbol{W} \cdot H_l^s e^{-is\lambda} d\mu d\lambda, \tag{13}$$

上式中 $\mu = \sin\varphi$。

把（11）式代入（7）式，同乘 $H_l^{s*} e^{-is\lambda}$，沿球面积分，注意到 Hough 函数是算子 F 的特征解以及 Hough 函数的正交性，于是从（7）式可得：

$$\frac{dw_l^s}{dt} + i(\sigma_l^s)_0 w_l^s + i\sum_{l=1}^{\infty} b_{l'l}^s w_l^s = 0, \qquad l = 1,2,3,\cdots. \tag{14}$$

其中：

$$b_{l'l} = \int_{-1}^{1} \Big\{ (\hat{U}_l^s \hat{U}_{l'}^s + \hat{V}_l^s \hat{V}_{l'}^s + \hat{Z}_l^s \hat{Z}_{l'}^s) + 2\bar{\bar{\omega}}\sin\varphi(\hat{V}_l^s \hat{U}_{l'}^s + \hat{U}_l^s \hat{V}_{l'}^s) \\ - \cos\varphi \frac{d\bar{\bar{\omega}}}{d\varphi} \hat{V}_l^s \hat{U}_{l'}^s - r\frac{d\bar{h}_0}{d\varphi} \hat{V}_l^s \hat{Z}_{l'}^s + \frac{(\bar{h}_0 - 1)}{\cos\varphi} \Big[s\hat{U}_l^s - \frac{d}{d\varphi}(\hat{V}_l^s \cos\varphi) \Big] \hat{Z}_{l'}^s \Big\} d\mu, \tag{15}$$

σ_0 是 Laplace 潮汐方程的特征频率。记 $W_l^s = \alpha_l^s + i\beta_l^s$，则（14）可写成

$$\begin{cases} \dfrac{d\alpha_l^s}{dt} - (\sigma_{l'}^s)_0 \beta_l^s - \sum_l^L b_{l'l}^s \beta_l^s = 0, \\ \dfrac{d\beta_l^s}{dt} + (\sigma_{l'}^s)_0 \alpha_l^s + \sum_l^R b_{l'l}^s \alpha_l^s = 0, \end{cases} \tag{16}$$

令

$$X = (\alpha_1^s, \alpha_2^s, \cdots, \alpha_L^s, \beta_1^s, \beta_2^s, \cdots, \beta_L^s), \tag{17}$$

$$D = \begin{bmatrix} 0 & -E \\ E & 0 \end{bmatrix}, \tag{18}$$

$$E = \begin{bmatrix} (\sigma_1^s)_0 + b_{11}^s & b_{12}^s & b_{13}^s & \cdots & b_{1L}^s \\ b_{21}^s & (\sigma_2^s)_0 + b_{22}^s & b_{23}^s & \cdots & b_{2L}^s \\ \cdots & \cdots & \cdots & \cdots & \cdots \\ b_{L1}^s & b_{L2}^s & b_{L3}^s & \cdots & (\sigma_L^s)_0 + b_{LL}^s \end{bmatrix}. \tag{19}$$

这样，可得谱系数方程如下：

$$\frac{dX}{dt} + DX = 0, \tag{20}$$

研究扰动的发展变化归根到底就是求解方程（20）。

3.2 数值方法

把方程（20）离散化，即

$$\frac{X^{n+1} - X^n}{\Delta t} + D\frac{X^{(n+1)} + X^n}{2} = 0, \tag{21}$$

则可得 X 的迭代方程如下：

$$X^{n+1} = \left(1 + \frac{\Delta t D}{2}\right)^{-1}\left(1 - \frac{\Delta t D}{2}\right)X^n. \tag{22}$$

在计算过程中，沿经圈的积分用 64 点高斯积分，模式初值也都插到高斯格点上；截断波数至 $s=16$，截断模态数至 $L=60$。在 $s \geq 1$ 时，30 个对称模为 LEG = 0, 2, 4, …, 16，LR = 1, 3, 5, …, 23，LWG = 0, 2, 4, …, 16；30 个反对称模为 LEG = 1, 3, …, 17，LR = 0, 2, 4, …, 24，LWG = 1, 3, 5, …, 17；在 $s=0$ 时，成对选取重力波 l_g，$l_g^* = 0, 1, 2, …, 35$；LR = −1, 1, 2, …, 23 也是 60 个模态。

在时间积分方案中，步长取 $\Delta t = 0.25$（相应的有量纲量约 0.5 h），每积分一步都计算了流场、位势场和散度场。

3.3 扰动模式初值

由于平流层爆发性增温发生在中高纬，用 1979 年 2 月 22 日 10 hPa 上 30°N 以北地区的地转偏差作为模式的初值，因此，初始流场为中心位于（60°E，60°N）和（90°W，60°N）的两个反气旋环流。

南半球的初始扰动都取为 0，即 $u' = v' = h' = 0$。计算结果表明，在积分过程中这种情形不随时间变化，流场和位势场都保持为 0，以后我们只画北半球区域扰动场的演变。

4 平流层地转适应的数值模拟

利用上述数值模式和数值方法，并利用模式的初始场进行积分至 50 小时，即积分 100 步，从而得到初始场作用下的北半球平流层地转适应的变化情况。

图 3 是把图 2 所示的初始扰动流场按 Hough 函数展开后再按（11）式求和所得，可以看到，图 3 所示的流场与图 2 所示的初始场是一致的，而位势场都是零值。这可以说明在本模式数值解中应用 Hough 函数是正确的。

图 4a~e 分别是模式积分为 0.5、1.0、2.0、25.0 和 50.0 h 后的 10 hPa 扰动流场、位势场的分布情形。从图 4a 可以看到，扰动流场很少变化，仍然保持为两个反气旋性结构，而扰动位势场略发生了改变，在（55°E，55°N）和（75°W，60°N）出现了两个位势高值区，这两个区域正好是反气旋性地转偏差的大致位置。这说明位势场已表现出向流场调整的特点。

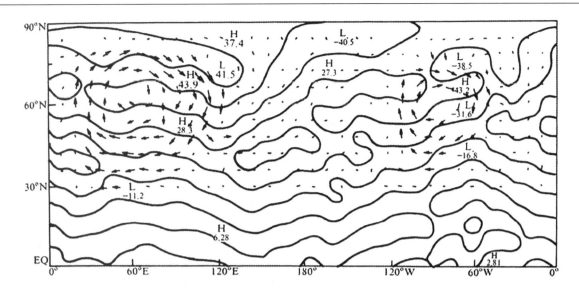

图 3 积分起步（t=0 h）时的流场、高度场（实线，单位：gpm）的分布

图 4b 是积分到第 2 步（约 1 h）的 10 hPa 上扰动流场和位势场的分布情形。可以看到，在图 4b 中的两个位势高值区加强，已经发展成为高压，中心值分别已达 361 gpm 和 402 gpm，而流型仍然是少变的。

图 4c 是模式积分到 2.0 h 时的 10 hPa 面上扰动流场与高度场分布。可以看到，扰动流型仍然少变，但在扰动高度场已经有两个很强的高压，中心值分别为 664 gpm 和 756 gpm，这与初始场非地转流场相配置；并且，模式积分到 4.0 h 时扰动流场和扰动高度场与图 4c 相类似，扰动流场和位势高度场已经较好地满足地转关系。

然而，当模式积分 4.0 h 以后，在模式的积分过程中扰动流场和位势高度场的这种地转关系配置少变（图略）；当模式积分至 25.0 h（见图 4d），在位势场的高压有些减弱；并且，当模式积分到 50.0 h，流型已经发生大的改变，不再是前面所述的两个反气旋性的结构了，高度场也被多个高低压中心所代替。在这过程中，虽然流场和位势场发生了大的改变，但两者之间仍然保持着较好的地转关系。

从上面的积分结果可知，本次地转适应大致在 2~4h 内完成，是一个快过程；并且此次地转适应过程中流场是少变的，主要是高度场向流场进行调整，以达到地转适应；另一方面，如果注意到每张图中单位矢量所代表的大小，可以发现，流型虽然少变，但流场也呈减小趋势，这就是说，也存在这流场向气压场的适应，因此，适应过程是相互的。

5 地转适应过程中散度场的演变

正如引言中所述，在地转适应中，散度场的变化主要反映了惯性重力波的活动（参见文献[9,14]）。因此，为了分析在这次平流层地转适应过程中惯性重力波的活动情况，我们利用散度场的演变来研究平流层地转适应过程中惯性重力波的活动情况。从图 5a~e 中所示的散度场的变化可以看到；与每个反气旋性流场相对应，总有一对流辐散辐

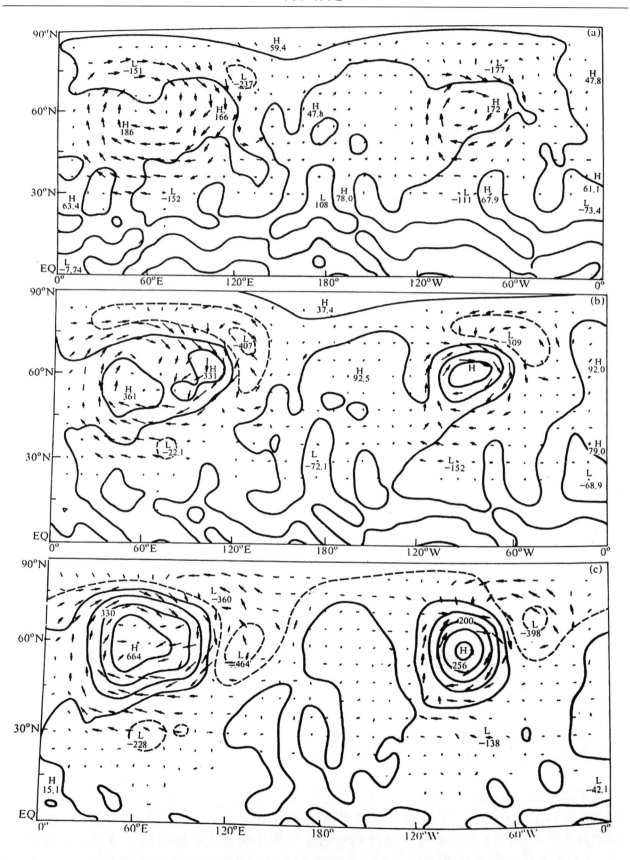

图4 模式积分时扰动流场和扰动高度场（实线，单位：gpm）的分布
(a) 积分 0.5 h; (b) 积分 1.0 h; (c) 积分 2.0 h;

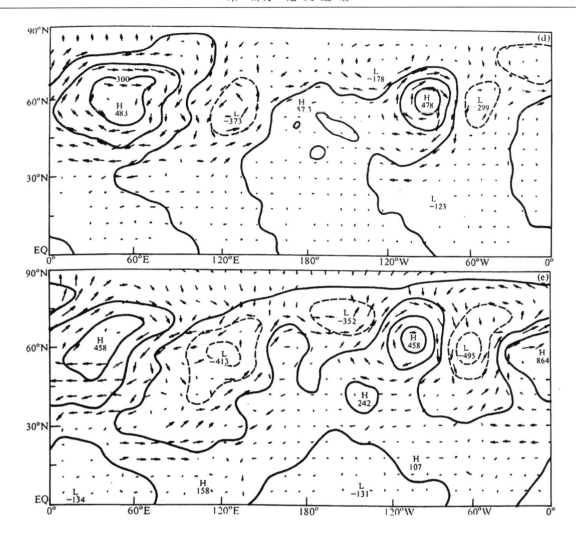

图 4（续）　模式积分时扰动流场和扰动高度场（实线，单位：gpm）的分布
(d) 积分 25.0 h；(e) 积分 50.0 h

合很强的区域；并且，在积分的前几步，散度场各极值中心（辐散辐合中心）有明显的移动，其中心量值都是逐渐加大的（图中散度都乘了 10^5），表明在地转适应过程中，辐散辐合是加强的，这说明惯性重力波活动加强。然而，在积分到 25.0 h 和 50.0 h，散度场值逐渐减小，如图 5d 中散度很小，只剩下零线，这说明地转适应完成后，散度量值减小，惯性重力波活动减弱。

为了进一步讨论在地转适应过程中散度随时间的演变，我们作了散度沿纬向—时间剖面图（见图 6a 和 b）。从图 6 可以看到，在高纬地区散度的极值区随时间向东移动，这说明惯性重力波激发出来后向东传播；而在中纬地区惯性重力波的传播特征不如高纬度地区明显，大体上是东传的，有些区域也有向西传的特征。

图 7 是模式积分过程中散度沿 50°E 和 75°W（强烈的非地转运动位于这两个经度）的纬度—时间剖面图。从图 7 很明显地看出，在地转适应过程中惯性重力波活动是向南传播的。

图 5 模式积分时散度场的分布（单位：10^{-5} s^{-1}）
(a) 积分 0.5 h；(b) 积分 1.0 h；(c) 积分 2.0 h；

图 5（续） 模式积分时散度场的分布（单位：$10^{-5}\,s^{-1}$）
(d) 积分 25.0 h；(e) 积分 50.0 h

从图 6 和图 7 可以看到，在这次平流层地转适应过程中，流场散度演变有两个很重要的特征：其一是散度在模式积分到 3 h 左右达到极值，然后量值减小；其二是当模式积分到 10 h 后，散度场差不多都为零值。这说明由于地转偏差，惯性重力波激发在地转适应过程开始到 3 h 前最为强烈。观测事实表明，在平流层爆发性增温期间，经常可观测到周期为几小时大振幅的温度波动，这是由于一种周期为 2～4 h 的大振幅惯性重力波所造成[15]。因此，本数值试验与观测事实相符。并且，由于惯性重力波能量频散，当模式积分到 10 h 后，散度急剧减小，这时流场与气压场处于地转平衡关系。这表明单从气压场与流场的配置来看，这次平流层地转适应的时间是 2～4 h，而从散度场的演变来看，要使惯性重力波能量很小，大约需要 5～10 h 左右。

6 结论与讨论

本文利用 Hough 函数构造一个线性正压全球谱模式，并以 1979 年 2 月 22 日平流

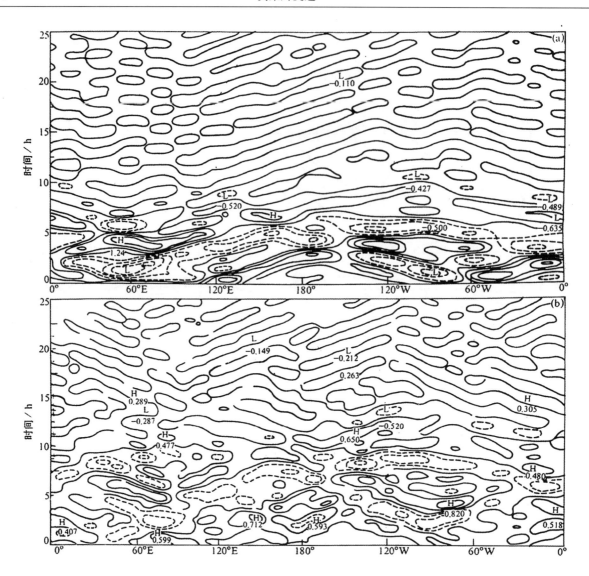

图 6 模式积分过程中散度的经度-时间剖面图（单位：10^{-5} s^{-1}）
(a) 沿 75°N；(b) 沿 45°N

层爆发性增温时 10 hPa 的实际很强的地转偏差作为初始场，模拟了平流层爆发性增温时扰动的地转适应过程，并从散度场的变化讨论了平流层爆发性增温时惯性重力波的激发和传播。从模式模拟的结果表明：由于平流层爆发性增温时，伴随着大范围的非地转运动，从而产生地转适应过程。在这过程中气压场与风场互相调整，经过 2~4 h 它们之间达到地转关系。并且，在这地转适应过程中散度场发生强烈变化，当模式积分到 3 h 时，散度场达到最大，之后慢慢衰减，这说明在这适应过程中所激发的惯性重力波的活动是很剧烈的。因此，平流层爆发性增温期间强烈的系统性的非地转运动是平流层惯性重力波的一个重要源区。

本研究表明，在平流层发生爆发性增温时，若没有对流层行星波继续上传，由于平流层强烈非地转运动所激发的惯性重力波的能量频散，只要经过 2~4 h 左右气压场与流场互相调整，从而很快达到地转平衡，并且，散度场经过 10 h 左右就很快被削减。

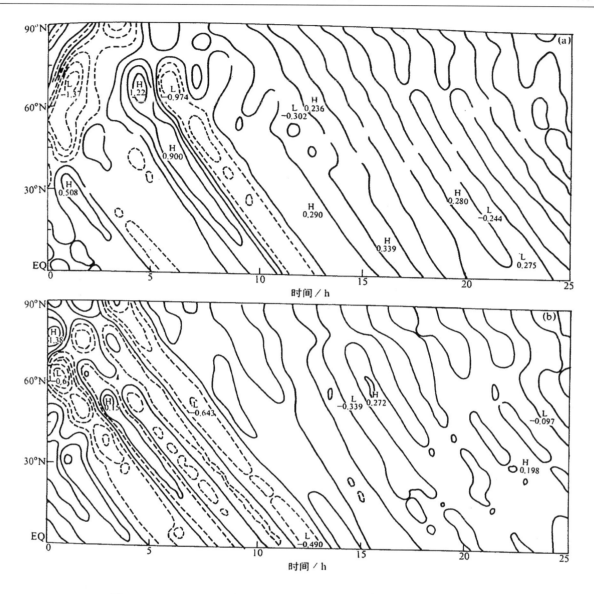

图 7 模式积分过程中散度的纬度时间剖面图 （单位：10^{-5} s^{-1}）
(a) 沿 50°E；(b) 沿 75°W

然而，由于平流层爆发性增温期间，对流层行星波可能在数天期间不断上传到平流层，因此，在平流层爆发性增温期间，平流层地转适应过程并不是通过一次过程就进行完毕，而可能是连续进行数天。一旦对流层行星波不能上传到平流层，爆发性增温也就很快停止了，则平流层爆发性增温时非地转运动在 2～4 h 之后达到地转平衡。

从本文研究结果可以看到，在平流层大尺度地转适应过程中气压场向风场适应。叶笃正和李麦村[9]阐述了在对流层大尺度运动中地转适应过程往往是风场向气压场适应，并且，他们指出在高层（平流层）大尺度运动的地转适应过程与对流层是不同的，往往是动力性的。这可能是由于平流层冬季风场变化很大，而气压场变化往往比较小，因此，在地转适应过程中往往是动力性的，即气压场向风场适应。因此，本文的结果证实了他们理论上所分析的结论。

致谢: 作者撰稿时, 与叶笃正、曾庆存院士进行了多次讨论, 他们对本文提出了许多宝贵意见, 在此谨表示感谢。

参 考 文 献

1. Rossby, C. G., On the mutual adjustment of pressure and velocity disturbances in centrain simple current systems. I, *J. Mar. Res.*, 1937, **1**, 15~27.
2. Rossby, C. G., On the mutual adjustment of pressure and velocity disturnabces in centrain simple current systems. II, *J. Mar. Res.*, 1938, **2**, 239~263.
3. Cahn, A., An investigation of the free oscillation of a simple current system, *J. Meteor.*, 1945, **2**, 113~119.
4. Charney, J. G., The dynamics of long waves in a baroclinic westerly current, *J. Meteor.*, 1947, **4**, 135~163.
5. Yeh. T. C., On the formation of quasi-geostrophic motion in the atmosphere, *J. Metoer. Soc. Japan*, the 75th Anniversary Volume, 1957, 130~134.
6. 曾庆存, 扰动特性对大气适应过程的影响和测风资料的使用问题, 气象学报, 1963, **33**(1), 37~50.
7. 曾庆存, 大气中的适应过程和发展过程 (一) 物理分析和线性理论, 气象学报, 1963, **33**(2), 163~174.
8. 曾庆存, 大气中的适应过程和发展过程 (二) 非线性问题, 气象学报, 1963, **33**(3), 281~289.
9. 叶笃正、李麦村, 大气运动中的适应问题, 北京: 科学出版社, 1965, 125pp.
10. Wiin-Nielsen, A., On geostrophic adjustment on the sphere, *Beitr. Phys. Atmos.*, 1976, **49**, 254~271.
11. Janic, Z. and A. Wiin-Nielsen, On geostrophic adjustment and numerical procedures in a rotating fluid, *J. Atmos. Sci.*, 1977, **34**, 297~310.
12. Walterscheid, R. L. and F. J. Bouecher, Jr., A simple model of the transient response of the thermosphere to impulsive forcing, *J. Atmos. Sci.*, 1984, **41**, 1062~1072.
13. Fritts, D. C. and Luo, Z., Gravity wave excitation by geostrophic adjustment of the jet stream. I: two-dimensional forcing, *J. Atmos. Sci.*, 1992, **49**, 681~697.
14. Matsuda, Y. and H. Takayama, Evolution of disturbance and geostrophic adjustment on the sphere, *J. Meteor. Soc. Japan*, 1989, **67**, 949~965.
15. Heath. D, F. et al., Observations of the global structure of the stratosphere and mesosphere with sounding rockets and with remote sensing techniques from satellite, *Structure and Dynamics of the Upper Atmosphere*, Amsterdam: Elsevier Scientific Publishing Company, 1974, 131~198.
16. 陈金中、黄荣辉, 中层大气重力波的一种激发机制及其数值模拟 I. 非地转不稳定和波结构, 大气科学, 1995, **19**(5), 554~562.
17. 黄荣辉、邹捍, 球面斜压大气中上传行星波与纬向平均气流的相互作用, 大气科学, 1989, **13**(4), 383~392.
18. Kasahara, A., Normal modes of ultralong waves in the atmosphere, *Mon. Wea. Rev.*, 1976, **104**, 669~690.
19. Kasahara, A., Numerical integration of the global barotropic primitive equation with Hough harmonic expansions, *J. Atmos. Sci.*, 1977, **34**, 687~701.
20. 董双林、吴津生、王宗皓, 哈佛函数的计算和应用: (I) 计算, 气象学报, 1981, **39**(1), 1~9.
21. Longuet-Higgins, M. S., The eigenfunctions of Laplace's tidal equations over a sphere, *Philos. Trans. Roy. Soc. London*, 1968, **A262**, 511~627.

附录　Hough 函数

Hough 函数是 Laplace 潮汐方程的特征解。虽然 Hough 函数不是有基流时线性正压浅水波方程的特征解，但对于相同数量的模式变量，Hough 函数展开法是一种比球函数更有效的工具，利用 Hough 函数展开法可以很方便地把快波（重力波）和漫波（涡漩波）分开；另外，Hough 函数展开法中三个要素 u、v、h 只有一个共同的谱系数，这意味着三个要素之间有着一种内在的动力学约束，物理意义更明显。

Laplace 潮汐方程组如下：

$$\frac{\partial u}{\partial t} - fv = -\frac{\partial \Phi}{a\cos\varphi \partial \lambda}, \tag{A1}$$

$$\frac{\partial v}{\partial t} + fu = -\frac{\partial \Phi}{a\partial \varphi}, \tag{A2}$$

$$\frac{\partial h}{\partial t} + \frac{h_e}{a\cos\varphi}\left(\frac{\partial u}{\partial \lambda} + \frac{\partial v\cos\varphi}{\partial \varphi}\right) = 0. \tag{A3}$$

上述方程组中的 u、v 分别是纬向、经向速度，$\Phi = gh$ 为位势，方程组的特征解即是 Hough 函数。

引入流函数 Ψ 和速度势 χ，则

$$u = \frac{1}{a\cos\varphi}\frac{\partial \chi}{\partial \lambda} - \frac{1}{a}\frac{\partial \Psi}{\partial \varphi}, \qquad v = \frac{1}{a}\frac{\partial \chi}{\partial \varphi} + \frac{1}{a\cos\varphi}\frac{\partial \Psi}{\partial \lambda}. \tag{A4}$$

这样就有涡度 $\zeta = \nabla^2 \Psi$，而散度 $D = \nabla^2 \chi$，并设

$$\begin{bmatrix} \chi \\ \Psi \\ h \end{bmatrix} = \begin{bmatrix} \dfrac{gh_e}{2\Omega}\hat{\chi}_{(\varphi)} \\ \dfrac{gh_e}{2\Omega}\hat{\Psi}_{(\varphi)} \\ h_e \hat{Z}_{(\varphi)} \end{bmatrix} e^{i(s\lambda - \nu t)}, \tag{A5}$$

记 $\mu = \sin\varphi$，$m = \cos\varphi\dfrac{\partial}{\partial \mu} = (1-\mu^2)\dfrac{\partial}{\partial \mu}$，$\sigma = \dfrac{\nu}{2\Omega}$，其中 σ 是无量纲频率，s 是波数，代入方程（A1）

~（A3），可得如下方程组：

$$(\sigma\nabla^2 - s)(i\hat{\chi}) + (\mu\nabla^2 + m)\hat{\Psi} = \nabla^2 \hat{Z}, \tag{A6}$$

$$(\sigma\nabla^2 - s)\hat{\Psi} + (\mu\nabla^2 + m)(i\hat{\chi}) = 0, \tag{A7}$$

$$\sigma\hat{Z} = -\frac{1}{\varepsilon}\nabla^2(i\hat{\chi}), \tag{A8}$$

其中 ε 是 Lamb 数，$\varepsilon = 4a^2\Omega^2 / gh_e$，$\nabla^2$ 是 Laplace 算符，

$$\nabla^2 = \frac{d}{d\mu}\left[(1 - \mu^2)\frac{d}{d\mu}\right] - \frac{s^2}{1 - \mu^2}.$$

无量纲振幅 $\hat{\chi}$，$\hat{\Psi}$，$\hat{Z}$ 只是 μ 的函数，可按连带 Legendre 函数展开，

$$\begin{bmatrix} \hat{\chi} \\ \hat{\Psi} \\ \hat{Z} \end{bmatrix} = \sum_{n=s}^{\infty} \begin{bmatrix} iA_n^s \\ B_n^s \\ C_n^s \end{bmatrix} p_n^s(\mu). \tag{A9}$$

把方程（A8）带入方程（A6）与（A7），并利用标准化 Legendre 函数的正交性，可得

$$A_n^s[n(n+1)\sigma - s] - B_{n+1}^s n(n+2)\sqrt{\frac{(n+1)^2 - s^2}{4(n+1)^2 - 1}}$$
$$- B_{n-1}^s(n-1)(n+1)\sqrt{\frac{n^2 - s^2}{4n^2 - 1}} - C_n^s n(n+2) = 0, \tag{A10}$$

$$B_n^s[n(n+1)\sigma - s] - A_{n+1}^s n(n+2)\sqrt{\frac{(n+1)^2 - s^2}{4(n+1)^2 - 1}} - A_{n-1}^s(n-1)(n+1)\sqrt{\frac{n^2 - s^2}{4n^2 - 1}}$$
$$- \frac{n(n+1)}{\varepsilon} A_n^s - \sigma C_n^s = 0. \tag{A11}$$

1 $s \geq 1$ 的情形

方程（A10）和（A11）据 Legendre 函数的奇偶性可分为下列对称情形和反对称情形，并令

$$\begin{cases} X = (A_s^s, B_{s+1}^s, C_{s+2}^s, A_{s+2}^s, B_{s+3}^s, C_{s+2}^s, \cdots), \\ Y = (B_s^s, A_{s+1}^s, C_{s+1}^s, B_{s+2}^s, A_{s+3}^s, C_{s+3}^s, \cdots), \end{cases} \tag{A12}$$

并令

$$d_n = \frac{n+1}{n}\sqrt{\frac{n^2 - s^2}{4n^2 - 1}}, \qquad e_n = \frac{n}{n+1}\sqrt{\frac{(n+1)^2 - s^2}{4(n+1)^2 - 1}},$$

$$K_n = -\frac{S}{n(n+1)}, \qquad r_n = -\frac{n(n+1)}{\varepsilon}.$$

这样，就有下面两种情形：
(1) 当在对称情形，则有

$$(A - \sigma I)X = 0, \tag{A13}$$

其中

$$A = \begin{bmatrix} K_s & d_{s+1} & -1 & 0 & 0 & 0 & \cdots \\ e_s & K_{s+1} & 0 & e_{s+2} & 0 & 0 & \cdots \\ r_s & 0 & 0 & 0 & 0 & 0 & \cdots \\ 0 & e_{s+1} & 0 & K_{s+2} & d_{s+3} & -1 & \cdots \\ 0 & 0 & 0 & e_{s+2} & K_{s+3} & 0 & \cdots \\ 0 & 0 & 0 & r_{s+2} & 0 & 0 & \cdots \\ \cdots & \cdots & \cdots & \cdots & \cdots & \cdots & \\ \cdots & \cdots & \cdots & \cdots & \cdots & \cdots & \end{bmatrix}.$$

(2) 当在反对称情形，则有

$$(B - \sigma I)Y = 0, \tag{A14}$$

$$B = \begin{bmatrix} K_s & d_{s+1} & 0 & 0 & 0 & 0 & \cdots \\ e_s & K_{s+1} & -1 & e_{s+2} & 0 & 0 & \cdots \\ 0 & r_s & 0 & 0 & 0 & 0 & \cdots \\ 0 & e_{s+1} & 0 & K_{s+1} & 0 & -1 & \cdots \\ 0 & 0 & 0 & e_{s+2} & -1 & d_{s+4} & \cdots \\ 0 & 0 & 0 & 0 & 0 & 0 & \cdots \\ \cdots & \cdots & \cdots & \cdots & \cdots & \cdots & \\ \cdots & \cdots & \cdots & \cdots & \cdots & \cdots & \end{bmatrix}.$$

A、B 分别是所得到的对称情形和反对称情形的特征矩阵，并设

$$\begin{bmatrix} u \\ iv \\ h \end{bmatrix} = \begin{bmatrix} \sqrt{gh_e}\,\hat{U}_{(\mu)} \\ \sqrt{gh_e}\,\hat{V}_{(\mu)} \\ \hat{Z}_{(\mu)} \end{bmatrix}, \tag{A15}$$

代入（A4）式，可得：

$$\hat{U} = \frac{\gamma(is\hat{\chi} - m\hat{\Psi})}{\cos\varphi}, \tag{A16}$$

$$\hat{V} = \frac{i\gamma(is\hat{\Psi} + m\hat{\chi})}{\cos\varphi}, \tag{A17}$$

$$\frac{d\hat{V}\cos\varphi}{d\varphi} = \frac{i\gamma(ism\hat{\Psi} + m^2\hat{\chi})}{\cos\varphi}, \tag{A18}$$

其中 $\gamma = \varepsilon^{-1/2}$。

由（A12）和（A13）求解可得到 X、Y，即得到（A9）式中的展开系数。X、Y 的量级是任意的，可按每个模态能量相同进行归一化，然后按（A9）式和（A16）、（A17）式便可求得对称情形和反对称情形下的 Hough 函数了。这样，可记：

$$H_l^s = \begin{bmatrix} \hat{U}_l^s \\ -i\hat{V}_l^s \\ \hat{Z}_l^s \end{bmatrix}. \tag{A19}$$

（A19）式中 H_l^s 即为 Hough 矢量函数、$H_l^s e^{is\lambda}$ 为 Hough 函数，其中 l 是模态经圈指数，s 为波数。

这样，由（A12）和（A13）式可求出 X、Y，并从（A9）式以及（A16）～（A18）式求出 Hough 函数的矢量函数，并从 $H_l^s e^{is\lambda}$ 求出对应于某一波数 s 的 Hough 函数的平面结构。

一般以 LWG 和 LEG 分别代表西传和东传重力波的经圈模指数，LR 代表 Rossby 波的经圈指数，并且，第一个西传重力波（对应于 LWG=0）为 Rossby-重力混合波，第一个东传重力波（对应于 LEG=0）为 Kelvin 波。

2 $s=0$ 的情形

在波数 $s=0$ 时，运动与 λ 无关，代表纬向平均运动。此时，惯性重力波模态可按上面办法构造，东传和西传惯性重力波成对出现，频率大小相等，符号相反，相应的特征函数成复共轭。由于涡旋波频率都为 0，已无法再按 QR 算法来构造。$s=0$，$\sigma=0$ 时，（A6）～（A8）式可退化为地转关系，因而，此时的 Rossby 模亦称为地转模。

第二部分　自述·追忆·致辞

我的自述
——家贫更加激励我努力拼搏

我一直很想写一写我是如何从一个农村的放牛娃成长为一名中国科学院院士的,但由于没有时间,一直未能成文。20世纪90年代初借学部办公室约院士自述稿之机,我才把我永远不能忘却的求学经历简要自述一下。此稿在院士自述稿的基础上做了适当补充,特别补充了20世纪90年代初到21世纪初的研究经历。

一、穷苦童年勤学习

我出生在福建省惠安县一个非常穷苦的农民家里,祖祖辈辈地无一垄,房无片瓦,父亲靠给人家当雇工或长工来养活全家。我出生后,父亲没有工作,只得带领全家四处流浪,后来在前黄村落了户,给一老人(我称曾祖母)当后代,继承了一点土地。那年代,饭都吃不饱,根本谈不上上学。新中国成立前,我家祖祖辈辈没有人上过学。1949年家乡解放了,我家分得土地,我也能上学了。

那时,我家兄弟姐妹多,只靠父亲一人劳动,家庭生活十分困难。有几次我几乎要停学,邻居婶妈坚持劝我母亲,应该让我继续上学;若交不起学费,她帮着出。听完她的劝说,母亲咬着牙坚持让我上学。由于没有钱买铅笔,另一邻居婶妈给我一支她儿子用剩的铅笔,我十分珍惜它,每次削笔千注意万注意,怕把它削坏,这支铅笔我用了近两个学期。之后,我就用经济富裕家庭子弟用剩下的笔头写字。我一放学就帮家里去地里放牛,边放牛边读书,每学期我成绩均名列班上前茅。就这样我只用了四年时间读完六年小学。

二、艰难求学难忘怀

1953年考初中时,由于考分高,我被位于县城的惠安一中(是一所老学校)录取了。但惠安一中离家太远,而离我家较近的山腰镇刚办一所中学(惠安二中),我向当时惠安一中负责新生报到的老师要求转到惠安二中念书,那位老师立即答应,他说:"有的是学生要从惠安二中转到惠安一中来念书,你这可是从好的学校换到差的学校,你要想好"。我向他说明我家的情况,他答应帮我换学校,过了几天,就让我到惠安二中报到。当时惠安二中刚建校,还没有校舍,只有一座破旧的锦山庙和一个临时的工棚当作我们的教室;没有电,暗弱的煤油灯是我们的照明;没有运动场,田埂空地是我们的操场。

由于没有钱交伙食费,每星期我自己从家挑二十斤白薯和三十多斤柴草,步行约五里到学校,由学校负责给我们这些穷苦子弟蒸白薯。由于顿顿吃白薯,用大酱当菜,吃完饭后烧心难忍。为了攒钱让我交注册费,母亲舍不得让兄弟姐妹吃一个鸡蛋,把攒下的鸡蛋拿去卖,给我当作新学期的注册费。而我自己也尽量利用寒暑假,边放牛边捡人家收成后漏在地里的白薯和花生,积累起来去卖,凑足每学期的注册费。然而,在这艰苦的环境里,我和同学们却孜孜不倦,勤奋学习。我在初中阶段,由于老师们的辛勤教导,培养了我对数学、物理的兴趣,年年取得好成绩。

由于惠安二中刚建校不久,没有高中,1956年我以优异的成绩考上惠安一中。惠安一中学生多,经济富裕家庭出身的子弟交得起伙食费就由学校提供饭菜,交不起伙食费的学生自己从家带粮食和副食。这样我每周星期日必须挑三十斤干柴草和二十多斤白薯,以及一小罐自制的豆酱和咸菜当副食,到离我家乡四十里的学校,学校负责帮我们蒸白薯。一下课,同学们尽快去食堂认领自己的蒸罐,吃完饭,又把白薯洗干净切好再放到蒸罐。看到经济富裕家庭出身的同学吃白米饭,心里很想能吃一顿白米饭该多好。由于家庭困难,我衣衫褴褛,再冷的冬天也只穿两件破烂的冬衣,直到高中毕业前,我还从没穿过一双买的鞋,一双木拖鞋伴我度过了中学时代。上学用的文具只有靠有时辅导经济富裕家庭的同学的功课得几分钱劳务费来购买。然而，生活的艰辛并没有使我退却，反而更激励我勤奋读书，我的学习成绩一直是班上第一名，在年级(有五个班)名列前茅，很多老师也很喜欢我。至今无论我走到哪里，这段艰辛的求学经历也难以忘怀。

三、喜悦迈入大学门

1959年，我高中毕业了。在高考前，需填升学志愿，我一个农村出身的学生，父母亲和亲属都没上过学，不知道该填什么学校。班主任刘老师主动找我，帮助我填升学志愿表，他依据我的学习成绩，第一志愿填北京大学。高考我们必须到泉州市去考，在临考前一天我与几位同学从惠安一中步行六十里到泉州。在高考期间三天，学校负责我们的伙食，四菜一汤，有肉、鱼、青菜，比较富裕家庭出身的同学因考试紧张吃不下，我和几位经济条件差的同学猛吃三天，我觉得那是我从出生以来吃得最好的三天，至今还记忆犹新。

我以优异的成绩考上北京大学地球物理系。接到北京大学录取通知书，母亲又高兴，又忧愁，几天几夜睡不着，家里几分钱都没有，去北京的路费至少需要20元，儿子一件新衣服也没有，一双鞋也没有，这怎么办？看到母亲流泪，我也十分着急。有一天我抱着试试看的心态，到惠安一中把录取通知书让班主任刘老师看，刘老师为我高兴。我把上学的困难向刘老师说了，他马上说："你不要愁，我立即去向许惜今校长反映"。很快许校长批准补助我20元作为去上学的路费。我兴高采烈地拿着这宝贵的20元回家。母亲又向村里生产队借了20元，给我做一套新衣服和买一双鞋。母亲为我做衣服时边做边流泪，真是"慈母手中线，游子身上衣，临行密密缝，意恐迟迟归"。这20元是我能够到北京求学的路费。我至今还十分感激刘老师，1965年大学毕业时回母校，想亲自感谢他对我的关心，可惜他不在学校，据说，他已回老家了，我深感遗憾。

1959年8月底，我带着一个麻袋作行李袋，一根小扁担和家里唯一的一床没有破洞的被子，穿着我一生中第一次买的鞋，带着学校补助给我的20元钱，与另一位同学第一次坐上汽车到厦门。在厦门用了17.3元买了一张到北京的火车票，坐上北上的火车，经江西的鹰潭、浙江的杭州、上海等车站转车到江苏的南京，当时还没有南京长江大桥，从南京轮渡到浦口，再从浦口坐车经济南车站、天津车站转车到北京。由于第一次坐火车，不熟悉，在济南转错车，转到去青岛的车，后再回济南，在天津又转错车，转到去沈阳的车，后又回天津，再坐上开往北京的车，历经十一天，终于来到北京。临行时，母亲给我蒸了几斤米糕，一路做干粮，由于车上温度高，到上海时米糕全发霉了，只好倒掉；另一位同学带的干粮没有变质，分了一些给我，这样在车上花了0.7元买一点饭，到了北京还剩2元钱。我带着这2元钱、一个麻袋、一条被子和一根小扁担，心里充满喜悦，忘记旅途的困倦，迈入了北京大学这一引导我认识现代科学的大学校门。

四、努力学好基础课

我一个乡巴佬走进美丽的北京大学校园，像刘姥姥进大观园一样，对一切都感到新鲜，都觉得那么好。我只带了一张小草席，到宿舍看到同学床上垫着褥子，觉得很奇怪。新买的鞋舍不得穿，光着脚在校园走，后来学校规定不让光脚在校园走，这才不得不穿上在老家买的新鞋。党和学校有关部门对我十分关心，给我最高助学金，到冬天还补助给我棉衣和褥子，学校和班主任经常问寒问暖。

刚进入大学，对大学上课没有固定教室很不习惯，由于我身材瘦小，下课换教室经常落在同学后面，没有跟上，曾几次没找到教室，有时上大课(如物理课)只好坐在教室后排。经过几次，我吸取教训，与同一宿舍城市出身的同学约定好，一下课我紧跟这位同学去要上课的教室，这样，再也不会找不到教室。大学学习我还用我中学时代的学习方法，上课前，提前预习，上课时注意听老师讲课，记好笔记，课后要复习，这样我学习成绩在班上位于前列。我特别喜欢数学、物理课程，在北京大学六年间系统学习了分析数学、复变函数、数理统计与概率论、特殊函数、数学物理方程、计算数学等课程；物理课，有普通物理、原子物理、光学、理论力学、流体力学等基础课。学好这些基础课程，为我学好专业课程奠定了基础。每到寒暑假，由于家乡路途遥远，没有钱买车票回家乡，就留在学校。因为我写字整齐，利用寒暑假给学校印刷厂刻教材的蜡版，1 天可得 1 元劳务费，这样一个暑假可得约 30 元劳务费，不仅自己可增添衣服，而且可为回家乡准备路费。我利用这得来的劳务费 6 年中回了两次老家。

学校对我生活和学习的关心，使我暗下决心，只有努力学习才能报答党和人民对我的培养，报答父母亲对我养育之恩。是这样一个纯朴的信念使我勤奋学习。在北京大学地球物理系，我学的是气象专业，在六年的学习生活中，我脑海中一直在思索一个问题，如何用现代的数学、物理概念来解释千变万化的天气与气候。

五、步入研究新旅程

1965 年，在我毕业前，班主任挑了我和班上几位学习成绩好的同学考研究生，并让我报考中国科学院叶笃正先生指导的研究生。当我考上研究生，看到同学分配到有关工作单位心里很羡慕，向班主任王晓林老师要求，不上研究生要去工作。王老师劝我还是去读研究生好，我只好听王老师的话，到中国科学院地球物理研究所读研究生，在叶笃正教授指导下从事大气动力学的研究。这里我要特别感谢我大学的班主任王老师，是他的指引，我才上了中国科学院研究生院。

当我学完研究生基础课程之后，满怀希望走进研究所时，"文化大革命"爆发了，我的学业和科研也就中断了。1968 年按规定，与其他的研究生和去苏联留学刚回国的留学生一起下放到农场劳动一年多。从农场回到研究所后，我幸运地参加了我国卫星气象学的研究，从此步入了大气科学的研究旅程。

1978 年秋经过考试，我作为第一批留学生被派到日本留学，因联系导师出了一点偏差，延至 1979 年才到东京大学地球物理系气象学科留学，在我的导师叶笃正先生的推荐下，在国际著名大气动力学家与数值模拟预报专家岸保勘三郎教授指导下从事大气动力学研究。我深知这个研究室历史悠久，设备先进，图书资料充足，并拥有世界上最先进的计算机，曾培养了许多国际上著名的大气动力学和数值模拟专家，深感能到这个研究室深造的机会难得。在岸保勘三郎教授的指引下，我进行了既是当时国际大气科学研究领域的前沿，又是十分艰难的研究课题——准定常行星波动力学的研究。这是一个关系到两周以上大气环流和短期气候变化的重要动力学研究课题。为了尽量利用东京大学先进的计算机来研究这个动力学问题，我每天早出晚归，工作达 14 小时以上。我从观测事实出发，应用波在缓变媒质中传播的理论，设计了一个 34 层数值模式，并利用这个模式研究了地球大气准定常行星波的形成以及三维空间传播特征，提出准定常行星波在地球大气中传播的两支波导理论。由于我在准定常行星波动力学方面做出有创新性的研究，获得了东京大学理学博士学位。

六、潜心研究出成果

从日本回国后，我一方面把所从事的行星波动力学研究继续下去，把准定常行星波动力学的研究成果应用到关于我国短期气候变化的机理与预测研究；另一方面在研究生院开设"高等大气动力学"课程，之后又开设"地球流体动力学"课程，培养有一定特色的研究生。当时我们一家四口挤居在筒子楼一间 14 平方米的房间，两个孩子晚上都要在家里做作业。我每天吃完晚饭，就到研究所去看书，做研究工作，待孩子熟睡后，才回来"接班"，不管刮风下雨，从不间断。为了把准定常行星波动力学理论应用到东亚大气环流与短期气候变化机制的研究，我放弃了节假日休息。有的学生看到我这样生活，与我开玩笑说："黄老师，你这样生活有什么意思？"我笑着说"能这样生活，我觉得很幸福"。其实，我有我的志趣和信念，我认为与 x, y 打交道，其乐无穷，搞科学研究是我最大的乐趣，在科学园地里耕耘，可以忘掉一切。我认为，不要把搞科研仅仅看成是一种谋生手段，没有志向和兴趣是搞不出什么创新性的研究的。

1985 年我晋升为研究所研究员，之后又成为博士生导师，获得"国家级有突出贡献中青年科学家"称号，并获得了全国"五一"劳动奖章。说实话，对名誉我是很淡泊的，我认为最重要的是为科学的发展扎扎实实作点贡献,为国家、为人民解决一点实际科学问题。鉴于我国夏季旱涝气候灾害的严重性,我开始研究热带太平洋暖池对东亚夏季风系统变化影响的研究,在对我国旱涝灾害变化特征和成因作深入的分析基础上,指出了西太平洋暖池热状态及上空对流活动对东亚大气环流与我国短期气候变化起重要作用及东亚/太平洋型遥相关的机理;提出不同热带太平洋 ENSO 的演变阶段对我国夏季旱涝灾害有不同影响及其物理机制,并把这种理论用于我国夏季旱涝预测,为我国旱涝的季度与超季度预报提出了一个新思路。从 1987 年起先后曾获多项国家和中国科学院科技奖。1991 年我被增选为中国科学院学部委员(后改为院士),在学部对学部委员候选人最后一轮评审之前,我应东京大学的邀请赴日本东京大学气候研究中心当客座教授。由于当时我还没有自己的 e-mail,评审结果公布之后,我并不知道自己被选为中国科学院学部委员,后来在东京大学图书馆看中文报纸,无意中看到新增中国科学院学部委员名单中有我。说实话,我并没有觉得自己当了学部委员有什么了不起。我在科研上能够做出些许成绩应该特别感谢我的两位恩师——叶笃正先生与岸保勘三郎先生对我的培养,是这两位恩师教我如何去发现和探索新的科学问题,如何以严谨的科学态度去论证和解释所发现的问题,是这两位恩师用宽厚的肩膀顶托我向科学高处攀登。这里我还要特别感谢我的夫人张锦英,她任劳任怨,把全部家务和教育子女的任务担负起来,无微不至地关心我的生活和身体。我的成绩有她非常大的功劳。

七、努力完成新使命

1992 年我从日本东京大学气候研究中心回国,回国后除了担负着许多学术事务、社会工作与行政事务外,还主持了一些关于灾害性气候预测方面的重大研究项目,特别是 1997~2003 年主持了《国家重点基础研究发展规划》第一批项目"我国重大气候灾害的形成机理和预测理论研究"。除了研究工作外,这期间还培养了多名博士生,形成了有一定特色、结构合理的研究梯队。如何带领这支队伍做出具有特色的创新研究,为国家大气科学的发展做出更大的贡献,这是摆在我前进道路上的新使命,我要以更大的努力去为我国大气科学的发展而奋斗。

从 20 世纪 70 年代起,随着全球气候变化研究的兴起,国际大气科学研究进入新的发展阶段。大气科学不仅从大气内部的动力、热力过程来研究大气环流的变化,而且从大气与海洋、陆面、冰雪等相互作用来研究大气环流与气候的变化,把大气与海洋、陆面、冰雪等看成一个整体,即气候系统。因此,我和我的学生们从热带太平洋和印度洋海-气相互作用、西北干旱和半干旱区陆-气相互作用、青藏高原的积雪来研究东亚季风系统的变异成因及其动力学机理,为我国开展短期气候预测提供有科学依据的预测因子。为了更好地开展关于东亚季风系统的研究,从 1996 起我还积极组织开展了中、日、韩关于东亚季风的合作研究,把我国关于东亚季风的研究及时与国际交流。

从 1995 年到 2014 年我又担任《大气科学》刊物的主编,在稿件的审查方面花费了大量精力,在我和编委会成员以及编辑部的努力下,《大气科学》已成为国内大气科学学科重要的刊物之一。为了推动海峡两岸青年大气科学学者的学术交流,从 1996 年起我和我国台湾有关学者积极组织海峡两岸青年大气科学学者学术交流研讨会,促进了两岸青年大气科学研究的学术交流和友谊。

鉴于西北太平洋台风是我国严重的天气灾害,以及开展影响我国台风季度预测的重要性,从 2007 年起,我和研究团队开展了西北太平洋热带气旋和台风气候学研究。通过几年研究,揭示了西太平洋和南海热带气旋和台风生成的几类大尺度环境场,以及年际、年代际变化特征,特别是提出西北太平洋季风槽和热带波动对热带气旋和台风生成的动力作用及其机理;并且指出了季风槽区对流层下层的强辐合和对流层上层的强辐散以及对流层下层风场的经向切变容易使热带的罗斯贝重力波(MRG 波)变成热带低压型波动(TD 波)的动力过程,因而是西北太平洋热带气旋和台风生成的关键动力因子,为开展西北太平洋台风活动和登陆我国台风的季度预测提供了科学依据。并且,为了更好地开展西北太平洋台风活动气候学研究,2010 年到 2016 年我又积极组织和开展了海峡两岸关于西北太平洋台风气候学的合作研究,促进了海峡两岸关于西北太平洋热带气旋和台风气候学研究的发展。

为了今后我国大气科学的更大发展,必须研讨我国大气科学和全球气候变化的研究发展战略,从 2011

年起到 2016 年我和有关学者主持和组织了我国大气科学和全球气候变化研究发展战略研究。此研究系统地回顾了近百年来国内外大气科学和全球气候变化的发展历程和现状、当前学科的发展动向和今后的发展趋势，提出了今后 10 年我国大气科学和全球气候变化研究八个重大科学问题以及需重点发展的研究领域，并且提出了今后应采取的重大战略措施。

总之，从 1959 年进入北京大学地球物理系气象专业学习，毕业之后一直在中国科学院大气物理研究所做科学研究至今(2019 年)整六十年。回顾这六十载春秋，我遵从父母亲的教导，"清清白白做人，老老实实干事"，一直勤恳学习和工作，从不敢怠慢，以做研究为乐，以培养学生为乐；从不与人争名、争利，也从不计较待遇；虽做出些许成绩，但从无张扬，与人为善，助人为乐；现虽积劳成疾，但还以乐观心态，盼望能够早日康复，再与学生一道从事研究工作。为表达我的心情，特作如下打油诗一首：

潜心研究六十载，

功名利禄视尘埃。

勤恳一生无悔怨，

健康乐观盼未来。

黄荣辉

于 2019 年写

缅怀我的恩师

——叶笃正先生的学术成就和治学精神

2019 年

引　言

本文是我在 2016 年 2 月中国科学院大气物理研究所召开的"纪念叶笃正先生诞辰百周年暨学术思想研讨会"上讲话稿经过整理而写成。在那会上，我作为叶先生亲自指导的学生之一，向大家简单回顾叶先生七十余载从事大气科学和全球气候变化研究的丰功伟绩，以及对他认真、严谨治学精神的些许体会。此文是为缅怀我的恩师—叶笃正先生的科学成就和治学精神，传承先生的治学风范而写。

叶先生是我国现代大气科学奠基人之一、中国科学院院士、国际著名气象学家，开创我国现代气象之先河。叶笃正先生不仅对我国和国际大气科学和全球气候变化研究发展做出了重大贡献，而且他严谨、认真的科学精神，是我们学习之楷模。

下面简单回顾叶先生对大气科学和全球气候变化研究的重大贡献。

一、创立了大气长波的频散理论

叶笃正先生早在 60 多年前作为芝加哥学派成员之一，提出了罗斯贝波(即 Rossby 波)的频散理论。他指出，在西风带上罗斯贝所提出的大气长波(或称 Rossby 波)可能有比它的相速度大的群速度($C_g>C$)，因此，在中高纬度西风带上大气长波的能量可能早于扰动本身向西风气流的下游传播。这可以很好说明了在中高纬度某地区上空西风带上游有一波槽存在，此扰动的群速度 C_g 有可能比它的相速度 C 大，因此，此扰动的能量有可能早于此扰动本身传播到此地区的下游上空，这样在该地区的下游上空就会产生新的扰动。因此，在此大的扰动尚未到来之前，在它的下游地区上空就会有新的小扰动出现。这就是天气学或动力气象学上称为"上游效应"，已成为现代大气动力气象的经典理论。

叶先生所提出的大气长波的频散理论，不仅几十年来广泛应用于短期和中期天气预报，而且推动了行星波动力学研究的发展。特别是这一理论推动了准定常行星波在球面大气的二维传播特征的研究，为解释大气环流异常遥相关的机理研究提供了科学基础；此外，这一理论还推动了准定常行星波在球面大气的三维传播规律的研究，为对流层与平流层相互作用过程及其机理的研究奠定了理论基础。

二、创立了东亚大气环流理论

(1) 提出东亚大气环流突变理论

20 世纪 50 年代中期，鉴于我国天气预报的需要，叶先生和他的合作者研究了东亚大气环流。叶先生与我国著名气象学家顾震潮先生、陶诗言先生及杨鉴初先生等合作在《Tellus》杂志上发表了 3 篇重要论文，揭示了冬季和夏季东亚上空平均大气环流的动力和热力结构及年际变化特征，引起了国际上的关注。之后，叶先生和陶诗言先生、李麦村先生发现，东亚和北美大气环流在过渡季节(春季、秋季)有急剧变化的现象。在这两个地区，大气环流的变化是有阶段性的突变，而不是逐渐、平稳地过渡的；特别是他们指出，东亚大气环流从冬季到夏季的季节变化是在突变中完成的，这种行星尺度的突变使东亚夏季风暴发，这一理论比国际上的相关研究(20 世纪 80 年代，Krishamurti 与 Ramanathan 以及 McBride 分别对印度夏季风和澳大利亚夏季风的研究中指出这一突变)早 20 多年。这一发现不仅对东亚大气环流研究的发展有重要科学意义，而且在东亚地区的天气预报中有广泛的应用价值。

(2) 提出准定常行星波的形成机制

20世纪50年代，在Rossby提出大气长波理论之后，准定常行星波(特别是东亚大槽)的形成机理成为令人感兴趣的问题。Charney和Eliassen提出地形强迫作用形成了准定常行星波；另一方面，Smagorinsky提出非绝热加热是准定常行星波形成的原因之一。叶先生和朱抱真先生提出，地形强迫作用和非绝热加热共同作用是准定常行星波形成的重要机制。这一理论发表在Tellus上，至今还被广泛引用，已被国际公认为准定常行星波形成的正确机制。

(3) 提出阻塞高压的演变过程及其机理

北半球的阻塞高压对东亚地区冬季的寒潮和夏季的梅雨锋维持有重要影响。20世纪50年代末，叶先生和合作者系统研究了北半球冬季阻塞高压的形成、发展、维持和崩溃过程。他们所著的《北半球冬季阻塞形势的研究》书中阐述的理论至今仍在我国天气预报中广泛运用。

他的研究团队关于东亚和北半球大气环流的特征、形成和变异机理的许多研究成果总结整理在《大气环流的若干基本问题》书中。在此书，叶先生和合作者详细地讨论了北半球大气环流的观测特性和基本要素、大气中的动力过程，特别是系统讨论了准地转运动和大气长波能量、角动量传输、热平衡、水汽输送、急流的形成和维持、西风带中槽脊的形成、长波不稳定等大气环流的基本问题。因此，这本书被视为关于大气环流动力学的最早出版物之一，并被翻译成俄语出版。

三、提出大气运动适应过程的尺度理论

20世纪50年代，叶先生深入研究了大气运动的适应理论。在大气运动中究竟是气压场还是风场为主导呢？这是50年代大气动力学的一个基本问题。叶先生深入研究之后于1957年提出了自己的观点：大尺度运动仍是以气压场的变化为主导，较小尺度的大气运动则是以风场的变化为主导，从而赋予了大气准地转运动受尺度影响的概念。曾庆存先生从数学上严格证明，当大气扰动的水平尺度大于临界尺度(即Rossby变形半径)时，风场向气压场适应，然而，当大气扰动尺度小于临界尺度时，气压场向风场适应，这一理论不仅被视为是地转适应理论中的重大突破，而且推动了大气大尺度环流动力学的深入研究，并开辟了中、小尺度环流系统动力学研究的新途径。

20世纪70年代末至80年代初，他又与李麦村先生共同提出，在大气各种空间尺度的系统生成与发展中都有三个不同时间尺度的变化阶段。叶先生与李麦村先生合著的《大气运动的适应问题》一书中，系统地总结了关于大气适应理论，深入讨论了大气地转适应中的物理特性、正压大气和斜压大气中的地转适应过程、流体静力学适应过程、大尺度与中尺度大气运动中风场和气压场适应过程等。

四、开创了青藏高原气象学研究

20世纪50年代至80年代，叶先生率领一批当年的青年科研人员，开展了青藏高原及对周围地区天气气候影响的研究，分析了青藏高原对亚洲季风和东亚大气环流及天气气候变异的动力、热力作用，从而开创了青藏高原气象学的研究。在20世纪60年代初期，叶先生就指出，青藏高原在夏季是一个巨大热源，在冬季是一个冷源。叶先生首先发现，当西风带自西向东越过青藏高原时，受青藏高原的影响，西风带会分为两支：一支绕过青藏高原西北侧向东流去，称为北支西风急流；另一支沿着青藏高原南侧，越过云贵高原经长江下游至日本，并在对流层上层存在一支西风急流，又称为南支西风急流。此外，叶先生等研究还指出，当南支西风急流越过青藏高原南缘时，受到青藏高原的加热作用，该急流中心高度会明显降低，这一发现引起了当时国内外学者的广泛关注。

叶先生等还深入研究了夏季青藏高原热源及其对亚洲季风和东亚大气环流的影响。20世纪70年代以

后，叶先生等利用当时很少的观测资料，对青藏高原热力状况、环流变化、高原上对流系统的作用、青藏高原在全球环流形成和变异中的重要性、青藏高原大型垂直流场等问题做了大量研究。叶先生等指出，夏季从高原上升的气流可以在遥远的地区下沉，导致高原与遥远地区有重要的遥相关作用；在春夏秋季节，由于青藏高原对大气加热作用，使得位于高原周围的大气环流产生重要变化。此外，叶先生等气象学家还研究了夏季由青藏高原热源引起的强对流小系统与大尺度天气系统的非线性相互作用，指出了夏季高原上空的小尺度对流活动对高原大尺度环流的维持起重要作用。

叶先生关于青藏高原对东亚和全球大气环流影响的研究作了系统总结。早在 1958 年，他与顾震潮先生、陶诗言先生、杨鉴初先生等合写的《西藏高原气象学》一书，是当时国内外唯一的西藏高原气象学专著。1979 年他又与高由禧先生合著了《青藏高原气象学》一书，全面系统地总结了这一领域的研究成果，被国际上公认为这一领域的经典著作。

五、开拓全球气候变化理论

20 世纪 80 年代初，叶先生担任中国科学院副院长，他不仅在中国科学院领导了地球科学的研究活动，而且开始发展新的研究领域。在叶先生等科学家的倡导下，我国兴起了全球变化的研究热潮。该研究把地球的各个部分(大气、海洋、冰雪、陆地、生物、人类活动)作为一个整体，研究其中各个部分过程的相互作用，从而进行以气候变化为主导的全球环境演变研究。

他不仅作为 JSC-WCRP 的成员参与了 WCRP 的拟定，也作为 IGBP 特别委员会成员参与了拟定 IGBP。他作为 WCRP 和 IGBP 中国委员会的主席，领导并组织了我国气候系统动力学和全球变化研究。他和他的研究团队对 IGBP 的发展做出了重大贡献，特别是提出了 IGBP 中一些新的概念。由于他的努力，全球气候变化研究在我国特别活跃。

他和他的团队研究指出，中国中部地区存在千年时间尺度、百年时间尺度和年代际时间尺度的突变，并指出从 20 世纪 60 年代中期开始，从萨赫勒东部到中国北部地带突然变得比以前干；在第一次全球变化大会(1984)，叶先生就提出了一个新的概念：地球表面过渡地域的气候和生态系统在全球变化中的重要性，这些区域对全球气候变化比较敏感，在这些区域环境变化的强信号比其他区域早些时候表现出来。此外，从 20 世纪 90 年代起，叶先生号召开展应对全球气候变化的适应研究，并提出了"人类有序活动"的概念。由于叶先生的倡导和努力，我国关于全球变化特别是全球气候变化的研究在国际上占有重要地位。

叶先生的一生是为现代大气科学和全球气候变化研究努力奋斗的一生。他在国际上首先创立了大气长波的频散理论和东亚大气环流理论，不仅为大气动力学理论的发展做出了卓越贡献，而且为天气预报和短期气候预测奠定了理论基础；他提出大气运动适应过程的尺度理论，开辟了中、小尺度环流系统动力学研究的新途径；他在国际上开创了青藏高原气象学的研究，提出了青藏高原对亚洲季风(特别是东亚季风)形成和变异的热力和动力作用；他还提出了全球气候变化的适应理论，推动了全球气候变化研究的发展。中国科学院前院长卢嘉锡先生对叶先生的丰功伟绩做出了高度评价，在叶先生八十大寿的贺词中写道：

> 叶茂根深东亚环流结硕果，
> 　学笃风正全球变化创新篇。

由于叶先生对大气科学发展的丰功伟绩和重大贡献，他先后获国家自然科学奖一等奖、二等奖、三等奖，多次获中国科学院自然科学奖，并获 1995 年陈嘉庚地球科学奖，何梁何利基金科学与技术成就奖。1981 年当选芬兰科学院外籍院士，1982 年当选英国皇家气象学会荣誉会员，1990 年当选美国气象学会荣誉会员。特别是他以卓越的学术成就荣获了世界气象组织 WMO 奖和国家最高科学技术奖，铭记史册。

六、科学精神之楷模

我们今天在此隆重纪念叶笃正先生诞辰百周年，不仅仅是由于叶先生在学术上做出了彪炳史册的研究，是我国气象学的先河,更重要的是他的科学精神,是我们学习的楷模。

(1) 学习叶先生的爱国精神

新中国成立后,叶先生毅然放弃国外优越的工作条件和优厚的生活待遇,冲破阻力,与家人于1950年10月回到祖国。回国后,叶先生被任命为中国科学院地球物理研究所北京工作站主任,在赵九章先生领导下,与顾震潮先生和陶诗言先生等在原1928年建立的中央研究院气象研究所基础上筹建了中国科学院地球物理研究所气象学研究室。叶先生满怀建设新中国的喜悦,开始了创建新中国气象研究的艰苦工作。

1966年叶先生参与创建了中国科学院大气物理研究所,1978年任所长,之后一直担任名誉所长。叶先生以认真的科学精神、严谨的科学作风、宽广的胸怀带领和团结我国大气科学研究团体取得一个又一个的创新研究成果,使大气物理研究所不断发展、壮大。即便是在病重期间,他仍然关心着研究所的发展和大气科学的研究。今天,叶先生参与创建的中国科学院大气物理研究所已发展成为拥有数百名研究人员和数百名博士硕士研究生,几乎包括大气科学所有分支学科的国际知名的综合性研究机构。

(2) 学习叶先生勇于创新的科学精神

叶先生以他敏锐的科学洞察力,勇于创新的精神,把我国的气象研究带到国际前沿。在20世纪50年代大气环流的形成和维持是国际大气科学研究的前沿,叶先生在我国率先领导他的团队开展了东亚大气环流理论、地转适应理论和青藏高原气象学的研究,提出了许多国际领先的大气科学理论,使我国有关大气动力学和大气环流理论的研究位于国际前列。在70年代末80年代初,大气科学的发展进入了一个新的阶段,大气科学研究不仅仅从大气内部动力、热力过程来研究,而且还从大气与海洋、大气与陆地、冰雪圈等的相互作用来研究大气环流和气候变化的成因。叶先生及时主持召开了我国大气科学发展研讨会,会上他前瞻性地提出了开展气候变化研究的重要性,从此之后,我国气候变化研究迅速发展。今天我国关于气候变化研究蓬勃发展与当年叶先生所指出的研究方向是分不开的。

叶先生不仅具有敏锐的科学洞察力和勇于创新的科学精神,而且努力倡导我国大气科学研究与国际的交流和合作,推进中国的大气科学研究走在世界的前列。叶先生组织领导了大量的国际学术活动,参与了许多国际组织工作,曾代表中国参与了世界气候研究计划(WCRP)和国际地圈生物圈计划(IGBP)的筹划与制定。1982年至1988年,他任国际科学联盟理事会(ICSU)和世界气象组织(WMO)联合科学委员会(JSC)委员,并在1983~1987年任国际大气物理和气象协会(IAMAP)执行委员。1987~1995年任国际大地测量和地球物理联合会(IUGG)执行局成员,1987~1999年任国际地圈生物圈计划科学委员会(IGBP)委员。在国际组织任职期间,叶先生做了大量工作,促进了我国环境科学和地球科学研究与国际交流和合作,并且他介绍和推荐我国许多科学家到国际学术组织任职,使我国大气科学在国际上有关研究计划的制定有一定话语权。

(3) 学习叶先生认真严谨的科学作风

叶先生一贯提倡认真、严谨的科学精神,他不仅以身作则,而且对学生在学术上要求严格,注重培养学生具有扎实基础和严谨的学风。他经常教导学生,做学问要"认真、认真、再认真",写文章要"谨慎、谨慎、再谨慎"。并且,他鼓励学生要独立思考,提出与导师不同的学术观点。

(4) 学习叶先生理论联系实际的科学态度

叶先生的研究总是根据国家建设和防灾减灾的实际需要开展。从20世纪50年代起,他根据我国天气预报和气候预测的需求,及时带领研究团队开展了东亚大气环流理论和青藏高原气象学的研究。他不仅重视理论研究,而且重视天气预报和气候预测的实际业务,他对我国气象局的工作十分关心,做了大量指导性工作。在20世纪90年代他根据我国旱涝灾害的严重性,带领研究团队做了我国长江流域、黄河流域旱涝规律与成因研究，为我国的旱涝气候预测提供了科学依据。叶先生在 86 岁高龄时还亲自到我国典型干旱区观测场进行实地考察。

叶先生一贯强调理论研究要与天气气候预报的实际需要相结合，重视研究部门和天气预报业务部门互相协作。他担任中国气象学会理事长多年，团结高校、业务单位和科研机构，开展了很多学术交流活动，对气象学会的发展做出重大贡献。并且，他对高等院校的青年教师和广大气象台预报员也十分关心。叶先生把他获得的各项国内外科学奖金基本上都贡献给大气物理研究所作为鼓励青年人才成长的研究奖金。

(5) 学习叶先生重视人才培养

叶先生非常重视培养人才，他以"上善若水"的崇高品德，像清澈的泉水，滋养着一批又一批优秀大气科学人才茁壮成长。叶先生在培养人才方面做了大量工作，他曾在北京大学、中国科学技术大学等学校任教多年，担任我所学位委员会主任多年，每次他都亲自参加研究生的开题报告，不仅时刻关心着学生的研究，强调研究生研究的前沿性和基础性，特别重视大气环流和气候变化的过程和机理研究，而且他非常关心学生的生活，经常嘘寒问暖，亦师亦父。因此，他培养了一大批优秀大气科学人才，其中许多人已成为国内外享有声誉的学者。

叶先生离开我们已多年了，斯人已去，精神长存，文章不朽，成就永在。让我们继承发扬先生认真、严谨的科学作风和孜孜不倦追求科学真理的精神。

难忘师生情
——缅怀岸保勘三郎先生

2012 年

引　　言

每一位有些许成就的学者，除了自己的勤奋努力外，还离不开老师含辛茹苦的培养。我在日本东京大学留学的日子是我一生成长过程中难以忘怀的一段经历。这期间我的恩师岸保勘三郎教授对我辛勤指导，用他宽厚的肩膀顶托我向大气动力学高处攀登。此文是为缅怀我的恩师——岸保勘三郎先生而写。

一、科学前沿细指引

1978 年经过考试和培训，我作为第一批赴日留学生派到日本留学，但由于指导教师的联系出了一点偏差，1979 年我才到东京大学理学部气象学科留学，经我硕士导师叶笃正先生的推荐，在国际著名学者岸保勘三郎教授指导下从事大气动力学研究。这个研究室有悠久的历史，曾培养出许多国际一流的大气动力学和数值模拟专家，对国际大气科学的发展做出许多重大贡献；并且，东京大学拥有充足的图书资料和当时最先进的电子计算机，我深深感到能到这个研究室留学是很荣幸的。

岸保勘三郎教授不仅是国际著名的大气动力学和数值天气预报专家，担负许多当时国际大气科学学术组织和日本国内气象学科学术组织的重要职务，而且他还是一位深受中国气象学者敬重的气象学家。早在 1957 年他就访华，为我国培训了一批数值天气预报专家。他的慈祥、宽容以及对学生认真培养的态度，给我留下深刻印象，深深铭记在我的脑海里，至今都难于忘怀。

我到日本后的第二天，按规定到东京大学理学部报到，办理手续之后在岸保先生秘书的带领下到气象学科研究室。我第一次到这个研究室，岸保先生在百忙之中亲自带我到东京大学地球物理系有关学科的研究室参观，随后他主持了研究室全体会议，欢迎我到这个研究室留学。在欢迎会之后两天，他让我到他的办公室，在介绍这个研究室的基本情况之后，他就问我"你喜欢动力学研究，还是喜欢数值模式研究？"，我回答"我喜欢大气动力学研究"。于是岸保先生向我提出"准定常行星波形成、传播与异常是大气环流演变和短期气候变异的关键问题之一，是当前国际大气动力学的研究前沿，但比较难，你若不怕困难，这个问题很值得研究，你不妨试试"。当时我因对准定常行星波动力学一无所知，故没有马上回答。回到我自己的办公室，想来想去也拿不定主意，下班后回到宿舍，一夜难眠。我想虽然在北京大学地球物理系天气动力学专业读了六年大学本科，之后在中国科学院又读了三年研究生，但后期却遇上了"文化大革命"，下放农场劳动一年多，回所后，又让我研究气象卫星红外遥感反演和遥感通道选择，行星波动力学我从来没有接触过，要答应做这个问题研究，肯定会遇到很多困难，工作肯定很紧张。可是要与岸保先生的助手（相当于助理教授）一起研究数值模式，我更是没有基础，充其量也是岸保先生助手的助手。经过再三考虑，特别我想起我去日本临行时，叶笃正先生对我嘱咐"此次你能到日本留学，机会难得，一定要做出一个有创新的研究"。于是我选择了当时大气动力学前沿问题之一——准定常行星波的形成、传播与异常作为研究课题。过了一周之后，岸保先生又让我到他的办公室，在问我到日本的一些生活情况之后，他就问我"今后研究课题你想好了没有？"，我回答"想好了，我想研究准定常行星波动力学问题"。听了我的回答，岸保先生很高兴，并说"很好，不过你今后会很辛苦"。于是他就给我介绍了好几篇关于准定常行星波动力学的研究文献给我看，我用了一个月左右时间，把岸保先生给我推荐的参考文献读完。这些文献使我对准定常行星波动力学的前沿问题有一定了解和兴趣。正是岸保先生的指引，使我开始向准定常行星波动力学的大门迈进。

二、研究过程常指导

这个研究室研究气氛很浓,他们有一个传统,研究工作再忙每周都要安排一个下午师生在一起的"轮讲会",把各人研究情况互相交流。第一次安排我在"轮讲会"轮讲,我十分紧张,我就讲了我对今后准定常行星波动力学研究的一些想法,我说"很多科学家研究地形和热源对准定常行星波形成的作用都是用二层模式,但由于准定常行星波会垂直传播,因此,必须要用在大气垂直方向有足够分辨率的模式才能正确描述准定常行星波的特征,这样可以更好地研究地形和热源对准定常行星波形成的作用"。我讲完之后,岸保先生肯定我这个想法,国际著名的大气动力学家松野太郎先生(当时他是这个研究室副教授,在岸保先生退休后,他继任这个研究室教授,之后为日本全球气候变化研究所所长、日本气象学会理事长)也肯定了我的看法。

有了想法,关键是努力去实现这个想法。经过两个月数值模式推演和设计之后,我碰到第一个难题是利用大型计算机的计算程序问题。当时东京大学计算机所用是 FORTRAN 计算语言,输入方式是卡片输入,我在国内虽会几种计算语言,如手编、符号、AGO 等语言,偏偏不会 FORTRAN 语言,因此我只能从头学 FORTRAN 计算语言。经过一段时间,我学会了 FORTRAN 计算语言。但之后,又遇到第二个难题,就是把 FORTRAN 语言一句一句打到卡片上,在 20 世纪 70 年代末把计算程序输入到计算机还不是现在的终端键盘输入,而是用卡片输入,它比当时国内用纸带输入更快、更方便。由于我所设计的数值模式比较复杂,计算程序量很大,因此,所用卡片有几千张,这样每天打卡、对卡要花费很多时间,并且几千张卡只要有一张卡中一个数码打错,计算就无法进行。为此,我调试这个模式的计算程序整整花了四个月时间。这四个月中我睡不着、吃不香,有时程序在计算机上通不了,彻夜难眠,不知程序错在哪里,有时是"踏遍青山无觅处,得来全不费功夫",经常睡一小会儿觉,忽然想到在某段程序有错,一大早连早饭也顾不上吃就赶到研究室,修改程序卡片。岸保先生尽管很忙,每周总要抽出时间与我讨论一次,他经常鼓励我"世界上没有一个复杂的程序一次就编成,在计算机上都要经过无数次调试,总是步步逼近,今天调试程序通不过,明天、后天,…总有一天会通过,计算结果会出来"。事实如岸保先生所说,我用了四个月时间终于把 34 层数值模型的程序调试通过,得到第一个计算结果,计算出北半球实际地形和热源强迫所产生的准定常行星波的振幅与位相分布。我在研究室师生的"轮讲会"上讲了这些计算结果,听我讲完之后,岸保先生一方面肯定了我的研究结果,另一方面他又指出在大气中准定常行星波形成方面还有许多问题需要进一步研究。在岸保先生的指导下,经过反复数值试验,我计算出比较符合实际的北半球冬季地形和热源强迫产生的准定常行星波振幅和位相的分布,以及几种地形和热源强迫所产生准定常行星波分布的差异。

岸保先生为我已得到初步的研究成果感到十分高兴,并看到我已步入准定常行星波动力学的研究大门,故他就让我自己愿意研究什么就计算什么,只是提示我可以利用此数值模型研究准定常行星波动力学的许多问题,并形象地比喻说:"你设计的数值模型只是意味你把炒菜的锅灶准备好,今后炒什么菜就看你想吃什么菜"。于是我利用此 34 层数值模型不仅研究了北半球冬、夏季各种地形和热源强迫所产生的准定常行星波的分布,而且还应用此模式再结合波在缓变媒质中传播理论,在前人研究的基础上,研究了地球大气准定常行星波在地球大气三维空间中的传播特征和规律,提出了地球大气准定常行星波在三维空间传播的两支波导理论,以及准定常行星波变异的成因等。这些研究结果在 20 世纪 80 年代初就发表在《Journal of the Meteorological Society of Japan》和《中国科学》刊物上。正是岸保先生对我耐心指导和鼓励,我才能做出这些有创新性的研究成果。

三、依依不舍师生情

在东京大学留学的日子里,我与岸保先生建立了深厚的师生情谊。他不仅是我学术上的恩师,而且在生活上也十分关心我。岸保先生得知我是公费留学生,国家支付的生活费用不多,他就经常让我去日本一些大学参观和参加会议,研究室支付差旅费,这样可以节省一些费用。他经常问我在日本生活是否习惯,问寒问暖,每逢年节,他都要请我上他家一起过节,热情招待。

由于我在地球大气中准定常行星波的形成、传播与异常做出有创新的研究,日本东京大学授予我理学博士学位。在博士论文答辩之后,我回国了,不久岸保先生也从东京大学退休了,但他仍像以前一样,十分关心我的研究,把国际上大气动力学新的研究动态及时写信让我知道;并且,每次我去东京参加一些学术会议,他都到会议学术报告厅去听我的报告,有一次我参加在筑波召开的"国际大气环流和热带气象学术研讨会",我的报告安排在下午的后半段,当时他身体不适,我劝他回东京,但他仍坚持听完我的报告才回东京。到了90年代初,岸保先生得了脑血栓,行走不利索,必须拿拐杖,我每次去日本参加会议,都要去拜访我的恩师。当恩师得知我要去拜访他,他和我的师娘都十分高兴,准备了丰盛的食品;并且每次柱着拐杖坚持到电车站接我,我要回旅馆时,他也坚持一定送我到车站,我再三劝他回去,他都不肯移步,直到电车开走了,他仍拄着拐杖站在站台上望着向远处开走的电车。那种依依不舍师生情至今深深留在我的脑海里,他那慈祥、和蔼可亲的形象经常浮现在我眼前。

四、后 记

岸保先生在 2011 年冬季由于脑血栓加重不幸去世。2012 年日本气象学会举行岸保先生追思会,从世界各地来了许多学者,挤满了会议厅,会上只安排了四位学者讲话,除了会议主持人做追思会的主旨讲话(讲述了岸保先生的生平和在学术上的重大贡献)外,日本著名气象学家、岸保先生老朋友增田先生讲述岸保先生对日本气象学会改革和发展做出的重大贡献;国际著名气象学家松野教授追思了岸保先生对国际大气动力学和数值天气预报发展的丰功伟绩;我作为岸保先生的学生代表在会上除了追思岸保先生对中日气象合作和交流做出的重大贡献外,还追思了岸保先生对我的培养和师生情谊。此文就是我在 2012 年日本气象学会举办的岸保先生追思会所讲的部分内容。我想岸保先生在另一世界并不寂寞,因为有许多的日本和国际上的学者在怀念着他,这其中有受过他精心栽培的我对他不尽的思念。

上善若水
——忆许惜今校长
2011 年

许惜今校长是我成长过程中不能忘怀的一位好校长。他与原惠安二中老校长曾锦章先生启蒙我走上求知之路，使我从一个穷苦的放牛娃成长为一名中国科学院院士。对于他们的恩情，我无论何时走到国内外任何一个科学讲坛上，都不能忘怀，他们的教诲正是时时刻刻激励我努力奋斗的动力。

许惜今校长在惠安一中任校长，虽只有短暂的二年多，但他砥砺耕耘，励精图治，把惠安一中搞得有声有色，创造了惠安一中鼎盛时期。他积极组织学生们勤工俭学，并言传身教，以身示范与学生们一起劳动。他尊师重教，把各方面老师凝聚在他的周围，特别是把一些著名老师团结在一起，如物理课的连家瑶老师、数学课的张跃辉老师、化学课的陈淑媛老师、生物课的孙昆化老师、语文课的曾毓贤老师等，这些老师不仅是我县最著名的老师，而且在当时泉州专区（现称泉州市）也是鼎鼎有名。在20世纪50年代后期，惠安一中高中部老师的阵容可以说是母校自创立到当时最强大的。这些老师具有丰富的教学经验，他们讲课生动，听这些老师的讲课简直是一种享受，50多年前这些老师的教学场面至今还深深铭刻在我脑海里。更令人钦佩的是在当时大跃进的年代中，许惜今校长顶着各种压力，狠抓教学，把教学放在最重要的地位。他不仅亲自任班主任，而且经常组织观摩教学，帮助年轻老师改进教学方法。在晚自修时，他还经常到各班去看看学生们自习情况，了解作业情况和老师们的教学情况。正是由于他的尊师重教，使惠安一中的教学质量迅速提高。在1958年和1959年高考方面取得显著的好成绩，如高15组参加高考近300人，有200多人考上不同大学，其中有几位同学还考上了北大、清华、科大等名校。因此，许校长在惠安一中的时期是惠安一中鼎盛的时期之一。

许校长虽逝世已35年了，但他的精神永在。他为国家、为人民不计个人得失、刚正不阿、直言不讳、鞠躬尽瘁、死而后已的崇高精神永远是我们学习的榜样。一段人生，愈多磨难，愈能呈现人的矢志不移，愈能呈现人的坚韧不拔。许校长一生坎坷，经历了许多起伏和磨难，这些起伏展示了他不计个人得失、刚正不阿的崇高精神；这些磨难呈现了他鞠躬尽瘁、矢志不移的伟大品格。

古人用"上善若水"来形容伟人的高尚品格，水滋养着万物而从不求回报。许校长的品格如清澈的泉水一样，他用智慧、才干和精神创造了惠安一中的鼎盛时期，滋养了惠安一中的一大批学生，但他没有求过一丝的回报。他教书育人，硕果累累，几经沉浮鞠躬尽瘁，最后无怨无悔地走完了他平凡而伟大的一生。在纪念许校长诞辰九十周年之际，抚今追昔，感触甚多，特作如下对联，以表达我对许校长的缅怀之情：

春风化雨桃李芬芳育英才，
鞠躬尽瘁为人师表存风范。

在纪念谢义炳先生诞辰百周年会上的讲话

2017 年

今天我们聚集一堂，隆重纪念谢先生诞辰百周年，这是因为谢先生不仅是我国伟大的现代大气科学教育家，而且是一位国内外著名的气象学家。谢先生是我国现代大气科学教育的开拓者之一，他在我国大气科学学科建设、人才培养、大气科学研究的发展等方面做出了卓越贡献。在他建议并参与组织领导下，北京大学物理系气象专业发展成拥有大气物理、气象、空间物理、天体物理、地球物理等完整的地球和行星物理的教育体系，为国家培养一大批大气科学、空间科学、地球物理和天体物理的优秀人才。这些人才在科学研究、教学和业务系统中发挥重要作用。据我不完全统计，地球物理系培养出包括大气科学、地球物理、天体物理、空间物理学科近 20 位两院院士，其中大气科学领域就有 12 位两院院士。

谢先生不仅在人才培养方面做出重大贡献，而且对我国大气科学的发展提出了具有前瞻性、战略性的长远目标。在 20 世纪 90 年代初，他发起并亲自组织了第一、二次我国大气科学发展战略研讨会。这两次研讨会不仅详细分析了国内外大气科学的发展动向，而且对我国大气科学的发展目标、应研究的主要科学问题及应采取的措施进行了热烈讨论。在他病重期间，还时刻关心我国大气科学发展战略研究，他让我和刘适达老师一定要把大气科学学科发展战略研讨会继续下去。在国家自然科学基金委员会地球科学部、中国科学院地学部、有关大学的支持下，大气科学学科发展战略研讨会在谢先生仙逝之后又召开了 4 次。这些研讨会的召开，对促进我国大气科学的发展起到重要作用。当年在研讨会上谢先生所提出的我国大气科学重大研究方向和目标，对我国大气科学的发展起着重要作用。

谢先生也是我国现代大气环流理论和天气动力学研究开拓者之一，他在全球大气环流理论、湿斜压大气动力学、寒潮、锋面、台风和暴雨等领域作出许多开创性的研究。今天我们纪念谢先生诞辰百周年，要学习谢先生"求实创新"的科学精神，谢先生一贯提倡在科学研究上不仅要瞄准国际前沿，而且要针对我国天气气候的实际科学问题，他关于东北低压的演变和锋面结构、我国夏季暴雨发生的环流特征及动力学、热带辐合带(ITCZ)不稳定及台风的生成等方面的研究论文，不仅成为这些研究领域的经典性著作，而且为我国天气预报提供了很有效的科学依据。

今天我们纪念谢先生诞辰百周年，要学习谢先生"坚持真理"的科学精神，早在 20 世纪 70 年代和 80 年代谢先生提出湿空气动力学，系统地研究了湿空气在天气气候系统演变中的重要性，当时有一些学者对此有不同看法，产生了争论，谢先生始终坚持湿空气动力学的正确性。今天随着全球气候变化研究的深入，湿空气在全球气候变化过程中的重要作用已成为全球气候变化中一个前沿研究课题。

谢先生已离开我们 22 年了，斯人已去，文章永在，精神永存！让我们继续发扬谢先生求实创新，坚持真理的科学精神。

在福建省泉港二中（原惠安二中）五十周年校庆庆典上的贺词

2002 年

今天，我满怀喜悦的心情来参加母校建校五十周年庆典活动。首先请允许我代表原惠安二中校友向母校表示祝贺。

首先，贺母校五十春秋教书育人结硕果。在这五十个春秋，母校经历了风风雨雨，从只有一座破旧老庙当教室的初级中学，变成了一所拥有崭新的现代化教学大楼、三级达标的完全中学。此时此刻，不禁使我回想起往事。忆当年，煤油灯下同学们挤在这座破旧庙里孜孜不倦地学习；看如今，楼上楼下电灯电话。忆当年，在台湾国民党敌机的轰隆轰炸时，同学们在地下防空洞中的琅琅读书声；看如今，只听到同学们在整洁的教室中英语的朗读声和电脑的按键声。忆当年，衣衫褴褛、光着脚的同学们在操场迎着凛冽的寒风跑操；看如今，脚穿耐克鞋、腿穿牛仔裤的同学们在校园潇洒地漫步。这是多么大的变化，这个变化是时代的变化，是国家经济的变化。然而，如今这么好的学习环境决不能忘记建校初期的艰辛，母校建校初期正是在这样艰苦的教学环境中培育了数百上千的数理化成绩突出的人才。如今我们惠安二中的校友不仅遍布全省各地，而且在全国及世界五大洲也能见到惠安二中校友的足迹，他们正在为我国的科技、教育和经济建设事业作出优异成绩，或是为国际科技的发展而拼搏，这是值得我们惠安二中师生为之自豪的。

在庆贺母校建校五十周年时，我们不能忘记曾锦章校长、孙祖恩教务主任、郭端修、吴再兴、吴明发、吴炳仁、连凤辉等老师。正是这批老师在艰苦的条件下把惠安二中办起来，把我们这批穷孩子培养成才，他们正是遵循伟大教育家陶行知先生所倡导的"捧着一颗心来，不带半根草去"的精神。此时此刻，我们应该感谢这些老师为母校的创建作出重大贡献，缅怀他们的丰功伟绩。

我是母校建校第二年入学（初二组）的学生，时光似箭，日月如梭，我离开母校已整整 46 年了，真是"少小离家老大回，乡音未改鬓毛衰"。今天我借母校庆祝建校五十周年之机，感谢母校的诸位老师对我的培养和教导。我能够从一个穷苦家庭出身的孩子，成长为一名中国科学院院士，这完全归功于母校的培养和各位老师的教导。正是在母校这段艰辛的学习生活，激励着我对科学的探索，无论我在中国科学院研究岗位上，或是在国内外科学讲台上，我都没有忘怀这段艰辛的生活。没有惠安二中老师启蒙我对数理化的钻研，今天我就不可能在大气科学领域做出些许创新的研究，也不可能在科学道路上留下一点足迹。

最后，在这大好的日子里，祝母校素质教育谱新篇，新世纪是泉港二中在素质教育再展雄风的世纪，新世纪是我们泉港二中再创辉煌的世纪。我作为母校的一名学子，愿意与泉港二中的师生一道，为提高我们泉港二中的教学质量、为提高我们泉港二中的声誉而奋斗。

"海峡两岸大气科学研究生学术研讨会"总结

2002 年

"海峡两岸大气科学研究生学术研讨会"于 2002 年 12 月 19 日至 20 日在台湾大学凝态中心举行。来自海峡两岸大气科学领域有关大学和研究所的师生计 80 余人参加了此次研讨会。台湾"中央"大学刘兆汉校长、台湾大学陈泰然教务长以及周仲岛、王作台等教授出席了此次研讨会。由中国科学院大气物理研究所、北京大学大气科学系、南京大学大气科学系、南京气象学院、中国气象科学研究院、中山大学大气科学系以及中国科技大学地球和空间科学系师生组成的大陆师生代表团参加了此会。现就参加此次会议的感想和体会总结如下。

一、两岸大气科学研究交流之必要

由于两岸大气科学师生面临共同的天气、气候和环境的问题，因此两岸师生在暴雨、台风等天气灾害、旱涝等气候灾害、大气环流、气候系统动力学和数值模式、大气化学与污染、雷达探测和大气遥感、大气物理和人工影响天气以及全球变化等领域已作出很好的研究。为了交流这些成果，两岸大气科学界的师生遵照"创新、交流、修学、敦谊"之共识，已分别在两岸风光秀丽之地联合举行了多次两岸大气科学研究生学术研讨会，互相交流了两岸师生在上述诸领域所取得的研究进展。在前几次研讨会成功举行的基础上，在陈泰然教务长的关照下，周仲岛教授的精心组织和主持下，这次海峡两岸大气科学研究生学术研讨会成功地在台湾大学举行，本次研讨会无论学术报告所涉及的内容和学术水平比前几次研讨会又前进了一步。

从此次研讨会所报告的内容看，两岸师生所从事的研究课题不仅处于国际前沿，而且研究课题紧紧与两岸经济和社会发展所急需解决的天气气候问题相联系。两岸老师们和同学们普遍反映，此次研讨会学术水平比较高，学术气氛浓厚，讨论热烈。在研讨过程中，青年学者不仅展现了自己的研究成就，相互启发，共同探讨未知的课题，而且也领略到老师们严谨的治学精神和很高的学术造诣。许多青年学者反映，此次研讨会上提出了不少新的学术观点，闻之令人耳目一新，特别在台风的生成和发展机制、暴雨与中尺度系统的生成和发展过程、季风变异与旱涝灾害、气候变化和数值模拟等方面的研究取得很好的进展，这些研究给两岸青年学者留下深刻印象。

二、两岸大气科学师生互相学习之必要

大陆代表团利用研讨会召开之前的 16、17 日两天先后参观了台湾"中央"大学、台湾大学和中国文化大学的校园和图书馆，以及这些大学的大气科学系。师生们每到一处，均受到各有关大学大气科学系师生的热情欢迎。大陆师生不仅领略到我国台湾有关大学校园的整洁，而且通过座谈和互相介绍，对我国台湾各有关大学的大气科学系的教学、科研、设备有了直观的了解。两岸师生互相促膝谈及教学心得和学习情况，相互学习。在参观访问中，同学们研修学问，开阔眼界，寓教于乐。

三、两岸师生在交流、修学过程中加深了相互了解与友谊

在此次研讨会过程中，两岸师生不仅在学术上交流研究成果，共同探索大气科学的未知领域，而且通过交流与修学，互相了解，增进友谊。无论在研讨会上的休息时间或在参观座谈中，师生们抓紧时间互相留言，探讨合作研究的课题和路径。特别在研讨会之后赴日月潭、高雄等地参观时，两岸师生进行联欢，同唱一首歌，这些不仅加强两岸师生之间的合作，而且大大增进了两岸同学之间的友谊。在高雄临别时，师生们都互相留下联系地址，为今后对共同感兴趣的大气科学问题展开合作研究做准备。

总之，此次研讨会两岸师生遵照"创新、交流、修学、敦谊"之原则，不仅顺利完成了学术交流这一目的，而且建立了在学术交流基础上的紧密联系和协作；并且，这种交流不仅仅是体现在研讨会上，而且将会延伸到今后的科研和工作中。因此，研讨会不仅为两岸广大青年学者的学术交流、展现自己研究成果提供平台，而且为海峡两岸的青年学者的修学和友谊架构桥梁。两岸师生都希望海峡两岸大气科学的学术交流能更加密切，海峡两岸大气科学研究生学术研讨会能一届一届延续下去。

《福建省惠安一中高十五组同学毕业五十周年纪念册》序言

2009 年

　　在参加母校惠安一中创办九十周年庆典时,我曾倡议,在 2009 年母校九十三周年校庆时正是我们高中毕业五十周年,应该好好纪念和庆祝一番。作为这次纪念活动的珍贵礼物,《福建省惠安一中高十五组同学毕业五十周年纪念册》已编辑出版了。它展示了我们惠安一中高十五组同学高中毕业五十年来的奋斗历程和多姿多彩的人生轨迹,是一册有史、有实、有趣、可忆、可思、可亲的妙作佳篇,表达了我们的共同心愿。

　　母校惠安一中是我们人生旅途中的重要一站,这里留下了我们青春的足迹、辛勤的汗水和纯真的友谊。三年同窗共学,我们烙上惠安一中精神的印记,塑造了我们今生的思想、品行、情操和理想,夯实了日后各位同学所从事事业的基础,练就了为人民服务的本领。在毕业时我们深知,未来生活的征程是从这里起航的,我们取得的点滴进步和些许业绩都渗透着母校付出的辛劳和期望。我们深深眷恋着在校时的美好时光,忘不了"和平屋""民主楼"同学们日日夜夜攻读的身影,忘不了在操场上争创体育达标的飒爽英姿,忘不了同窗学友真诚友爱的纯真笑脸,更忘不了当年老师的谆谆教诲。正如同学诗中所云:那段经历让人久久难以忘怀,因为那里有我们青春年华留下的脚印,有我们激情时代存留的记忆……。

　　半个世纪过去了,而今我们重又聚首,早已不复往日童颜,正如唐人贺知章诗中所云"少小离家老大回,乡音未改鬓毛衰,儿童相见不相识,笑问客从何处来"。岁月如歌,我们每个学子都是这首歌中的一个音符,共同谱写了这五十年不平凡的人生交响乐章。

　　五十年弹指一挥间,在这期间先后有六十多位同学离我们而去,我们没有忘记他们,在这里谨向他们表示沉痛哀悼!向他们的亲属致以深切问候!

　　祝愿同学们身体健康,并愿《福建省惠安一中高十五组同学毕业五十周年纪念册》能相伴左右,时常拾起我们青春留下的欢歌笑语。

海峡两岸青年大气科学学术研讨会的开幕词

2009 年

各位老师、各位同学：上午好！

自从 1996 年两岸青年大气科学家首次相聚于山东威海市以来，至今已有 13 年了，两岸大气科学师生先后在台北、威海、张家界、成都、昆明等地召开了多届青年大气科学学术研讨会。今天我们又相聚在塞上江南的银川市，召开新一届两岸青年大气科学研讨会，在此祝贺此会顺利召开。

这个序列会是由中国科学院大气物理研究所和中国文化大学共同发起，并联合两岸有关院校和有关科研机构共同组织和召开的高水准青年大气科学学术盛会。13 年来，这个序列会议遵循"创新、交流、修学、敦谊"之原则，不仅为两岸大气科学领域优秀研究生和年轻研究人员提供了学术交流和修学的良好机会，而且也促进了两岸年轻的大气科学学者的联谊活动，加强了彼此的沟通和理解，增进了之间的友谊。两岸学子在这短暂的会议期间，在这学术百花园中得以濡染芬芳与智慧，发挥聪明才智。两岸青年学者不仅展现了自己的研究成就，相互启发，共同探讨未知，提出了不少新的学术观点，而且领略到老师们严谨的治学精神和学术造诣。此外，在会后参观旅游中大家不仅领略祖国的美好河山，而且还继续讨论共同感兴趣的科学问题，研修学问，开阔眼界，寓教于乐。

回顾这 13 年来历届会议研讨所涉及的大气科学领域十分广泛，研讨的科学问题，包括了全球变暖、气候动力学、气候预测、季风和大气环流、台风和暴雨、中小尺度环流与灾害性天气、大气物理、大气探测与遥感、大气环境、大气化学与边界层物理、地球流体力学等领域，几乎涉及大气科学各个领域的前沿课题。这充分显示出两岸学子以严谨的学术作风，孜孜以求地探索着大气科学的前沿问题。

我作为两岸青年大气科学学术研讨会的最早倡议人之一，协助刘广英教授主持了历届会议的筹备和组织工作。当年也许可算是中年学者，现已成为老年学者，完成了历史使命，因此，从这一届之后我不再参与会议的具体组织工作。此届两岸青年大气科学学术研讨会许多会务都由年轻一代来筹备组织。"桐花万里丹山路，雏凤清于老凤声"。我相信此届研讨会将比以往几届办得更好。虽然我不再做具体组织工作，但我一定会继续关心两岸青年学者的学术交流。

"乘风破浪会有时，直挂云帆济沧海"。当前大气科学正要迎接一个飞跃发展期的到来，我们两岸青年学子一定要不失时机把握这个机遇，求实创新，不懈努力，加强交流，共同迎接即将到来的大气科学飞跃发展期。

最后祝会议取得成功！

《泉港区第二中学(原惠安二中)校志》序言

2009 年

古云："盛世修志，志亦盛事"。

母校泉港二中(前惠安二中)编修的《泉港区第二中学校志》在编纂人员的辛勤笔耕下今已问世，欣然拜读，它是一册有史、有实、有趣的妙作佳篇。

忆往昔，五十七年前在母校艰难求学的身影不禁浮现在我的眼前，思绪万端。破旧的锦山庙是我们唯一的教室，暗弱的煤油灯是我们唯一的照明，田埂空地是我们晨练的操场。然而，在这样的环境中，同学们却孜孜不倦，勤奋读书，琅琅的读书声萦绕校园，争创成为三好学生的飒爽英姿到处可见，同学们在这样的青涩磨练中不断进取，几度风雨，几多奋斗。这破旧的校舍里留下了我和同学们青春年华的脚印，这里给我们留下的美好记忆至今也难以忘怀。

再回首，时光流转，五十七年过去了，弹指一挥间。泉港二中蹒跚起步，历经风雨，艰难成长，现今旧貌换新颜，高大宽敞的教学楼，设施先进的科技楼，舒适清洁的宿舍楼，一个现代祥和的校园呈现在眼前。让人更感欣慰的是，现已发展为一所省二级达标的高中学校，它将再次起航，身披更加灿烂的朝阳。

一条流河，愈多起伏，愈能展示它的奔腾咆哮，愈能展示它的雄浑壮阔。

一册校志，愈多曲折，愈能书写她的辉煌灿烂，愈能书写她的多姿多彩。

一段人生，愈多磨难，愈能呈现你的矢志不渝，愈能呈现你的坚韧不拔。

不思量，自难忘。母校泉港二中在我们心中将永远铭记，因为她是我们人生旅途中的重要一站，在这里我们烙上了泉港二中(原惠安二中)精神的印记，铸造了我们今生的思想、品行、情操和理想的雏形，夯实了今生我们所从事事业的基础。我和同学们现在的些许业绩都渗透着母校付出的辛劳和期望，是母校的清泉滋养了我们的成长。正因为如此，泉港二中的发展历程深深地烙在我们心中，融入我们的血液，泉港二中的精神时刻激励着我们努力前进。五十七年来，许多老师用辛勤汗水浇灌我们这些穷学子成长，为泉港二中从一座破旧宫庙发展到今日现代化的名校献出了他们的美丽人生，他们的精神正是伟大教育家陶行知先生所说"捧着一颗心来，不带半棵草去"的奉献精神。今日我们不能忘记他们，我想编修此志也能寄托我们对他们的哀思。

先贤后学，如能从这本校志中读出一点人生的感悟，也就无愧于母校的教育与培养，更不违编修此志的宗旨了。

在"首届大气科学学科建设与人才培养研讨会"上的讲话

——发展大气科学 培养创新人才

2011 年

研究生教育是培养国家高层次人才的重要途径，是实施"科教兴国"伟大战略的重要任务。当今世界，经济和综合国力的竞争实质上是科学技术的竞争，是人才的竞争，培养高层次人才是当前我国大气科学教育的重大战略目标。青年硕士、博士生不仅是当前我国大气科学科研的主力军的一部分，而且是我国大气科学在第一线从事知识创新研究的生力军。因此，培养好具有创新能力的青年硕士、博士，不仅是每位教师的重要任务，也是国家赋予我们大气科学每位博士生导师的历史使命。

最近，中国科学院地学部把"大气科学与全球气候变化重大科学问题"作为当前我国学科发展战略研究一个重要项目。这个项目旨在回顾国际和我国近百年来(特别是最近 60 年)大气科学发展历程及中国科学家对国际大气科学和全球气候变化研究发展的贡献；分析当前国际大气科学和全球气候变化研究发展的特征、动向和趋势；提炼我国当前大气科学和全球气候变化研究的重大科学问题，从而提出符合我国发展需求的重大战略研究方向及应采取相应的发展措施。在此会上提出我国应从大气科学大国走向大气科学强国的口号。

然而，目前我国大气科学人才培养方面离此目标还有一定距离，我国从当前在读博士生人数而论，已近千人，与美国持平，但我国大气科学人才培养方面还缺乏创新；各有关院校专业设置太窄，大气科学招生数太少，教学还缺乏基础性和前沿性，课程多而杂，并且各有关院校专业千篇一律，缺失特色，特别是大气探测、大气物理、中层大气有关专业设置还不多，人才缺乏。这些都要求我们在座各位共同努力，把我国大气科学人才培养搞上去。下面谈谈有关发展大气科学、培养创新人才方面的一些看法和点滴体会。

一、首要的是扩展大气科学二级学科，多培养致力于大气科学研究的学生

第一，大气科学经过百年的发展，现已成为一门现代的综合性学科，它在地学就现代科学而论具有引领性；并且，随着我国经济的快速发展，不仅在天气预报和短期气候变化预测需要人才，而且在应对全球气候变化、人工影响天气、气象卫星和雷达探测、风能和太阳能的利用、环境保护等方面急需大批大气科学人才。但是，现在我们大气科学只有 2 个二级学科，大大局限了我们硕士、博士的招生，无法满足上述各项事业发展的需求。相比之下，我们兄弟学科海洋科学拥有 7 个二级学科、地质学则有 10 多个二级学科。因此，大气科学首要的任务是把大气科学二级学科扩展，把大气物理与大气环境分开，而且应增设地球流体力学、气候系统和大气探测和气象工程等二级学科，尽快使大气科学教育和人才培养跟上国家经济发展。

第二，当前有些大学大气科学的招生规模和老师队伍比起 20 世纪 60 年代不仅没有发展，而且还在萎缩，有的名校在 50 和 60 年代教授达 20 多名，招生数达近百名，而现在教授加副教授的编制也不过 10 多名。有的学校由于合并，原本很有名的大气科学专业逐渐消失。量是质的基础，没有量，谈什么质。当前在高校的老师负担很大，他们有很重的教书任务，还要承担很重的科研任务，还要求做出创新性研究，说实在话，这是不好做到的。因此，在一些名校应尽快扩大大气科学教师队伍，设置一定附属研究中心、研究院等，以便来扩大师资队伍，扩大招生规模。

第三，大家知道，大气科学是一门既艰苦、待遇又低的学科，因此，很多学生对大气科学没兴趣，不愿从事大气科学研究，一些名校进入大气科学学习的本科生很多不是第一志愿，加上从事大气科学研究的学生去欧美一些国家留学比较容易，致使大批学生出国留学并留在国外，这造成 20 世纪 80 年代后期到 90

年代国内大气科学研究生的生源紧缺，从新世纪初开始，这种局面虽已扭转，但一些名校的学生报考大气科学研究生仍不令人满意，这可能造成目前国内大气科学高层次人才缺乏的原因之一。

因此，如何吸引、凝聚全国大气科学的本科学生，特别是那些有作为的学生到大气科学来读研究生还是存在问题。在此，我呼吁大家应不厌其烦地讲述我国大气科学在国民经济建设、社会发展和环境保护的重要性，以及中国大气科学研究在世界的重要性及其地位，以鼓舞同学们热爱大气科学。叶笃正先生在解放初期国家最需要大气科学的科研人才时，毅然放弃在美国的学术地位和优厚的生活待遇毅然回国，开拓了中国现代大气科学的科学研究，培育了一大批国家现代大气科学的高层次人才，他是我们学习的楷模。

二、重要的是大气科学课程设置要不断创新，激励学生对大气科学的兴趣

当前国家学科领域非常需要具有高素质的高层次人才，我认为，当前我国大气科学的有关院校和科研院所不仅必须而且完全有能力培养出大气科学的高层次人才，这就要求我们大气科学本科和研究生教学要不断创新。

第一，大气科学从19世纪地理学科的一个分支学科发展成为今天已拥有多个分支学科的一门综合性现代学科。在它的发展中提出了许多重要理论，有些理论对整个自然科学有重要贡献，如非线性动力学，这些理论不断推动着大气科学的发展。但我国有些高校的课程几十年一贯制，不能把当前新发展的分支学科及时向学生介绍，因此，我国大气科学的课程设计要不断创新。我们大气科学的课程设计，既要强调基础性又要有一定前沿性，如地球系统、非线性动力学要适当介绍一些，去掉一些冗余的课程。

第二，大气科学的基础课程要不断创新，现在我国高校推行PPT教学，有的老师平常很忙，上课前把PPT拿到教室一放，效果如同函授教学；有的学校大气科学的本科生还不知道20世纪60年代就发现的赤道大气波动。当前，大气科学迅速发展，这就要求我国大气科学研究生主要课程不仅具有基础性，而且要不断创新，要把一些大气科学的新进展、新理论及时增加到教学内容中去，要不断吸收国际上一些先进国家的大气科学教学内容。

第三，为了激励学生们对大气科学的兴趣，培养一批有创新精神的大气科学高层次人才，作为博士生导师，首先必须把课程不断创新，把课教得活泼、生动。科研是很重要，但讲课也很重要，因此，尽管大家学术事务和行政工作繁忙，一定要从事一定的教学任务。当前不仅要提倡和鼓励一些大气科学院士、专家给有关院校和科研机构的研究生讲课，而且院校之间以及科研机构和院校之间的讲课老师要互相流动，互相交流，争取把我国大气科学某些主要课程凝炼成国际大气科学的优秀课程或一流课程。

第四，营造一个和谐、宽松而有浓厚的科研气氛的科研团队对于激励学生对大气科学研究的兴趣是很重要。现在是大科学时代，不是以前师傅带徒弟的小作坊科研时代，科研项目不是只凭个人的灵感能完成，要靠一个有合作精神的团队来完成，特别是大气科学，更是这样。大气科学不仅需要研究气候系统中大气、海洋、陆面、冰雪变化及其它们的相互作用，而且还得研究这些相互作用对天气气候变化的影响，这是一个庞大而复杂的研究任务，没有一个通力合作的团队是完成不了的。要完成这个庞大气候系统的研究，就必须要求有的学生从大气环流变化来研究，有的学生从海洋热力变化来研究，有的学生从陆面来研究，有的学生从冰雪变化来进行研究，有的学生从事观测资料分析，有的学生从事动力诊断研究，有的学生从事数值模拟研究。在这团体中同学们互相讨论、互相合作、互相配合、互相帮助，形成一个和谐、生动活泼的研究团体。

三、关键的是培养学生对研究的进取，要有创新的科学精神

培养什么样的高层次人才才能适应21世纪的要求？许多导师都在谈论和探讨这个问题。有的专家认为，21世纪高层次人才是具有创新思想全面发展的人才；有的专家认为，21世纪需要的高层次人才是能认识和掌握自然界和社会发展规律并对国家和社会有无私奉献精神的人才。总之，21世纪需要的高层次人才是高素质、具有创新能力、全面发展的人才。要造就这样高层次的人才不是一件容易的事。下面，我想谈

谈几点初浅看法：

第一，要求导师必须站得高、看得远，不仅要学生学好基础课程和专业课程，重视文献阅读，而且重视培养研究生对于实验、观测资料的分析、机理的研究以及利用计算机进行数值模拟等创新能力。导师在研究生的研究过程中要始终强调创新，要求学生不要在文献堆中找问题，既要查阅文献，又不要被文献束缚住，选题不仅要立足于国家需求，而且更要注意学科的前沿性和国际的发展趋势；并且，研究生的研究课题不要仅仅局限于完成国家重大科研项目，而是要强调具有一定学科的基础性、前沿性和创新性。当前我国大气科学不少博士研究生的研究课题存在着只是完成导师所承担的科研项目，缺乏研究的基础性、前沿性和创新性，更谈不上对大气科学将来一些学科发展方向前沿科学问题的探讨。此外，大气科学的院校和研究机构要高度重视博士研究生开题报告制度，把博士研究生开题报告看成是高层次创新人才培养重要的一环，要认真地对博士研究生学位论文研究是否具有前沿性、创新性进行把关。

第二，要培养高层次、有创新能力大气科学人才还必须把学生放到国内外有关的学术研讨会去磨炼。为了培养学生的创新能力，要利用国内、国际大气科学学术会议，研究项目全体会议以及暑假举办的青年大气科学学术研讨会等召开之机，鼓励学生们去参加这些会议，并发表论文，与国内外有关学者进行认真讨论。通过参加这些会议，使学生能认识到国内外本领域的研究进展、发展趋势，从而提高他们的创新能力。因此，利用广泛的国内外学术交流，给研究生们提供良好的学习机会，促进研究生的创新能力的培养。

第三，为了要培养学生对科学的认真态度和治学的严谨精神，对学生的论文要严格把关：一是本研究组学术沙龙会上要认可；二是在全国或项目学术讨论会上要认可；三是国际学术讨论会上要认可。经过严格把关后的论文投到有关 SCI 收录的刊物上审稿都较容易通过。并且，学生在文章中不要动不动就用"发现"等大词汇，在文章中"表明"或"指出"大气中一些新现象、新机理就已足矣。各位老师尽管各种事务繁忙，对于学生的论文，都应修改多遍，认真把关。要培养学生对科学的认真态度和治学的严谨精神就必须从一个个小课题的具体研究做起，要从一篇篇文章的撰写、修改、发表做起，持之以恒，从不松懈，只有这样，才能形成治学严谨的作风。另外，要求博士生的文章不要动不动就签上导师的名字，要锻炼学生独立发表论文的能力，只有这样，才能逐渐锻炼学生自己去发展一个分支学科。

第四，导师对研究生成才负有义不容辞的责任，导师不仅是研究生学术的指路人，也是研究生灵魂的工程师，既要教书又要育人。导师不仅自己要治学严谨、作风正派、为人师表，而且对学生在治学、道德和团结协作精神也要提出严格要求。应该说，最近几年来，国家对科研加大了投入，通过许多重大科研规划的出台，使我国科研水平有很大提高，国际影响也不断扩大了。但是，也应该看到，当前很多老师为了要任务，申请课题，或是参加种种评审会、学术会议等，不得不成为"空中飞人"，加上各种晋升、课题评估和结题验收都需要论文，这样，许多论文要靠学生来完成。为了应付课题评估、结题验收，许多老师不得不追着学生要论文，这就容易使学生养成浮躁作风。这种现象在别的科学领域存在，也在我们大气科学存在。现在，认真搞观测、细致分析数据的人少了，而从网络上下载一个模式，让计算机转出一个结果管它符合不符合实际的人多了；刻苦钻研、探讨物理机制的人少了，而从文献上引用外国人的思路重复做一些研究管它符合不符合中国实际的人多了；对课题研究孜孜以求的人少了，而过分夸大结果的人多了。这些作风是不利于大气科学的发展。因此，当前在大气科学研究生培养上，不仅要培养学生有创新能力，而且要培养学生对科学的认真态度和治学的严谨精神。

总之，为适应新世纪我国经济和社会快速发展的需求，特别是为适应国家在防灾减灾和应对全球气候变化方面对高素质、高层次大气科学创新人才的需求，我们应尽快把我国大气科学的教育和人才培养搞上去。当前，一批批对科学有志向、有兴趣、对研究有进取的高层次大气科学年青创新人才正在涌现，他们继承老一辈科学家正在为我国大气科学的发展做出重要贡献，"桐花万里丹山路，雏凤清于老凤声"。让我们紧紧把握住这大好时机，为国家培养和造就更多有志向的大气科学的高层次创新人才，完成国家赋予我们神圣的历史使命。"长风破浪会有时，直挂云帆济沧海"，希望大家为把我国从大气科学大国变成大气科学强国而努力奋斗！

海峡两岸台风暴雨合作研究成果交流会的开幕词
2011 年

今天我们在风景秀丽的张家界召开海峡两岸关于全球变暖对西北太平洋台风活动和登陆中国台风的影响及其机理专家研讨会。这次会议是在海峡两岸关于台风暴雨合作项目共同资助下召开的。

正如大家所知道，西北太平洋热带气旋和台风给海峡两岸带来严重天气灾害，每年大约有 7~8 个台风登陆我国，给两岸带来数百亿元严重的经济损失和数百人的人员伤亡，因此，开展西北太平洋热带气旋和台风活动季内、年际和年代际变化的气候学研究，不仅具有重要的科学意义，而且对于开展台风季节预报也具有重要的应用价值。海峡两岸气象学家遵照"交流、创新、敦谊"之宗旨，在国家自然科学基金委员会地区合作基金和台湾李国鼎基金会的支持下，开展了对西北太平洋热带气旋和台风活动的合作研究，经过两年多的合作研究，取得了丰硕成果。这次会议的目的是交流近两年来合作研究的部分成果，并商讨进一步合作的可能性及三年合作研究成果的总结。

这次会议我们很荣幸地邀请到国际著名气象学家、前日本 Frontier 研究所长松野太郎教授以及德国知名气象学家 Hans-Graf 教授给会议作 Special Lecture；并且也请 Hawaii 大学的朱宝信教授和李天明教授以及台湾大学隋中兴教授给会议做特邀报告；这次研讨会得到中国科学院港澳合作办公室的资助和中国科学院大气物理研究所的大力支持，让我们表示感谢。由于此次会议是海峡两岸科学讨论会，并为了便于交流，此次会议语言除了 Special Lecture 用英语外，特邀报告和一般学术交流均用中文。

张家界分布着全球有名的喀斯特地貌，是闻名世界的风景区，希望各位专家在会议之余能领略祖国美好河山。

最后祝会议圆满成功，希望大家在张家界度过美好时光。

在中山大学大气科学系建系五十周年庆典会上的致词

2011 年

金秋的羊城，万紫千红。在这个美好的季节，我们在这里隆重聚会，庆祝中山大学大气科学系建系五十周年，这不仅是中山大学大气科学系发展的一件大事，而且也是全国大气科学发展的大事。今天，全国大气科学各有关大学院所领导都来祝贺！在这里我谨向中山大学大气科学系全体老师、同学们和中山大学广大校友，表示衷心的祝贺！

五十前的今天，中山大学气象学专业创办，他是我国华南地区唯一的大气科学学科人才培养单位。自从 1961 年创建以来，中山大学大气科学系坚持以人才培养为根本任务，多年来，一直致力于培养高素质、高层次、多样化的人才，已形成本科、硕士、博士和博士后完整的大气科学人才培养体系，为中国气象事业的发展和中国经济建设造就了一大批杰出人才；并且，中山大学大气科学系长期坚持以热带气象学和季风、大气环境学为主要科学研究和办学方向，保持鲜明的学科特色和区域性特征，在季风、热带大气环流、数值模拟、大气环境和应用气象等方面取得重要研究成果。这些成果在国内大气科学界有较高的学术地位，在国际上也有较大影响，因此可以说，中山大学大气科学系的五十年是培育英才的五十年，是探求科学真理的五十年。

五十年弹指一挥间，回顾昨天，中山大学大气科学系成果累累，成于大气；展望未来，中山大学大气科学系再创辉煌，名扬天下。最后，我祝福中山大学大气科学系的教学事业如珠江之水奔流不息，中山大学大气科学系的科学研究如南岭之巅攀登不止！

庆贺福建省泉港二中(原惠安二中)建校六十周年的贺词

2012 年

泉港第二中学：

泉港金秋，温暖如春，在这美好的季节，我们迎来了母校的六十华诞。在此谨向泉港二中建校六十周年表示衷心的祝贺！并向曾经培养过我的老师送上真诚的祝福和崇高的敬意！向全校师生和广大校友表示诚挚的问候！

泉港二中(原惠安二中)自1952年创建以来，砥砺耕耘六十载，励精图治一甲子，今天成为一所省二级达标初、高中齐全的中学。忆往昔，五十九年前在母校艰难求学的身影不禁浮现在我的眼前，师长们的谆谆教导和同学们的挑灯夜读的场面历历在目，思绪万端。当年破旧的锦山宫是我们唯一的教室，暗弱的煤油灯是我们唯一的照明，田埂空地是我们晨练唯一的操场。然而，在这样的环境中，同学们却孜孜不倦勤奋读书，琅琅的读书声萦绕校园，争创"三好"生的飒爽英姿到处可见。同学们在这样的青涩磨练中勤奋读书，不断进取，几度风雨，几多奋斗，大家的学习成绩不亚于当时的名校。这破旧的校舍里留下了我和同学们青春年华的脚印，给我们留下了难忘回忆；这里是我们人生旅途中重要的一站，在这里烙上了泉港二中(原惠安二中)精神的印记；这里塑造了我们今生思想、品行、情操和理想的雏形，夯实了我们今生所从事事业的基础。我无论何时走到国内外任何一个科学讲坛上，都不曾忘怀这段经历，这段经历成为激励我努力奋斗的动力。

时光流转，光阴荏苒，六十年过去了，弹指一挥间。当年惠安二中蹒跚起步，历经风雨，艰难成长，如今泉港二中旧貌换新颜，高大宽敞的教学楼，设施先进的科技楼，舒适整洁的宿舍楼，校园到处充满着学习、创新、祥和的气氛。建校六十年来，母校正是始终秉承"文明、勤奋、求实、创新"的校训，为国家和社会培养了三万多名初、高中毕业生，得到了社会的广泛赞誉。

回顾昨天，泉港二中教书育人硕果累累，成于勤奋；展望未来，泉港二中素质教育再创辉煌，名扬全市。"长风破浪会有时，直挂云帆济沧海"。我祝愿母校的人才培养如闽江之水奔流不息，母校的教学质量如武夷之巅攀登不止。

"东亚能量和水分循环及其与季风相互作用国际研讨会"开幕词

2012 年

"东亚能量和水分循环及其与季风相互作用国际研讨会"今天在著名的历史文化名城敦煌如期召开了。首先，请允许我代表会议组织委员会热烈欢迎各位学者在百忙之中出席本次会议。本次会议由973项目"全球变暖背景下东亚能量和水分循环变异及其对我国极端气候的影响"第二、第四、第五课题、国家自然科学基金委员会重点基金项目"青藏高原地区地气相互作用和陆面物理过程模式创建研究"、全球变化专项项目"青藏高原气候系统变化及其对东亚区域的影响与机制研究"、国家自然科学基金委员会杰出青年基金项目"青藏高原地气相互作用观测与卫星遥感应用"、"东亚冬季风系统变异及其内动力学机理研究"、中国科学院知识创新工程项目"全球变暖背景下我国冬季冰冻雨雪灾害发生的年际变化及其机理研究"等6个项目和课题共同组织的，并得到这些项目和课题的资助。在此，我谨代表会议组织委员会对这些项目和课题的资助表示感谢！

大家知道，气候系统中的能量和水分循环是影响天气和气候异常重要的物理过程，认识这些过程对于理解天气变化和气候变异都具有重要意义。大气圈与其他圈层之间进行能量和水分交换主要是通过下垫面进行的，而下垫面类型的不同影响着海气和陆气间的能量和水分循环过程。东亚地区具有复杂而典型的下垫面，既包括戈壁、沙漠、黄土高原、草原等干旱、半干旱区，又有东部的海洋和季风湿润区，还有全球最高的大地形——青藏高原。一方面，这些不同类型下垫面上的能量和水分循环对东亚典型的季风气候变异有重要影响；另一方面，东亚季风的变异也会影响这些下垫面状况，从而影响着东亚地区的能量和水分循环。这种相互作用的过程决定了东亚地区天气气候的变异。因此，本次会议的目的是：第一，分析和讨论东亚季风气候系统变异特征、过程及其与下垫面水分和能量循环的关系；第二，分析和讨论中国西北地区和中亚干旱区陆-气相互作用特征及其对东亚季风系统的影响和关联。本次研讨会将通过30余个报告交流上述项目的研究成果，并探讨东亚地区典型下垫面的能量和水分循环过程及其与东亚季风系统的相互作用。

来自国内外12个研究机构和有关院校计50余名学者参加本次研讨会。这次会议我们还邀请到来自美国、日本、印度、中国香港等地的部分专家，特别是我们邀请到著名的季风研究专家美国C.P. Chang教授、美国NOAA的Yang Song博士、日本气象研究所的Kusunoki博士、香港中文大学的吴仁广教授等在本次会议作Special Lecture和邀请报告。为了使会议交流和讨论更方便，会议将采用中文和英文并用的形式，除了在Special Lecture用英文外，其他可用中文演讲和讨论。

敦煌是古代"丝绸之路"的重镇，它是一座具有几千年历史的世界文化名城，这里是典型的干旱区，并有著名的雅丹地貌。可借用唐代著名诗人王之涣所写的凉州词"黄河远上白云间，一片孤城万仞山。羌笛何须怨杨柳，春风不度玉门关。"来形容敦煌地区的自然景观。希望各位学者在会议之余能领略中国西部的美丽沙漠和戈壁。

最后，祝各位学者在敦煌愉快，预祝会议圆满成功！

在"海峡两岸台风和暴雨合作研究成果发表研讨会"上的致词

2012 年

今天很高兴能参加在福州召开的"海峡两岸台风和暴雨合作研究成果发表研讨会"。正如大家所知道，台风和暴雨是我们两岸共同遭受的严重天气灾害，西北太平洋热带气旋和台风给海峡两岸带来严重损失，每年大约有7～8个台风登陆我国，给两岸带来数百亿严重的经济损失和数百人的人员伤亡，如2006年7月12日在福建霞浦登陆的台风"碧丽斯"，2009年9月7日在我国台湾南部登陆的台风"莫拉克"给两岸带来巨大的经济损失和重大人员伤亡。两岸开展台风和暴雨合作研究，不仅对于台风和暴雨的演变特征、结构、强度、路径以及强降水和强风等天气学认识的提高以及西太平洋热带气旋和台风活动及登陆中国地点的季节内、年际和年代际变异的气候学研究具有重要科学意义，而且对于开展台风和暴雨的预报、预警和季度预测也具有实际的应用价值。

正因为如此重要，海峡两岸气象学家遵照"交流、创新、敦谊"之宗旨，从2010年起，在国家自然科学基金委员会地区合作基金和台湾李国鼎基金会的支持和资助下，两岸关于台风和暴雨的合作研究的各个领域都取得了丰硕成果。这次会议，我们两岸同仁不仅可以在一起很好总结、交流近三年的合作研究成果，而且可以商讨进一步合作研究的可能性。我所负责的台风气候学方面的项目也想利用这次机会与兄弟项目交流近两年多合作研究的部分成果，并可向兄弟项目学习更多知识。

这次会议在福州召开，对我来说更是一次好机会，福建是一个人杰地灵的地方，福建与台湾有千丝万缕的地缘、人缘、血缘关系。福建是我老家，是生我养我的地方，从小就亲身体验到台风和暴雨所带来的严重灾害，"少小离家老大回，乡音未改鬓毛衰"，到老了很想回家乡，能在两岸关于暴雨和台风合作研究中为家乡做一点事。

借此机会，感谢北京大学张庆红教授和福建气象局为此次会议的召开做了大量前期准备工作。最后祝会议圆满成功，在中华民族传统节日中秋佳节来临之际，祝大家节日愉快、合家欢乐！

在"第三海洋研究所海洋-大气化学与全球变化重点实验室成立十周年庆典会"上的致词

2012 年

海洋是地球气候系统最重要组成部分之一，它吸收经过大气层太阳辐射的大量能量，并反馈于大气，影响大气的运动及气候变化；并且，海洋还吸收人类活动大量排放的 CO_2 等温室气体，减缓了温室效应。然而，海洋也因此受到了全球变暖明显的影响，如海平面上升、海水温度升高、海水酸化及其生态影响等，这些已成为当今人类社会面临的重大问题。鉴于全球变化与海洋-大气化学过程及海洋相互作用的重要关系，国家海洋局于 2002 年 8 月在第三海洋研究所成立了海洋-大气化学与全球变化重点实验室。此实验室主要通过海洋-大气系统的观测实验、理论分析和数值模拟等手段，研究在全球气候变暖背景下我国近海和邻近海以及极地海洋-大气系统物理和化学过程的变化特征和机理，及其对我国海洋环境和气候变化的影响。

十年来，我作为此重点实验室学术委员会主任，亲自见证了实验室研究的进步和发展。此实验室根据国家需求与国际海洋科学发展前沿，围绕我国海洋环境与全球气候变化等重要问题，立足于中国近海及邻近海域，以极地和大洋为科研考察的重点，形成了一个颇具特色的重点实验室。这个实验室在海洋-大气生物地球化学循环、全球变化区域响应与适应对策、极区海洋和大气化学、同位素海洋化学和海洋核放射监测技术和应对气候变化的海藻生物技术等研究方向取得了丰硕的科研成果。实验室在《Science》《Journal of Geophysical Research》《Remote Sensing of Environment》《中国科学》《海洋学报》《大气科学》等国内外有影响力的刊物上发表了大量的学术论文和多本专著。最近，实验室为了更好地服务于国家经济建设和社会发展的需求，促进与有关研究机构的交流及进一步发展，总结了这十年来的研究成果，汇编了庆贺实验室成立十周年的三辑论文选集：《海洋大气化学与海洋生物地球化学过程研究》《海洋同位素、气溶胶化学与海洋技术研发》《海洋气候动力学与海洋环境研究》。这些都反映了此实验室在十年间，对我国近海和邻近海以及极地海洋-大气系统物理和化学过程的变化特征和机理及其对我国海洋环境和气候变化的影响做了大量的创新性研究工作，所得到的成果是喜人的。并且，此实验室的人才培养也取得很大成功。

总之，回顾昨天，实验室研究成果累累，成于创新；展望未来，实验室人才荟萃，名扬学界。希望实验室今后进一步对有关海洋-大气化学与全球变化进行深入的研究，抓紧人才培养，提高实验室服务于我国应对气候变化科技支撑的能力，争取早日成为国内外有影响力的海洋-大气化学与全球变化重点实验室。最后，我祝愿实验室的人才培养如九龙江之水奔流不息，实验室的科学研究如武夷山之巅攀登不止。

附录 论文目录

1. 黄荣辉, 刘瑞芝, 周凤仙, 袁重光, 曾庆存. 1977. 大气中二氧化碳红外吸收带透过率的计算. 大气科学, 1(3): 188~198.
2. 黄荣辉. 1979. 卫星资料在数值预报中的应用与四维分析. 气象科学, (1): 51~57.
3. 黄荣辉, 袁重光. 1980. 红外遥测气温垂直分布中几种方法的比较及误差分析. 大气科学, 4(1): 49~57.
4. 黄荣辉. 1980. 关于四维分析中一些问题的初步试验, 全国第二次数值预报论文集. 北京: 科学出版社, 169~174.
5. Huang Ronghui, and Gambo K. 1981. The response of a model atmosphere in middle latitude to forcing by topography and stationary heat sources. J. Meteor. Soc. Japan, 59(2): 220~237.
6. 黄荣辉, 李荣凤. 1981. 大气气温和风场垂直分布的统计特征及其在设计数值天气预报模式中的应用. 大气科学, 5(3): 300~309.
7. Huang Ronghui, and Gambo K. 1982. The response of a hemispheric multi-level model atmosphere to forcing by topography and stationary heat sources. Part I, Forcing by topography. J. Meteor. Soc. Japan, 60(1): 78~92.
8. Huang Ronghui, and Gambo K. 1982. The response of a hemispheric multi-level model to foring by topography and stationary heat sources. Part II, Forcing by stationary heat sources and forcing by topography and stationary heat sources. J. Meteor. Soc. Japan, 60(1): 93~108.
9. 黄荣辉, 岸保勘三郎. 1983. 关于冬季北半球定常行星波传播另一波导的研究. 中国科学(B辑), (10): 940~950.
10. Huang Ronghui, and Gambo K. 1983. The response of a hemispheric multi-level model atmosphere to forcing by topography and stationary heat sources in summer. J. Meteor. Soc. Japan. 61(4): 495~509.
11. 黄荣辉. 1983. 冬季格陵兰高原对北半球定常行星波形成的作用. 大气科学, 7(4): 393~402.
12. 黄荣辉. 1984. 球面大气定常行星波的波作用守恒方程与用波作用通量所表征的定常行星波传播波导. 中国科学(B辑), (8): 766~775.
13. 黄荣辉. 1984. 冬季北半球地形与定常热源强迫所产生的定常行星波及其动量通量、热量通量的计算. 气象学报, 40(1): 1~10.
14. 黄荣辉. 1984. 冬季北半球高纬度环流对定常行星波的动力作用. 全国寒潮中期预报文集, 北京: 北京大学出版社, 152~163.
15. 黄荣辉. 1984. 冬季热源位置与热源的南北宽度对定常行星波的影响. 大气科学, 8(2): 117~125.
16. Huang Ronghui. 1984. The characteristics of the forced stationary planetary wave propagations in the Northern Hemispheric summer. Adv. Atmos. Sci., 1(1): 84~94.
17. 黄荣辉. 1984. 冬、夏平流层环流的数值模拟. 高原气象, 3(3): 1~13.
18. 黄荣辉, 曾庆存, 杨大升. 1984. 几十年来关于大气环流与大尺度动力学的研究进展与2000年研究展望. 气象科技, (5): 1~7.
19. 黄荣辉. 1984. 一个诊断基本气流中波动的新物理量——Eliassen-Palm通量. 力学进展, 14(2): 175~181.
20. Huang Ronghui. 1985. The simulation of three-dimensional teleconnection in the summer circulation over the Northern Hemisphere. Adv. Atmos. Sci., 2(1): 81~92.
21. 黄荣辉. 1985. 夏季青藏高原与落基山脉对北半球定常行星波形成的动力作用. 大气科学, 9(3): 243~250.
22. 黄荣辉. 1986. 夏季西藏高原对北半球定常行星波形成的热力作用. 大气科学, 10(1): 1~8.

23. 黄荣辉. 1985. 夏季青藏高原热源异常对北半球大气环流异常的作用. 气象学报, 43(2): 208~220.
24. 黄荣辉. 1985. 夏季青藏高原对于南亚平均季风环流形成与维持的热力作用. 热带气象, 1(1): 1~8.
25. 黄荣辉. 1985. 平流层与中间层大气动力学的研究. 大气科学, 9(4): 413~422.
26. 黄荣辉. 1985. 冬季低纬度热源异常在北半球对流层大气环流异常中的作用. 气象学报, 43(4): 410~423.
27. 黄荣辉. 1986. 大气中定常行星波的Eliassen-Palm通量与波折射指数的关系. 大气科学, 10(2): 145~153.
28. 黄荣辉. 1986. 大气行星尺度运动的动力特征. 大气科学, 10(4): 348~356.
29. 黄荣辉. 1986. 冬季低纬度热源异常对北半球大气环流影响的物理机制. 中国科学(B辑), (1): 91~103.
30. Huang Ronghui, and Yan Bangliang. 1987. The physical effects of topography and heat sources on the formation and maintenance of the summer monsoon over Asia. Adv. Atmos. Sci., 4(1): 13~23.
31. 黄荣辉. 1987. 与数值预报模式的系统误差有关的几个行星波动力学问题. 气象学报, 45(1): 6~13.
32. Huang Ronghui, and Li Weijing. 1987. Influence of the heat source anomaly over the western tropical Pacific on the subtropical high over East Asia. Proc. International Conference on the General Circulation of East Asia, April 10-15, 1987, Chengdu, China, 40~51.
33. 黄荣辉, 李维京. 1988. 夏季热带西太平洋上空的热源异常对东亚上空副热带高压的影响及其物理机制. 大气科学, 12(SI): 107~116.
34. 黄荣辉. 1988. 涡旋粘性对夏季副热带准定常行星尺度环流的影响. 气象学报, 46(1): 9~19.
35. 黄荣辉, 严邦良. 1988. 热带基本气流对冬季北半球中高纬度准定常行星波的影响. 气象学报, 46(2): 154~163.
36. Huang Ronghui, and Lu Li. 1989. Numerical simulation of the relationship between the anomaly of subtropical high over East Asia and the convective activities in the western tropical Pacific. Adv. Atmos. Sci., 6(2): 202~214.
37. 黄荣辉, 邹捍. 1989. 球面斜压大气中上传行星波与纬向平均气流的相互作用. 大气科学, 13(4): 383~392.
38. 黄荣辉. 1989. 关于大气环流遥相关及低频振荡的研究进展与问题. 大气科学, 13(4): 480~490.
39. Huang Ronghui, and Wu Yifang. 1989. The influence of ENSO on the summer climate change in China and its mechanisms. Adv. Atmos. Sci., 6(1): 21~32.
40. 黄荣辉, 严邦良. 1989. 一个描述河陆风变化的数学模式及其数值试验. 大气科学, 13(1): 11~21.
41. Huang Ronghui. 1989. The influence of eddy viscosity on the subtropical planetary-scale circulation in summer. Acta Meteorologica Sinica, 3(2): 145~155.
42. 黄荣辉. 1990. 引起我国夏季旱涝的东亚大气环流异常遥相关及其物理机制的研究. 大气科学(竺可桢诞辰百周年纪念刊), 14(1): 108~117.
43. Huang Ronghui, and Wang Lianying. 1990. Relationship between the interannual variation of total ozone in the Northern Hemisphere and the QBO of basic flow in the tropical stratosphere. Adv. Atmos. Sci., 7(1): 47~56.
44. 黄荣辉. 1990. ENSO事件及热带海-气相互作用动力学的最近研究进展. 大气科学, 14(2): 234~242.
45. 黄荣辉, 张庆云. 1990. 华北降水的年代际变化及其对经济的影响, 水资源问题研讨会论文集. 北京: 水利电力出版社, 95~101.
46. 黄荣辉, 杨广基, 吴仪芳, 周家斌, 陈烈庭, 张庆云. 1990. 综合长期预报方法及对旱涝季度与超季度预报试验. 大气科学(竺可桢诞辰百周年纪念刊), 14(1): 26~31.
47. 黄荣辉. 1990. 关于气候灾害及其成因与预测研究. 地球科学进展, 5(5): 59~63.
48. 黄荣辉. 1990. 热带西太平洋暖池热状态与对流活动对我国夏季降水年际变化的影响及其在旱涝季度预报中应用, 长期天气预报论文集. 北京: 气象出版社, 38~52.

49. 黄荣辉, 严邦良, 岸保勘三郎. 1991. 基本气流在 ENSO 对北半球冬季大气环流影响中的作用. 大气科学, 15(1): 44~54.

50. Huang Ronghi. 1992. The East Asia/Pacific pattern teleconnection of summer circulation and climate anomaly in East Asia. Acta Meteorologica Sinica, 6(1): 25~37.

51. Huang Ronghui, and Sun Fengying. 1992. Impacts of the tropical western Pacific on the East Asian summer monsoon. J. Meteor. Soc. Japan, 70(1B): 243~256.

52. 黄荣辉. 1992. 夏季大气环流异常的东亚太平洋型遥相关及其在我国旱涝预测中的应用. 长期天气预报和日地关系研究, 章基嘉, 黄荣辉主编. 北京: 海洋出版社, 6~19.

53. Huang Ronghui, Yin Baoyu, and Liu Aidi. 1992. Intraseasonal variability of the East Asian summer monsoon and its association with the convective activities in the tropical western Pacific. Climate Variability, Edited by Ye Duzheng et al., Beijing: China Meteorological Press, 134~155.

54. 黄荣辉. 1992. 近40年我国夏季旱涝变化及其成因初探. 气候变化若干问题研究, LASG Monog., No.2. 北京: 气象出版社, 14~29.

55. 黄荣辉, 严邦良. 1993. 用线性化全球原始方程谱模式研究地形强迫行星波垂直传播特征. 大气科学, 17(3): 257~267.

56. 黄荣辉, 孙凤英. 1994. 热带西太平洋暖池的热状态及其上空的对流活动对东亚夏季气候异常的影响. 大气科学, 18(2): 141~151.

57. Huang Ronghui. 1994. Interaction between the 30-60 day oscillation, the Walker circulation and the convective activities in the tropical western Pacific and their relations to the interannual oscillation. Adv. Atmos. Sci., 11(3): 367~384.

58. 黄荣辉, 孙凤英. 1994. 热带西太平洋暖池上空对流活动对东亚夏季季节内变化的影响. 大气科学, 18(4): 456~465.

59. 黄荣辉, 傅云飞, 臧晓云. 1996. 亚洲季风与 ENSO 循环的相互作用. 气候与环境研究, 1(1): 38~54.

60. Huang Ronghui, and Chen Wen. 1996. The dispersions and propagations of planetary wave in the spherical atmosphere. From Atmospheric Circulation to Global Change. Edited by IAP, Beijing: China Meteorological Press, 197~214.

61. 黄荣辉, 徐飞亚. 1996. 大气科学研究的一个前沿领域-气候系统动力学与气候预测. 地球科学进展, 11(3): 231~237.

62. 黄荣辉. 1996. 现代大气科学基础研究发展趋势、特点与前沿研究课题. 现代大气科学前沿与展望, 国家自然科学基金委员会等编. 北京: 气象出版社, 7~11.

63. 黄荣辉, 郭其蕴. 1996. 我国的气候灾害及其对国民经济的影响. 科学中国人, 8: 41~43.

64. 黄荣辉. 1997. 我国灾害气候的主要成因与预测. 科学中国人, ZI.: 16~18.

65. 黄荣辉等. 1997. 综合旱涝季度预报方法及对1991~1995年夏季季度与超季度预测试验. 气候与环境研究, 2(1): 1~15.

66. 黄荣辉. 1997. 我国的气候灾害及其成因. 科学, 50(2): 33~37.

67. 黄荣辉. 1997. 我国的气候灾害特征、成因及其预测研究. 百名院士科技系列报告集, 周光召、朱光亚主编, 下册. 北京: 科学出版社, 327~350.

68. Huang R H, Li X, Yuan C G, Lu R Y, Moon S E, Kim B J. 1997. Seasonal prediction experiments of the summer droughts and floods during the early 1990's in East Asia with numerical models. Adv. Atmos. Sci., 15(4): 433~446.

69. 黄荣辉, 张人禾. 1997. ENSO 循环与东亚季风环流相互作用过程的诊断研究. 赵九章纪念文集, 叶笃正主编. 北京: 科学出版社, 93~109.

70. Huang Ronghui, Zang Xiaoyun, Zhang Renhe, and Chen Jilong. 1998. The westerly anomalies over the tropical Pacific and their dynamical effect on the ENSO cycles during 1980~1994. Adv. Atmos. Sci., 15(2): 135~151.

71. Huang Ronghui, Ren Baohua, and Huang Gang. 1998. Further investigation on the impact of the tropical western Pacific on the East Asian summer monsoon. 研究丛书 No.3, CCSR. Tokyo University.

72. 黄荣辉, 张振洲, 黄刚, 任保华. 1998. 夏季东亚季风区水汽输送特征及其与南亚季风区水汽输送的差别. 大气科学, 22(4): 460~469.

73. 黄荣辉. 1998. 关于东亚季风的研究, 东亚季风和中国暴雨, 中国科学院大气物理研究所编. 北京: 气象出版社, 127~136.

74. 黄荣辉, 徐予红, 王鹏飞, 周连童. 1998. 1998 年夏长江流域特大洪涝特征及其成因探讨. 气候与环境研究, 3(4): 300~313.

75. 黄荣辉, 黄刚, 任保华. 1999. 东亚夏季风的研究进展及其需进一步研究的问题. 大气科学(特刊), 23(2): 129~141.

76. 黄荣辉, 徐予红, 周连童. 1999. 我国夏季降水的年代际变化及华北干旱化趋势. 高原气象, 18(4): 465~476.

77. 黄荣辉, 张人禾, 严邦良. 1999. 关于东亚气候系统年际变化研究进展及其需进一步研究的问题. 中国基础研究, (2): 66~74.

78. 黄荣辉. 2000. 20 世纪大气环流与大尺度动力学研究进展与回顾. 21 世纪初大气科学回顾与展望, 国家自然科学基金委员会地球科学等编. 北京: 气象出版社, 89~97.

79. 黄荣辉, 章国材, 陆则慰. 2000. 21 世纪大气科学前沿与展望. 21 世纪初大气科学回顾与展望, 国家自然科学基金委员会地球科学部等编. 北京: 气象出版社, 3~9.

80. Huang Ronghui, Zhang Renhe, and Zhang Qingyun. 2000. The 1997/98 ENSO cycle and its impact on summer climate anomalies in East Asia. Adv. Atmos. Sci., 17(3): 348~362.

81. 黄荣辉. 2001. 关于我国重大气候灾害的形成机理和预测理论研究进展. 中国基础研究, (8): 4~8.

82. Huang Ronghui, Wu Bingyi, Sung-Gil Hong, and Jai-Ho Oh. 2001. Sensitivity of numerical simulations of the East Asian summer monsoon rainfall and circulation to different cumulus parameterization schemes. Adv. Atmos. Sci., 18(1): 23~41.

83. 黄荣辉, 张人禾, 严邦良. 2001. 热带西太平洋纬向风异常对 ENSO 循环的动力作用. 中国科学(D 辑), 31(8): 697~704.

84. 黄荣辉, 陈金中. 2002. 平流层球面大气地转适应过程和惯性重力波的激发. 大气科学, 26(3): 289~306.

85. 黄荣辉, 周连童. 2002. 我国重大气候灾害特征、形成机理和预测研究. 自然灾害学报. 11(1): 1~9.

86. 黄荣辉, 陈文. 2002. 关于亚洲季风与 ENSO 循环相互作用研究最近的进展. 气候与环境研究. 7(2): 146~159.

87. Huang Ronghui. 2002. Studies on the variability and prediction of East Asian summer monsoon. Korean J. Atmos. Sci., 5(1): 1~12.

88. 黄荣辉, 陈文, 丁一汇, 李崇银. 2003. 关于季风动力学及季风与 ENSO 循环相互作用的研究. 大气科学, 27(4): 484~502.

89. Huang Ronghui, Zhou Liantong, and Chen Wen. 2003. The progresses of recent studies on the variabilities of the East Asian monsoon and their causes. Adv. Atmos. Sci., 20(1): 55~69.

90. 黄荣辉, 陈际龙, 周连童, 张庆云. 2003. 关于中国重大气候灾害与东亚气候系统之间关系的研究. 大气科学(纪念特刊), 27(4): 770~787.

91. 黄荣辉. 2003. 中国的重大气候灾害及其成因. 科学的前沿(中国科学院研究生院演讲录 第四辑), 余翔林编. 北京: 科学出版社, 69~83.

92. Huang Ronghui, Huang Gang, and Wei Zhigang. 2004. Climate variations of the summer monsoon over China. East Asian Monsoon, Chang C P, Ed., World Scientific Publishing Co. Pte.Ltd., 213~270.

93. Huang Ronghui, Chen Wen, and Zhang Renhe. 2004. Recent advances in studies of the interaction between the East Asian winter and summer monsoons and ENSO cycle. Adv. Atmos. Sci., 21(3): 407~424.

94. 黄荣辉, 顾雷, 徐予红, 张启龙, 吴尚森, 曹杰. 2005. 东亚夏季风爆发和北进的年际变化特征及其与热带西太平洋热力状态的关系. 大气科学, 29(1): 20~36.

95. 黄荣辉. 2006. 我国重大气候灾害的形成机理和预测理论研究. 地球科学进展. 21(6): 564~575.

96. 黄荣辉, 陈际龙, 黄刚, 张启龙. 2006. 中国东部夏季降水的准两年周期振荡及其成因. 大气科学, 30(4): 545~560.

97. 黄荣辉, 蔡榕硕, 陈际龙, 周连童. 2006. 我国旱涝气候灾害的年代际变化及其与东亚气候系统变化的关系. 大气科学, 30(5): 730~743.

98. 黄荣辉, 韦志刚, 李锁锁, 周连童. 2006. 黄河上游和源区气候、水文的年代际变化及其对华北水资源的影响. 气候与环境研究, 11(3): 245~258.

99. Huang Ronghui, and Wang Lin. 2006. Interdecadal variation of Asian winter monsoon and its association with the planetary wave activity. Proc. Symposium on Asian Monsoon, Kuala Lumpur, Malaysia, 126.

100. Huang Ronghui, Gu Lei, Zhou Liantong, and Wu Shangsen. 2006. Impact of thermal state of the tropical western Pacific on onset date and process of the South China Sea summer monsoon. Adv. Atmos. Sci., 23(6): 909~924.

101. 黄荣辉, 陆则慰, 范蔚茗, 等. 2006. 20世纪大气科学的发展成就和21世纪初大气科学的前沿科学问题. 国家自然科学基金委员会地球科学部等编, 北京: 气象出版社, 11~23.

102. 黄荣辉, 陈光华. 2007. 西北太平洋热带气旋移动路径的年际变化及其机理研究. 气象学报, 65(5): 683~694.

103. 黄荣辉, 魏科, 陈际龙, 陈文. 2007. 东亚2005年和2006年冬季风异常及其与准定常行星波活动的关系. 大气科学, 31(6): 1033~1048.

104. 黄荣辉. 2007. 呵护我们的家园——地球大气. 科学名家讲座(第六辑), 北京青少年科技俱乐部活动委员会编. 哈尔滨: 哈尔滨出版社, 79~137.

105. Huang Ronghui, Chen Jilong, and Huang Gang. 2007. Characteristics and variations of the East Asian monsoon system and its impacts on climate disasters in China. Adv. Atmos. Sci., 24(6): 993~1023.

106. 黄荣辉, 顾雷, 陈际龙, 黄刚. 2008. 东亚季风系统的时空变化及其对我国气候异常影响的最近研究进展. 大气科学, 32(4): 691~719.

107. 黄荣辉, 陈际龙. 2010. 我国东、西部夏季水汽输送特征及其差异. 大气科学, 34(6): 1035~1045.

108. 黄荣辉, 王磊. 2010. 台风在我国登陆地点的年际变化及其与夏季东亚/太平洋型遥相关的关系. 大气科学, 34(5): 853~864.

109. 黄荣辉, 陈际龙, 刘永. 2011. 我国东部夏季降水异常主模态的年代际变化及其与东亚水汽输送的关系. 大气科学, 35(4): 589~606.

110. 黄荣辉, 刘永, 王磊, 王林. 2012. 从2009年秋到2010年春我国西南地区严重干旱成因分析. 大气科学, 36(3): 443~457.

111. 黄荣辉, 陈栋, 刘永. 2012. 中国长江流域洪涝灾害和持续性暴雨的发生特征及成因. 成都信息工程学院学报, 27(1): 1~19.

112. Huang Ronghui, Chen Jilong, Wang Lin, and Lin Zhongda. 2012. Characteristics, processes, and causes of the spatio-temporal variations of the East Asian monsoon system. Adv. Atmos. Sci., 29(6): 910~942.

113. 黄荣辉, 周德刚, 陈文, 周连童, 韦志刚, 张强, 高晓清, 卫国安, 侯旭宏. 2013. 关于中国西北干旱区陆-气相互作用及其对气候影响研究的最近进展. 大气科学, 37(2): 189~210.

114. 黄荣辉, 刘永, 冯涛. 2013. 20 世纪 90 年代末中国东部夏季降水和环流的年代际变化特征及其内动力成因. 科学通报, 58(8): 617~628.

115. 黄荣辉, 刘永, 皇甫静亮, 冯涛. 2014. 20 世纪 90 年代末东亚冬季风年代际变化特征及其内动力成因. 大气科学, 38(4): 627~644.

116. 黄荣辉, 皇甫静亮, 刘永, 杜振彩, 陈国森, 陈文, 陆日宇. 2016. 从 Rossby 波能量频散理论到准定常行星波动力学研究的发展. 大气科学, 40(1): 3~21.

117. 黄荣辉, 皇甫静亮, 武亮, 冯涛, 陈光华. 2016. 关于西北太平洋季风槽年际和年代际变异及其对热带气旋生成影响和机理的研究. 热带气象学报, 32(6): 767~785.

118. 黄荣辉, 皇甫静亮, 刘永, 冯涛, 武亮, 陈际龙, 王磊. 2016. 西太平洋暖池对西北太平洋季风槽和台风活动影响过程及其机理的最近研究进展. 大气科学, 40(5): 877~896.

119. Huang Ronghui, Liu Yong, Du Zhencai, Chen Jilong, and Huangfu Jingliang. 2017. Differences and links between the East Asian and South Asian summer monsoon systems: Characteristics and variability. Adv. Atmos. Sci., 34(10): 1204~1218.

120. 黄荣辉, 陈文, 魏科, 王林, 皇甫静亮. 2018. 平流层大气动力学及其与对流层大气相互作用的研究: 进展与问题. 大气科学, 42(3): 463~487.